AF767141

Computation, Optimization, and Machine Learning in Seismology

Advanced Textbook Series

1. **Unconventional Hydrocarbon Resources: Techniques for Reservoir Engineering Analysis**
   Reza Barati and Mustafa M. Alhubail
2. **Geomorphology and Natural Hazards: Understanding Landscape Change for Disaster Mitigation**
   Tim R. Davies, Oliver Korup, and John J. Clague
3. **Remote Sensing Physics: An Introduction to Observing Earth from Space**
   Rick Chapman and Richard Gasparovic
4. **Geology and Mineralogy of Gemstones**
   David Turner and Lee A. Groat
5. **Data Analysis for the Geosciences: Essentials of Uncertainty, Comparison, and Visualization**
   Michael W. Liemohn
6. **Earth System Geophysics**
   Steven R. Dickman
7. **Earth's Natural Hazards and Disasters**
   Bethany D. Hinga
8. **Computation, Optimization, and Machine Learning in Seismology**
   Subhashis Mallick

Advanced Textbook 8

Computation, Optimization, and Machine Learning in Seismology

Subhashis Mallick
University of Wyoming, USA

This work is a co-publication of the American Geophysical Union and John Wiley and Sons, Inc.

AGU
ADVANCING EARTH AND SPACE SCIENCES

WILEY

This edition first published 2026

Published by John Wiley & Sons, Inc., Hoboken, New Jersey.

Published under the aegis of the AGU Publications Committee

Matthew Giampoala, Vice President, Publications
Steven A. Hauck, II, Chair, Publications Committee
For details about the American Geophysical Union, visit us at www.agu.org.

Wiley also publishes its books in a variety of electronic formats and by print-on-demand. Some content that appears in standard print versions of this book may not be available in other formats. For more information about Wiley products, visit our web site at www.wiley.com.

The manufacturer's authorized representative according to the EU General Product Safety Regulation is Wiley-VCH GmbH, Boschstr. 12, 69469 Weinheim, Germany, e-mail: Product_Safety@wiley.com.

Registered Office
John Wiley & Sons, Inc., 111 River Street, Hoboken, NJ 07030, USA
Editorial Office
111 River Street, Hoboken, NJ 07030, USA
For details of our global editorial offices, customer services, and more information about Wiley products visit us at www.wiley.com.

Library of Congress Cataloging-in-Publication Data applied for:

Paperback ISBN: 9781119654469

Cover Design: Wiley
Cover Image: © reklamlar/Getty Images, © bartvdd/Getty Images

Set in 9.5/12.5pt STIXTwoText by Straive, Chennai, India

To my wife Gopa for her constant support and encouragement.

Contents

Preface

Seismology is a very important geophysical exploration tool that has been utilized by Earth scientists for many years. The analysis of naturally occurring earthquakes and the understanding that the Earth is composed of the crust, mantle, and core dates back to the 1800s. The introduction of exploration seismology in the early twentieth century, particularly in the context of oil and gas exploration, marked a transformative expansion in the field of seismology. Finally, the advent of the digital era in the mid-1960s marked the birth of modern seismology. Employing digital technology to locate earthquakes around the globe became the foundational basis for contemporary concepts of plate tectonics, leading to the emergence of a new discipline—*computational seismology*.

The primary objective of computational seismology is to theoretically simulate seismic waves on a computer to create subsurface Earth models and align the computed responses with field observations to ascertain the actual subsurface rock properties. In the initial phase of the digital era, when computing technology was serial, this process was executed manually; that is, one would begin with an "initial" or "guess" model, compute theoretical responses, and manually match those responses with field observations. Subsequently, based on this match or mismatch, the model would be modified and recomputed, continuing the iterative process until a satisfactory match was achieved.

Between 1980 and 2000, advancements in computational techniques helped automate this matching procedure, leading to the emergence of another new field: seismic inversion or optimization. The application of inversion or optimization has since become routine in nearly every domain of geophysics. Recently, novel computing technologies based on artificial intelligence, broadly categorized as *machine learning*, have been developed. The applications of machine learning in geophysics are extensive, including geological interpretation of seismic sections via pattern recognition, signal-to-noise ratio enhancement in data, seismic depth imaging, and beyond. Indeed, machine learning can significantly reduce the computational resources required for large-scale seismic optimization problems.

The advancements in computational technology, coupled with recent developments in data science and artificial intelligence methodologies, enable the resolution of highly complex problems concerning the recorded seismic data. Modern seismic methods have progressed from the analysis of stacked data to prestack data, encompassing complex applications such as waveform-based inversion, depth imaging, and, most recently, the application of machine learning to resolve imaging and inverse problems. These contemporary applications necessitate (1) a thorough understanding of seismic wave theory and its efficient implementation on computers; (2) the integration of seismic theory with optimization theory to estimate subsurface geological structures, lithologies, and fluid properties; and (3) the utilization of recent advancements in machine learning to facilitate efficient implementations of these optimization methods in high-performance computing environments.

However, through my experience teaching both undergraduate and graduate seismology courses for several decades, I have observed that many students lack an understanding of the fundamentals. To address this gap, I have made earnest efforts to create this textbook, titled *Computation, Optimization, and Machine Learning in Seismology*, which adopts a comprehensive approach to seismology, encompassing fundamental concepts to the current state-of-the-art. This book provides an overview of the theory of seismic wave propagation and various methods for computing forward synthetic seismograms, before delving into optimization, inverse methods, and machine learning. The primary motivation for this textbook is to introduce students to seismology and expectations for the future of the field. Although primarily intended for senior undergraduate and graduate students, this textbook may also serve as a reference for academic and research professionals.

Following Chapter 1 (the introduction), Chapter 2 discusses the mathematics necessary for the subsequent chapters. However, a supplementary mathematical background in differential and integral calculus, partial derivatives, and linear algebra will be beneficial for maximizing the understanding gained from this book. Following this mathematical foundation, Chapter 3 addresses the fundamentals of linearized elasticity, while Chapters 4 and 5 explore the details of wave propagation, including various methods for computing synthetic seismic responses. Chapter 6 discusses the key concepts of function optimization in single and multiple dimensions, while Chapters 7, 8, and 9 sequentially examine local, global, and machine learning methodologies for solving seismic inverse problems. Finally, Chapter 10 considers the future of seismic inversion and provides concluding remarks. To reinforce the learning outcomes of each chapter, numerous examples of key concepts in practice are included, along with homework exercises at the end of Chapters 1–9.

Ideally, this book should be taught as a two-semester course on seismic wave theory and optimization, with Chapters 1–5 covered in the first semester and Chapters 6–10 in the second semester. Alternatively, this book may be utilized for a one-semester course on seismic inverse theory, with a primary emphasis on Chapters 6–10, while Chapters 1–5 serve as background reading.

Throughout my career, I have been fortunate to associate with many eminent individuals who have influenced and motivated me to write this book. Although the list is extensive, I must acknowledge a select few. First, I am deeply indebted to my graduate advisor, Neil Frazer, for introducing me to the field of seismology and for teaching some excellent courses. In fact, I have taken the liberty of incorporating many of his course notes into this book. My collaboration with Mrinal K. Sen as a colleague, fellow graduate student, and long-time friend, with whom I frequently discuss research, has been an invaluable resource. I am also grateful to all my graduate students, particularly Amit Padhi, Tao Li, Mohit Ayani, and Lingxiao Jia, whose research contributions comprise significant elements of this book. Discussions with former colleagues Steve Holbrook and Pejman Tahmesebi, as well as my current colleagues Po Chen, Dario Grana, Ken Dueker, Mike Cheadle, Vladimir Alvarado, and Morteza Dezam, have also been instrumental in writing various sections of this book. Amit Padhi and Soman Chacko reviewed the first three chapters and provided numerous constructive criticisms. Neena Ganjoo, Noel McGlinchey, and Keerthana Govindarajan from Wiley, as well as Jenny Lunn from the American Geophysical Union, were exceptionally supportive throughout this writing project, and bi-weekly meetings with Taka'aki Taira (UC Berkeley) were crucial in shaping the book into its present form. I would also like to express my gratitude to my two sons, Auritra and Abhik, and my daughter-in-law, Christie, for their encouragement. Finally, I am profoundly indebted to my wife, Gopa, for her motivation, patience, and unwavering support throughout this journey.

Subhashis Mallick
University of Wyoming, USA

Availability Statement

Synthetic seismograms shown in Figures 2.3 and 2.4 were generated using real well-log data, RSU#1 from the Rock-Springs Uplift (RSU), Wyoming, USA (Pafeng et al., 2017), and the proprietary reflectivity modeling software ANIVEC, developed by Mallick and Frazer (1987, 1988, 1990, 1991).

The input required to generate Figures 3.10 and 3.14 is provided in the table associated with Problem 3.12, and the plots shown were generated with codes that used the equations for phase velocity, group velocity, and the iterative group-to-phase transformation flow provided in Figure 3.18. The synthetic data shown in Figure 3.15b were generated from the ANIVEC modeling software (Mallick and Frazer, 1987, 1988, 1990, 1991), and the input model is shown in Figure 3.15a. The observed velocity as a function of the observed angle (Figure 3.16b) was generated using the code written for picking the time-varying reflection events in Figure 3.15b and the procedure shown in Figure 3.16a. Figure 3.17 was generated from the group-to-phase transformation flow of Figure 3.18.

The real well-log data shown in Figure 4.1 are the RSU#1 well-log data. The synthetic seismograms shown in Figure 4.18b are computed using the ANIVEC synthetic seismogram computation software (Mallick and Frazer, 1987, 1988, 1990, 1991).

Several other figures were generated by the author, who can be contacted with any queries.

References

Mallick, S., and Frazer, L.N., 1987, Practical aspects of reflectivity modeling, *Geophysics*, *52*, 1355–1364, https://doi.org/10.1190/1.1442248.

Mallick, S., and Frazer, L.N., 1988, Rapid computation of multioffset vertical seismic profile synthetic seismograms for layered media, *Geophysics*, *53*, 479–491, https://doi.org/10.1190/1.1442479.

Mallick, S., and Frazer, L.N., 1990, Computation of synthetic seismograms for stratified azimuthally anisotropic media, *Journal of Geophysical Research*, *95*, 8513–8526, https://doi.org/10.1029/JB095iB06p08513.

Mallick, S., and Frazer, L.N., 1991, Reflection/transmission coefficients and azimuthal anisotropy in marine seismic studies, *Geophysical Journal International*, *105*, 241–252, https://doi.org/10.1111/j.1365-246X.1991.tb03459.x.

Pafeng, J., Mallick, S., and Sharma, H., 2017, Prestack waveform inversion of three-dimensional seismic data – an example from the Rock Springs Uplift, *Wyoming, USA*, *Geophysics*, *82*, B1–B12, https://doi.org/10.1190/GEO2016-0079.1.

About the Companion Website

This book is accompanied by a companion website for instructors.

www.wiley.com/go/Mallick_1stedition

The website includes solutions to problems and figure PPTs.

1

Introduction to Key Concepts in Seismic Inversion and Elastic Wave Theory

1.1 Background

Geophysical inversion is a well-established technique that yields quantitative descriptions of the earth's subsurface. Inversion of measurements of earthquake waves, for example magnetic, electrical, and gravity fields, has long been used to understand the macroscopic structure of the earth. Seismic inversion is routinely used in the oil and gas industry for exploration risk assessment and for reservoir characterization. The purpose of this book is to familiarize readers with the theory behind various approaches to inversion and their common application. Our focus is on seismic applications of inversion, including methods that utilize machine learning (ML) and artificial intelligence (AI) that are now being increasingly applied in all fields of geophysics.

1.2 Seismology—A Historical Perspective

Broadly speaking, seismology is the study of the mechanical vibrations of the earth. These vibrations are generated either from natural sources such as earthquakes or from the artificial (human-generated) sources such as from the use of explosives or vibrators. Elastic waves associated with these vibrations propagate within the earth and are recorded as seismograms on various measuring devices, the most common of which are "seismometers" or "geophones." Analyzing these seismograms to estimate the elastic properties of the earth, and relating these to subsurface lithology, fluids, in situ stress fields, etc., is the primary objective of seismology.

Based on the source generating the mechanical vibrations, seismology is broadly classified into (1) *earthquake* or *passive-source* seismology and (2) *exploration* or *active-source* seismology. Irrespective of a passive or an active source, the fundamental mathematical theory for both is the propagation of the elastic waves. This book focuses on elastic wave theory and how it is used to decipher the earth's internal structure and physical properties from recorded seismograms. From a historical perspective, however, it is important to briefly introduce the above two broad categories of seismology and their respective applications.

1.2.1 Earthquake Seismology

The study of earthquakes dates back thousands of years, a comprehensive review of which can be found in Ben-Menahem (1995). Earthquakes are caused by a sudden release of energy within the subsurface, primarily due to the release of elastic strain. Analysis of the seismograms generated from earthquakes that are recorded by permanent seismometers deployed around the globe enables the determination of their location, magnitude, focal mechanism,

Computation, Optimization, and Machine Learning in Seismology, Advanced Textbook 8, First Edition. Subhashis Mallick.

and the subsurface structure of the earth. Additionally, earthquake seismology plays a major role in developing our understanding of plate tectonics. In earthquake seismology, the relative distances between the permanent seismic sensors are large (on the order of hundreds of kilometers) and the temporal frequencies of the elastic waves generated from the earthquakes are low (typically less than 1 Hz). Consequently, the vertical and lateral resolution of the earth structure obtained from the analysis of these seismograms is low (tens to hundreds of kilometers). The depth of penetration of these waves is, however, high, and therefore earthquake seismology is useful for studying the structure of earth's deep interior. A very good account of the science of earthquake seismology can be found in Bath (1979), Stein and Wysession (2002), Shearer (2009), among others.

Finding the structure of the earth's deep interior, developing the concepts of plate tectonics, etc., are the fundamental focus of earthquake seismology. In the past few decades, the oil and gas industry has made an interesting adaptation of the concepts of earthquake seismology in a technique known as "microseismic monitoring." Where hydrocarbon-bearing reservoirs are impermeable, the reservoir fluids are squeezed out by fracturing under great pressure. In microseismic monitoring, the microearthquakes (microseisms) resulting from such fracturing are recorded by borehole seismic sensors and analyzed to characterize the distribution, extent, size, and nature of the induced fractures. This helps to optimize well placement and design for efficient drainage of the reservoir hydrocarbons (Grechka and Heigl, 2017). Microseismic monitoring is also used in carbon dioxide sequestration and steam injection related applications.

1.2.2 Exploration Seismology

Exploration seismology evolved from earthquake seismology. Here, the source generating elastic waves is artificially excited. Compared with the earthquake seismology, in exploration seismology the seismic sensors (usually geophones on land and hydrophones in marine environments) are placed at relatively short distances apart (50 m or even less), and the temporal frequencies of the elastic waves are high with a dominant frequency between 20 and 50 Hz. The lateral and vertical resolution of the subsurface structures typically resolved from exploration seismology is therefore higher than earthquake seismology. However, because high frequencies rapidly attenuate with depth, the depth of penetration in exploration seismology is much lower than for earthquake seismology, and its use is limited to finding detailed structures at shallow depths, usually less than 10 km.

In exploration seismology, artificial sources are placed at or the near surface of the earth or within a borehole. On land, the seismic sensors (usually geophones) are placed either at the surface or within a borehole. In marine environments on the other hand, these sensors (usually hydrophones) are placed either near the water surface or at the water bottom, or within a borehole below the ocean floor. Irrespective of the land or marine environment, the elastic or acoustic wavefields, excited at the source locations propagate within the earth layers. The propagating wavefield, characterized by a temporal seismic wavelet, is reflected at boundaries between subsurface layers of contrasting elastic or acoustic impedance. The sequence of reflections is recorded as seismograms at the receiver (sensor) locations. Analyzing these recorded seismograms to interpret subsurface structure and rock properties is the focus of exploration seismology (Dobrin and Savit, 1988; Telford et al., 1990). Of all the different types of geophysical exploration methods, exploration seismology is by far the most commonly used and the most successful in terms of accuracy, depth of penetration, and subsurface resolution (Selley and Sonnenberg, 2014).

1.3 Mathematical Foundations of Seismology

Following the discovery of Hooke's law of linearized elasticity in 1660, there have been many advances in the mathematics of elastic wave theory, a comprehensive account of which can be found in the classic textbook written by Love (1892, reprinted 1944), and in more recently published textbooks by Aki and Richards (2002) and Chapman (2004).

Before the digital era, early applications of seismology were limited to the analysis of wave arrival travel times in analog records of seismic data. With the advent of computers these travel-time-based analyses gradually evolved into more quantitative analyses of the recorded seismic waveforms (Chapman, 2004).

With current advances in high-performance computing, seismology has developed into a more mature science. Complex partial differential equations that describe the propagation of seismic wave fields can now be numerically solved in reasonable timeframes. This, in turn, allowed going beyond the travel-time-based qualitative analyses into quantitative analyses using sophisticated optimization techniques. More recently, the use of ML methods, a subset of AI, has been introduced in all aspects of seismology for these analyses. The primary focus of this book is to provide an overview of these quantitative methods.

1.4 Seismic Inversion

Seismic optimization is a subset of a broad category, called seismic inversion. Therefore, before understanding the meaning of optimization, it is first necessary to understand the meaning of inversion by explaining some fundamental concepts.

1.4.1 The Meaning of Inversion

Many applications in science, engineering, economics, etc., require finding an optimal model that satisfies a given set or sets of observations. Finding such an optimal model from a set or sets of observations is called inversion. To understand what is meant by inversion, consider the scenario shown in Figure 1.1.

As shown in Figure 1.1, if a person kicks a soccer ball so that it travels with a speed v, the distance d the ball will travel in a time t is given by

$$d = vt. \tag{1.1}$$

To obtain the speed v with which the ball travels, we can measure the distance d that it travels. Additionally, we can also use a stopwatch and measure the time t that the ball takes to travel the distance d. Thus, given the set of observation $\{d, t\}$, it is straightforward to find the speed v as

$$v = \frac{d}{t}. \tag{1.2}$$

The above example is a simple form of inversion in which the "physics of the problem" given as $d = vt$ is inverted to find the model (v) from a set of observations $\{d, t\}$.

Note that the above inverse problem has the following elements:

1) A set of observations, also called the "observed data." In the above example, this is $\{d, t\}$.
2) The physics of the problem. In the above example, it is $d = vt$.

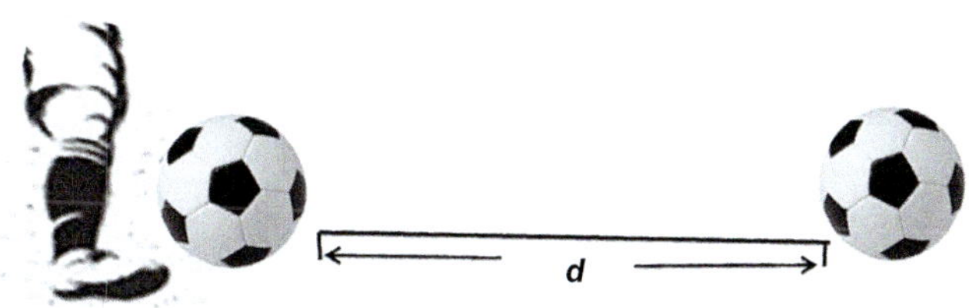

Figure 1.1 Example of a person kicking a soccer ball.

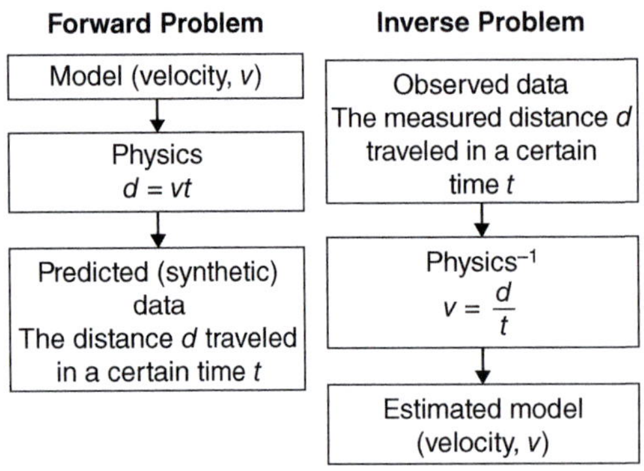

Figure 1.2 The forward and inverse problems with reference to the soccer-ball example of Figure 1.1.

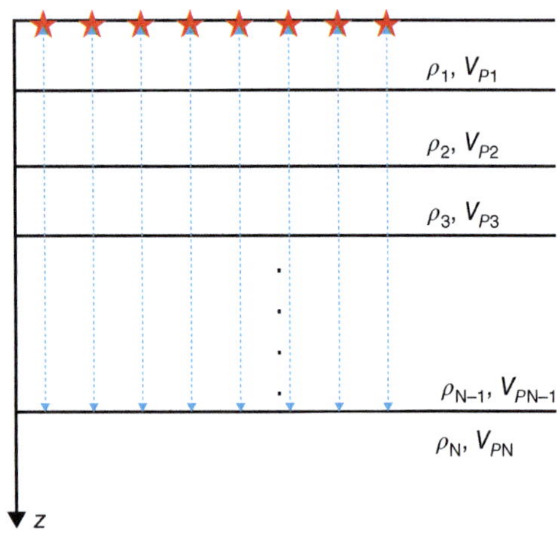

Figure 1.3 A schematic representation of an N-layered earth model. For any layer i, the P-wave velocity and density are represented respectively as V_{P_i} and ρ_i. The seismic waves originating from each surface source location travel vertically, get reflected from the different interfaces, and are recorded at the receiver locations that coincide with the source locations. The source and receiver locations are denoted by red asterisks.

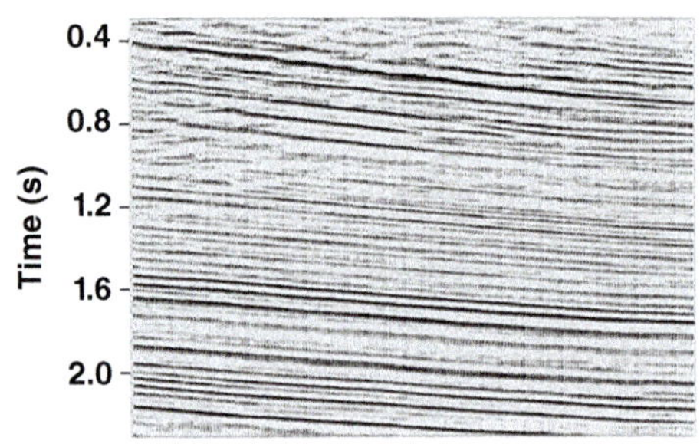

Figure 1.4 Example of poststack seismic data. Events represent normal-incidence reflections from interfaces between different subsurface layers recorded in two-way travel time.

Consequently, the inversion has two distinct components: (1) the forward problem and (2) the inverse problem, as shown in Figure 1.2 with reference to the soccer-ball example.

Given a model (the speed v in the soccer-ball example) and the physics $d = vt$, the forward problem is computing the data, i.e., the distance traveled in the given time. Because here the data are computed assuming a speed v, they are called the synthetic or predicted data. Given this forward problem, the corresponding inverse problem is to estimate the model (i.e., the speed v) from the set of observation. To estimate the model, the forward problem must be inverted. For the soccer-ball example, such an inverse physics is simply $v = \frac{d}{t}$, and the solution to the problem is straightforward. For geophysics including seismic, the problem is, however, not as simple and requires further elaboration.

1.4.2 Seismic Problems

Going past the soccer-ball example, let us now consider some seismic examples where direct inversion of physics can be similarly performed. Our first such example is poststack or normal-incidence seismic inversion, as illustrated in Figures 1.3–1.5. In this example, we consider a subsurface model of N layers, with P-wave velocities $V_{P1}, V_{P2}, \dots, V_{PN}$ and densities $\rho_1, \rho_2, \dots, \rho_N$ (Figure 1.3).

Poststack seismic data are a record of normal-incidence reflections from interfaces between layers in the subsurface, as a function of two-way time (Figure 1.4). For an interface between two given layers, say k and $k+1$, the normal-incidence reflection coefficient R_k is given as (Aki and Richards, 2002)

$$R_k = \frac{\Delta Z_k}{2\overline{Z}_k} \tag{1.3}$$

with

$$\Delta Z_k = Z_{k+1} - Z_k = \rho_{k+1} V_{Pk+1} - \rho_k V_{Pk} \tag{1.4}$$

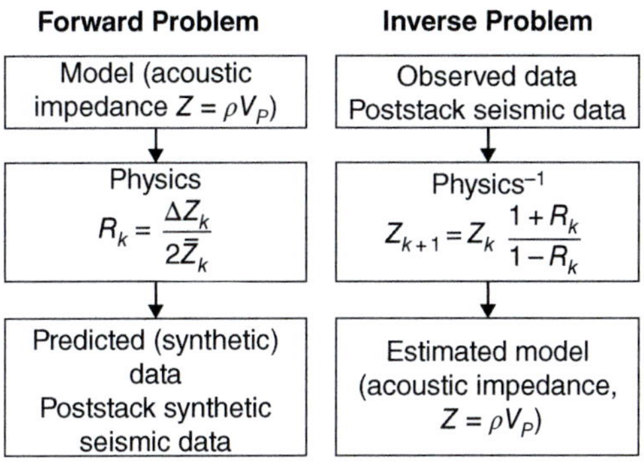

Figure 1.5 Forward and inverse problems for poststack seismic inversion.

and

$$\bar{Z}_k = \frac{Z_{k+1} + Z_k}{2} = \frac{\rho_{k+1} V_{Pk+1} + \rho_k V_{Pk}}{2}. \tag{1.5}$$

Poststack seismic inversion uses the physics of the forward problem given by equation 1.3 to estimate the acoustic impedance (Z_{k+1}) from poststack seismic data. From equations 1.3–1.5, we can write

$$Z_{k+1} = Z_k \frac{1 + R_k}{1 - R_k}. \tag{1.6}$$

Poststack inversion estimates the acoustic impedance using the recursive relation given by equation 1.6. Assuming that the seismic data are processed and calibrated such that reflection amplitudes correspond to normal-incidence reflection coefficients (equation 1.3), equation 1.6 is recursively used from the top to down for estimating the acoustic impedances of all layers. From a prior knowledge of $Z1$ (acoustic impedance of the top layer) and $R1$ (the reflection coefficient of the first interface recorded on the seismic data), equation 1.6 is used to estimate $Z2$. From $Z2$ and $R2$, $Z3$ is then estimated, and so on until we reach the end of the data. When this operation is performed for all receiver locations, the result is an acoustic impedance section (2D seismic data) or volume (3D seismic data). It must, however, be noted that although the basic physics for poststack seismic inversion is given by equations 1.3–1.6, the process is not as simple. First, the seismic data must be carefully processed such that reflection amplitudes are representative of normal-incidence reflection coefficients. This is achieved by calibrating the seismic data to well data via well-to-seismic ties. This in turn allows for the estimation of the seismic wavelet and the deconvolution of the seismic data with the estimated wavelet such that the data amplitudes represent reflection coefficients. The background acoustic impedance trend over the entire seismic data volume must also be estimated using well data. Because these are beyond the scope of this book, we do not address them here. Readers may refer to Oldenburg et al. (1983) for computational details of poststack inversion. Figure 1.5 shows a schematic illustration of the forward and inverse problems for poststack seismic inversion.

Let us now move from poststack to a class of prestack seismic inversion, popularly known as amplitude-variation-with-angle (AVA) inversion. In the seismic industry, AVA inversion is routinely used for subsurface characterization (Russell, 1988; Hampson, 1991). As illustrated in Figure 1.6, for prestack data the source and receiver locations are not coincident, and consequently, the recorded seismic reflections are functions of the source-to-receiver offset (or the angle of incidence to be specific) and time. Recorded traces from different source and receiver locations are sorted into the common midpoint (CMP) domain as shown in Figure 1.6 such that their midpoints are common, and there is a one-to-one relationship between offset distance and the angle of incidence at each subsurface layer interface. This yields a group of traces called a CMP gather (leftmost panel, Figure 1.7), in which reflections from different interfaces in the subsurface are represented at each CMP location as a function of both source-to-receiver offset and time. For AVA inversion, these offset domain seismic data are first compensated for the geometrical spreading loss and corrected for the normal moveout (NMO). Geometrical spreading compensation (GSC) is correction applied to the loss of the seismic data amplitudes for the spherical divergence,

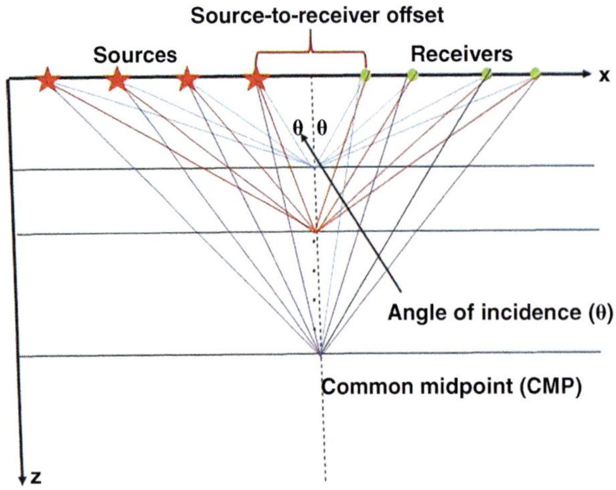

Figure 1.6 Illustration of CMP domain sorting for prestack seismic data.

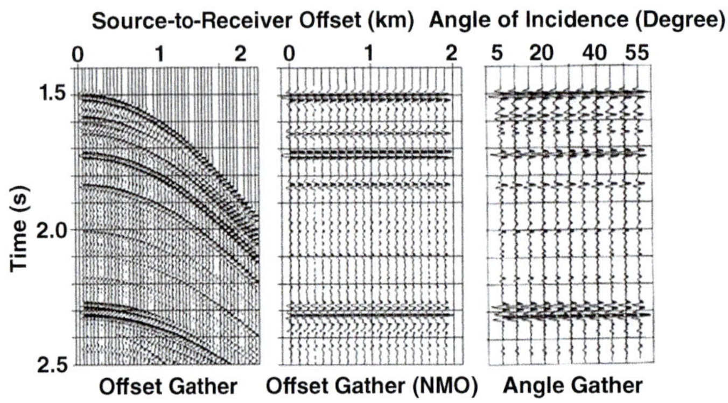

Figure 1.7 CMP-sorted offset domain prestack seismic data without and with NMO correction, and the corresponding computed angle gather.

and NMO correction is the correction that is applied to the prestack seismic data to their corresponding zero-offset two-way times. The procedure for both GSC and NMO corrections can be found in any seismic data processing book (for example, see Yilmaz, 1989). The spreading and NMO-corrected offset domain prestack (CMP) gather corresponding to the gather in the leftmost panel in Figure 1.7 is illustrated in the middle panel of the same figure. Because AVA inversion works in the angle-of-incidence domain, the next step in AVA inversion is to convert the NMO-corrected offset domain prestack data into the angle domain. This is performed by an offset-to-angle transformation as described by Todd and Backus (1985), Resnick (1993), and Mukhopadhyay and Mallick (2011). Offset-to-angle transformation converts the seismic data such that reflections in each trace in the resultant angle gather correspond to fixed incidence angles (θ) and time instead of fixed offsets and time, as illustrated in the rightmost panel of Figure 1.7.

Reflection amplitudes in the angle-domain prestack seismic data are assumed to represent plane-wave reflection coefficients for the given angle of incidence. AVA inversion uses the plane-wave reflection coefficient formula as the "physics of the forward problem." While the exact expression of the plane P-wave reflection coefficient is complex (Zoeppritz, 1919; Mallick, 1993; Aki and Richards, 2002), approximations to it suitable for practical application come from the pioneering work by Bortfeld (1961) and later from Richards and Frasier (1976), Shuey (1985), among others. These approximate formulations are sufficiently accurate for small to moderate

(~40–45°) incidence angles, and where contrasts between elastic properties across layer interfaces are not large.

Consider an interface separating two media with different elastic properties. Also consider that the P- and S-wave velocity and density of the upper medium are given respectively as V_{P1}, V_{S1}, and ρ_1, and those for the lower medium are given as V_{P2}, V_{S2}, and ρ_2. The plane wave P–P reflection coefficient R_{PP} as a function of the average angle of incidence θ (i.e., the average of the incidence and the transmission angles) can be expressed approximately as (Aki and Richards, 2002)

$$R_{PP} \approx A + B\sin^2\theta + C\sin^2\theta\tan^2\theta. \tag{1.7}$$

In equation 1.7, A is the AVA intercept,

$$A = \frac{1}{2}\left(\frac{\Delta V_P}{V_P} + \frac{\Delta\rho}{\rho}\right), \tag{1.8}$$

B is the AVA gradient,

$$B = \frac{1}{2}\frac{\Delta V_P}{V_P} - 2\left(\frac{V_S}{V_P}\right)^2\left(2\frac{\Delta V_S}{V_S} + \frac{\Delta\rho}{\rho}\right), \tag{1.9}$$

and C is the AVA curvature,

$$C = \frac{1}{2}\frac{\Delta V_P}{V_P}. \tag{1.10}$$

Also, in equations 1.8–1.10, ΔV_P, ΔV_S, and $\Delta\rho$ are the contrasts, given as

$$\Delta V_P = V_{P2} - V_{P1}, \tag{1.11}$$

$$\Delta V_S = V_{S2} - V_{S1}, \tag{1.12}$$

$$\Delta\rho = \rho_2 - \rho_1, \tag{1.13}$$

and V_P, V_S, and ρ are the average properties, given as

$$V_P = \frac{V_{P1} + V_{P2}}{2}, \tag{1.14}$$

$$V_S = \frac{V_{S1} + V_{S2}}{2}, \tag{1.15}$$

and

$$\rho = \frac{\rho_1 + \rho_2}{2}. \tag{1.16}$$

Before formulating the AVA inverse problem, let us first investigate the accuracy of the approximate reflection coefficient formula given by equation 1.7 in which A, B, and C are defined by equations 1.8–1.16. This is demonstrated by a simple example shown in Figure 1.8. Figure 1.8a on the left represents an interface between two elastic solids where V_{P1}, V_{S1}, ρ_1, V_{P2}, V_{S2}, and ρ_2 have the same meaning as above with the units for V_{P1}, V_{S1}, V_{P2}, and V_{S2} being in meters/second (m/s) and those for ρ_1 and ρ_2 being in grams/cubic-centimeter (g/cm^3). Figure 1.8b on the right is the computed reflection coefficient for the interface using both exact reflection coefficient the interface (solid curve) and approximate reflection coefficient formula (dashed curve), with the reflection coefficient (R_{PP}) as the vertical axis and $\sin^2\theta$ as the horizontal axis. Note that the exact and approximate reflection coefficients are very close to one another up to $\sin^2\theta = 0.75$ which corresponds to an incidence angle $\theta = 60°$, demonstrating the fact that the approximate formula of equation 1.7 is reasonably accurate. The intercept of the reflection coefficient curve on the vertical ($\theta = 0°$; $\sin^2\theta = 0$) axis is the AVA intercept A. It can be also seen from Figure 1.8b that for relatively small incidence angles ($\theta \leq 30°$ or $\sin^2\theta \leq 0.25$), R_{PP} is almost linear with $\sin^2\theta$ and the AVA gradient B is the slope of this linear behavior and is given as $\Delta R_{PP}/\Delta(\sin^2\theta)$. Finally, for $\theta > 40°$, $\sin^2\theta > 0.41$, the measure of the departure from the linear behavior of R_{PP} versus $\sin^2\theta$ is dictated by the AVA curvature C. The objective of the AVA inversion is to fit equation 1.7 to the reflection coefficients, estimate A, B, C, and then use equations 1.8–1.16 and other information to estimate V_P, V_S, and ρ, which are discussed later.

Equations 1.7 through 1.16 provide the fundamental basis for the "physics of the forward problem" for AVA inversion. Consider the angle-domain seismic data shown in Figure 1.9, where the reflection amplitudes for the first reflection in the data are picked along the red dashed line. If there are a total of n incidence angles, assume

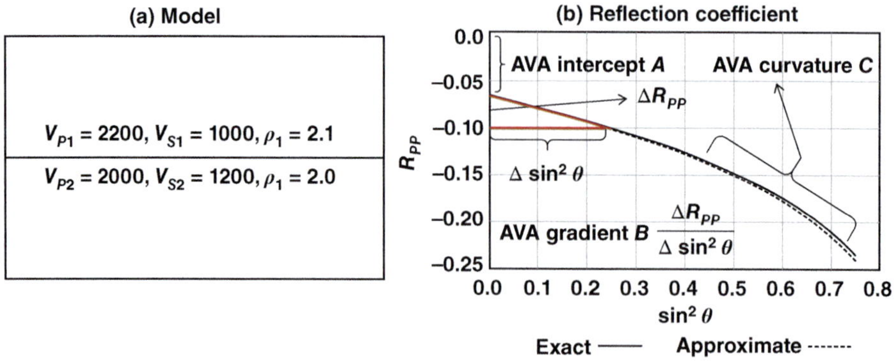

Figure 1.8 Example of the plane P-wave reflection coefficient R_{PP}. (a) The model showing the interface between two isotropic elastic layers with different properties. V_{P1}, V_{P2}, V_{S1}, and V_{S2} are in m/s while ρ_1 and ρ_2 are in g/cm^3. (b) Computed R_{PP} for the model shown in (a) using the exact formula (solid line) and the approximate formula (dashed line) given by equations 1.7–1.16 demonstrating that the approximate formula is very close to the exact formula. The AVA intercept (A) is value of the intercept of the reflection coefficient curve in the vertical (R_{PP}) axis. For incidence angles less than 30°, R_{PP} varies linearly with $\sin^2\theta$, and the slope of this liner behavior is the AVA gradient (B). Finally, the departure from the linear behavior for large incidence angles is controlled by the AVA curvature (C).

that these reflection amplitudes are given as $d_{11}, d_{21}, d_{31}, \dots, d_{n1}$. If these n incidence angles are represented by $\theta_1, \theta_2, \theta_3, \dots, \theta_n$, then assuming that the seismic data are processed to preserve amplitudes, $d_{11}, d_{21}, d_{31}, \dots, d_{n1}$ represent the reflection coefficients for angles $\theta_1, \theta_2, \theta_3, \dots, \theta_n$. So, from the AVA formula (equation 1.7), we can write

$$\begin{aligned} d_{11} &= A_1 + B_1 \sin^2\theta_1 + C_1 \sin^2\theta_1 \tan^2\theta_1, \\ d_{21} &= A_1 + B_1 \sin^2\theta_2 + C_1 \sin^2\theta_2 \tan^2\theta_2, \\ &\vdots \\ d_{n1} &= A_1 + B_1 \sin^2\theta_n + C_1 \sin^2\theta_n \tan^2\theta_n, \end{aligned} \tag{1.17}$$

with A_1, B_1, and C_1 being the AVA intercept, gradient, and curvature corresponding to the first reflection marked by the red dashed line in Figure 1.9.

Equation 1.17 can be written as

$$\begin{pmatrix} d_{11} \\ d_{21} \\ \vdots \\ d_{n1} \end{pmatrix} = \begin{pmatrix} 1 & \sin^2\theta_1 & \sin^2\theta_1 \tan^2\theta_1 \\ 1 & \sin^2\theta_2 & \sin^2\theta_2 \tan^2\theta_2 \\ \vdots & \vdots & \vdots \\ 1 & \sin^2\theta_n & \sin^2\theta_n \tan^2\theta_n \end{pmatrix} \begin{pmatrix} A_1 \\ B_1 \\ C_1 \end{pmatrix}. \tag{1.18}$$

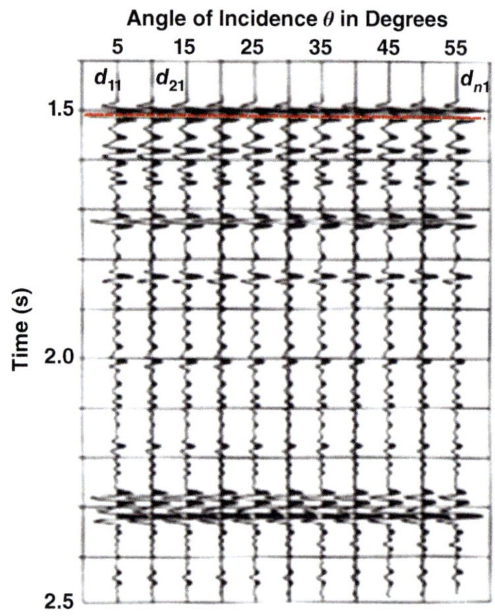

Figure 1.9 Illustration of the AVA inversion problem.

Equation 1.18 represents the seismic data at a given time sample for all angles of incidence. Thus, if there are a total of m data samples, we can easily extend equation 1.18 to represent all data samples (that is all reflection on the data) and write

$$\mathbf{d} = \mathbf{Gm}, \tag{1.19}$$

with

$$\mathbf{d} = \begin{pmatrix} d_{11} & d_{12} & \cdots & d_{1m} \\ d_{21} & d_{22} & \cdots & d_{2m} \\ \vdots & \vdots & \cdots & \vdots \\ d_{n1} & d_{n2} & \cdots & d_{nm} \end{pmatrix}, \tag{1.20}$$

$$\mathbf{G} = \begin{pmatrix} 1 & \sin^2\theta_1 & \sin^2\theta_1 \tan^2\theta_1 \\ 1 & \sin^2\theta_2 & \sin^2\theta_2 \tan^2\theta_2 \\ \vdots & \vdots & \vdots \\ 1 & \sin^2\theta_n & \sin^2\theta_n \tan^2\theta_n \end{pmatrix}, \tag{1.21}$$

and

$$\mathbf{m} = \begin{pmatrix} A_1 & A_2 & \cdots & A_m \\ B_1 & B_2 & \cdots & B_m \\ C_1 & C_2 & \cdots & C_m \end{pmatrix}. \tag{1.22}$$

Each column of the $3 \times m$ matrix $\mathbf{m}$ (equation 1.22) represents the AVA intercept, gradient, and curvature of the given time sample of the seismic data. Equation 1.19 shows a classical least-squares inversion problem, and the estimated model $\mathbf{m}^{est}$ can be obtained as

$$\mathbf{m}^{est} = (\mathbf{G}^T\mathbf{G})^{-1}\mathbf{G}^T\mathbf{d}. \tag{1.23}$$

In equation 1.23, the superscript T represents a matrix transpose and the superscript -1 represents a matrix inverse.

While equation 1.23 provides a pure least-squares solution for the AVA inversion, in practice inverting the matrix on the right-hand side of equation 1.23 can be unstable and is mitigated in a damped least-squares approach by adding a small term on the diagonal of the matrix to be inverted. This damped least-squares solution is given as

$$\mathbf{m}^{est} = (\mathbf{G}^T\mathbf{G} + \varepsilon^2\mathbf{I})^{-1}\mathbf{G}^T\mathbf{d}. \tag{1.24}$$

In equation 1.24, $\mathbf{I}$ is the identity matrix of the same size of $\mathbf{G}^T\mathbf{G}$ and ε is a small number suitably chosen to stabilize matrix inversion. Note that the model $\mathbf{m}$ consists of the AVA intercept, gradient, and curvature for each time (i.e., depth) sample. Therefore, once $\mathbf{m}$ is estimated, equations 1.9–1.16 can be used to estimate the subsurface model of V_P, V_S, and ρ. A schematic illustration of the forward and inverse problem for the AVA inversion is shown in Figure 1.10.

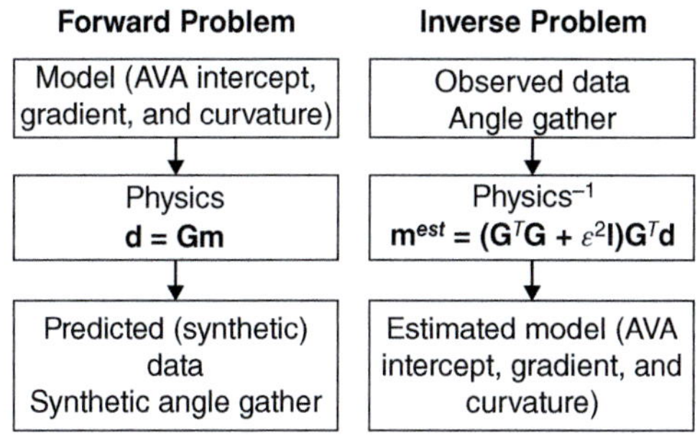

Figure 1.10 The forward and inverse problems for the AVA inversion.

1.4.3 Operator-Based and Model-Based Inversions—The Concept of Optimization

A variety of geophysical inversions such as poststack and prestack AVA-based seismic inversions can be performed by directly inverting the physics of the forward problem, like the soccer-ball example. These methods invert the underlying physics and apply the inverse operator to the data to estimate the model and are therefore called the *operator-based* inversions. Even though such operator-based inversions are straightforward to implement, they are usually not stable. Geophysical data are seldom noise-free, and applying operator-based inversions to noisy data makes the inversion unstable and inaccurate (Singh et al., 1989; Ziolkowski et al., 1989). Additionally, while such operator-based methods can be applied to solve problems like post-stack and AVA inversions, they cannot be used to solve more complex inversion problems. For example, although AVA inversion is the current state of the art in the seismic industry, the fundamental theory behind AVA inversion assumes that the recorded prestack seismic data contain only primary P wave reflections and are not contaminated by complex effects of wave propagation. In a modeling study, Mallick (2007) demonstrated that such an assumption may be true for small

(<25–30°) incidence angles. For Incidence angles >25–30°, complex wave effects, such as interbed multiple reflections, mode conversions, ray bending, etc., seriously affect the P-wave reflections, and the assumptions made in AVA inversion become increasingly invalid. Even at small incidence angles, assuming multiple uncontaminated primary reflections is seldom true in hard-rock and carbonate environments with strong impedance contrasts. The seismic industry is aware of this limitation, and AVA inversion results are always interpreted in conjunction with well data and other constraining geological, petrophysical, and rock-physics information (Mallick and Adhikari, 2015). Given these long-recognized limitations of AVA inversion, and with the increasing availability of high-performance computers, the industry is now moving toward wave-equation-based methods for inverting prestack seismic data. The forward and the corresponding inverse problems for such wave-equation-based inversions are illustrated in Figure 1.11.

Inverting the physics in the wave-equation-based method is not as simple and straightforward as in poststack or AVA inversions. It requires starting with an initial or guess model, using the physics of the forward problem to compute the synthetic data, matching this with the observed data, and iteratively modifying the model until the synthetic data computed from the model matches the observed data to a user-defined accuracy. This class of inversions is based on the *optimization methods*, which, in the context of seismology, is called *model-based inversion*. The primary objective of this book is to discuss different flavors of model-based inversion methods in seismology. Strictly speaking, the model-based inversions are a subset of the optimization methods. In this book, however, the terms optimization methods and model-based inversions are interchangeably used.

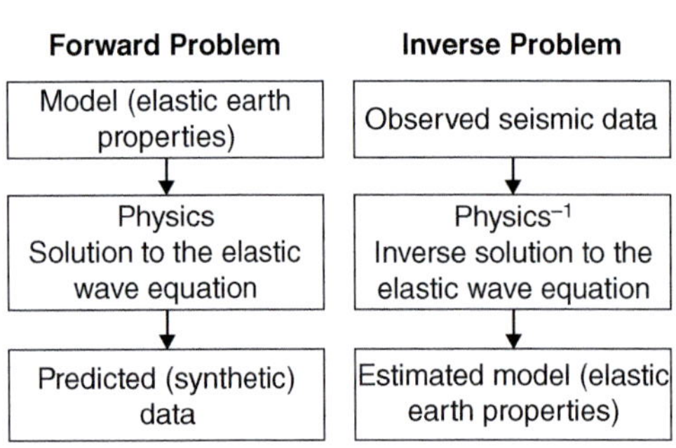

Figure 1.11 The forward and inverse problems for wave-equation-based inversion.

1.4.4 Fundamental Concepts of the Optimization Method (Model-Based Inversion)

Starting again with the soccer-ball example (Figure 1.1), we now try to solve the inverse problem differently and develop the fundamental concepts of optimization. Reiterating the forward problem: if a person kicks a soccer ball with a speed v, the distance d the ball will travel in a time t is given by $d = vt$. Given the set of observations $\{d, t\}$, and directly using the physics $d = vt$, instead of the physics^{-1} $v = d/t$, we will now solve the problem. Starting with an initial guess of the speed, say v', we use the physics to compute the "synthetic data" as

$$d' = v't. \tag{1.25}$$

Of course, the calculated distance d' is not going to be the same as the observed distance d unless we are very lucky in guessing $v' = v$. We now define a function f as

$$f = 1 - \frac{|d - d'|}{d + d'}. \tag{1.26}$$

Note that f is 1 only when $d' = d$, else it is less than 1. In optimization or model-based inversion, we keep choosing different values of v', use equation 1.25 to compute d' and then use equation 1.26 to compute f. Because f attains a maximum value of 1 at the true value of v, by plotting the computed f for different v' and choosing the value of v' where f attains a maximum, we obtain the solution to the optimization problem, as illustrated in Figure 1.12.

In the example shown in Figure 1.12, the problem is cast as maximizing f. It is also possible to differently define f and cast the optimization as a minimization problem,

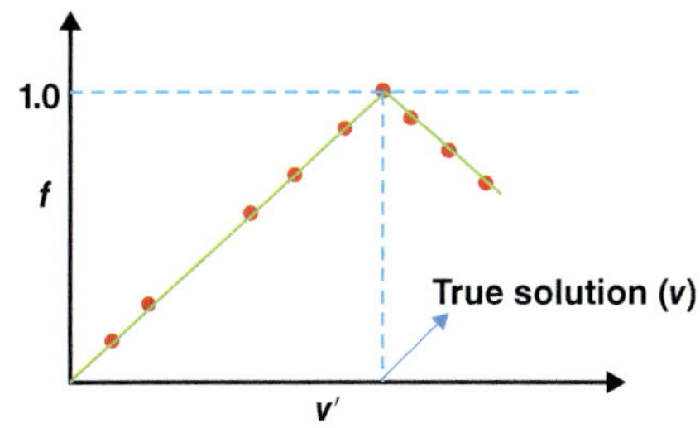

Figure 1.12 Illustration of the model-based inversion (optimization) solution with reference to the soccer-ball example.

which is left as an exercise at the end of this chapter.

In optimization problems, the function f is called the *objective function* or simply the *objective*, and depending upon how it is defined, an optimization is equivalent to finding the maximum or the minimum of the objective.

1.5 Model, Data, and Objective Spaces

There are three distinct components in all optimization problems: (1) the model, (2) the data, and (3) the objective. The model and the objective are related to one another via the data and the underlying physics of the forward problem. In optimization, the model or the decision space is usually denoted as **X**. Although in the soccer-ball example above, the model is a scalar-valued quantity, in general it is not. In seismic inverse problems under an isotropic elastic assumption for example, the model consists of the P-wave velocity (V_P), S-wave velocity (V_S), and density (ρ) values at each depth sample in the subsurface. For an anisotropic elastic inversion, the model will contain anisotropic parameters in addition to V_P, V_S, and ρ. For electrical, magnetic, and gravity inverse problems, the model will consist respectively of electrical resistivities, magnetic susceptibilities, and densities. We thus define the model or decision space as

$$\mathbf{X} = [\mathbf{x}_1, \mathbf{x}_2, \cdots, \mathbf{x}_N]^T, \tag{1.27}$$

in which each $\mathbf{x}_i, i = 1, 2, \ldots, N$ is a vector-valued quantity (V_P, V_S, ρ, …, etc.).

Similarly, the data space is defined as

$$\mathbf{D} = [\mathbf{d}_1, \mathbf{d}_2, \ldots, \mathbf{d}_J]^T, \tag{1.28}$$

with each $\mathbf{d}_j, j = 1, 2, \ldots, J$ being a vector-valued quantity representing the observed data.

Finally, the objective space is defined to be

$$\mathbf{Y} = [y_1, y_2, \ldots, y_M]^T, \tag{1.29}$$

in which each $y_k, k = 1, 2, \ldots, M$, is a scalar-valued quantity representing the objective of each data component. Although at the beginning, we deal with single-objective optimization where the objective space consists of a single value, at the end we discuss multi-objective optimizations (MOO) where the objective space has multiple values. Additionally, note that in all optimizations there is a direct mapping of the model (decision) space onto the objective space $\mathbf{X} \Rightarrow \mathbf{Y}$ via the physics of the forward problem and the data space **D**. For the soccer-ball example above, this mapping is performed via equations 1.25 and 1.26.

1.6 Different Flavors of Optimization

In the soccer-ball example above, we formulated the optimization as follows:

1) Start with an initial guess model (speed v').
2) Compute the synthetic data using the physics of the forward problem, $d' = v't$.
3) Compute the objective, $f = 1 - \frac{|d-d'|}{d+d'}$.
4) Modify v' and repeat steps 1 through 3 until the maxima of the objective f is obtained as shown in Figure 1.11.

All optimization methods follow steps 1 through 4, but we have not yet discussed how to guess the initial model and take the subsequent steps to modify the model to drive the objective to its maximum (or minimum). For the soccer-ball example, we may exhaustively search the model space by keeping on guessing the model v', computing the forward problem, and computing the objective until we find the optimum point. For seismic optimization, however, this is not as simple. The primary reason for this is that the forward problem is much more complex than $d' = v't$, and computing the predicted data and evaluating the objective is computationally challenging for most seismic and other geophysical optimization problems. Additionally, the model space in these problems is so large that it is impossible to exhaustively search the entire model space like we can do for the soccer-ball example. Consequently, starting with a good initial model and finding the optimum point with as few objective evaluations as possible is a primary goal for these optimization methods.

1.6.1 Local (Gradient-Based) Optimization

Casting the optimization as maximizing the objective and restricting the analysis to just a single model parameter, one way of choosing the initial model and iteratively modifying it until the final (true) model is estimated is illustrated in Figure 1.13. As discussed in detail later, in many geophysical optimization problems it is usually possible to use various constraints and estimate an initial model that is relatively close to the true model. Such an initial model is represented by the black circle in the objective space in Figure 1.13. Starting with this initial model, we compute the objective f. In addition, we also compute

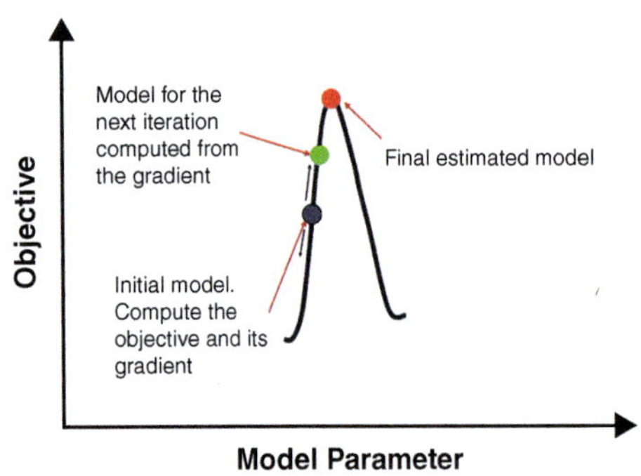

Figure 1.13 Illustration of the gradient-based optimization in one dimension.

the gradient of the objective, i.e., we compute $df/dm \approx \Delta f/\Delta m$ for a suitably chosen step size Δm. Assuming that the shape of the objective function is approximately linear within the chosen step size from a starting model, say m_0, model m_1 for the next iteration would then be

$$m_1 = m_0 + \left(\frac{\Delta f}{\Delta m}\right)_{m=m_0} \Delta m. \tag{1.30}$$

This new model is shown as a green circle in Figure 1.13. From m_1, we then obtain m_2 and continue the process until the true model (m_{true}) shown as a red circle in Figure 1.13 is estimated.

The method illustrated above is an effective optimization approach. However, such methods assume that the objective has a well-defined and unique peak (or trough if cast as minimization). It also assumes that the shape of the objective is approximately known such that the step size (Δm) can be suitably chosen. Because the model space is searched by these methods using the gradient of the objective, they are called gradient-based optimization methods. Additionally, because these methods tend to converge to the peak or trough closest to the initial model, they are also called "local optimizations" (Tarantola, 2005). Local optimizations have been widely used in seismic inverse problems such as depth imaging (Sava and Hill, 2009), AVA inversion (Russell, 1988; Hampson, 1991, among others), waveform inversion (Pratt, 1999; Sen and Roy, 2003; Guasch et al., 2012; Liu et al., 2012; Warner et al., 2013, among others), etc. Due to their computational efficiency, they are still the most popular methods for solving geophysical (including seismic) inverse problems. We discuss different flavors of local optimization methods in Chapters 6 and 7.

1.6.2 Global Optimization

Although gradient-based optimizations are compute-efficient, for many geophysical optimizations, the shape of the objective function is not as simple as illustrated in Figure 1.13, and it may look like the one shown in Figure 1.14.

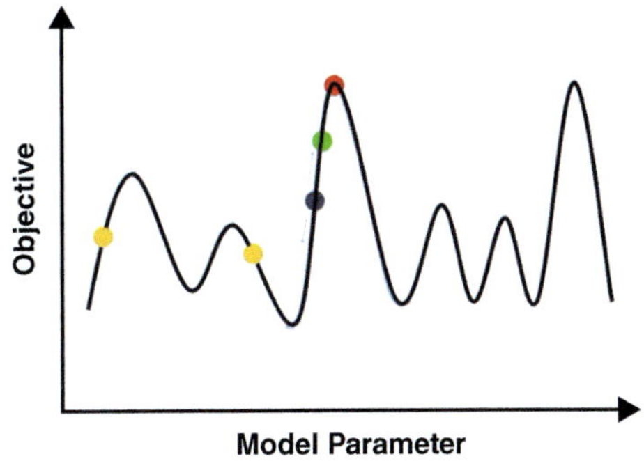

Figure 1.14 Illustration of a scenario where a local-gradient-based optimization may fail to converge.

In such a case, choosing an initial model like the ones represented by the yellow circles in Figure 1.14 would cause convergence to an incorrect solution.

To circumvent the issue of convergence to local optima, another class of optimizations has been developed which does not make any assumption on the shape of the objective function. These methods are based on random search with some directivity in the model space as opposed to the local methods, and they belong to the class of global optimization.

In global optimization, a random set of initial models within a user-defined search window is generated and their respective objectives are evaluated as represented by the black circles in Figure 1.15. After generating this initial set of models and computing their objectives, a selection criterion, shown as the green dashed line in Figure 1.15, is set as follows:

1) The models whose objectives are above the green line will be chosen with a high probability in the next step of iteration.
2) The models whose objectives are below the green line will be chosen with a low probability in the next iteration.

Like local optimization, here also we have cast the problem as a maximization problem. The above selection process is likely to choose more models above the green line and less below it. After this process, an interpolation is made between the selected models to generate new models, shown as red circles in Figure 1.16.

Once these new models are generated and their objectives are evaluated, a new selection criterion is chosen which is above the previous selection criterion and is shown as the red dashed line in Figure 1.16. Models are again selected using this new selection criterion, and the selected models are again interpolated to generate another set of new models (Figure 1.17). This process is repeated multiple times to exhaustively sample the model space such that the entire objective is evaluated (Figure 1.18). Note that these methods do not make any assumption of the shape of the objective function, nor do they depend upon the choice of the initial models. They attempt to evaluate the entire objective and are therefore called "global optimization"

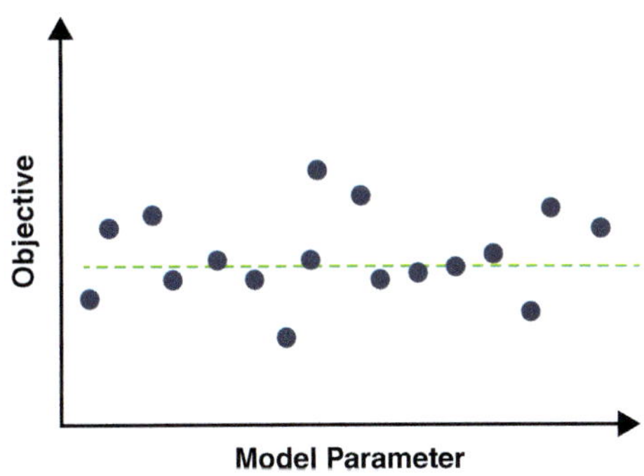

Figure 1.15 Illustration of global optimization—step 1: generation of the initial set and the choice of the selection criterion.

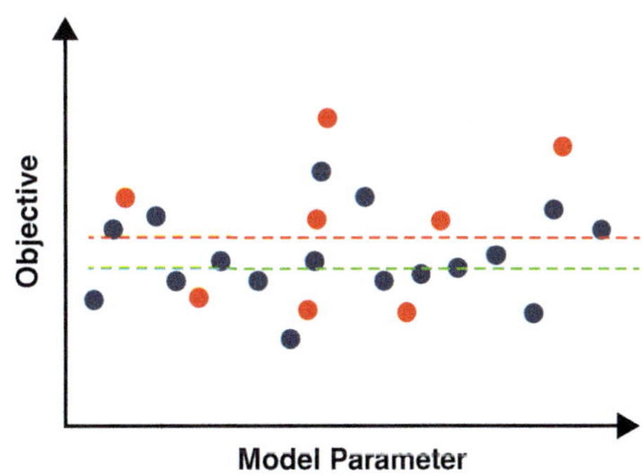

Figure 1.16 Illustration of global optimization—step 2: generation of the new models.

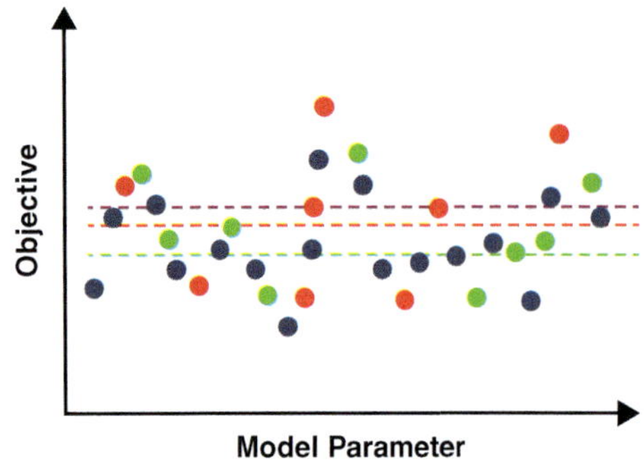

Figure 1.17 Illustration of global optimization—step 3: generation of another set of models, a process which iteratively continues until the model space is exhaustively searched.

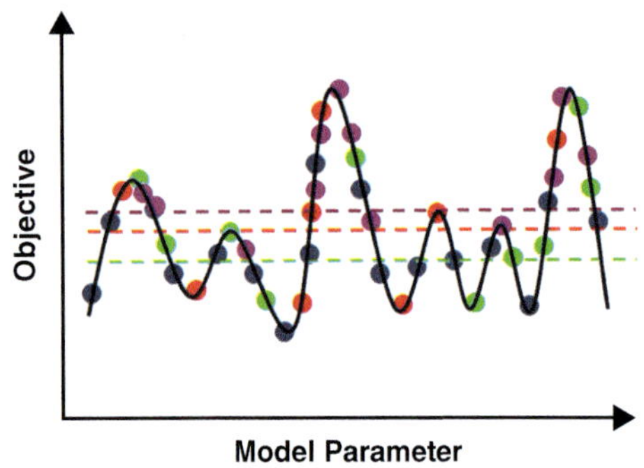

Figure 1.18 Illustration of global optimization and evaluation of the objective at the final step.

methods. It must be noted that although these global optimizations are elegant for solving highly nonlinear inverse problems, they are much more computationally demanding than their local counterparts.

Because of computational complexity, global optimization methods were not as popular as the local methods before the introduction of high-performance parallel computers. In recent years, however, there have been increasing applications of these global methods for solving complex geophysical optimization problems (see, for example, Sen and Stoffa, 1991, 1992, 1996; Stoffa and Sen 1991; Mallick, 1995, 1999; Mallick and Adhikari, 2015; Pafeng et al., 2017; Jia et al., 2021; Mallick and Chakraborty, 2022, among others). We discuss different global optimization methods in Chapter 8.

1.6.3 Machine-Learning-Based Optimization

At present, gradient-based and global methods are popular for seismic optimization. Current advances in AI, specifically ML, a subset of AI, however, made tremendous impacts in all fields of applied science and engineering. In seismology, applications of AI technology are not new and have been used in the past to solve a variety of problems such as seismic trace editing, first break picking, statics correction (Taner et al., 1988; McCormack, 1990; McCormack et al. 1993; Khan, 2007), geohazard analysis (Lau et al., 2002), seismic facies classification (Zhao et al., 2015), etc. With advances in computing technology, these AI-based methods are now becoming viable tools to solve large-scale seismic optimization problems.

Like operator-based methods, AI/ML methods tend to map the data space **D** directly onto the model or the decision space **X**. However, the operator for this mapping is estimated via optimization by training (learning). Figure 1.19, for example, shows a neural-network-based ML in which the training process consists of four distinct elements:

1) The input layer, which consists of the input (observed) data.
2) Hidden layers, which take the observed data from the input layer and perform internal calculations using the universal approximation theorem (Gelenbe et al., 1999) to estimate the model. The universal approximation theorem expresses any continuous function as a linear sum of its variables (i.e., the observed data from the input layer) with different activation functions and weights (Hajian and Styles, 2018).
3) The output layer, which contains the desired model.
4) The iterative optimization loop, which compares the estimated model from the output layer with the true model, modifies the weights, and feeds back to the

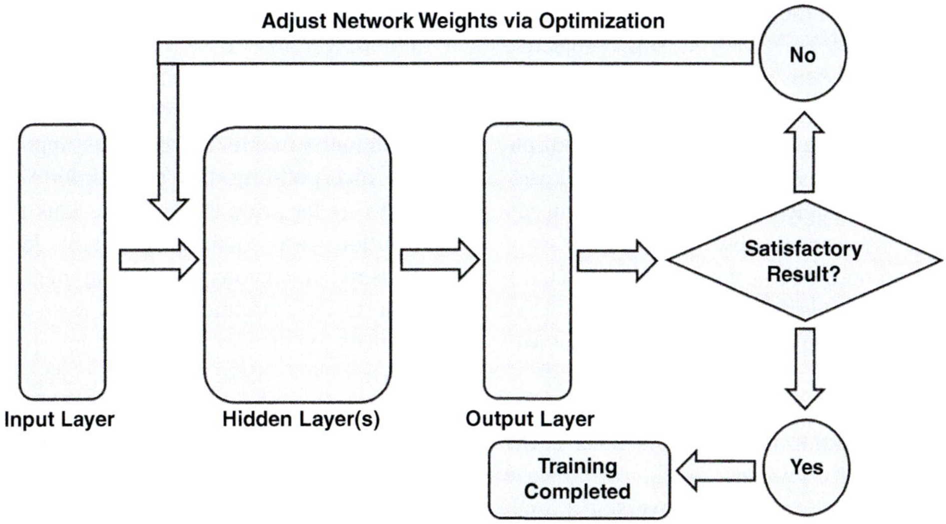

Figure 1.19 An overview of the machine-learning-based optimization using neural networks.

hidden layers for recomputing a new model and comparing. Once the model output from the network is satisfactory, the training/learning is assumed complete.

The training/learning phase is an optimization process, and both supervised and unsupervised (physics-informed) training have been used for seismic inversion (Biswas et al., 2019). Typically, a small subset of data is used for training. Once the training is completed and the optimum weights are estimated, they are applied to the rest of the data. The process is much more computationally efficient than the classical optimization methods. But depending upon the type of the problem, the training can be computationally challenging. For seismic trace editing, first break picking, statics correction, etc., an architecture with a few hidden layers is sufficient and the training process is efficient. Large-scale seismic inversion problems, however, require a deep-learning architecture with a cascade of hidden layers, which can take a lot of computation time. Therefore, the applications of AI/ML to solve large-scale seismic inversion problems is still immature. However, as computing technology improves, AI/ML could become the future for solving these problems. Note that AI/ML attempts to find an inverse operator to directly map the observed data onto the model (decision) space via optimization, which is not analogous to the optimization used in local or global optimization. In Chapter 9, we discuss different flavors of AI/ML for solving seismic optimization problems, and in Chapter 10 we discuss how AI/ML can revolutionize the future of seismic and other geophysical inversion methods.

1.6.4 Single and Multi-objective Optimization

Traditionally, optimization methods tend to define a single objective and attempt to find its minimum or maximum, i.e., the objective space **Y** in equation 1.29 is considered to have a single component. Many optimization or inverse problems, are, however, multi-objective. In inverting multi-component seismic data for example, there is one objective for each data component. If seismic data are jointly inverted with other geophysical data such as gravity, electrical, and magnetic data, it is a MOO problem.

Traditionally, multi-objective problems are dealt with by defining a single objective as a weighted sum of the objectives of different data components, followed by a single-objective optimization (Du and MacGregor, 2010; Karaulis et al., 2012). Multi-objective problems are, however, highly nonlinear with multiple solutions (Pareto-optimal solutions), and casting them as a single-objective optimization can provide only a small subset of this entire set of the Pareto-optimal solutions (Deb et al., 2002). In MOO, the objective is assumed as a vector, and all its components are simultaneously optimized.

In geophysics, MOO has been successfully used for estimating an anisotropic model of the lithosphere (Kozlovskaya et al., 2007), wave equation migration velocity inversion (Singh et al., 2008), earthquake focal mechanisms (Heyburn and Fox, 2010), and waveform inversion of multicomponent seismic data (Padhi and Mallick, 2013, 2014; Li and Mallick, 2015; Li et al., 2016). In AI/ML, MOO is called multi-task learning (MTL), and some applications of MTL can be found in Igel (2005), Jin and Sendhoff (2008), Drugan (2015), Parisi et al. (2016), Sener and Koltun (2018), Wu et al. (2019), Ross et al. (2019), Zhang and Alkhalifah (2019), among others. We discuss the details of the MOO methods in Chapter 8 and MTL in Chapters 9 and 10.

1.7 Bayesian Approach to Inversion/Optimization

The Bayesian approach to inversion/optimization is named after Thomas Bayes, who was one of the eighteenth century pioneers in probability theory. As discussed later, although the Bayesian approach is fundamentally different from the classical approaches discussed above, in general, it results in the same solution (Aster et al., 2018).

The primary difference between the classical and Bayesian approaches is in the nature of the solution of the optimization problem. As previously described, in the classical approach there is an unknown model, say $\mathbf{m}_{true}$, that we aim to find. In the Bayesian approach on the contrary, the model $\mathbf{m}$ is assumed to be a random variable, and therefore, its solution is a probability distribution of the model parameters. Because the model is assumed to be a random variable, the Bayesian approach allows incorporating the prior information about $\mathbf{m}$ as its prior distribution. This prior distribution can come in different forms. For seismic inversions as an example, it is possible to use wireline P- and S-wave sonic and density log measurements and interpolate and extrapolate these between geologically interpreted horizons in poststack seismic data to provide an initial or guess model over the entire seismic dataset (Figure 1.20). For most seismic optimization problems, we discuss in this book, this is our preferred method for the initial model generation.

Once the initial model is obtained, the prior distributions for each model parameter can be defined in many ways. For example, we can assume a Gaussian distribution around each model parameter at each location as the prior. Alternatively, we may also define a search window around each model parameter shown in Figure 1.20, based on the geological knowledge of the area and assuming equal (uniform) probability for the random models within the search window. Irrespective of the choice of the prior distribution, a Bayesian approach would essentially start from this user-defined prior distribution and provide the posterior probability density function (PDF). The PDF generated from the Bayesian approach can, in turn, be used to estimate the maximum likelihood model. In addition, the PDF can also be used to quantify the uncertainty associated with the estimation of each model parameter.

The mathematical formulations of the Bayesian approach are discussed in Chapter 8.

1.8 Summary and Organization of the Book

Although many geophysical inverse problems can be directly solved using an operator-based inversion, model-based or optimization

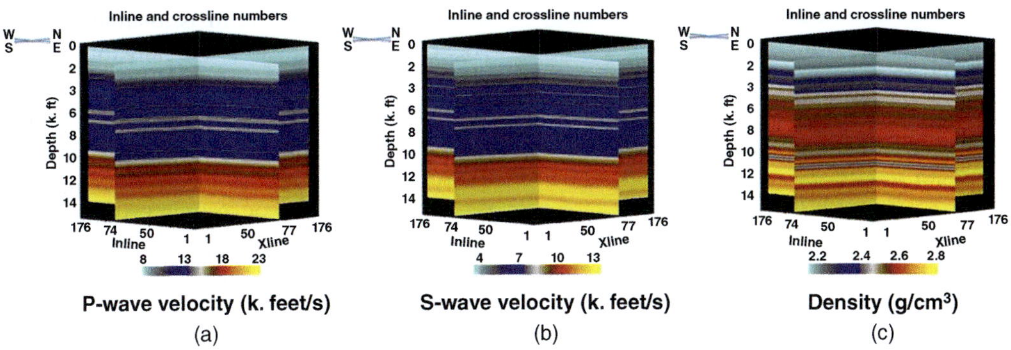

Figure 1.20 An illustration of generating an initial or guess model for the seismic inverse problem. Pafeng et al. (2017)/with permission of GeoScienceWorld.

methods are, in general, preferred over the operator-based inversions. In these latter optimization methods, the objective is defined as the match or mismatch between the observed and the synthetic (or predicted) data. Before optimization, the model (decision), data, and the objective spaces must be clearly defined, and the mapping of the model space onto the objective space (or spaces) must be established for the objective evaluations. All optimization methods deal with either maximization or minimization of the objective depending upon whether the problem is cast as a maximization or a minimization problem. Optimization methods may be local or global. While the local optimization methods are compute-efficient, they may sometimes fail, especially when the inverse problem is non-unique, meaning that that there are multiple minima or maxima in the objective function, and the starting model is far from the true (desired) model. Global methods can handle non-uniqueness because they do not depend upon the choice of the starting model, but they are much more computationally demanding than the local methods. Current advances of ML/AI methods show these to have the potential for being the future of optimization. Finally, the Bayesian inversion approach assumes the model parameters to be random variables, which, in turn, allows estimating their posterior PDFs and quantifying uncertainty in the model parameter estimates.

This book is organized as follows: Chapter 1 (current chapter) provides a brief overview, Chapter 2 provides some mathematical background, Chapters 3–5 discuss elastic wave theory, Chapters 6–9 discuss various optimization methods, and finally Chapter 10 provides the directions for future developments and concluding remarks.

1.9 Exercises

Problem 1.1 Consider an interface between two layers with P-wave velocity and density of the upper layer being V_{P1} and ρ_1 and those for the lower layer being V_{P2} and ρ_2. Then, according to equation 1.3, the normal-incidence reflection coefficient, i.e., the reflection coefficient for the incidence-angle $\theta = 0°$, is

$$R_0 = \frac{\rho_2 V_{P2} - \rho_1 V_{P1}}{\rho_2 V_{P2} + \rho_1 V_{P1}}.$$

Now, look at the AVA equation (equation 1.7) that provides the reflection coefficient for any arbitrary incidence-angle θ in terms of the AVA intercept A, AVA gradient B, and AVA curvature C. Here, the AVA intercept A must be the reflection coefficient at $\theta = 0°$. However, in equation 1.8, A is defined as

$$A = \frac{1}{2}\left(\frac{\Delta V_P}{V_P} + \frac{\Delta \rho}{\rho}\right),$$

where

$$\Delta V_P = V_{P2} - V_{P1}; \quad \Delta \rho = \rho_2 - \rho_1;$$
$$V_P = \frac{V_{P2} + V_{P1}}{2}; \quad \rho = \frac{\rho_2 + \rho_1}{2}.$$

For small contrasts in the P-wave velocity and density across the interface, show that the R_0 and A are equivalent.

Problem 1.2 The AVA gradient B is given in equation 1.9 as

$$B = \frac{1}{2}\frac{\Delta V_P}{V_P} - 2\left(\frac{V_S}{V_P}\right)^2\left(2\frac{\Delta V_S}{V_S} - \frac{\Delta\rho}{\rho}\right),$$

where $\Delta V_S = V_{S2} - V_{S1}$; $V_S = \frac{V_{S2}+V_{S1}}{2}$, and the other parameters are as defined in Problem 1.1.

Show that the above expression is equivalent to

$$B = \frac{1}{2}\frac{\Delta V_P}{V_P} - \frac{2\Delta\mu}{\rho V_P^2},$$

where $\Delta\mu$ is the rigidity modulus contrast across the interface, given as

$$\Delta\mu = \rho_2 V_{S2}^2 - \rho_1 V_{S1}^2.$$

Problem 1.3 If the S-to-P-wave velocity ratio (V_S/V_P) of an elastic medium is γ, then the Poisson's ratio (υ) of the medium is given by

$$\upsilon = \frac{1-2\gamma^2}{2(1-\gamma^2)}.$$

Starting with the expression for the AVA gradient B in terms of the rigidity modulus, provided above in Problem 1.2, express B in terms of the average Poisson's ratio and Poisson's ratio contrast across the interface.

Problem 1.4 AVA inversion provides the estimate of the AVA intercept A, AVA gradient B, and the AVA curvature C. From these estimates, how do you estimate the P- and S-wave velocity and density contrasts ($\Delta V_P/V_P$, $\Delta V_S/V_S$, $\Delta\rho/\rho$)? Do you need to make some additional assumptions?

Problem 1.5 Derive the least-squares AVA inversion equation 1.23 from equation 1.19.

Problem 1.6 In the soccer-ball example, we formulated the inversion or optimization solution by guessing different velocities (v'), computing the forward problem $d' = v't$, and then computing the objective. Note that here we use the observed time (t) and compute different distances d' for different $v'$and match with the measured distance d in computing the objective. How would you cast the same optimization by using the measured distance d, computing different times, say t', for different v', and matching with the measured time t to compute the objective? Explain in detail. Cast the problem as minimization instead of maximization.

References

Aki, K., and Richards, P.G., 2002, *Quantitative Seismology*, Second Edn, University Science Books, ISBN 978-1891389634.

Aster, R.C., Borchers, B., and Thurber, C.H., 2018, *Parameter Estimation and Inverse Problems*, Third Edn, Elsevier, ISBN 978-0128046517.

Bath, M., 1979, *Introduction to Seismology*, Second, Revised Edn, Birkhäuser Basel, Basel ISBN 978-3034852838.

Ben-Menahem, A., 1995, A concise history of mainstream seismology: origins, legacy, and perspectives, *Bulletin of the Seismological Society of America*, *85*(4), 1202–1225.

Biswas, R., Sen, M.K., Das, V., and Mukerji, T., 2019, Prestack and poststack inversion using a physics-guided convolutional neural network, *Interpretation*, *7*, SE161–SE174, https://doi.org/10.1190/INT-2018-0236.1.

Bortfeld, R., 1961, Approximation to the reflection and transmission coefficients of plane longitudinal and transverse waves, *Geophysical Prospecting*, *9*, 485–502, https://doi.org/10.1111/j.1365-2478.1961.tb01670.x.

Chapman, C.H., 2004, *Fundamentals of Seismic Wave Propagation*, Cambridge University Press, ISBN 978-0521815383.

Deb, K., Pratap, A., Agarwal, A., and Meyarivan, T., 2002, A fast and elitist multi-objective genetic algorithm: NSGA-II, *IEEE Transactions on Evolutionary Computations*, *6*, 181–197, https://doi.org/10.1109/4235.996017.

Dobrin, M.B., and Savit, C.H., 1988, *Introduction to Geophysical Prospecting*, Fourth Edn, McGraw-Hill, ISBN 978-0070171961.

Drugan, M.M., 2015, Multi-objective optimization perspectives on reinforcement learning algorithms using reward vectors, ESANN 2015 Proceedings, European symposium on Artificial Neural Networks, Computational Intelligence and Machine Learning. Bruges (Belgium), 22–24 April 2015.

Du, A., and MacGregor, L.M., 2010, Reservoir characterization from joint inversion of marine CSEM and seismic AVA data using Genetic Algorithms: a case study based on the Luva gas field, SEG Technical Program Expanded Abstracts, https://doi.org/10.1190/1.3513888.

Gelenbe, E., Mao, Z.H., and Li, Y.D., 1999, Function approximation with spiked random networks, *IEEE Transactions on Neural Networks*, *10*, 3–9, https://doi.org/10.1109/72.737488.

Grechka, V., and Heigl, W.M., 2017, *Microseismic Monitoring*, Society of Exploration Geophysicists, ISBN 978-1560803478.

Guasch, L., Warner, M., Nangoo, T., J. Morgan, Umpleby, A., Stekl, I., and Shah, N., 2012, Elastic 3D full-waveform inversion, SEG Technical Program Expanded Abstracts, https://doi.org/10.1190/segam2012-1239.

Hajian, A., and Styles, P., 2018, *Application of Soft Computing and Intelligent Methods in Geophysics*, Springer, ISBN: 978-3319665313.

Hampson, D., 1991, AVO inversion, theory and practice, *The Leading Edge*, *10*, 39–42, https://doi.org/10.1190/1.1436820.

Heyburn, R., and Fox, B., 2010, Multi-objective analysis of body and surface waves from the Market Rasen (UK) earthquake, *Geophysical Journal International*, *181*, 532–544, https://doi.org/10.1111/j.1365-246X.2010.04523.x.

Igel, C., 2005, Multi-objective model selection for support vector machines, in C.A. Coello Coello, A. Hernández Aguirre, and E. Zitzler, eds., *Evolutionary Multi-Criterion Optimization. EMO 2005*, Lecture Notes in Computer Science, Vol. *3410*, Springer, Berlin, Heidelberg https://doi.org/10.1007/978-3-540-31880-4_37.

Jia, L., Mallick, S., and Wang, C., 2021, Data-driven prestack waveform inversion using genetic algorithm: methodology and examples, *Interpretation*, *9*, T1069–T1088, https://doi.org/10.1190/INT-2020-0201.1.

Jin, Y., and Sendhoff, B., 2008, Pareto-based multiobjective machine learning: an overview and case studies, *IEEE Transactions on Systems, Man and Cybernetics, Part C: Applications and Reviews*, *38*, 397–415, https://doi.org/10.1109/TSMCC.2008.919172.

Karaulis, M., Revil, A., Zhang, D., and Werkema, D.D., 2012, Time-lapse joint inversion of crosswell DC resistivity and seismic data: a numerical investigation, *Geophysics*, *77*, D141–D157, https://doi.org/10.1190/geo2012-0011.1.

Khan, K.A., 2007, Artificial-intelligence-based automated first break picking and quality control, SEG Technical Program Expanded Abstracts, https://doi.org/10.1190/1.2792598.

Kozlovskaya, E., Vecsey, L., Plomerová, J., and Raita, T., 2007, Joint inversion of multiple data types with the use of multiobjective optimization: problem formulation and application to the seismic anisotropy investigations, *Geophysical Journal International*, *171*, 761–779, https://doi.org/10.1111/j.1365-246X.2007.03540.x.

Lau, A., Gonzalez, A., Mallick, S., and Gillespie, D., 2002, Waveform gather inversion and attribute-guided interpolation: a two-step approach to log prediction, *The Leading Edge*, *21*, 1024–1027, https://doi.org/10.1190/1.1518440.

Li, T., and Mallick, S., 2015, Multicomponent, multi-azimuth pre-stack seismic waveform inversion for azimuthally anisotropic media using a parallel and computationally efficient non-dominated sorting genetic algorithm,

Geophysical Journal International, *200*, 1136–1154, https://doi.org/10.1093/gji/ggu445.

Li, T., Mallick, S., Tamimi, N., and Davis, T., 2016, Inversion of wide-azimuth multi-component vertical seismic profile data for anisotropic subsurface properties, SEG Technical Program Expanded Abstracts, https://doi.org/10.1190/segam2016-13875028.1.

Liu, F., Guasch, L., Morton, S.A., Warner, M., Umpleby, A., Meng, Z., Fairhead, S., and Checkles, S., 2012, 3-D time-domain full waveform inversion of a Vallhall OBC dataset, SEG Technical Program Expanded Abstracts, https://doi.org/10.1190/segam2012-1105.1.

Love, A.E.H., 1944, *A Treatise on the Mathematical Theory of Elasticity*, Fourth Edn, Dover Publications, ISBN 978-1107618091.

Mallick, S., 1993, A simple approximation to the P-wave reflection coefficient and its implication in the inversion of amplitude variation with offset data, *Geophysics*, *58*, 544–552, https://doi.org/10.1190/1.1443437.

Mallick, S., 1995, Model-based inversion of amplitude-variations-with-offset data using a genetic algorithm, *Geophysics*, *60*, 939–954, https://doi.org/10.1190/1.1443860.

Mallick, S., 1999, Case history: some practical aspects of prestack waveform inversion using a genetic algorithm: an example from the East Texas woodbine gas sand, *Geophysics*, *64*, 326–336, https://doi.org/10.1190/1.1444538.

Mallick, S., 2007, Amplitude-variation-with-offset, elastic impedance, and wave-equation synthetics – a modeling study, *Geophysics*, *72*, C1–C7, https://doi.org/10.1190/1.2387108.

Mallick, S., and Adhikari, S., 2015, Amplitude-variation-with-offset and prestack waveform inversion: a direct comparison using a real data example from the Rock Springs uplift, Wyoming, USA, *Geophysics*, *80*, B45–B59, https://doi.org/10.1190/GEO2014-0233.1.

Mallick, S., and Chakraborty, D., 2022, Prediction of the ocean water sound speeds via attribute-guided seismic waveform inversion, *Geophysics*, *87*, U67–U79, https://doi.org/10.1190/geo2021-0442.1.

McCormack, M.D., 1990, Seismic trace editing and first break picking using neural networks, SEG Technical Program Expanded Abstracts, https://doi.org/10.1190/1.1890185.

McCormack, M.D., Zaucha, D.E., and Dushek, D.W., 1993, First-break refraction event picking and seismic data trace editing using neural networks, *Geophysics*, *58*, 67–78, https://doi.org/10.1190/1.1443352.

Mukhopadhyay, P.K., and Mallick, S., 2011, An accurate ray-based offset-to-angle transform from normal moveout uncorrected multicomponent data in a transversely isotropic medium with vertical symmetry axis, *Geophysics*, *76*, C41–C51.

Oldenburg, D.W., Scheuer, T., and Levy, S., 1983, Recovery of acoustic impedance from reflection seismogram, *Geophysics*, *48*, 1318–1337, https://doi.org/10.1190/1.1441413.

Padhi, A., and Mallick, S., 2013, Accurate estimation of density from the inversion of multicomponent prestack seismic waveform data using a nondominated sorting genetic algorithm, *The Leading Edge*, *32*, 94–98, https://doi.org/10.1190/tle32010094.1.

Padhi, A., and Mallick, S., 2014, Multicomponent pre-stack seismic waveform inversion in transversely isotropic media using a non-dominated sorting genetic algorithm, *Geophysical Journal International*, *196*, 1600–1618, https://doi.org/10.1093/gji/ggt460.

Pafeng, J., Mallick, S., and Sharma, H., 2017, Prestack waveform inversion of three-dimensional seismic data – an example from the Rock Springs uplift, Wyoming, USA, *Geophysics*, *82*, B1–B12, https://doi.org/10.1190/GEO2016-0079.1.

Parisi, S., Pirotta, M., and Restelli, M., 2016, Multi-objective reinforcement learning through continuous pareto manifold approximation, *Journal of Artificial Intelligence Research*, *57*, 187–227, https://doi.org/10.1613/jair.4961.

Pratt, R.G., 1999, Seismic waveform inversion in the frequency domain. Part 1: theory and verification in a physical scale model, *Geophysics, 64*, 888–901, https://doi.org/10.1190/1.1444597.

Resnick, J., 1993, Seismic data processing for AVO and AVA analysis, in J.P. Castagna, and M.M. Backus, eds., *Offset dependent reflectivity: Theory and practice of AVO and AVA analysis: SEG*, 175–189, Society of Exploration Geophysicists.

Richards, P.G., and Frasier, C.W., 1976, Scattering of elastic waves from depth-dependent inhomogeneties, *Geophysics, 41*, 441–458, https://doi.org/10.1190/1.1440625.9.

Ross, Z.E., Yue, Y., Meier, M., Hauksson, E., and Heaton, T.H., 2019, PhaseLink: a deep learning approach to seismic phase association, *Journal of Geophysical Research: Solid Earth, 124*, 856–869, https://doi.org/10.1029/2018JB016674.

Russell, B.H., 1988, Introduction to Seismic Inversion Methods: S. N. Domenico Series Editor, SEG, Course Notes Series, 2.

Sava, P., and Hill, S.J., 2009, Overview and classification of wavefield seismic imaging methods, *The Leading Edge, 28*, 170–183, https://doi.org/10.1190/1.3086052.

Selley, R.C., and Sonnenberg, S.A., 2014, *Elements of Petroleum Geology*, Elsevier, ISBN 978-0123860316.

Sen, M.K., and Roy, I.G., 2003, Computation of differential seismograms and iteration adaptive regularization in prestack waveform inversion, *Geophysics, 68*, 2026–2039, https://doi.org/10.1190/1.1635056.

Sen, M.K., and Stoffa, P.L., 1991, Nonlinear one-dimensional seismic waveform inversion using simulated annealing, *Geophysics, 56*, 1624–1638, https://doi.org/10.1190/1.1442973.

Sen, M.K., and Stoffa, P.L., 1992, Rapid sampling of model space using genetic algorithms: examples from seismic waveform inversion, *Geophysical Journal International, 108*, 281–292, https://doi.org/10.1111/j.1365-246X.1992.tb00857.x.

Sen, M.K., and Stoffa, P.L., 1996, Bayesian inference, Gibbs's sampler and uncertainty estimation in geophysical inversion, *Geophysical Prospecting, 44*, 313–350, https://doi.org/10.1111/j.1365-2478.1996.tb00152.x.

Sener, O. and Koltun, V., 2018, Multi-Task Learning as Multi-Objective Optimization, 32nd Conference on Neural Information Processing Systems (NeurIPS 2018), Montréal, Canada, https://doi.org/10.48550/arXiv.1810.04650.

Shearer, P.M., 2009, *Introduction to Seismology*, Second Edn, Cambridge University Press, ISBN 978-052170842-5.

Shuey, R.T., 1985, A simplification of the Zoeppritz equations, *Geophysics, 50*, 609–614, https://doi.org/10.1190/1.1441936.

Singh, S.C., West, G.F., Bregman, N.D., and Chapman, C.H., 1989, Full waveform inversion of reflection data, *Journal of Geophysical Research, 94*, 1777–1794, https://doi.org/10.1029/JB094iB02p01777.

Singh, V.P., Duquet, B., Legér, M., and Schoenauer, M., 2008, Automatic wave-equation migration velocity inversion using multiobjective evolutionary algorithms, *Geophysics, 73*, VE61–VE73, https://doi.org/10.1190/1.2966008.

Stein, S., and Wysession, M., 2002, *An Introduction to Seismology, Earthquakes and Earth Structure*, Wiley-Blackwell, ISBN 978-0865420786.

Stoffa, P.L., and Sen, M.K., 1991, Nonlinear multiparameter optimization using genetic algorithms: inversion of plane-wave seismograms, *Geophysics, 56*, 1794–1810, https://doi.org/10.1190/1.1442992.

Taner, T.M., Lu, L., and Baysal, E., 1988, Unified method for 2-D and 3-D refraction statics with first break picking by supervised learning, SEG Technical Program Expanded Abstracts, https://doi.org/10.1190/1.1892424.

Tarantola, A., 2005, *Inverse Problem Theory and Methods for Model Parameter Estimation*, Society for Industrial and Applied Mathematics, ISBN 978-0898715729.

Telford, W.M., Geldart, M.P., and Sheriff, R.E., 1990, *Applied Geophysics*, Second Edn,

Cambridge University Press, ISBN 978-0521339384.

Todd, C.P., and Backus, M.M., 1985, Offset-dependent reflectivity in a structural context, SEG Technical Program Expanded Abstracts, 586–588, https://doi.org/10.1190/1.1892805.

Warner, M., Ratcliffe, A., Nangoo, T., Morgan, J., Umpleby, A., Shah, N., Vinje, V., Štekl, I., Guasch, L., Win, C., Conroy, G., and Bertrand, A., 2013, Anisotropic 3D full-waveform inversion, *Geophysics*, *78*, R59–R80, https://doi.org/10.1190/geo2012-0338.1.

Wu, X., Liang, L., Shi, Y., Geng, Z., and Fomel, S., 2019, Multitask learning for local seismic image processing: fault detection, structure-oriented smoothing with edge-preserving, and seismic normal estimation by using a single convolutional neural network, *Geophysical Journal International*, *219*, 2097–2109, https://doi.org/10.1093/gji/ggz418.

Yilmaz, Ö., 1989, *Seismic Data Processing*, Society of Exploration Geophysicists, ISBN 978-1560800941.

Zhang, Z., and Alkhalifah, T., 2019, High-resolution reservoir characterization using deep learning aided elastic full-waveform inversion: the North Sea field data example, *Geophysics*, *84*, R741–R751, https://doi.org/10.1190/segam2019-3198055.1.

Zhao, T., Jayaram, V., Roy, A., and Marfurt, K.J., 2015, A comparison of classification techniques for seismic facies recognition, *Interpretation*, *3*, SAE29–SAE58, https://doi.org/10.1190/INT-2015-0044.1.

Ziolkowski, A., Fokkema, J.T., Koster, K., Confurius, A., and Vanboom, R., 1989, Inversion of common mid-point seismic data, in P.L. Stoffa, ed., *Tau-P: A Plane Wave Approach to the Analysis of Seismic Data*, 141–176, Kluwer Academic Publishers, ISBN 978-9401068840.

Zoeppritz K., 1919, Erdbebenwellen VIII B, Über Reflexion ad Durchgang seismischer wellen durch Unstetigkeitsflachen, Göttinger Nach. 66–84.

2

Mathematical Background

A comprehensive review of some mathematical concepts is essential to understand the fundamentals of seismic wave propagation, which is needed for optimization and machine-learning applications. In this chapter, we review three such concepts:

1) Fourier series and Fourier integrals.
2) Partial differential equations (PDEs).
3) Fundamentals of tensors.

In addition to the above three concepts, other mathematical concepts that are useful in seismology are discussed later in the subsequent chapters.

2.1 Fourier Series and Fourier Integrals

We frequently use the Fourier series and Fourier integrals in our discussion of the seismic wave propagation. We therefore provide below some basic concepts, which are later used in Chapters 3–5. For additional details on the Fourier series and integrals, see Sommerfeld (1949).

2.1.1 Fourier Series

Let $f(x)$ be an arbitrary function, given in the interval $-\pi \leq x \leq +\pi$. This function may, for example, be a seismogram. We now wish to approximate this function as the sum of $2n+1$ trigonometric terms:

$$\begin{aligned} S_n(x) = A_0 + A_1 \cos x + A_2 \cos 2x + \cdots \\ + A_n \cos nx + B_1 \sin x \\ + B_2 \sin 2x + \cdots + B_n \sin nx. \end{aligned} \tag{2.1}$$

We approximate this sum using the principle of least squares. So, by denoting the error $\varepsilon_n(x) = f(x) - S_n(x)$, we now define the mean square error as

$$M = \frac{1}{2\pi} \int_{-\pi}^{+\pi} \varepsilon_n^2 dx, \tag{2.2}$$

and attempt to reduce M to a minimum and find the coefficients A_k and B_k. The requirement that M must be a minimum leads to

$$\begin{aligned} -\frac{\partial M}{\partial A_k} = \frac{1}{\pi} \int_{-\pi}^{+\pi} [f(x) - S_n(x)] \cos kx dx = 0, \\ k = 0, 1, 2, \ldots n, \\ -\frac{\partial M}{\partial B_k} = \frac{1}{\pi} \int_{-\pi}^{+\pi} [f(x) - S_n(x)] \sin kx dx = 0, \\ k = 1, 2, \ldots n. \end{aligned} \tag{2.3}$$

Equation 2.3 represents a set of $2n+1$ equations to determine the $2n+1$ coefficients A_k and B_k. Note that according to equation 2.3, the evaluation of each coefficient A_k and B_k is independent of one another and is not recursively connected. To verify this, recall the well-known orthogonality relations that exist among the trigonometric functions,

$$\int_{-\pi}^{+\pi} \cos kx \sin lx dx = 0 \tag{2.4}$$

and

$$\left.\begin{array}{l} \displaystyle\int_{-\pi}^{+\pi} \cos kx \cos lx dx \\ \displaystyle\int_{-\pi}^{+\pi} \sin kx \sin lx dx \end{array}\right\} = 0, k \neq l. \tag{2.5}$$

Note that for $k = l = 0$, the first integral in equation 2.5 is 2π. Both integrals shown in equation 2.5 can be combined as

$$\frac{1}{2\pi}\int_{-\pi}^{+\pi} \cos kx \cos lx dx = \frac{1}{2\pi}\int_{-\pi}^{+\pi} \sin kx \sin lx dx = \delta_{kl}. \tag{2.6}$$

In equation 2.6, δ_{kl} is the Kronecker delta

$$\delta_{kl}\begin{cases} = 0, & k \neq l \\ = 1, & k = l \end{cases}. \tag{2.7}$$

Also, for $k = l = 0$, we have

$$\frac{1}{2\pi}\int_{-\pi}^{+\pi} \cos kx \cos lx dx = \frac{1}{2\pi}\int_{-\pi}^{+\pi} dx = 1. \tag{2.8}$$

Now, if we substitute equations 2.4, 2.6, and 2.8 into equation 2.3, then in the integrals with S_n all terms except the kth term vanish, and we directly obtain the Fourier representation of the coefficients as

$$A_0 = \frac{1}{2\pi}\int_{-\pi}^{+\pi} f(x) dx \tag{2.9}$$

and

$$\left.\begin{array}{l} A_k = \displaystyle\frac{1}{\pi}\int_{-\pi}^{+\pi} f(x) \cos kx dx \\ B_k = \displaystyle\frac{1}{\pi}\int_{-\pi}^{+\pi} f(x) \sin kx dx \end{array}\right\}, k > 0. \tag{2.10}$$

Accuracy of the approximation, given by equations 2.1, increases with the number of constants A_k and B_k at our disposal, or, in other words, it increases with our choice of n. However, because A_k and B_k for $k < n$ are independent of n, previously calculated A_k and B_k remain unchanged if we increase the number of terms from n to $n+1$, and only the new coefficients A_{n+1} and B_{n+1} need to be newly evaluated. The coefficients A_k and B_k once found are final. From equation 2.10, it can be readily seen that if $f(x)$ is an even function, i.e., if $f(x) = f(-x)$, all B_k vanish. Similarly, for odd functions, i.e., for $f(x) = -f(-x)$, all A_k vanish. Therefore, the even functions can be represented as a pure cosine series and the odd functions as a pure sine series.

In the Fourier representation, there is nothing to prevent us from letting the number of terms n to grow to infinity, or, in other words, go to the limit $n \to \infty$, and let the finite series, represented by equation 2.1, grow into an infinite Fourier series. However, before dealing with such an infinite series, it is first important to discuss the completeness of the system of functions used here as the basis. It can be seen that if any of the terms in the Fourier series, say the kth cosine term, were omitted, the function $f(x)$ cannot be described by the remaining terms with arbitrary accuracy. Even if we pass to the limit $n \to \infty$, a finite error $A_k \cos kx$ would remain in the representation of $f(x)$. Let us, for example, consider the case that we want to express $\cos nx$ by an incomplete series of all cosine terms with $k < n$ and $k > n$. Because of the orthogonality, all A_k would then vanish and thus the error in the representation of $\cos nx$ would be $\cos nx$ itself. One would not of course disturb the regularity of the trigonometric functions by omitting one term. But this leads to what is called the relation of completeness in the representation of continuous functions. This relation of completeness is, in fact, contained in the basis of the system of least squares and needs little more elaboration. This elaboration, in turn, would lead us to the Fourier integrals.

What we learn from mathematics is that a system of functions, say $\varphi_1, \varphi_2, \ldots, \varphi_k, \ldots$, is complete, if for every continuous function $f(x)$, the mean error according to the principle of least squares (equations 2.2 and 2.3) goes to zero as $n \to \infty$. In equation 2.1, we assumed this system of functions to be sinusoidal. Here, we extend it to any system of functions φ which are orthogonal and normalized to 1, i.e.,

$$\int \varphi_k \varphi_l dx = 0; \quad \int \varphi_k^2 dx = 1. \tag{2.11}$$

Equation 2.11 implies that

$$A_k = \int f(x)\varphi_k(x)dx. \tag{2.12}$$

If, in equations 2.11 and 2.12, we let the lower and upper limits of the integral to be a and b so that the length of the expansion is $b - a$, then, from the least-squares equation 2.2, we can write

$$\begin{aligned}(b-a)M &= \int_a^b \left[f(x) - \sum_{k=0}^{n} A_k \varphi_k \right]^2 dx \\ &= \int_a^b f^2(x)dx - 2\sum_{k=0}^{n} A_k \\ &\times \int_a^b f(x)\varphi_k(x)dx + \sum_{k=0}^{n} A_k^2.\end{aligned} \tag{2.13}$$

Now, equation 2.12 implies that the middle term on the right-hand side of equation 2.13 is twice the third term, but with an opposite sign. Therefore,

$$(b-a)M = \int_a^b [f(x)]^2 dx - \sum_{k=0}^{n} A_k^2. \tag{2.14}$$

So, for the squared error $M \to 0$ as $n \to \infty$, we must have

$$\sum_{k=0}^{\infty} A_k^2 = \int_a^b [f(x)]^2 dx. \tag{2.15}$$

Equation 2.15 is the mathematical formulation of the relation of completeness in representing any continuous function $f(x)$ within an arbitrary interval $\langle a, b \rangle$ using a set of normalized orthogonal functions φ (equation 2.11).

Having described the relation of completeness for any set of normalized orthogonal functions, we now revisit the Fourier coefficients A_k and B_k defined in equations 2.9 and 2.10 and express them in more useful forms for computer implementations. In the following, we do this for an infinite Fourier series. However, the methods we develop is also valid for a truncated Fourier series.

Starting from equation 2.1, replacing $S_n(x)$ by $f(x)$, using equations 2.9 and 2.10, and replacing the variables of integrations by ξ instead of x, we first write

$$\begin{aligned}f(x) &= \frac{1}{2\pi}\int f(\xi)d\xi + \frac{1}{\pi}\sum_{k=1}^{\infty}\int f(\xi)\cos k\xi d\xi \\ &\quad \cdot \cos kx + \frac{1}{\pi}\sum_{k=1}^{\infty}\int f(\xi)\sin k\xi d\xi \cdot \sin kx \\ &= \frac{1}{2\pi}\int f(\xi)d\xi + \frac{1}{\pi}\sum_{k=1}^{\infty}\int f(\xi)\cos k(x-\xi)d\xi \\ &= \frac{1}{2\pi}\left\{\int f(\xi d\xi + \sum_{k=1}^{\infty}\left(\int f(\xi)e^{ik(x-\xi)}d\xi \right.\right. \\ &\quad \left.\left. + \int f(\xi)e^{-ik(x-\xi)}d\xi\right)\right\}.\end{aligned} \tag{2.16}$$

In equation 2.16, we omitted the lower and upper limits of integrations, and they respectively are $-\pi$ and $+\pi$.

Note that in the last term on the right-hand side of equation 2.16, we can consider the summation of positive k in $e^{-ik(x-\xi)}$ to be the summation for the corresponding negative values of k in $e^{-ik(x-\xi)}$. We therefore replace the term by

$$\sum_{k=-1}^{-\infty} f(\xi)e^{ik(x-\xi)}d\xi = \sum_{k=-\infty}^{-1} f(\xi)e^{ik(x-\xi)}d\xi. \tag{2.17}$$

This, in turn, leads equation 2.16 to

$$f(x) = \frac{1}{2\pi}\sum_{k=-\infty}^{+\infty}\int f(\xi)e^{ik(x-\xi)}d\xi. \tag{2.18}$$

We finally introduce the Complex Fourier coefficients C_k and write equation 2.18 as

$$f(x) = \sum_{k=-\infty}^{+\infty} C_k e^{ikx}, \tag{2.19}$$

with

$$C_k = \frac{1}{2\pi} \int f(\xi) e^{-ik\xi} d\xi. \tag{2.20}$$

It can be readily seen that the relation between the previously defined Fourier coefficients A_k and B_k, defined in equations 2.9 and 2.10, and our newly introduced complex Fourier coefficient C_k (equation 2.20) to be

$$C_k = \begin{cases} \frac{1}{2}(A_k - iB_k), & k > 0, \\ \frac{1}{2}(A_{|k|} + iB_{|k|}), & k < 0, \end{cases}$$

$$C_0 = A_0. \tag{2.21}$$

By replacing real Fourier coefficients A_k and B_k with the complex Fourier coefficient C_k, the complex representation of equations 2.19 and 2.20 is much simpler than the real representation and allows us to go to the Fourier integrals.

2.1.2 Fourier Integrals

In the Fourier representation, both real and complex, the function $f(x)$ was defined in the interval $-\pi \le x \le +\pi$. We now want to define the function $f(z)$ for an arbitrary interval $-a \le x \le +a$. To do so, we first make the substitution

$$x = \frac{\pi z}{a}. \tag{2.22}$$

The above substitution transforms equations 2.9 and 2.10 into

$$A_0 = \frac{1}{2a} \int_{-a}^{+a} f(z) dz \tag{2.23}$$

and

$$\left.\begin{matrix} A_k \\ B_k \end{matrix}\right\} = \frac{1}{a} \int_{-a}^{+a} f(z) \left.\begin{matrix} \cos \\ \sin \end{matrix}\right\} \left(\frac{\pi k z}{a}\right) dz. \tag{2.24}$$

Or, for the complex representation, given by equations 2.19–2.21, we obtain

$$f(z) = \sum_{k=-\infty}^{+\infty} C_k e^{i\frac{\pi}{a}kz}, \tag{2.25}$$

with

$$C_k = \frac{1}{2a} \int_{-a}^{+a} f(\xi) e^{-i\frac{\pi}{a}k\xi} d\xi. \tag{2.26}$$

We now make a in equations 2.25 and 2.26 very large so that the sequence of values $\omega_k = \frac{\pi}{a}k$ in equation 2.25 becomes dense. Replacing ω_k by ω, we write the difference between the consecutive values of ω as

$$d\omega = \frac{\pi}{a}, \quad \frac{1}{a} = \frac{d\omega}{\pi}. \tag{2.27}$$

And, from equation 2.27 and from the definition of $\omega = \frac{\pi}{a}k$, we write equation 2.26 as

$$C_k = \frac{d\omega}{2\pi} \int_{-a}^{+a} f(\xi) e^{-i\omega\xi} d\xi. \tag{2.28}$$

Now, as a gets large, we can replace the series summation in equation 2.26 by an integration. For the time being, let us avoid saying the limits of the integral are $\pm\infty$ and use $\pm\Omega$ instead and express equation 2.25 as

$$f(x) = \lim_{\Omega\to\infty} \lim_{a\to\infty} \frac{1}{2\pi} \int_{-\Omega}^{+\Omega} e^{i\omega x} d\omega \times \int_{-a}^{+a} f(\xi) e^{-i\omega\xi} d\xi. \tag{2.29}$$

Note that in the definition of $f(z)$ in equation 2.25 the variable z is arbitrary, and we replaced it with x in equation 2.29. Finally, by carrying out the limits, we write equation 2.29 as

$$f(x) = \frac{1}{2\pi} \int_{-\infty}^{+\infty} d\omega \int_{-\infty}^{+\infty} f(\xi) e^{i\omega(x-\xi)} d\xi. \tag{2.30}$$

Equation 2.30 is the complex form of the Fourier integral. Now

$$e^{i\omega(x-\xi)} = \cos[\omega(x-\xi)] + i \sin[\omega(x-\xi)]. \tag{2.31}$$

Because sine is an odd function of ω and cosine is an even function of ω, if we insert equation 2.31 into equation 2.30, we obtain

$$f(x) = \frac{1}{\pi}\int_{0}^{+\infty} d\omega \int_{0}^{+\infty} f(\xi)\cos[\omega(x-\xi)]d\xi. \tag{2.32}$$

Equation 2.32 is the real form of the Fourier integral. Again, because $\cos[\omega(x-\xi)] = \cos\omega x \cos\omega\xi + \sin\omega x \sin\omega\xi$, we can also express equation 2.32 as

$$f(x) = \int_{0}^{\infty} a(\omega)\cos\omega x d\omega + \int_{0}^{\infty} b(\omega)\sin\omega x d\omega, \tag{2.33}$$

where

$$a(\omega) = \frac{1}{\pi}\int_{-\infty}^{+\infty} f(\xi)\cos\omega\xi d\xi,$$

$$b(\omega) = \frac{1}{\pi}\int_{-\infty}^{+\infty} f(\xi)\sin\omega\xi d\xi. \tag{2.34}$$

Note that in the Fourier integral representation of equations 2.32–2.34, $b(\omega)$ must vanish if $f(x)$ is an even function and $a(\omega)$ must vanish if $f(x)$ is an odd function. This, in turn, leads to a pure cosine or a pure sine integral. One or the other can be produced when $f(x)$ is given for the values of $x \geq 0$ by continuing $f(x)$ either as an even or an odd function on the negative side. Thus, for an even continuation, we write

$$f(x) = \int_{0}^{\infty} a(\omega)\cos\omega x d\omega,$$

$$a(\omega) = \frac{2}{\pi}\int_{0}^{\infty} f(\xi)\sin\omega\xi d\xi. \tag{2.35}$$

And, for an odd continuation, we write

$$f(x) = \int_{0}^{\infty} b(\omega)\sin\omega x d\omega,$$

$$b(\omega) = \frac{2}{\pi}\int_{0}^{\infty} f(\xi)\sin\omega\xi d\xi. \tag{2.36}$$

The pure cosine or pure sine representations as above are useful for some specific problems. For the time being, let us assume the variable x to be the time coordinate t. Then, in equation 2.33, we have an arbitrary process $f(t)$ in time expressed as its harmonic components in frequency ω. In Fourier integrals, we are concerned with a continuous spectrum ranging over all frequencies, while in Fourier series, we are concerned with the discrete spectrum consisting of a fundamental tone plus harmonic overtones. In our notation, the amplitude of the spectrum is given by

$$c(\omega) = \sqrt{[a(\omega)]^2 + [b(\omega)]^2}, \tag{2.37}$$

and the phase $\gamma(\omega)$ is given by

$$\gamma(\omega) = \tan^{-1}\frac{b(\omega)}{a(\omega)}, \tag{2.38}$$

and the relation between these quantities is best expressed as

$$c(\omega)e^{i\gamma(\omega)} = a(\omega) + ib(\omega). \tag{2.39}$$

Note that the Fourier integral uses both the amplitude and phase spectrum.

2.1.3 Fourier Transforms

Using the variable t in place of x, we can write the complex form of the Fourier integral (equation 2.30) as two equations

$$f(t) = \frac{1}{\sqrt{2\pi}}\int_{-\infty}^{+\infty} \varphi(\omega)e^{i\omega t}d\omega,$$

$$\varphi(\omega) = \frac{1}{\sqrt{2\pi}}\int_{-\infty}^{+\infty} f(t)e^{-i\omega t}dt. \tag{2.40}$$

Note that combining both equations shown in equation 2.40, we obtain equation 2.30. Disregard the variable of integration in the second equation of equation 2.40. The function $\varphi(\omega)$ is equivalent to $a(\omega) + ib(\omega)$ in equation 2.39 and contains both amplitude and phase of the oscillating process $f(t)$. Also note that the functions $f(t)$ and $\varphi(\omega)$ are reciprocals of each other such that the one is determined by the other via similar integral equations. These functions $f(t)$ and $\varphi(\omega)$ are called Fourier transforms of each other or the Fourier transform pairs. Also, out of the two equations shown in equation 2.40, the first equation is called the forward Fourier transform and the second is called the inverse Fourier transform. Although we describe the Fourier transforms using equation 2.40, their many equivalent forms exist. The integration components for these forms are the same, but they are differently normalized. Below, we list a few of such forms of the forward and inverse Fourier transform pairs:

$$f(t) = \int_{-\infty}^{\infty} \phi(\omega) e^{i\omega t} d\omega,$$
$$\phi(\omega) = \frac{1}{2\pi} \int_{-\infty}^{\infty} f(t) e^{-i\omega t} dt. \tag{2.41}$$

$$f(t) = \int_{-\infty}^{\infty} \widetilde{f}(\omega) e^{i\omega t} d\omega,$$
$$\widetilde{f}(\omega) = \frac{1}{2\pi} \int_{-\infty}^{\infty} f(t) e^{-i\omega t} dt. \tag{2.42}$$

$$f(t) = \int_{-\infty}^{\infty} f(\omega) e^{i\omega t} d\omega,$$
$$f(\omega) = \frac{1}{2\pi} \int_{-\infty}^{\infty} f(t) e^{-i\omega t} dt. \tag{2.43}$$

So far, we have shown Fourier transforms in single dimensions. Its extension to multiple dimensions is straightforward using multiple integral transforms, which we discuss later while discussing the seismic wavefield propagation in three dimensions.

Fourier transforms have numerous practical applications in seismology. It is therefore worthwhile to discuss some of the important practical issues with the implementation of the Fourier transformations in digital computers. Here, we do not discuss how to write a code to do Fourier transforms. There are many excellent textbooks (for example, the classic textbook by Press et al., 1989) that provide how to write such a code. Additionally, most modern high-level computer languages have Fourier transform as a built-in function, and it is not necessary to write such a code from scratch.

Irrespective of our preferences of writing our own code or using one built-in within high-level computer languages, they have one thing in common. The Fourier transform pairs, given by equation 2.40 or any of its equivalent forms (equations 2.41–2.43), are for continuous functions of infinite length. However, our functions are sampled at discrete sample points and are of finite length. We may therefore ask ourselves the following questions:

1) How do we compute Fourier transforms for discretely sampled functions?
2) What are the consequences for these functions being of finite length?

Here, we concentrate on computing the inverse Fourier transform:

$$S(t) = \frac{1}{2\pi} \int_{-\infty}^{\infty} \widetilde{S}(\omega) e^{-i\omega t} d\omega. \tag{2.44}$$

Extension of the concepts we discuss is obvious for the forward transform as they both are of identical nature. We specifically discuss the following three topics:

1) Discretization or sampling.
2) Anti-aliasing.
3) Windowing.

Additional details on these issues and other practical aspects of numerical implementation of Fourier transforms can be found in Mallick and Frazer (1987).

Discretization (Sampling)

Using equation 2.44, we compute the time-domain signal $S(t)$ by computing the frequency domain signal $\widetilde{S}(\omega)$ and then inverse Fourier transforming it to time. As shown later in Chapter 5, computing seismic response for discrete frequencies and then Fourier transforming to time domain is a common method for computing forward synthetic seismograms. Therefore, the issues we are discussing here are of fundamental importance.

To implement equation 2.44 in the computer, the integral on the right-hand side of this equation must be replaced by a summation. In addition, the lower and upper limits of the sum must be replaced from $-\infty$ to $+\infty$ to a finite sum, say from $-\Omega$ to $+\Omega$. For now, let us first replace the integral to a sum, leaving its lower and upper limits unchanged and write equation 2.44 as

$$S(t) \approx \frac{\Delta\omega}{2\pi} \sum_{n=-\infty}^{\infty} e^{-in(\Delta\omega)t} \widetilde{S}(n\Delta\omega). \quad (2.45)$$

In equation 2.45, $\Delta\omega$ is the sampling interval with which $\widetilde{S}(\omega)$ is sampled in the frequency (ω) domain. Now, what is the consequence of replacing the integration of a continuous function with a sum of a function discretely sampled with the sampling interval $\Delta\omega$?

Note that the right-hand side of equation 2.45 can be written as

$$\frac{\Delta\omega}{2\pi} \sum_{n=-\infty}^{\infty} e^{-in(\Delta\omega)t} \widetilde{S}(n\Delta\omega)$$

$$= \frac{1}{2\pi} \int_{-\infty}^{\infty} d\omega \widetilde{S}(\omega) e^{-i\omega t}$$

$$\times \left[\Delta\omega \sum_{n=-\infty}^{\infty} \delta(\omega - n\Delta\omega) \right]. \quad (2.46)$$

And, if we compare the right-hand side of equation 2.46 with the integral equation of the inverse Fourier transform (equation 2.44), we immediately see that discretizing a continuous function in the Fourier transform is equivalent to multiplying the original integral with a picket fence or a Dirac comb that contains delta functions at regular intervals of $\Delta\omega$. This Dirac comb can be mathematically expressed by the term within the square brackets on the right-hand side of equation 2.46 and is pictorially represented by Figure 2.1.

Without any proof, we now state a useful formula, known as the Poisson summation formula, which is valid for Fourier transform pairs. Further details including the proof of this formula can be found in Zygmund (1968) and Pinsky (2002).

Poisson summation formula: Consider the Fourier transform pair:

$$\widetilde{\varphi}(\omega) = \int_{-\infty}^{\infty} \varphi(t) e^{i\omega t} dt. \quad (2.47)$$

Then,

$$\sum_{n=-\infty}^{\infty} \varphi(t + nT) = \frac{1}{T} \sum_{n=-\infty}^{\infty} e^{-2n\pi i \frac{t}{T}}$$

$$\times \widetilde{\varphi}\left(\frac{2n\pi}{T}\right). \quad (2.48)$$

Therefore, if we set $T = 2\pi/\Delta\omega$, then the Poisson summation formula (equation 2.48) becomes

$$\sum_{n=-\infty}^{\infty} \varphi(t + nT) = \frac{\Delta\omega}{2\pi} \sum_{n=-\infty}^{\infty} e^{-in(\Delta\omega)t}$$

$$\times \widetilde{\varphi}(n\Delta\omega). \quad (2.49)$$

And using equation 2.49, we rewrite equation 2.45 as

$$S(t) \approx \frac{\Delta\omega}{2\pi} \sum_{n=-\infty}^{\infty} e^{-in(\Delta\omega)t} \widetilde{S}(n\Delta\omega)$$

$$= \sum_{n=-\infty}^{\infty} S(t + nT). \quad (2.50)$$

Equation 2.50 allows us to summarize the consequence of discretizing the Fourier transform integral and is illustrated in Figure 2.2. Consider that we want to synthesize a time-domain signal of length $T = T_S$. In doing so, also assume that we choose the sampling interval $\Delta\omega < 2\pi/T_S$, for example we choose $\Delta\omega = 2\pi/T_0$ (see Figure 2.2). Then, according to equation 2.50, we will not obtain the part of the signal beyond $T = T_0$, which is shown as dashed line in Figure 2.2. This part of the signal, in fact, will be wrapped around in the front from $t = 0$ and will interfere with the signal computed for $0 \geq t \leq T_0$. However, if we choose $\Delta\omega \leq 2\pi/T_s$, we will obtain the entire signal over the desired interval $0 \geq t \leq T_S$.

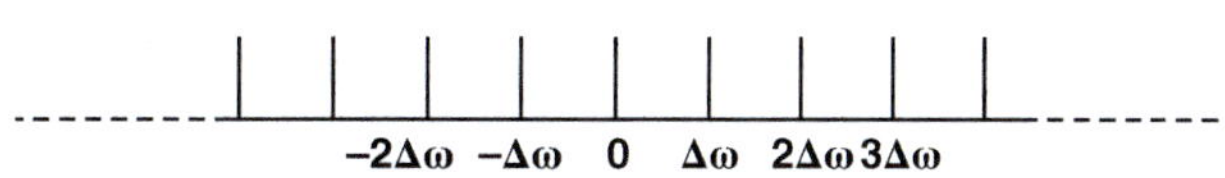

Figure 2.1 Example of a Dirac comb or picket fence, representing spikes (delta functions) at regular intervals of $\Delta\omega$.

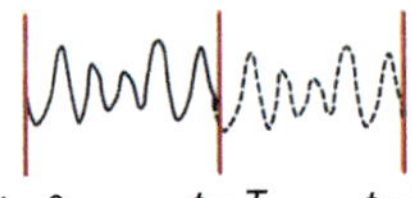

Figure 2.2 Illustration of the consequence of discretizing.

One must wonder that to synthesize a time-domain signal of a desired length T_s, if we choose $\Delta\omega \leq 2\pi/T_s$, then everything should be fine, and so would there be a point to these discussions? We use Figure 2.3 to demonstrate the reason. Consider that we are interested only in the first 1 s of the seismic data in solving a seismic inverse problem. As discussed in Chapter 1, we start from an initial model and compute the synthetic data to match with the observed data. We compute synthetic seismic data using a frequency–wavenumber approach, where we first compute them in the frequency–wavenumber domain and convert them into time and space using Fourier transforms. We discuss details of seismogram computations later in Chapter 5. Because we are only interested in the first 1 s, i.e., $T_S = 1$, we set $\Delta\omega = \frac{2\pi}{T_s} = 2\pi$ for the synthetic computation, which is shown in Figure 2.3a. Note that all events inside the red curve belong to times >1 s and are aliased (wrapped around) in the front. Of course, if we set T_s to 2 s and therefore use $\Delta\omega = \pi$, as shown in Figure 2.3b, these aliased events move to their correct times. For clarity, some events (A, B, C, D, E, and F) are marked in both seismograms shown in Figure 2.3. Note that changing $\Delta\omega$ from 2π to π doubles the computation time. And, as we previously discussed in Chapter 1, a seismic inversion/optimization requires multiple iterations of seismogram computations. Therefore, if we need to invert only the first 1 s of the seismic data, it is desirable to find ways to set $\Delta\omega$ to 2π instead of π which does not generate the wrapped-around events shown in Figure 2.3a. This, in turn, can save a lot of computation time, which we discuss next.

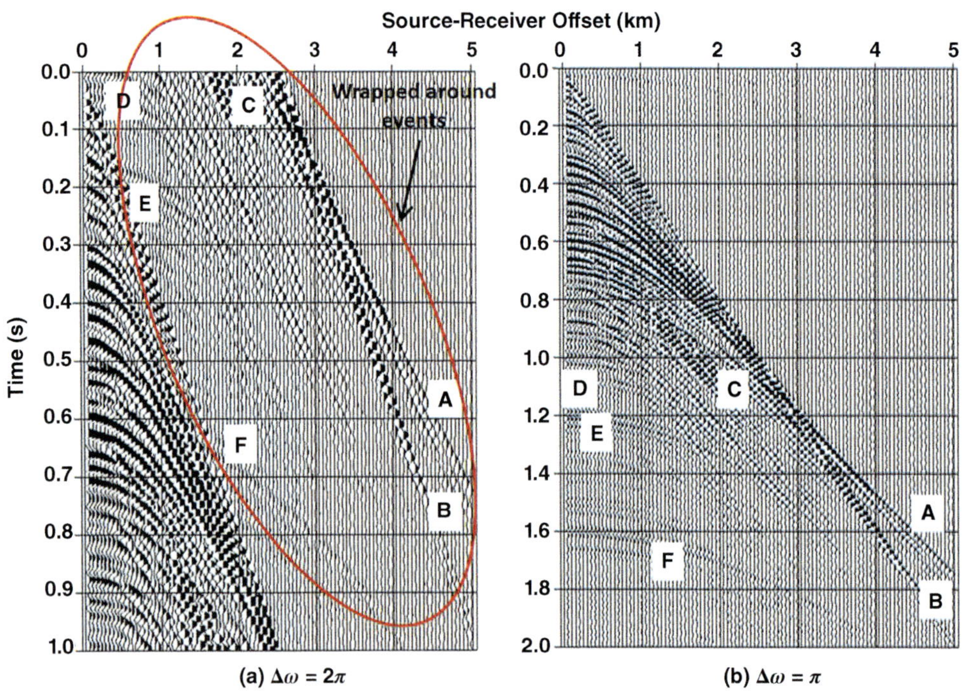

Figure 2.3 Illustration of the wraparound (a) or aliasing (b) caused from discretization of the Fourier transform integral to a summation for a synthetic seismogram computation.

Anti-aliasing

As seen for the desired time window T_S, setting $\Delta\omega > 2\pi/T_S$ causes the later parts of the signal to be aliased. This is because by approximating an integral by summation, we synthesize $\sum_{n=-\infty}^{+\infty} S(t+nT_S)$, as shown in equation 2.50. Now, what happens if we attempt to synthesize $e^{-at}S(t)$ instead of $S(t)$ for a suitably chosen value of a? From equation 2.50, the effect of discretization would then give

$$\sum_{n=-\infty}^{\infty} S(t+nT_S)e^{\{-a(t+nT_S)\}}$$
$$= \sum_{n=-\infty}^{-1} S(t+nT_S)e^{(-anT_S)} + S(t)e^{-at}$$
$$+ \sum_{n=1}^{\infty} S(t+nT_S)e^{(-anT_S)}. \tag{2.51}$$

Because the seismic signal is causal and there are no events before $t = 0$, the first term on the right-hand side of equation 2.51 is zero. The second term is our signal $S(t)$ that we want to synthesize but decayed by a factor e^{-at}. Finally, the third term is the aliased or wrapped-around signal which is decayed by a factor of e^{-aT_S} or higher with T_S being our desired time window length. Therefore, after Fourier transforming the frequency domain signal to time domain, if we multiply the computed signal for each time sample t by e^{+at}, we will obtain our desired signal. But the wrapped-around components of the signal will remain decayed. For example, we want e^{-aT_S} to be a equal to $1/D$, meaning that we want the wrapped-around signals to be decayed at least by a factor of D. This leads to

$$a = \frac{\ln(D)}{T_S}. \tag{2.52}$$

For example, if we want the wrapped-around signals to be decayed by a factor of 50, all we need to do is set $a = \ln(50)/T_S$. But the question that remains to be answered is how do we compute $e^{-at}S(t)$ instead of $S(t)$? Considering that $S(t)$ and $\widetilde{S}(\omega)$ are Fourier transform pairs, then from equation 2.42, the forward Fourier transform of $e^{-at}S(t)$ can be written as

$$\mathfrak{F}_t(e^{-at}S(t)) = \int_{-\infty}^{\infty} dt e^{i\omega t}e^{-at}S(t)$$
$$= \int_{-\infty}^{\infty} dt e^{i(\omega+ia)t}S(t) = \widetilde{S}(\omega+ia). \tag{2.53}$$

Therefore, the inverse Fourier transform of $\widetilde{S}(\omega+ia)$ is $e^{-at}S(t)$. So, if we make the frequencies complex by adding an imaginary component equal to a, then the corresponding time-domain signal of the computed frequency domain signal is $e^{-at}S(t)$. From equation 2.52, we already know how to choose a, and therefore we know how to attenuate the undesired aliased or wrapped-around parts of the signal.

Figure 2.4 shows the result of the same synthetic seismogram computation as the one shown in Figure 2.3a using complex frequencies with $D = 50$. There are no undesired wrapped-around arrivals in Figure 2.4. This, in

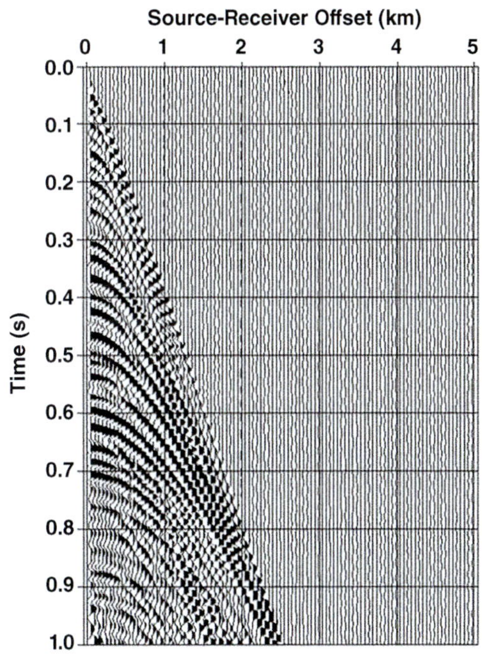

Figure 2.4 Same as Figure 2.3a but applying anti-aliasing with the parameter D defined in equation 2.52 set to a value of 50. Note that all wrapped-around events have now disappeared.

turn, indicates that replacing real frequencies with complex frequencies is a powerful technique for removing undesired aliased events in the computed seismograms.

Windowing

So far, we have thoroughly discussed the issue of aliasing caused by discretization of the Fourier transform integrals and approximating it as a summation. We also described how the undesired aliased events can be removed using complex frequencies. However, we kept our summation as infinite (see equation 2.45). In practice, this infinite sum must be truncated to a finite sum, say from $-\Omega$ to $+\Omega$. Truncating an infinite sum into a finite one has the effect of windowing, which we now consider. Without any loss of generality, here we discuss the truncation of an integral instead of a sum. We also use the well-known "*convolution theorem*," which we state below, without proof. For further details on the convolution theorem, including its proof, see any book on Fourier analysis or digital signal processing (for example, Zygmund, 1968; Robinson and Treitel, 2000; Pinsky, 2002, etc.).

Convolution Theorem: Assume that $f(t) \leftrightarrow \tilde{f}(\omega)$ and $g(t) \leftrightarrow \tilde{g}(\omega)$ are Fourier transform pairs. Then,

$$f(t) * g(t) \leftrightarrow \tilde{f}(\omega)\tilde{g}(\omega). \tag{2.54}$$

The symbol "*" in equation 2.54 represents convolution and given as

$$f(t) * g(t) = \int f(t)g(t-\tau)d\tau. \tag{2.55}$$

The convolution theorem therefore states that convolution in time domain is multiplication in the frequency domain and vice versa.

Now, note that truncating our integral to $-\Omega$ to $+\Omega$ from $-\infty$ to $+\infty$ is equivalent to

$$\frac{1}{2\pi}\int_{-\Omega}^{\Omega} d\omega e^{-i\omega t}\tilde{S}(\omega) = \frac{1}{2\pi}\int_{-\infty}^{\infty} d\omega e^{-i\omega t} B \times \left(\frac{\omega}{\Omega}\right)\tilde{S}(\omega), \tag{2.56}$$

where

$$B\left(\frac{\omega}{\Omega}\right) = \begin{cases} 1, & \left|\frac{\omega}{\Omega}\right| < 1 \\ 1/2, & \left|\frac{\omega}{\Omega}\right| = 1. \\ 0, & \left|\frac{\omega}{\Omega}\right| > 1 \end{cases} \tag{2.57}$$

Therefore, using the convolution theorem, if $S(t) \leftrightarrow \tilde{S}(\omega)$ are Fourier transform pairs, equation 2.56 implies

$$\frac{1}{2\pi}\int_{-\Omega}^{\Omega} d\omega e^{-i\omega t}\tilde{S}(\omega) = S(t) * \mathfrak{F}_t^{-1}\left[B\left(\frac{\omega}{\Omega}\right)\right]. \tag{2.58}$$

In equation 2.58, the symbol $\mathfrak{F}_t^{-1}$ represents an inverse Fourier transform. But it can be readily seen that

$$\mathfrak{F}_t^{-1}\left[B\left(\frac{\omega}{\Omega}\right)\right] = \frac{\sin(\Omega t)}{\pi t}. \tag{2.59}$$

Therefore,

$$\frac{1}{2\pi}\int_{-\Omega}^{\Omega} d\omega e^{-i\omega t}\tilde{S}(\omega) = S(t) * \frac{\sin(\Omega t)}{\pi t}. \tag{2.60}$$

The resulting effect of equation 2.60 is that the signal is characterized by a central peak of finite width π/Ω with sidelobes on either side. These sidelobes can be reduced by filtering. For example, if the source wavelet is available, we can multiply the source wavelet spectrum with $\tilde{S}(\omega)$ before computing the truncated integral on the right-hand side of equation 2.58. Another good way of performing such filtering is by using a Hanning window and given as (see Mallick and Frazer, 1987)

$$W_H\left(\frac{\omega}{\Omega}\right) = \begin{cases} 0.5\left(1+\cos\left(\frac{\omega\pi}{\Omega}\right)\right), & |\omega| \le \Omega \\ 0, & |\omega| > \Omega \end{cases}. \tag{2.61}$$

The inverse Fourier transform of the Hanning window has much smaller sidelobes than $\sin(\Omega t/\pi t)$ which improves the quality of the computed time-domain wavelets.

2.2 Partial Differential Equations

Solution to the seismic wave propagation problems requires knowledge of PDEs. Here, we provide some fundamentals of PDEs, which we later use in the subsequent chapters. Discussions here are sufficient for understanding the seismic wave propagation problems, and for further details on the PDEs, refer to any textbook on Mathematical Physics, for example Menzel (1961), Dettman (1969), and Dennery and Krzywicki (1967).

2.2.1 How Do the Simplest Partial Differential Equations Arise?

We start with the potential equation

$$\nabla^2 u = 0 \tag{2.62}$$

or

$$\nabla^2 u = -(4\pi)\rho, \tag{2.63}$$

where $\rho =$ density

Equations 2.62 and 2.63 are known in the "theory of gravitation" as the expression of the field-action approach as opposed to the action-at-a-distance approach of Newton, and the Laplace Operator ∇^2 is defined in Cartesian coordinates as

$$\nabla^2 = \nabla \cdot \nabla, \tag{2.64}$$

which in a right-handed Cartesian coordinate system, with x and y being the horizontal axes and z being the vertical axis, positive downward (Figure 2.5) is written as

$$\nabla^2 = \nabla \cdot \nabla = \frac{\partial^2}{\partial x^2} + \frac{\partial^2}{\partial y^2} + \frac{\partial^2}{\partial z^2}. \tag{2.65}$$

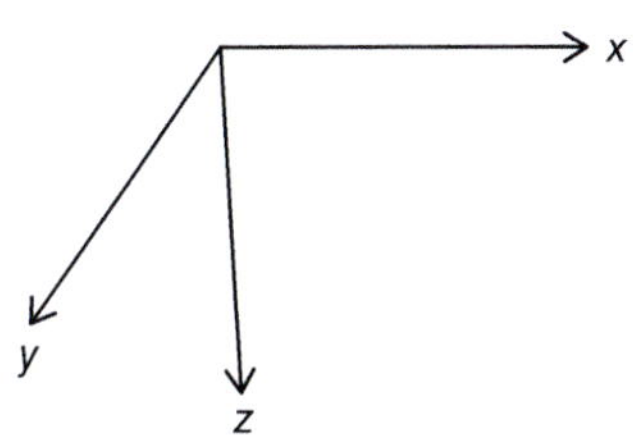

Figure 2.5 A Cartesian coordinate system.

The same equations are fundamental to the electrostatic and magnetic fields with equation 2.62 being valid for an empty space and equation 2.63 in the presence of a source of density ρ. The factor 4π in equation 2.63 is deliberately put inside the parenthesis because it can be removed by a suitable choice of units. Equation 2.62 also appears in the hydrodynamics of incompressible fluids with u being the velocity potential. We also mention here that the two-dimensional potential equation

$$\frac{\partial^2 u}{\partial x^2} + \frac{\partial^2 u}{\partial y^2} = 0 \tag{2.66}$$

is the basis of the "Riemannian function theory," which we may characterize as the "Field theory" of the analytic functions of type $f(x + iy)$.

Another well-known PDE is the wave equation

$$\nabla^2 u = \frac{1}{c^2}\frac{\partial^2 u}{\partial t^2}. \tag{2.67}$$

Equation 2.67 is fundamental to acoustics where c is the speed of sound. It is also fundamental to electrodynamics of variable fields where c is the speed of light. In the special theory of relativity, we may also write equation 2.67 as

$$\Delta^2 u = 0, \tag{2.68}$$

with

$$\Delta^2 = \sum_{k=1}^{4} \frac{\partial^2}{\partial x_k^2}, \tag{2.69}$$

by introducing a fourth coordinate $x_4 = ict$, in addition to the spatial coordinates x_1 (or x), x_2 (or y), and x_3 (or z). All the PDEs discussed above (equations 2.62, 2.63, 2.66–2.68) are similar in nature and belong to the "Type-1 partial differential equations."

In the general theory of elasticity, we have as a special case the differential equation for the transverse vibration of a thin disc and given as

$$\nabla^2(\nabla^2 u) = -\frac{1}{c^2}\frac{\partial^2 u}{\partial t^2}, \tag{2.70}$$

in which

$$\nabla^2(\nabla^2) = \frac{\partial^4}{\partial x^4} + 2\frac{\partial^4}{\partial x^2 \partial y^2} + \frac{\partial^4}{\partial y^4}. \tag{2.71}$$

Analogously, the differential equation for an oscillating elastic rod is given by

$$\frac{\partial^4 u}{\partial x^4} = -\frac{1}{c^2}\frac{\partial^2 u}{\partial t^2}. \tag{2.72}$$

The class of equations represented by equations 2.70 and 2.72 are different from those of Type-1 and are called "Type-2 partial differential equations."

A third type of PDEs arises to represent the states of equilibrium, which we introduce here in connection with the heat conduction (equalization of the energy differences). We must, however, point out that diffusion (equalization of material densities), fluid friction (equalization of the impulse differences), or pure electric conduction (equalization of the differences of potential) satisfy the same equation.

Let $\mathbf{G}$ be a vector of the magnitude and direction of heat flow, and assume that the initial point P is surrounded by an element of volume $d\tau$. Then, the outflow of heat energy from $d\tau$ per unit time is $\nabla \cdot (\mathrm{G}d\tau)$. A decrease per unit of time in the amount of heat Q in $d\tau$, which we denote by $-\partial Q/\partial r$, corresponds to this. Therefore,

$$\nabla \cdot \mathbf{G}d\tau = -\frac{\partial Q}{\partial t}. \tag{2.73}$$

Our heat conductor is here considered to be a rigid body so that we can neglect expansion, and the heat content is the same as the energy content. Now every increase dQ in heat causes an increase in the temperature of $d\tau$, and every decrease $-dQ$ in heat causes a decrease in the temperature of $d\tau$. Denoting the temperature as u, we then have

$$\begin{aligned} dQ &= cdmdu, \\ dm &= \rho d\tau. \end{aligned} \tag{2.74}$$

In equation 2.74, c is the specific heat and ρ is the material density. Using equation 2.74, we write equation 2.73 as

$$\nabla \cdot \mathbf{G} = -c\rho\frac{\partial u}{\partial t}. \tag{2.75}$$

We now apply Fourier's law of thermal conduction, which, in an isotropic medium, relates $\mathbf{G}$ and u as

$$\mathbf{G} = -\kappa \nabla u, \tag{2.76}$$

where κ is the thermal conductivity of the material. Using equation 2.76, we can write equation 2.75 as

$$\begin{aligned} \nabla^2 u &= \frac{1}{k}\frac{\partial u}{\partial t}, \\ k &= \frac{\kappa}{c\rho}. \end{aligned} \tag{2.77}$$

The parameter k in equation 2.77 is the temperature conductivity. Fourier's law (equation 2.77) was adapted to the case of diffusion in physiology, where u stands for the concentration of dissolved matter in the solvent, $\mathbf{G}$ is the material flow of the dissolved matter, and k is diffusion coefficient. For inner friction of an incompressible fluid, k is the kinematic viscosity, and equation 2.77 is the Navier–Stokes equation of laminar flow. The analog to Fourier's law in electricity is Ohm's law where u stands for the potential, $\mathbf{G}$ for the specific electric current, and k for the specific resistance of the conductor.

Schrödinger's equation, encountered in wave mechanics, formally belongs to the same equation type

$$\nabla^2 u = \frac{2m}{i\hbar}\frac{\partial u}{\partial t},$$

where $\hbar$ is the Planck's constant divided by 2π and m is the mass of the particle. However, because the real constant k in equation 2.77 is replaced in the above equation by a complex constant $i\hbar/2m$, Schrödinger's equation becomes an oscillation, not an equalization process. We can see this in the passage to the case of periodicity in time, if we set

$$\begin{aligned} u &= \psi e^{-i\omega t}, \\ \omega &= \frac{W}{\hbar}, \end{aligned} \tag{2.78}$$

with W being the energy of the state, then the Schrödinger's equation becomes

$$\begin{aligned} \nabla^2 \psi + C\psi &= 0, \\ C &= \frac{2m}{\hbar^2}W. \end{aligned} \tag{2.79}$$

And, starting from the wave equation (equation 2.67) and setting $u = \Psi \exp(-i\omega t)$, we obtain equation 2.78. Therefore, the diffusion equation with a complex constant is an oscillation equation rather than a diffusion equation. The diffusion equation (equation 2.77) belongs to the "Type-3 partial differential equations."

From the above, we note a family of resemblance among the PDEs, which stems from the invariance under rotation and translation. The differential operator of second order implied by this invariance is just the Laplace ∇^2. In the case of space-time invariance of relativity, this is replaced by a four-dimensional Laplace operator Δ^2. The equations that we discussed so far are PDEs, and they form the basis of present-day physics, including seismic wave propagation.

2.2.2 Elliptic, Hyperbolic, and Parabolic Partial Differential Equations: Theory of Characteristics

We restrict ourselves to the case of two independent variables x and y, in which case the most general form of a linear PDE of second order is

$$L(u) = A\frac{\partial^2 u}{\partial x^2} + 2B\frac{\partial^2 u}{\partial x \partial y} + C\frac{\partial^2 u}{\partial y^2} + D\frac{\partial u}{\partial x} + E\frac{\partial u}{\partial y} + F = 0. \tag{2.80}$$

In equation 2.80, A, B, C, D, E, and F are functions of x and y. A more general form of equation 2.80 is

$$A\frac{\partial^2 u}{\partial x^2} + 2B\frac{\partial^2 u}{\partial x \partial y} + C\frac{\partial^2 u}{\partial y^2} = \Phi\left(u, \frac{\partial u}{\partial x}, \frac{\partial u}{\partial y}, x, y\right), \tag{2.81}$$

in which Φ need not be linear in any of its variables, as they are in equation 2.80. We now discuss the conditions for solvability of equation 2.81.

Let Γ be a given curve in the xy plane along which both u and the derivative $\partial u/\partial n$ of u in the direction of the normal n are prescribed. Given these initial conditions, we now discuss whether a solution of equation 2.81 exists. To do so, we define

$$p = \frac{\partial u}{\partial x}, \quad q = \frac{\partial u}{\partial y},$$
$$r = \frac{\partial^2 u}{\partial x^2}, \quad s = \frac{\partial^2 u}{\partial x \partial y}, \quad t = \frac{\partial^2 u}{\partial y^2}. \tag{2.82}$$

And write equation 2.81 as

$$Ar + 2Bs + Ct = \Phi. \tag{2.83}$$

Furthermore, the following relations are valid in general, and therefore hold on Γ

$$dp = rdx + sdy \tag{2.84}$$

and

$$dq = sdx + tdy. \tag{2.85}$$

Now, because p and q are known on Γ, equations 2.83–2.85 are a set of three linear equations with its determinant being

$$\Delta = \begin{vmatrix} A & 2B & C \\ dx & dy & 0 \\ 0 & dx & dy \end{vmatrix} = Ady^2 - 2Bdxdy + Cdx^2. \tag{2.86}$$

Consequently, r, s, and t can be calculated from equations 2.83–2.85 only when the determinant Δ is not zero. This, in turn, is equivalent to finding $Ady^2 - 2Bdxdy + Cdx^2 = 0$ or

$$\left(\frac{dy}{dx}\right) = \begin{cases} \dfrac{B + \sqrt{B^2 - AC}}{A} \\ \dfrac{B - \sqrt{B^2 - AC}}{A} \end{cases}. \tag{2.87}$$

Equation 2.87 provides two sets of curves along which the determinant Δ vanishes where the solutions to the PDE is not possible. These curves are called the "characteristic curves" or "characteristics" and are given by

$$\alpha(x, y) = -\left(\frac{B + \sqrt{B^2 - AC}}{A}\right)x + y = \text{const} \tag{2.88}$$

and

$$\beta(x, y) = -\left(\frac{B - \sqrt{B^2 - AC}}{A}\right)x + y = \text{const.} \tag{2.89}$$

Now, referring to our original equation 2.81, we now consider the following three cases:

1) $B^2 - AC < 0$
2) $B^2 - AC > 0$
3) $B^2 - AC = 0$

Case 1, $B^2 - AC < 0$: By substituting

$$v = \frac{\alpha(x,y) + \beta(x,y)}{2},$$
$$w = \frac{\alpha(x,y) - \beta(x,y)}{2i}, \tag{2.90}$$

equation 2.81 takes the form

$$\left(\frac{AC - B^2}{A}\right)\left(\frac{\partial^2 u}{\partial v^2} + \frac{\partial^2 u}{\partial w^2}\right) = \Phi'\left(u, v, w, \frac{\partial u}{\partial v}, \frac{\partial u}{\partial w}\right). \tag{2.91}$$

The curves represented by equation 2.91 are elliptic in which the characteristics are complex conjugates. This family of PDEs is called "elliptic partial differential equations."

Case 2, $B^2 - AC > 0$: By substituting

$$v = \frac{\alpha(x,y) + \beta(x,y)}{2},$$
$$w = \frac{\alpha(x,y) - \beta(x,y)}{2}, \tag{2.92}$$

equation 2.81 becomes

$$\left(\frac{AC - B^2}{A}\right)\left(\frac{\partial^2 u}{\partial v^2} - \frac{\partial^2 u}{\partial w^2}\right) = \Phi'\left(u, v, w, \frac{\partial u}{\partial v}, \frac{\partial u}{\partial w}\right). \tag{2.93}$$

Equation 2.93 represents the "hyperbolic partial differential equations" in which the characteristics form two distinct families.

Case 3, $B^2 - AC = 0$: By substituting

$$v = \alpha(x,y) = \beta(x,y), w = x, \tag{2.94}$$

equation 2.81 becomes

$$A\frac{\partial^2 u}{\partial w^2} = \Phi'\left(u, v, w, \frac{\partial u}{\partial v}, \frac{\partial u}{\partial w}\right), \tag{2.95}$$

representing "parabolic partial differential equations" in which only one real family of characteristics exists.

Therefore, to summarize, the PDE, given by equation 2.81, is elliptic when $B^2 - AC < 0$, hyperbolic when $B^2 - AC > 0$, and parabolic when $B^2 - AC = 0$.

2.2.3 Simple Examples of the Partial Differential Equations

In Section 2.1.1, we described the PDEs as type 1 (equations 2.62, 2.63, 2.66, 2.68), type 2 (equations 2.20 and 2.72), and type 3 (equation 2.77). Also, starting from the general form, in Section 2.2.2 we defined the elliptic, hyperbolic, and parabolic types of PDEs. Here, we attempt to combine both sections using some simple examples.

Example 1: The wave equation: For simplicity, consider the one-dimensional wave equation, which, from equation 2.67, can be written as

$$\frac{\partial^2 u}{\partial x^2} - \frac{1}{c^2}\frac{\partial^2 u}{\partial t^2} = 0, \tag{2.96}$$

where x is the spatial coordinate and t is the time coordinate. Equation 2.96 is of hyperbolic type with the characteristics given as $\alpha(x, y) = x + ct = \text{const}$ and $\beta(x, y) = x - ct = \text{const}$. By using α and β as the independent variables, we can write equation 2.96 as

$$\frac{\partial^2 u}{\partial\alpha\partial\beta} = 0. \tag{2.97}$$

The most general solution to equation 2.97 is

$$u = g(\alpha) + h(\beta) \equiv g(x + ct) + h(x - ct), \tag{2.98}$$

where g and h are arbitrary differentiable functions of x and t. Let us now prescribe the boundary condition for $t = 0$ (initial condition) as

$$\left.\begin{aligned} u|_{t=0} &= a(x) \\ \left.\frac{\partial u}{\partial t}\right|_{t=0} &= b(x) \end{aligned}\right\}, \ x_1 \le x \le x_2. \tag{2.99}$$

We can then write from equation 2.98

$$g(x) + h(x) = a(x) \tag{2.100}$$

and

$$c\frac{dg(x)}{dx} - c\frac{dh(x)}{dx} = b(x). \tag{2.101}$$

From equation 2.101, we obtain

$$g(x) - h(x) = \frac{1}{c}\int_{x_1}^{x} b(x')dx' + \text{const}. \tag{2.102}$$

And, by combining equations 2.100 and 2.102, we write

$$g(x) = \frac{1}{2}a(x) + \frac{1}{2c}\int_{x_1}^{x} b(x')dx' + \text{const} \tag{2.103}$$

and

$$h(x) = \frac{1}{2}a(x) - \frac{1}{2c}\int_{x_1}^{x} b(x')dx' + \text{const.} \tag{2.104}$$

And, by combining equations 2.98 and 2.104, we write

$$u(x,t) = \frac{1}{2} \times \left\{ a(x+ct) + a(x-ct) + \frac{1}{c}\int_{x-ct}^{x+ct} b(x')dx' \right\}. \tag{2.105}$$

Because $a(x)$ and $b(x)$ are defined in the interval $[x_1, x_2]$, equation 2.105 is meaningful when $x_1 \leq x \pm ct \leq x_2$, which, as shown in Figure 2.6, represents a rectangular region in the $x-t$ plane.

When $x_1 \to \infty$ and $x_2 \to \infty$, i.e., when the spatial domain is infinite, the rectangle covers the entire $x-t$ plane, and the solution is determined everywhere for any time. However, if the initial conditions are specified along a finite segment of the x axis, one needs additional information to find solution everywhere. This information may consist of having boundary values for all times, for example $u(x_1, t) = \xi(t)$; $u(x_2, t) = \zeta(t)$. To simplify the discussion, let us consider a case where $x_1 = 0$ and $x_2 \to \infty$. It is then sufficient to give only one boundary condition: for example, $u(0, t) = 0$. We can then replace our problem with the one where the initial conditions are prescribed for all x where $-\infty \leq x \leq +\infty$ and write

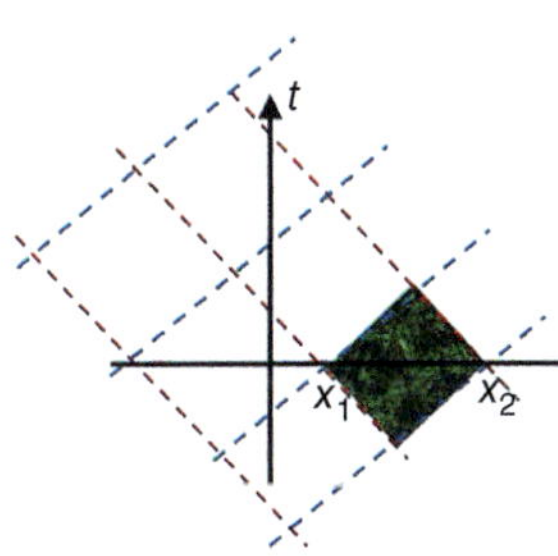

Figure 2.6 Domain of definition of equation 2.105 in the $x-t$ plane.

$$u|_{t=0} = A(x)$$

$$\left.\frac{\partial u}{\partial t}\right|_{t=0} = B(x) \tag{2.106}$$

where

$$A(x) = a(x)$$

$$B(x) = b(x) \tag{2.107}$$

for $x > 0$, and the values of $A(x)$ and $B(x)$ for $x < 0$ are, for the moment, unknown. The solution of the above is given by

$$u(x,t) = \frac{1}{2} \times \left\{ A(x+ct) + A(x-ct) + \frac{1}{c}\int_{x-ct}^{x+ct} B(x')dx' \right\}. \tag{2.108}$$

And, to satisfy the boundary condition $u(0, t) = 0$ we specified above, we must have

$$A(ct) + A(-ct) + \frac{1}{c}\int_{-ct}^{ct} B(x')dx' = 0. \tag{2.109}$$

Because the integral is an odd function of t, both A and B must be odd functions, i.e.,

$$A(x) = -A(-x); B(x) = -B(-x). \tag{2.110}$$

Equations 2.107 and 2.110 give A and B with solution given by equation 2.108.

Example 2: The diffusion equation: Restricting to one dimension, we write the diffusion equation (equation 2.77) as

$$\frac{\partial^2 u}{\partial x^2} - \frac{1}{a^2}\frac{\partial u}{\partial t} = 0. \tag{2.111}$$

Equation 2.111 is of parabolic type, and the characteristics are the lines $t = \text{const}$ in the $x-t$ plane. It is easy to see that by specifying the boundary conditions along a characteristic, one would not have a well-posed problem.

For example, suppose we specify the value of u at a given time $t = 0$, i.e.,

$$u(x, 0) = b(x). \tag{2.112}$$

Equation 2.112 determines $\partial^2 u/\partial t^2$ for $t = 0$, which in turn determines the initial value of $\partial u/\partial t$ because from the equation itself we get

$$\left.\frac{\partial u}{\partial t}\right|_{t=0} = a^2 \frac{d^2 b(x)}{dx^2}. \tag{2.113}$$

It is thus evident that we no longer have the freedom to specify $\partial u/\partial t]_{t=0}$, and if the boundary condition is imposed along the line $t = 0$, they would overdetermine the problem. We consider an example that shows to what extent a single initial condition completely determines the solution. Assume that

$$u|_{t=0} = \begin{cases} 1, & x > 0 \\ 0, & x < 0 \end{cases}. \tag{2.114}$$

We also note that both the differential equation and the initial condition are invariant under a scale transformation

$$x \to sx$$
$$t \to s^2 t. \tag{2.115}$$

for an arbitrary s. Therefore, the solution must also be invariant under this transformation $u(x, t) = u(sx, s^2 t)$. For $t > 0$, we put

$$s = \frac{1}{\sqrt{t}} \tag{2.116}$$

and find that $u(x, t)$ is, in fact, a function of a single variable $y = x/\sqrt{t}$, in terms of which equation 2.111 becomes

$$\frac{d^2 u}{dy^2} + \frac{y}{2a^2}\frac{du}{dy} = 0, \tag{2.117}$$

with

$$u(-\infty) = 0; \quad u(+\infty) = 1. \tag{2.118}$$

Note that one solution to equation 2.117 is

$$u_1 = \text{const}. \tag{2.119}$$

To obtain the second solution, we use a method called the method of variation of constants. Consider a general second-order homogeneous differential equation

$$a(x)\frac{d^2 u}{dx^2} + b(x)\frac{du}{dx} + c(x)u = 0; \quad a(x) \neq 0. \tag{2.120}$$

If $u_1(x)$ is one solution of equation 2.120, then we write the second solution as

$$u_2(x) = u_1(x)h(x), \tag{2.121}$$

in which $h(x)$ must be determined. Inserting equation 2.121 into equation 2.120 we obtain

$$a(x)\frac{d^2(u_1 h)}{dx^2} + b(x)\frac{d(u_1 h)}{dx} + c(x)(u_1 h) = 0, \tag{2.122}$$

which, after some algebra, gives the differential equation for $h(x)$ as

$$\frac{d}{dx}\ln\left(\frac{dh(x)}{dx}\right) = -p(x) - 2\frac{d}{dx}\ln\{u_1(x)\}. \tag{2.123}$$

In equation 2.123, $p(x)$ is defined as

$$p(x) = \frac{b(x)}{a(x)}$$
$$a(x) \neq 0. \tag{2.124}$$

Equation 2.123 can be integrated to give

$$\frac{dh(x)}{dx} = \frac{\text{const}}{u_1^2(x)} \exp\left(-\int_{x_0}^{x} p(x')dx'\right), \tag{2.125}$$

where x_0 is an arbitrary initial point. Integrating equation 2.124 further to obtain $h(x)$ and then using equation 2.121, we obtain the second solution $u_2(x)$ as

$$u_2(x) = (\text{const})u_1(x)\int_{x_0}^{x} \frac{dx''}{u_1^2(x'')} \exp \times \left(-\int_{x_0}^{x''} p(x')dx'\right). \tag{2.126}$$

Now, for our case

$$p(y) = \frac{b(y)}{a(y)} = \frac{y}{2a^2}. \tag{2.127}$$

Therefore,

$$u(y) = (\text{const})\int_{y_0}^{y} dy' e^{-y'^2/4a^2}. \tag{2.128}$$

To obtain the constants (k), we use the boundary conditions to write

$$u(-\infty) = k \int_{y_0}^{-\infty} dy e^{-y^2/4a^2} = 0$$
$$u(+\infty) = k \int_{y_0}^{+\infty} dy e^{-y^2/4a^2} = 1. \quad (2.129)$$

Both equations shown in equation 2.129 can be combined into a single equation

$$k \int_{-\infty}^{+\infty} dy e^{-y^2/4a^2} = 1. \quad (2.130)$$

From equation 2.130, we obtain

$$k = \frac{2a}{\sqrt{\pi}}. \quad (2.131)$$

Plugging in the value of k for the constant into equation 2.128, we obtain

$$u(y) = \frac{2a}{\sqrt{\pi}} \int_{-\infty}^{y} dy' e^{-y'^2/4a^2}. \quad (2.132)$$

And, finally, using equation 2.116, we obtain

$$u(x,t) = \frac{2a}{\sqrt{\pi}} \int_{-\infty}^{x/\sqrt{t}} dy e^{-y^2/4a^2}; \quad t > 0. \quad (2.133)$$

Equation 2.133 is the solution to the diffusion equation for $t > 0$. It is important to note that there is no solution to the diffusion equation for $t < 0$, satisfying the boundary condition, given by equation 2.114. By setting $s = 1/\sqrt{-t}$, we can get the diffusion equation as

$$\frac{d^2u}{dy^2} - \frac{y}{2a^2}\frac{du}{dy} = 0, \quad (2.134)$$

leading to the solution of type

$$u(y) = (\text{const}) \int_{y_0}^{y} dy' e^{y'^2/4a^2}. \quad (2.135)$$

But it does not satisfy the boundary conditions.

Example-3: Two-dimensional Laplace Equation: Let a function of two variables $u(x, y)$ satisfy the two-dimensional Laplace equation and given as

$$\frac{\partial^2 u}{\partial x^2} + \frac{\partial^2 u}{\partial y^2} = 0, \quad (2.136)$$

within a closed curve C. This is an elliptic equation, with $A=1$, $B=0$, and $C=1$ (see quotation 2.81). Additionally, we have only imaginary solutions, and there are no characteristic curves on the $x-y$ plane. We therefore need to use quite different arguments compared to those developed in the preceding two examples. First, assume that our problem has two solutions, u_1 and u_2. Then, their difference $u_1 - u_2$ must also satisfy the Laplace equation within the domain of C and vanish on C. Note that if a function satisfies Laplace equation throughout the region enclosed by C, it may be considered in that region as the real and imaginary parts of a certain analytic function of $z = x + iy$. To see this, let $u(x, y)$ and $v(x, y)$ be the real and imaginary parts of an analytic function $f(z)$ in the complex plane, i.e.,

$$f(z) = u(x,y) + iv(x,y). \quad (2.137)$$

We assume that at any point z in the complex plane, first-order derivatives of the functions $u(x, y)$ and $v(x, y)$ with respect to x and y are defined or in other words

$$\frac{df}{dz} = \lim_{\Delta z \to 0} \frac{f(z+\Delta z) - f(z)}{\Delta z}$$
$$= \lim_{\substack{\Delta x \to 0 \\ \Delta y \to 0}} \frac{u(x+\Delta x, y+\Delta y) - u(x,y) + i\{v(x+\Delta x, y+\Delta y) - v(x,y)\}}{\Delta x + i\Delta y}. \quad (2.138)$$

We now impose the condition that the right-hand side of equation 2.138 should give the same result irrespective of the order $\Delta x \to 0$ or $\Delta y \to 0$ is taken. Thus, by setting $\Delta y \to 0$, then $\Delta x \to 0$ gives

$$\frac{df}{dz} = \frac{\partial u(x,y)}{\partial x} + i\frac{\partial v(x,y)}{\partial x}. \quad (2.139)$$

And, reversing the order leads to

$$\frac{df}{dz} = -i\frac{\partial u(x,y)}{\partial y} + \frac{\partial v(x,y)}{\partial y}. \quad (2.140)$$

Therefore, from equations 2.139 and 2.140, we must have

$$\frac{\partial u(x,y)}{\partial x} = \frac{\partial v(x,y)}{\partial y},$$
$$\frac{\partial u(x,y)}{\partial y} = -\frac{\partial v(x,y)}{\partial x}. \qquad (2.141)$$

Equation 2.141 is known as the "Cauchy–Reimann" condition. We can also readily verify the following from the Cauchy–Reimann condition

$$\left.\begin{aligned} &\frac{\partial^2 u}{\partial x^2} + \frac{\partial^2 u}{\partial y^2} \\ &\frac{\partial v}{\partial x^2} + \frac{\partial^2 v}{\partial y^2} \end{aligned}\right\} = 0, \qquad (2.142)$$

Thus, the real and complex components of an analytic function satisfy the Laplace equation and are called "harmonic functions."

2.2.4 Adjoint Differential Forms

Recall our most general form of linear PDE of second order, given as $L(u) = 0$, shown in equation 2.80. We now introduce the concept of the differential form $M(v)$, which is adjoint to $L(u)$. The adjoint is defined such that the expression $vL(u) - uM(v)$ can be generally integrable and expressed as

$$vL(u) - uM(v) = \frac{\partial X}{\partial x} + \frac{\partial Y}{\partial y}. \qquad (2.143)$$

The problem is to determine M as a function of v and X, Y as functions of both u and v. To do this, we use the following identities:

$$vA\frac{\partial^2 u}{\partial x^2} - u\frac{\partial^2 Av}{\partial x^2} = \frac{\partial}{\partial x}\left(Av\frac{\partial u}{\partial x} - u\frac{\partial Av}{\partial x}\right), \ldots \qquad (2.144)$$

$$vB\frac{\partial^2 u}{\partial x\partial y} - u\frac{\partial^2 Bv}{\partial x\partial y} = \frac{\partial}{\partial x}\left(vB\frac{\partial u}{\partial y}\right) - \frac{\partial}{\partial y}\left(u\frac{\partial Bv}{\partial x}\right) = \ldots \qquad (2.145)$$

and

$$Dv\frac{\partial u}{\partial x} - u\frac{\partial}{\partial x}(-Dv) = \frac{\partial}{\partial x}(Duv), \ldots \qquad (2.146)$$

Three dots (...) in equations 2.144 and 2.146 indicate that they remain valid when we replace x by y and A, D by C, E. And the three dots in equation 2.145 indicate that we may use the symmetric expressions obtained from interchanging x and y. From this, we obtain

$$M(v) = \frac{\partial^2 Av}{\partial x^2} + 2\frac{\partial^2 Bv}{\partial x\partial y} + \frac{\partial^2 Cv}{\partial y^2} - \frac{\partial Dv}{\partial x} - \frac{\partial Ev}{\partial y} + Fv, \qquad (2.147)$$

$$X = A\left(v\frac{\partial u}{\partial x} - u\frac{\partial v}{\partial x}\right) + B\left(v\frac{\partial u}{\partial y} - u\frac{\partial v}{\partial y}\right) + \left(D - \frac{\partial A}{\partial x} - \frac{\partial B}{\partial y}\right)uv, \qquad (2.148)$$

and

$$Y = B\left(v\frac{\partial u}{\partial x} - u\frac{\partial v}{\partial x}\right) + C\left(v\frac{\partial u}{\partial y} - u\frac{\partial v}{\partial y}\right) + \left(E - \frac{\partial B}{\partial x} - \frac{\partial C}{\partial y}\right)uv. \qquad (2.149)$$

We can readily verify that L and M are reciprocal, i.e., $L(v)$ is the adjoint differential form to $M(u)$. Of particular interest in mathematical physics, including seismic wave propagation, are those differential expressions for which $L(u) = M(u)$. They are called "self-adjoint," and it can be readily verified that the conditions of self-adjointness is

$$\frac{\partial A}{\partial x} + \frac{\partial B}{\partial y} = D, \frac{\partial B}{\partial x} + \frac{\partial C}{\partial y} = E. \qquad (2.150)$$

2.3 Fundamentals of Tensor Algebra and Tensor Calculus

We previously stated in Chapter 1 that the fundamental basis for the linearized elastic wave theory is the Hooke's law. We discuss Hooke's law in Chapter 3 where we define the second-order strain and stress tensors and linearly relate them with one another using a fourth-order elastic stiffness/compliance tensor. Consequently, a good understanding of the tensors and the related mathematics is important before discussing seismic wave propagation.

Notions of tensors are best understood in invariant notation with non-orthogonal curvilinear coordinates. Mathematical details of such tensor algebra and calculus can be found in Backus (1967) and Fedorov (1968), and here we provide an outline.

2.3.1 System of Coordinates

Applications of mathematical and numerical methods to the geometrical and mechanical phenomena are formulated using a system of reference or a system of coordinates. Such a system allows a relation to be established between the numbers and points in space. For a three-dimensional Euclidean space $\mathfrak{R}^3$ for example, a system of coordinates attaches each point in space with a triplet of numbers, called the coordinates of the point.

In physics, many types of coordinate systems such as the *Cartesian*, *Cylindrical*, and *Spherical* coordinate systems are popular and are used to solve different types of problems. We introduce these different types later, and here we use the *curvilinear* coordinate system. It becomes evident later that all different types of coordinate systems are subsets of the curvilinear system, which we introduce here.

Figure 2.7 shows the illustration of the curvilinear coordinate system in $\mathfrak{R}^3$, which we use to develop the concept of the tensors. First, we define the origin of our system as the point "O" and associate the origin with the triplet of numbers (0, 0, 0). From the origin, we define three curves, which we denote as x^1, x^2, and x^3, shown with solid black in Figure 2.7. We designate the curves x^1, x^2, and x^3 to be the coordinate axes. By defining the origin and the coordinate axes, we can now relate any point P in the three-dimensional space with the triplet of numbers (a^1, a^2, a^3), as the measures along the three coordinate axes, shown with red dashed lines in Figure 2.7. We call this triplet (a^1, a^2, a^3) to be the coordinate of the point P. Note that the measures a^1, a^2, and a^3 can be distance, angle, or anything else. Also note that at any arbitrary point P in space, we can define the coordinate axes as curves parallel to x^1, x^2, and x^3 from that point. The curves along which any two of the given coordinates are constants are called the coordinate curves. For example, the curve along which x^2 and x^3 are constant is defined as the x^1 coordinate curve. Because in defining our coordinates we did not assume any orthogonality between them, the tangents to the coordinate curves, drawn from any arbitrary point P, are not coplanar and form a triad, which in general are not orthogonal. We therefore call our system of reference as *non-orthogonal curvilinear* coordinates. Here, we develop the tensor analysis using this system and show later that all other known coordinate systems are just special cases of this general system.

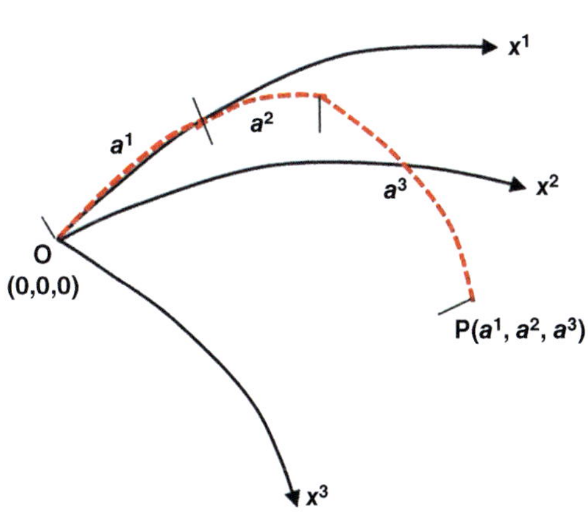

Figure 2.7 Illustration of the curvilinear system of coordinates.

2.3.2 What Are Tensors?

In tensor algebra, *tensor* refers to a generalized term given to a quantity. In an n-dimensional Euclidean space $\mathfrak{R}^n$, a scalar as a is defined as a tensor of order zero, a vector $\vec{\mathbf{v}}$ is a tensor of order one, $\vec{\vec{\mathbf{T}}}$ is a tensor of order two, and so on. In $\mathfrak{R}^n$, a tensor of order k maps onto n^k distinct components. Thus, a scalar maps onto $\mathfrak{R}^n \to \mathfrak{R}$, a vector maps onto $\mathfrak{R}^n \to \mathfrak{R}^n$, a tensor of order two maps $\mathfrak{R}^n \to \mathfrak{R}^n \times \mathfrak{R}^n$, a tensor of order three maps $\mathfrak{R}^n \to \mathfrak{R}^n \times \mathfrak{R}^n \times \mathfrak{R}^n$, and so on. Restricting to a three-dimensional Euclidean space $\mathfrak{R}^3$, we thus have $3^0 = 1$ component for a scalar a, $3^1 = 3$ components for a vector $\vec{\mathbf{v}}$, $3^2 = 9$ components for a second-order tensor $\vec{\vec{\mathbf{T}}}$, $3^3 = 27$ components for a third-order tensor $\vec{\vec{\vec{\Gamma}}}$, $3^4 = 81$ components for a fourth-order tensor $\vec{\vec{\vec{\vec{C}}}}$, and so on.

In the following, we omit using the arrows and designate the tensors of order greater than one with boldface symbols such as **v**, **T**, **Γ**, **C**, and explicitly mention its order. And for the tensors or order zero (scalars), we use non-boldface symbols such as a, b, c, etc.

2.3.3 Basis Vectors

Consider two adjacent points in $\Re^3$, M (x^1, x^2, x^3) and M_1 $(x^1 + dx^1, x^2 + dx^2, x^3 + dx^3)$, shown in Figure 2.8. If O is the origin of the coordinate system, then the directed linear segment $\overline{OM}$ is independent of the choice of the coordinates and is represented by the position vector **r** as

$$\mathbf{r} = \mathbf{r}(x^1, x^2, x^3). \tag{2.151}$$

From the *chain rule* of the elementary partial differential calculus, we can then write

$$\begin{aligned} d\mathbf{r} &= \frac{\partial \mathbf{r}}{\partial x^1}dx^1 + \frac{\partial \mathbf{r}}{\partial x^2}dx^2 + \frac{\partial \mathbf{r}}{\partial x^3}dx^3 \\ &= \sum_{i=1}^{3} \frac{\partial \mathbf{r}}{\partial x^i}dx^i = \sum_{i=1}^{3} dx^i \partial_i(\mathbf{r}). \end{aligned} \tag{2.152}$$

From here on, we omit the summation sign and assume that the repeated indices, appearing once as a superscript and then as a subscript or vice versa, are always summed over the index. In addition, we also assume that $\partial/\partial x^i \equiv \partial_i$. Thus, we write equation 2.152 as

$$d\mathbf{r} = dx^i \partial_i(\mathbf{r}). \tag{2.153}$$

Note that $d\mathbf{r}$ defined by equation 2.153 is represented in Figure 2.8 as the directed linear segment $\overline{MM_1}$ and is called the *displacement vector*. The vectors $\partial_i(\mathbf{r})$ in $\Re^3$ form three ordered points that are directed along the tangent to x^i coordinate axes at the point M under consideration. In Figure 2.8, $\partial_1(\mathbf{r})$, $\partial_2(\mathbf{r})$, and $\partial_3(\mathbf{r})$ are shown as blue arrows. In fact, the triplet of vectors represented as $\partial_i(\mathbf{r})$, in conjunction with the corresponding components along x^i, determines any directed segment **v** from the point M. Therefore, $\partial_i(\mathbf{r})$ forms the basis of defining any directed object from any arbitrary point M and is considered as the *basis vectors* in the space. Please also note that, in general, the basis vectors are not unit vectors. However, when dx^i in equation 2.153 equals the length of the arc along the x^i coordinate curve, then the basis vectors $\partial_i(\mathbf{r})$ are also *unit*

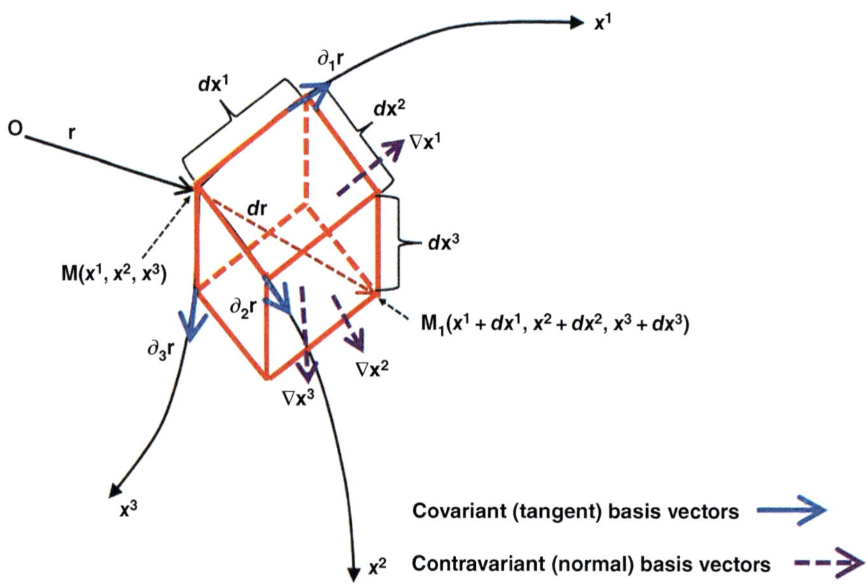

Figure 2.8 Illustration to define the basis vectors in curvilinear coordinate system.

vectors. Proof of this is straightforward and left as an exercise.

To summarize, basis vectors are fundamental vectors in a curvilinear system by which any vector from a given point in space can be expressed. They are directed along the tangent to the coordinate curves drawn from the given point and therefore can vary from point to point. In general, they are not unit vectors, but when the given coordinate is the measure of arc length, the basis vectors are unit vectors.

2.3.4 The Gradient Operator and the Covariant and Contravariant Basis Vectors

The rate of change of any quantity (scalar/vector/tensor) is called the gradient of that quantity and is directed along the direction along which the rate of change is maximum. Referring to Figure 2.8, let Q be any field defined in $\mathfrak{R}^3$. Also, let M be any point with coordinates (x^1, x^2, x^3) and M_1 be another point, infinitesimally close to M with coordinates $(x^1 + dx^1, x^2 + dx^2, x^3 + dx^3)$, so that the directed segment $\overline{MM_1}$ forms the displacement vector $d\mathbf{r}$. If Q is the value of the field at point M and $Q + dQ$ is the value of the field at M_1, then from the definition of the gradient we write

$$dQ = d\mathbf{r} \cdot \nabla Q. \tag{2.154}$$

Now, Q is a field in space, i.e., $Q = Q(x^1, x^2, x^3)$. Therefore,

$$dQ = dx^i \partial_i Q. \tag{2.155}$$

Because Q is arbitrary, we set x^i for Q in equation 2.154 and write

$$dx^i = d\mathbf{r} \cdot \nabla x^i. \tag{2.156}$$

Then, by setting dx^i from equation 2.156 into equation 2.155, we obtain

$$dQ = (d\mathbf{r} \cdot \nabla x^i)\partial_i Q$$

or

$$dQ = d\mathbf{r} \cdot (\nabla x^i \partial_i Q). \tag{2.157}$$

By comparing equations 2.154 and 2.157, we obtain

$$\nabla Q = (\nabla x^i \partial_i)Q. \tag{2.158}$$

Therefore, the gradient operator ∇ is defined in curvilinear coordinates as

$$\nabla = \nabla x^i \partial_i. \tag{2.159}$$

Now, in equation 2.153, we defined the displacement vector as $d\mathbf{r} = dx^j \partial_j(\mathbf{r})$. Substituting this into equation 2.156, we obtain

$$dx^i = [dx^j \partial_j(\mathbf{r})] \cdot \nabla x^i$$

or

$$\nabla \mathrm{x}^{\mathrm{i}} \cdot \partial_{\mathrm{j}}(\vec{\mathbf{r}}) = \frac{dx^{\mathrm{i}}}{dx^{\mathrm{j}}} = \delta^{\mathrm{i}}_{\mathrm{j}}. \tag{2.160}$$

Therefore, ∇x^i and $\partial_i(\mathbf{r})$ are mutually inverse of each other. Like $\partial_i(\mathbf{r})$, ∇x^i also forms a triplet of vectors through which any vector $\mathbf{A}$ can be defined as

$$\mathbf{A} = A^i \partial_i(\mathbf{r}) = A_i \nabla \mathrm{x}^i. \tag{2.161}$$

As we previously noted, $\partial_i(\mathbf{r})$ is directed along the tangent of the x^i axis at the given point M. From the definition of the gradient on the other hand, ∇x^i is directed along the outward normal to the $x^j - x^k$ surface at point M where $i \neq j$, $i \neq k$, and $j \neq k$, Thus, in $\mathfrak{R}^3$, ∇x^1 is along the normal to the $x^2 - x^3$ surface; ∇x^2 is along the normal to the $x^3 - x^1$ surface; and ∇x^3 is along the normal to the $x^1 - x^2$ surface. In Figure 2.8, ∇x^1, ∇x^2, and ∇x^3 are shown with purple arrows. Because of this geometrical relation, $\partial_i(\mathbf{r})$ are called the *tangent* or *covariant* basis vectors, and ∇x^i are called the *normal* or *contravariant* basis vectors. From the definition of the vector $\mathbf{A}$, shown in equation 2.161, A^i are the *contravariant* components of $\mathbf{A}$, and A_i are the *covariant* components of $\mathbf{A}$. Therefore, if a vector $\mathbf{A}$ is defined at any point in space, it can be defined either in terms of its contravariant component and covariant basis vectors or its covariant component and contravariant basis vectors as defined in equation 2.161.

Now, for any vector, say $\mathbf{v}$, we can write

$$\partial_i(\mathbf{r}) \cdot \mathbf{v} = \partial_i(\mathbf{r}) \cdot (v_j \nabla x^j) = v_j(\partial_i(\mathbf{r}) \cdot \nabla x^j)$$
$$= v_j \delta^j_i = v_i.$$

Therefore,

$$v_i = \partial_i(\mathbf{r}) \cdot \mathbf{v}. \tag{2.162}$$

Similarly,

$$v^i = \nabla x^i \cdot \mathbf{v}. \tag{2.163}$$

The derivation of equation 2.163 is straightforward. Equations 2.162 and 2.163 are the basis for obtaining the covariant and contravariant components.

2.3.5 Concept of Tensors

We pointed out earlier in Section 2.3.2 that a tensor is a general term given to a quantity, and we are now ready to develop the concept of tensors. A tensor of order (rank) n denotes a scalar-valued function of n vector variables. For example, a second-order tensor $\mathbf{T}$ can be defined as

$$T(\mathbf{u}, \mathbf{v}) = \mathbf{T} : \mathbf{uv}, \tag{2.164}$$

where $\mathbf{u}$ and $\mathbf{v}$ are vectors and the symbol ":" is a double-dot product. Similarly, a third-order tensor $\mathbf{S}$ is defined as

$$S(\mathbf{u}, \mathbf{v}, \mathbf{w}) = \mathbf{S} \vdots \mathbf{uvw}, \tag{2.165}$$

with $\mathbf{u}, \mathbf{v}, \mathbf{w}$ being three vectors and the symbol $\vdots$ representing a triple dot product. Now,

$$\begin{aligned} S(\mathbf{u}, \mathbf{v}, \mathbf{w}) &= S(u^i\partial_i(\mathbf{r}), v^j\partial_j(\mathbf{r}), w^k\partial_k(\vec{\mathbf{r}})) \\ &= \mathbf{S} \vdots (\partial_i(\mathbf{r})\partial_j(\mathbf{r})\partial_k(\mathbf{r}))u^iv^jw^k \\ &= S_{ijk}u^iv^jw^k. \end{aligned} \tag{2.166}$$

Thus, by analogy with the vectors, or the tensors of order one, we may write

$$\mathbf{S} = S_{ijk}\nabla x^i\nabla x^j\nabla x^k. \tag{2.167}$$

And, by analogy with vectors, S_{ijk} are the covariant components of $\mathbf{S}$. We can similarly write

$$\begin{aligned} S(\mathbf{u}, \mathbf{v}, \mathbf{w}) &= S(u_i\nabla x^i, v_j\nabla x^j, w_k\nabla x^k) \\ &= \mathbf{S} \vdots (u_i\nabla x^i v_j\nabla x^j w_k\nabla x^k) \\ &= S^{ijk}u_iv_jw_k, \end{aligned} \tag{2.168}$$

which, in turn, leads to

$$\mathbf{S} = S^{ijk}\partial_i(\mathbf{r})\partial_j(\mathbf{r})\partial_k(\mathbf{r}), \tag{2.169}$$

where S^{ijk} are the contravariant components of $\mathbf{S}$. Note that the summation conventions of the indices repeating as subscript and superscript or as superscript and subscript are implied in equations 2.167 and 2.168. Apart from the covariant and contravariant components, for tensors of rank greater than one, there can also be mixed components, such as

$$\mathbf{S} = S_{i\ k}^{\ j}\nabla x^i\partial_j(\mathbf{r})\nabla x^k. \tag{2.170}$$

Therefore, a tensor is defined as an invariant object of kind

$$\mathbf{T} = T_{\alpha\beta....}^{\quad\ \ \gamma\delta....}\nabla x^\alpha\nabla x^\beta....\partial_\gamma(\vec{\mathbf{r}})\partial_\delta(\vec{\mathbf{r}})...., \tag{2.171}$$

in which the summation is carried out over the indices $\alpha, \beta, \ldots, \gamma, \delta, \ldots$. Each term in the product $\nabla x^\alpha\nabla x^\beta \cdots \partial_\gamma(\mathbf{r})\partial_\delta(\mathbf{r}) \cdots$ has a definite order, and every term is considered as linearly independent with respect to the other terms in the sum. The object $\mathbf{T}$ is independent of the choice of the coordinate system although its components do depend upon such choice. In general, by interchanging any two upper or any two lower indices in the components of a tensor will give another tensor with components different from the components of the original tensor. However, if after interchanging any two lower or any two upper indices in the components of a tensor the tensor does not change, then the tensor is called *symmetric* with respect to these indices. And, if such interchanged indices differ only in sign, the tensor is then called *antisymmetric* with respect to those indices.

2.3.6 The Identity Tensor

The identity tensor, usually denoted as $\mathbf{I}$, is a second-order tensor and is defined by the property that for any vector $\mathbf{v}$ we have

$$\mathbf{v} \cdot \mathbf{I} = \mathbf{I} \cdot \mathbf{v} = \mathbf{v}. \tag{2.172}$$

Components of $\mathbf{I}$ are given as

$$g^{ij} = \mathbf{I} : \nabla x^i\nabla x^j = \nabla x^i \cdot \nabla x^j. \tag{2.173}$$

Equation 2.173 gives the contravariant components of $\mathbf{I}$. We can also express the contravariant and other mixed components as

$$\left.\begin{aligned} g_{ij} &= \partial_i(\mathbf{r})\cdot\partial_j(\mathbf{r}),\\ g^i{}_j &= \nabla x^i\cdot\partial_j(\vec{\mathbf{r}}) = \delta^i{}_j,\\ g_i{}^j &= \partial_i(\vec{\mathbf{r}})\cdot\nabla x^j = \delta_i{}^j. \end{aligned}\right\} \tag{2.174}$$

Note from above that the identity tensor is a symmetric tensor and given as

$$\begin{aligned} \mathbf{I} &= g^{ij}\partial_i(\mathbf{r})\partial_j(\mathbf{r}) = g_{ij}\nabla x^i\nabla x^j\\ &= g^i{}_j\partial_i(\mathbf{r})\nabla x^j\\ &= g_i^j\nabla x^i\partial_j(\mathbf{r}). \end{aligned} \tag{2.175}$$

Also note that

$$\begin{aligned} \mathbf{I} &= g_i^j\nabla x^i\partial_j(\mathbf{r}) = \delta_i^j\nabla x^i\partial_j(\mathbf{r})\\ &= \nabla x^i\partial_i(\mathbf{r}) = \nabla\mathbf{r}. \end{aligned} \tag{2.176}$$

Now, from the definition of the identity tensor,

$$\begin{aligned} \nabla x^i &= \nabla x^i\cdot\mathbf{I} = \nabla x^i\cdot(\nabla x^j\partial_j(\mathbf{r}))\\ &= (\nabla x^i\cdot\nabla x^j)\partial_j(\mathbf{r}) = g^{ij}\partial_j(\mathbf{r}). \end{aligned} \tag{2.177}$$

We can similarly show that

$$\partial_i(\mathbf{r}) = g_{ij}\nabla x^j. \tag{2.178}$$

Now, for any vector $\mathbf{v}$, its contravariant components v^i are given by equation 2.163 as $v^i = \nabla x^i.\mathbf{v}$. But $\nabla x^i.\mathbf{v} = \mathbf{v}.\nabla x^i = (v_j\nabla x^j).\nabla x^i = v_j g^{ij}$. Therefore,

$$v^i = g^{ij}v_j. \tag{2.179}$$

Similarly, we can show that

$$v_i = g_{ij}v^j. \tag{2.180}$$

Equations 2.177–2.180 are the index raising and lowering rule for obtaining contravariant basis vectors or components from covariant basis vectors or components and vice versa from the components of the identity tensor. We can further extend this index raising and lowering rule to a tensor of any arbitrary order. For example,

$$\left.\begin{aligned} T_{ij\ldots}{}^{klm\ldots} &= g^{\alpha i}T^{\alpha}{}_{j\ldots}{}^{klm\ldots}\\ T_{ij\ldots}{}^{klm\ldots} &= g_{\alpha k}T_{ij\ldots\alpha}{}^{lm\ldots} \end{aligned}\right\}. \tag{2.181}$$

2.3.7 Elements of Tensor Algebra

Now that we have defined the basis vectors and tensors, we are ready to define the algebraic operations such as addition and multiplication with tensors.

Addition of Tensors

Addition of tensors are defined only for the tensors of the same rank. To obtain the components of a tensor $\mathbf{T}$, which is the sum of two tensors, say $\mathbf{T}'$ and $\mathbf{T}''$, it is necessary to carry out the operation of raising and lowering of indices to bring individual tensors to the same index structure. After this, the components of the sum are simply the sums of the corresponding components, and the rank of the summed tensor is same as the rank of the individual tensors that are summed.

Multiplication of a Tensor by a Scalar

Multiplication of a tensor by a scalar a gives another tensor of the same rank all whose components are multiplied by a.

Multiplication of Tensors

Let $\mathbf{T}'$ is a fourth-order tensor and given as

$$\mathbf{T}' = T'_{ij}{}^{kl}\nabla x^i\nabla x^j\partial_k(\mathbf{r})\partial_l(\mathbf{r}) \tag{2.182}$$

and $\mathbf{T}''$ be a second-order tensor and given as

$$\mathbf{T}'' = T''_s{}^w\nabla x^s\partial_w(\mathbf{r}). \tag{2.183}$$

It is then possible to form a new tensor by multiplication of the components of the above two tensors. The product is another tensor of rank equal to the sum of ranks of the individual tensors. Thus, if $\mathbf{T} = \mathbf{T}'\mathbf{T}''$, then $\mathbf{T}$ is a sixth-order tensor and given as

$$\begin{aligned} \mathbf{T} = \mathbf{T}'\mathbf{T}'' &= T'_{ij}{}^{kl}T''_s{}^w\nabla x^i\nabla x^j\partial_k(\mathbf{r})\partial_l(\mathbf{r})\\ &\times\nabla x^s\partial_w(\mathbf{r}) = T_{ij}{}^{kl}{}_s{}^w\nabla x^i\nabla x^j\partial_k(\mathbf{r})\\ &\times\partial_l(\mathbf{r})\nabla x^s\partial_w(\mathbf{r}). \end{aligned} \tag{2.184}$$

From equation 2.184, we can write the components of the new tensor $\mathbf{T}$ as

$$T_{ij}{}^{kl}{}_s{}^w = T'_{ij}{}^{kl}T''_s{}^w. \tag{2.185}$$

It must be noted that the resultant tensor depends upon the order of multiplication. Multiplication of two tensors can be extended easily to the product of any number of tensors. We also note here that in the product, summation is implied with respect to all indices $1, 2, \ldots, n$, where n is the number of dimensions.

The Dot Product Between Tensors

As discussed above, we have shown that the product between a fourth-order tensor $\mathbf{T}'$ and a second-order tensor $\mathbf{T}''$ is a sixth-order tensor $\mathbf{T}$. This can easily be extended to the fact that the product between any two tensors of order m and n is a tensor of order $m+n$. The product between a tensor and a scalar, which we discussed above in Section "Multiplication of Tensors," is, in fact, the same. Because a scalar is a tensor of order zero, the product of a scalar and a tensor of order m is a tensor of order m.

Besides the plain product between two tensors of orders m and n resulting in a tensor of higher or same order $(m+n)$, there are also other products between two tensors. From two given tensors, it is possible to obtain new tensors of rank smaller than the sum of their ranks by using the invariant operation of scalar multiplication between any pair of triplets of base vectors, corresponding to a given index. For example, using the same fourth-order tensor $\mathbf{T}'$ and the second-order tensor $\mathbf{T}''$, we can obtain a second-order tensor $\mathbf{T}^*$ using the following double-scalar or double-dot product:

$$\begin{aligned}\mathbf{T}^* = \mathbf{T}' : \mathbf{T}'' &= [T'^{\,kl}_{ij}\nabla x^i \nabla x^j \partial_k(\mathbf{r})\partial_1(\mathbf{r})] : [T''^{\,w}_{s}\nabla x^s \partial_w(\mathbf{r})]\\ &= T'^{\,kl}_{ij} T''^{\,w}_{s}(\nabla x^i \cdot \nabla x^s)(\nabla x^j \cdot \partial_w(\mathbf{r}))\partial_k(\mathbf{r})\partial_l(\mathbf{r})\\ &= T'^{\,kl}_{ij} T''^{\,w}_{s} g^{is}\delta^j{}_w \partial_k(\mathbf{r})\partial_l(\mathbf{r})\\ &= T'^{\,kl}_{ij} T''^{\,j}_{s} g^{is}\partial_k(\mathbf{r})\partial_l(\mathbf{r})\\ &= T'^{\,kl}_{ij} T''^{\,ij}\partial_k(\mathbf{r})\partial_l(\mathbf{r}) = T^{*kl}\partial_k(\mathbf{r})\partial_l(\mathbf{r}). \end{aligned} \quad (2.186)$$

Scalar product of any tensor can also be carried out with the identity tensor. For example, if $\mathbf{T}$ is a second-order tensor, given as $\mathbf{T} = T^{ij}\partial_i(\mathbf{r})\partial_j(\mathbf{r})$, then

$$\begin{aligned}\mathbf{T}\cdot\mathbf{I} &= [T^{ij}\partial_i(\mathbf{r})\partial_j(\mathbf{r})]\cdot[\delta_l{}^k \nabla x^l \partial_k(\mathbf{r})]\\ &= T^{ij}\partial_i(\mathbf{r})\cdot\partial_k(\mathbf{r})\nabla x^k \partial_j(\mathbf{r})\\ &= T^{ij}g_{ik}\nabla x^k \partial_j(\mathbf{r}) = T_k{}^j \nabla x^k \partial_j(\mathbf{r}). \end{aligned} \quad (2.187)$$

Similarly,

$$\begin{aligned}\mathbf{T}\cdot\mathbf{I}\cdot\mathbf{I} &= T_k{}^j \nabla x^k \partial_j(\mathbf{r})\cdot[\delta^m{}_n \partial_m(\mathbf{r})\nabla x^n]\\ &= T_k{}^j[\partial_j(\mathbf{r})\cdot\partial_m(\mathbf{r})]\nabla x^k \nabla x^m\\ &= T_k{}^j g_{jm}\nabla x^k \nabla x^m\\ &= T_{km}\nabla x^k \nabla x^m. \end{aligned} \quad (2.188)$$

Thus, raising and lowering indices reduce to repeated scalar multiplication with the identity tensor. However, instead of using one index of the identity tensor used in equations 2.187 and 2.188, if we use both indices and compute a double-dot product, we obtain a tensor of lower rank. For example, let $\mathbf{T}$ be a fourth-order tensor given as

$$\mathbf{T} = T_{mn}{}^{rs}\nabla x^m \nabla x^n \partial_r(\mathbf{r})\partial_s(\mathbf{r}). \quad (2.189)$$

Then,

$$\mathbf{T}^* = \mathbf{T} : \mathbf{I} = [T_{mn}{}^{rs}\nabla x^m \nabla x^n \partial_r(\mathbf{r})\partial_s(\mathbf{r})] : [\delta^i{}_j \partial_i(\mathbf{r})\nabla x^j], \quad (2.190)$$

which, after some algebra leads to

$$\begin{aligned}\mathbf{T}^* = \mathbf{T} : \mathbf{I} &= T_{mi}{}^{is}\nabla x^m \partial_s(\mathbf{r})\\ &= T^*{}_m{}^s \nabla x^m \partial_s(\mathbf{r}). \end{aligned} \quad (2.191)$$

The invariant operation of transformation of fourth-order tensor $\mathbf{T}$ into a second-order tensor $\mathbf{T}^*$ shown above is produced by summing its components on the n and t when $n = r$ and is called *contraction*. Obviously, contraction reduces a tensor of rank n to another tensor of rank $n-2$. Additionally, it is also evident that contraction is possible only for the tensors with rank $n \geq 2$.

The Vector (Cross) Product

The vector or cross product is useful in rotations. To define such product, we restrict to the three-dimensional Euclidean space $\Re^3$ in a curvilinear coordinate system $\{x^1, x^2, x^3\}$ such that the basis vectors $\partial_1(\mathbf{r})$, $\partial_2(\mathbf{r})$, $\partial_3(\mathbf{r})$ and $\nabla x^1, \nabla x^2, \nabla x^3$ form a right-handed system (Figure 2.8). Tensors are independent of either a left- or a right-handed system. To start our discussion, we first define the third-order tensor $\mathbf{J}$ as

$$\mathbf{J} = J^{ijk}\partial_i(\mathbf{r})\partial_j(\mathbf{r})\partial_k(\mathbf{r}) = J_{ijk}\nabla x^i \nabla x^j \nabla x^k. \quad (2.192)$$

Therefore, the components of **J** can be defined as

$$\left.\begin{aligned} J^{ijk} &= \mathbf{J} \vdots \nabla x^i \nabla x^j \nabla x^k \\ &= \nabla x^i \cdot \nabla x^j \times \nabla x^k \\ \mathrm{J}_{\mathrm{ijk}} &= \mathbf{J} \vdots \partial_i(\mathbf{r})\partial_j(\mathbf{r})\partial_k(\mathbf{r}) \\ &= \partial_i(\mathbf{r}) \cdot \partial_j(\mathbf{r}) \times \partial_k(\mathbf{r}) \end{aligned}\right\}. \tag{2.193}$$

Now, from the knowledge of scalar triple product in a right-handed system, we write

$$\begin{aligned} &\partial_1(\mathbf{r}) \cdot \partial_2(\mathbf{r}) \times \partial_3(\mathbf{r}) = \partial_2(\mathbf{r}) \cdot \partial_3(\mathbf{r}) \times \partial_1(\mathbf{r}) \\ &\quad = \partial_3(\mathbf{r}) \cdot \partial_1(\mathbf{r}) \times \partial_2(\mathbf{r}) \\ &\quad = -\partial_2(\mathbf{r}) \cdot \partial_1(\mathbf{r}) \times \partial_3(\mathbf{r}) \\ &\quad = -\partial_1(\mathbf{r}) \cdot \partial_3(\mathbf{r}) \times \partial_2(\mathbf{r}) \\ &\quad = -\partial_3(\mathbf{r}) \cdot \partial_2(\mathbf{r}) \times \partial_1(\mathbf{r}). \end{aligned} \tag{2.194}$$

Therefore, $J_{123} = J_{231} = J_{312} = -J_{213} = -J_{132} = -J_{321}$. Thus,

$$J_{ijk} = \partial_i(\mathbf{r}) \cdot \partial_j(\mathbf{r}) \times \partial_k(\mathbf{r})\varepsilon_{ijk}. \tag{2.195}$$

Similarly,

$$J^{ijk} = \nabla x^i \cdot \nabla x^j \times \nabla x^k \varepsilon^{ijk}. \tag{2.196}$$

The symbols ε_{ijk} and ε^{ijk} in equations 2.195 are defined such that they are +1 where the indices (i, j, k) are in the order 123, 231, and 312, are −1 where they are in the order 231, 132, and 321, and are zero when any two or all the indices are repeated. Now, from the definition of the identity tensor $\mathbf{I} = \nabla x^i \partial_i(\mathbf{r}) = \partial_i(\mathbf{r}) \nabla x^i$ and equation 2.195, it is straightforward to show that

$$\left.\begin{aligned} \nabla x^1 &= \frac{\partial_2(\mathbf{r}) \times \partial_3(\mathbf{r})}{\partial_1(\mathbf{r}) \cdot \partial_2(\mathbf{r}) \times \partial_3(\mathbf{r})} \\ \nabla x^2 &= \frac{\partial_3(\mathbf{r}) \times \partial_1(\mathbf{r})}{\partial_2(\mathbf{r}) \cdot \partial_3(\mathbf{r}) \times \partial_1(\mathbf{r})} \\ \nabla x^3 &= \frac{\partial_1(\mathbf{r}) \times \partial_2(\mathbf{r})}{\partial_3(\mathbf{r}) \cdot \partial_1(\mathbf{r}) \times \partial_2(\mathbf{r})} \end{aligned}\right\}. \tag{2.197}$$

Equation 2.196 and the definition of the identity tensor similarly provide

$$\left.\begin{aligned} \partial_1(\mathbf{r}) &= \frac{\nabla x^2 \times \nabla x^3}{\nabla x^1 \cdot \nabla x^2 \times \nabla x^3} \\ \partial_2(\mathbf{r}) &= \frac{\nabla x^3 \times \nabla x^1}{\nabla x^2 \cdot \nabla x^3 \times \nabla x^1} \\ \partial_3(\mathbf{r}) &= \frac{\nabla x^1 \times \nabla x^2}{\nabla x^3 \cdot \nabla x^1 \times \nabla x^2} \end{aligned}\right\}. \tag{2.198}$$

Now from equation 2.194, the denominators of all equations in equation 2.197 are equal. Denoting this denominator as J, we can then write

$$\begin{aligned} \nabla x^1 \cdot \nabla x^2 \times \nabla x^3 &= J^{-3}[\{\partial_2(\mathbf{r}) \times \partial_3(\mathbf{r})\} \\ &\quad \cdot \{\partial_3(\mathbf{r}) \times \partial_1(\mathbf{r})\} \times \{\partial_1(\mathbf{r}) \times \partial_2(\mathbf{r})\}]. \end{aligned} \tag{2.199}$$

And, from the vector identity

$$\mathbf{a} \times \mathbf{b} \times \mathbf{c} = (\mathbf{a} \cdot \mathbf{c})\mathbf{b} - (\mathbf{b} \cdot \mathbf{c})\mathbf{a}, \tag{2.200}$$

for any three vectors **a**, **b**, **c**, we write equation 2.199, after some algebra, as

$$\nabla x^1 \cdot \nabla x^2 \times \nabla x^3 = J^{-1}. \tag{2.201}$$

From equation 2.201, in conjunction with equations 2.195 and 2.196, we now readily see that J^{ijk} and J_{ijk} are inverse of each other for $i \neq j \neq k$.

We now establish the general expressions if the cross products between any two basis vectors $\nabla x^i, \nabla x^j$ or $\partial_i(\mathbf{r}), \partial_j(\mathbf{r})$ is known, then the cross products between the vectors and tensors can easily be computed:

$$\begin{aligned} \nabla x^i \times \nabla x^j &= \mathbf{I} \cdot \nabla x^i \times \nabla x^j \\ &= (\partial_k(\mathbf{r})\nabla x^k) \cdot (\nabla x^i \times \nabla x^j) \\ &= \partial_k(\mathbf{r})\nabla x^k \cdot \nabla x^i \times \nabla x^j \\ &= \partial_k(\mathbf{r})J^{kij} = \mathbf{I} \cdot \partial_k(\mathbf{r})J^{kij} \\ &= \nabla x^l \partial_l(\mathbf{r}) \cdot \partial_k(\mathbf{r})J^{kij} \\ &= \nabla x^l g_{lk} J^{kij} = \nabla x^l \mathrm{J}_l{}^{ij} \\ &= \nabla x^k J_k{}^{ij}. \end{aligned} \tag{2.202}$$

But, from the knowledge of scalar triple product

$$\begin{aligned} J_k{}^{ij} &= \partial_k(\mathbf{r}) \cdot \nabla x^i \times \nabla x^j \\ &= \nabla x^i \cdot \nabla x^j \times \partial_k(\mathbf{r}) = J_{ij}{}^k. \end{aligned} \tag{2.203}$$

Using equation 2.203, we thus write equation 2.202 as

$$\nabla x^i \times \nabla x^j = \nabla x^k J^{ij}{}_k. \tag{2.204}$$

In a similar fashion, it can be shown that

$$\nabla x^i \times \nabla x^j = \partial_k(\mathbf{r})J^{ijk}. \tag{2.205}$$

And, from equations 2.196 and 2.201, we have

$$J^{ijk} = \nabla x^i \cdot \nabla x^j \times \nabla x^k \varepsilon^{ijk} = J^{-1}\varepsilon^{ijk}. \tag{2.206}$$

From equations 2.204–2.206, we obtain

$$\begin{aligned}\nabla x^i \times \nabla x^j &= \nabla x^k J^{ij}{}_k = \partial_k(\mathbf{r}) J^{ijk} \\ &= J^{-1}\varepsilon^{ijk}\partial_k(\mathbf{r}). \end{aligned} \tag{2.207}$$

Similarly, we can also show that

$$\begin{aligned}\partial_i(\mathbf{r}) \times \partial_j(\mathbf{r}) &= \partial_k(\mathbf{r}) J_{ij}{}^k = \nabla x^k J_{ijk} \\ &= J\varepsilon_{ijk}\nabla x^k. \end{aligned} \tag{2.208}$$

Equations 2.207 and 2.208 provide the formulas for cross products of the contravariant and covariant basis vectors. These equations, however, have the term J, which contains the cross product of the basis vectors (equations 2.194 and 2.201). Therefore, unless we find a way to express J in some other way, equations 2.207 and 2.208 are of little use. Now, the covariant components of the identity tensor can be expressed as a matrix

$$\mathbf{g_{ij}} = \begin{pmatrix} g_{11} & g_{12} & g_{13} \\ g_{21} & g_{22} & g_{23} \\ g_{31} & g_{32} & g_{33} \end{pmatrix}. \tag{2.209}$$

Note that the matrix $\mathbf{g_{ij}}$ defined in equation 2.209 is symmetric and its each component can be obtained from equation 2.174. It is also important to point out that although symmetric, $\mathbf{g_{ij}}$ is not an identity matrix for general curvilinear coordinate systems. This is also true for the matrix $\mathbf{g^{ij}}$ with all elements being the contravariant components of the identity matrix. However, from equation 2.174, the matrices written as the mixed components of the identity tensor are identity matrices in any coordinate system.

Using equation 2.174 for each component of the matrix $\mathbf{g_{ij}}$, it can be shown that the determinant of this matrix is given as

$$J = [\det(\mathbf{g_{ij}})]^{\frac{1}{2}} = g. \tag{2.210}$$

We discuss the determinant in the next subsection. Equation 2.210, in conjunction with equations 2.207 and 2.208, provides the way to obtain the cross products of the basis vectors. And, as we previously mentioned, once the cross products between the basis vectors are defined, they can be defined for the tensors.

Second-Order Tensors

Because the second-order tensors in $\Re^3$ can be written as 3×3 matrices (see equation 2.209), we can use the matrix algebra to define the following useful operations with them.

Determinant of a Second-Order Tensor: If $\mathbf{A}$ is a second-order tensor, then we denote the determinant of $\mathbf{A}$ as $\det(\mathbf{A})$ and denote as

$$\det(\mathbf{A}) = A^1{}_i A^2{}_j A^3{}_k \varepsilon^{ijk} = A_1{}^i A_2{}^j A_3{}^k \varepsilon_{ijk}. \tag{2.211}$$

The determinant of a second-order tensor is invariant, i.e., it does not change with the choice of the coordinate system being used to define the tensor. Sometimes the notations $\det(\mathbf{A_{ij}})$ or $\det(\mathbf{A^{ij}})$ are used to denote the determinant of the matrix with the covariant components A_{ij} or the contravariant components A^{ij} of the tensor $\mathbf{A}$ expressed as

$$\left.\begin{aligned}\det(\mathbf{A_{ij}}) &= A_{i1}A_{j2}A_{k3}\varepsilon^{ijk} \\ \det(\mathbf{A^{ij}}) &= A^{i1}A^{j2}A^{k3}\varepsilon_{ijk}\end{aligned}\right\}. \tag{2.212}$$

We must, however, point out that

$$\begin{aligned}\det(\mathbf{A}) &= \det(\mathbf{A^i{}_j}) = \det(\mathbf{A_i{}^j}) \\ &\neq \det(\mathbf{A_{ij}}) \neq \det(\mathbf{A^{ij}}) \end{aligned} \tag{2.213}$$

for a general curvilinear coordinate system. Expanding equation 2.211 in terms of the basis vectors and from the definition of ε_{ijk}, it can be shown that

$$\det(\mathbf{A}) = \frac{(\partial_1(\mathbf{r}) \cdot \mathbf{A}) \cdot (\partial_2(\mathbf{r}) \cdot \mathbf{A}) \times (\partial_3(\mathbf{r}) \cdot \mathbf{A})}{\partial_1(\mathbf{r}) \cdot \partial_2(\mathbf{r}) \times \partial_3(\mathbf{r})}. \tag{2.214}$$

On the right-hand side of the equation 2.214, the numerator denotes the new volume element under the tensor mapping $\mathbf{A}$. The denominator on the right-hand side of equation 2.214 on the other hand is the original volume element J we previously introduced in our discission of the cross product. Therefore, $\det(\mathbf{A})$ is the change in volume element due to the operator $\mathbf{A}$ and is independent of the coordinate system used to define its components.

Transpose of a Second-Order Tensor: If $\mathbf{A}$ is a second-order tensor, then $\mathbf{A}^T$ is another second-order tensor called **A** transpose or the transpose of **A**, which has the property that for any two vectors **u** and **v**, we have

$$\mathbf{u} \cdot \mathbf{A} \cdot \mathbf{v} = \mathbf{v} \cdot \mathbf{A}^{\mathrm{T}} \cdot \mathbf{u}. \tag{2.215}$$

Note that if we represent **A** as $\mathbf{A} = A_{ij} \nabla x^i \nabla x^j$, then $\mathbf{A}^T = A_{ji} \nabla x^i \nabla x^j = A_{ij} \nabla x^j \nabla x^i$. Therefore, if

$$\mathbf{A} = \sum_{ij} \mathbf{u}_i \mathbf{v}_j, \tag{2.216}$$

then

$$\mathbf{A}^T = \sum_{ji} \mathbf{v}_j \mathbf{u}_i. \tag{2.217}$$

Also

$$\left.\begin{aligned} A_{ij} &= \partial_i(\mathbf{r}) \cdot \mathbf{A} \cdot \partial_j(\mathbf{r}) = \partial_j(\mathbf{r}) \cdot \mathbf{A}^T \cdot \partial_i(\mathbf{r}) = A_{ji} \\ A_i^{\ j} &= \partial_i(\mathbf{r}) \cdot \mathbf{A} \cdot \nabla x^j = \nabla x^j \cdot \mathbf{A}^T \cdot \partial_i(\mathbf{r}) = A^j_{\ i} \\ A^i_{\ j} &= \nabla x^i \cdot \mathbf{A} \cdot \partial_j(\mathbf{r}) = \partial_j(\mathbf{r}) \cdot \mathbf{A}^T \cdot \nabla x^i = A_j^{\ i} \\ A^{ij} &= \nabla x^i \cdot \mathbf{A} \cdot \nabla x^j = \nabla x^j \cdot \mathbf{A}^T \cdot \nabla x^i = A^{ji} \end{aligned}\right\}. \tag{2.218}$$

Trace of a Second-Order Tensor: Let **A** be a second-order tensor defined in $\Re^3$ as

$$\mathbf{A} = \sum_{ij} \mathbf{u}_i \mathbf{v}_j, \tag{2.219}$$

where **u** and **v** are vectors. Then, the trace of **A** is denoted as

$$\mathrm{tr}(\mathbf{A}) = \sum_{ij} \mathbf{u}_i \cdot \mathbf{v}_j. \tag{2.220}$$

As an example, for any vector **u**, $\nabla.\ \mathbf{u} = \mathrm{tr}(\nabla \mathbf{u})$. The operation $\nabla.\ \mathbf{u}$ is called the divergence of the vector **u**, which we introduce in the next section. In general, if $\mathbf{A} = A_{ij} \nabla x^i \nabla x^j$, then

$$\begin{aligned} \mathrm{tr}(\mathbf{A}) &= A_{ij} \nabla x^i \cdot \nabla x^j = A_{ij} g^{ij} = A^{ij} g_{ij} \\ &= A_i^{\ i} = A^i_{\ i}. \end{aligned} \tag{2.221}$$

Note that like determinant, the trace of a second-order tensor is invariant, i.e., it is independent of the choice of the coordinate system.

Inverse of a Second-Order Tensor: If **A** and **B** are second-order tensors such that $\mathbf{A} \cdot \mathbf{B} = \mathbf{B} = \mathbf{A} = \mathbf{I}$, with **I** being the identity tensor, then **A** is the inverse of **B** and **B** is the inverse of **B**, and we write

$$\mathbf{A} = \mathbf{B}^{-1}; \quad \mathbf{B} = \mathbf{A}^{-1}. \tag{2.222}$$

It must be noted that $\mathbf{A}^{-1}$ exists only when $\det(\mathbf{A}) \neq 0$. If $\det(\mathbf{A}) = 0$, then **A** is said to be singular. If **A** is non-singular, i.e., it has an inverse, then

$$\mathbf{u} \cdot \mathbf{A} \cdot \mathbf{v} \neq 0, \tag{2.223}$$

for all vectors **u**, $\mathbf{v} \neq \mathbf{0}$. If $\mathbf{u} \cdot \mathbf{A} \cdot \mathbf{v} > 0$ for all **u**, $\mathbf{v} \neq \mathbf{0}$, then **A** is said to be positive-definite.

2.3.8 Elements of Tensor Calculus

In the preceding sections, we discussed the concept of tensors, their representation using the basis vectors, elementary tensor algebra, and useful matrix operations for the second-order tensors. We are now ready to discuss elementary tensor calculus, and discuss the following important topics, which will be useful later:

1) Differentiation of tensors.
2) The divergence operator.
3) The curl operator.
4) Differentiation of basis vectors.
5) Change of coordinates.

Differentiation of Tensors

If **T** is a second-order tensor, then $\nabla \mathbf{T} = \nabla x^i \partial_i \mathbf{T}$ is a third-order tensor, which can be expressed as a combination of the triad $\nabla x^i, \nabla x^j, \nabla x^k$ or $\partial_i(\mathbf{r}), \partial_j(\mathbf{r}), \partial_k(\mathbf{r})$, or any combination of them. We define coefficients of such an expansion as $D_i T_{jk}, D^i T^{jk}$, etc. Therefore,

$$\begin{aligned} \nabla \mathbf{T} &= (D_i T_{jk}) \nabla x^i \nabla x^j \nabla x^k, \\ D_i T_{jk} &= \nabla \mathbf{T} \ \vdots\ \partial_i(\mathbf{r}) \partial_j(\mathbf{r}) \partial_k(\mathbf{r}). \end{aligned} \tag{2.224}$$

$$\begin{aligned} \nabla \mathbf{T} &= (D_i T_j^{\ k}) \nabla x^i \nabla x^j \partial_k(\mathbf{r}), \\ D_i T_j^{\ k} &= \nabla \mathbf{T} \ \vdots\ \partial_i(\mathbf{r}) \partial_j(\mathbf{r}) \nabla x^k. \end{aligned} \tag{2.225}$$

$$\begin{aligned} \nabla \mathbf{T} &= (D^i T^{jk}) \partial_i(\mathbf{r}) \partial_j(\mathbf{r}) \partial_k(\mathbf{r}), \\ D^i T^{jk} &= \nabla \mathbf{T} \ \vdots\ \nabla x^i \nabla x^j \nabla x^k. \end{aligned} \tag{2.226}$$

Now, for any scalar Q, vectors $\mathbf{u}$, $\mathbf{v}$, and second-order tensor $\boldsymbol{\tau}$, it can be readily verified that

$$\left.\begin{aligned} D_i(Qu_j) &= (D_iQ)u_j + Q(D_iu_j) \\ D_i(u_jv_k) &= (D_iu_j)v_k + u_j(D_iv_k) \\ D_i(\tau_{kl}v_m) &= (D_i\tau_{kl})v_m + \tau_{kl}(D_iv_m) \end{aligned}\right\}. \tag{2.227}$$

Therefore, for any scalar Q, $D_iQ = \partial_iQ$; for any vector $\mathbf{u}$, $D_iu_j = \nabla\mathbf{u}:\partial_i(\mathbf{r})\partial_j(\mathbf{r})$; for any second-order tensor $\boldsymbol{\tau}$, $D_i\tau_{kl} = \nabla\boldsymbol{\tau}\vdots\partial_i(\mathbf{r})\partial_k(\mathbf{r})\partial_l(\mathbf{r})$; and so on. Again, if $\boldsymbol{\tau}$ and $\mathbf{M}$ are second-order tensors, then their product $\boldsymbol{\tau}\mathbf{M}$ is clearly a fourth-order tensor, i.e.,

$$\boldsymbol{\tau}\mathbf{M} = \tau_{ij}M^l{}_m\nabla x^i\nabla x^j\partial_l(\mathbf{r})\nabla x^m, \tag{2.228}$$

with

$$\left.\begin{aligned} \boldsymbol{\tau} &= \tau_{ij}\nabla x^i\nabla x^j \\ \mathbf{M} &= M^l{}_m\partial_l(\mathbf{r})\nabla x^m \end{aligned}\right\}. \tag{2.229}$$

Therefore, for two second-order tensors, we have

$$\begin{aligned} \nabla(\boldsymbol{\tau}\mathbf{M}) = D_i(\tau_{jk}M^l{}_m)&\nabla x^i\nabla x^j\nabla x^k \\ &\times\partial_l(\mathbf{r})\nabla x^m, \end{aligned} \tag{2.230}$$

where

$$\begin{aligned} D_i(\tau_{jk}M^l{}_m) = \nabla(\boldsymbol{\tau}\overset{\rightrightarrows}{\mathbf{M}})\ \vdots\vdots\ &\partial_i(\mathbf{r})\partial_j(\mathbf{r}) \\ &\times\partial_k(\mathbf{r})\nabla x^l\partial_m(\mathbf{r}). \end{aligned} \tag{2.231}$$

Using the definition of the gradient operator $\nabla = \nabla x^n\partial_n$, equation 2.231, after some algebra, leads to

$$D_i(\tau_{jk}M^l{}_m) = D_i(\tau_{jk})M^l{}_m + \tau_{jk}D_i(M^l{}_m). \tag{2.232}$$

The Divergence Operator

Divergence is obtained by "dotting into each other" the first two members of each "polyad" of the gradient. For example, if $\mathbf{u}$ is a vector, then

$$\nabla\mathbf{u} = D_iu_j\nabla x^i\nabla x^j \tag{2.233}$$

and

$$\nabla\cdot\mathbf{u} = (D_iu_j)\nabla x^i\cdot\nabla x^j = D_iu_jg^{ij} = D_iu^i. \tag{2.234}$$

We can similarly show that

$$\nabla\cdot\mathbf{u} = D^iu_i. \tag{2.235}$$

And, as we mentioned in the preceding section, $\nabla.\ \mathbf{u}$ is the trace of the square matrix formed by the components of $\nabla\mathbf{u}$. For a second-order tensor $\mathbf{T}$, $\nabla\mathbf{T} = D_iT_{jk}\nabla x^i\nabla x^j\nabla x^k$. Therefore,

$$\begin{aligned} \nabla\cdot\mathbf{T} &= D_i(T_{jk})\nabla x^i\cdot\nabla x^j\nabla x^k \\ &= D_i(T_{jk})g^{ij}\nabla x^k \\ &= D_iT^i{}_k\nabla x^k. \end{aligned} \tag{2.236}$$

We can get the divergence of the tensors of order >2 in a similar fashion.

The Curl Operator

Curl of vectors and tensors is obtained by "crossing into each other" the first two members of each "polyad" of the gradient, for example, if $\mathbf{u}$ is a vector so that $\nabla\mathbf{u} = D_iu_j\nabla x^i\nabla x^j$, then

$$\begin{aligned} \nabla\times\mathbf{u} &= D_iu_j\nabla x^i\times\nabla x^i = D_iu_jJ^{ij}{}_k\nabla x^k \\ &= D_iu_jJ^{ijk}\partial_k(\mathbf{r}) = D_iu_jJ^{-1}\varepsilon^{ijk}\partial_k(\mathbf{r}). \end{aligned} \tag{2.237}$$

Similarly, from $\nabla\mathbf{u} = D^iu^j\partial_i(\mathbf{r})\partial_j(\mathbf{r})$, we have

$$\begin{aligned} \nabla\times\mathbf{u} &= D^iu^j\partial_i(\mathbf{r})\times\partial_j(\mathbf{r}) = D^iu^jJ_{ijk}\nabla x^k \\ &= D^iu^jJ^{-1}\varepsilon_{ijk}\partial_k(\mathbf{r}). \end{aligned} \tag{2.238}$$

For a second-order tensor $\mathbf{T}$, we have

$$\nabla\mathbf{T} = D_iT_{jk}\nabla x^i\nabla x^j\nabla x^k. \tag{2.239}$$

Therefore,

$$\begin{aligned} \nabla\times\mathbf{T} &= D_iT_{jk}\nabla x^i\times\nabla x^j\nabla x^k \\ &= D_iT_{jk}J^{ijm}\partial_m(\mathbf{r})\nabla x^k \\ &= D_iT_{jk}J^{ij}{}_n\nabla x^n\nabla x^k. \end{aligned} \tag{2.240}$$

But $\nabla = \nabla x^i\partial_i$, therefore

$$\begin{aligned} \nabla\mathbf{T} &= \nabla x^i\partial_i(T_{jk}\nabla x^j\nabla x^k) \\ &= \partial_i(T_{jk})\nabla x^i\nabla x^j\nabla x^k + T_{jk}\nabla x^i\partial_i(\nabla x^j)\nabla x^k \\ &\quad + T_{jk}\nabla x^i\nabla x^j\partial_i(\nabla x^k). \end{aligned} \tag{2.241}$$

Comparison of equations 2.240 and 2.241 implies that there must be an expression for D_i in terms of the spatial coordinates. For this, we must first know what happens to the basis vectors when differentiated with respect to spatial coordinates, which we discuss next.

Differentiation of the Basis Vectors

In the preceding subsections, we defined the gradient, divergence, and curl operators on tensors. In defining them, we used the operators D_i and D^i. But we are yet to define how to get these operators, and here we attempt to do so by discussing the differentiation of the basis vectors with respect to the spatial coordinate axes. As before, we restrict to the general curvilinear coordinates and develop these formalisms. In the following subsection, we discuss the change of coordinates, demonstrating that our discussions using the curvilinear coordinates provide an easy and convenient way to move between different coordinate systems.

From our preceding discussions, we have seen that the derivative of a vector by a scalar variable is also a vector which can be represented by a sum of three vectors parallel to the basis vectors. In a general curvilinear coordinate system, the covariant or tangent basis vectors $\partial_i(\mathbf{r})$, $\partial_j(\mathbf{r})$, $\partial_k(\mathbf{r})$ or their contravariant or normal conjugates $\nabla x^i, \nabla x^j, \nabla x^k$ depend upon the position of the point space. By displacing the point, these basis vectors also change. This spatial change in the basis vectors is characterized by the values of the derivatives such as $\partial_i[\partial_j(\mathbf{r})]$ or $\partial_i(\nabla x^j)$. These derivatives are also vectors that can be represented by a sum of three vectors parallel to the basis vectors. Therefore, a relation of the following type must hold:

$$\partial_i(\partial_j(\mathbf{r})) = \Gamma^k_{i\,j}\partial_k(\mathbf{r}) = \Gamma_{kij}\nabla x^k. \quad (2.242)$$

In equation 2.141, Γ^k_{ij} and Γ_{kij} are the symbols of connection, known as the *Christoffel symbols*. The first one (Γ^k_{ij}) is the *Christoffel symbol of first kind* and the second (Γ_{kij}) is the *Christoffel symbol of second kind*. Now

$$\partial_i(\partial_j(\mathbf{r})) = \frac{\partial^2(\mathbf{r})}{\partial x^i \partial x^j} = \Gamma_i{}^k{}_j\partial_k(\vec{\mathbf{r}}) = \Gamma_{kij}\nabla x^k. \quad (2.243)$$

However, $\partial_i[\partial_j(\mathbf{r})] = \partial_j[\partial_i(\mathbf{r})]$, and so, we must have

$$\Gamma_i{}^k{}_j = \Gamma_j{}^k{}_i, \quad (2.244)$$

and

$$\Gamma_{kij} = \Gamma_{kji}. \quad (2.245)$$

Again, from equation 2.242, it follows that

$$\left.\begin{aligned} \Gamma_i{}^k{}_j &= \partial_i(\partial_j(\mathbf{r}))\cdot\nabla x^k \\ \Gamma_{kij} &= \partial_i(\partial_j(\mathbf{r}))\cdot\partial_k(\mathbf{r}) \end{aligned}\right\}. \quad (2.246)$$

Now, we have previously shown that $\nabla x^k = g^{kl}\partial_l(\mathbf{r})$ in equation 2.177. Therefore,

$$\begin{aligned} \Gamma_i{}^k{}_j &= \partial_i(\partial_j(\mathbf{r}))\cdot\nabla x^k = \partial_i(\partial_j(\mathbf{r}))\cdot\partial_l(\mathbf{r})g^{kl} \\ &= \Gamma_{lij}g^{kl}. \end{aligned} \quad (2.247)$$

And from equation 2.178, we have $\partial_k(\mathbf{r}) = g_{kl}\nabla x^l$, and therefore

$$\begin{aligned} \Gamma_{kij} &= \partial_i(\partial_j(\mathbf{r}))\cdot\partial_k(\mathbf{r}) = \partial_i(\partial_j(\mathbf{r}))\cdot\nabla x^l g_{kl} \\ &= \Gamma_i{}^l{}_j g_{kl}. \end{aligned} \quad (2.248)$$

Now, from the second equation of equation 2.246, we obtain, after some algebra

$$\Gamma_{kij} = \partial_i g_{jk} - \partial_j(\mathbf{r})\cdot\partial_i(\partial_k(\mathbf{r})) \quad (2.249)$$

and

$$\Gamma_{kji} = \partial_j g_{ik} - \partial_i(\mathbf{r})\cdot\partial_j(\partial_k(\mathbf{r})). \quad (2.250)$$

But, according to equation 2.245 $\Gamma_{kij} = \Gamma_{kji}$. Therefore,

$$\Gamma_{kij} = \partial_j g_{ik} - \partial_i(\mathbf{r})\cdot\partial_j(\partial_k(\mathbf{r})). \quad (2.251)$$

Adding equations 2.249 and 2.251 gives

$$\begin{aligned} 2\Gamma_{kij} &= \partial_i(g_{jk}) + \partial_j(g_{ik}) \\ &\quad - [\partial_j(\mathbf{r})\cdot\partial_i(\partial_k(\mathbf{r})) + \partial_i(\mathbf{r})\cdot\partial_j(\partial_k(\mathbf{r}))]. \end{aligned} \quad (2.252)$$

But

$$\begin{aligned} \partial_k g_{ij} &= \partial_k(\partial_i(\mathbf{r})\cdot\partial_j(\mathbf{r})) \\ &= \partial_k(\partial_i(\mathbf{r}))\cdot\partial_j(\mathbf{r}) + \partial_i(\mathbf{r})\cdot\partial_k(\partial_j(\vec{\mathbf{r}})) \end{aligned} \quad (2.253)$$

and

$$\begin{aligned} \partial_k(\partial_i(\mathbf{r})) &= \frac{\partial}{\partial x^k}\left(\frac{\partial(\mathbf{r})}{\partial x^i}\right) = \frac{\partial}{\partial x^i}\left(\frac{\partial(\mathbf{r})}{\partial x^k}\right) \\ &= \partial_i(\partial_k(\mathbf{r})). \end{aligned} \quad (2.254)$$

Equations 2.253 therefore leads to

$$\partial_k g_{ij} = \partial_j(\mathbf{r})\cdot\partial_i(\partial_k(\mathbf{r})) + \partial_i(\mathbf{r})\cdot\partial_j(\partial_k(\mathbf{r})). \quad (2.255)$$

Using equation 2.255, we simplify equation 2.252 as

$$2\Gamma_{kij} = \partial_i(g_{jk}) + \partial_j(g_{ik}) - \partial_k(g_{ij}). \quad (2.256)$$

Finally, from equations 2.155 and 2.247, we obtain the expressions for the Christoffel symbol of first and second kinds as

$$\left.\begin{aligned} \Gamma_{kij} &= \frac{1}{2}(\partial_i(g_{kj}) + \partial_j(g_{ik}) - \partial_k(g_{ij})) \\ \Gamma_{i\ j}^{\ k} &= g^{kl}\Gamma_{lij} \\ &= \frac{1}{2}g^{kl}(\partial_i(g_{lj}) + \partial_j(g_{il}) - \partial_l(g_{ij})) \end{aligned}\right\}. \quad (2.257)$$

Note that in equation 2.257 we expressed the Christoffel symbols in terms of the components of the identity tensor **I** and their derivatives, which are easy to calculate. We can now derive some useful formula using the Christoffel symbol.

In equation 2.174, we have seen that $\partial_k(\mathbf{r}).\nabla x^j$ is constant in space and is given as the Kronecker delta function, being one when $k=j$, and otherwise zero. Therefore, its derivative with respect to the spatial coordinates is zero, which leads to

$$-\Gamma_{i\ k}^{\ j} = \partial_k(\mathbf{r}) \cdot (\partial_i \nabla x^j). \quad (2.258)$$

Now, for any vector, $\mathbf{v} = v_i \nabla x^i = v^i \partial_k(\mathbf{r})$. We can therefore write equation 2.258 as

$$\partial_i(\nabla x^j) = -\Gamma_{i\ k}^{\ j}\nabla x^k. \quad (2.259)$$

And, from our definition of the gradient operator (equation 2.159), $\nabla(\nabla x^j) = \nabla x^i \partial_i(\nabla x^j)$. Therefore, equation 2.259 leads us to

$$\nabla(\nabla x^j) = -\Gamma_{ik}^{j}\nabla x^i \nabla x^k. \quad (2.260)$$

The definition of the Christoffel symbols (equation 2.257), in conjunction with equation 2.260, allows us to derive the following explicit forms for any scalar Q, vector **v**, and second-order tensor **T**. Equivalent expressions for tensors of order >2 can be similarly derived

$$\left.\begin{aligned} D_i Q &= \partial_i Q \\ D_i v_j &= -\Gamma_{ij}^{k} v_k + \partial_i v_j \\ D_i v^j &= \Gamma_{ik}^{j} v^k + \partial_i v^j \\ D_i T_{jk} &= \partial_i T_{jk} - \mathrm{T}_{lk}\Gamma_{ij}^{l} - \mathrm{T}_{mj}\Gamma_{i\ k}^{m} \\ D_i T_j^{\ k} &= \partial_i T_j^{\ k} - T_l^{\ k}\Gamma_{ij}^{l} + T_j^{\ m}\Gamma_{i\ m}^{\ k} \end{aligned}\right\}. \quad (2.261)$$

Deriving equation 2.261 is left as an exercise. Note that equation 2.261 allows us to write the gradient, divergence, and curl operators in explicit forms. Additionally, using equation 2.261, we can also readily verify that for the identity tensor

$$\nabla \mathbf{I} = \nabla\nabla(\mathbf{r}) = \mathbf{0}. \quad (2.262)$$

Equation 2.262 implies that the identity tensor is spatially invariant. Deriving equation 2.262 is also left as an exercise.

Change of Coordinates

We restrict to a coordinate system in $\mathcal{R}^3$ with covariant (tangent) basis vectors $\{\partial_1(\mathbf{r}), \partial_2(\mathbf{r}), \partial_3(\mathbf{r})\}$ and the contravariant (normal) basis vectors $\{\nabla x^1, \nabla x^2, \nabla x^3\}$. We now move to a new coordinate system $\{\bar{x}^1, \bar{x}^2, \bar{x}^3\}$ with tangent basis vectors $\{\partial_{\bar{1}}(\mathbf{r}), \partial_{\bar{2}}(\mathbf{r}), \partial_{\bar{3}}(\mathbf{r})\}$ and normal basis vectors $\{\nabla\bar{x}^1, \nabla\bar{x}^2, \nabla\bar{x}^3\}$. Our objective here is to find the relationship between these two coordinate systems. Now, from the definition of the gradient operator (equation 2.159) for any scalar Q, we can write ∇Q in both coordinate systems as

$$\nabla Q = \nabla x^j \partial_j(\mathbf{r}) \quad (2.263)$$

and

$$\nabla Q = \nabla \bar{x}^j \partial_{\bar{j}} Q. \quad (2.264)$$

Similarly, the chain rule for the derivatives in both coordinates can be written as

$$\begin{aligned} dQ &= \frac{\partial Q}{\partial x^1}dx^1 + \frac{\partial Q}{\partial x^2}dx^2 + \frac{\partial Q}{\partial x^3}dx^3 \\ &= dx^j \partial_j Q \end{aligned} \quad (2.265)$$

and

$$\begin{aligned} dQ &= \frac{\partial Q}{\partial \bar{x}^1}d\bar{x}^1 + \frac{\partial Q}{\partial \bar{x}^2}d\bar{x}^2 + \frac{\partial Q}{\partial \bar{x}^3}d\bar{x}^3 \\ &= d\bar{x}^j \partial_{\bar{j}} Q. \end{aligned} \quad (2.266)$$

Equations 2.263–2.266 are the basis for obtaining the expressions for the coordinate transformation. We first develop this for the general curvilinear coordinates and then move on to other familiar coordinate systems such as the Cartesian coordinates, cylindrical coordinates, and spherical coordinates.

Curvilinear Coordinates: We substitute x^i for Q into equation 2.264, which gives

$$\nabla x^i = \nabla \bar{x}^j \partial_{\bar{j}} x^i. \tag{2.267}$$

Now, we let $Q = \bar{x}^j$ in equation 2.265 to obtain

$$d\bar{x}^j = dx^i \partial_i \bar{x}^j = dx^i \frac{\partial \bar{x}^j}{\partial x^i}. \tag{2.268}$$

Now, letting $\mathbf{r}$ for Q in equations 2.265 and 2.266, we get

$$d\mathbf{r} = dx^i \partial_i(\mathbf{r}) = d\bar{x}^j \partial_{\bar{j}}(\mathbf{r}). \tag{2.269}$$

Combining equations 2.268 and 2.269 gives

$$dx^i \partial_i(\mathbf{r}) = dx^i \partial_i \bar{x}^j \partial_{\bar{j}}(\mathbf{r}). \tag{2.270}$$

Finally, equations 2.267 and 2.270 provide the relationship between the contravariant and covariant basis vectors between the two curvilinear coordinates as

$$\left.\begin{aligned} \nabla x^i &= \nabla \bar{x}^j \frac{\partial x^i}{\partial \bar{x}^j} \\ \partial_i(\mathbf{r}) &= \partial_{\bar{j}}(\mathbf{r}) \frac{\partial \bar{x}^j}{\partial x^i} \end{aligned}\right\}. \tag{2.271}$$

Once the relationship between the basis vectors for the two coordinate system is established by equation 2.271, transforming the components of any tensor from one system to the other is straightforward.

Before moving on to the other coordinates, let us first review the covariant (tangent) and contravariant (normal) basis vectors in curvilinear systems.

In Figure 2.9, we show the right-handed, curvilinear coordinate system $\{x^1, x^2, x^3\}$ in $\Re^3$ from any arbitrary point P in space. The covariant basis vectors $\{\partial_1(\mathbf{r}), \partial_2(\mathbf{r}), \partial_3(\mathbf{r})\}$ are then respectively directed along the tangent to the x^1, x^2, and x^3 axes, drawn from the Point P. The basis vector $\partial_1(\mathbf{r})$ for example is shown as a red arrow in Figure 2.8. The contravariant basis vectors ∇x^i, $i=1, 2, 3$ on the other hand are directed along the outward normal to the infinitesimal surface drawn from the point P along the $x^j - x^k$ axes, where $i \neq j$, $i \neq k$, and $j \neq k$. The direction of the contravariant basis vector ∇x^1 from the point P is illustrated with a cyan arrow in Figure 2.9. Note that

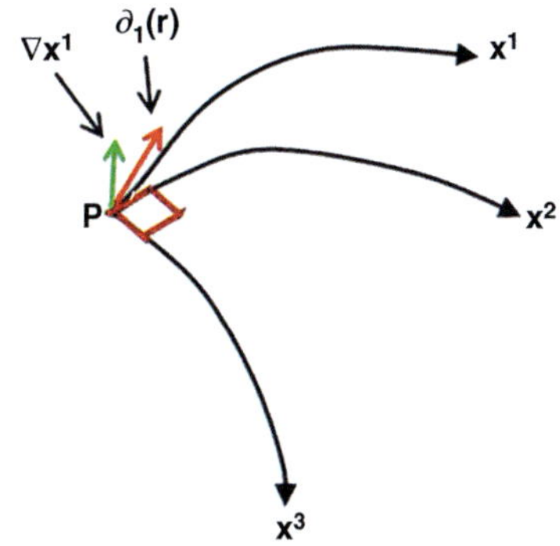

Figure 2.9 Geometrical representations of the covariant (tangent) and contravariant (normal) basis vectors.

$\partial_i(\mathbf{r}) \neq \nabla x^i$ for a general curvilinear coordinate system.

Cartesian Coordinates: The Cartesian coordinate system in $\Re^3$ is the coordinate system where the coordinate axes $\{x^1, x^2, x^3\}$ are mutually orthogonal straight lines. They are also designated as $\{x, y, z\}$. As will be evident soon, the superscript–subscript convention, which we introduced for the curvilinear systems, is no longer necessary for the Cartesian system, and the Cartesian coordinate axes are also referred as $\{x_1, x_2, x_3\}$.

We use the right-handed Cartesian coordinate system, which, in geophysics, we usually assume that the positive x^1 (or x or x_1) axis directed along north, positive x^2 (or y or x_2) axis directed along east, and the positive x^3 (or z or x_3) axis directed vertically downward in depth. Choosing the origin of the coordinate system to be the point O in space with coordinates (0, 0, 0), the coordinates for any arbitrary point P in space are obtained by measuring the distance of the point along the three axes (Figure 2.10). From the geometrical representation of the tangent and normal basis vectors, given in Figure 2.9, we can readily see that for the Cartesian system they are same, i.e., $\partial_i(\mathbf{r}) = \nabla x^i$. Or to be explicit, we have, for a Cartesian coordinate system,

$$\partial_1(\mathbf{r}) = \nabla x^1; \;\; \partial_2(\mathbf{r}) = \nabla x^2; \;\; \partial_3(\mathbf{r}) = \nabla x^3. \tag{2.272}$$

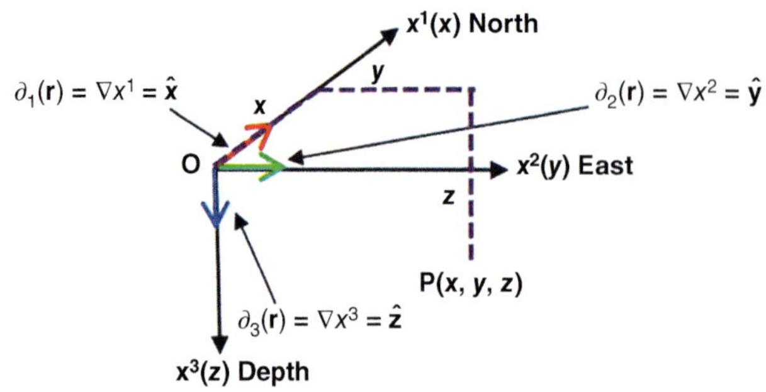

Figure 2.10 Illustration of the right-handed Cartesian coordinate system. Note that the covariant and contravariant basis vectors are the same and are unit vectors along the three coordinate axes.

Because covariant and contravariant basis vectors are the same, the subscript and superscript convention is no longer required in the Cartesian system, and we refer the three coordinate axes as $\{x, y, z\}$ or $\{x_1, x_2, x_3\}$. Additionally, because the measures of the coordinates along the axes are distances, the basis vectors are also unit vectors along the three axes, i.e.,

$$\left.\begin{aligned} \partial_1(\mathbf{r}) &= \nabla x^1 = \hat{\mathbf{x}} = \hat{\mathbf{x}}_1 \\ \partial_2(\mathbf{r}) &= \nabla x^2 = \hat{\mathbf{y}} = \hat{\mathbf{x}}_2 \\ \partial_3(\mathbf{r}) &= \nabla x^3 = \hat{\mathbf{z}} = \hat{\mathbf{x}}_3 \end{aligned}\right\}, \tag{2.273}$$

with

$$\partial_i(\mathbf{r}) \cdot \partial_j(\mathbf{r}) = \delta_{ij}, \tag{2.274}$$

where δ_{ij} is the Kronecker delta and given as

$$\delta_{ij} = \begin{cases} 0; & i \neq j \\ 1; & i = j \end{cases}. \tag{2.275}$$

It can be readily verified that the gradient (∇) operator for the Cartesian system is given as

$$\nabla = \hat{\mathbf{x}}\frac{\partial}{\partial x} + \hat{\mathbf{y}}\frac{\partial}{\partial y} + \hat{\mathbf{z}}\frac{\partial}{\partial z} = \hat{\mathbf{x}}_1\frac{\partial}{\partial x_1} + \hat{\mathbf{x}}_2\frac{\partial}{\partial x_2} + \hat{\mathbf{x}}_3\frac{\partial}{\partial x_3}. \tag{2.276}$$

Equation 2.276 allows obtaining the gradient of any tensor, divergence of any tensor of order >1, and curl between any tensors of order >1. For example, the Laplacian operator ∇^2, introduced in Section 2.2, can be expressed in the Cartesian coordinate system as

$$\nabla^2 = \nabla \cdot \nabla = \frac{\partial^2}{\partial x^2} + \frac{\partial^2}{\partial y^2} + \frac{\partial^2}{\partial z^2}. \tag{2.277}$$

Cylindrical Coordinate System: Cylindrical coordinates of a point in space are given as the radial distance r, azimuth θ, and vertical distance z. Therefore, as shown in Figure 2.11, by setting the origin of the coordinates to be the point O with coordinates (0, 0, 0), the coordinates (r, θ, η) for any point in space in a right-handed cylindrical coordinates are obtained by stepping a radial distance r at an angle θ, clockwise from the positive x-axis, and then a distance η along the positive z-axis.

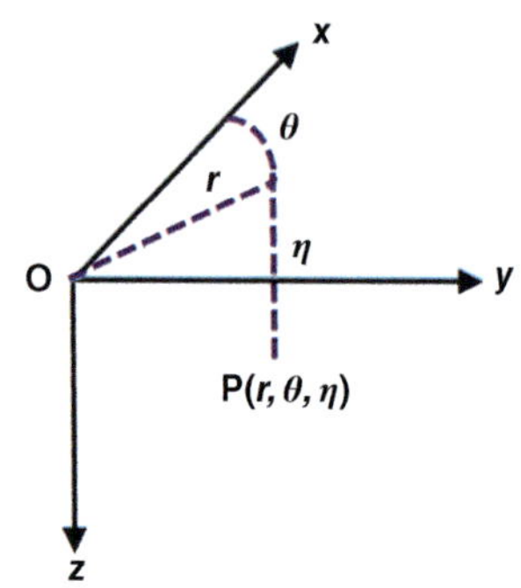

Figure 2.11 Illustration of a right-handed cylindrical coordinate system with reference to a right-handed Cartesian coordinate system.

If the point P shown with the coordinates (x, y, z) in Figure 2.10 and with the coordinates (r, θ, η) in Figure 2.11 is the same, then we can readily verify that

$$\left.\begin{aligned} x &= r\cos\theta \\ y &= r\sin\theta \\ z &= \eta \end{aligned}\right\}. \tag{2.278}$$

Then, using the general relation between the basis vectors between two different coordinate systems, we previously derived (equation 2.271), we can easily derive the covariant and contravariant basis vectors for the cylindrical system as

$$\left.\begin{aligned} \partial_1(\mathbf{r}) &= \hat{\mathbf{r}} \\ \partial_2(\mathbf{r}) &= r\hat{\boldsymbol{\theta}} \\ \partial_3(\mathbf{r}) &= \hat{\mathbf{z}} \end{aligned}\right\} \tag{2.279}$$

and

$$\left.\begin{aligned} \nabla x^1 &= \hat{\mathbf{r}} \\ \nabla x^2 &= \tfrac{1}{r}\hat{\boldsymbol{\theta}} \\ \nabla x^3 &= \hat{\mathbf{z}} \end{aligned}\right\}. \tag{2.280}$$

In equations 2.279 and 2.280, $\hat{\mathbf{r}}, \hat{\boldsymbol{\theta}}$, and $\hat{\mathbf{z}}$ are the unit vectors along the three coordinate axes for the cylindrical system. It is important to note that for the cylindrical system the basis vectors along the r and z coordinate axes are unit vectors because the coordinates along these axes are given as the distance. But for the θ coordinate axis where the coordinate is given as the angle, the basis vectors are not unit vectors.

From equation 2.280, we can also define the gradient and the Laplacian operators in the cylindrical coordinates as

$$\nabla = \hat{\mathbf{r}}\frac{\partial}{\partial r} + \frac{\hat{\boldsymbol{\theta}}}{r}\frac{\partial}{\partial \theta} + \hat{\mathbf{z}}\frac{\partial}{\partial z} \tag{2.281}$$

and

$$\nabla^2 = \frac{\partial^2}{\partial r^2} + \frac{1}{r^2}\frac{\partial^2}{\partial \theta^2} + \frac{\partial^2}{\partial z^2}. \tag{2.282}$$

Spherical Coordinate System: As shown in Figure 2.12, the spherical coordinate system is defined by the radial distance r, colatitude θ, and azimuth φ. Therefore, by setting the origin of the coordinate system to be O (0,0,0), the coordinates (r, θ, φ) for any arbitrary point P in space are obtained by the radial distance r, colatitude θ given as the angle, measured clockwise from the positive z-axis, and the azimuthal angle φ, measured clockwise from the positive x-axis.

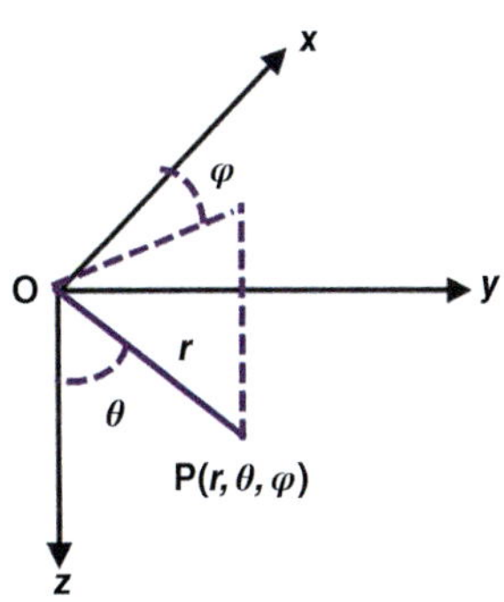

Figure 2.12 Illustration of the right-handed spherical coordinate system.

Like cylindrical coordinate system, if we start from the Cartesian coordinates as the reference and use change of coordinates equations (equation 2.271), the covariant and contravariant basis vectors in the spherical coordinates can be derived as

$$\left.\begin{aligned} \partial_1(\mathbf{r}) &= \hat{\mathbf{r}} \\ \partial_2(\mathbf{r}) &= r\hat{\boldsymbol{\theta}} \\ \partial_3(\mathbf{r}) &= r\sin\theta\hat{\boldsymbol{\varphi}} \end{aligned}\right\} \tag{2.283}$$

and

$$\left.\begin{aligned} \nabla x^1 &= \hat{\mathbf{r}} \\ \nabla x^2 &= \frac{1}{r}\hat{\boldsymbol{\theta}} \\ \nabla x^3 &= \tfrac{1}{r\sin\theta}\hat{\boldsymbol{\varphi}} \end{aligned}\right\}. \tag{2.284}$$

Like the cylindrical coordinates, $\hat{\mathbf{r}}$, $\hat{\boldsymbol{\theta}}$, and $\hat{\boldsymbol{\varphi}}$ in equations 2.283 and 2.284 are the unit vectors along the three coordinate axes. Note that for spherical coordinates the basis vectors are unit vectors only along the r coordinate axis.

Form equation 2.284, the gradient and Laplacian operators in spherical coordinates are given as

$$\nabla = \hat{\mathbf{r}}\frac{\partial}{\partial r} + \frac{\hat{\boldsymbol{\theta}}}{r}\frac{\partial}{\partial \theta} + \frac{\hat{\boldsymbol{\varphi}}}{r\sin\theta}\frac{\partial}{\partial \phi} \tag{2.285}$$

and

$$\nabla^2 = \frac{\partial^2}{\partial r^2} + \frac{1}{r^2}\frac{\partial^2}{\partial \theta^2} + \frac{1}{r^2\sin^2\theta}\frac{\partial^2}{\partial \phi^2}. \tag{2.286}$$

2.3.9 Useful Theorems in Tensor Calculus

We conclude discussions on tensors with two important theorems in tensor calculus: (1) divergence theorem (also known as the Gauss' theorem and Ostrogradsky's theorem) and (2) Stokes' theorem, which is frequently used in seismology. Details of both these

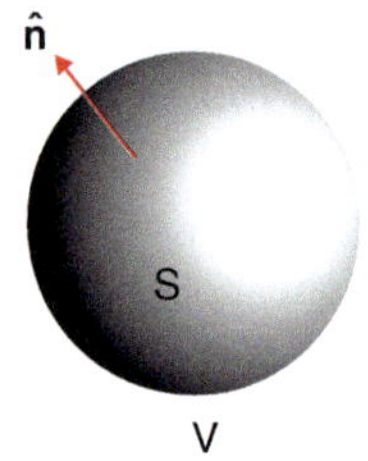

Figure 2.13 An arbitrary volume V in $\Re^3$ enclosed by an outer surface S on which the unit outward normal vector $\hat{\mathbf{n}}$ is defined everywhere.

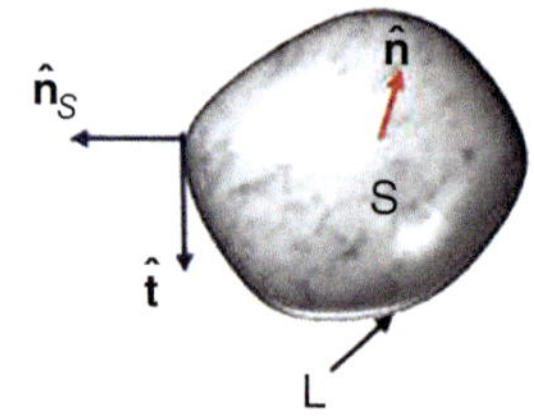

Figure 2.14 A surface S enclosed by a line L. Outward unit normal vector $\hat{\mathbf{n}}$ is defined everywhere on S, and the unit normal and tangent vectors $\hat{\mathbf{n}}_S$ and $\hat{\mathbf{t}}$ are defined everywhere on L.

theorems can be found in Kreyszig et al. (2011), and here we just provide them without proof.

Divergence Theorem: Let V be an arbitrary volume in $\Re^3$ enclosed by an outer surface S, with unit outward normal vector $\hat{\mathbf{n}}$ defined on every point on S (Figure 2.13).

If $\mathbf{T}$ is a tensor field defined everywhere within the volume including its outer surface, then the divergence theorem is given as

$$\left.\begin{aligned} \int_V \nabla \mathbf{T} dV &= \int_S \hat{\mathbf{n}} \mathbf{T} ds \\ \int_V (\nabla \cdot \mathbf{T}) dV &= \int_S (\hat{\mathbf{n}} \cdot \mathbf{T}) ds \\ \int_V (\nabla \times \mathbf{T}) dV &= \int_S (\hat{\mathbf{n}} \times \mathbf{T}) ds \end{aligned}\right\}. \tag{2.287}$$

Stokes's Theorem: Stokes's theorem is the divergence theorem for a surface in $\Re^3$. Consider a surface S enclosed by a line L with outward unit normal vector $\hat{\mathbf{n}}$ defined on S and the unit normal and tangent vectors ($\hat{\mathbf{n}}_S$ and $\hat{\mathbf{t}}$) defined on L (Figure 2.14).

If $\mathbf{T}$ is a tensor field defined everywhere in S, including L, then the Stokes' theorem is given as

$$\left.\begin{aligned} \int_S \nabla_S \mathbf{T} dS &= \oint_L \hat{\mathbf{n}}_S \mathbf{T} dl \\ \int_S (\hat{\mathbf{n}} \times \nabla_S \mathbf{T}) dS &= \oint_L (\hat{\mathbf{t}} \mathbf{T}) dl \\ \int_S \hat{\mathbf{n}} \cdot (\nabla_S \mathbf{T}) dS &= \oint_L (\hat{\mathbf{t}} \cdot \mathbf{T}) dl \end{aligned}\right\}. \tag{2.288}$$

2.4 Chapter Summary

In this chapter, we discussed some important mathematical concepts that will be useful in discussing the subsequent chapters. We first started our discussion on the Fourier series and Fourier integrals. In these discussions, we started from the discrete Fourier series, generalized it to continuous Fourier integrals, and defined the forward and inverse Fourier transforms. We also discussed different practical aspects of the numerical implementation of the Fourier transforms on the digital computers. We then moved on to PDEs and briefly discussed the parabolic, hyperbolic, and elliptic PDEs. Finally, we concluded the chapter with discussions on the fundamentals of tensors, tensor algebra, and tensor calculus.

2.5 Exercises

Problem 2.1 Starting from the complex representation of the Fourier integral, given by equation 2.30, derive the Fourier transform pairs, given by equation 2.40.

Problem 2.2 Equation 2.52 shows that to attenuate the wrapped-around seismic signal to be attenuated by a factor $1/D$ in synthesizing $f(t)e^{-aT_s}$ instead of $f(t)$, where T_s is the time length of the seismic data, we must set $a = \ln(D)/T_S$. Verify this to be true.

Problem 2.3 We discussed that truncating the infinite integral in the inverse Fourier transform to a finite one between $-\Omega$ and $+\Omega$ is equivalent to multiplying the original frequency domain signal $\widetilde{S}(\omega)$ with $B\left(\frac{\omega}{\Omega}\right)$, given by equation 2.57.

1) Using the value of $|\Omega|$ to be 256π, compute $B\left(\frac{\omega}{\Omega}\right)$ for ω ranging between -256π and $+256\pi$ at regular intervals of $\Delta\omega = \pi/2$. Note that the angular frequency ω (radians/second) is related to the circular frequency f in Hz (cycles/second) as $\omega = 2\pi f$. So, in this step you compute the function between the frequencies -128 and $+128$ Hz at intervals of $\Delta f = \frac{1}{4} = 0.25$ Hz, with a total of 1024 frequency samples.
2) Give a time shift of 2 s to the computed data in step 1 by multiplying each sample by $e^{2i\omega}$.
3) Inverse Fourier transform the data computed in step 2 using either your own code or a built-in routine in any high-level computer language. Because Δf was ¼ Hz, the 1024 time-domain samples after Fourier transform will be 4 s long with time sampling interval of 4/1024 s.
4) Display the time-domain signal from step 3.

Problem 2.4 Repeat problem 2.3 using the Hanning window, defined by equation 2.61.

Problem 2.5 Repeat problem 2.3 or 2.4 using a Ricker wavelet, given as $R(f) = \frac{2}{\sqrt{\pi}}\frac{f^2}{f_C^3}e^{-\left(\frac{f}{f_C}\right)^2}$, where f is the frequency in Hz and f_C is the central or peak frequency in Hz. Use $f_C = 64$ Hz in your calculations.

Problem 2.6 Starting from equation 2.122, derive equation 2.123.

Problem 2.7 Derive equation 2.131 from 2.130.

Problem 2.8 Starting from the Cauchy–Reimann condition (equation 2.141), verify if the conditions of the harmonic functions, given by equation 2.142, are correct.

Problem 2.9 If the adjoint of the differential form $L(u)$ is $M(v)$ and satisfies equation 2.143, show that $L(v)$ is the adjoint differential form to $M(u)$.

Problem 2.10 Derive equation 2.150.

Problem 2.11 Derive the components of the identity tensor, given by equation 2.174

Problem 2.12 Derive equations 2.196 and 2.197.

Problem 2.13 Prove the vector identity, given by the equation 2.200.

Problem 2.14 Provide step-by-step derivations for obtaining the covariant and contravariant basis vectors for a cylindrical coordinate system, given by equations 2.279 and 2.280.

Problem 2.15 Provide step-by-step derivations for obtaining the covariant and contravariant basis vectors for a spherical coordinate system, given by equations 2.283 and 2.284.

References

Backus, G.E., 1967, Converting vector and tensor equations to scalar equations in spherical coordinates, *Geophysical Journal International*, *13*, 71–102, https://doi.org/10.1111/j.1365-246X.1967.tb02147.x.

Dennery, P., and Krzywicki, A., 1967, *Mathematics for Physicists*, Harper & Row, ISBN 0-486-69193-4.

Dettman, J.W., 1969, *Mathematical Methods in Physics and Engineering*, McGraw-Hill, ISBN 0-486-65649-7.

Fedorov, F.I., 1968, *Theory of Elastic Waves in Crystals*, Springer, ISBN 978-1475712773.

Kreyszig, E., Kreyszig, H., and Norminton, E.J., 2011, *Advanced Engineering Mathematics*, 10th Edn, John Wiley and Sons, ISBN 978-0470458365.

Mallick, S., and Frazer, L.N., 1987, Practical aspects of reflectivity modeling, *Geophysics*, *52*, 1355–1364, https://doi.org/10.1190/1.1442248.

Mallick, S., and Frazer, L.N., 1988, Rapid computation of multioffset vertical seismic profile synthetic seismograms for layered media, *Geophysics*, *53*, 479–491, https://doi.org/10.1190/1.1442479.

Mallick, S., and Frazer, L.N., 1990, Computation of synthetic seismograms for stratified azimuthally anisotropic media, *Journal of Geophysical Research*, *95*, 8513–8526, https://doi.org/10.1029/JB095iB06p08513.

Mallick, S., and Frazer, L.N., 1991, Reflection/transmission coefficients and azimuthal anisotropy in marine seismic studies, *Geophysical Journal International*, *105*, 241–252, https://doi.org/10.1111/j.1365-246X.1991.tb03459.x.

Menzel, D.H., 1961, *Mathematical Physics*, Dover Publications, ISBN 0-486-60056-4.

Pafeng, J., Mallick, S., and Sharma, H., 2017, Prestack waveform inversion of three-dimensional seismic data – an example from the Rock Springs Uplift, Wyoming, USA, *Geophysics*, *82*, B1–B12, https://doi.org/10.1190/GEO2016-0079.1.

Pinsky, M., 2002, *Introduction to Fourier Analysis and Wavelets*, Brooks Cole, ISBN 978-0534376604.

Press, W.H., Flannery, B.P., Teulolsky, S.A., and Vetterling, W., 1989, *Numerical Recipes The Art of Scientific Computing (FORTRAN Version)*, Cambridge University Press, ISBN 978-0521308113.

Robinson, E.A., and Treitel, S., 2000, *Geophysical Signal Analysis*, Society of Exploration Geophysicists, ISBN 978-1560801047.

Sommerfeld, A., 1949, *Partial Differential Equations in Physics*, Academic Press, ISBN 0-126546584.

Zygmund, A., 1968, *Trigonometric Series*, Second Edn, Cambridge University Press, (published 1988), ISBN 978-0521358859.

3

Fundamentals of the Linearized Elastic Wave Theory

3.1 Introduction

As we discussed in Chapter 1, an important element to seismic optimization is computing synthetic or predicted data. These synthetic data, when matched with the observed data, allow us to compute the objective and cast our optimization, either as minimization or a maximization. Computing synthetic data, in turn, requires discussion of the theory of elasticity and the propagation of elastic waves, which we discuss in this chapter (Chapter 3) and the next two chapters (Chapters 4 and 5). In this chapter, we review the fundamental theory of linearized elasticity, derive the elastodynamic wave equation, and solve the elastodynamic wave equation for a homogeneous medium. After discussing the elastic waves homogeneous media, in Chapters 4 and 5, we move on to the elastic wave propagation in a heterogeneous earth and provide different flavors of computing synthetic seismic data.

In our discussion, we will use the right-handed Cartesian system (Figure 2.10) as our reference. As we discussed in Chapter 2, in a Cartesian system, there is no difference between the covariant and contravariant basis vectors, and the basis vectors are unit vectors along the three coordinate axes (see equation 2.272). Additionally, we will not use any superscript and assume the three coordinate axes as x (or x_1), y (or x_2), and z (or x_3) and interchangeably use $x \leftrightarrow x_1$, $y \leftrightarrow x_2$, and $z \leftrightarrow x_3$ as the coordinate axes.

3.2 The Stress Tensor and Traction

Consider a volume V in three-dimensional (3D) Euclidean space $\Re^3$, enclosed by a surface S. Also assume that at any point within the surface S, we can uniquely define a unit normal vector $\hat{\mathbf{n}}$ pointing outward (Figure 3.1).

If this volume is subject to a force, every surface that is enclosed within the volume will have a force. Such a force on any arbitrary surface can then be expressed as three components along the three coordinate axes $\{x, y, z\}$. Let us then consider an infinitesimal volume element along the three coordinate axes and look at the forces that are acting on each surface of this volume element.

As shown in Figure 3.2, forces acting on each surface (plane) of this infinitesimal volume element can be represented by three-component vectors. For example, in the $x_1 - x_2$ (or $x - y$) plane, the components of this vector are $[\tau_{31}, \tau_{32}, \tau_{33}]^T$ or $[\tau_{zx}, \tau_{zy}, \tau_{zz}]^T$. These vectors acting on all six surfaces can be combined as a 3×3 matrix

$$\boldsymbol{\tau} = \begin{pmatrix} \tau_{11} & \tau_{12} & \tau_{13} \\ \tau_{21} & \tau_{22} & \tau_{23} \\ \tau_{31} & \tau_{32} & \tau_{33} \end{pmatrix}, \tag{3.1}$$

or as

$$\boldsymbol{\tau} = \begin{pmatrix} \tau_{xx} & \tau_{xy} & \tau_{xz} \\ \tau_{yx} & \tau_{yy} & \tau_{yz} \\ \tau_{zx} & \tau_{zy} & \tau_{zz} \end{pmatrix}. \tag{3.2}$$

Computation, Optimization, and Machine Learning in Seismology, Advanced Textbook 8, First Edition. Subhashis Mallick.

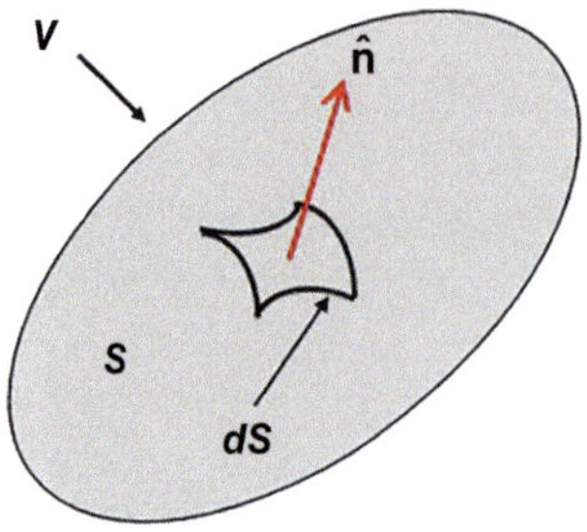

Figure 3.1 Illustration of a volume V in three-dimensional Euclidean space $\Re^3$, enclosed by surface S such that the outward unit normal vector $\hat{\mathbf{n}}$ can be defined everywhere on S.

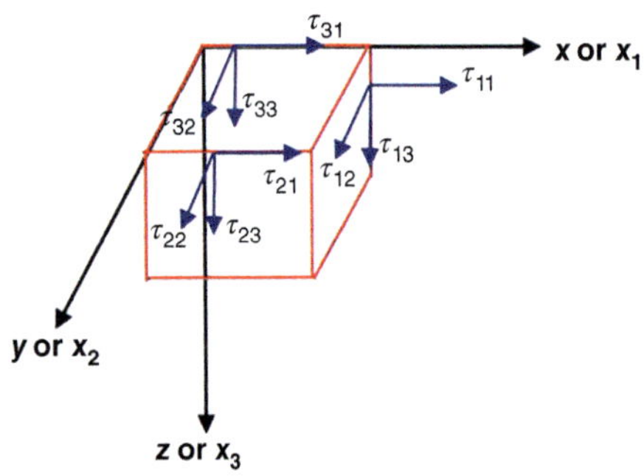

Figure 3.2 Analysis of the forces acting on an infinitesimal volume element when a volume is subject to force.

From our discussion on Chapter 2, we immediately identify $\boldsymbol{\tau}$ as a second-order tensor with nine distinct components, which can be written as

$$\begin{aligned}\boldsymbol{\tau} = {} & \tau_{11}\hat{\mathbf{x}}_1\hat{\mathbf{x}}_1 + \tau_{12}\hat{\mathbf{x}}_1\hat{\mathbf{x}}_2 + \tau_{13}\hat{\mathbf{x}}_1\hat{\mathbf{x}}_3 \\ & + \tau_{21}\hat{\mathbf{x}}_2\hat{\mathbf{x}}_1 + \tau_{22}\hat{\mathbf{x}}_2\hat{\mathbf{x}}_2 + \tau_{23}\hat{\mathbf{x}}_2\hat{\mathbf{x}}_3 \\ & + \tau_{31}\hat{\mathbf{x}}_3\hat{\mathbf{x}}_1 + \tau_{32}\hat{\mathbf{x}}_3\hat{\mathbf{x}}_2 + \tau_{33}\hat{\mathbf{x}}_3\hat{\mathbf{x}}_3. \quad (3.3)\end{aligned}$$

In Chapter 2, we introduced the summation convention for indices appearing as superscript and subscript or vice versa. Because we do not use superscript in Cartesian system, we extend the summation convention for repeated indices. Using this convention, we write equation 3.3 as

$$\boldsymbol{\tau} = \tau_{ij}\hat{\mathbf{x}}_i\hat{\mathbf{x}}_j. \quad (3.4)$$

Note from Figure 3.2, for the infinitesimal volume element to remain in space without rotation or translation, the component of the opposing forces on the element must equal or we must have $\tau_{ij} = \tau_{ji}$. This condition, in turn, implies that the stress tensor must be symmetric. Therefore, only six out of nine components of the stress tensor are distinct, and the matrix representation of the stress tensor becomes

$$\boldsymbol{\tau} = \begin{pmatrix} \tau_{11} & \tau_{12} & \tau_{13} \\ \tau_{12} & \tau_{22} & \tau_{23} \\ \tau_{13} & \tau_{23} & \tau_{33} \end{pmatrix} = \begin{pmatrix} \tau_{xx} & \tau_{xy} & \tau_{xz} \\ \tau_{xy} & \tau_{yy} & \tau_{yz} \\ \tau_{xz} & \tau_{yz} & \tau_{zz} \end{pmatrix}. \quad (3.5)$$

Also from Figure 3.2, the force acting on the $x_1 - x_2$ (or $x - y$) plane is given by $\tau_{31}\hat{\mathbf{x}}_1 + \tau_{32}\hat{\mathbf{x}}_2 + \tau_{33}\hat{\mathbf{x}}_3$ or $\tau_{zx}\hat{\mathbf{x}} + \tau_{zy}\hat{\mathbf{y}} + \tau_{zz}\hat{\mathbf{z}}$, and from equation 3.3 or 3.4, we can immediately see this force to be $\hat{\mathbf{x}}_3 \cdot \boldsymbol{\tau}$ or $\hat{\mathbf{z}} \cdot \boldsymbol{\tau}$. The force acting on the $x_1 - x_2$ plane is thus obtained from the dot product of the unit normal vector to the plane with the stress tensor and is called traction. We can similarly derive the expressions for the traction acting on all surfaces of the infinitesimal volume element shown in Figure 3.2. We can further extend this concept to any arbitrary volume V enclosed by surface S. If the volume V is subject to a force, then the traction acting at any point within any surface S enclosing V (see Figure 3.1) is a vector and is given by

$$\mathbf{t} = \hat{\mathbf{n}} \cdot \boldsymbol{\tau}. \quad (3.6)$$

In equation 3.6, $\hat{\mathbf{n}}$ is the outward unit normal vector to the surface S from the given point (see Figure 3.1).

3.3 Strain (Deformation) Tensor

In Section 3.2, we considered an elastic medium (continuum) being subject to an external force and defined the second-order symmetric stress tensor within the medium and the traction vector acting on any surface. We now consider how the elastic medium reacts when subject to such force.

As illustrated in Figure 3.3, when a medium is subject to an external force or a load, the medium responds to the load by deformation.

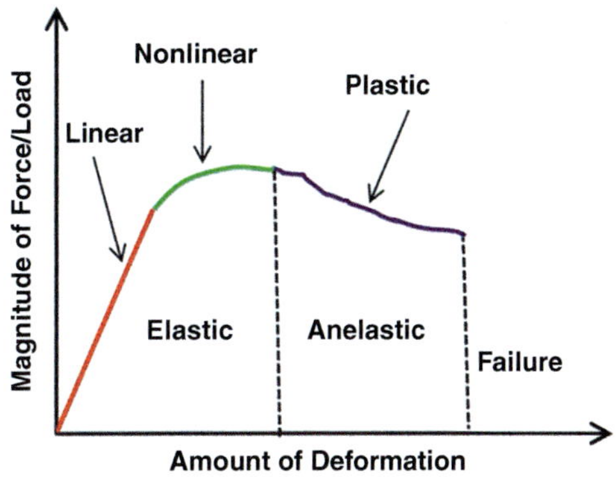

Figure 3.3 Behavior of an elastic continuum when subject to an external force.

For example, if an elastic medium such as an iron bar is pulled from both sides with a force, it responds by stretching and the original shape is changed. If we take examples from geology, a sedimentary rock, originally deposited as highly porous and permeable, is compacted and loses porosity and permeability upon increasing burial due to loading from the overlying newly deposited sediments. In plate tectonics, when the rigid tectonic plates subside at the subduction zones, they deform the surrounding rock formations. Initial stage of this deformation, when the amount of load and the corresponding amount of deformation is small, the relationship between them is linear, represented by the red line in Figure 3.3. With increasing amount of load, this linear relation gradually becomes a nonlinear relation, represented by the green curve in Figure 3.3. The regime covered by both linear and nonlinear relations is elastic where the medium returns to its original shape upon removal of the external force or load. However, when the force/load increases further, the medium enters an anelastic regime where it tends to flow plastically and cannot return to its original shape when the load is removed. In plate tectonics for example, this situation occurs when the subducting plates cause folding structures in the sediments through which they subside. Increasing overburden load from the hot spot volcanisms at places like the Hawaiian Islands makes permanent flexural bending in the underlying lithosphere. Finally, when the amount of the external force or load is very high, the medium can no longer deform and fails via fracturing. In the subsurface of the earth, such fracturing and faulting due to the subducting plates is the primary mechanism for earthquakes. To summarize, there are three distinct phases in the behavior of an elastic continuum in response to an external force (Fedorov, 1968; Achenbach, 1973):

1) *Phase 1—elastic phase*: For small to moderate forces or loads, the magnitude of deformation is initially linearly and then nonlinearly related to the applied force. The deformation in this phase is reversible, meaning that when the external force is removed, the elastic continuum reverts to its original shape.
2) *Phase 2—anelastic phase*: As the magnitude of the external force or load increases further, the deformation is anelastic, and no longer reversible and characterized by plastic or ductile flow.
3) *Phase 3—failure phase*: When the external force is very large, and the medium can no longer withstand the force and breaks apart via fracturing or brittle failure.

In this book, we deal with far-field seismic wavefields where a linear relation is a reasonable assumption (Fedorov, 1968; Achenbach, 1973; Auld, 1973; Aki and Richards, 2002; Chapman, 2004). We therefore primarily focus on the linear component of phase 1, represented by the red line in Figure 3.3. Nonlinear elasticity, represented by the green curve in Figure 3.3, is important for studying the behavior of the elastic waves near the source, for example, in analyzing the rupture propagation in a tectonically active area such as the San-Andreas fault. It is however important to point out the anelastic and failure phases in context with the microseismic monitoring from induced seismicity due to fracturing. To induce fracturing and enhance fluid flow in impermeable reservoirs, it is important that the rock is brittle such that the length of the anelastic zone is minimal, requiring the

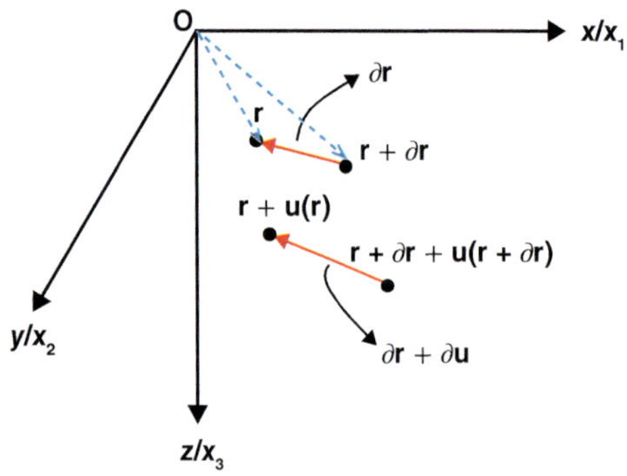

Figure 3.4 Deformation within an elastic continuum when subject to an external force.

analysis of the rock behavior in the anelastic and failure phases. Here, we focus on the elastic zone where the external force is small, and the behavior of the medium is linear.

To quantify deformation, consider two adjacent points in an elastic continuum (medium) with position vectors $\mathbf{r}$ and $\mathbf{r}+\delta\mathbf{r}$ (Figure 3.4). Now, in the event of any external force, the elastic continuum adjusts itself by deforming its original shape via a displacement of each point within the continuum to a different position. This displacement is, of course, a vector, which we call the displacement vector and denote it by $\mathbf{u}$. Therefore, after deformation, the positions of the original points with position vectors $\mathbf{r}$ and $\mathbf{r}+\delta\mathbf{r}$ are now $\mathbf{r}+\mathbf{u}(\mathbf{r})$ and $\mathbf{r}+\delta\mathbf{r}+\mathbf{u}(\mathbf{r}+\delta\mathbf{r})$, as shown in Figure 3.4. Because the difference between the original points is $\delta\mathbf{r}$ and $\mathbf{u}$ is the displacement vector (see Figure 3.4), the difference between the two points after deformation is $\delta\mathbf{r}+\delta\mathbf{u}$, which is also shown in Figure 3.4. Alternatively, this difference $\delta\mathbf{r}+\delta\mathbf{u}$ may also be written as the difference between the two position vectors after deformation or, in other words,

$$\delta\mathbf{r}+\delta\mathbf{u}=\mathbf{r}+\delta\mathbf{r}+\mathbf{u}(\mathbf{r}+\delta\mathbf{r})-[\mathbf{r}+\mathbf{u}(\mathbf{r})]. \tag{3.7}$$

Because $\delta\mathbf{r}$ is small, we expand $\mathbf{u}(\mathbf{r}+\delta\mathbf{r})$ as a Taylor series and take only the first term to write

$$\mathbf{u}(\mathbf{r}+\delta\mathbf{r})=\mathbf{u}(\mathbf{r})+(\delta\mathbf{r}\cdot\nabla)\mathbf{u}(\mathbf{r}), \tag{3.8}$$

which leads equation 3.7 to

$$\delta\mathbf{u}=\delta\mathbf{r}\cdot\nabla\mathbf{u}. \tag{3.9}$$

In equation 3.9, we replaced $\mathbf{u}(\mathbf{r})$ with $\mathbf{u}$. Now, we may write

$$\nabla\mathbf{u}=\frac{1}{2}[\{\nabla\mathbf{u}+(\nabla\mathbf{u})^T\}+\{\nabla\mathbf{u}-(\nabla\mathbf{u})^T\}], \tag{3.10}$$

leading equation 3.9 to

$$\delta\mathbf{u}=\delta\mathbf{r}\cdot\left[\frac{1}{2}\{\nabla\mathbf{u}+(\nabla\mathbf{u})^{\mathbf{T}}\}+\frac{1}{2}\{\nabla\mathbf{u}-(\nabla\mathbf{u})^{\mathbf{T}}\}\right]. \tag{3.11}$$

We now use the vector identity $\mathbf{a}\times\mathbf{b}\times\mathbf{c}=\mathbf{c}\cdot(\mathbf{ab})-\mathbf{c}\cdot(\mathbf{ba})$ and write

$$\delta\mathbf{r}\cdot\{\nabla\mathbf{u}-(\nabla\mathbf{u})^{\mathbf{T}}\}=\nabla\times\mathbf{u}\times\delta\mathbf{r}. \tag{3.12}$$

From equation 3.12, the second term on the right-hand side of equation 3.11 is a pure rotation. And, the first term, given as

$$\boldsymbol{\varepsilon}=\frac{1}{2}[\nabla\mathbf{u}+(\nabla\mathbf{u})^T], \tag{3.13}$$

is a symmetric second-order tensor representing the elastic deformation of the medium and is called the *strain* or *deformation* tensor. Like stress tensor, the strain tensor can be written in component notation as

$$\boldsymbol{\varepsilon}=\varepsilon_{ij}\hat{\mathbf{x}}_i\hat{\mathbf{x}}_j \tag{3.14}$$

or as a matrix

$$\boldsymbol{\varepsilon}=\begin{pmatrix}\varepsilon_{11} & \varepsilon_{12} & \varepsilon_{13}\\ \varepsilon_{12} & \varepsilon_{22} & \varepsilon_{23}\\ \varepsilon_{13} & \varepsilon_{23} & \varepsilon_{33}\end{pmatrix}=\begin{pmatrix}\varepsilon_{xx} & \varepsilon_{xy} & \varepsilon_{xz}\\ \varepsilon_{xy} & \varepsilon_{yy} & \varepsilon_{yz}\\ \varepsilon_{xz} & \varepsilon_{yz} & \varepsilon_{zz}\end{pmatrix}. \tag{3.15}$$

3.4 Static Relation—The First Fundamental Equation in Elasticity

The first fundamental equation in elasticity is derived from a law of physics that states that the force F required to extend (or compress) a spring by some distance x is linearly proportional to x, i.e., $F=kx$, where k is a constant. This fundamental law of physics is known as the Hooke's law, named after the seventeenth century British physicist Robert Hooke, who stated this law in 1676 and later published the proof of this law in 1678. With relation to the deformation versus force diagram shown in Figure 3.3, we can clearly see that Hooke's law relates to the elastic regime where the deformation and force are linearly related. Therefore, in terms of the stress and strain tensors, the generalized from of Hooke's law is given as (Aki and Richards, 2002)

$$\boldsymbol{\tau} = \mathbf{c} : \boldsymbol{\varepsilon}, \tag{3.16}$$

where $\boldsymbol{\tau}$ is the second-order stress tensor, $\boldsymbol{\varepsilon}$ is the second-order strain tensor, $\mathbf{c}$ is the fourth-order elastic-stiffness tensor, and the symbol ":" is a double dot product between $\mathbf{c}$ and $\boldsymbol{\varepsilon}$. In component notation and using the summation convention over the repeated indices, we can also write equation 3.16 as

$$\tau_{ij} = c_{ijkl}\varepsilon_{kl}. \tag{3.17}$$

Note that the components of c_{ijkl} are the constants of proportionality for the linear relationship between the different components of the stress and strain tensors, which we may explicitly write. Now, as a fourth-order tensor, the elastic-stiffness tensor $\mathbf{c}$ has $3^4 = 81$ components. However, we already showed that both stress and strain tensors are symmetric, i.e., $\tau_{ij} = \tau_{ji}$ and $\varepsilon_{kl} = \varepsilon_{lk}$. These symmetry properties of both stress and strain tensors imply $c_{ijkl} = c_{jikl} = c_{ijlk} = c_{jilk}$, which, in turn, reduces the number of independent components of the elastic-stiffness tensor to 36, and 6 independent components of the stress and tensors can be linearly related via the following matrix equation:

$$\begin{pmatrix} \tau_{11} \\ \tau_{22} \\ \tau_{33} \\ \tau_{23} \\ \tau_{31} \\ \tau_{12} \end{pmatrix} = \begin{pmatrix} c_{1111} & c_{1122} & c_{1133} & c_{1123} & c_{1131} & c_{1112} \\ c_{2211} & c_{2222} & c_{2233} & c_{2223} & c_{2231} & c_{2212} \\ c_{3311} & c_{3322} & c_{3333} & c_{3323} & c_{3331} & c_{3312} \\ c_{2311} & c_{2322} & c_{2333} & c_{2323} & c_{2331} & c_{2312} \\ c_{3111} & c_{3122} & c_{3133} & c_{3123} & c_{3131} & c_{3112} \\ c_{1211} & c_{1222} & c_{1233} & c_{1223} & c_{1231} & c_{1212} \end{pmatrix} \begin{pmatrix} \varepsilon_{11} \\ \varepsilon_{22} \\ \varepsilon_{33} \\ 2\varepsilon_{23} \\ 2\varepsilon_{31} \\ 2\varepsilon_{12} \end{pmatrix}. \tag{3.18}$$

Finally, from the thermodynamic argument of the existence of the strain energy function, it can be proved that $c_{ijkl} = c_{klij}$ (Auld, 1973; Aki and Richards, 2002). This, in turn, implies that the 6×6 matrix on the right-hand side of equation 3.18 is symmetric with 21 independent components. Replacing c_{ijkl} with C_{IJ}, we finally write equation 3.18 as

$$\mathbf{T} = \mathbf{CE}, \tag{3.19}$$

where

$$\begin{aligned} \mathbf{T} &= [\tau_{11}, \tau_{22}, \tau_{33}, \tau_{23}, \tau_{31}, \tau_{12}]^T \\ &= [\tau_{xx}, \tau_{yy}, \tau_{zz}, \tau_{yx}, \tau_{zx}, \tau_{xy}]^T, \end{aligned} \tag{3.20}$$

$$\begin{aligned} \mathbf{E} &= [\varepsilon_{11}, \varepsilon_{22}, \varepsilon_{33}, 2\varepsilon_{23}, 2\varepsilon_{31}, 2\varepsilon_{12}]^T \\ &= [\varepsilon_{xx}, \varepsilon_{yy}, \varepsilon_{zz}, 2\varepsilon_{yz}, 2\varepsilon_{zx}, 2\varepsilon_{xy}]^T, \end{aligned} \tag{3.21}$$

and $\mathbf{C}$ is the 6×6 symmetric elastic-stiffness matrix and given as

$$\mathbf{C} = \begin{pmatrix} C_{11} & C_{12} & C_{13} & C_{14} & C_{15} & C_{16} \\ C_{12} & C_{22} & C_{23} & C_{24} & C_{25} & C_{26} \\ C_{13} & C_{23} & C_{33} & C_{34} & C_{35} & C_{36} \\ C_{14} & C_{24} & C_{34} & C_{44} & C_{45} & C_{46} \\ C_{15} & C_{25} & C_{35} & C_{45} & C_{55} & C_{56} \\ C_{16} & C_{26} & C_{36} & C_{46} & C_{56} & C_{66} \end{pmatrix}. \tag{3.22}$$

By comparing equation 3.22 with equation 3.18, the correspondence between the components c_{ijkl} of the stiffness tensor and C_{IJ} of our newly defined stiffness matrix $\mathbf{C}$ can be directly found. From now on, we will use equation 3.19 for the generalized Hooke's law and consider this to be the *constitutive relation* or the first fundamental equation in the elasticity theory. It is, however, important to note that it is sometimes convenient to express equation 3.19 as its inverse

$$\mathbf{E} = \mathbf{ST}, \tag{3.23}$$

where $\mathbf{S} = \mathbf{C}^{-1}$ is the symmetric 6×6 *elastic compliance matrix.*

3.4.1 Orthogonal Transformation of the Elastic Stiffness Matrix

Equation 3.19 is a simplified matrix formulation of the Hooke's law, as shown in equations 3.16 and 3.17, where we used the symmetry properties of the fourth-order elastic-stiffness tensor $\mathbf{c}$ and expressed it as an elastic-stiffness matrix $\mathbf{C}$. However, we know from Chapter 2 that the components of a tensor vary with the choice of the coordinate system. Therefore, if the stiffness tensor **c** is defined for a right-handed Cartesian coordinate system, as shown in Figures 2.10, 3.2, or 3.4, and we move to a different coordinate system, the components of **c** will also vary in general. And, because the components of the elastic-stiffness matrix **C** are from the original stiffness tensor **c** (see equations 3.18 and 3.22), **C** must also vary for such coordinate transformation. Here, we consider one such simple transformation of one Cartesian reference system to another, which is called the orthogonal transformation.

We start in two dimensions (Figure 3.5a), and let $\{x_1, x_2\}$ be the coordinate system in which any arbitrary vector **v** can be written as $\mathbf{v} = v_1\hat{\mathbf{x}}_1 + v_2\hat{\mathbf{x}}_2$. We now rotate the axes by an arbitrary angle θ to a new system $\{x'_1, x'_2\}$, as shown in Figure 3.5a. Assume that the same vector is represented in the new system as $\mathbf{v} = v'_1\hat{\mathbf{x}}'_1 + v'_2\hat{\mathbf{x}}'_2$. We can then readily derive that the components of **v** in the primed system are related to those in the unprimed system through a transformation matrix, whose components are the direction cosines of the primes system to the unprimed system, i.e.,

$$\begin{pmatrix} v'_1 \\ v'_2 \end{pmatrix} = \begin{pmatrix} \cos(x'_1, x_1) & \cos(x'_1, x_2) \\ \cos(x'_2, x_1) & \cos(x'_2, x_2) \end{pmatrix} \begin{pmatrix} v_1 \\ v_2 \end{pmatrix} = \begin{pmatrix} a_{11} & a_{12} \\ a_{21} & a_{22} \end{pmatrix} \begin{pmatrix} v_1 \\ v_2 \end{pmatrix} = \begin{pmatrix} \cos\theta & \sin\theta \\ -\sin\theta & \cos\theta \end{pmatrix} \begin{pmatrix} v_1 \\ v_2 \end{pmatrix}. \quad (3.24)$$

Equation 3.24 implies that the primed system is related to the unprimed system via a transformation matrix **a** given as

$$\mathbf{a} = \begin{pmatrix} \cos\theta & \sin\theta \\ -\sin\theta & \cos\theta \end{pmatrix}. \quad (3.25)$$

Because the vector **v** is arbitrary, we can invert equation 3.24 to write

$$\begin{pmatrix} v_1 \\ v_2 \end{pmatrix} = \mathbf{a}^{-1} \begin{pmatrix} v'_1 \\ v'_2 \end{pmatrix}, \quad (3.26)$$

where

$$\mathbf{a}^{-1} = \begin{pmatrix} \cos(x_1, x'_1) & \cos(x_1, x'_2) \\ \cos(x_2, x'_1) & \cos(x_2, x'_2) \end{pmatrix} = \begin{pmatrix} \cos\theta & -\sin\theta \\ \sin\theta & \cos\theta \end{pmatrix}. \quad (3.27)$$

And, from equations 3.25 and 3.27, we verify that $\mathbf{a}^{-1} = \mathbf{a}^T$, and the transformation represented by Figure 3.5 is an *orthogonal transformation*.

Such an orthogonal transformation can be readily extended to three dimensions (Figure 3.5b) as

$$\begin{pmatrix} v'_1 \\ v'_2 \\ v'_3 \end{pmatrix} = \begin{pmatrix} a_{11} & a_{12} & a_{13} \\ a_{21} & a_{22} & a_{23} \\ a_{31} & a_{32} & a_{33} \end{pmatrix} \begin{pmatrix} v_1 \\ v_2 \\ v_3 \end{pmatrix} \quad (3.28)$$

and

$$\begin{pmatrix} v_1 \\ v_2 \\ v_3 \end{pmatrix} = \mathbf{a}^{-1} \begin{pmatrix} v'_1 \\ v'_2 \\ v'_3 \end{pmatrix} = \mathbf{a}^{\mathrm{T}} \begin{pmatrix} v'_1 \\ v'_2 \\ v'_3 \end{pmatrix}. \quad (3.29)$$

Like we did in two dimensions, for 3D transformation for angles φ and θ about the x_1 and x_3 axes (Figure 3.5b), the components of the transformation matrix **a** can be readily

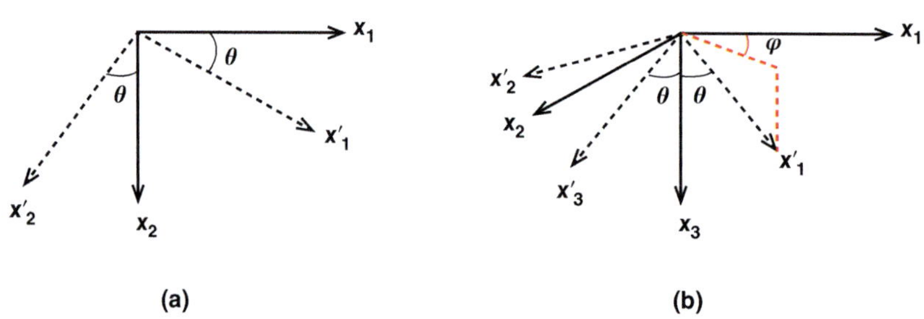

Figure 3.5 Orthogonal transformation in (a) two dimensions and (b) three dimensions.

derived as

$$\mathbf{a} = \begin{pmatrix} \cos(x_1, x_1') & \cos(x_1, x_2') & \cos(x_1, x_3') \\ \cos(x_2, x_1') & \cos(x_2, x_2') & \cos(x_2, x_3') \\ \cos(x_3, x_1') & \cos(x_3, x_2') & \cos(x_3, x_3') \end{pmatrix}. \tag{3.30}$$

We derive the components of the transformation matrix **a** later. Given the relationships shown by equations 3.28 and 3.29, we can further verify the following for second-, third-, and fourth-order tensors:

$$\left.\begin{aligned} A'_{ij} &= a_{ik}a_{jl}A_{kl} \\ B'_{ijk} &= a_{ip}a_{jq}a_{kr}A_{pqr} \\ C'_{ijkl} &= a_{ip}a_{jq}a_{kr}a_{ls}C_{pqrs} \end{aligned}\right\}. \tag{3.31}$$

And we can similarly obtain expressions for tensors of order >4.

We now go to the constitutive relation (equation 3.19) and derive expressions of the elastic-stiffness matrix **C**, given by equation 3.22 under orthogonal transformations. We start by explicitly writing our constitutive relation in matrix form as

$$\begin{pmatrix} \tau_{11} \\ \tau_{22} \\ \tau_{33} \\ \tau_{23} \\ \tau_{31} \\ \tau_{12} \end{pmatrix} = \begin{pmatrix} C_{11} & C_{12} & C_{13} & C_{14} & C_{15} & C_{16} \\ C_{12} & C_{22} & C_{23} & C_{24} & C_{25} & C_{26} \\ C_{13} & C_{23} & C_{33} & C_{34} & C_{35} & C_{36} \\ C_{14} & C_{24} & C_{34} & C_{44} & C_{45} & C_{46} \\ C_{15} & C_{25} & C_{35} & C_{45} & C_{55} & C_{56} \\ C_{16} & C_{26} & C_{36} & C_{46} & C_{56} & C_{66} \end{pmatrix} \begin{pmatrix} \varepsilon_{11} \\ \varepsilon_{22} \\ \varepsilon_{33} \\ 2\varepsilon_{23} \\ 2\varepsilon_{31} \\ 2\varepsilon_{12} \end{pmatrix}.$$

In the above equation, the components τ_{ij} and ε_{ij} are the components of two second-order tensors $\boldsymbol{\tau}$ and $\boldsymbol{\varepsilon}$. For orthogonal transformations of second-order tensors (equation 3.31), we thus write $\tau'_{ij} = a_{ik}a_{jl}\tau_{kl}$ and $\varepsilon'_{ij} = a_{ik}a_{jl}\varepsilon_{kl}$. Using this, we can explicitly write the orthogonal transformation of the stress tensor components as

$$\begin{pmatrix} \tau'_{11} \\ \tau'_{22} \\ \tau'_{33} \\ \tau'_{23} \\ \tau'_{31} \\ \tau'_{12} \end{pmatrix} = \mathbf{M} \begin{pmatrix} \tau_{11} \\ \tau_{22} \\ \tau_{33} \\ \tau_{23} \\ \tau_{31} \\ \tau_{12} \end{pmatrix}, \tag{3.32}$$

with

$$\mathbf{M} = \begin{pmatrix} a_{11}^2 & a_{12}^2 & a_{13}^2 & 2a_{12}a_{13} & 2a_{13}a_{11} & 2a_{11}a_{12} \\ a_{21}^2 & a_{22}^2 & a_{23}^2 & 2a_{22}a_{23} & 2a_{23}a_{21} & 2a_{21}a_{22} \\ a_{31}^2 & a_{32}^2 & a_{33}^2 & 2a_{32}a_{33} & 2a_{33}a_{31} & 2a_{31}a_{32} \\ a_{21}a_{31} & a_{22}a_{32} & a_{23}a_{33} & a_{23}a_{32} + a_{22}a_{33} & a_{23}a_{31} + a_{21}a_{33} & a_{21}a_{32} + a_{22}a_{31} \\ a_{31}a_{11} & a_{32}a_{12} & a_{33}a_{13} & a_{33}a_{12} + a_{32}a_{13} & a_{33}a_{11} + a_{31}a_{13} & a_{31}a_{12} + a_{32}a_{11} \\ a_{11}a_{21} & a_{12}a_{22} & a_{13}a_{23} & a_{13}a_{22} + a_{12}a_{23} & a_{13}a_{21} + a_{11}a_{23} & u_{11}a_{22} + a_{12}a_{21} \end{pmatrix} \tag{3.33}$$

And those for the strain tensor components as

$$\begin{pmatrix} \varepsilon'_{11} \\ \varepsilon'_{22} \\ \varepsilon'_{33} \\ 2\varepsilon'_{23} \\ 2\varepsilon'_{31} \\ 2\varepsilon'_{12} \end{pmatrix} = \mathbf{N} \begin{pmatrix} \varepsilon_{11} \\ \varepsilon_{22} \\ \varepsilon_{33} \\ 2\varepsilon_{23} \\ 2\varepsilon_{31} \\ 2\varepsilon_{12} \end{pmatrix}, \tag{3.34}$$

where

$$\mathbf{N} = \begin{pmatrix} a_{11}^2 & a_{12}^2 & a_{13}^2 & a_{12}a_{13} & a_{13}a_{11} & a_{11}a_{12} \\ a_{21}^2 & a_{22}^2 & a_{23}^2 & a_{22}a_{23} & a_{23}a_{21} & a_{21}a_{22} \\ a_{31}^2 & a_{32}^2 & a_{33}^2 & a_{32}a_{33} & a_{33}a_{31} & a_{31}a_{32} \\ 2a_{21}a_{31} & 2a_{22}a_{32} & 2a_{23}a_{33} & a_{23}a_{32}+a_{22}a_{33} & a_{23}a_{31}+a_{21}a_{33} & a_{21}a_{32}+a_{22}a_{31} \\ 2a_{31}a_{11} & 2a_{32}a_{12} & 2a_{33}a_{13} & a_{33}a_{12}+a_{32}a_{13} & a_{33}a_{11}+a_{31}a_{13} & a_{31}a_{12}+a_{32}a_{11} \\ 2a_{11}a_{21} & 2a_{12}a_{22} & 2a_{13}a_{23} & a_{13}a_{22}+a_{12}a_{23} & a_{13}a_{21}+a_{11}a_{23} & a_{11}a_{22}+a_{12}a_{21} \end{pmatrix}. \tag{3.35}$$

From equations 3.33 and 3.34, we see that the transformation matrix **N** for the strain tensor is like the matrix **M** for the stress tensor except for a shift of a factor of 2 from the lower left-hand corner to the upper right-hand corner. The matrices **M** and **N** are called respectively the *Bond stress and strain transformation matrices* (Auld, 1973).

Now, to our constitutive relation given as $\mathbf{T} = \mathbf{CE}$ (equation 3.19), we apply the Bond stress transformation $\mathbf{T}' = \mathbf{MT} = \mathbf{MCE}$. We then apply the Bond strain transformation $\mathbf{E}' = \mathbf{NE}$ or $\mathbf{E} = \mathbf{N}^{-1}\mathbf{E}'$. This gives

$$\mathbf{T}' = \mathbf{MT} = \mathbf{MCE} = \mathbf{MCN}^{-1}\mathbf{E}'. \tag{3.36}$$

From equation 3.36, we now have the expression for the elastic-stiffness matrix under orthogonal transformation as

$$\mathbf{C}' = \mathbf{MCN}^{-1}. \tag{3.37}$$

From equation 3.37, it appears that the transformation of the coefficient (stiffness) matrix for coordinate rotation (orthogonal transformation) requires inverting the 6×6 Bond strain transformation matrix **N**. However, it is straightforward to show that $\mathbf{N}^{-1} = \mathbf{M}^T$, which leads to the equation for the expression for the elastic-stiffness matrix for orthogonal transformation as

$$\mathbf{C}' = \mathbf{MCM}^T. \tag{3.38}$$

The proof for $\mathbf{N}^{-1} = \mathbf{M}^T$ is left as an exercise. It can also be shown that the expression for the Bond transformation for the elastic compliance matrix **S** defined in equation 3.23 is

$$\mathbf{S}' = \mathbf{NSN}^T, \tag{3.39}$$

where **N** is the Bond strain transformation matrix defined in equation 3.35.

Having defined the Bond stress and strain transformation matrices, we now generalize the rotation of the elastic-stiffness matrix under orthogonal transformations.

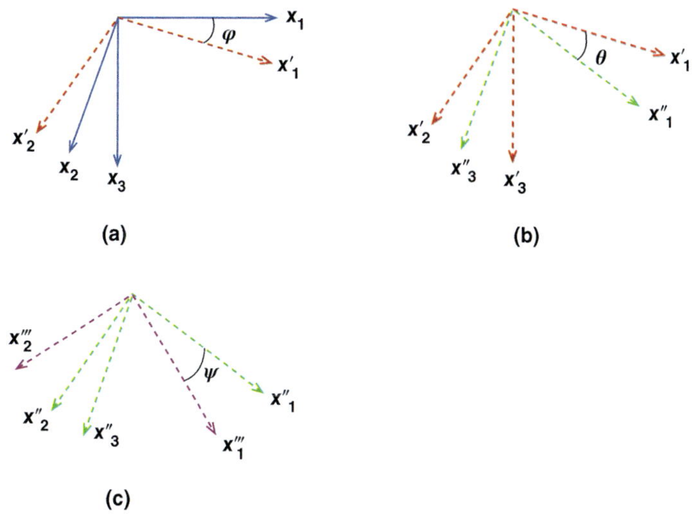

Figure 3.6 Euler rotation: (a) step 1, (b) step 2, and (c) step 3.

Any general rotation of rectangular coordinates can be performed by applying successive rotations about different coordinate axes. A standard method of doing this is known as *Euler rotation* (Auld, 1973), as demonstrated in Figure 3.6. We assume the original Cartesian coordinate axes as $\{x_1, x_2, x_3\}$, as shown in Figure 3.6a, in reference to which the elastic-stiffness matrix **C** is defined by equation 3.22. In Euler rotation, we define the elastic-stiffness matrix in a different Cartesian system via three successive orthogonal transformation steps as explained below.

Step 1: In step 1, we rotate clockwise by an angle φ with respect to the x_3 coordinate axis. As shown by the red dashed lines in Figure 3.6a, we move into a different coordinate system $[x_1', x_2', x_3']$ after this rotation where x_1' and x_2' are rotated by an angle φ with respect to the original coordinates (x_1, x_2), but x_3' remains the same as the original coordinate x_3.

The transformation matrix **a** (equation 3.30) for this rotation then becomes

$$\mathbf{a} = \begin{pmatrix} \cos(x_1', x_1) & \cos(x_1', x_2) & \cos(x_1', x_3) \\ \cos(x_2', x_1) & \cos(x_2', x_2) & \cos(x_2', x_3) \\ \cos(x_3', x_1) & \cos(x_3', x_2) & \cos(x_3', x_3) \end{pmatrix}$$

$$= \begin{pmatrix} \cos\varphi & \sin\varphi & 0 \\ -\sin\varphi & \cos\varphi & 0 \\ 0 & 0 & 1 \end{pmatrix}. \tag{3.40}$$

And, from equation 3.33, we derive the Bond stress transformation matrix for this rotation as

$$\mathbf{M}_\varphi = \begin{pmatrix} \cos^2\varphi & \sin^2\varphi & 0 & 0 & 0 & \sin 2\varphi \\ \sin^2\varphi & \cos^2\varphi & 0 & 0 & 0 & -\sin 2\varphi \\ 0 & 0 & 1 & 0 & 0 & 0 \\ 0 & 0 & 0 & \cos\varphi & -\sin\varphi & 0 \\ 0 & 0 & 0 & \sin\varphi & \cos\varphi & 0 \\ -\dfrac{\sin 2\varphi}{2} & \dfrac{\sin 2\varphi}{2} & 0 & 0 & 0 & \cos 2\varphi \end{pmatrix}. \tag{3.41}$$

Finally, from equation 3.38, we write the elastic-stiffness matrix for the new coordinate system as

$$\mathbf{C}_\varphi = \mathbf{M}_\varphi \mathbf{C} \mathbf{M}_\varphi^{\mathrm{T}}. \tag{3.42}$$

Step 2: We now rotate clockwise by an angle θ with respect to the x_2' axis from step 1 to go to another system $\{x_1'', x_2'', x_3''\}$. As shown in Figure 3.6b, x_1'' and x_3'' coordinate axes (green dashed lines in Figure 3.6b) will be at an angle θ with respect to x_1' and x_3' coordinate axes (dashed red lines), but the x_2'' coordinate axis will be the same as the $x_1'x_1'x_1'x_2'$ axis. Again, from equation 3.30, we obtain the transformation matrix **a** for this rotation

$$\mathbf{a} = \begin{pmatrix} \cos(x_1'', x_1') & \cos(x_1'', x_2') & \cos(x_1'', x_3') \\ \cos(x_2', x_1') & \cos(x_2', x_2') & \cos(x_2', x_3') \\ \cos(x_3'', x_1') & \cos(x_3'', x_2') & \cos(x_3'', x_3') \end{pmatrix}$$

$$= \begin{pmatrix} \cos\theta & 0 & -\sin\theta \\ 0 & 1 & 0 \\ \sin\theta & 0 & \cos\theta \end{pmatrix}, \tag{3.43}$$

then, from equation 3.33, we obtain the Bond stress transformation matrix

$$\mathbf{M}_\theta = \begin{pmatrix} \cos^2\theta & 0 & \sin^2\theta & 0 & -\sin 2\theta & 0 \\ 0 & 1 & 0 & 0 & 0 & 0 \\ \sin^2\theta & 0 & \cos^2\theta & 0 & \sin 2\theta & 0 \\ 0 & 0 & 0 & \cos\theta & 0 & \sin\theta \\ \frac{\sin 2\theta}{2} & 0 & -\frac{\sin 2\theta}{2} & 0 & \cos 2\theta & 0 \\ 0 & 0 & 0 & -\sin\theta & 0 & \cos\theta \end{pmatrix}, \tag{3.44}$$

and, from equations 3.38 and 3.42, we obtain the elastic-stiffness matrix for the new coordinate system as

$$\mathbf{C}_\theta = \mathbf{M}_\theta \mathbf{C}_\varphi \mathbf{M}_\theta^{\mathrm{T}} = \mathbf{M}_\theta \mathbf{M}_\varphi \mathbf{C} \mathbf{M}_\varphi^{\mathrm{T}} \mathbf{M}_\theta^{\mathrm{T}}. \tag{3.45}$$

Step 3: In step 3, we now clockwise rotate by an angle ψ about the x_3'' axis from step 2. As shown in Figure 3.6c, the new horizontal coordinate axes (x_1''', x_2'''), shown in purple, will be at an angle ψ with respect to the x_1'' and x_2'' axes (in green) from the previous rotation, and the x_3''' axis will be the same as the x_3'' axis. The transformation matrix **a** for this rotation is the same as that for step 1 (equation 3.40) with φ replaced by ψ from which we obtain the Bond stress transformation matrix $\mathbf{M}_\psi$ for step 3 rotation as

$$\mathbf{M}_\psi = \begin{pmatrix} \cos^2\psi & \sin^2\psi & 0 & 0 & 0 & \sin 2\psi \\ \sin^2\psi & \cos^2\psi & 0 & 0 & 0 & -\sin 2\psi \\ 0 & 0 & 1 & 0 & 0 & 0 \\ 0 & 0 & 0 & \cos\psi & -\sin\psi & 0 \\ 0 & 0 & 0 & \sin\psi & \cos\psi & 0 \\ -\frac{\sin 2\psi}{2} & \frac{\sin 2\psi}{2} & 0 & 0 & 0 & \cos 2\psi \end{pmatrix}. \tag{3.46}$$

Finally, the stiffness matrix after step 3 is given as

$$\begin{aligned} \mathbf{C}_\psi &= \mathbf{M}_\psi \mathbf{C}_\theta \mathbf{M}_\psi^{\mathrm{T}} = \mathbf{M}_\psi \mathbf{M}_\theta \mathbf{M}_\varphi \mathbf{C} \mathbf{M}_\varphi^{\mathrm{T}} \mathbf{M}_\theta^{\mathrm{T}} \mathbf{M}_\psi^{\mathrm{T}} \\ &= (\mathbf{M}_\psi \mathbf{M}_\theta \mathbf{M}_\varphi) \mathbf{C} (\mathbf{M}_\psi \mathbf{M}_\theta \mathbf{M}_\varphi)^{\mathrm{T}}. \end{aligned} \tag{3.47}$$

3.4.2 Elastic Symmetries

The elastic-stiffness matrix with 21 independent components (equation 3.33) is the most generalized form of an anisotropic elastic medium, known as the *triclinic* medium (Fedorov, 1968; Auld, 1973). Triclinic medium is, however, rare, and for most media of geological interest, the stiffness matrices take much simpler forms. These forms are derived from imposing the symmetry properties of the medium using Euler rotations that we discussed above. Although there are many elastic symmetries, those of geological interest are (1) isotropic, (2) hexagonal, and (3) orthorhombic symmetries, which we discuss below.

Isotropic Elastic Medium: An isotropic elastic medium is given by the property that stiffness matrix is invariant under any orthogonal transformation and is represented by two independent constants as

$$\mathbf{C} = \begin{pmatrix} C_{11} & C_{11}-2C_{66} & C_{11}-2C_{66} & 0 & 0 & 0 \\ C_{11}-2C_{66} & C_{11} & C_{11}-2C_{66} & 0 & 0 & 0 \\ C_{11}-2C_{66} & C_{11}-2C_{66} & C_{11} & 0 & 0 & 0 \\ 0 & 0 & 0 & C_{66} & 0 & 0 \\ 0 & 0 & 0 & 0 & C_{66} & 0 \\ 0 & 0 & 0 & 0 & 0 & C_{66} \end{pmatrix}$$
$$= \begin{pmatrix} \lambda+2\mu & \lambda & \lambda & 0 & 0 & 0 \\ \lambda & \lambda+2\mu & \lambda & 0 & 0 & 0 \\ \lambda & \lambda & \lambda+2\mu & 0 & 0 & 0 \\ 0 & 0 & 0 & \mu & 0 & 0 \\ 0 & 0 & 0 & 0 & \mu & 0 \\ 0 & 0 & 0 & 0 & 0 & \mu \end{pmatrix}. \tag{3.48}$$

As evident from equation 3.48, the two elastic constants for the isotropic symmetry are C_{11} and C_{66} or λ and μ. The parameters λ and μ are known as the *Lamé parameters* (Achenbach, 1973; Aki and Richards, 2002). An acoustic medium is a special case for an isotropic symmetry which is valid for fluids where $C_{66} = \mu = 0$ and the medium is characterized by just one constant, $C_{11} = \lambda$.

By defining the second fundamental equation of elasticity, known as the *momentum equation*, combining it with equation 3.19 to write the *elastodynamic equation*, and solving it for homogeneous medium, it will be shown later in this chapter that an isotropic elastic system is characterized by two waves types: (1) a pure longitudinal wave, called the primary or P-wave propagating with a constant velocity $\mathbf{V}_P$ in any direction with magnitude

$$V_P = \sqrt{\frac{C_{11}}{\rho}} = \sqrt{\frac{\lambda + 2\mu}{\rho}}, \tag{3.49}$$

and a pure transverse wave, called the secondary or S-wave propagating with a constant velocity $\mathbf{V}_S$ in any direction with magnitude

$$V_s = \sqrt{\frac{C_{66}}{\rho}} = \sqrt{\frac{\mu}{\rho}}. \tag{3.50}$$

In equations 3.49 and 3.50, ρ is the density of the medium. In seismology, we try to estimate the physical properties such as V_P, V_S, and ρ from recorded seismic data, not the elastic constants or Lamé parameters. It is therefore sometimes convenient to express the elastic-stiffness matrix $\mathbf{C}$ in equation 3.48 in terms of these parameters by setting $C_{11} = \lambda + 2\mu = \rho V_P^2$ and $C_{44} = \mu = \rho V_S^2$.

Hexagonal Elastic Medium: When the stiffness matrix is invariant under rotation about one coordinate axis, it is known as the hexagonal elastic medium. For example, if we impose the constraint that the stiffness matrix is invariant to any arbitrary rotation about the vertical (x_3) axis, i.e., the symmetry axis is vertical, it can be shown that

$$\mathbf{C} = \begin{pmatrix} C_{11} & C_{11} - 2C_{66} & C_{13} & 0 & 0 & 0 \\ C_{11} - 2C_{66} & C_{11} & C_{13} & 0 & 0 & 0 \\ C_{13} & C_{13} & C_{33} & 0 & 0 & 0 \\ 0 & 0 & 0 & C_{44} & 0 & 0 \\ 0 & 0 & 0 & 0 & C_{44} & 0 \\ 0 & 0 & 0 & 0 & 0 & C_{66} \end{pmatrix}. \tag{3.51}$$

From equation 3.51, we can see that there are only five elastic constants in a hexagonal medium, C_{11}, C_{13}, C_{33}, C_{44}, and C_{66}. In exploration seismology literature, the hexagonally symmetric medium represented by equation 3.51 is also called a transversely isotropic medium with a vertical symmetry axis or a VTI medium.

From the solution of the elastodynamic wave equation in a homogeneous VTI medium, it is shown later in the chapter that such a medium is characterized by three types of waves: (1) P wave, (2) fast S wave (S_1 wave), and (3) slow S wave (S_2 wave). P and S_1 waves are pure longitudinal and transverse waves only along certain directions and therefore they are also called quasi-P (qP) and quasi-S_1 (qS_1) waves. The S_2 wave, however, is a pure transverse wave along all propagation directions. The vertical (x_3) axis being the axis of symmetry, the P, S_1, and S_2 waves propagate with the same phase velocity along all directions in the horizontal ($x_1 - x_2$) plane, but they vary vertically. For a VTI medium with the elastic-stiffness coefficients given by equation 3.51 and density ρ, the P-, S_1-, and S_2-wave phase velocities in the vertical (x_3) direction are given respectively as

$$V_{P0} = \sqrt{\frac{C_{33}}{\rho}},$$
$$V_{S0} = \sqrt{\frac{C_{55}}{\rho}} = \sqrt{\frac{C_{44}}{\rho}},$$
$$V_{S20} = \sqrt{\frac{C_{55}}{\rho}} = \sqrt{\frac{C_{44}}{\rho}}. \tag{3.52}$$

And, in the horizontal ($x_1 - x_2$ plane) they are

$$V_{Phor} = \sqrt{\frac{C_{11}}{\rho}},$$
$$V_{Shor} = \sqrt{\frac{C_{55}}{\rho}} = \sqrt{\frac{C_{44}}{\rho}},$$
$$V_{S2hor} = \sqrt{\frac{C_{66}}{\rho.}} \tag{3.53}$$

When a VTI medium is rotated 90° about any horizontal axis, the symmetry axis of the medium becomes horizontal, and the medium is called a transversely isotropic medium with a horizontal symmetry axis or an HTI medium. For example, assuming the Euler angles $\varphi = 0°$, $\theta = 90°$, and $\psi = 0°$, and setting the Bond stress transformation matrix $\mathbf{M}_\theta$ (equation 3.44) for $\theta = 90°$, we obtain the elastic-stiffness matrix for an HTI medium as

$$\mathbf{C}_{HTI} = \begin{pmatrix} C_{33} & C_{13} & C_{13} & 0 & 0 & 0 \\ C_{13} & C_{11} & C_{11} - 2C_{66} & 0 & 0 & 0 \\ C_{13} & C_{11} - 2C_{66} & C_{11} & 0 & 0 & 0 \\ 0 & 0 & 0 & C_{66} & 0 & 0 \\ 0 & 0 & 0 & 0 & C_{44} & 0 \\ 0 & 0 & 0 & 0 & 0 & C_{44} \end{pmatrix}. \tag{3.54}$$

The HTI medium with stiffness matrix represented by equation 3.54 has the x_2 axis as the symmetry and the P, S_1, and S_2 wave phase velocities do not vary in the $x_2 - x_3$ plane, but they do in all other planes. We may further rotate $\mathbf{C}_{HII}$ of equation 3.54 with $\varphi = 90°, \theta = 0°$, and $\psi = 0°$, or $\varphi = 0°$, $\theta = 0°$, and $\psi = 90°$, and define a different HTI medium with x_1 symmetry axis. Furthermore, if we rotate the VTI medium about the x_3 axis with an angle other than 90°, we obtain a transversely isotropic medium with a tilted symmetry axis or a TTI medium.

Like isotropic medium, it is sometimes convenient to express the stiffness coefficients for hexagonal medium in terms of the quantities we measure from seismic data. Thomsen (1986), for example, defines a VTI medium in terms of the vertical P- and S-wave velocities V_{P0} and V_{S0}, given in equation 3.52 and three dimensionless parameters

$$\begin{aligned} \varepsilon &= \frac{C_{11} - C_{33}}{2C_{33}}, \\ \delta &= \frac{(C_{13} + C_{44})^2 - (C_{33} - C_{44})^2}{2C_{33}(C_{33} - C_{44})}, \\ \gamma &= \frac{C_{66} - C_{55}}{2C_{55}} = \frac{C_{66} - C_{44}}{2C_{44}}. \end{aligned} \tag{3.55}$$

In exploration geophysics, V_{P0}, V_{S0}, ε, δ, and γ are called the *Thomsen parameters*. Note that the dimensionless scalar parameter ε in equation 3.55 is not the same as $\boldsymbol{\varepsilon}$ which we defined earlier to denote the second-order strain tensor. We discuss the physical interpretation of the Thomsen parameters later in the chapter.

Orthorhombic Elastic Medium: An orthorhombic elastic medium is represented by nine independent elastic constants, and the stiffness matrix is given as

$$\mathbf{C} = \begin{pmatrix} C_{11} & C_{12} & C_{13} & 0 & 0 & 0 \\ C_{12} & C_{22} & C_{23} & 0 & 0 & 0 \\ C_{13} & C_{23} & C_{33} & 0 & 0 & 0 \\ 0 & 0 & 0 & C_{44} & 0 & 0 \\ 0 & 0 & 0 & 0 & C_{55} & 0 \\ 0 & 0 & 0 & 0 & 0 & C_{66} \end{pmatrix}. \tag{3.56}$$

An orthorhombic (ORT) medium is therefore represented by nine independent constants, C_{11}, C_{22}, C_{33}, C_{12}, C_{13}, C_{23}, C_{44}, C_{55}, and C_{66}. In an ORT medium, P wave is pure longitudinal only along the three coordinate axes. If the density of such a medium is ρ, then the P-wave phase velocities along the x_1, x_2, and x_3 axes are given respectively as

$$\begin{aligned} V_{Px1} &= \sqrt{\frac{C_{11}}{\rho}}, \\ V_{Px2} &= \sqrt{\frac{C_{22}}{\rho}}, \\ V_{P0} &= \sqrt{\frac{C_{33}}{\rho}}. \end{aligned} \tag{3.57}$$

The fast and slow S waves, S_1 and S_2, are also pure transverse modes only along the x_1, x_2, and x_3 axes and their phase velocities are given respectively as

$$\begin{aligned} V_{Sx1} &= \sqrt{\frac{C_{44}}{\rho}}, \\ V_{Sx2} &= \sqrt{\frac{C_{66}}{\rho}}, \\ V_{S0} &= \sqrt{\frac{C_{55}}{\rho}}. \end{aligned} \tag{3.58}$$

Like hexagonally symmetric medium, in exploration geophysics, Tsvankin (1997) represented an orthorhombically symmetric medium by V_{P0}, V_{S0}, and seven-dimensional parameters $\varepsilon^{(1)}$, $\varepsilon^{(2)}$, $\delta^{(1)}$, $\delta^{(2)}$, $\delta^{(3)}$, $\gamma^{(1)}$, and $\gamma^{(2)}$, which are given as

$$
\begin{aligned}
\varepsilon^{(1)} &= \frac{C_{22} - C_{33}}{2C_{33}}, \\
\varepsilon^{(2)} &= \frac{C_{11} - C_{33}}{2C_{33}}, \\
\delta^{(1)} &= \frac{(C_{23} + C_{44})^2 - (C_{33} - C_{44})^2}{2C_{33}(C_{33} - C_{44})}, \\
\delta^{(2)} &= \frac{(C_{13} + C_{55})^2 - (C_{33} - C_{55})^2}{2C_{33}(C_{33} - C_{55})}, \\
\delta^{(3)} &= \frac{(C_{12} + C_{66})^2 - (C_{11} - C_{66})^2}{2C_{11}(C_{11} - C_{66})}, \\
\gamma^{(1)} &= \frac{C_{66} - C_{55}}{2C_{55}}, \\
\gamma^{(2)} &= \frac{C_{66} - C_{44}}{2C_{44}}.
\end{aligned} \tag{3.59}
$$

We refer the above representation of an ORT medium as the *Tsvankin–Thomsen* notation.

The ORT medium defined by equation 3.56 with reference to the x_1, x_2, and x_3 coordinate axes can be arbitrarily rotated using Euler angles φ, θ, and ψ via the Bond transformation matrices.

3.4.3 Geological Interpretation of the Elastic Symmetries—The Concept of an Equivalent Medium

Different elastic symmetries, such as isotropic, hexagonal, orthorhombic, etc., are derived from the symmetry properties of the crystals (Fedorov, 1968; Auld, 1973). When the individual rock-forming minerals are preferentially aligned, the seismic properties of the rock can be anisotropic. Earth's mantle, for example, is primarily composed of olivine, which in the upper mantle is characterized by orthorhombic anisotropy. At the spreading centers, these orthorhombically symmetric crystals tend to preferentially align along the direction of the spreading, and quantifying such an anisotropic behavior is one parameter used to infer the past- and present-day plate motions (Schaeffer

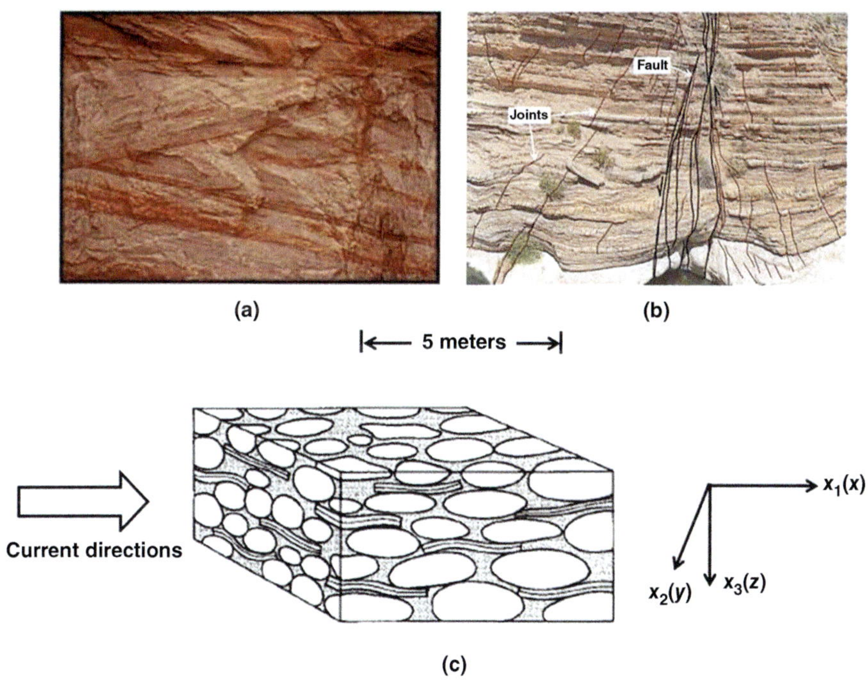

Figure 3.7 (a) Cross-bedded sedimentary rock (URL: GemLand/ https://www.gemland.com/sedona/sandstone.htm/last accessed October 4, 2024). (b) Naturally fractured sedimentary rock (from Michelena et al., 2019/Society of Exploration Geophysicists). The scale for both (a) and (b) are shown at the bottom. (c) Deposition of elongated sediments in relation to the direction of the current (from Selly and Sonnenberg, 2014/with permission of Elsevier).

et al., 2016). Anisotropy of rocks, caused by such preferential alignment of the individual anisotropic rock-forming minerals, is called *inherent anisotropy*.

The presence of such inherent anisotropy is not uncommon in the deep interior of the earth. For most sedimentary rocks encountered in the shallow crustal exploration, the rock-forming minerals are usually randomly deposited and behave isotopically. These individual rocks are, however, much thinner than the seismic wavelength. Figure 3.7a, for example, shows a cross-bedded sediment showing thin sedimentary rocks. Assuming an average sedimentary rock P-wave velocity of 2500 m/s and a dominant frequency in seismic data of 35 Hz, the dominant seismic wavelength is approximately 70 m. It is thus impossible for the seismic data to resolve the individual layers, and it would find an average property, typically on the order of a quarter of the dominant wavelength (~18 m). In finding such average property, the seismic waves, however, will encounter different elastic stiffness between perpendicular and parallel to the layering of the beds, and therefore the rock will behave like an anisotropic medium. In addition to thin layering, the rocks are often fractured. As shown in Figure 3.7b, spacing of these fractures may be much smaller than the dominant seismic wavelength, which also leads to an apparent anisotropic behavior in the observed seismic data. Elongated sediments tend to be deposited with their longest and intermediate axes respectively parallel and perpendicular to the horizontal (x_1 and x_2) and the shortest axis along the vertical (x_3) to the paleocurrent direction by which they were deposited (Figure 3.7c), which also leads to an apparent anisotropic behavior. In contrast to inherent anisotropy, the forms of apparent anisotropy shown in Figure 3.7 are called *induced anisotropy*, which, in turn, leads to the concept of an *equivalent medium*. We use Figure 3.8 to understand the fundamental concepts of an equivalent medium by applying an external force (stress) and analyzing the corresponding deformation (strain) within the medium in reaction to this force.

Figure 3.8a shows a uniform (homogeneous) medium. For example, it may represent a sedimentary rock in which the individual rock-forming mineral grains are randomly oriented, and the entire medium is uniform and homogeneous. In the real world, clean sandstone for example may be the representative of such a medium. Now, if we pull the medium with some force along the x_1 or x axis, shown by the solid black arrows, the medium will be stretched along the x axis, and squeezed along $x_2(y)$ and $x_3(z)$ axes. Considering the medium is perfectly uniform, the amount of squeezing in y and z axes will be the same. Thus, from the analysis of the externally applied stress and the corresponding strain, and assuming a linear relation between the stress strain, we immediately see that there are two constants involved for relating the stress with strain: (1) the ratio between the amount of elongation along the x axis and the magnitude of the applied external force, and (2) the ratio between the squeeze along the y or z axis and the magnitude of the external force. Additionally, because the medium is uniform, these constants will be the same, if we pull the medium along the y or z axis, shown respectively by the red and green arrows in Figure 3.8a. We therefore conclude that a medium shown in Figure 3.8a can be represented by only two elastic constants, and from our previous discussion, it is an isotropic elastic medium.

We now consider a medium represented by Figure 3.8b where the medium consists of thin layers, which are individually isotropic. Now, like Figure 3.8a, if we pull this medium with a certain force along the x axis (black arrows), the medium will be stretched along the x axis and squeezed along y and z axes. However, because the y axis is parallel to the bedding plane, while the z axis is perpendicular to it, the amount of squeezing along these two directions will be different. Consequently, to relate the applied force along the x axis with the corresponding deformations along x, y, and z, we need three elastic constants relating the applied force to (1) the elongation along x, (2) squeeze along y, and (3) squeeze along z. Additionally, because the y axis is also parallel

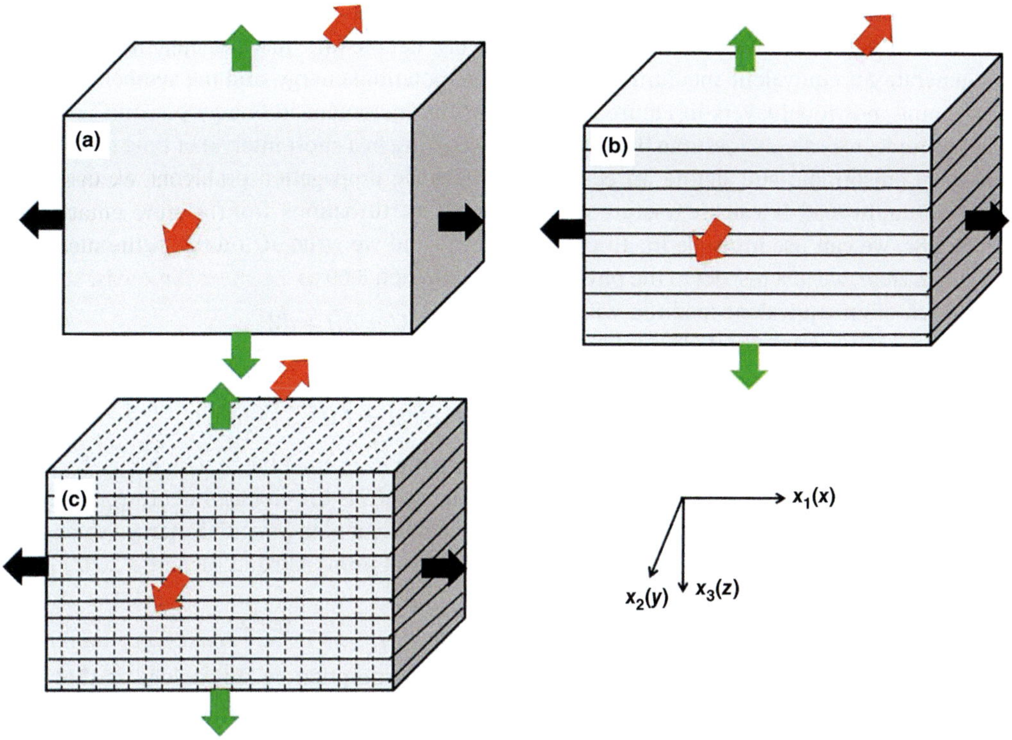

Figure 3.8 Some common forms of rocks encountered in geological materials. (a) Isotropic medium, (b) thinly layered isotropic medium, and (c) thinly layered isotropic medium with fractures. The right-handed Cartesian coordinate system is also shown at the bottom right-hand corner.

to the bedding plane, applying a force along *y* (red arrows in Figure 3.8b) and relating it to the deformation along the three axes can be described by the same three elastic constants. If we now apply an external force along the *z* axis (green arrows), the medium will stretch along *z* and squeeze along *x* and *y*. Also, because *z* is perpendicular to the bedding plane, the amount of stretch will be different from that when the medium is pulled parallel to the bedding. Also, the amount of squeeze along *x* and *y* will be the same, but different from those when the force is applied horizontally. Therefore, to relate the applied force along the *z* axis with the corresponding deformations along *x*, *y*, and *z*, we need two additional elastic constants relating the applied force to (4) the elongation along *z* and (5) squeeze along *x* and *y*. In long wavelength limit, a medium with thin layers that are individual isotropic requires five elastic constants, which from our previous discussion is equivalent to a hexagonally symmetric medium. Also note that the vertical or *z* axis is the axis of symmetry for such a medium, and therefore, it is specifically a VTI medium. By defining a VTI medium with horizontal layers, we can also arbitrarily rotate such medium using Euler angles to represent any cross-bedded structure shown in Figure 3.7a.

To the thinly layered medium, we now add a set of vertical fractures, represented by dashed lines in Figure 3.8c, and apply force in three directions and analyze the corresponding deformations. By using the same arguments as above, we can immediately verify that applying force in each coordinate direction and relating them to the deformations along all three direction require three independent elastic constants, i.e., a total of nine independent elastic constants, which, in other words, is equivalent to an orthorhombic medium.

The three examples shown in Figure 3.8 can be extended for many other forms. For example, introducing vertical fractures in an isotropic medium leads to an HTI medium. We can further use Euler angles to tilt the fractures and generate an equivalent medium. Instead of assuming individual layers in Figure 3.8b,c to be isotropic, we can also assume them to be arbitrarily anisotropic and define an equivalent medium. Instead of a single fracture set of Figure 3.8c, we can use multiple fracture sets and so on. Here, we just restrict to the physical description of an equivalent medium without any mathematical details and refer to Backus (1962), Schoenberg (1983), Schoenberg and Douma (1988), and Schoenberg and Sayers (1995) for further details.

3.5 Strain Energy Function and the Positive-Definite Conditions

In Chapter 1, we discussed the principles of model-based inversion or optimization methods. For seismic problems, these methods start from an initial guess model (or models) and use certain criteria to walk in the model space until the synthetic or predicted data computed from the model (or models) satisfactorily matches with the observed data. During this walk in the model space, it is mandatory to choose physically meaningful models, or, in other words, we must impose constraints such that the model parameters such as V_P, V_S, ρ, ε, δ, γ, etc., are within their feasible limits, otherwise computing synthetic data will not be possible. The way to impose such constraints are derived from definition of the *strain energy function* and *positive-definite conditions* in the theory of elasticity. Therefore, before moving onto the second fundamental equation in elasticity and deriving the elastodynamic equation, it is important to briefly discuss this topic.

Strain Energy Function: According to the first law of thermodynamics, the total energy in an elastic body occupying a volume V must be balanced, or, in other words, we must have

$$\delta M + \delta H = \int_V (\delta K + \delta U)dV. \tag{3.60}$$

In equation 3.60, M, H, K, and U respectively denote the mechanical, heat, kinetic, and potential energy, and the symbol δ stands for the increment in energy per unit volume, occurring in a short interval of time dt. In seismic wave propagation problems, we consider small perturbations from a state of equilibrium, and we write δU on the right-hand side of equation 3.60 as

$$\delta U = dH + dW, \tag{3.61}$$

with

$$dH = TdE \tag{3.62}$$

being the heat function, in which T is the temperature, dE is the change in entropy, and dW is the increment in potential energy, which can be derived from (Aki and Richards, 2002)

$$\frac{\partial W}{\partial \varepsilon_{ij}} = \tau_{ij} = c_{ijkl}\varepsilon_{kl}. \tag{3.63}$$

The assumption of seismology is that the process of deformation in a medium from the passage of an elastic wave is adiabatic, i.e., there is no gain or loss of heat during the passage of seismic wave because the time constant of thermal diffusion in rocks is much longer than the passage of waves. Thus, entropy is constant and $dE = 0$ and therefore $dH = 0$. So, by integrating equation 3.63, we obtain

$$W = \frac{1}{2}\tau_{ij}\varepsilon_{ij} = c_{ijkl}\varepsilon_{kl}\tau_{ij}. \tag{3.64}$$

W defined above is called the *strain energy function*. Exact derivation of equation 3.64 is given in Aki and Richards (2002) and therefore not repeated here.

Positive-Definite Condition: Consider an elastic continuum in an undeformed state, i.e., in a state of zero strain. This state corresponds to a static equilibrium with minimum strain energy. Now, if the medium is deformed by applying an external force, the strain energy must always increase. This, in turn, imposes constraints on the elastic constants that all principal determinants in the fourth-order elastic-stiffness tensor c_{ijkl} must be positive and greater than zero. Or the determinant det$|\mathbf{C}|$ of the 6×6 elastic-stiffness matrix $\mathbf{C}$ (equation 3.22) must be greater than zero.

Imposing the positive-definite condition of $\det|\mathbf{C}| > 0$ to a VTI medium with stiffness matrix, given by equation 3.51, leads to the following constraints (Auld, 1973):

$$\left.\begin{array}{l} C_{11} \geq |C_{11} - 2C_{66}| \\ C_{44} \geq 0 \\ (C_{11} - C_{66})C_{33} \geq C_{13}^2 \end{array}\right\}. \tag{3.65}$$

We can further use equation 3.65 and show that for an isotropic medium we must have

$$\left.\begin{array}{l} V_P^2 \geq |V_P^2 - 2V_S^2| \\ V_S \geq 0 \\ V_P > V_S \end{array}\right\}. \tag{3.66}$$

Constraints satisfying the positive-definite conditions for the orthorhombic and other symmetries can be similarly obtained.

3.6 Dynamic Relation—Second Fundamental Equation in Elasticity

Consider a closed volume V, bounded by the surface S on which the outward unit normal vector $\hat{\mathbf{n}}$ is defined on every point (Figure 2.13). Now assume that the volume is subjected to a time-variant external force with the body force density (force per unit volume) $\mathbf{f}(\mathbf{r}, t)$, where $\mathbf{r}$ is the position vector (equation 2.151) for any point within the volume, and t is the time. This external force will lead to an existence of the stress tensor $\boldsymbol{\tau}$ within the volume V with the traction, defined on any point on the surface S as $\mathbf{T}(\hat{\mathbf{n}}) = \hat{\mathbf{n}} \cdot \boldsymbol{\tau}$. If this externally applied force induces a space- and time-variant particle displacement field $\mathbf{u}(\mathbf{r}, t)$ within V, then by applying Newton's second law of motion of equating the rate of change of momentum due to this particle displacement field to the forces acting on the volume V and the surface S, we write

$$\frac{\partial}{\partial t}\int_V \rho\dot{\mathbf{u}}dV = \int_V \mathbf{f}dV + \int_S \mathbf{T}(\hat{\mathbf{n}})dS. \tag{3.67}$$

In equation 3.67 $\dot{\mathbf{u}} = \frac{\partial \boldsymbol{u}}{\partial t}$ and $\rho = \rho(\mathbf{r})$ is the density (mass per unit volume) field within V.

Now, recalling $\mathbf{T}(\hat{\mathbf{n}}) = \hat{\mathbf{n}} \cdot \boldsymbol{\tau}$ and applying the divergence theorem (equation 2.286), we convert the surface integral on the right-hand side of equation 3.67 to a volume integral

$$\int_S \mathbf{T}(\hat{\mathbf{n}})dS = \int_S \hat{\mathbf{n}} \cdot \boldsymbol{\tau}dS = \int_V \nabla \cdot \boldsymbol{\tau}dV. \tag{3.68}$$

Finally, combining equations 3.67 and 3.68, we write

$$\rho\ddot{\mathbf{u}} = \nabla \cdot \boldsymbol{\tau} + \mathbf{f}, \tag{3.69}$$

where $\ddot{\mathbf{u}} = \frac{\partial^2 \mathbf{u}}{\partial \mathbf{u}^2}$. Equation 3.69 can be Fourier transformed into the frequency domain as

$$-\rho\omega^2\mathbf{u} = \nabla \cdot \boldsymbol{\tau} + \mathbf{f}, \tag{3.70}$$

in which ω is the frequency in radians per second. Equations 3.69 and 3.70 are respectively the time and frequency domain versions of the second fundamental equation of elasticity and are called the *momentum equations.*

3.7 Elastodynamic Equation

In seismology, the body force vector $\mathbf{f}$ is localized to a small region and the rest of the region is source free with $\mathbf{f} = \mathbf{0}$, where the momentum equations can be written as

$$\left.\begin{array}{l} \rho\ddot{\mathbf{u}} = \nabla \cdot \boldsymbol{\tau} \\ -\rho\omega^2\mathbf{u} = \nabla \cdot \boldsymbol{\tau} \end{array}\right\}. \tag{3.71}$$

We restrict to the right-handed Cartesian coordinate system (Figures 3.2 and 3.4), apply divergence operator from Chapter 2, and use Hooke's law (equation 3.19) and the expression of the strain tensor in terms of the displacement vector (equation 3.13), to write equation 3.71, after some algebra as

$$\left.\begin{array}{l} \nabla_S\mathbf{C}(\nabla_S)^T\mathbf{u} = \rho\ddot{\mathbf{u}} \\ \nabla_S\mathbf{C}(\nabla_S)^T\mathbf{u} = -\rho\omega^2\mathbf{u} \end{array}\right\}, \tag{3.72}$$

with $\mathbf{C}$ being the 6×6 stiffness matrix (equation 3.22) and ∇_S defined as

$$\nabla_S = \begin{pmatrix} \partial_1 & 0 & 0 & 0 & \partial_3 & \partial_2 \\ 0 & \partial_2 & 0 & \partial_3 & 0 & \partial_1 \\ 0 & 0 & \partial_3 & \partial_2 & \partial_1 & 0 \end{pmatrix} = \begin{pmatrix} \partial_x & 0 & 0 & 0 & \partial_z & \partial_y \\ 0 & \partial_y & 0 & \partial_z & 0 & \partial_x \\ 0 & 0 & \partial_z & \partial_y & \partial_x & 0 \end{pmatrix}. \tag{3.73}$$

Equation 3.72 is the *elastodynamic* equation in both time and frequency domains. Note that in deriving equation 3.72, we did not make any assumption about the symmetry of the medium. Equation 3.72 is therefore valid for any inhomogeneous, anisotropic medium in three dimensions. Any forward computation of wave-equation-based synthetic seismic data, used in the model-based inversion methods, starts from this elastodynamic equation.

3.8 Solution of the Elastodynamic Equation in Homogeneous Elastic Medium

As illustrated in Figure 3.9, assume a plane wave propagating in a homogeneous medium along an arbitrary direction with the unit vector $\hat{\boldsymbol{l}}$ given as

$$\hat{\mathbf{l}} = \hat{\mathbf{x}}_1 l_1 + \hat{\mathbf{x}}_2 l_2 + \hat{\mathbf{x}}_3 l_3 = \hat{\mathbf{x}}_1 \sin\theta\cos\varphi + \hat{\mathbf{x}}_2 \sin\theta\sin\varphi + \hat{\mathbf{x}}_3 \cos\theta. \quad (3.74)$$

Such a plane wave will have fields proportional to $e^{i\omega(|\mathbf{p}|\hat{\boldsymbol{l}}.\mathbf{r}-t)}$, where $\mathbf{r}$ is the position vector, t is time, and $|\mathbf{p}|$ is the slowness or the inverse velocity such that the slowness vector $\mathbf{p}$ along $\hat{\boldsymbol{l}}$ is defined as

$$\mathbf{p} = |\mathbf{p}|\sin\theta\cos\varphi\hat{\mathbf{x}}_1 + |\mathbf{p}|\sin\theta\sin\varphi\hat{\mathbf{x}}_2 + |\mathbf{p}|\cos\theta\hat{\mathbf{x}}_3. \quad (3.75)$$

From here onward we will replace $|\mathbf{p}|$ with just p. We Fourier transform the frequency domain elastodynamic equation (the bottom equation of equation 3.72) from spatial coordinates x_1, x_2, and x_3 into the wave numbers $k_1 = \omega p l_1 = \omega p \sin\theta\cos\varphi, k_2 = \omega p l_2 = \omega p \sin\theta\sin\varphi$, and $k_3 = \omega p l_3 = \omega\rho\cos\theta$. This, in turn, replaces the spatial derivatives ∂_1, ∂_2, and ∂_3 with $i\omega p l_1 = i\omega p\sin\theta\cos\varphi, i\omega p l_2 = i\omega p \sin\theta\sin\varphi$, and $i\omega p l_3 = i\omega p\cos\theta$ such that ∇_S defined in equation 3.73 becomes

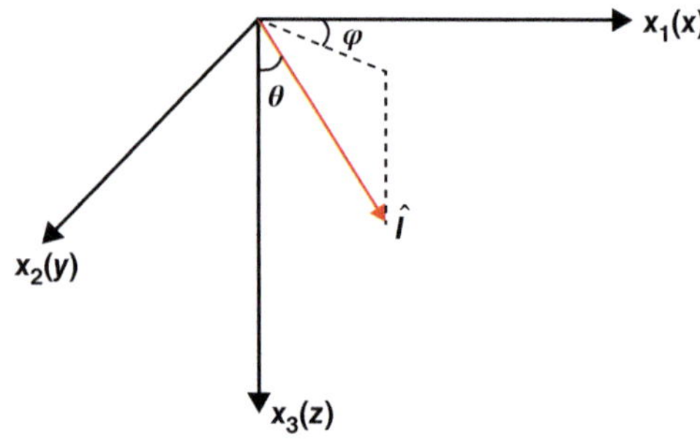

Figure 3.9 Illustration of the propagation of a plane wave along an arbitrary direction, associated with the unit vector $\hat{\mathbf{l}}$.

$$\nabla_S = i\omega p \begin{pmatrix} l_1 & 0 & 0 & 0 & l_3 & l_2 \\ 0 & l_2 & 0 & l_3 & 0 & l_1 \\ 0 & 0 & l_3 & l_2 & l_1 & 0 \end{pmatrix}. \quad (3.76)$$

Substituting ∇_S from equation 3.76 into the elastodynamic equation leads to

$$(p^2\boldsymbol{\Gamma} - \rho\mathbf{I}) \cdot \mathbf{u} = \mathbf{0}, \quad (3.77)$$

in which $\mathbf{u} = [u_1,\ u_2,\ u_3]^T = [u_x,\ u_y,\ u_z]^T$ is the displacement vector with its components being the components of the particle displacement along the three coordinate axes, ρ is the density, and $\mathbf{I}$ is the 3×3 identity matrix. The matrix $\boldsymbol{\Gamma}$ in equation 3.77 is a symmetric 3×3 matrix of the form

$$\boldsymbol{\Gamma} = \begin{pmatrix} \alpha & \delta & \upsilon \\ \delta & \beta & \zeta \\ \upsilon & \zeta & \gamma \end{pmatrix}, \quad (3.78)$$

with

$$\left.\begin{aligned}
\alpha &= C_{11}l_1^2 + C_{66}l_2^2 + C_{55}l_3^2 + 2C_{56}l_2l_3 + 2C_{15}l_3l_1 + 2C_{16}l_1l_2 \\
\beta &= C_{66}l_1^2 + C_{22}l_2^2 + C_{44}l_3^2 + 2C_{24}l_2l_3 + 2C_{46}l_3l_1 + 2C_{26}l_1l_2 \\
\gamma &= C_{55}l_1^2 + C_{44}l_2^2 + C_{33}l_3^2 + 2C_{34}l_2l_3 + 2C_{35}l_3l_1 + 2C_{45}l_1l_2 \\
\delta &= C_{16}l_1^2 + C_{26}l_2^2 + C_{45}l_3^2 + (C_{46} + C_{25})l_2l_3 + (C_{14} + C_{56})l_3l_1 + (C_{12} + C_{66})l_1l_2 \\
\upsilon &= C_{15}l_1^2 + C_{46}l_2^2 + C_{35}l_3^2 + (C_{45} + C_{36})l_2l_3 + (C_{13} + C_{55})l_3l_1 + (C_{14} + C_{56})l_1l_2 \\
\zeta &= C_{56}l_1^2 + C_{24}l_2^2 + C_{34}l_3^2 + (C_{44} + C_{23})l_2l_3 + (C_{36} + C_{45})l_3l_1 + (C_{25} + C_{46})l_1l_2
\end{aligned}\right\}. \quad (3.79)$$

Equation 3.77 is called the *Christoffel equation* and the matrix $\boldsymbol{\Gamma}$ is called the *Christoffel matrix*. A solution of the Christoffel equation can be obtained by setting the determinant $|p^2\boldsymbol{\Gamma} - \rho\mathbf{I}|$ to zero, which, from equation 3.78, leads to

$$(\alpha\beta\gamma + 2\delta\zeta\upsilon - \alpha\zeta^2 - \beta\upsilon^2 - \gamma\delta^2)p^6 + \{\rho(\delta^2 + \zeta^2 + \upsilon^2 - \alpha\beta - \beta\gamma - \gamma\alpha)\}p^4 + \rho^2(\alpha + \beta + \gamma)p^2 - \rho^3 = 0. \tag{3.80}$$

Because the determinant equation is cubic in p^2, there are three solutions (eigenvalues) in p^2, which, for a general anisotropic elastic medium, correspond to the up- and downgoing quasi-P (qP) waves and the fast and slow quasi-S (qS_1 and qS_2) waves. Plugging in these eigenvalues into the Christoffel equation gives the eigenvectors, providing the direction of the particle motion vector $\mathbf{u}$. A solution to the Christoffel equation along different directions (i.e., for different values of the azimuth φ and colatitude θ shown in Figure 3.9) is a surface in three dimensions and is called the *slowness surface* for different wavefields.

For any general anisotropic system, we can numerically solve the determinant equation 3.80 and obtain the slowness surface and the corresponding polarization directions (particle motions) for the qP, qS_1, and qS_2 wavefields. For isotropic, hexagonal, and orthorhombic symmetries, which are mostly common in most of the crust and upper mantle, the Christoffel equation can be analytically solved. For the isotropic symmetry, the solution is straightforward and left as an exercise. In the following, we consider the hexagonal and orthorhombic symmetries.

3.8.1 Solution of the Christoffel Equation for a Hexagonally Symmetric Medium

A hexagonally symmetric elastic medium is characterized by the existence of a symmetry axis such that properties do not vary in the plane perpendicular to this axis. In VTI medium for example with the elastic-stiffness matrix given by equation 3.51, where the symmetry axis is vertical (x_3 of the z axis), the properties do not vary in the horizontal ($x_1 - x_2$ or $x - y$) plane. Therefore, if we consider wave propagation along the $x_1 - x_3$ ($x - y$) plane by setting the azimuth φ in Figure 3.9 to 0° such that $l_1 = \sin\theta$, $l_2 = 0$, and $l_3 = \cos\theta$, and solve the Christoffel equation, we have the solution for all directions. Furthermore, these VTI solutions can be arbitrarily rotated to find solutions for the HTI or TTI. We therefore consider a VTI medium and obtain the solution of the Christoffel equation along an arbitrary direction in the vertical $x_1 - x_3$ plane, oriented at an angle θ with respect to the vertical (x_3) axis. By setting the displacement vector as

$$\mathbf{u} = [u_1, u_3, u_3]^T, \tag{3.81}$$

Christoffel equation (equation 3.77) for the VTI medium in the $x_1 - x_3$ plane becomes

$$\begin{pmatrix} p^2(C_{11}l_1^2 + C_{44}l_3^2) - \rho & p^2(C_{13} + C_{44})l_3l_1 & 0 \\ p^2(C_{13} + C_{44})l_3l_1 & p^2(C_{44}l_1^2 + C_{33}l_3^2) - \rho & 0 \\ 0 & 0 & p^2(C_{66}l_1^2 + C_{44}l_3^2) - \rho \end{pmatrix} \begin{pmatrix} u_1 \\ u_3 \\ u_2 \end{pmatrix} = 0. \tag{3.82}$$

Equation 3.82 is equivalent to two equations, which are, after setting $l_1 = \sin\theta$ and $l_3 = \cos\theta$, one matrix equation

$$\begin{pmatrix} p^2(C_{11}\sin^2\theta + C_{44}\cos^2\theta) - \rho & p^2(C_{13} + C_{44})\sin\theta\cos\theta \\ p^2(C_{13} + C_{44})\sin\theta\cos\theta & p^2(C_{44}\sin^2\theta + C_{33}\cos^2\theta) - \rho \end{pmatrix} \begin{pmatrix} u_1 \\ u_3 \end{pmatrix} = 0, \tag{3.83}$$

and a scalar equation

$$[p^2(C_{66}\sin^2\theta + C_{44}\cos^2\theta) - \rho]u_2 = 0. \tag{3.84}$$

Solution to equation 3.84 is

$$p = \sqrt{\frac{\rho}{C_{66}\sin^2\theta + C_{44}\cos^2\theta}}. \tag{3.85}$$

From equation 3.84, it is evident that the polarization (particle motion) for this wavefield is along the x_2 axis, meaning that the solution is a pure transverse wave. This wavefield solution is the slow S-wave solution, which is also called the S_2 or horizontally polarized S wave or simply SH wave. The phase velocity of this wavefield is the inverse of the slowness, i.e.,

$$V_{S2} = \sqrt{\frac{C_{66}\sin^2\theta + C_{44}\cos^2\theta}{\rho}}. \tag{3.86}$$

In the vertical ($\theta = 0°$) and horizontal ($\theta = 90°$), the phase velocities are

$$\left.\begin{aligned} V_{S20} &= \sqrt{\frac{C_{44}}{\rho}} \\ V_{S2hor} &= \sqrt{\frac{C_{66}}{\rho}} \end{aligned}\right\}, \tag{3.87}$$

which we previously introduced (see equations 3.52 and 3.53).

Now, setting the determinant of the matrix on the left-hand side of equation 3.83 to zero gives

$$\begin{aligned} &p^4[(C_{11}\sin^2\theta + C_{44}\cos^2\theta)(C_{44}\sin^2\theta + C_{33}\cos^2\theta) - (C_{13} + C_{44})^2\sin^2\theta\cos^2\theta] \\ &\quad - p^2(C_{11}\sin^2\theta + C_{44}\cos^2\theta + C_{44}\sin^2\theta + C_{33}\cos^2\theta)\rho + \rho^2 = 0, \end{aligned} \tag{3.88}$$

with the solutions

$$p = (2\rho)^{\frac{1}{2}}\left[\begin{array}{c} C_{11}\sin^2\theta + C_{33}\cos^2\theta + C_{44} \\ \hline \pm\sqrt{\{(C_{11} - C_{44})\sin^2\theta + (C_{44} - C_{33})\cos^2\theta\}^2 + (C_{13} + C_{44})^2\sin^2 2\theta} \end{array}\right]^{-\frac{1}{2}}. \tag{3.89}$$

And, from equation 3.83, the particle motions for both solutions given by equation 3.89 are polarized in the $x_1 - x_3$ plane. Out of the two solutions, the one with the plus (+) sign before the second term within the square bracket on the right-hand corresponds to the P wavefield, and the one with the minus (−) sign corresponds to the S_1 or the vertically polarized S wavefield (SV). It can also be seen that the particle motions for both these wavefields are mutually orthogonal, but neither pure longitudinal nor pure transverse along any arbitrary direction. However, along certain directions, for example along the horizontal (x_1) or vertical (x_3) axes, P wavefield is pure longitudinal and S_1 wavefield is pure transverse. The phase velocities for P and S_1 are the inverse of their respective slowness, which, from equation 3.89, can be readily derived as

$$\left.\begin{aligned} V_P &= \sqrt{\frac{C_{11}\sin^2\theta + C_{33}\cos^2\theta + C_{44} + \sqrt{\{(C_{11} - C_{44})\sin^2\theta + (C_{44} - C_{33})\cos^2\theta\}^2 + (C_{13} + C_{44})^2\sin^2 2\theta}}{2\rho}} \\ V_{S1} &= \sqrt{\frac{C_{11}\sin^2\theta + C_{33}\cos^2\theta + C_{44} - \sqrt{\{(C_{11} - C_{44})\sin^2\theta + (C_{44} - C_{33})\cos^2\theta\}^2 + (C_{13} + C_{44})^2\sin^2 2\theta}}{2\rho}} \end{aligned}\right\}. \tag{3.90}$$

Again, for pure vertical ($\theta = 0°$) and pure horizontal ($\theta = 90°$) propagations, we obtain from equation 3.90

$$\left.\begin{aligned} V_{P0} &= \sqrt{\frac{C_{33}}{\rho}} \\ V_{S0} &= \sqrt{\frac{C_{44}}{\rho}} \\ V_{Phor} &= \sqrt{\frac{C_{11}}{\rho}} \\ V_{Shor} &= \sqrt{\frac{C_{44}}{\rho}} \end{aligned}\right\}, \qquad (3.91)$$

which we previously listed (equations 3.52 and 3.53) when we introduced the hexagonal symmetry. Because S_1 is polarized in the vertical ($x_1 - x_3$) plane and S_2 is polarized along the x_2 axis, which is perpendicular to the $x_1 - x_3$ plane and horizontally oriented, from now on we will designate the S_1 wave as the vertically polarized S wave or SV wave and S_2 as the horizontally polarized S wave or SH wave. Accordingly, we will also use the notations V_{SV} and V_{SH} for the phase velocities of these two waves instead of V_{S1} and V_{S2}, which we used in equations 3.90 and 3.86.

We now use the Thomsen parameters V_{P0}, V_{S0}, ε, δ, and γ, which we previously introduced in equations 3.52 and 3.52. We can readily see that the SH (S_2) wave phase velocity (equation 3.86) can be given as

$$V_{SH} = V_{S0}\sqrt{1 + 2\gamma \sin^2\theta}. \qquad (3.92)$$

Expressing V_{SH} in Thomsen notations V_{S0} and γ by equation 3.92 is more convenient than expressing it in terms of elastic coefficients C_{44} and C_{66} provided in equation 3.86. V_{S0} is the phase velocity in the vertical direction which we can measure from seismic data. Also, the SH wave phase velocity in the horizontal direction is $\sqrt{C_{66}/\rho}$ (see equation 3.87), and thus $\gamma = (C_{66} - C_{44})/2C_{44}$ can be considered as the magnitude of the SH-wave anisotropy, which we can also estimate from seismic data. Equation 3.92 is therefore more convenient to use than equation 3.86. Equation 3.92 is the exact expression for V_{SH} in Thomsen notation. From this exact expression, Thomsen (1986) approximated the square root on the right-hand side of equation 3.92 as a truncated Taylor series and obtained a *weak anisotropic* expression as

$$V_{SH} \approx V_{S0}(1 + \gamma \sin^2\theta). \qquad (3.93)$$

Now, for the P-SV waves, we combine both equations shown in equation 3.90 and write

$$V^2 = \frac{1}{2\rho}[C_{11}\sin^2\theta + C_{33}\cos^2\theta + C_{44} \pm D(\theta)], \qquad (3.94)$$

with

$$D(\theta) = \sqrt{\begin{aligned}&\{(C_{11} - C_{44})\sin^2\theta + (C_{44} - C_{33})\cos^2\theta\}^2 \\ &+ (C_{13} + C_{44})^2 \sin^2 2\theta\end{aligned}}. \qquad (3.95)$$

A plus (+) and a minus (−) sign before $D(\theta)$ on the right-hand side of equation 3.94 respectively represent the P- and SV-wave phase velocities. Now, introducing Thomsen parameters V_{P0}, V_{S0}, ε, and δ, equations for the P- and SV-wave phase velocities V_P and V_{SV} can be written from equations 3.94 and 3.95, after some algebra as

$$\left.\begin{aligned} V_P^2 &= V_{P0}^2[1 + \varepsilon \sin^2\theta + D^*(\theta)] \\ V_{SV}^2 &= V_{S0}^2\left[1 + \frac{V_{P0}^2}{V_{S0}^2}\varepsilon \sin^2\theta - \frac{V_{P0}^2}{V_{S0}^2}D^*(\theta)\right] \end{aligned}\right\}, \qquad (3.96)$$

with

$$D^*(\theta) = \frac{1}{2}\zeta\left[\left\{1 + \frac{4\delta^*}{\zeta^2}\sin^2\theta\cos^2\theta\right.\right. \\ \left.\left. + \frac{4(\zeta + \varepsilon)\varepsilon}{\zeta^2}\sin^4\theta\right\}^{\frac{1}{2}} - 1\right]. \qquad (3.97)$$

The parameters ζ and δ^* in equation 3.97 are given as

$$\left.\begin{aligned} \zeta &= 1 - \frac{V_{P0}^2}{V_{S0}^2} \\ \delta^* &= \zeta(2\delta - \varepsilon) \end{aligned}\right\}. \qquad (3.98)$$

Equation 3.96 in conjunction with equations 3.97 and 3.98 provide the exact expressions for the P- and SV-wave phase velocities for the VTI symmetry. By approximating the square root inside the square bracket on the right-hand side of equation 3.97, Thomsen (1986) expressed $D^*(\theta)$ as

$$D(\theta)^* \approx (2\delta - \varepsilon)\sin^2\theta\cos^2\theta + \varepsilon\sin^4\theta, \tag{3.99}$$

using which the *weak-anisotropic* approximations for the P- and SV-wave phase velocities can be derived as

$$\left.\begin{aligned} V_P &\approx V_{P0}(1 + \delta\sin^2\theta\cos^2\theta + \varepsilon\sin^4\theta) \\ V_{SV} &\approx V_{S0}\left[1 + \frac{V_{P0}^2}{V_{S0}^2}(\varepsilon - \delta)\sin^2\theta\cos^2\theta\right] \end{aligned}\right\}. \tag{3.100}$$

Figure 3.10 provides the P-wave phase velocity (V_P) as a function of the phase angle (θ) using both exact formula and weakly anisotropic approximate formula, given respectively by equations 3.96 and 3.100 for two formations: (1) Taylor Sandstone and (2) Green River Shale. The anisotropic parameters V_{P0}, V_{S0}, ε, δ, and γ for both these formations are given in Thomsen (1986) and in the table provided in Problem 3.12. Note from Figure 3.10, for weak P-wave anisotropy ($\varepsilon = 0.11$, $\delta = -0.035$), the exact and approximate P-wave phase velocities for Taylor Sandstone are nearly equal for almost all phase angles between vertical ($\theta = 0°$) and horizontal ($\theta = 90°$) propagations, but for propagation (phase) angles above 80°, there are some minor differences. For moderate values of P-wave anisotropy ($\varepsilon = 0.195$, $\delta = -0.22$) of the Green River Shale on the other hand, the exact and approximate P-wave phase velocities differ for phase angles above 20°, but they both follow the same trend of variations. The exact and weak-anisotropic phase velocities V_{SH} and V_{SV} also be similarly computed and compared, which are left as exercises.

3.8.2 Solution of the Christoffel Equation for an Orthorhombically Symmetric Medium

Elastic-stiffness matrix for orthorhombic symmetry (equation 3.56) is like that of the hexagonal symmetry (equation 3.51) except that for orthorhombic symmetry $C_{11} \neq C_{22} \neq C_{33}$, $C_{44} \neq C_{55}$, $C_{12} \neq C_{11} - 2C_{66}$, and $C_{23} \neq C_{13}$ such that there are nine independent elastic constants instead of five. Previously, we designated such an orthorhombic medium to be equivalent to a set of vertical fractures in a stack of mirobedded layers in which each of the microbeds is individually isotropic (Figure 3.8c). Using a right-handed Cartesian

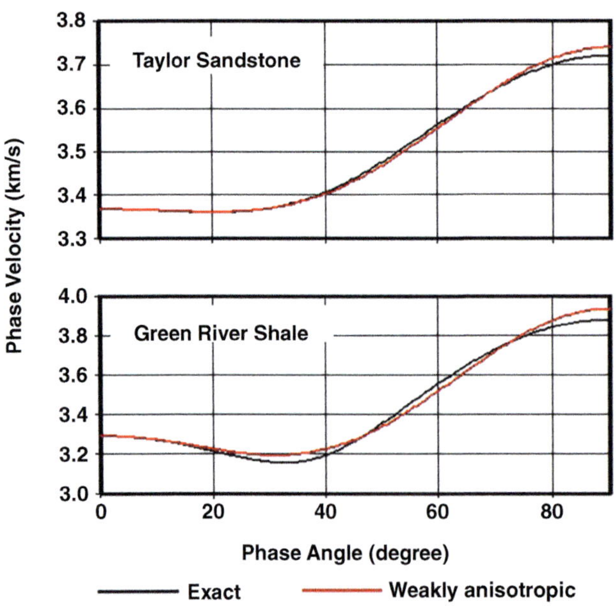

Figure 3.10 Exact (black) and weak-anisotropic (red) P-wave phase velocity V_P as a function of the phase angle θ for Taylor Sandstone (top) and Green River Shale (bottom).

coordinate system such that the microbeds are parallel to the horizontal ($x_1 - x_2$) plane and the fractures are parallel to the $x_2 - x_3$ plane as shown in Figure 3.8, we can immediately see that all three coordinate planes ($x_1 - x_2$, $x_2 - x_3$, and $x_3 - x_1$) are symmetry planes.

For any arbitrary direction of propagation in 3D with $l_1 = \sin\theta\cos\varphi$, $l_2 = \sin\theta\sin\varphi$, and $l_3 = \cos\theta$ (see Figure 3.9 and equation 3.74), the components of Christoffel matrix (equation 3.79) are

$$\left.\begin{aligned} \alpha &= C_{11}l_1^2 + C_{66}l_2^2 + C_{55}l_3^2 \\ \beta &= C_{66}l_1^2 + C_{22}l_2^2 + C_{44}l_3^2 \\ \gamma &= C_{55}l_1^2 + C_{44}l_2^2 + C_{33}l_3^2 \\ \delta &= (C_{12} + C_{66})l_1 l_2 \\ \upsilon &= C_{13}l_3 l_1 \\ \zeta &= (C_{44} + C_{23})l_2 l_3 \end{aligned}\right\}, \tag{3.101}$$

using which a solution to the cubic equation in p^2 (equation 3.80) can be numerically obtained, and the P-SV- and SH-wave slowness and the corresponding phase velocities can be computed. It can also be verified that, in general, none of these wave modes are pure longitudinal or transverse. Therefore, for orthorhombic symmetry, these should be considered as the quasi-P (qP) wave and the fast and slow quasi-S (qS_1, qS_2) waves. In the $x_1 - x_2$, $x_2 - x_3$, and $x_1 - x_3$ symmetry planes, however, these wavefields behave like the hexagonally symmetric medium where the Christoffel equation decouples into 2×2 P-SV system of equations and a scalar SH equation, for which analytical solutions can be obtained. In the vertical ($x_1 - x_3$) plane for example, $l_1 = \sin\theta$, $l_2 = 0$, and $l_3 = \cos\theta$. Therefore, for propagation in the $x_1 - x_3$ plane, equation 3.101 simplifies to

$$\left.\begin{aligned} \alpha &= C_{11}\sin^2\theta + C_{55}\cos^2\theta \\ \beta &= C_{66}\sin^2\theta + C_{44}\cos^2\theta \\ \gamma &= C_{55}\sin^2\theta + C_{33}\cos^2\theta \\ \delta &= 0 \\ \upsilon &= C_{13}\cos\theta\sin\theta \\ \zeta &= 0 \end{aligned}\right\}. \tag{3.102}$$

Therefore, the Christoffel equation (equation 3.77) becomes

$$\begin{pmatrix} p^2(C_{11}\sin^2\theta + C_{55}\cos^2\theta) - \rho & 0 & p^2(C_{13} + C_{55})\cos\theta\sin\theta \\ 0 & p^2(C_{66}\sin^2\theta + C_{44}\cos^2\theta) - \rho & 0 \\ p^2(C_{13} + C_{55})\cos\theta\sin\theta & 0 & p^2(C_{55}\sin^2\theta + C_{33}\cos^2\theta) - \rho \end{pmatrix} \begin{pmatrix} u_1 \\ u_2 \\ u_3 \end{pmatrix} = 0. \tag{3.103}$$

Like hexagonal symmetry, if we rearrange the displacement vector $\mathbf{u}$ from $[u_1, u_2, u_3]^T$ to $[u_1, u_3, u_2]^T$, equation 3.103 decouples into a 2×2 matrix equation

$$\begin{pmatrix} p^2(C_{11}\sin^2\theta + C_{55}\cos^2\theta) - \rho & p^2(C_{13} + C_{55})\cos\theta\sin\theta \\ p^2(C_{13} + C_{55})\cos\theta\sin\theta & p^2(C_{55}\sin^2\theta + C_{33}\cos^2\theta) - \rho \end{pmatrix} \begin{pmatrix} u_1 \\ u_3 \end{pmatrix} = 0 \tag{3.104}$$

and a scalar equation

$$[p^2(C_{66}\sin^2\theta + C_{44}\cos^2\theta) - \rho]u_2 = 0. \tag{3.105}$$

Equation 3.105 represents a pure shear motion with the wavefield propagation in the $x_1 - x_3$ plane and the particle displacement polarized vertically along the x_2 axis. This solution is therefore

SH wave with the phase velocity given as

$$V_{SH} = \sqrt{\frac{C_{66}\sin^2\theta + C_{44}\cos^2\theta}{\rho}}. \tag{3.106}$$

For vertical and horizontal propagations ($\theta = 0°$ and $90°$), it follows from equation 3.106 that the SH-wave phase velocities are respectively $\sqrt{C_{44}/\rho}$ and $\sqrt{C_{66}/\rho}$.

Comparing equation 3.104 with equation 3.83, we can readily write the phase velocity for the P-SV system for an orthorhombic medium in the $x_1 - x_3$ plane as

$$V^2 = \frac{1}{2\rho}[C_{11}\sin^2\theta + C_{33}\cos^2\theta + C_{55} \pm E(\theta)], \tag{3.107}$$

with

$$E(\theta) = \sqrt{\begin{array}{l}\{(C_{11} - C_{55})\sin^2\theta \\ +(C_{55} - C_{33})\cos^2\theta\}^2 \\ +(C_{13} + C_{55})^2\sin^2 2\theta\end{array}}. \tag{3.108}$$

Like equation 3.94, which we previously derived for the hexagonal medium, the plus (+) and minus (–) signs before $E(\theta)$ on the right-hand side of equation 3.107 respectively represent the P- and SV-wave phase velocities. In addition, we must also note that the particle displacement polarization directions of these P- and SV-wave modes are mutually orthogonal, but they are not pure longitudinal or transverse models in any arbitrary direction. For pure vertical and horizontal propagations (that is for phase angles $\theta = 0°$ and $90°$), they are pure modes, and phase velocities for which, from equations 3.107 and 3.108, can be derived as

$$\left.\begin{array}{l} V_{P_{Vert}} = \sqrt{\dfrac{C_{33}}{\rho}};\ V_{P_{Hor}} = \sqrt{\dfrac{C_{11}}{\rho}}; \\ V_{SV_{Vert}} = V_{SV_{Hor}} = \sqrt{\dfrac{C_{55}}{\rho}}. \end{array}\right\} \tag{3.109}$$

In other symmetry planes ($x_2 - x_3$ and $x_1 - x_2$), analytical solutions for P, SV, and SH can similarly be obtained, which, in fact, can be directly written from the solutions in the $x_1 - x_3$ plane by using the appropriate stiffness constants. In the $x_2 - x_3$ plane for example, we can obtain the solutions from those provided above in the $x_1 - x_3$ plane (equations 3.106–3.108) just by the following replacements and interchanges (Tsvankin, 1997):

$$C_{11} \rightarrow C_{22};\ C_{55} \leftrightarrow C_{44};\ C_{13} \rightarrow C_{23}. \tag{3.110}$$

In equation 3.110, the $\rightarrow$ means a replacement and $\leftrightarrow$ represents an interchange. The solutions in the $x_1 - x_2$ plane can also similarly be obtained.

Thus, for an orthorhombic medium, a general solution of the Christoffel equation must be numerically obtained. However, analytical solutions exist in all three symmetry planes. These analytical solutions are like those for the hexagonally symmetric medium. Because the properties for a hexagonally symmetric medium are invariant perpendicular to the symmetry axis, once we obtain the solution in any plane containing the symmetry axis, we have the solution in all directions. In a VTI medium for example, if we have solutions in the vertical ($x_1 - x_3$) plane, we obtain solutions in all directions. For orthorhombic symmetry, solutions are different for each symmetry plane. To provide Thomsen notations, such as V_{P0}, V_{S0}, ε, δ, and γ, to represent an orthorhombically symmetric medium, Tsvankin (1997) introduced the Tsvankin–Thomsen parameters V_{P0}, V_{S0}, $\varepsilon^{(1)}$, $\varepsilon^{(2)}$, $\delta^{(1)}$, $\delta^{(2)}$, $\delta^{(3)}$, $\gamma^{(1)}$, and $\gamma^{(2)}$, provided in equations 3.58 and 3.59. Like Thomsen parameters defined for a VTI medium, Tsvankin–Thomsen parameters also have physical meanings. The parameters V_{P0} and V_{S0} are the vertical (along the x_3 axis) P- and SV-wave velocities. The Thomsen parameter ε in two vertical planes ($x_2 - x_3$ and $x_1 - x_3$) are respectively represented by $\varepsilon^{(1)}$ and $\varepsilon^{(2)}$ and are related to the amount of P-wave anisotropy (relative difference in the P-wave phase velocities between horizontal and vertical) in those planes. The parameters $\delta^{(1)}$, $\delta^{(2)}$, and $\delta^{(3)}$ are equivalent to the Thomsen parameter δ in the $x_2 - x_3$, $x_1 - x_3$, and $x_1 - x_2$ planes and control the variations of both P- and SV-wave phase velocities as functions of phase angle in those planes. Finally, $\gamma^{(1)}$ and $\gamma^{(2)}$ represent the Thomsen parameter γ in the $x_2 - x_3$ and $x_1 - x_3$ planes and are related to the SH-wave anisotropy in those planes. If $\varepsilon^{(1)} = \varepsilon^{(2)} = \varepsilon$, $\delta^{(1)} = \delta^{(2)} = \delta$, $\delta^{(3)} = 0$, and $\gamma^{(2)} = \gamma^2 = \gamma$, the orthorhombic medium is a hexagonal medium. Using these Tsvankin–Thomsen parameters, exact and approximate (weakly anisotropic) phase velocities for P-, SV-, and SH-waves in the symmetry planes can be obtained like the VTI medium. For details, see Tsvankin (1997).

3.9 Ray (Group) Angle and Ray (Group) Velocity

Computing slowness along direction $\hat{\boldsymbol{l}} = \sin\theta\cos\varphi\hat{\mathbf{x}}_1 + \sin\theta\sin\varphi\hat{\mathbf{x}}_2 + \cos\theta\hat{\mathbf{x}}_3$ for different values of θ and φ is a surface in three dimensions, which is called the *slowness surface*. By taking the reciprocal of the computed slowness along these directions is also a surface, known as the *phase velocity* surface. The P, SV, and SH waves with constant phase and amplitudes proportional to $e^{i\omega(|\mathbf{p}|\hat{l}.\mathbf{r}-t)}$ propagate with different values of slowness (or phase velocity) along these directions. While these wavefields propagate along these directions, the energy propagates along the direction parallel to the outward normal to the slowness surface. In an isotropic medium, where the wavefields propagate with the same slowness in all directions, the slowness surface is spherical in shape, and the directions of the phase and energy propagations are identical. For any anisotropic medium, however, the normal is not parallel to the propagation direction. Consequently, the energy propagates differently and forms a different surface in three dimensions, known as the *energy* or *ray* or the *group velocity surface*. Therefore, for an anisotropic medium, the phase slowness or phase velocity, associated with the phase angle θ and azimuth φ, for propagation direction $\hat{\boldsymbol{l}}$ (Figure 3.9) is associated with the corresponding group velocity $\mathbf{V}_g$ propagating along a different direction with group angle ϕ and azimuth ξ.

3.9.1 Mathematical Formulation of the Group and Phase Directions for an Elastic Medium with Arbitrary Anisotropy

We use Figure 3.11 to elaborate our discussion on the phase and group velocity. Assuming

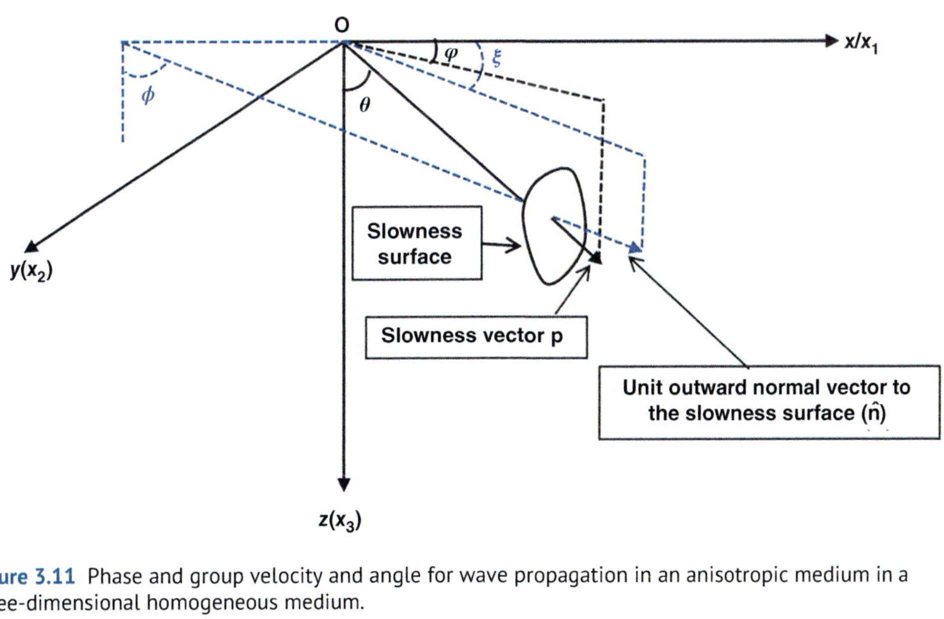

Figure 3.11 Phase and group velocity and angle for wave propagation in an anisotropic medium in a three-dimensional homogeneous medium.

a right-handed Cartesian coordinate system, the solid black arrow in Figure 3.11 is the slowness vector $\mathbf{p}$, propagating along the direction $\hat{\boldsymbol{l}} = \sin\theta\cos\varphi\hat{\mathbf{x}}_1 + \sin\theta\sin\varphi\hat{\mathbf{x}}_2 + \cos\theta\hat{\mathbf{x}}_3$ with azimuth φ and phase angle θ, and the shaded region is the slowness surface in the neighborhood of this slowness vector. The unit outward normal vector $\hat{\mathbf{n}}$ to the slowness surface is shown with the dashed blue arrow in Figure 3.11. In general, the direction of this unit normal vector is different from the slowness vector with colatitude ϕ and azimuth ξ, and the group velocity vector $\mathbf{V}_g$ is directed along this normal and has the magnitude equal to the component of the phase velocity vector (inverse of the slowness vector) along the normal (Musgrave, 1970; Auld, 1973). Or, in other words,

$$\mathbf{V}_g = \frac{\hat{\mathbf{n}}}{\mathbf{p}\cdot\hat{\mathbf{n}}} = (\hat{\mathbf{n}}\cdot\mathbf{V}_{ph})\hat{\mathbf{n}}, \qquad (3.111)$$

in which $\mathbf{V}_{ph}$ is the phase velocity vector for P, SV, or SH wavefields. Equation 3.111 is the fundamental relation between the group velocity and slowness (phase velocity) vectors and can be used to numerically obtain the group velocity vector from the slowness vector, as illustrated below.

As illustrated in Figure 3.12, we compute the slowness along a given direction $\{\theta, \varphi\}$ and around its neighborhood $\{\theta+\Delta\theta, \varphi-\Delta\varphi\}$, $\{\theta+\Delta\theta, \varphi+\Delta\varphi\}$, $\{\theta-\Delta\theta, \varphi+\Delta\varphi\}$, and $\{\theta-\Delta\theta, \varphi-\Delta\varphi\}$ so that they form a portion of the slowness surface about $\{\theta, \varphi\}$. In Figure 3.12, $\mathbf{p}(\theta, \varphi)$ is shown from point P as the solid black arrow. The slowness values $\mathbf{p}(\theta+\Delta\theta, \varphi-\Delta\varphi)$, $\mathbf{p}(\theta+\Delta\theta, \varphi+\Delta\varphi)$, $\mathbf{p}(\theta-\Delta\theta, \varphi+\Delta\varphi)$, and $\mathbf{p}(\theta-\Delta\theta, \varphi-\Delta\varphi)$ on the other hand are shown as solid red arrows from points P_1, P_2, P_3, and P_4 in Figure 3.12. Finally, the portion of the slowness surface formed by joining the points P_1, P_2, P_3, and P_4 is shown as dashed red curves in Figure 3.12. Now, from elementary vector algebra, discussed in Chapter 2, the four vectors $\mathbf{t}_1$, $\mathbf{t}_2$, $\mathbf{t}_3$, and $\mathbf{t}_4$, shown as solid blue arrows in Figure 3.12, can be written as

$$\left.\begin{aligned}
\mathbf{t}_1 &= \mathbf{p}(\theta+\Delta\theta, \varphi+\Delta\varphi) - \mathbf{p}(\theta+\Delta\theta, \varphi-\Delta\varphi)\\
\mathbf{t}_2 &= \mathbf{p}(\theta+\Delta\theta, \varphi+\Delta\varphi) - \mathbf{p}(\theta-\Delta\theta, \varphi+\Delta\varphi)\\
\mathbf{t}_3 &= \mathbf{p}(\theta-\Delta\theta, \varphi+\Delta\varphi) - \mathbf{p}(\theta-\Delta\theta, \varphi-\Delta\varphi)\\
\mathbf{t}_4 &= \mathbf{p}(\theta-\Delta\theta, \varphi-\Delta\varphi) - \mathbf{p}(\theta+\Delta\theta, \varphi-\Delta\varphi)
\end{aligned}\right\}. \qquad (3.112)$$

If we choose $\Delta\theta$ and $\Delta\varphi$ to be sufficiently small, then these four vectors $\mathbf{t}_1$, $\mathbf{t}_2$, $\mathbf{t}_3$, and $\mathbf{t}_4$ will coincide with the dashed red curves and lie on the slowness surface. Consequently, using the concept of the cross product of vectors, discussed in Chapter 4, the outward unit normal vectors $\hat{\mathbf{n}}_1, \hat{\mathbf{n}}_2, \hat{\mathbf{n}}_3$, and $\hat{\mathbf{n}}_4$ to the slowness surface at points P_1, P_2, P_3, and P_4

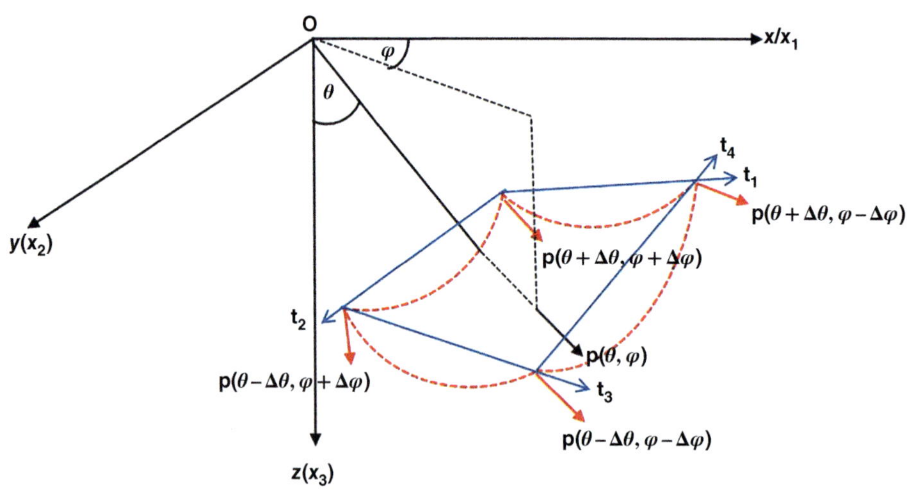

Figure 3.12 Illustration of numerical computation of the group velocity vector for an anisotropic medium.

can be written as

$$\left.\begin{aligned} \hat{\mathbf{n}}_1 &= \frac{\mathbf{t}_1 \times \mathbf{t}_4}{|\mathbf{t}_1 \times \mathbf{t}_4|} \\ \hat{\mathbf{n}}_2 &= \frac{\mathbf{t}_2 \times \mathbf{t}_1}{|\mathbf{t}_2 \times \mathbf{t}_1|} \\ \hat{\mathbf{n}}_3 &= \frac{\mathbf{t}_2 \times \mathbf{t}_3}{|\mathbf{t}_2 \times \mathbf{t}_3|} \\ \hat{\mathbf{n}}_4 &= \frac{\mathbf{t}_3 \times \mathbf{t}_4}{|\mathbf{t}_3 \times \mathbf{t}_4|} \end{aligned}\right\}. \tag{3.113}$$

In equation 3.113, $|\mathbf{v}|$ represents the magnitude of the vector $\mathbf{v}$. For small $\Delta\theta$ and $\Delta\varphi$, the average of $\hat{\mathbf{n}}_1, \hat{\mathbf{n}}_2, \hat{\mathbf{n}}_3$, and $\hat{\mathbf{n}}_4$ is the unit outward normal $\hat{\mathbf{n}}$ at point P corresponding to the slowness $\mathbf{p}(\theta, \varphi)$. Once $\hat{\mathbf{n}}$ is obtained, we can use equation 3.111 to obtain $\mathbf{V}_g$, and from $\mathbf{V}_g$, we can obtain the group angle ϕ and azimuth ξ (Figure 3.11).

3.9.2 Analytical Expressions for Group Velocity and Angle for Specific Symmetries

For an elastic medium with arbitrary anisotropy, we can solve the Christoffel equation (equation 3.77) to compute $\mathbf{p}(\theta, \varphi)$, $\mathbf{p}(\theta+\Delta\theta, \varphi-\Delta\varphi)$, $\mathbf{p}(\theta+\Delta\theta, \varphi+\Delta\varphi)$, $\mathbf{p}(\theta-\Delta\theta, \varphi+\Delta\varphi)$, and $\mathbf{p}(\theta-\Delta\theta, \varphi-\Delta\varphi)$ for any $\{\theta, \varphi\}$ pair and suitably chosen small values of $\Delta\theta$ and $\Delta\varphi$ and compute $\mathbf{V}_g$ from equations 3.111–3.113, and by repeating this for different values of θ and φ, we can get the slowness (or phase velocity) and group velocity surfaces. However, as we previously noted, for a hexagonally symmetric medium the P- and SV-wave particle motions are restricted within the plane containing the symmetry axis, and the SH-wave motion is restricted in the direction perpendicular to this plane. For wave propagation along the symmetry ($x_1 - x_2$, $x_2 - x_3$, and $x_3 - x_1$) planes in an orthorhombically symmetric medium, the wavefields behave in a similar fashion. In these situations, the slowness (or phase velocity) is independent of the azimuth φ and functions of only the phase angle θ. We also previously seen that the slowness (and therefore phase velocity) as a function of θ under such situation can be analytically computed (for example, equations 3.93 and 3.96–3.98 for a VTI medium and equations 3.106–3.108 for an orthorhombic medium with the wavefield propagating in the $x_1 - x_3$ plane). Because of this azimuthal independence, the group velocity vector ($\mathbf{V}_g$) and therefore the group angle (ϕ) can also be analytically computed. To elaborate on this analytic computation, we use a VTI medium. Results for propagation in HTI and TTI, and for an orthorhombic medium propagation in the symmetry planes, can be similarly obtained.

Figure 3.13 shows an illustration of the seismic wavefield propagation in the vertical

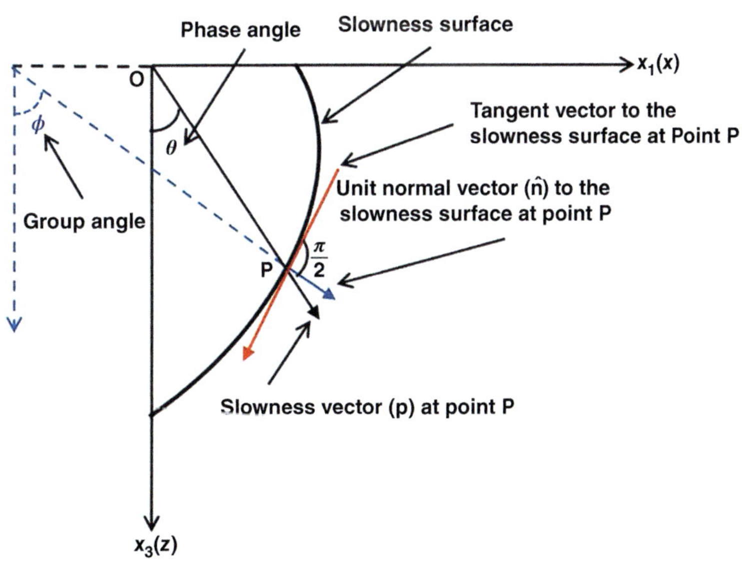

Figure 3.13 The group velocity and group angle as a function of the phase velocity (slowness) and phase angle for an elastic medium with VTI symmetry.

$(x_1 - x_3)$ plane with the slowness surface (or strictly a two-dimensional (2D) section of the slowness surface in the $x_1 - x_3$ plane) as the solid black curve and the slowness vector **p** at a point P on the slowness surface with phase angle θ as the solid black arrow. From elementary vector algebra, we write the slowness vector **p** at point P as

$$\mathbf{p} = |\mathbf{p}|(\hat{\mathbf{x}}_1 \sin\theta + \hat{\mathbf{x}}_3 \cos\theta), \tag{3.114}$$

in which $|\mathbf{p}|$ is the magnitude of **p** and $\hat{\mathbf{x}}_1$ and $\hat{\mathbf{x}}_3$ are the unit vectors along the x_1 and x_3 coordinate axes. Again, from elementary differential geometry, the tangent vector **t** at the point P on the slowness surface and shown as the solid red arrow in Figure 3.13 is given as

$$\mathbf{t} = \frac{d\mathbf{p}}{d\theta}. \tag{3.115}$$

And, by substituting **p** from equation 3.114 into equation 3.115, we obtain

$$\mathbf{t} = \hat{\mathbf{x}}_1\left(|\mathbf{p}|\cos\theta + \sin\theta\frac{d\mathbf{p}}{d\theta}\right) - \hat{\mathbf{x}}_3\left(|\mathbf{p}|\sin\theta - \cos\theta\frac{d\mathbf{p}}{d\theta}\right). \tag{3.116}$$

Now, by definition, the phase velocity vector $\mathbf{V}_{ph}$ is directed along the slowness vector with magnitude equal to the reciprocal of the magnitude of the slowness vector or, in other words,

$$\mathbf{V}_{ph} = \frac{\hat{\mathbf{p}}}{|\mathbf{p}|}, \tag{3.117}$$

where $\hat{\mathbf{p}}$ is the unit vector along the direction of the slowness. Using equation 3.117, we can then write equation 3.116 as

$$\mathbf{t} = \frac{1}{|\mathbf{V}_{ph}|^2}\left[\left\{|\mathbf{V}_{ph}|\cos\theta - \sin\theta\frac{d|\mathbf{V}_{ph}|}{d\theta}\right\} \times\hat{\mathbf{x}}_1 - \left\{|\mathbf{V}_{ph}|\sin\theta + \cos\theta\frac{d|\mathbf{V}_{ph}|}{d\theta}\right\}\hat{\mathbf{x}}_3\right]. \tag{3.118}$$

In equation 3.118 $|\mathbf{V}_{ph}|$ means the magnitude of the phase velocity vector. When $|\mathbf{V}_{ph}|$ in equation 3.118 is replaced by the P-, SV-, or SH-wave phase velocities, given by equations 3.96 and 3.92, **t** becomes tangent vector at point P with phase angle θ for the slowness surfaces of the respective wavefields.

Note that the magnitude of **t** is given as

$$|\mathbf{t}| = \sqrt{|\mathbf{V}_{ph}|^2 + \left(\frac{d|\mathbf{V}_{ph}|}{d\theta}\right)^2}. \tag{3.119}$$

The unit tangent vector $\hat{\mathbf{t}}$ is therefore given as $\hat{\mathbf{t}} = \mathbf{t}/|\mathbf{t}|$. Finally, the unit normal vector at point P, shown as the blue arrow in Figure 3.13, is perpendicular to the tangent vector, which from equations 3.118 and 3.119 can be derived as

$$\hat{\mathbf{n}} = \frac{1}{|\mathbf{t}|}\left[\left\{|\mathbf{V}_{ph}|\sin\theta + \cos\theta\frac{d|\mathbf{V}_{ph}|}{d\theta}\right\} \times\hat{\mathbf{x}}_1 + \left\{|\mathbf{V}_{ph}|\cos\theta - \sin\theta\frac{d|\mathbf{V}_{ph}|}{d\theta}\right\}\hat{\mathbf{x}}_3\right]. \tag{3.120}$$

Thus, from the definition of the group velocity given by equation 3.111, we obtain, after some straightforward algebra,

$$\mathbf{V}_g = \left[\left\{|\mathbf{V}_{ph}|\sin\theta + \cos\theta\frac{d|\mathbf{V}_{ph}|}{d\theta}\right\} \times\hat{\mathbf{x}}_1 + \left\{|\mathbf{V}_{ph}|\cos\theta - \sin\theta\frac{d|\mathbf{V}_{ph}|}{d\theta}\right\}\hat{\mathbf{x}}_3\right]. \tag{3.121}$$

From equation 3.121, we can also derive the magnitude of the group velocity vector and is given by

$$|\mathbf{V}_g| = \sqrt{|\mathbf{V}_{ph}|^2 + \left(\frac{d|\mathbf{V}_{ph}|}{d\theta}\right)^2}, \tag{3.122}$$

and the group angle ϕ can be obtained from

$$\tan\phi = \frac{\tan\theta + \dfrac{1}{|\mathbf{V}_{ph}|}\dfrac{d|\mathbf{V}_{ph}|}{d\theta}}{1 - \dfrac{\tan\theta}{\mathbf{V}_{ph}}\dfrac{d|\mathbf{V}_{ph}|}{d\theta}}. \tag{3.123}$$

Therefore, if we have the analytical expression of the derivative of the phase velocity with respect to the phase angle, we then have the analytical expression of the group velocity vector (equations 3.121) and thus its magnitude (equation 3.122) and the group angle (equation 3.123). For the SH wave with the phase velocity given by equation 3.92, its derivative with respect to the phase angle is straightforward, and therefore the analytical expression for the group velocity for SH waves is easy to obtain and left as an exercise. Computing this derivative for P and SV waves is, however, not straightforward and requires further elaboration.

From equation 3.96, we write

$$\left.\begin{aligned} V_P &= V_{P0}\sqrt{1+\varepsilon\sin^2\theta + D^*(\theta)} \\ V_{SV} &= V_{So}\sqrt{1+\frac{V_{P0}^2}{V_{S0}^2}\varepsilon\sin^2\theta - \frac{V_{P0}^2}{V_{S0}^2}D^*(\theta)} \end{aligned}\right\}, \tag{3.124}$$

in which $D^*(\theta)$ is defined in equation 3.97. Therefore,

$$\frac{dV_P}{d\theta} = \frac{1}{2}\frac{V_{P0}^2}{V_P}\left[\varepsilon\sin 2\theta + \frac{dD^*(\theta)}{d\theta}\right] \tag{3.125}$$

and

$$\frac{dV_{SV}}{d\theta} = \frac{1}{2}\frac{V_{P0}^2}{V_{SV}}\left[\varepsilon\sin 2\theta - \frac{dD^*(\theta)}{d\theta}\right]. \tag{3.126}$$

From equations 3.125 and 3.126, it is evident that if we can obtain an analytic expression for the derivative of $D^*(\theta)$ with respect to the phase angle, we obtain the analytic expressions for $dV_P/d\theta$ and $dV_{SV}/d\theta$, and if we replace $|\mathbf{V}_{ph}|$ with V_P or V_{SV} in equations 3.121–3.123, we then have the analytic expressions for the group velocity vector ($\mathbf{V}_g$), its magnitude ($|\mathbf{V}_g|$) and the group angle (ϕ) for both P and SV waves.

To obtain the derivative of $D^*(\theta)$ with respect to the phase angle θ, we modify the right-hand side of equation 3.97 and write, after some algebra,

$$D^*(\theta) = \frac{1}{2} \times \zeta\left\{1+\frac{\delta^*}{\zeta^2}(\sin 2\theta)^2 + \frac{4(\zeta+\varepsilon)\varepsilon}{\zeta^2}\sin^4\theta\right\}^{\frac{1}{2}} - \frac{1}{2}\zeta. \tag{3.127}$$

Therefore,

$$\frac{dD^*(\theta)}{d\theta} = \frac{1}{2}\frac{\zeta}{\left\{1+\frac{\delta^*}{\zeta^2}(\sin 2\theta)^2 + \frac{4(\zeta+\varepsilon)\varepsilon}{\zeta^2}\sin^4\theta\right\}^{\frac{1}{2}}} \times \left\{4\frac{\delta^*}{\zeta^2}\sin 2\theta\cos 2\theta + 16\frac{(\zeta+\varepsilon)\varepsilon}{\zeta^2} \times \sin\theta\cos\theta\sin^2\theta\right\} \tag{3.128}$$

From equation 3.127, we can write the denominator on the right-hand side of equation 3.128 as, after some algebra,

$$\left\{1+\frac{\delta^*}{\zeta^2}(\sin 2\theta)^2 + \frac{4(\zeta+\varepsilon)\varepsilon}{\zeta^2}\sin^4\theta\right\}^{\frac{1}{2}} = \frac{2D^*(\theta)+\zeta}{\zeta}. \tag{3.129}$$

Finally, combining equations 3.128 and 3.129, we have

$$\frac{dD^*(\theta)}{d\theta} = \frac{2\sin 2\theta}{2D^*(\theta)+\zeta} \times [\delta^* + 2\{(\zeta+\varepsilon)\varepsilon - \delta^*\}\sin^2\theta]. \tag{3.130}$$

In equation 3.130, the parameter ζ is defined in terms of the vertical P- and S-wave velocities (V_{P0}, V_{S0}), and the parameter δ^* is defined in terms of ζ and the Thomsen parameters ε and δ in equation 3.98. Equation 3.130 is therefore the exact analytic representation of the derivative of $D^*(\theta)$ with respect to the phase angle θ. Plugging in the right-hand side of equation 3.130 for this derivative in equations 3.125 and 3.126 therefore gives the analytic expressions of the derivatives of V_P and V_{SV} with respect to θ. Finally, using these derivatives and the corresponding phase velocities for $|\mathbf{V}_{ph}|$ in equations 3.121–3.123 provides the analytic expressions for P- and SV-wave group velocities and angles.

By ignoring the derivative term on the right-hand side of equation 3.122 so that $|\mathbf{V}_{ph}| \approx |\mathbf{V}_g|$ and using the weakly anisotropic P- SV- and SH-wave phase velocities, provided in equations 3.100 and 3.93, Thomsen (1986) provides the *weakly anisotropic* equivalents for the group angles as

$$\left.\begin{array}{l}\tan\phi_P \approx \tan\theta_P[1+2\delta+4(\varepsilon-\delta)\sin^2\theta_P]\\ \tan\phi_{SV} \approx \tan\theta_{SV}\left[1+2\frac{V_{P0}^2}{V_{S0}^2}(\varepsilon-\delta)(1-2\sin^2\theta_{SV})\right]\\ \tan\phi_{SH} \approx \tan\theta_{SH}[1+2\gamma]\end{array}\right\}. \tag{3.131}$$

In equation 3.131 ϕ_P, ϕ_{SV}, and ϕ_{SH} are the group angles and θ_P, θ_{SV}, and θ_{SH} are the phase angles for P, SV, and SH wavefields. The derivation of equation 3.131 is straightforward and left as an exercise.

For wavefields propagating in the symmetry planes in an orthorhombic medium, derivations of the exact and weakly anisotropic group velocities and group angles are identical to that of the VTI medium besides the fact that they contain additional parameters; see Tsvankin (1997) for details.

Using the same examples of the Taylor Sandstone and Green River Shale, P-wave phase velocities as functions of the phase angle are previously shown in Figure 3.10, in Figure 3.14 we show the dependence between the P-wave phase and group angles using both the exact and weakly anisotropic formulas. Comparing the phase velocities (Figure 3.10) and the phase angle versus group angle (Figure 3.14), we observe the following:

1) For weak anisotropy like Taylor Sandstone, the exact and approximate phase velocities are about the same except for $\theta > 80°$. However, the exact and approximate phase versus relations are different (although small) beyond 10°.
2) For moderate anisotropy like Green River Shale, the exact and approximate phase velocities, although different beyond 20°, are still sufficiently close. For phase versus group angles, however, they are substantially different, except for very small (<5°) and very large (>85°) values of the phase angles. For example, at phase angle $\theta = 20°$, the approximate and exact group angles are respectively 17° and 29°. For $\theta = 40°$, they are respectively 48° and 59°. These numbers are substantially different.

Here, we provided both exact and weakly anisotropic approximations for historical reasons. The results shown in Figures 3.10 and 3.14, however, suggest that keeping in consideration of developments in computing technology over the past decades, we completely do away with the weakly anisotropic approximations and use only the exact formulations.

3.9.3 Importance of the Group and Phase—An Optimization Problem of Practical Importance

In the above section, we thoroughly discussed the concepts of the phase and group

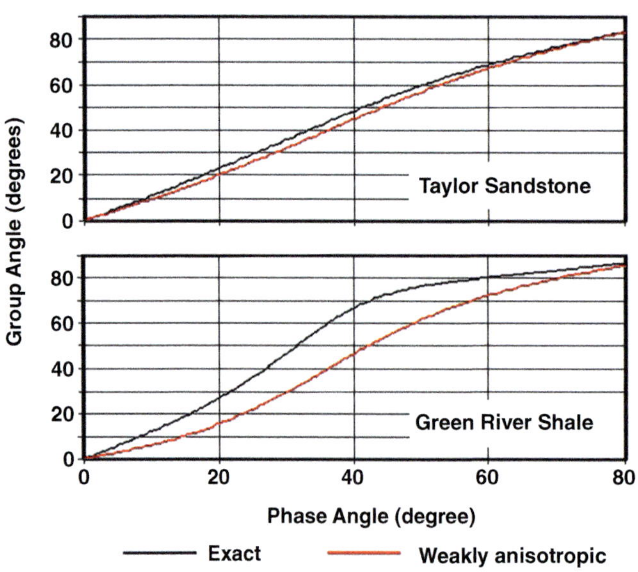

Figure 3.14 Relationship between the P-wave phase and group angles for Taylor Sandstone (top) and Green River Shale (bottom). The black curves are computed using the exact formula, and the red curves are computed using the weakly anisotropic formula.

velocities and angles for wave propagation in an anisotropic medium and provided ways how to compute them numerically for general anisotropy and analytically for specific symmetries. Aim for seismology is to decipher subsurface lithological, fluid, and structural properties from the acquired seismic data. One such fundamental property is the velocity of propagation. Seismic data are acquired in time-offset domain. From our previous discussions, however, the velocity of propagation is a function of angle (phase or group), not offset. Therefore, to analyze how the velocity of propagation varies with angle in an anisotropic medium, it is necessary to convert the offsets to angles. Additionally, if we compare between the phase and group velocities, the dependence of the phase velocity as a function of the phase angle is much simpler than that between the group velocity and group angle. Going beyond just the velocities, one popular method for subsurface characterization is the amplitude-variation-with-angle (AVA) analysis. In AVA analysis, we calibrate the seismic reflection amplitudes on the recorded seismic data to the angle-dependent reflection coefficients. These angle-dependent reflection coefficients are, in turn, functions of the phase angle, not group angle. It is thus important that when we convert the offsets in the recorded seismic data into angles, they are phase angles, not group angles.

For now, we stay simple and consider a model shown Figure 3.15a with a VTI medium underlain by an isotropic half-space. We choose the properties of the VTI medium to be that of the Mesaverde Clay Shale, which we obtained from Thomsen (1986). Using this model, we compute the synthetic seismic response and show it in Figure 3.15b. The P–P reflection from the VTI/isotropic layer interface is shown in Figure 3.15b, from which we can pick the travel time t for any source–receiver offset x as shown by the red lines.

Picking the travel times for all offsets in the synthetic data shown in Figure 3.15b will provide the variation of t as a function of x.

For the simple model, shown in Figure 3.15a, we can easily use these $x - t$ data and compute the velocity and angle and is illustrated in Figure 3.16. From Figure 3.16a, we can relate the incidence-angle α with the source–receiver offset x and the reflector depth h as

$$\alpha = \tan^{-1}\left(\frac{x}{2h}\right). \tag{3.132}$$

Now, if $V_{obs}(\alpha)$ is the observed velocity for this angle α, then from Figure 3.16a, we write

$$V_{obs}(\alpha) = \frac{2\sqrt{\left(\frac{x}{2}\right)^2 + h^2}}{t}, \tag{3.133}$$

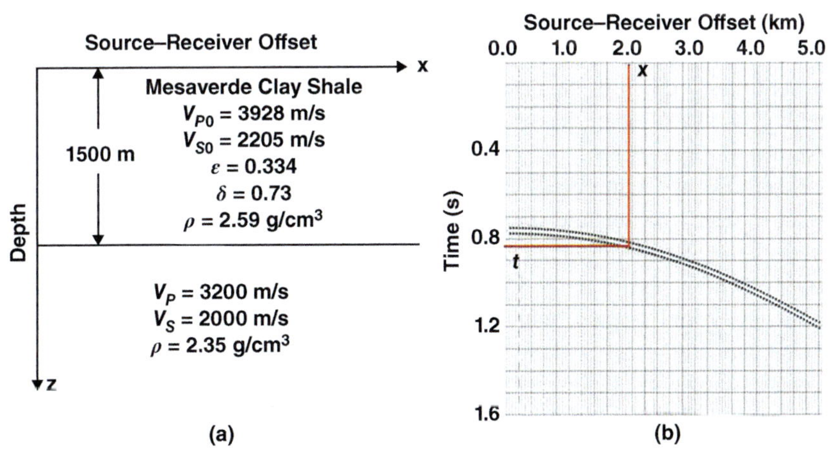

Figure 3.15 (a) A model with a VTI layer overlain by an isotropic half-space. (b) Synthetic seismic response showing the vertical component of the particle motion computed using the model shown in (a). The red lines show how at any point of the reflection event the source–receiver offset (x) and the corresponding two-way travel time (t) can be picked.

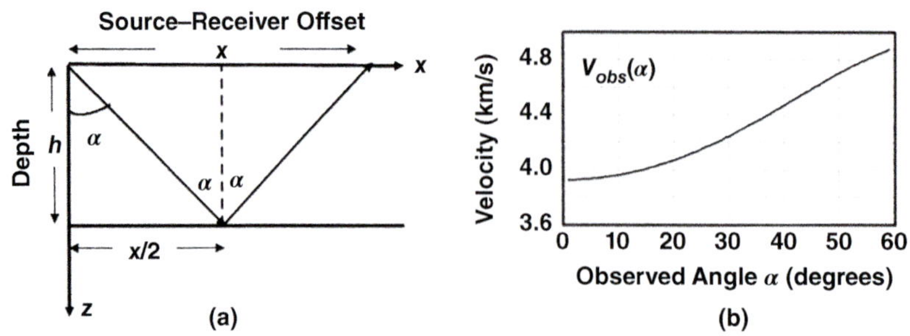

Figure 3.16 Illustration of how the angle-dependent velocity can be computed from the x–t data obtained from the synthetic seismogram shown in Figure 3.15b. (a) The depth-domain model showing how the angle of incidence (α) is related to the source–receiver offset (x) and the reflector depth (h). (b) Variation of the velocity as a function of the observed angle, computed from the $x - t$ data picked from the synthetic seismogram shown in Figure 3.15b.

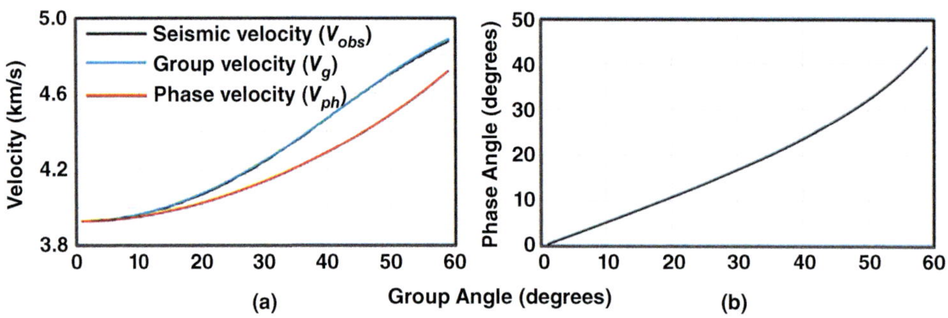

Figure 3.17 (a) Angle-dependent velocities. The black curve is identical to V_{obs} shown in Figure 3.16b. The red curve is computed by assuming that the observe angle α from Figure 3.16 is the phase angle θ and using the exact P-wave phase velocity formula. The cyan curve, on the other hand, is computed assuming that the observed angle α is the group angle ϕ and computing the corresponding group velocities. (b) Relation between the phase and group angles for the Mesaverde Clay Shale layer that was used to compute the synthetic data shown in Figure 3.15b.

where t is the travel time, picked from the data for the source–receiver offset x. If we repeat this for all source–receiver offsets and their corresponding travel times, we obtain the observed velocity field V_{obs} as a function of the observed angle α, as shown in Figure 3.16b.

Note that because we computed α and $V_{obs}(\alpha)$ explicitly from the seismic data, we have referred them as the observed angle and velocity. We now use Figure 3.17 to illustrate what are α and $V_{obs}(\alpha)$, which we can directly obtain from the seismic data. The black curve, shown in Figure 3.17a, is identical to $V_{obs}(\alpha)$ of Figure 3.16b. By assuming that the observed angles α are phase angles, we use equations 3.96 and 3.97 and compute the P-wave phase velocities and show them as the red curve in Figure 3.17a. Finally, we assume the observe angles to be group angles and compute the corresponding group velocities, which we show as cyan in Figure 3.17a. From Figure 3.17a, it is thus evident that the angle (α) and the corresponding velocities $V_{obs}(\alpha)$ are the group angles and group velocities or, in other words, $\alpha \equiv \phi$ and $V_{obs}(\alpha) \equiv |\mathbf{V}_g(\phi)|$.

Therefore, the angle-dependent velocities obtainable from the observed seismic data are the group velocities as functions of group angles. If we ignore this fundamental fact and assume that the angle-dependent velocities obtainable from travel times are phase velocities as functions of phase angles, we introduce error in our estimations of the subsurface properties. As shown in Figure 3.17b, the difference between the phase and the group angles could be substantial. For a group angle (ϕ) of 20°, the corresponding phase angle (θ) is 10°. For $\phi = 40°$, $\theta \approx 22°$, and for $\phi = 50°$, $\theta \approx 32°$. Thus, to correctly characterize the subsurface properties, the group angles and velocities obtainable from the observed seismic data must be converted to their corresponding phase velocities and angles.

From our previous discussions, given the phase velocity as a function of phase angle, it is straightforward to compute the corresponding group velocities and group angles. For any arbitrary anisotropy, we can compute them numerically using equations 3.111–3.113. And, for specific symmetries, we can compute them analytically using equations 3.121–3.130. However, from the group velocities as functions of the group angles that are available from seismic data, computing the corresponding phase velocities as functions of phase angles is not straightforward and must be iteratively obtained via an optimization. Figure 3.18 shows an illustration of such an optimization. Restricting to a layer and a half-space example of Figure 3.15 where we can directly obtain the observed angle α and the observed velocity $V_{obs}(\alpha)$, as shown in Figure 3.16b, the steps for this optimization are the following:

1) We set the observed angle α to the group angle ϕ and the observed velocity $V_{obs}(\alpha)$ to the group velocity $|\mathbf{V}_g|(\phi)$.
2) We assume that the phase angle θ is equal to the group angle ϕ.
3) We compute the phase velocity $|\mathbf{V}_{ph}|(\theta)$, and from θ and $|\mathbf{V}_{ph}|(\theta)$, we compute the corresponding group angles ϕ_{comp} and the group velocities $|\mathbf{V}_g^{comp}|(\phi_{comp})$.
4) We compare if the computed group angle ϕ_{comp} and the corresponding group velocity $|\mathbf{V}_g^{comp}|(\phi_{comp})$ are sufficiently close to α and $V_{obs}(\alpha)$.
   a) If no, we modify the phase angle θ and go to step 3.
   b) If yes, we assume that we solved the problem, output results, and exit.

We used the above iterative steps to show that the observed angle α is the group angle ϕ, the observed velocity $V_{obs}(\alpha)$ is the group velocity $|\mathbf{V}_g|(\phi)$, and the difference between the group and phase angles ϕ and θ is shown in Figure 3.17. The process to iteratively update

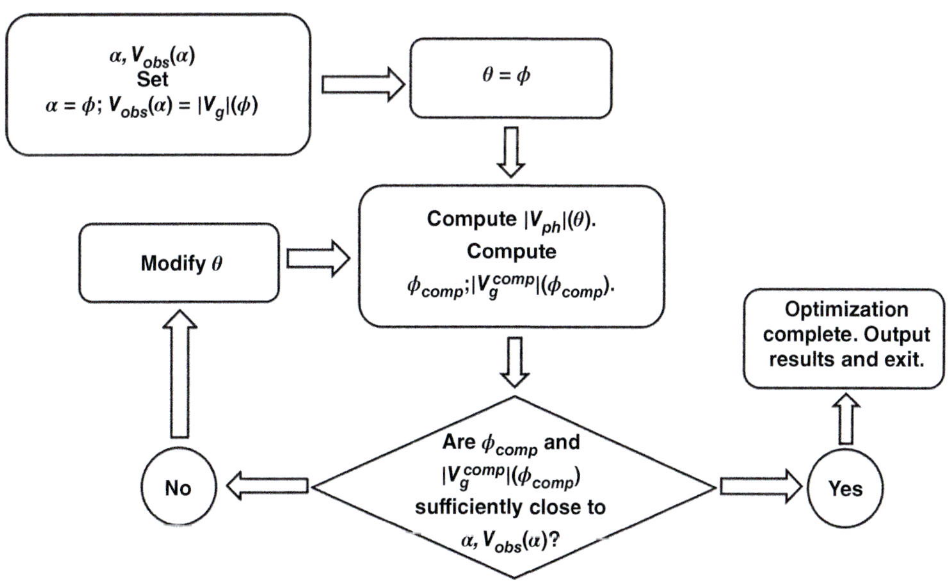

Figure 3.18 Illustration of an optimization procedure to compute the phase velocity as a function of phase angle from the group velocity as a function of group angle.

the phase angle θ (steps 3 and 4) can be performed using a classical optimization or a machine-learning approach, discussion of which we postpone to later chapters.

3.10 Radiation Patterns from Seismic Sources

In deriving the Christoffel equation (equation 3.77), we started from the elastodynamic equation in a source-free region (equation 3.72) and specialized to plane waves propagating along an arbitrary direction with unit vector $\hat{\boldsymbol{l}} = \sin\theta\cos\varphi\hat{\mathbf{x}}_1 + \sin\theta\sin\varphi\hat{\mathbf{x}}_2 + \cos\theta\hat{\mathbf{x}}_3$ (Figure 3.9). In earth, seismic waves are generated by excitations from sources from localized regions, which can be considered as *point sources*. In earthquake seismology, these sources are natural earthquake sources, and for exploration seismology, they are artificial sources like an explosion or a vibration. We now discuss the radiation patterns from these sources in a homogeneous elastic medium. For simplicity, we restrict to the wave propagation from point sources in a homogeneous isotropic elastic medium and postpone the discussion of the wave propagation in an anisotropic medium for later.

3.10.1 The Laplacian Operator and Its Inverse

In Chapter 2, we introduced the Laplacian operator $\nabla^2 = \nabla \cdot \nabla$ in Cartesian, cylindrical, and spherical coordinate systems (see equations 2.276, 2.281, and 2.285). We also observed in Chapter 2 that this Laplacian operator is fundamental to the partial differentials of all kinds, and that governing the seismic wave propagation, which is specifically a hyperbolic type of partial differential equation, is no exception. Therefore, before discussing the radiation patters, it is first important to look at ∇^2 in detail and introduce the inverse of the Laplacian operator, which we denote as ∇^{-2}.

For any function Q, defined in 3D Euclidean space $\Re^3$ which vanishes faster than $1/|\mathbf{r}|^{2+\varepsilon}$ as $|\mathbf{r}| \to \infty$, where $\mathbf{r}$ is the position vector in $\Re^3$ and ε is a small number, we define

$$[\nabla^{-2}Q](\mathbf{r}) = \int_{\Re^3} K(\mathbf{r},\mathbf{r}')Q(\mathbf{r}')dV', \quad (3.134)$$

where

$$K(\mathbf{r},\mathbf{r}') = -\frac{1}{4\pi|\mathbf{r}-\mathbf{r}'|}. \quad (3.135)$$

Note that from equations 3.134 and 3.135, it follows that

$$\nabla^2[\nabla^{-2}Q](\mathbf{r}) \equiv \nabla^2 \int_{\Re^3} K(\mathbf{r},\mathbf{r}')Q(\mathbf{r}')dV' = \int_{\Re^3} [\nabla^2 K(\mathbf{r},\mathbf{r}')Q(\mathbf{r}')dV']. \quad (3.136)$$

Therefore, if we can show that $\nabla^2 K(\mathbf{r},\mathbf{r}') = \delta(\mathbf{r}-\mathbf{r}')$ with $\delta(\mathbf{r}-\mathbf{r}')$ being the delta function, then equation 3.136 leads to

$$\nabla^2[\nabla^{-2}Q](\mathbf{r}) = \int_{\Re^3} \delta(\mathbf{r}-\mathbf{r}')Q(\mathbf{r}')dV' = Q(\mathbf{r}), \quad (3.137)$$

then we establish that ∇^{-2} defined by equations 3.134 and 3.135 is the inverse of the Laplacian operator. Because both $K(\mathbf{r},\mathbf{r}')$ and $\delta(\mathbf{r}-\mathbf{r}')$ are functions of $\mathbf{r}$ and $\mathbf{r}'$, without loss of generality, we can set $\mathbf{r}' = \mathbf{0}$ and write

$$K(\mathbf{r},\mathbf{r}') = K(\mathbf{r},\mathbf{0}) = -\frac{1}{4\pi|\mathbf{r}|} \quad (3.138)$$

and

$$\delta(\mathbf{r}-\mathbf{r}') = \delta(\mathbf{r}). \quad (3.139)$$

We now use the definition of ∇^2 in spherical coordinates (equation 2.285) and recall that $K(\mathbf{r},\mathbf{r}') = K(\mathbf{r},\mathbf{0})$ is independent of the colatitude θ and azimuth φ (Figure 2.12), we write

$$\nabla^2 K(\mathbf{r},\mathbf{0}) = \nabla^2\left(-\frac{1}{4\pi|\mathbf{r}|}\right) = \frac{1}{r^2}\partial_r\left[r^2\partial_r\left(-\frac{1}{4\pi|\mathbf{r}|}\right)\right] = 0; \quad \mathbf{r} \neq \mathbf{0}. \quad (3.140)$$

Equation 3.140 therefore implies that $\nabla^2 K(\mathbf{r}, \mathbf{0})$ is zero everywhere except at the origin. Equivalently, $K(\mathbf{r}, \mathbf{r}')$ is zero everywhere except at $\mathbf{r} = \mathbf{r}'$. However, to verify if this is really a delta function, we must show that its integral over a volume, irrespective of how small the volume is, must be equal to 1.

In Figure 3.19, we define an infinitesimal sphere about the origin with radius ε and bounding surface S_ε. Note that we can always define such a sphere about the origin irrespective of how small ε is. Also, the unit outward normal vector $\hat{\mathbf{n}}$ to this sphere would be equal to $\hat{\mathbf{r}}$, the unit vector along the radial direction (see Figure 3.19). Now

$$\begin{aligned}\int_V \nabla^2\left(-\frac{1}{4\pi|\mathbf{r}|}\right) dV &= \int_\varepsilon \nabla \cdot \nabla\left(-\frac{1}{4\pi r}\right) dV \\ &= \int_{\partial S_\varepsilon} \hat{\mathbf{n}} \cdot \nabla\left(-\frac{1}{4\pi r}\right) dA \\ &= \int_{\partial S_\varepsilon} \hat{\mathbf{r}} \cdot \hat{\mathbf{r}}\partial_r\left(-\frac{1}{4\pi r}\right) dA \\ &= \int_{\partial S_\varepsilon} \frac{1}{4\pi r^2} dA = \left.\frac{1}{4\pi r^2}\right|_{r=\varepsilon} \int_{\partial S_\varepsilon} dA \\ &= \frac{1}{4\pi''^2} 4\pi''^2 = 1. \end{aligned} \tag{3.141}$$

Equation 3.141 proves that $\nabla^2 K(\mathbf{r}, \mathbf{r}') = \delta(\mathbf{r} - \mathbf{r}')$. Therefore, the operator ∇^{-2}, defined by equations 3.134 and 3.135, is the true inverse of the Laplacian operator ∇^2. Three important properties of ∇^{-2} are that, for any tensor $\mathbf{Q}$, we can write

$$\left.\begin{aligned} \nabla \times \nabla^{-2}\mathbf{Q} &= \nabla^{-2}(\nabla \times \mathbf{Q}) \\ \nabla \cdot \nabla^{-2}\mathbf{Q} &= \nabla^{-2}(\nabla \cdot \mathbf{Q}) \\ \nabla^{-2}(\partial_i\partial_j\mathbf{Q}) &= \partial_i\partial_j(\nabla^{-2}\mathbf{Q}) \end{aligned}\right\}. \tag{3.142}$$

Proofs of the above properties are straightforward and left as an exercise.

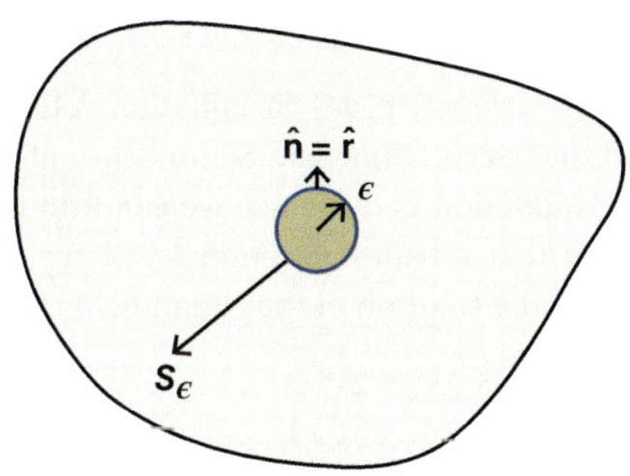

Figure 3.19 An infinitesimally small spherical volume about the origin with radius ε and the bounding surface S_ε.

3.10.2 Helmholtz Representation Theorem

Helmholtz representation theorem, also known as the fundamental theorem in vector calculus, is stated as follows.

If $\mathbf{f}$ is a vector field in $\Re^3$ which vanishes faster than $1/|\mathbf{r}|^{2+\varepsilon}$ as $\mathbf{r} \to \infty$ with $\mathbf{r}$ and ε as previously defined, then there exists a scalar φ and a vector $\boldsymbol{\psi}$ such that

$$\mathbf{f} = \nabla\varphi - \nabla \times \boldsymbol{\psi}, \tag{3.143}$$

with

$$\nabla \cdot \boldsymbol{\psi} = 0. \tag{3.144}$$

To prove Helmholtz representation theorem, we set

$$\left.\begin{aligned} \varphi &= \nabla^{-2}(\nabla \cdot \mathbf{f}) \\ \boldsymbol{\psi} &= \nabla^{-2}(\nabla \times \mathbf{f}) \end{aligned}\right\}. \tag{3.145}$$

Then, using the commutative properties of ∇^{-2}, given in equation 3.142, we write

$$\left.\begin{aligned} \varphi &= \nabla^{-2}(\nabla \cdot \mathbf{f}) = \nabla \cdot (\nabla^{-2}\mathbf{f}) \\ \boldsymbol{\psi} &= \nabla^{-2}(\nabla \times \mathbf{f}) = \nabla \times (\nabla^{-2}\mathbf{f}) \end{aligned}\right\}. \tag{3.146}$$

Therefore,

$$\begin{aligned} \nabla\varphi - \nabla \times \boldsymbol{\psi} &= \nabla\nabla \cdot (\nabla^{-2}\mathbf{f}) - \nabla \times \nabla \times (\nabla^{-2}\mathbf{f}) \\ &= (\nabla\nabla \cdot - \nabla \times \nabla\times)(\nabla^{-2}\mathbf{f}) \\ &= \nabla^2(\nabla^{-2}\mathbf{f}) = \mathbf{f}. \end{aligned} \tag{3.147}$$

And

$$\nabla \cdot \boldsymbol{\Psi} = \nabla \cdot \nabla \times (\nabla^{-2}\mathbf{f}) = 0. \tag{3.148}$$

We have therefore proved the Helmholtz representation theorem. From Helmholtz representation theorem, it also follows that for a vector field $\mathbf{f}$ defined in $\Re^3$ which vanishes faster than $1/|\mathbf{r}|^{2+\varepsilon}$ as $\mathbf{r} \to \infty$, and if $\nabla \cdot \mathbf{f} = 0$ and $\nabla \times \mathbf{f} = \mathbf{0}$, then $\mathbf{f}$ is zero everywhere in $\Re^3$. This follows directly from the fact that if we use Helmholtz representation theorem (equation 3.143) to represent $\mathbf{f} = \nabla\varphi - \nabla\times\boldsymbol{\psi}$, in which φ and $\boldsymbol{\psi}$ are defined in equation 3.145, then if $\nabla\cdot\mathbf{f} = 0$ and $\nabla\times\mathbf{f} = \mathbf{0}$, then $\mathbf{f}$ must be zero.

3.10.3 Momentum Equation for Isotropic Elastic System

Restricting to isotropic elastic system with the elastic-stiffness matrix $\mathbf{C}$ given by equation 3.48, the time and frequency domain versions of the momentum equation (equations 3.69 and 3.70) respectively become

$$\rho\ddot{\mathbf{u}} = (\lambda + 2\mu)\nabla(\nabla\cdot\mathbf{u}) - \mu\nabla\times\nabla\times\mathbf{u} + \mathbf{f}, \tag{3.149}$$

and

$$-\rho\omega^2\mathbf{u} = (\lambda + 2\mu)\nabla(\nabla\cdot\mathbf{u}) - \mu\nabla\times\nabla\times\mathbf{u} + \mathbf{f}. \tag{3.150}$$

Proofs of equations 3.148 and 3.149 are not difficult and left as an exercise. We restrict here to the frequency domain version of the momentum equation and rewrite equation 3.150 as

$$\mathbf{u} = -\frac{(\lambda + 2\mu)}{\rho\omega^2}\nabla(\nabla\cdot\mathbf{u}) + \frac{\mu}{\rho\omega^2}\nabla\times\nabla\times\mathbf{u} - \frac{\mathbf{f}}{\rho\omega^2}. \tag{3.151}$$

We now use the Helmholtz representation theorem to write the source term $\mathbf{f}$ in equation 3.150 as $\mathbf{f} = \nabla\varphi - \nabla\times\boldsymbol{\psi}$, in which φ and $\boldsymbol{\psi}$ are defined in equation 3.146 such that $\nabla.\,\boldsymbol{\psi} = 0$. We also define

$$\left.\begin{aligned} P &= -\frac{(\lambda + 2\mu)}{\rho\omega^2}\nabla\cdot\mathbf{u} - \frac{1}{\rho\omega^2}\varphi \\ \mathbf{S} &= -\frac{\mu}{\rho\omega^2}\nabla\times\mathbf{u} - \frac{1}{\rho\omega^2}\boldsymbol{\Psi} \end{aligned}\right\}. \tag{3.152}$$

Using equation 3.152, we then write the momentum equation (equation 3.151) as

$$\mathbf{u} = \nabla P - \nabla\times\mathbf{S}, \tag{3.153}$$

in which

$$\nabla\cdot\mathbf{S} = 0. \tag{3.154}$$

Using equations 3.153 and 3.154, and from the definition of $\nabla^2 = \nabla\,.\,\nabla$, we can write from equation 3.152

$$\left.\begin{aligned} \left(\nabla^2 + \frac{\rho\omega^2}{(\lambda + 2\mu)}\right)P &= -\frac{1}{(\lambda + 2\mu)}\varphi \\ \left(\nabla^2 + \frac{\rho\omega^2}{\mu}\right)\mathbf{S} &= -\frac{1}{\mu}\boldsymbol{\Psi} \end{aligned}\right\}. \tag{3.155}$$

And by comparing equation 3.155 with the one-dimensional wave equation (equation 2.96), which we previously introduced in Chapter 2, we can immediately identify equation 3.155 to be the frequency domain representation of the 3D inhomogeneous wave equation with P and $\mathbf{S}$ respectively propagating with speeds α and β and given as

$$\left.\begin{aligned} \alpha &= \sqrt{\frac{\lambda + 2\mu}{\rho}} \\ \beta &= \sqrt{\frac{\mu}{\rho}} \end{aligned}\right\}. \tag{3.156}$$

And, from our previous discussions (for example, equations 3.49 and 3.50), we recognize that $\alpha = V_P$ is the P-wave velocity and $\beta = V_S$ is the S-wave velocity. Also note from our previous discussions in Chapter 2, both equations shown in equation 3.154 are hyperbolic type of partial differential equations.

3.10.4 Green's Function for Hyperbolic Partial Differential Equations

To solve the wave equations with non-zero source terms, represented by equation 3.155, it is convenient to first solve

$$(\nabla^2 + k^2)G(\mathbf{r}, \mathbf{r}') = \delta(\mathbf{r} - \mathbf{r}'). \tag{3.157}$$

The function $G(\mathbf{r}, \mathbf{r}')$ in equation 3.157 is called the *Green's Function*. Significance of the Green's function is that once we can find this function that satisfies equation 3.157, we can then find the solution of the equation of type

$$(\nabla^2 + k^2)\Phi = \chi, \tag{3.158}$$

which is equivalent to both equations shown in equation 3.155 as

$$\Phi(\mathbf{r}) = \int_{\Re^3} G(\mathbf{r}, \mathbf{r}')\chi(\mathbf{r}')dV'. \tag{3.159}$$

We can prove that equation 3.159 to be correct by direct substitution

$$(\nabla^2 + k^2)\Phi = \int_{\Re^3} (\nabla^2 + k^2)G(\mathbf{r}, \mathbf{r}')\chi(\mathbf{r}')dV'$$

$$= \int_{\Re^3} \delta(\mathbf{r} - \mathbf{r}')\chi(\mathbf{r}')dV' = \chi(\mathbf{r}). \quad (3.160)$$

Green's Theorem: The solution of $(\nabla^2 + k^2)G(\mathbf{r}, \mathbf{r}') = \delta(\mathbf{r} - \mathbf{r}')$, which satisfies the radiation condition that the energy flux is outward from the source and no energy is radiating toward the source, is given by

$$G(\mathbf{r}, \mathbf{r}') = -\frac{e^{ik|\mathbf{r}-\mathbf{r}'|}}{4\pi|\mathbf{r} - \mathbf{r}'|}. \quad (3.161)$$

Proof of the Green's Theorem: Without any loss of generality, we set the source to be at the origin, i.e., we set $\mathbf{r}' = \mathbf{0}$. To prove the Green's theorem, we must then show that

$$(\nabla^2 + k^2)G(\mathbf{r}) = (\nabla^2 + k^2)\left(-\frac{e^{ikr}}{4\pi r}\right) = 0, \quad (3.162)$$

when $r \neq 0$ and

$$\int_V (\nabla^2 + k^2)\left(-\frac{e^{ikr}}{4\pi r}\right) dV = 1, \quad (3.163)$$

with V being any closed volume with the origin as the interior point. Proof of equation 3.162 is straightforward and follows

$$(\nabla^2 + k^2)\left(-\frac{e^{ikr}}{4\pi r}\right) = \frac{1}{r^2}\partial_r\left(r^2\partial_r\left(-\frac{e^{ikr}}{4\pi r}\right)\right)$$

$$-k^2\frac{e^{ikr}}{4\pi r} = \frac{1}{r^2}\partial_r\left(-\frac{r^2}{4\pi}\left(\frac{ikre^{ikr} - e^{ikr}}{r^2}\right)\right)$$

$$-k^2\frac{e^{ikr}}{4\pi r} = k^2\frac{e^{ikr}}{4\pi r} - k^2\frac{e^{ikr}}{4\pi r} = 0, \quad (3.164)$$

when $r \neq 0$. To prove equation 3.163, note that we have already proved that

$$(\nabla^2 + k^2)\left(\frac{e^{ikr}}{4\pi r}\right)$$

is zero everywhere except at the origin. Therefore, the integral of this quantity over any volume containing the origin is the same as the integral of this quantity over any other volume containing the origin. We can then use a convenient volume, a sphere S_ε of radius ε with its center at the origin. We therefore write

$$\int_V (\nabla^2 + k^2)\left(-\frac{e^{ikr}}{4\pi r}\right) dV = \int_{S_\varepsilon} \nabla^2\left(-\frac{e^{ikr}}{4\pi r}\right) dV$$

$$+ \int_{S_\varepsilon} k^2\left(-\frac{e^{ikr}}{4\pi r}\right) dV$$

$$= \int_{\partial S_\varepsilon} \hat{\mathbf{r}} \cdot \nabla\left(-\frac{e^{ikr}}{4\pi r}\right) dA + \int_{S_\varepsilon} k^2\left(-\frac{e^{ikr}}{4\pi r}\right) dV$$

$$= \int_{\partial S_\varepsilon} \hat{\mathbf{r}} \cdot \hat{\mathbf{r}}\partial_r\left(-\frac{e^{ikr}}{4\pi r}\right) dA - k^2\int_0^\varepsilon \frac{e^{ikr}}{4\pi r}(4\pi r^2)dr$$

$$= -\frac{1}{4\pi}\int_0^{2\pi}\int_0^{\pi} \frac{ikre^{ikr} - e^{ikr}}{r^2}(\varepsilon^2 \sin\theta d\theta d\phi) - k^2$$

$$\times \int_0^\varepsilon re^{ikr}dr = 1, \quad (3.165)$$

which proves equation 3.163, and thus we give the complete proof of Green's Theorem. Now that we know how to obtain the Green's function for the inhomogeneous wave equation (equation 3.155), we are ready to solve the wave equation and obtain the P- and S-wave radiation patterns from seismic sources. In seismology, we are generally interested in point sources, meaning that the sources located at a single point in $\Re^3$ and without any loss of generality, we will assume this point to be at the origin. In the following, we separately discuss radiation patterns from three different source types: (1) a uniform (explosive) source, (2) a point double couple, and (3) a point force.

3.10.5 Radiation Patterns from a Uniform (Explosive) Point Source

To represent a uniform source, we set the force term $\mathbf{f}$ in equation 3.151 as

$$\mathbf{f} = -g(\omega)\nabla\delta(\mathbf{r}). \quad (3.166)$$

Note that the gradient of the delta function, $\nabla\delta(\mathbf{r})$, on the right-hand side of equation 3.166 represents forces of uniform magnitudes, pointing to the origin from all directions, and therefore $-\nabla\delta(\mathbf{r})$ represents uniform forces along all directions pointed outward from the origin, which is equivalent to an explosion. Finally, $g(\omega)$ on the right-hand side of equation 3.166 is the time/frequency dependence of the source or the source wavelet spectrum. Equation 3.166 is therefore the mathematical representation of a point explosion from the origin, which we often encounter in exploration seismology, such as a dynamite explosion in the land or an air gun explosion in the marine environment.

From our Helmholtz decomposition of **f** into $\nabla\varphi - \nabla\times\boldsymbol{\psi}$, with φ and $\boldsymbol{\psi}$ being defined by equation 3.146, for an explosive source, we obtain

$$\left.\begin{aligned} \varphi &= \nabla^{-2}(\nabla\cdot\mathbf{f}) = \nabla^{-2}(\nabla\cdot(-g(\omega)\nabla\delta(\mathbf{r}))) \\ &= -g(\omega)\nabla^{-2}(\nabla^2\delta(\mathbf{r})) = -g(\omega)\delta(\mathbf{r}) \\ \boldsymbol{\psi} &= \nabla^{-2}(\nabla\times\mathbf{f}) = \nabla^{-2}(\nabla\times(-g(\omega)\nabla\delta(\mathbf{r}))) \\ &= \mathbf{0} \end{aligned}\right\}. \tag{3.167}$$

By substituting φ from equation 3.167, we obtain the equation for P waves (the first equation in equation 3.155) as

$$\left(\nabla^2 + \frac{\omega^2}{\alpha^2}\right)P = \frac{g(\omega)\delta(\mathbf{r})}{\lambda+2\mu}, \tag{3.168}$$

in which α is as defined in equation 3.156. Finally, from Green's theorem, the solution to equation 3.168 is given as

$$P = -\frac{g(\omega)e^{ik|\mathbf{r}|}}{(\lambda+2\mu)4\pi|\mathbf{r}|}. \tag{3.169}$$

We can also generalize the solution, shown by equation 3.169, to a point explosion located at position $\mathbf{r}'$ as

$$P = -\frac{g(\omega)e^{ik|\mathbf{r}-\mathbf{r}'|}}{(\lambda+2\mu)4\pi|\mathbf{r}-\mathbf{r}'|}. \tag{3.170}$$

Also, because $\boldsymbol{\psi} = \mathbf{0}$ (see equation 3.167), in an isotropic elastic medium, a point explosion does not generate any S wave.

To find the P-wave radiation pattern form point explosion, we use equation 3.153 and write

$$\mathbf{u} = \nabla P = -\frac{g(\omega)}{\lambda+2\mu}\hat{\mathbf{r}}\partial_r\left(\frac{e^{i\frac{\omega}{\alpha}r}}{4\pi r}\right)$$

$$= -\frac{g(\omega)\hat{\mathbf{r}}}{\lambda+2\mu}\left[\frac{i\frac{\omega}{\alpha}e^{i\frac{\omega}{\alpha}r}}{4\pi r} - \frac{e^{i\frac{\omega}{\alpha}r}}{4\pi r^2}\right]$$

$$= -\frac{\hat{\mathbf{r}}g(\omega)i\omega e^{i\omega r/\alpha}}{4\pi\rho\alpha^3 r} + \frac{\hat{\mathbf{r}}g(\omega)e^{i\omega r/\alpha}}{4\pi\rho\alpha^2 r^2}. \tag{3.171}$$

From the right-hand side of equation 3.171, we see that the first term decays with a factor $1/r$ and the second term decays with a factor of $1/r^2$ with distance r. The first term is called the *far-field* radiation pattern, which we mostly record in our seismic data. The second term is the near-field radiation pattern, which rapidly decays with distance of propagation. We explicitly write the far- and near-field terms on the right-hand side of equation 3.171 as

$$\left.\begin{aligned} \mathbf{u}^{ff} &= -\frac{\hat{\mathbf{r}}g(\omega)i\omega e^{i\omega r/\alpha}}{4\pi\rho\alpha^3 r} \\ \mathbf{u}^{nf} &= \frac{\hat{\mathbf{r}}g(\omega)e^{i\omega r/\alpha}}{4\pi\rho\alpha^2 r^2} \end{aligned}\right\}. \tag{3.172}$$

Now, the frequency domain representation of the point explosion (equation 3.166) can be represented in time domain as

$$\mathbf{f} = -g(t)\nabla\delta(\mathbf{r}), \tag{3.173}$$

with $g(t)$ being the source wavelet in time domain. Assume that $g(t)$ is a heavyside step function $H(t)$ and given as

$$g(t) = H(t) = \begin{cases} 0; & t\le 0 \\ 1; & t>0 \end{cases}. \tag{3.174}$$

Then, in frequency domain, we have

$$g(\omega) = \pi\delta(\omega) - \frac{1}{i\omega}$$

or

$$i\omega g(\omega) = -1 + i\pi\omega\delta(\omega). \tag{3.175}$$

Equation 3.175 implies that in the far field $i\omega g(\omega) = -1$, which allows us to write the far-field term $\mathbf{u}^{ff}$ in equation 3.172 as

$$\mathbf{u}^{ff} = \frac{\hat{\mathbf{r}}e^{i\omega r/\alpha}}{4\pi\rho\alpha^3 r}. \tag{3.176}$$

And, because by definition

$$\delta\left(t-\frac{r}{\alpha}\right) = \frac{1}{2\pi}\int_{-\infty}^{\infty} e^{-i\omega(r-r')}d\omega,$$

equation 3.176 can be represented in time domain as

$$\mathbf{u}^{ff} = \frac{\hat{\mathbf{r}}\delta\left(t - \frac{r}{\alpha}\right)}{4\pi\rho\alpha^3 r}. \quad (3.177)$$

3.10.6 Radiation Patterns from a Point Double Couple or a Moment-Tensor Source

We now generalize the time-domain representation of a point explosion (equation 3.173) as

$$\mathbf{f} = -\mathbf{M}(t) \cdot \nabla\delta(\mathbf{r}). \quad (3.178)$$

In equation 3.178, $\mathbf{M}(t)$ is a symmetric second-order tensor, known as the *moment tensor*. The moment tensor is used to represent an earthquake source and is written as (Aki and Richards, 2002)

$$\mathbf{M}(t) = M_0(\hat{\mathbf{n}}\hat{\mathbf{e}} + \hat{\mathbf{e}}\hat{\mathbf{n}})H(t). \quad (3.179)$$

In equation 3.179, $H(t)$ is the heavyside step function, defined by equation 3.174; $\hat{\mathbf{n}}$ is the unit vector, normal to the fault plane; and $\hat{\mathbf{e}}$ is the unit vector in the direction of the fault motion causing the earthquake. Note that $\hat{\mathbf{n}} \cdot \hat{\mathbf{e}} = 0$. Finally, M_0 in equation 3.179 is the seismic moment and given as

$$M_0 = \langle \mu A \Delta s \rangle = \int_A \mu \Delta s dA = \langle \mu \Delta s \rangle A. \quad (3.180)$$

In equation 3.180, μ is the rigidity modulus of the rock, Δs is the displacement along the fault, A is the area of the fault, and for any property P, $\langle P \rangle$ is the average. Note that the components of $\mathbf{M}$ are moment (force times the distance) and that is why it is called the moment tensor. Moment tensor can be obtained from the earthquake focal mechanism solutions, which can be found in Aki and Richards (2002).

We use Figure 3.20 to understand the physical meaning of a moment-tensor source. We assume that our Cartesian system of coordinate to be the triplet $\{x, y, z\}$ and the fault motion is parallel to the y coordinate axis. In this situation, $\hat{\mathbf{n}} = \hat{\mathbf{x}}$ and $\hat{\mathbf{e}} = \hat{\mathbf{y}}$. Therefore, from equation 3.198

$$\mathbf{M} = M_0(\hat{\mathbf{x}}\hat{\mathbf{y}} + \hat{\mathbf{y}}\hat{\mathbf{x}})H(t). \quad (3.181)$$

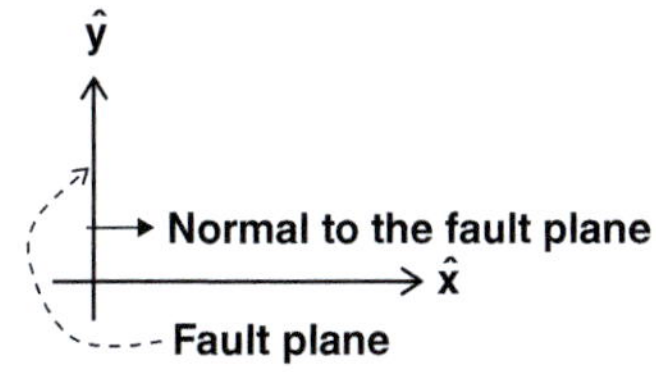

Figure 3.20 Illustration of a fault in a two-dimensional (x–y) plane with motion parallel to the y axis and the direction normal to the fault along the x axis.

Therefore, we write the source term (equation 3.178) as

$$\mathbf{f} = -M_0(\hat{\mathbf{x}}\hat{\mathbf{y}} + \hat{\mathbf{y}}\hat{\mathbf{x}})H(t) \cdot \nabla\delta(\mathbf{r}). \quad (3.182)$$

But

$$\nabla\delta(\mathbf{r}) = \hat{\mathbf{x}}\delta'(x)\delta(y)\delta(z) + \hat{\mathbf{y}}\delta(x)\delta'(y)\delta(z) + \hat{\mathbf{z}}\delta(x)\delta(y)\delta'(z), \quad (3.183)$$

where $\delta'(x), \delta'(y)$, and $\delta'(z)$ respectively indicate the derivatives of the delta functions $\delta(x)$, $\delta(y)$, and $\delta(z)$ with respect to the x, y, and z coordinate axes. Using the right-hand side of equation 3.183 for $\nabla\delta(\mathbf{r})$ into equation 3.182, gives, after some simple algebra,

$$\mathbf{f} = M_0 H(t)\delta(z)[\hat{\mathbf{x}}\delta(x)\{-\delta'(y)\} + \hat{\mathbf{y}}\{-\delta'(x)\}\delta(y)]. \quad (3.184)$$

We use Figure 3.21 to explain the physical meaning of equation 3.184. Figure 3.21a shows the delta function $\hat{\mathbf{x}}\delta(x)$, represented by a spike along the x coordinate axis at the origin (note that the x axis is shown along the vertical direction and the y axis is shown along the horizontal direction in this figure). Figure 3.21b shows the pictorial representation of the first term, $-\hat{\mathbf{x}}\delta(x)\delta'(y)$ on the right-hand side of equation 3.184. The derivative $\delta'(y)$ forms a couple with opposing forces along the direction of the x axis immediately left (O^-) and right (O^+) forming a point couple with a net clockwise rotation, as shown in Figure 3.21c. Figure 3.21d–f explains the second term $-\hat{\mathbf{y}}\delta(y)\delta'(x)$ of the right-hand side of equation 3.183, forming another point couple with a net counterclockwise rotation as shown in Figure 3.21f. Because we do not expect any net rotation

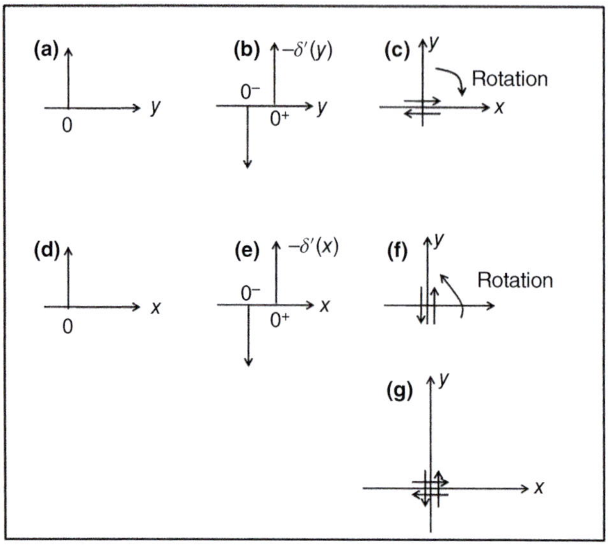

Figure 3.21 Pictorial explanation of equation 3.183. (a) The delta function $\hat{\mathbf{x}}\delta(x)$, represented by a spike along the x axis at the origin. (b) Representation of the first term, $-\hat{\mathbf{x}}\delta(x)\delta'(y)$ on the right-hand side of equation 3.183, showing that it forms a couple about the origin, with 0^+ and 0^- representing the positions immediately right and immediately left from the origin. (c) The point couple represented by the first term $-\hat{\mathbf{x}}\delta(x)\delta'(y)$ on the right-hand side of equation 3.183 with the direction of the net rotation shown by an arrow. Note that the y axis is shown along the horizontal direction in (a) and (b) but along the vertical direction in (c). Parts (d)–(f) are the same as (a)–(c) but for representing the second term $-\hat{\mathbf{y}}\delta(y)\delta'(x)$ on the right-hand side of equation 3.183. Part (g) shows that the net effect of the first and second terms on the right-hand side of equation 3.183 is a point double couple.

from these two terms, the clockwise rotation from the first term must balance the counterclockwise rotation from the second term, leading the moment-tensor source to a *point double couple*, shown in Figure 3.21g. These point double-couple seismic sources have been traditionally used in the field of earthquake seismology. With the introduction of microseismic monitoring, which we briefly discussed in Chapter 1, the same point double-couple sources are now used by the seismic industry in applications related to characterizing hydraulic fracture propagation for unconventional and geothermal reservoir exploration, for monitoring the carbon-sequestered reservoirs to mitigate global warming, etc.

Before discussing radiation patterns from moment tensor or point double-couple sources, it is important to clarify one additional point. In deriving equation 3.184, we started with Figure 3.20 with fault motion parallel to the y coordinate axis and the normal along the x coordinate axis. Instead, if we started with the fault motion parallel to the x coordinate with normal along the y coordinate axis by setting $\hat{\mathbf{n}} = \hat{\mathbf{y}}$ and $\hat{\mathbf{e}} = \hat{\mathbf{x}}$ and followed the same procedure, we would still obtain equation 3.184. This is the classical problem of the fault-plane ambiguity, which we should always keep in mind. The term M_0 given by equation 3.180 approximates the true seismic moment. The true seismic moment resulting from a rupture and faulting is represented by a polynomial, and equation 3.184 contains only the first two terms of this polynomial, and the resulting moment-tensor representation of equation 3.179 is the centroid-moment-tensor (CMT) representation of an earthquake source (Backus and Mulcahy, 1976; McGuire et.al., 2001). It is well known that this CMT representation fails to specify which of the two nodal planes is the actual fault plane (Aki and Richards, 2002), which we demonstrate here. In the finite-moment-tensor (FMT) representation, one additional term of the seismic moment is included. By including this additional term, FMT resolves this classical fault-plane ambiguity problem by providing additional information such as the characteristic duration (Silver and Jordan, 1983), dimensions of the faulting, and the directivity of the vector of the fault slip (Ben-Menahem, 1961). Chen et al. (2005, 2010) demonstrated that the FMT can resolve the fault-plane ambiguity for small-to-medium-sized earthquakes.

Before deriving the radiation patterns from a point double-couple source, we must first point out a few important facts about the moment tensor. First, in the moment-tensor representation, given by equation 3.178, if the moment tensor $\mathbf{M}(t)$ is set to a second-order identity tensor $\mathbf{I}$, then the source is identical to a uniform source or a point explosion, the radiation patterns for which we already derived. Second, for any earthquake, the moment tensor, defined by equations 3.179 and 3.180, is always a symmetric tensor. Third, during the occurrence of an earthquake, we do not expect any net volume change, which, in turn, imposes an additional constraint on $\mathbf{M}(t)$, that is, its trace must be zero or

$$\mathrm{tr}[\mathbf{M}(t)] = 0. \tag{3.185}$$

In Chapter 2 (see equation 2.220), we have defined the trace of a second-order tensor. Finally, the moment tensor is position-independent, and so we must have

$$\left.\begin{aligned} \nabla \cdot [\mathbf{M}(t)] &= 0 \\ \nabla \times [\mathbf{M}(t)] &= \mathbf{0} \end{aligned}\right\}. \tag{3.186}$$

To the body force $\mathbf{f}$, represented by a moment tensor via equation 3.177, we now do the Helmholtz decomposition $\mathbf{f} = \nabla\varphi - \nabla\times\boldsymbol{\psi}$, with φ and $\boldsymbol{\psi}$ being defined by equation 3.145, and separately solve φ and $\boldsymbol{\psi}$.

The Component (φ) of the Moment-Tensor Source: The Helmholtz decomposition for the moment-tensor source gives, from equation 3.146 $\varphi = \nabla^{-2}(\nabla\cdot\mathbf{f}) = \nabla^{-2}\{\nabla\cdot(\nabla\varphi - \nabla\times\boldsymbol{\psi})\}$. But the Helmholtz decomposition implies that $\nabla\cdot\boldsymbol{\psi} = 0$ (equation 3.145). Now

$$\begin{aligned} \nabla\cdot\mathbf{f} &= \nabla\cdot[-\mathbf{M}(t)\cdot\nabla\delta(\mathbf{r})] \\ &= -[\nabla\cdot\mathbf{M}(t)\cdot\nabla\delta(\mathbf{r}) + \mathbf{M}(t) : \nabla\nabla\delta(\mathbf{r})]. \end{aligned}$$

And, from equation 3.186, we write, from above,

$$\nabla\cdot\mathbf{f} = -\mathbf{M}(t) : \nabla\nabla\delta(\mathbf{r}). \tag{3.187}$$

From our discussion of the tensors in Chapter 2, for any second-order tensor $\mathbf{A}$ and vector $\mathbf{b}$

$$\begin{aligned} \nabla\cdot(\mathbf{A}\cdot\mathbf{b}) &= \nabla x^i\cdot\partial_i(\mathbf{A}\cdot\mathbf{b}) \\ &= \nabla x^i\cdot(\partial_i\mathbf{A}\cdot\mathbf{b} + \mathbf{A}\cdot\partial_i\mathbf{b}) \\ &= \nabla x^i\cdot(\partial_i\mathbf{A})\cdot\mathbf{b} + \nabla x^i\cdot\mathbf{A}\cdot\partial_i\mathbf{b} \\ &= \nabla\cdot\mathbf{A}\cdot\mathbf{b} + \mathbf{A} : \nabla x^i\partial_i\mathbf{b} \\ &= \nabla\cdot\mathbf{A}\cdot\mathbf{b} + \mathbf{A} : \nabla\mathbf{b}. \end{aligned} \tag{3.188}$$

Therefore,

$$\begin{aligned} \varphi &= \nabla^{-2}(\nabla\cdot\mathbf{f}) = -\nabla^{-2}[\mathbf{M}(t) : \nabla\nabla\delta(\mathbf{r})] \\ &= -\mathbf{M}(t) : \nabla\nabla[\nabla^{-2}\delta(\mathbf{r})]. \end{aligned} \tag{3.189}$$

Finally, from the definition of ∇^{-2}, we obtain

$$\varphi = -\mathbf{M}(t) : \nabla\nabla\left(\frac{1}{4\pi|\mathbf{r}|}\right). \tag{3.190}$$

The Component ($\boldsymbol{\psi}$) of the Moment-Tensor Source: From equation 3.146, in the Helmholtz decomposition $\mathbf{f} = \nabla\varphi - \nabla\times\boldsymbol{\psi}$, $\boldsymbol{\psi} = \nabla^{-2}(\nabla\times\mathbf{f})$. For the moment-tensor source, we thus have

$$\begin{aligned} \boldsymbol{\Psi} &= \nabla^{-2}(\nabla\times\mathbf{f}) = \nabla^{-2}[\nabla\times\{-\mathbf{M}(t)\cdot\nabla\delta(\mathbf{r})\}] \\ &= -\nabla^{-2}[\nabla\times\mathbf{M}(t)\cdot\nabla\delta(\mathbf{r}) \\ &\quad - \{\mathbf{M}(t)\}^T \otimes \{\nabla\nabla\delta(\mathbf{r})\}^T] \\ &= -\nabla^{-2}[\mathbf{M}(t)\otimes\nabla\nabla\delta(\mathbf{r})] \\ &= -\mathbf{M}(t)\otimes[\nabla\nabla\{\nabla^{-2}\delta(\mathbf{r})\}]. \end{aligned} \tag{3.191}$$

The symbol $\otimes$ in equation 3.191 represents a dot–cross product of two tensors, or a dot product, followed by a cross product of two tensors. Using the invariant notation, we introduced in Chapter 2, assume that two second-order tensors $\mathbf{A}$ and $\mathbf{B}$ are represented as

$$\left.\begin{aligned} \mathbf{A} &= A_{ij}\nabla x^i\nabla x^i \\ \mathbf{B} &= B^{kl}\partial_k(\mathbf{r})\partial_l(\mathbf{r}) \end{aligned}\right\}. \tag{3.192}$$

Then, from the concepts, developed in Chapter 2,

$$\mathbf{A} \cdot \mathbf{B} = A_{ij}B^{jl}\nabla x^i \partial_l(\mathbf{r}), \tag{3.193}$$

and $\mathbf{A} \otimes \mathbf{B}$ is a vector given as

$$\mathbf{A} \otimes \mathbf{B} = A_{ij}B^{jl}\nabla x^i \times \partial_l(\mathbf{r}). \tag{3.194}$$

Finally, from the definition of ∇^{-2}, equation 3.191 leads to

$$\boldsymbol{\Psi} = -\mathbf{M}(t) \otimes \nabla\nabla\left(\frac{1}{4\pi|\mathbf{r}|}\right). \tag{3.195}$$

Equations 3.190 and 3.195 provide φ and ψ for source located at the origin. For source, located at $\mathbf{r} = \mathbf{r}'$, they can be generalized as

$$\left.\begin{aligned} \varphi &= -\mathbf{M}(t) : \nabla\nabla\left(\frac{1}{4\pi|\mathbf{r}-\mathbf{r}'|}\right) \\ \boldsymbol{\Psi} &= -\mathbf{M}(t) \otimes \nabla\nabla\left(\frac{1}{4\pi|\mathbf{r}-\mathbf{r}'|}\right) \end{aligned}\right\}. \tag{3.196}$$

And, from Green's theorem, the solution for P- and S-wave equations (equation 3.155) is

$$\begin{aligned} P(\mathbf{r}) &= -\int_{\Re^3} G_\alpha(\mathbf{r},\mathbf{r}')\frac{\varphi(\mathbf{r}')}{\lambda+2\mu}dV(\mathbf{r}') \\ &= \int_{\Re^3} \frac{e^{i(\omega/\alpha)|\mathbf{r}-\mathbf{r}'|}}{4\pi|\mathbf{r}-\mathbf{r}'|}\frac{\varphi(\mathbf{r}')}{\lambda+2\mu}dV(\mathbf{r}') \end{aligned} \tag{3.197}$$

and

$$\begin{aligned} \mathbf{S}(\mathbf{r}) &= -\int_{\Re^3} G_\beta(\mathbf{r},\mathbf{r}')\frac{\boldsymbol{\Psi}(\mathbf{r}')}{\mu}dV(\mathbf{r}') \\ &= \int_{\Re^3} \frac{e^{i(\omega/\beta)|\mathbf{r}-\mathbf{r}'|}}{4\pi|\mathbf{r}-\mathbf{r}'|}\frac{\boldsymbol{\Psi}(\mathbf{r}')}{\mu}dV(\mathbf{r}'). \end{aligned} \tag{3.198}$$

Using φ and ψ from equation 3.196 in equations 3.197 and 3.199, we obtain

$$\begin{aligned} P(\mathbf{r}) &= -\int_{\Re^3} G_\alpha(\mathbf{r},\mathbf{r}')\frac{\varphi(\mathbf{r}')}{\lambda+2\mu}dV(\mathbf{r}') \\ &= \int_{\Re^3} \frac{e^{i(\omega/\alpha)|\mathbf{r}-\mathbf{r}'|}}{4\pi|\mathbf{r}-\mathbf{r}'|}\frac{\varphi(\mathbf{r}')}{\lambda+2\mu}dV(\mathbf{r}') \\ &= \int_{\Re^3} \frac{G_\alpha(\mathbf{r},\mathbf{r}')}{\lambda+2\mu}\left[\mathbf{M} : \nabla'\nabla'\left(-\frac{1}{4\pi|\mathbf{r}'|}\right)\right] \\ &\times dV(\mathbf{r}') = -\frac{\mathbf{M}}{\lambda+2\mu} : \int_{\Re^3} G_\alpha(\mathbf{r},\mathbf{r}') \\ &\times \left[\nabla'\nabla'\left(-\frac{1}{4\pi|\mathbf{r}'|}\right)\right] dV(\mathbf{r}') \end{aligned} \tag{3.199}$$

and

$$\begin{aligned} \mathbf{S}(\mathbf{r}) &= -\int_{\Re^3} G_\beta(\mathbf{r},\mathbf{r}')\frac{\boldsymbol{\Psi}(\mathbf{r}')}{\mu}dV(\mathbf{r}') \\ &= \int_{\Re^3} \frac{e^{i(\omega/\beta)|\mathbf{r}-\mathbf{r}'|}}{4\pi|\mathbf{r}-\mathbf{r}'|}\frac{\boldsymbol{\Psi}(\mathbf{r}')}{\mu}dV(\mathbf{r}') \\ &= \int_{\Re^3} \frac{G_\beta(\mathbf{r},\mathbf{r}')}{\mu}\left[\mathbf{M} \otimes \nabla'\nabla'\left(-\frac{1}{4\pi|\mathbf{r}'|}\right)\right] \\ &\times dV(\mathbf{r}') = -\frac{\mathbf{M}}{\mu} \otimes \int_{\Re^3} G_\beta(\mathbf{r},\mathbf{r}') \\ &\times \left[\nabla'\nabla'\left(-\frac{1}{4\pi|\mathbf{r}'|}\right)\right] dV(\mathbf{r}'). \end{aligned} \tag{3.200}$$

It is apparent from equations 3.199 and 3.200 that if we evaluate the integral

$$\boldsymbol{\Phi} = \int_{\Re^3} G_v(\mathbf{r},\mathbf{r}')\left[\nabla'\nabla'\left(-\frac{1}{4\pi|\mathbf{r}'|}\right)\right] dV(\mathbf{r}'), \tag{3.201}$$

where the subscript v is either α or β, and apply it separately to $P(\mathbf{r})$ in equation 3.199 or to $\mathbf{S}(\mathbf{r})$ in equation 3.200, where $G_v(\mathbf{r}, \mathbf{r}')$ is the Green's function, which, from equation 3.161, is given as

$$G_v(\mathbf{r},\mathbf{r}') = -\frac{e^{ik|\mathbf{r}-\mathbf{r}'|}}{4\pi|\mathbf{r}-\mathbf{r}'|}; \quad k = \frac{\omega}{v}; \quad v = \begin{cases}\alpha \\ \beta\end{cases}, \tag{3.202}$$

we then solve the problem of the radiation patterns for P and S waves due to a point double-couple source.

The integral of equation 3.202 can be performed using thin spherical shells, which follows from the classical *Newton's Shell Theorem*. This theorem allows computational simplifications to the gravitational fields to the objects inside or outside of the spherically symmetric bodies. Here, we adopt this theorem in evaluating the integral shown in equation 3.201 in relation with computing the radiation patterns from the point double-couple sources.

As shown in Figure 3.22a, we define $\mathbf{r}_* = \mathbf{r} - \mathbf{r}'$ and perform our integral in thin

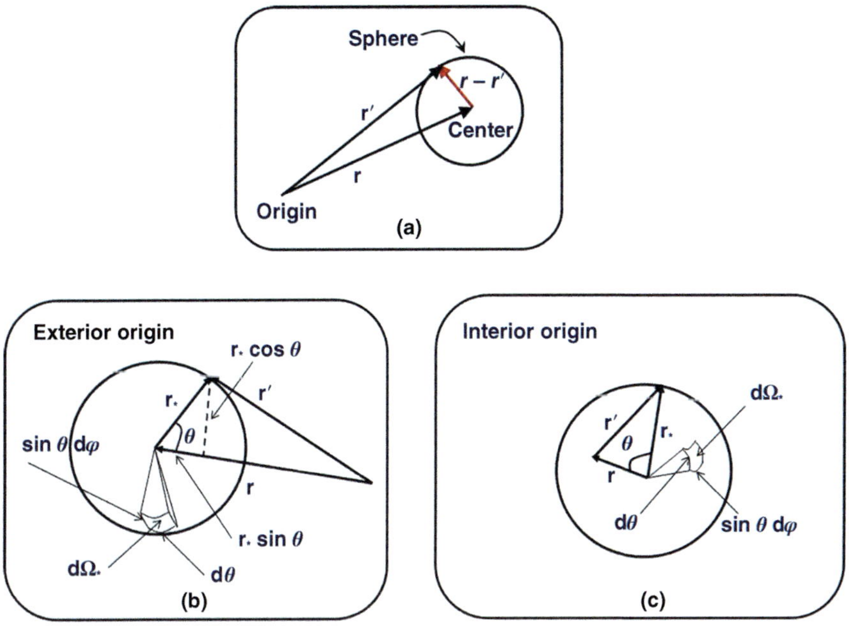

Figure 3.22 Illustrations for evaluating the volume integral, shown in equation 3.201; see the text for details.

spherical shells by letting

$$dV(\mathbf{r}') = |\mathbf{r}_*|^2 d|\mathbf{r}_*| d\Omega_* = r_*^2 dr_* d\Omega_*, \tag{3.203}$$

where

$$d\Omega_* = \sin\theta d\theta d\varphi \tag{3.204}$$

is the solid angle element in the surface of a unit sphere. This, in turn, leads equation 3.201 to

$$\boldsymbol{\Phi} = (4\pi)^{-2} \int_{r_*=0}^{\infty} dr_* r_*^2 \frac{e^{ikr_*}}{r_*} \int_{\Omega_*} d\Omega_* \nabla'\nabla' \left(\frac{1}{r'_*}\right). \tag{3.205}$$

Note that we replaced $G_v(\mathbf{r}, \mathbf{r}') = G_v(\mathbf{r}, \mathbf{r}_*)$ in equation 3.205 with the expression of the Green's function with $k = \omega/v$, where ω is the frequency and $v = \alpha$ or β for P and S waves. Newton's Shell Theorem says that the second integral on the right-hand side of equation 3.205 is given as

$$\int_{\Omega_*} d\Omega_* \nabla'\nabla' \left(\frac{1}{|\mathbf{r}'_*|}\right) = H(\mathbf{r} - \mathbf{r}_*)\nabla\nabla \int_{\Omega_*} \frac{d\Omega_*}{|\mathbf{r}'|}, \tag{3.206}$$

where $H(\mathbf{r} - \mathbf{r}_*)$ is the heavyside step function defined earlier in equation 3.174. Physically, equation 3.206 means that the gravitational potential (and all its derivatives) due to a uniform spherical shell of an arbitrary mass behaves as if all the mass is concentrated at the center of the shell and vanishes inside the shell. A complete proof of Newton's Shell Theorem is given in Aki and Richards (2002), and we do not repeat it here.

To use Newton's Shell Theorem in obtaining the radiation patterns from a point double-couple source, we state another theorem, also used by Newton which states that the integral on the right-hand side of equation 3.206 is given as

$$\int_{\Omega_*} \frac{d\Omega_*}{|\mathbf{r}'|} = 4\pi \left[\frac{H(\mathbf{r} - \mathbf{r}_*)}{|\mathbf{r}|} - \frac{H(\mathbf{r}_* - \mathbf{r})}{|\mathbf{r}_*|}\right]$$
$$= 4\pi \min\left(\frac{1}{|\mathbf{r}|}, \frac{1}{|\mathbf{r}_*|}\right), \tag{3.207}$$

in which $\min(a, b)$ stands for the minimum between a and b. Physical meaning of equation 3.207 is that the gravitational potential due to a uniform spherical shell of total mass M is equal to $M \min\left(\frac{1}{r}, \frac{1}{r_*}\right)$, where

r is the field point from the center of the shell and r_* is the radius of the shell.

To prove equation 3.207, we identify two scenarios, (1) with exterior origin (Figure 3.22b) and (2) with interior origin (Figure 3.22c), to write

$$\int_{\Omega_*} \frac{d\Omega_*}{r'} = \int_{\theta=0}^{\pi} d\theta \sin\theta \int_{\phi=0}^{2\pi} d\text{Œ} \frac{1}{r'}$$
$$= \int_{0}^{2\pi} d\text{Œ} \int_{0}^{\pi} \frac{d\theta \sin\theta}{r'}. \qquad (3.208)$$

In equation 3.208, we replaced $|\mathbf{r}'|$ with r'. Similarly, replacing $|\mathbf{r}|$ and $|\mathbf{r}_*|$ respectively with r and r_* and using the law of cosines to Figure 3.22b,c, we have $r\prime^2 = r^2 + r_*^2 - 2rr_* \cos\theta$, and hence, by differentiating both sides, $dr' = rr_* \sin\theta d\theta$, or

$$\frac{\sin\theta d\theta}{r'} = \frac{dr'}{rr_*}. \qquad (3.209)$$

It thus implies from equation 3.209 that we can evaluate the integral in equation 3.208 with limits of 0 and π for the angle θ using the appropriate change of variables for the two specific cases, identified in Figure 3.23.

Case 1—when the evaluation point is outside the sphere: Here $r > r_*$, represented by Figure 3.23a for the lower limit of θ (0), when $r' = r - r_*$, and by Figure 3.23b for the upper limit of θ (π), when $r' = r + r_*$. Consequently, using equation 3.209, we write

$$\int_0^{\pi} \frac{d\theta \sin\theta}{r'} = \int_{r-r_*}^{r+r_*} \frac{dr'}{rr_*}$$

which allows evaluating the integral of equation 3.208 as

$$\int_{\Omega_*} \frac{d\Omega_*}{r'} = \int_0^{2\pi} d\varphi \int_0^{\pi} \frac{d\theta \sin\theta}{r'} = 2\pi \int_{r-r_*}^{r+r_*} \frac{dr'}{rr_*}$$
$$= 2\pi \frac{(r+r_*) - (r-r_*)}{rr_*} = \frac{4\pi}{r}. \qquad (3.210)$$

Case 2—when the evaluation point is inside the sphere: This is represented by Figure 3.23c,d, with lower ($\theta = 0$) and upper ($\theta = \pi$) limits for r' respectively being $r_* - r$ and $r + r_*$. Therefore,

$$\int_0^{\pi} \frac{d\theta \sin\theta}{r'} = \int_{r_*-r}^{r_*+r} \frac{dr'}{rr_*}$$

which leads the evaluation of the integral of equation 3.208 to

$$\int_{\Omega_*} \frac{d\Omega_*}{r'} = \int_0^{2\pi} d\varphi \int_0^{\pi} \frac{d\theta \sin\theta}{r'} = 2\pi \int_{r_*-r}^{r_*+r} \frac{dr'}{rr_*}$$
$$= 2\pi \frac{(r_*+r) - (r_*-r)}{rr_*} = \frac{4\pi}{r_*}. \qquad (3.211)$$

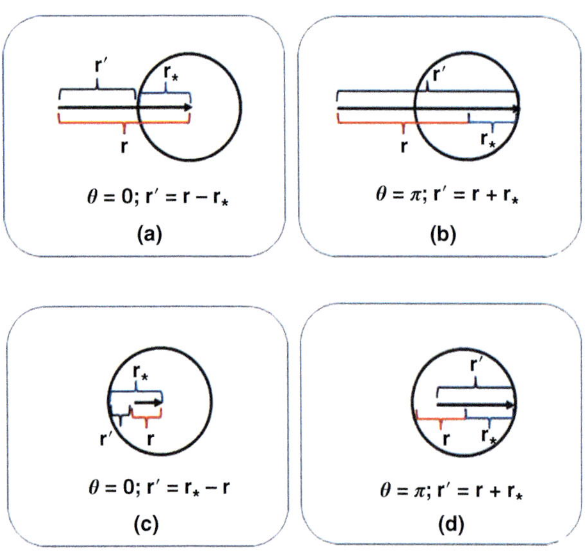

Figure 3.23 Figure related to the derivation of equation 3.207. (a) For the point outside the sphere and $\theta = 0$. (b) For the point outside the sphere and $\theta = \pi$. Parts (c) and (d) are the same as (a) and (b), but when the point is inside the sphere.

Equations 3.210 and 3.211 can be both combined as

$$\int_{\Omega_*} \frac{d\Omega_*}{r'} = \int_{\theta=0}^{\pi} d\theta \sin\theta \int_{\phi=0}^{2\pi} d\OE \frac{1}{r'}$$

$$= \int_0^{2\pi} d\OE \int_0^{\pi} \frac{d\theta \sin\theta}{r'}$$

$$= \begin{cases} \dfrac{4\pi}{r}; & r > r_* \\ \dfrac{4\pi}{r_*}; & r < r_* \end{cases}. \tag{3.212}$$

Equation 3.212 is equivalent to

$$\int_{\Omega_*} \frac{d\Omega_*}{r'} = 4\pi \min\left(\frac{1}{r}, \frac{1}{r_*}\right),$$

which leads to equation 3.207 and therefore the theorem is proved. Combining Newton's Shell Theorem (equation 3.206) with equation 3.207, which we just proved gives us

$$\int_{\Omega_*} d\Omega_* \nabla'\nabla'\left(\frac{1}{r'_*}\right) = H(r-r_*)\nabla\nabla \int_{\Omega_*} \frac{d\Omega_*}{r'}$$

$$= 4\pi H(r-r_*)\nabla\nabla \left[\frac{H(r-r_*)}{r} - \frac{H(r_*-r)}{r_*}\right]. \tag{3.213}$$

Using tensor algebra, we discussed in Chapter 2, we can write in spherical coordinates

$$\nabla\nabla[f(r)] = \widehat{\mathbf{r}}\widehat{\mathbf{r}}\left[f''(r) - \frac{1}{r}f'(r)\right] + \left[\frac{f'(r)}{r}\right]\mathbf{I}. \tag{3.214}$$

In equation 3.214, $f'(r)$ and $f''(r)$ are the first and second derivatives of f with respect to r and $\mathbf{I}$ is the identity tensor. Proof of equation 3.214 is left as an exercise.

We now use equation 3.124 to our evaluated integral, shown in equation 3.213, to further simplify it. First, note that

$$\partial_r\left[\frac{H(r-r_*)}{r}\right] = \partial_r H(r-r_*)\frac{1}{r}$$

$$+ H(r-r_*)\partial_r\left(\frac{1}{r}\right)$$

$$= \partial_r H(r-r_*)\frac{1}{r} - \frac{1}{r^2}H(r-r_*)$$

$$= \frac{\delta(r-r_*)}{r} - \frac{H(r-r_*)}{r^2}. \tag{3.215}$$

Similarly,

$$\partial_r\left[\frac{H(r_*-r)}{r_*}\right] = -\frac{\delta(r_*-r)}{r_*}. \tag{3.216}$$

Therefore, from equations 3.215 and 3.216, we have

$$\partial_r\left[\frac{H(r-r_*)}{r} + \frac{H(r_*-r)}{r_*}\right] = -\frac{H(r-r_*)}{r^2}. \tag{3.217}$$

Thus,

$$\partial_r^2\left[\frac{H(r-r_*)}{r} + \frac{H(r_*-r)}{r_*}\right] = -\partial_r\left[\frac{H(r-r_*)}{r^2}\right]$$

$$= -\frac{\delta(r-r_*)}{r^2} + \frac{2H(r-r_*)}{r^3}. \tag{3.218}$$

By using

$$f(r) = \left[\frac{H(r-r_*)}{r} - \frac{H(r_*-r)}{r_*}\right]$$

in equation 3.213 and using equations 3.214–3.218, we can write

$$\nabla\nabla[f(r)] = \nabla\nabla\left[\frac{H(r-r_*)}{r} - \frac{H(r_*-r)}{r_*}\right]$$

$$= \widehat{\mathbf{r}}\widehat{\mathbf{r}}\left[f''(r) - \frac{1}{r}f'(r)\right] + \left[\frac{f'(r)}{r}\right]\mathbf{I}$$

$$= \widehat{\mathbf{r}}\widehat{\mathbf{r}}\left[\frac{3H(r-r_*)}{r^3} - \frac{\delta(r-r_*)}{r^2}\right]$$

$$- \frac{H(r-r_*)}{r^3}\mathbf{I}. \tag{3.219}$$

Finally, from equation 3.219, equation 3.213 becomes

$$\int_{\Omega_*} d\Omega_* \nabla'\nabla'\left(\frac{1}{r'_*}\right) = H(r-r_*)\nabla\nabla$$

$$\int_{\Omega_*} \frac{d\Omega_*}{r'} = 4\pi H(r-r_*)\nabla\nabla$$

$$\times \left[\frac{H(r-r_*)}{r} - \frac{H(r_*-r)}{r_*}\right] = 4\pi H(r-r_*)$$

$$\times \left[\widehat{\mathbf{r}}\widehat{\mathbf{r}}\left\{\frac{3H(r-r_*)}{r^3} - \frac{\delta(r-r_*)}{r^2}\right\}\right.$$

$$\left. - \mathbf{I}\frac{H(r-r_*)}{r^3}\right]. \tag{3.220}$$

Also note that

$$\left.\begin{aligned} H(r-r_*)H(r-r_*) &= H(r-r_*) \\ H(r-r_*)\delta(r-r_*) &= \delta(r-r_*) \end{aligned}\right\}. \tag{3.221}$$

Therefore,

$$\int\limits_{\Omega_*} d\Omega_* \nabla'\nabla'\left(\frac{1}{r'_*}\right) = 4\pi\left[\hat{\mathbf{r}}\hat{\mathbf{r}}\left\{\frac{3H(r-r_*)}{r^3} - \frac{\delta(r-r_*)}{r^2}\right\} - \mathbf{I}\frac{H(r-r_*)}{r^3}\right]. \tag{3.222}$$

Using equation 3.222, we thus write the integral we want to evaluate (equation 3.205) as

$$\begin{aligned}
\boldsymbol{\Phi} &= (4\pi)^{-2}\int\limits_{r_*=0}^{\infty} dr_* r_*^2 e^{ikr_*}\int\limits_{\Omega_*} d\Omega_* \nabla'\nabla'\left(\frac{1}{r'_*}\right) \\
&= (4\pi)^{-2}\int\limits_{0}^{\infty} dr_* r_*^2 e^{ikr_*} 4\pi\left[\hat{\mathbf{r}}\hat{\mathbf{r}}\left\{\frac{3H(r-r_*)}{r^3} - \frac{\delta(r-r_*)}{r^2}\right\} - \mathbf{I}\frac{H(r-r_*)}{r^3}\right] \\
&= \frac{1}{4\pi}\left\{\begin{aligned} &\hat{\mathbf{r}}\hat{\mathbf{r}}\int_0^{\infty} dr_* r_*^2 e^{ikr_*}\frac{3H(r-r_*)}{r^3} - \hat{\mathbf{r}}\hat{\mathbf{r}}\int\limits_0^{\infty} dr_* r_*^2 e^{ikr_*}\frac{\delta(r-r_*)}{r^2} \\ &-\mathbf{I}\int_0^{\infty} dr_* r_*^2 e^{ikr_*}\frac{H(r-r_*)}{r^3} \end{aligned}\right\} \\
&= \frac{1}{4\pi}\left\{\hat{\mathbf{r}}\hat{\mathbf{r}}\frac{3}{r^3}\int\limits_0^{r} dr_* r_*^2 e^{ikr_*} - \hat{\mathbf{r}}\hat{\mathbf{r}}e^{ikr} - \mathbf{I}\int\limits_0^{r} dr_* r_*^2 e^{ikr_*}\right\} = \frac{1}{4\pi}\left\{(3\hat{\mathbf{r}}\hat{\mathbf{r}} - \mathbf{I})\frac{1}{r^3}\int\limits_0^{r} dr_* r_*^2 e^{ikr_*} - \hat{\mathbf{r}}\hat{\mathbf{r}}e^{ikr}\right\}.
\end{aligned} \tag{3.223}$$

Once $\boldsymbol{\Phi}$ is obtained, we can use equations 3.189 and 3.200 to write

$$\left.\begin{aligned} P &= \frac{\mathbf{M}}{\lambda+2\mu} : \boldsymbol{\Phi}; \quad k = \frac{\omega}{\alpha} \\ \mathbf{S} &= \frac{\mathbf{M}}{\mu}\otimes\boldsymbol{\Phi}; \quad k = \frac{\omega}{\beta} \end{aligned}\right\} \tag{3.224}$$

and obtain the radiation patterns from the displacement vector $\mathbf{u} = \nabla P - \nabla\times\mathbf{S}$. By carrying out the integral in equation 3.223 and then from equation 3.224, we can obtain both P-wave radiation pattern $\mathbf{u}_P = \nabla P$ and the S-wave radiation pattern $\mathbf{u}_s = \nabla\times\mathbf{S}$. Details of these derivations are given in Aki and Richards (2002), and below we write the far-field radiation patterns

$$\left.\begin{aligned} \mathbf{u}_P^{ff} &= \hat{\mathbf{r}}\frac{i\omega e^{i\omega|\mathbf{r}-\mathbf{r}_s|/\alpha}[\hat{\mathbf{r}}\cdot\mathbf{M}\cdot\hat{\mathbf{r}}]}{4\pi\rho\alpha^3|\mathbf{r}-\mathbf{r}_s|} \\ \mathbf{u}_S^{ff} &= \frac{i\omega e^{i\omega|\mathbf{r}-\mathbf{r}_s|/\beta}[(\hat{\boldsymbol{\theta}}\hat{\boldsymbol{\theta}}+\hat{\boldsymbol{\phi}}\hat{\boldsymbol{\phi}})\cdot\mathbf{M}\cdot\hat{\mathbf{r}}]}{4\pi\rho\beta^3|\mathbf{r}-\mathbf{r}_s|} \end{aligned}\right\}. \tag{3.225}$$

3.10.7 Radiation Patterns from a Point Force

For a point force, the body force vector **f** in the momentum equation (equation 3.151) is of the form

$$\mathbf{f} = \hat{\mathbf{a}}\delta(\mathbf{r}-\mathbf{r}_s). \tag{3.226}$$

In equation 3.226, we assume that the source is located at the origin and represents a force, oriented in an arbitrary direction with unit vector $\hat{\mathbf{a}}$. In exploration seismology, vibrator sources can be represented by such an equation. For example, in the Cartesian coordinate system, represented by $\{x_1, x_2, x_3\}$, if $\hat{\mathbf{a}} = \hat{\mathbf{x}}_3$, then equation 3.226 represents a vertical vibrator type of source. In general, a source wavelet term of type $g(\omega)$ is present on the right-hand side of equation 3.226, which we omit here for simplicity.

Using the same derivations, which were carried out for the uniform and double-couple sources, we can obtain the radiation patterns for the point force. We omit the detailed derivations here and just write the far-field radiation patterns for P and S waves in a spherical coordinates as

$$\left.\begin{aligned} \mathbf{u}_P^{ff} &= \frac{\hat{\mathbf{r}}(\hat{\mathbf{a}}\cdot\hat{\mathbf{r}})}{\rho\alpha^2}\frac{e^{i\omega r/\alpha}}{4\pi|\mathbf{r}-\mathbf{r}_s|} \\ \mathbf{u}_S^{ff} &= \frac{\hat{\mathbf{a}}\cdot(\widehat{\boldsymbol{\theta}}\widehat{\boldsymbol{\theta}}+\widehat{\boldsymbol{\phi}}\widehat{\boldsymbol{\phi}})}{\rho\beta^2}\frac{e^{i\omega r/\beta}}{4\pi|\mathbf{r}-\mathbf{r}_s|} \end{aligned}\right\}. \tag{3.227}$$

3.10.8 Summary of Radiation Patterns

In the above sections, we discussed three types of sources: (1) uniform (explosive), (2) force, and (3) double couple. In our discussion, we explicitly assumed that these sources are localized at a point and obtained expressions for both P- and S-wave radiation patterns at a distance $r = |\mathbf{r} - \mathbf{r}_s|$ away from the source location $\mathbf{r}_s$ in a 3D homogeneous isotropic elastic medium. All of them are, therefore, 3D radiation patterns from point sources. It is sometimes convenient to consider 2D sources where each of the above three source types are line sources instead of point sources. For example, computing synthetic seismograms for a 3D inhomogeneous elastic medium is computationally challenging. In many practical situations, replacing such a 3D medium with a 2D medium and computing seismograms for line sources is much less computationally demanding. These 2D radiation patterns are, however, different from 3D radiation patterns and require both amplitude and phase corrections to approximately match the real data,

Table 3.1 Far-field radiation patterns in 3D and 2D.

Dimension	3D	2D
Explosion $\mathbf{f} = -\nabla\delta(\mathbf{r}-\mathbf{r}_s)$	$\left.\begin{aligned}\mathbf{u}_P^{ff} &= \frac{i\omega}{\alpha}\frac{\hat{\mathbf{r}}}{\rho\alpha^2}\frac{e^{i\omega\frac{r}{\alpha}}}{4\pi r}\\ \mathbf{u}_S &= \mathbf{0}\end{aligned}\right\}$	$\left.\begin{aligned}\mathbf{u}_P^{ff} &= \frac{i\omega}{\alpha}\frac{\hat{\mathbf{r}}}{\rho\alpha^2}\left(\frac{\alpha}{8\pi\omega r}\right)^{1/2}e^{i\left(\omega\frac{r}{\alpha}+\frac{\pi}{4}\right)}\\ \mathbf{u}_S &= \mathbf{0}\end{aligned}\right\}$
Force $\mathbf{f} = \mathbf{a}\delta(\mathbf{r}-\mathbf{r}_s)$	$\left.\begin{aligned}\mathbf{u}_P^{ff} &= \hat{\mathbf{r}}\frac{\mathbf{a}\cdot\hat{\mathbf{r}}}{\rho\alpha^2}\frac{e^{i\omega\frac{r}{\alpha}}}{4\pi r}\\ \mathbf{u}_S^{ff} &= \frac{\mathbf{a}\cdot(\widehat{\boldsymbol{\theta}}\widehat{\boldsymbol{\theta}}+\widehat{\boldsymbol{\phi}}\widehat{\boldsymbol{\phi}})}{\rho\beta^2}\frac{e^{i\omega\frac{r}{\beta}}}{4\pi r}\end{aligned}\right\}$	$\left.\begin{aligned}\mathbf{u}_P^{ff} &= \hat{\mathbf{r}}\frac{\mathbf{a}\cdot\hat{\mathbf{r}}}{\rho\alpha^2}\left(\frac{\alpha}{8\pi\omega r}\right)^{1/2}e^{i\left(\omega\frac{r}{\alpha}+\frac{\pi}{4}\right)}\\ \mathbf{u}_S^{ff} &= \frac{\mathbf{a}\cdot(\widehat{\boldsymbol{\theta}}\widehat{\boldsymbol{\theta}}+\widehat{\mathbf{y}}\widehat{\mathbf{y}})}{\rho\beta^2}\left(\frac{\beta}{8\pi\omega r}\right)^{1/2}e^{i\left(\omega\frac{r}{\beta}+\frac{\pi}{4}\right)}\end{aligned}\right\}$
Earthquake $\mathbf{f} = -\mathbf{M}:\nabla\delta(\mathbf{r}-\mathbf{r}_s)$	$\left.\begin{aligned}\mathbf{u}_P^{ff} &= \frac{i\omega}{\alpha}\frac{\hat{\mathbf{r}}}{\rho\alpha^2}\hat{\mathbf{r}}\cdot\mathbf{M}\cdot\hat{\mathbf{r}}\frac{e^{i\omega\frac{r}{\alpha}}}{4\pi r}\\ \mathbf{u}_S^{ff} &= \frac{i\omega}{\beta}\frac{(\widehat{\boldsymbol{\theta}}\widehat{\boldsymbol{\theta}}+\widehat{\boldsymbol{\phi}}\widehat{\boldsymbol{\phi}})}{\rho\beta^2}\cdot\mathbf{M}\cdot\hat{\mathbf{r}}\frac{e^{i\omega\frac{r}{\beta}}}{4\pi r}\end{aligned}\right\}$	$\left.\begin{aligned}\mathbf{u}_P^{ff} &= \frac{i\omega}{\alpha}\frac{\hat{\mathbf{r}}}{\rho\alpha^2}\hat{\mathbf{r}}\cdot\mathbf{M}\cdot\hat{\mathbf{r}}\left(\frac{\alpha}{8\pi\omega r}\right)e^{i\left(\omega\frac{r}{\alpha}+\frac{\pi}{4}\right)}\\ \mathbf{u}_S^{ff} &= \frac{i\omega}{\beta}\frac{(\widehat{\boldsymbol{\theta}}\widehat{\boldsymbol{\theta}}+\widehat{\mathbf{y}}\widehat{\mathbf{y}})}{\rho\beta^2}\cdot\mathbf{M}\cdot\hat{\mathbf{r}}\left(\frac{\beta}{8\pi\omega r}\right)e^{i\left(\omega\frac{r}{\beta}+\frac{\pi}{4}\right)}\end{aligned}\right\}$

which are always 3D. We do not explicitly derive these 2D radiation patterns but provide both 3D and 2D far-field radiation patterns from different source types in an isotropic homogeneous medium in Table 3.1. We have already seen that in 3D the far-field amplitudes inversely decay with source–receiver distance r (equations 3.176, 3.224, and 3.226), and we list them all in the second column of Table 3.1. For 2D, listed in the third column of Table 3.1, the amplitudes decay inversely with $\sqrt{r}$. In addition, there is a $\pi/4$ phase shift in these 2D radiation patterns. Therefore, if 2D synthetic seismograms are computed instead of 3D in an optimization problem where these synthetic data are iteratively matched with the real data as we previously discussed in Chapter 1, it is necessary to approximately correct these 2D synthetic data prior to matching with real data. We discuss these issues further later in the book.

For further details on the radiation patterns, see Fedorov (1968), Achenbach (1973), and Aki and Richards (2002).

3.11 Chapter Summary

In this chapter, we discussed the fundamentals of linearized elastic wave theory. Starting from the definitions of stress and strain tensors, we first defined the first fundamental equation of elasticity, known as the Hooke's law. We defined different elastic symmetries and the concepts of an equivalent medium. We then introduced the second fundamental equation in elasticity, known as the momentum equation and combined it with the Hookes's law to formulate the elastodynamic equation. We then thoroughly investigated the elastodynamic equation to study wave propagation in a homogeneous medium and radiation pattern from seismic sources.

3.12 Exercises

Problem 3.1 Starting from equation 3.7, derive equation 3.9.

Problem 3.2 In deriving equation 3.12, we mentioned about the vector identity $\mathbf{a} \times \mathbf{b} \times \mathbf{c} = \mathbf{c} \cdot (\mathbf{ab}) - \mathbf{c} \cdot (\mathbf{ba})$. Verify that this identity is true.

Problem 3.3 Using the vector identity $\mathbf{a} \times \mathbf{b} \times \mathbf{c} = \mathbf{c} \cdot (\mathbf{ab}) - \mathbf{c} \cdot (\mathbf{ba})$, derive equation 3.12.

Problem 3.4 Starting from the linearized Hooke's law (equation 3.17) and noting the symmetry properties of the elastic-stiffness tensor ($c_{ijkl} = c_{jikl} = c_{ijlk} = c_{jilk}$), derive the matrix equation 3.18.

Problem 3.5 Given the properties of the orthogonal transformation of a vector in three dimensions (equations 3.28 and 3.29), derive equation 3.30.

Problem 3.6 Starting from the orthogonal transformation equations for the stress and strain tensors given as $\tau'_{ij} = a_{ik}a_{jl}\tau_{kl}$ and $\epsilon'_{ij} = a_{ik}a_{jl}\epsilon_{kl}$, derive equations 3.32 and 3.34 with the matrices $\mathbf{M}$ and $\mathbf{N}$ given respectively by equations 3.33 and 3.35.

Problem 3.7 Show that if the matrices $\mathbf{M}$ and $\mathbf{N}$ are the Bond stress and strain transformation matrices, given respectively by the equations 3.33 and 3.35, then $\mathbf{N}^{-1} = \mathbf{M}^T$.

Problem 3.8 Derive the expressions for the matrices $\mathbf{M}_\varphi$ and $\mathbf{M}_\theta$, given respectively in equation 3.44.

Problem 3.9 Starting with the VTI medium, with the elastic-stiffness matrix represented by equation 3.51, formulate the Bond transformation matrix for obtaining the HTI medium with x_1 as the symmetry axis. Provide the elastic-stiffness matrix for such medium.

Problem 3.10 Derive the expressions for the elastic-stiffness coefficients C_{11}, C_{13}, C_{33}, C_{44}, and C_{66} for the VTI medium in terms of the Thomsen parameters V_{P0}, V_{S0}, ε, δ, and γ.

Problem 3.11 Derive the expressions for elastic-stiffness coefficients C_{11}, C_{22}, C_{33}, C_{12}, C_{13}, C_{23}, C_{44}, C_{55}, and C_{66} for an orthorhombic medium in terms of the Tsvankin–Thomsen parameters V_{P0}, V_{S0}, $\varepsilon^{(1)}$, $\varepsilon^{(2)}$, $\delta^{(1)}$, $\delta^{(2)}$, $\delta^{(3)}$, $\gamma^{(1)}$, and $\gamma^{(2)}$ defined by the equations 3.57–3.59.

Problem 3.12 VTI medium properties for two rock formations, (1) Taylor Sandstone and (2) Green River Shale, are given in Thomsen (1986), which are shown in the below table:

Rock Formation	V_{PO} (m/s)	V_{SO} (m/s)	ε	δ	γ	ρ (g/cm^3)
Taylor Sandston	3368	1829	0.11	−0.035	0.255	2.5
Green River Shale	3392	1768	0.195	−0.22	0.18	2.075

Using your preferred language, write a computer code to do the following:

1) From the Thomsen notation, calculate the elastic-stiffness coefficient matrix for both Taylor Sandstone and Green River Shale and print them out.
2) Write the code for computing the Bond transformation matrices $\mathbf{M}_\varphi$, $\mathbf{M}_\theta$, and $\mathbf{M}_\psi$ (equations 3.41, 3.44, and 3.46).
3) Compute the rotated stiffness matrices and print them for Taylor Sandstone and Green River Shale for the following three sets of the Euler angles φ, θ, and ψ:

$$\varphi = 60°, \quad \theta = 0°, \quad \psi = 0°.$$
$$\varphi = 0°, \quad \theta = 90°, \quad \psi = 90°.$$
$$\varphi = 0°, \quad \theta = 30°, \quad \psi = 60°.$$

Problem 3.13 For the Green River Shale, use V_{P0}, V_{S0}, and ρ as they are given in Problem 3.12 and make them orthorhombic by setting the Tsvankin–Thomsen parameters as $\varepsilon^{(1)} = 0.165$, $\varepsilon^{(2)} = 0.215$, $\delta^{(1)} = -0.2$, $\delta^{(2)} = -0.22$, $\delta^{(3)} = -0.24$, $\gamma^{(1)} = 0.16$, $\gamma^{(2)} = 18$, and repeat Problem 3.12.

Problem 3.14 From the positive-definite conditions for the VTI medium (equation 3.65), specialize them to those for an isotropic medium (equation 3.66).

Problem 3.15 Starting from the momentum equation in a source-free region, given by equation 3.71, derive the elastodynamic equation (equation 3.72).

Problem 3.16 Starting from the elastodynamic equation given as $\nabla_S \mathbf{C}(\nabla_S)^T\mathbf{u} = -\rho\omega^2\mathbf{u}$ (equation 3.72) with the operator ∇_S given by equation 3.76, derive the Christoffel equation 3.76 with the components of the Christoffel matrix $\boldsymbol{\Gamma}$ given in equation 3.79. Also verify that the determinant $|p^2\boldsymbol{\Gamma} - \rho\mathbf{I}|$ is given by equation 3.80.

Problem 3.17 Starting from equation 3.88, derive the solution to the P- and SV-wave slowness, given by equation 3.89.

Problem 3.18 Starting from the P- and SV-wave phase velocities V_P and V_{SV}, given in equation 3.92 with $D(\theta)$, given in equation 3.93, obtain their equivalent forms in Thomsen notation, provided via equations 3.94–3.96.

Problem 3.19 Obtain the weak-anisotropic expressions for the phase velocities V_P, V_{SV}, and V_{SH} shown in equations 3.93 and 3.98.

Problem 3.20 Using any coding language of your preference, write a code for computing both exact and weakly anisotropic phase velocities V_P, V_{SV}, and V_{SH} as functions of the phase angle θ for a VTI medium. Use your code and compute the exact and weakly anisotropic $V_P(\theta)$ for Taylor Sandstone and Green River Shale, parameters for which are provided in the table shown in Problem 3.12 and reproduce Figure 3.10. Repeat this exercise for $V_{SV}(\theta)$ and $V_{SH}(\theta)$.

Problem 3.21 From the expression for the unit normal vector to the slowness surface for the VTI medium (equation 3.120) and the definition of the group velocity (equation 3.111), obtain the expression of the group velocity for the VTI medium, given by equation 3.121.

Problem 3.22 From the expression of the group velocity vector shown in equation 3.121, derive its magnitude (equation 3.122) and the group angle (equation 3.123).

Problem 3.23 Starting from the SH-wave phase velocity for VTI medium, given as $V_{SH} = V_{S0}\sqrt{1 + 2\gamma \sin^2\theta}$ (equation 3.92), and setting $\lceil \mathbf{V}_{ph} \rceil = V_{SH}$ in equations 3.122 and 3.123, obtain the magnitude of the group velocity $\lceil \mathbf{V}_g \rceil$ and the group angle ϕ.

Problem 3.24 From the expression of $D^*(\theta)$, shown in equation 3.97, verify that it is equivalent to equation 3.127. Then, derive its derivative shown in equation 3.130.

Problem 3.25 Derive all the equations shown in equation 3.131.

Problem 3.26 Extend the code written in Problem 3.19 to compute the exact and approximate group velocities and group angles and reproduce Figure 3.14.

Problem 3.27 Prove the commutative properties of ∇^{-2} given by equation 3.142.

Problem 3.28 Derive equation 3.214.

Problem 3.29 Derive equation 3.219.

References

Achenbach, J.D. (1973). *Wave Propagation in Elastic Solids*. North Holland Publishing Company ISBN 7204-0325-1.

Aki, K. and Richards, P.G. (2002). *Quantitative Seismology*, Seconde. University Science Books ISBN 0-978-1891389634.

Auld, B.A. (1973). *Acoustic Fields and Waves in Solids*, vol. *1*. John Wiley & Sons ISBN 0-471-03700-1.

Backus, G.E. (1962). Long-wave elastic anisotropy produced by horizontal layering. *Journal of Geophysical Research 67*: 4427–4420. https://doi.org/10.1029/JZ067i011p04427.

Backus, G.E. and Mulcahy, M. (1976). Moment tensors and other phenomenological descriptions of seismic sources-I. Continuous displacements. *Geophysical Journal International 46* (2): 341–361. https://doi.org/10.1111/j.1365-246X.1976.tb04162.x.

Ben-Menahem, A. (1961). Radiation of seismic surface-waves from moving sources. *Bulletin of the Seismological Society of America 51*: 403–435. https://doi.org/10.1029/JZ067i001p00345.

Chapman, C.H. (2004). *Fundamentals of Seismic Wave Propagation*. Cambridge University Press ISBN 978-0521815383.

Chen, P., Zhao, L., and Jordan, T.H. (2005). Finite moment tensor of the 3 September 2002 Yorba Linda Earthquake. *Bulletin of the Seismological Society of America 95* (3): 1170–1180, https://doi.org/. doi: 10.1785/0120040094.

Chen, P., Jordan, T.H., and Zhao, L. (2010). Resolving fault plane ambiguity for small earthquakes. *Geophysical Journal International 181*: 493–501, https://doi.org/. doi: 10.1111/j.1365-246X.2010.04515.x.

Fedorov, F.I. (1968). *Theory of Elastic Waves in Crystals*. Springer ISBN 978-1475712773.

Mallick, S. and Frazer, L.N. (1987). Practical aspects of reflectivity modeling. *Geophysics 52*: 1355–1364. https://doi.org/10.1190/1.1442248.

Mallick, S. and Frazer, L.N. (1988). Rapid computation of multioffset vertical seismic profile synthetic seismograms for layered media. *Geophysics 53*: 479–491. https://doi.org/10.1190/1.1442479.

Mallick, S. and Frazer, L.N. (1990). Computation of synthetic seismograms for stratified azimuthally anisotropic media. *Journal of Geophysical Research 95*: 8513–8526. https://doi.org/10.1029/JB095iB06p08513.

Mallick, S. and Frazer, L.N. (1991). Reflection/transmission coefficients and azimuthal anisotropy in marine seismic studies. *Geophysical Journal International 105*: 241–252. https://doi.org/10.1111/j.1365-246X.1991.tb03459.x.

McGuire, J.J., Zhao, L., and Jordan, T.H. (2001). Teleseismic inversion for the second-degree moments of earthquakes. *Geophysical Journal International 145*: 661–678. https://doi.org/10.1046/j.1365‐246x.2001.01414.x.

Michelena, R.J., Gilman, J.R., and Zahm, C.K. (2019). Seismic, geologic, geomechanics, and dynamic constraints in flow models of unconventional fractured reservoirs: Example from a south Texas field. *The Leading Edge 38* (2): 116–122. https://doi.org/10.1190/tle38020116.1.

Musgrave, M.J. (1970). *Crystal Acoustics: Introduction to the Study of Elastic Waves and Vibrations in Crystals*. Holden Day ISBN 978-0816262021.

Schaeffer, A.J., Lebedev, A., and Becker, T.W. (2016). Azimuthal seismic anisotropy in the Earth's upper mantle and the thickness of tectonic plates. *Geophysical Journal International 207*: 901–933. https://doi.org/10.1093/gji/ggw309.

Schoenberg, M. (1983). Reflection of elastic waves from periodically stratified media with interfacial slip. *Geophysical Prospecting 31*: 2650292. https://doi.org/10.1111/j.1365-2478.1983.tb01054.x.

Schoenberg, M. and Douma, J. (1988). Elastic wave propagation in media with parallel fractures and aligned cracks. *Geophysical Prospecting 36*: 571–590. https://doi.org/10.1111/j.1365‐2478.1988.tb02181.x.

Schoenberg, M. and Sayers, C.M. (1995). Seismic anisotropy of fractured rock. *Geophysics 60*: 204–211. https://doi.org/10.1190/1.1443748.

Selley, R.C. and Sonnenberg, S.A. (2014). *Elements of Petroleum Geology*. Elsevier ISBN 978-0123860316.

Silver, P. and Jordan, T.H. (1983). Total-moment spectra of fourteen large earthquakes. *Journal of Geophysical Research 88*: 3273–3293. https://doi.org/10.1029/JB088iB04p03273.

Thomsen, L.A. (1986). Weak elastic anisotropy. *Geophysics 51*: 1954–1966. https://doi.org/10.1190/1.1442051.

Tsvankin, I. (1997). Anisotropic parameters and P-wave velocity for orthorhombic media. *Geophysics 62*: 1292–1309. https://doi.org/10.1190/1.1444231.

4

Computation of Synthetic Seismograms in Inhomogeneous Medium: Approximate (Partial) Solutions

4.1 Introduction

In Chapter 3, we discussed propagation of elastic waves in a homogeneous medium. By defining the stress and strain tensors, we first formulated the linearized theory of elasticity, given as Hooke's law. We also looked at the symmetry properties to define different anisotropic elastic systems and discussed the concepts of orthogonal transformation and equivalent medium. Then, by combining Hooke's law with the equations of motion, we formulated the elastodynamic equation for an elastic continuum and discussed wave propagation in a homogeneous medium for different anisotropic systems and the radiation patterns from seismic sources in a homogeneous isotropic elastic medium.

Primary motivation behind Chapter 3 was to understand the propagation of elastic waves in a homogeneous elastic medium. However, the earth is not homogeneous and is characterized by inhomogeneity at various scales from continuous variations of elastic properties to sharp discontinuities separating different rock types. We use seismic data to estimate these subsurface rock properties, two essential components of which are (1) inversion and (2) rock physics. As we discussed in Chapter 1, to correctly handle noisy seismic data, it is necessary to use a model-based inversion or optimization method, which requires forward computation of the synthetic seismic response for a given model. Therefore, a thorough understanding of computing synthetic seismic response from a given earth model with depth varying elastic properties is important. In this chapter, we begin our discussion of computing synthetic seismic responses for such media. In this chapter, we strictly restrict to the approximate methods, which allow computing partial seismic responses and postpone discussion of computing complete responses to Chapter 5.

While inversion allows estimating the elastic properties like the P- and S-wave velocity, density, and other anisotropic properties, the properties that are of interest to a seismic interpreter are the rock properties, such as porosity, permeability, minerology, types of fluids contained within the rock pores, etc. This topic of mapping the elastic properties that are obtainable from seismic data onto the rock properties is rock physics. We postpone the discussion of the fundamentals of rock physics to the latter part of the book.

4.2 Fundamentals of Ray Theory and Computation of Partial Synthetic Seismic Response

Ray theory is based on the propagation of the energy along the ray paths, which are perpendicular to the wave fronts, the concept of which we already developed in Chapter 3 when

Computation, Optimization, and Machine Learning in Seismology, Advanced Textbook 8, First Edition. Subhashis Mallick.

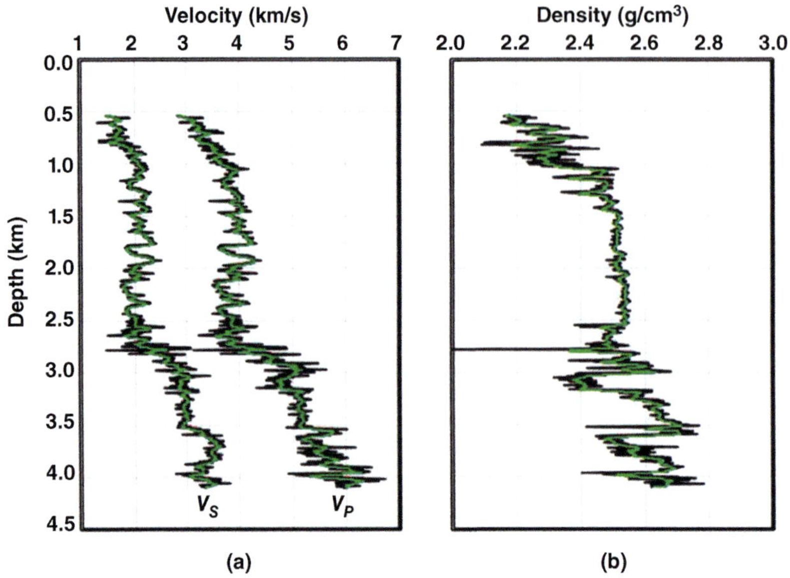

Figure 4.1 Real well-log measurements in black and their average properties in green. (a) V_P and V_S. (b) ρ.

we introduced the difference between the group and phase directions in an anisotropic elastic medium. We emphasize again, for the propagating wavefields, the wave fronts are the surfaces of constant phase, or, in other words, they are parallel to the slowness surface. The rays, on the other hand, are directed along the outward normal to the wave fronts. As the P and S waves propagate along the rays in an inhomogeneous medium, they come across discontinuities. Additionally, there are also regions on the earth, which can be described by systematic and continuous changes in the elastic properties. In global seismology, for example, both P- and S-wave velocities increase by almost a factor of two between the top and the bottom of the mantle. In exploration seismology, where we deal with much shallower regions of the earth's crust, we still encounter similar changes.

Figure 4.1, for example, shows a real well-log data with borehole measurements of the P- and S-wave velocity (V_P and V_S) and density (ρ) as functions of depth at a certain location. The black curves, shown in this figure, are the real measurements of these properties at 0.125 m depth intervals showing fine layering with varying V_P, V_S, and ρ. As we previously discussed in Chapter 3 in the context of an equivalent medium, propagating seismic waves do not see these thin individual layers, and the green curves in Figure 4.1 are computed by taking a moving average of the respective properties, which the seismic data are expected to see. As can be seen from these averaged values, in addition to sharp variations, there are also smooth variations of V_P, V_S, and ρ throughout the depth range of the well-log measurement.

These sharp discontinuities and more smooth and continuous variations of elastic properties distort the radiation patterns for both P and S wavefields, which we discussed for a homogeneous medium in Chapter 3. At sharp discontinuities, both P and S wavefields are reflected and transmitted, and at smooth discontinuities, the wavefield shapes are modified with propagation (Aki and Richards, 2002).

4.2.1 Snell's Law

Before we begin our discussion on ray theory and how to compute ray-theoretical synthetic seismic responses, it is first important to discuss one fundamental law in geometrical optics, known as Snell's law. Complete proof of Snell's law can be found in any standard book in optics (for example, Jenkins and White, 2001). Here, we state this law with reference to the seismic wave propagation.

Consider an interface separating two isotropic elastic media with different properties as illustrated in Figure 4.2. Also assume that the medium above the interface is characterized by the P-wave velocity α_1, S-wave velocity β_1, and density ρ_1, and the one below the interface is characterized by α_2, β_2, ρ_2. Now, if a plane P wave is incident on the interface, represented by the ray at an angle of incidence θ_i, it reflected P and S waves with angles of reflection θ_r and ϑ_r above the interface and transmitted P and S waves with transmission angles ξ_t and ζ_t below the interface (see Figure 4.2). The Snell's law says that the sine of the angle (incidence, reflection, and transmission) over the velocity for the incident, reflected, and transmitted ray paths is constant, which allows relating the angle of incidence θ_i to the angles of reflection (θ_r, ϑ_r) and transmission (ξ_t, ζ_t) as

$$\frac{\sin\theta_i}{\alpha_1} = \frac{\sin\theta_r}{\alpha_1} = \frac{\sin\vartheta_r}{\beta_1} = \frac{\sin\xi_t}{\alpha_2} = \frac{\sin\zeta_t}{\beta_2} = const. \tag{4.1}$$

We discuss the problem of reflection and transmission later in this chapter and their derivations in Chapter 5, and here we just state Snell's law that allows us to trace rays in a homogeneous elastic medium. The constant, given by equation 4.1 is called the *ray parameter* that is usually designated by the symbol p. Note that in our solution to the Christoffel equation, we designated p as the slowness, given as the inverse of the phase velocity ($p = 1/V_{ph}$). For an isotropic medium, there is no difference between the phase and group velocities and therefore $\alpha_1, \alpha_2, \beta_1, \beta_2$, shown in Figure 4.2 are the P- and S-wave phase velocities, and it is important to clarify the difference between the ray-parameter p represented by equation 4.1 with the slowness p we previously defined in Chapter 3.

Referring to a right-handed Cartesian coordinate system, shown in Figure 4.3, consider the wave propagating in a vertical plane with azimuth φ at a phase angle θ, shown by the red arrow in Figure 4.3. For a given $\{\theta, \varphi\}$ pair, the solution to the Christoffel equation gives the slowness vector given as

$$\mathbf{p} = \frac{1}{V_{ph}}(\hat{\mathbf{x}}\sin\theta\cos\varphi + \hat{\mathbf{y}}\sin\theta\sin\varphi + \hat{\mathbf{z}}\cos\theta). \tag{4.2}$$

Clearly, x, y, and z components of the slowness vector, shown in Figure 4.3, are given by

$$\left.\begin{aligned} p_x &= \frac{\sin\theta\cos\varphi}{V_{ph}} \\ p_y &= \frac{\sin\theta\sin\varphi}{V_{ph}} \\ p_z &= \frac{\cos\theta}{V_{ph}} \end{aligned}\right\}. \tag{4.3}$$

Additionally, as shown in Figure 4.3, the p_x and p_y combines to a total horizontal slowness vector, oriented at an angle φ with respect to

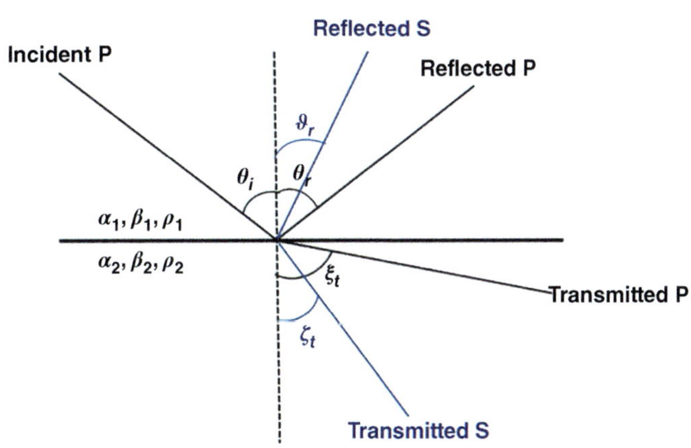

Figure 4.2 An interface separating two isotropic media with different elastic properties where an incident P wave generates reflected and transmitted P and S waves.

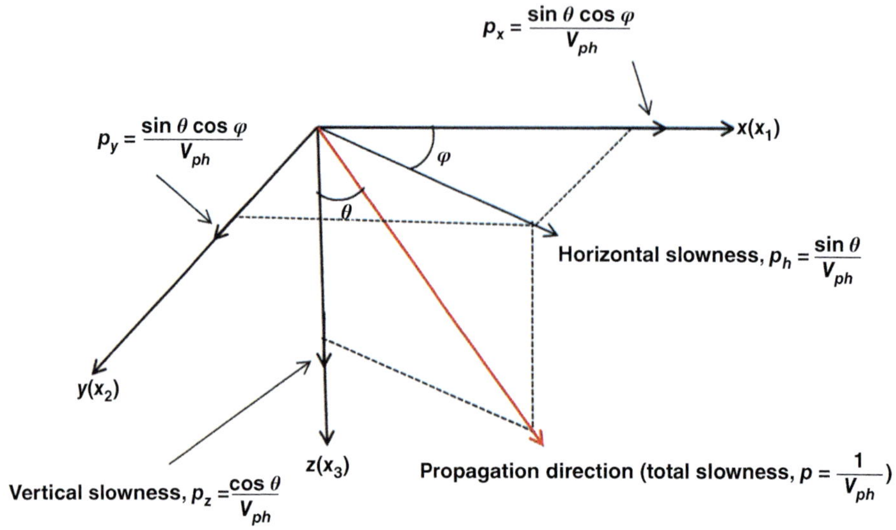

Figure 4.3 Relation between the total slowness and the ray parameter.

the *x* coordinate axis with magnitude

$$p_h = \sqrt{p_x^2 + p_y^2} = \frac{\sin\theta}{V_{ph}}. \tag{4.4}$$

Comparing equation 4.4 with Snell's law (equation 4.1), we immediately identify that the ray parameter is the horizontal slowness. Snell's law therefore states that the horizontal slowness for the ray path is always preserved. To be consistent with the seismology literature, we use the symbol p for the ray parameter, keeping in mind that it is the horizontal component of the slowness vector **p**, which we previously defined in relation with the Christoffel equation. We also generalize Snell's law for an arbitrary anisotropic medium and state that, for a seismic wave propagating with a phase velocity V_{ph} at phase angle θ, the ray parameter, defined as $p = \sin\theta / V_{ph}$, is constant.

4.2.2 Sign Convention for the Fourier Transforms in Seismology

In Chapter 3, we recognized that for a plane wave, propagating in an arbitrary direction outward with the unit vector $\hat{\mathbf{l}} = \sin\theta\cos\varphi\hat{\mathbf{x}} + \sin\theta\sin\varphi\hat{\mathbf{y}} + \cos\theta\hat{\mathbf{z}}$ (Figure 3.9, equation 3.74), the wavefield is proportional to $e^{i\omega(|\mathbf{p}|\hat{l}\cdot\mathbf{r}-t)}$, where **r** is the position vector, t is time, and $|\mathbf{p}|$ is the slowness vector. Now,

$$|\mathbf{p}|\hat{\mathbf{l}}\cdot\mathbf{r} = p_x x + p_y y + p_z z, \tag{4.5}$$

in which p_x, p_y, and p_z are defined by equation 4.3. Therefore,

$$e^{i\omega(|\mathbf{p}|\hat{\mathbf{l}}\cdot\mathbf{r}-t)} = e^{i\omega(p_x x + p_y y + p_z z - t)}. \tag{4.6}$$

Defining $k_x = \omega p_x$, $k_y = \omega p_y$, and $k_z = \omega p_z$, we can write equation 4.6 as

$$e^{i\omega(|\mathbf{p}|\hat{\mathbf{l}}\cdot\mathbf{r}-t)} = e^{i(k_x x + k_y y + k_z z - \omega t)}. \tag{4.7}$$

It is obvious from equation 4.7 that for a propagating wavefield in the outward direction away from the source, k_x, k_y, and k_z are spatial equivalents of the temporal frequency ω. Therefore, k_x, k_y, and k_z are called *spatial frequencies* or *wave numbers* in *x*, *y*, and *z* directions for a propagating wavefield in three dimensions. Also note from equation 4.7 that the sign of the wave numbers k_x, k_y, and k_z is opposite to that of ω. This, in turn, leads to the sign convention for the forward and inverse Fourier transforms in the spatial domain to be opposite to that of time domain for a propagating wavefield. In seismology, we acquire seismic data in space and time, which we may denote as $S(x, y, z, t)$. Therefore, if we define $S(k_x, k_y, k_x, \omega)$ as the seismic data in the frequency–wave number domain, then from the time–frequency domain Fourier transform pair $f(t) \leftrightarrow f(\omega)$ given by equation 2.43, we write the Fourier transform pair for $S(x, y, z, t) \leftrightarrow S(k_x, k_y, k_x, \omega)$ as

$$\left.\begin{aligned} S(x,y,z,t) &= \int_{-\infty}^{+\infty}\int_{-\infty}^{+\infty}\int_{-\infty}^{+\infty}\int_{-\infty}^{+\infty} S(k_x,k_y,k_z,\omega) \\ &\times e^{i(-k_x x-k_y y-k_z z+\omega t)}dk_x dk_y dk_z d\omega \\ S(k_x,k_y,k_z,\omega) &= \frac{1}{(2\pi)^4}\int_{-\infty}^{+\infty}\int_{-\infty}^{+\infty}\int_{-\infty}^{+\infty}\int_{-\infty}^{+\infty} S(x,y,z,t) \\ &\times e^{i(k_x x+k_y y+k_z z-\omega t)}dxdydzdt \end{aligned}\right\}. \tag{4.8}$$

We use the above sign convention for the temporal and spatial Fourier transforms in this book. In Chapter 5, we consider the computation of complete seismic response for a horizontally stratified medium, where we use $S(x, y, z, t) \leftrightarrow S(p_x, p_y, z, \omega)$ Fourier transform pair as

$$\left.\begin{aligned} S(x,y,z,t) &= \int_{-\infty}^{+\infty}\omega^2 d\omega\int_{-\infty}^{+\infty}dp_x\int_{-\infty}^{+\infty} \\ &\times S(p_x,p_y,z,\omega)e^{i\omega(-p_x x-p_y y+t)}dp_y \\ S(p_x,p_y,z,\omega) &= \frac{1}{(2\pi)^3}\int_{-\infty}^{+\infty}dt\int_{-\infty}^{+\infty}dx\int_{-\infty}^{+\infty} \\ &\times S(x,y,z,t)e^{i\omega(p_x x+p_y y-t)}dy \end{aligned}\right\}. \tag{4.9}$$

Note that in equation 4.9 we used the relations $k_x = \omega p_x$ and $k_y = \omega p_y$ and transformed the wave number variables k_x and k_y to ray-parameter (slowness) variables p_x and p_y.

4.2.3 Ray Tracing in a Smoothly Varying Laterally Homogeneous Medium

To account for how the inhomogeneity affects radiation patterns, we first want to derive the expressions for the amplitude, travel time, and distance functions for the particle motion **u** in a medium with smooth and continuous variations of the elastic properties. For simplicity, we restrict to an isotropic medium, where the P-wave velocity (α), S-wave velocity (β), and density (ρ) can be expressed as continuous functions of depth (z) only and are given as

$$\left.\begin{aligned} \alpha &= \alpha(z) \\ \beta &= \beta(z) \\ \rho &= \rho(z) \end{aligned}\right\}. \tag{4.10}$$

In addition, we also assume a point double-couple (moment-tensor) source, located at the position $\mathbf{r}_0$ with a body force density given as

$$\mathbf{f} = -\mathbf{M}\cdot\nabla\delta(\mathbf{r}-\mathbf{r}_0). \tag{4.11}$$

In a homogeneous medium, the wave fronts do not get distorted and therefore the rays are straight lines. For a medium with continuous change of the elastic properties, shown by equation 4.10 on the other hand, the wave front shapes will change with distance and hence the rays will, in general, be curves. In Figure 4.4, we show such a curve in purple. At the source location $\mathbf{r}_0$, we define the ray with the triplet $\{\hat{\boldsymbol{\eta}}_0, \hat{\boldsymbol{\theta}}_0, \hat{\boldsymbol{\varphi}}\}$, where $\hat{\boldsymbol{\eta}}_0$ is the unit vector along the radial direction and tangent to the ray at $\mathbf{r}_0$, $\hat{\boldsymbol{\theta}}_0$ is the unit vector along the colatitude, and $\hat{\boldsymbol{\varphi}}$ is the unit vector in the direction of the azimuth. In our discussion of the Christoffel equation for VTI medium in Chapter 2, the azimuth φ of the P, SV, and SH waves does not change with propagation. And, because an isotropic medium is a special case of a VTI medium where the dimensionless parameters ε, δ, and γ are zero, the azimuth of the ray in an isotropic medium also does not change with direction in a laterally invariant medium described by equation 4.10. Therefore, the ray direction at any point P away from the source location can be described by the unit vector triplet $\{\hat{\mathbf{t}}, \hat{\boldsymbol{\theta}}, \hat{\boldsymbol{\varphi}}\}$, as shown in Figure 4.4.

We now consider a unit focal sphere (see Figure 4.5a) and designate the points in this sphere by $\hat{\boldsymbol{\eta}}_0(\theta_0, \varphi)$. Thus, $\hat{\boldsymbol{\eta}}_0$ plays the role that the unit vector $\hat{\mathbf{r}}$ plays in a homogeneous medium. We also use the symbol p (ray parameter) to denote $\sin\theta/\alpha(z)$, which, according to Snell's law, remains constant along the direction of the ray path. Let us now consider a *ray tube*, generated from the source propagating P

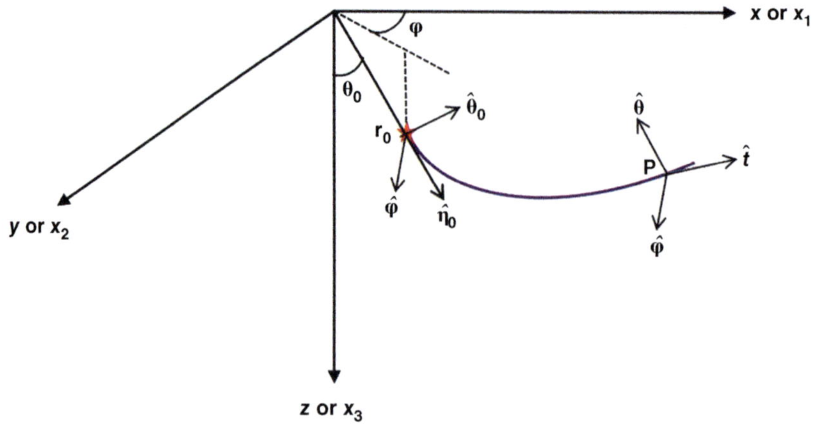

Figure 4.4 Ray path in a medium with continuous change of elastic properties, shown as a purple curve. At the source location $\mathbf{r}_0$, shown by a red asterisk, the ray direction is defined by the unit tangent vector $\hat{\boldsymbol{\eta}}_0$, unit colatitude vector $\hat{\boldsymbol{\theta}}_0$, and unit azimuth vector $\hat{\boldsymbol{\varphi}}$. For any other arbitrary point P in the ray path, they are respectively $\hat{\mathbf{t}}, \hat{\boldsymbol{\theta}}$, and $\hat{\boldsymbol{\varphi}}$.

waves, which, at any distance s away from the source, has a cross-sectional area $dA = dA(s)$, as shown in Figure 4.5a. Because this distance s is a function of the travel time (t), we write $dA = dA(t)$, which, as shown in Figure 4.5a, leaves the source with solid angle $d\Omega$ in direction $\hat{\boldsymbol{\eta}}_0$. From Figure 4.5b, this solid angle is given as

$$d\Omega = \sin\theta_0 d\theta_0 d\varphi. \tag{4.12}$$

The geometrical optics assumption requires that the energy flux to be conserved along the ray tube, as shown in Figure 4.5. The P-wave energy density in the ray tube is given by

$$E = \rho\omega^2 |\mathbf{u}_P|^2. \tag{4.13}$$

In equation 4.13, $\rho = \rho(z)$ is the density, ω is the temporal frequency, and $\mathbf{u}_P$ is the particle displacement vector for the P wave.

Therefore, the P-wave energy flux D, which is conserved along the ray tube, is given by

$$D = E\alpha dA. \tag{4.14}$$

Because this energy flux D is the same anywhere along the ray tube, the flux $D(r)$ at any arbitrary position is equal to the limit of the flux as we move toward the source. Therefore, we can write

$$D(r) = \lim_{r \to r_0} \{\rho(r)\omega^2 |\mathbf{u}(r)|^2 \alpha(r) dA(r)\}. \tag{4.15}$$

But, from Figure 4.5, we have

$$dA(r) = |\mathbf{r} - \mathbf{r}_0|^2 d\Omega. \tag{4.16}$$

And, from our discussion of the radiation patterns in Chapter 3 (see equation 3.224), we have

$$\mathbf{u}_P = \frac{i\omega e^{i\omega|\mathbf{r}-\mathbf{r}_0|\alpha} [\hat{\mathbf{t}}\hat{\boldsymbol{\eta}}_0\hat{\boldsymbol{\eta}}_0 : \mathbf{M}(\omega)]}{4\pi\rho(r)\alpha^3(r)|\mathbf{r} - \mathbf{r}_0|}. \tag{4.17}$$

Note that to obtain equation 4.17 from the far-field P-wave radiation pattern shown in equation 3.224, we replaced $\hat{\mathbf{r}}$ with $\hat{\boldsymbol{\eta}}_0$ and used the symmetry property of the moment tensor $\mathbf{M}(\omega)$. From equations 4.16 and 4.17, we thus write equation 4.15 as

$$D(\mathbf{r}) = \lim_{r \to r_0} \rho(r)\omega^2 \times \left[\frac{i\omega e^{i\omega|\mathbf{r}-\mathbf{r}_0|\alpha(r)} \{\hat{\mathbf{t}}\hat{\boldsymbol{\eta}}_0\hat{\boldsymbol{\eta}}_0 : \mathbf{M}(\omega)\}}{4\pi\rho(r)\alpha^3(r)|\mathbf{r} - \mathbf{r}_0|}\right]^2 \times \alpha(r)|\mathbf{r} - \mathbf{r}_0|^2 d\Omega. \tag{4.18}$$

Note that the term inside the square bracket on the right-hand side of equation 4.18 is the far-field P-wave displacement field for a moment-tensor source. We can also obtain the equivalent expressions for the S-wave field for the moment-tensor source by replacing $\mathbf{u}_P$ with $\mathbf{u}_S$ from equation 2.224. Similarly, we can obtain the equivalent expressions for other

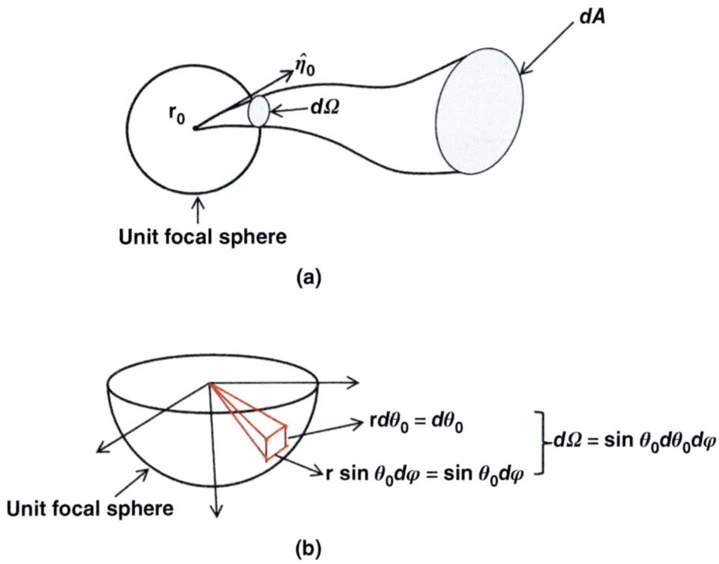

Figure 4.5 (a) Illustration of a ray tube with a cross-sectional area dA and a solid angle $d\Omega$ on the unit focal sphere around the source region. (b) Derivation of the solid angle $d\Omega = \sin\theta_0 d\theta_0 d\varphi$.

source types, such as a point explosion or a point force, by using the appropriate terms for the displacement field for those sources, given in Chapter 3. In doing all these, note that the displacement field used in ray theory is the far-field term. But as we move near the source by taking the limit in equation 4.18, we should strictly take the other terms, which is ignored in ray theory, and that is why the amplitudes predicted by ray theory are approximate, not exact.

Going back to equation 4.18, and taking the limit, we obtain

$$D(\mathbf{r}) = \frac{\omega^4 \left|\hat{\mathbf{t}}\hat{\boldsymbol{\eta}}_0\hat{\boldsymbol{\eta}}_0 : \mathbf{M}(\omega)\right|^2}{(4\pi)^2\rho_0^2\alpha_0^2} d\Omega. \qquad (4.19)$$

In equation 4.19, ρ_0 and α_0 respectively stand for the density and P-wave velocity at the location of the source at $\mathbf{r} = \mathbf{r}_0$. But, from equation 4.14 $D(\mathbf{r}) = E(\mathbf{r})\alpha(\mathbf{r})dA$. And, from equation 4.13, $E(\mathbf{r}) = \rho(\mathbf{r})\omega^2|\mathbf{u}_P(\mathbf{r})|^2$. Therefore,

$$D(\mathbf{r}) = \rho(\mathbf{r})\omega^2|\mathbf{u}_P(\mathbf{r})|^2\alpha(\mathbf{r})dA. \qquad (4.20)$$

Therefore,

$$|\mathbf{u}_P(\mathbf{r})|^2 = \frac{\omega^2|\hat{\mathbf{t}}\hat{\boldsymbol{\eta}}_0\hat{\boldsymbol{\eta}}_0 : \mathbf{M}(\omega)|^2}{(4\pi)^2\rho_0\alpha_0^5\alpha(\mathbf{r})(dA/d\Omega)}. \qquad (4.21)$$

By taking the square root considering that P-wave is longitudinal and multiplying by an appropriate phase lag, we solve for the P-wave displacement field $\mathbf{u}_P$ at a position $\mathbf{r}$ due to a source located at $\mathbf{r} = \mathbf{r}_0$

$$\mathbf{u}_P = \frac{i\omega\hat{\mathbf{t}}\hat{\boldsymbol{\eta}}\hat{\boldsymbol{\eta}} : \mathbf{M}(\omega)e^{i\omega T}}{4\pi\sqrt{\rho_0\rho(\mathbf{r})}\{\alpha_0^5\alpha(\mathbf{r})\}^{1/2}(dA/d\Omega)^{1/2}}. \qquad (4.22)$$

In equation 4.22, T is the P-wave travel time from source location ($\mathbf{r}_0$) to the location $\mathbf{r}$, that is

$$T = \int_{r_0}^{r} \frac{dS}{\alpha(\mathbf{r})} \qquad (4.23)$$

along the ray path. It can be readily seen that for a homogeneous medium, equation 4.22 leads to the far-field radiation pattern for the moment-tensor source given in equation 3.224.

Before moving further, it is first necessary to find a convenient expression for $d\Omega/dA$ in the denominator of equation 4.22.

As shown in Figure 4.6a, we introduce the function $\mathbf{X}(\hat{\boldsymbol{\eta}}_0, T)$ in $\Re^3$, defined by the requirements that for a fixed $\hat{\boldsymbol{\eta}}_0$ (which according to Figure 4.5a is a function of θ_0 and φ), and varying time T, $\mathbf{X}(\hat{\boldsymbol{\eta}}_0, T)$ is a ray, and for a fixed T and varying $\hat{\boldsymbol{\eta}}_0$, $\mathbf{X}(\hat{\boldsymbol{\eta}}_0, T)$ is a wave front. Therefore, for fixed $\hat{\boldsymbol{\eta}}_0$, any point lying

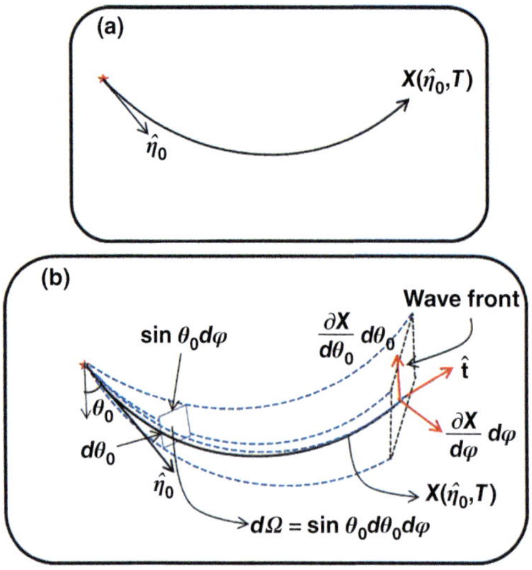

Figure 4.6 Illustration of the ray and wave front. (a) X as a function of a fixed $\hat{\boldsymbol{\eta}}_0$ and varying t. Different points of the function $\mathbf{X}$ for different times t forms a curve, which is the ray path. (b) Infinitesimal ray tube defined around the ray path, which leaves the source at time $t=0$ and propagates outward. At any fixed time t, these ray paths from a surface, shown as dashed black, is the wave front. This ray rube also forms a solid angle, shown in shaded gray, given as $d\Omega = \sin\theta_0 d\theta_0 d\varphi$. Note in (b) that $\hat{\mathbf{t}}$ is the unit tangent vector and T is time. The source location is represented by a red asterisk in both figures.

within $\mathbf{X}(\hat{\boldsymbol{\eta}}_0, T)$ is the point in space where the energy that left the source at time $T=0$ with tangent $\hat{\boldsymbol{\eta}}_0$ is now at time T (see Figure 4.6a). Considering the ray tube around this ray at any time T is the wave front (Figure 4.6b) with the cross-sectional area (see Figure 4.6b)

$$dA = \hat{\mathbf{t}} \cdot \left\{ \frac{\partial \mathbf{X}(\hat{\boldsymbol{\eta}}_0, T)}{\partial \theta_0} d\theta_0 \times \frac{\partial \mathbf{X}(\hat{\boldsymbol{\eta}}_0, T)}{\partial \varphi} d\varphi \right\}. \tag{4.24}$$

Figure 4.7 shows a two-dimensional (2D) cross-section of a ray and the corresponding ray tube generated from the source location at $\mathbf{r}_0$ (shown by a red asterisk) and arriving at a receiver location (shown by a brown circle) as a lateral distance Δ away from the source.

As shown in Figure 4.7, we use dA_1 to denote the cross-sectional area of the ray tube by the horizontal plane containing the receiver and dA to denote the cross-sectional area of the ray tube by the intersection of the wave front containing the receiver. From Figure 4.7, it is evident that

$$dA = dA_1 \hat{\mathbf{t}} \cdot \hat{\mathbf{z}}, \tag{4.25}$$

with $\hat{\mathbf{t}}$ and $\hat{\mathbf{z}}$ respectively being the unit vectors along the tangent to the ray at the receiver location and along the vertical direction (see Figure 4.7). Now, dA_1 in equation 4.25 is equivalent to dA we previously defined via Figure 4.6 and equation 4.24. Therefore,

$$dA = \hat{\mathbf{t}} \cdot \left(\frac{\partial \mathbf{X}}{\partial \theta_0} d\theta_0 \right) \times \left(\frac{\partial \mathbf{X}}{\partial \varphi} d\varphi \right). \tag{4.26}$$

We also note that the ray tube shown in Figure 4.7 is represented by the solid angle $d\Omega$ on focal sphere, given by equation 4.16.

In Figure 4.8, we elaborate Figure 4.7, showing the ray with unit tangent vectors at the source and receiver location oriented at angles θ_0 and θ_r with respect to the vertical (z) axis. From this figure, we readily express the first term on the right-hand side of equation 4.26 as

$$\left| \frac{\partial \mathbf{X}}{\partial \theta_0} d\theta_0 \right| = \left| \frac{\partial \mathbf{D}}{\partial \theta_0} d\theta_0 \right| \cos\psi. \tag{4.27}$$

In equation 4.27, $\mathbf{D}$ is the vector representing the lateral distance from receiver to source and pointing toward the source, and the angle ψ shown in Figure 4.8. In Figure 4.8, the segments $(\partial \mathbf{X}/\partial \theta_0)d\theta_0$ and $(\partial \mathbf{D}/\partial \theta_0)d\theta_0$ are represented by red and blue arrows. Because the ray is perpendicular to the wave front at the receiver location, it implies that

$$\psi = \pi - \theta_r$$

and therefore

$$\cos\psi = -\cos\theta_r. \tag{4.28}$$

Additionally, from Figure 4.8, we also have

$$\left| \frac{\partial \mathbf{X}}{\partial \varphi} d\varphi \right| = \mathbf{D} d\varphi. \tag{4.29}$$

From equations 4.16, 4.26–4.29, we therefore have

$$\frac{dA}{d\Omega} = -\frac{\left(\frac{\partial D}{\partial \theta_r} \right) \cos\theta_r D}{\sin\theta_r}. \tag{4.30}$$

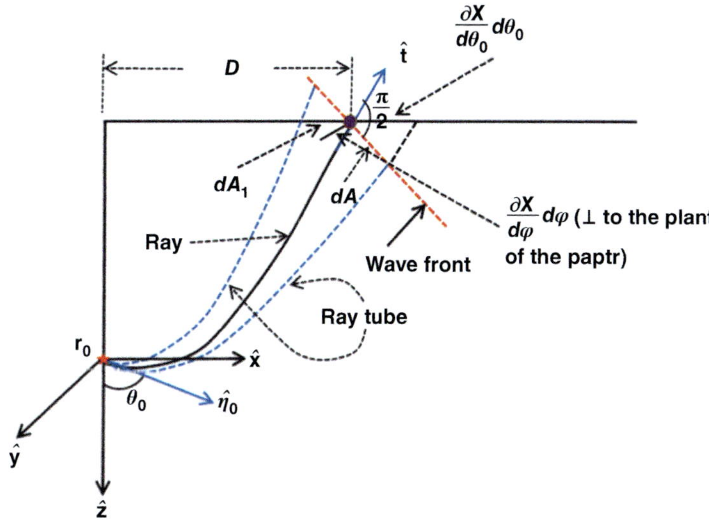

Figure 4.7 Two-dimensional cross-section of a ray and the corresponding ray tube from the source to receiver location. The source is represented by a red asterisk, and the receiver is represented by a brown circle. The lateral distance between the source and receiver is denoted by *D*.

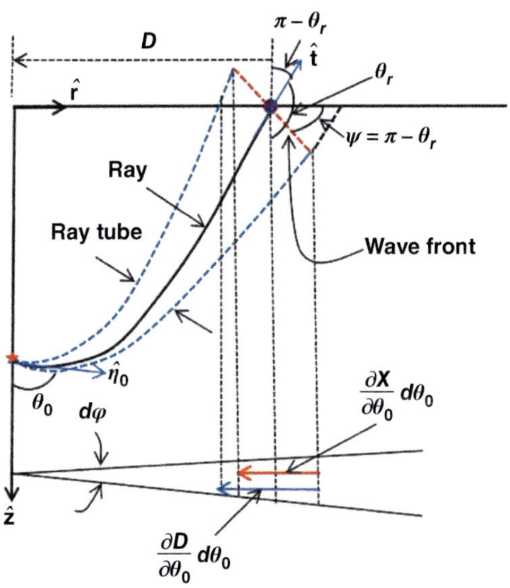

Figure 4.8 Same as Figure 4.7 but with elaboration of different angles and the line segments.

In equation 4.30, D is the magnitude of the vector **D**. Although equation 4.30 expresses $d\Omega/dA$ in terms of the lateral distance D and the angle θ_r, we need a more convenient expression for use in equation 4.22 and compute the displacement field. To do so, we use the ray-parameter p. Because this parameter is constant for the ray path, we must have $p = \sin\theta_r/\alpha_r = \sin\theta_0/\alpha_0$, where α_r and α_0 are the P-wave velocities at the receiver and source locations. Thus,

$$\left.\begin{aligned} \sin\theta_r &= p\alpha_r \\ \cos\theta_r &= \sqrt{1-p^2\alpha_r^2} \\ \sin\theta_0 &= p\alpha_0 \\ \cos\theta_0 &= \sqrt{1-p^2\alpha_0^2} \end{aligned}\right\}. \tag{4.31}$$

In terms of ray parameter, it is straightforward to show that

$$\frac{\partial D}{\partial \theta_0} = \frac{\partial D}{\partial p}\frac{\sqrt{1-p^2\alpha_0^2}}{\alpha_0}\operatorname{sgn}\left(\frac{\pi}{2}-\alpha_0\right), \tag{4.32}$$

using which we then have

$$\frac{dA}{d\Omega} = -\frac{(\partial_p D)\Delta q_r q_0 \operatorname{sgn}\left(\frac{\pi}{2}-\theta_0\right)}{p\alpha_0}, \tag{4.33}$$

where

$$\left.\begin{aligned} q_r &= \sqrt{\frac{1}{\alpha_r^2}-p^2} \\ q_0 &= \sqrt{\frac{1}{\alpha_0^2}-p^2} \\ \partial_p D &= \frac{\partial D}{\partial p} \end{aligned}\right\}. \tag{4.34}$$

Or

$$\left(\frac{dA}{d\Omega}\right)^{-1/2} = \frac{p^{1/2}\alpha_0^{1/2}}{\left\{\sqrt{\operatorname{sgn}\left(\theta_0 - \frac{\pi}{2}\right)\partial_p D \alpha_r^{1/2} q_r^{1/2} q_0^{1/2} D^{1/2}}\right\}}. \tag{4.35}$$

Note that we have not used any property of the P wave in deriving equation 4.35. Therefore, the same equation is applicable for S waves if we replace α_r with β_r and α_0 with β_0.

Using equation 4.35, we write from equation 4.16 the far-field ray-theoretical radiation for P waves at the receiver location as

$$\mathbf{u}_P(\mathbf{r}) \approx \frac{p^{1/2}[-i\omega\hat{\mathbf{t}}\hat{\boldsymbol{\eta}}_0\hat{\boldsymbol{\eta}}_0 : \mathbf{M}(\omega)]e^{i\omega\Upsilon(p)}}{4\pi\sqrt{\rho_r\rho_0}\alpha_r\alpha_0^3\sqrt{\operatorname{sgn}\left(\theta_0 - \frac{\pi}{2}\right)\partial_p D \times D^{1/2} q_0^{1/2} q_r^{1/2}}}. \tag{4.36}$$

In equation 4.36, $\Upsilon(p)$ is the total P-wave travel time from the source to the receiver. The ray-theoretical approximation for the motion due to S waves can be similarly obtained as

$$\mathbf{u}_S(\mathbf{r}) = \frac{p^{1/2}[-i\omega(\hat{\boldsymbol{\theta}}_0\hat{\boldsymbol{\theta}}_0 + \hat{\boldsymbol{\varphi}}\hat{\boldsymbol{\varphi}}) : \mathbf{M}(\omega)]e^{i\omega\Upsilon(p)}}{4\pi\sqrt{\rho_r\rho_0}\beta_r\beta_0^3\sqrt{\operatorname{sgn}\left(\theta_0 - \frac{\pi}{2}\right)\partial_p D \times D^{1/2} q_0^{1/2} q_r^{1/2}}}. \tag{4.37}$$

In contrast with equation 4.36, in equation 4.37, q_0 and q_r are in terms of S-wave velocities,

$$\left.\begin{aligned} q_0 &= \sqrt{\frac{1}{\beta_0^2} - p^2} \\ q_r &= \sqrt{\frac{1}{\beta_r^2} - p^2} \end{aligned}\right\}, \tag{4.38}$$

where β_0 and β_r are the S-wave velocities at the source and receiver locations. Additionally, $\Upsilon(p)$ in equation 4.37 is the total S-wave travel time from the source to the receiver location.

We must emphasize the following for equations 4.36 and 4.37:

1) P- and S-wave velocity and density, defined by equation 4.10, are smooth.
2) Source and receiver locations are many wavelengths apart from one another.
3) Neither P ray nor the S ray has a turning point at the source or receiver location.
4) Receiver location is not on a caustic (focusing). At the caustic $dA \to 0$, and therefore $\partial_p\Delta \to 0$, both equations 4.36 and 4.37 break down.

Additionally, we also note that $\mathbf{u}_P$ and $\mathbf{u}_S$, given by equations 4.36 and 4.37, are asymptotic, not exact.

4.2.4 Travel Time and Distance

In equations 4.36 and 4.37, D and Υ respectively represent the total lateral distance between the source and receiver and total travel time from source to receiver. We now generalize this for any arbitrary point P lying in the ray path at a depth Z and a lateral distance Δ from the source, for which the travel time from the source is T (Figure 4.9). Instead of being specific to any wave type, we assume that the wave speed is $c(z)$, where $c = \alpha$ for P waves and $c = \beta$ for S waves.

Referring to Figure 4.9, if dT is the travel time along the infinitesimal arc ds around any point P, then

$$dT = ds/c. \tag{4.39}$$

But

$$ds = \sqrt{(d\Delta)^2 + (dz)^2}. \tag{4.40}$$

Also, from Figure 4.9,

$$\frac{\hat{\mathbf{t}}}{c} = p\hat{\mathbf{x}} + q\operatorname{sgn}\left(\theta_0 - \frac{\pi}{2}\right)\hat{\mathbf{z}}, \tag{4.41}$$

where

$$\left.\begin{aligned} p &= \frac{\sin\theta}{c} \\ q &= \frac{\cos\theta}{c} = \sqrt{\frac{1}{c^2} - p^2} \end{aligned}\right\}. \tag{4.42}$$

Therefore,

$$\hat{\mathbf{t}} = \sin\theta\hat{\mathbf{x}} + \operatorname{sgn}\left(\theta_0 - \frac{\pi}{2}\right)\hat{\mathbf{z}} \tag{4.43}$$

and

$$\left.\begin{aligned} dz &= ds\hat{\mathbf{t}}\cdot\hat{\mathbf{z}} \\ d\Delta &= ds\hat{\mathbf{t}}\cdot\hat{\mathbf{x}} \end{aligned}\right\}. \tag{4.44}$$

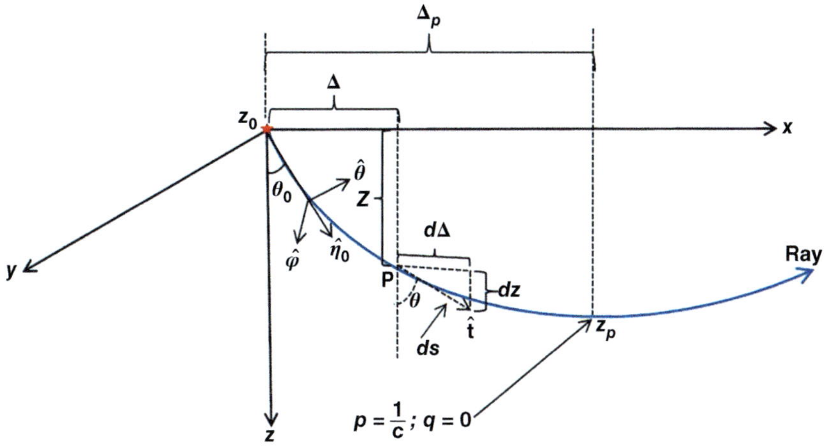

Figure 4.9 Calculation of time versus distance relationship for the propagating ray in a medium with continuous change of velocity with depth given by $c = c(z)$.

From the above relations, we write equation 4.39 as

$$dT = \frac{dz}{qc^2}$$

or, by integrating,

$$T = \int \frac{dz}{qc^2}. \tag{4.45}$$

Similarly, we can also obtain

$$\Delta = \int \frac{pdz}{q}. \tag{4.46}$$

By setting the limits of the integrals in equations 4.45 and 4.46 from the source depth z_0 to depth Z of the point P, or by providing the limits of the integration in equations 4.45 and 4.46, we can obtain the travel time T and the lateral distance Δ. Instead of doing so, we carry out these integrations from z_0 to z_p such that at z_p, $c = p^{-1}$ or $q = 0$, and write equations 4.45 and 4.46 as

$$\left.\begin{aligned} T(z,p) &= \int_{z_0}^{z_p} \frac{dz'}{q(z')c^2(z')} \\ \Delta(z,p) &= \int_{z_0}^{z_p} \frac{pdz'}{q(z')} \end{aligned}\right\}. \tag{4.47}$$

From equation 4.42, it is obvious that when p^{-1} or $q = 0$, θ must be equal to $\pi/2$. Therefore, z_p is the depth where the ray becomes horizontal beyond which the ray must turn (see Figure 4.9). This depth z_p is therefore called the *turning point* of the ray. Note that if $c(z)$ increases with depth, no ray can penetrate deeper than z_p. However, if there are low velocity zones where rays bend in the reverse direction than the one shown in Figure 4.9, rays may penetrate beyond z_p. Therefore, if there are more than one turning point in a medium, there must be low velocity zones between them.

From here onward, we will use z_s and z_r to respectively represent the source and receiver depths. In a medium where $c(z)$ continuously increases with depth, we define z_p to be the turning point. Thus, from Figure 4.10, the total travel time Υ and the total lateral distance D from source to receiver location are given by

$$\left.\begin{aligned} \Upsilon &= T(z_r,p) + \text{sgn}\left(\frac{\pi}{2} - \theta_s\right) T(z_s,p) \\ D &= \Delta(z_r,p) + \text{sgn}\left(\frac{\pi}{2} - \theta_s\right) \Delta(z_s,p) \end{aligned}\right\}. \tag{4.48}$$

In equation 4.48, $T(z_r, p)$, $T(z_s, p)$, $\Delta(z_r, p)$, and $\Delta(z_s, p)$ can be obtained from equation 4.47 by setting the appropriate limits of integration. In the example shown in Figure 4.10, $\text{sgn}\left(\frac{\pi}{2} - \theta_s\right)$ is +1, and thus in equation 4.48 we add the time and distances from z_r to z_p to those from z_s to z_p. However, if $\text{sgn}\left(\frac{\pi}{2} - \theta_s\right)$ is

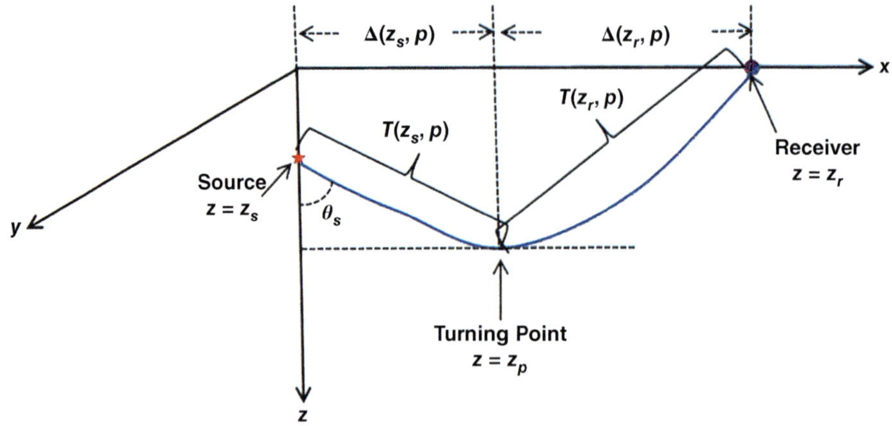

Figure 4.10 Calculation of the total travel time Υ and lateral distance D from source to receiver location for a medium with continuous increase of velocity with depth.

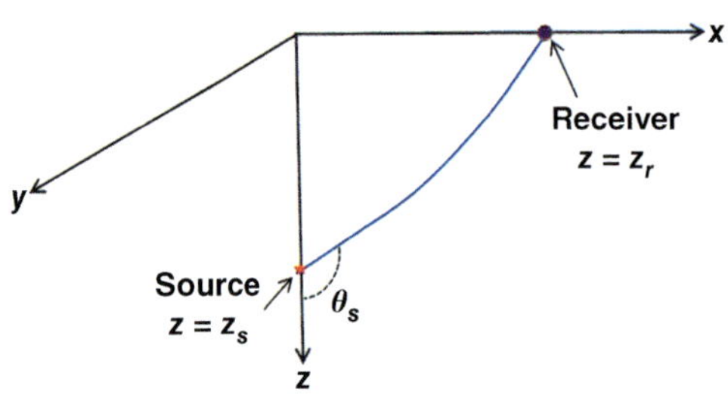

Figure 4.11 Situation where the source and receiver terms must be subtracted to compute total travel time and lateral distance.

−1 as shown in Figure 4.11, we must subtract (z_s, z_p) from (z_r, z_p) in equation 4.48.

We now introduce a new parameter $\tau(p, z)$ and given as

$$\tau(p,z) = T(p,z) - p(p,z), \tag{4.49}$$

which, using equation 4.47, we write as

$$\tau(p,z) = \int_z^{z_p} \frac{dz'}{q(z')c^2(z')} - p\int_z^{z_p} \frac{pdz'}{q(z')}$$

$$= \int_z^{z_p} \frac{dz'}{q(z')}\left\{\frac{1}{c^2(z')} - p^2\right\}. \tag{4.50}$$

Using the definition of q (equation 4.42), equation 4.50 simplifies to

$$\tau(z,p) = \int_z^{z_p} q(z)dz. \tag{4.51}$$

Again, from equations 4.42 and 4.47, it is straightforward to show that

$$\left.\begin{aligned} \partial_p\left(\frac{1}{q}\right) &= \frac{p}{q^3} \\ \partial_p T &= \int_z^{z_p} \frac{pdz'}{c^2(z')q^3(z')} \\ \partial_p \Delta &= \frac{1}{p}\partial_p T \end{aligned}\right\}. \tag{4.52}$$

In equation 4.52, $\partial_p = \partial/\partial p$. From equations 4.52 and 4.48, we also obtain

$$\partial_p \Upsilon = p\partial_p D.$$

Or

$$p = \frac{\partial_p \Upsilon}{\partial_p D} = \frac{dT}{d\Delta} \tag{4.53}$$

is the slope of the travel time curve. Additionally, it is also straightforward to show that

$$\frac{d^2T}{d\Delta^2} = \frac{1}{\partial_p \Delta}. \tag{4.54}$$

Note that the right-hand side of equation 4.54 is in the geometrical spreading term of the far-field radiation patterns for P and S waves (equations 4.36 and 4.37). Therefore, the amplitudes of the P and S waves is proportional to $\sqrt{d^2T/d\Delta^2}$.

4.2.5 A Practical Example—Linearly Varying Velocity with Depth

In the above section, we developed all equations necessary to compute ray-theoretical seismograms for a medium with continuous change of properties with depth shown by equation 4.10. We will now consider an example where velocity is linearly dependent on depth as

$$c(z) = c_0 + c_1 z.$$

Such a linear increase is demonstrated in Figure 4.12a. Although Figure 4.12a shows the velocity linearly increasing with depth, the expressions we derive below are valid for velocities linearly increasing or decreasing with depth. Deriving expressions for such a linear velocity gradient is useful because any arbitrary velocity as a function of depth can be always approximated as piecewise linear velocity functions (Figure 4.12b).

From equation 4.47, we obtain, after some algebra, the P-wave ray-theoretical amplitude for linear depth function as

$$u_P^{ff}(r) = \frac{\sqrt{p}\{-i\omega\hat{\mathbf{t}}\hat{\boldsymbol{\eta}}_0\hat{\boldsymbol{\eta}}_0 : \mathbf{M}(\omega)\}e^{i\omega\Upsilon(p)}}{4\pi\sqrt{\rho_s\rho_r\alpha_s\alpha_r}\sqrt{\frac{1}{c_1 p^2}\left\{\frac{q_r}{\alpha_s}\operatorname{sgn}\left(\frac{\pi}{2}-\theta_s\right)\frac{q_s}{\alpha_r}\right\}}}. \quad (4.55)$$

In equation 4.55, ρ_s and ρ_r are the densities at the source and receiver locations, α_s and α_r are the P-wave velocities at the source and receiver locations, $q_s = \sqrt{\frac{1}{\alpha_s^2} - p^2}$, and $q_r = \sqrt{\frac{1}{\alpha_r^2} - p^2}$. The S-wave amplitudes can be similarly derived. For details, see Aki and Richards (2002), Červený et al. (1978), and Kline and Kay (1966).

4.2.6 Reflection and Transmission Problem

So far, we discussed calculating ray-theoretical amplitudes for media with elastic properties smoothly varying with depth. In the subsurface, elastic properties on the other hand not only smoothly vary with depth, but also there are regions where there are sharp discontinuities separating media with varying elastic properties. These sharp discontinuities cause reflection and transmission from the interface separating two media with sharp discontinuities. We previously discussed this reflection and transmission in the context of Snell's law where we considered an incident P wave in an interface that generates the reflected and transmitted P and S waves (Figure 4.2), which are controlled by the reflection and transmission coefficients. We discuss how to compute the reflection and transmission coefficients in Chapter 5, and

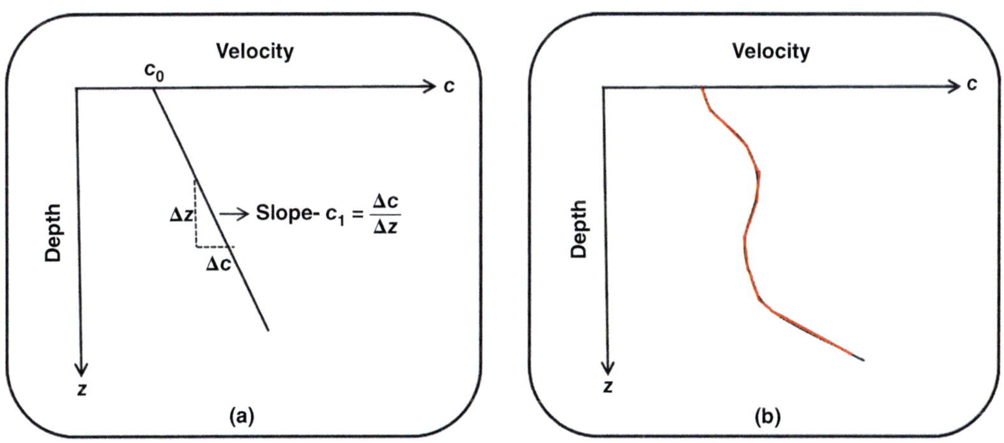

Figure 4.12 (a) Linearly varying velocity with depth having an intercept c_0 and slope c_1. (b) An arbitrary velocity function of depth (black), approximated by piecewise linear velocity functions (red).

here we just review some fundamentals and provide explicit and approximate expressions for reflection and transmission coefficients for an interface separating two isotropic solids.

As shown in Figure 4.13, consider an interface, located at depth $z = z_i$ separating two media with different elastic properties. Clearly, there is a discontinuity at depth z_i, and to represent this discontinuity, we use z_i^- and z_i^+ to the medium just above and just below the interface. As shown in Figure 4.13a, consider that a downward propagating wavefield represented by the vector $\mathbf{v}_d(z_i^-)$ is incident from above the interface. This wavefield will generate an upward propagating reflected wavefield $\mathbf{v}_u(z_i^-)$ above the interface and a downward propagating transmitted wavefield $\mathbf{v}_d(z_i^+)$. In general, all these wavefields are three-component vectors. For a general anisotropic medium, they are the amplitudes of P, S_1, and S_2 wave amplitudes. For an isotropic medium, which we discuss here, they are P, SV, and SH wave amplitudes. Note that these wavefields are not the components of the particle displacement vector **u**. In Chapter 3, we discussed Helmholtz decomposition of $\mathbf{u} = \nabla P - \nabla \times \mathbf{S}$ in which both P and $\mathbf{S}$ satisfy wave equation (equation 3.151), and the components of P, SV, and SH wavefields from this Helmholtz decomposition.

Having defined the downward incident wavefield $\mathbf{v}_d(z_i^-)$, generating the reflected and transmitted wavefields $\mathbf{v}_u(z_i^-)$ and $\mathbf{v}_d(z_i^+)$, we now define the downward reflection coefficient matrix $\mathbf{R}_d$ and the downward transmission coefficient matrix $\mathbf{T}_d$ such that

$$\left.\begin{array}{l} \mathbf{v}_u(z_i^-) = \mathbf{R}_d \mathbf{v}_d(z_i^-) \\ \mathbf{v}_d(z_i^+) = \mathbf{T}_d \mathbf{v}_d(z_i^-) \end{array}\right\}. \tag{4.56}$$

Clearly, $\mathbf{R}_d$ and $\mathbf{T}_d$ are 3×3 matrices. For a general anisotropic medium, their components are P-to-P (PP) P-to-S1 ($PS1$), P-to-S2 ($PS2$), S1-to-P ($S1P$), S1-to-S1 ($S1S1$), S1-to-S2 ($S1S2$), S2-to-P ($S2P$), S2-to-S1 ($S1S2$), and S2-to-S2 ($S2S2$) reflection and transmission coefficients, using which both equations shown in equation 4.56 can be written as

$$\begin{pmatrix} v_u^P(z_i^-) \\ v_u^{S1}(z_i^-) \\ v_u^{S2}(z_i^-) \end{pmatrix} = \begin{pmatrix} R_d^{PP} & R_d^{PS1} & R_d^{PS2} \\ R_d^{S1P} & R_d^{S1S1} & R_d^{S1S2} \\ R_d^{S2P} & R_d^{S2S1} & R_d^{S2S2} \end{pmatrix} \begin{pmatrix} v_d^P(z_i^-) \\ v_d^{S1}(z_i^-) \\ v_d^{S2}(z_i^-) \end{pmatrix} \tag{4.57}$$

and

$$\begin{pmatrix} v_d^P(z_i^+) \\ v_d^{S1}(z_i^+) \\ v_d^{S2}(z_i^+) \end{pmatrix} = \begin{pmatrix} T_d^{PP} & T_d^{PS1} & T_d^{PS2} \\ T_d^{S1P} & T_d^{S1S1} & T_d^{S1S2} \\ T_d^{S2P} & T_d^{S2S1} & T_d^{S2S2} \end{pmatrix} \begin{pmatrix} v_d^P(z_i^-) \\ v_d^{S1}(z_i^-) \\ v_d^{S2}(z_i^-) \end{pmatrix}. \tag{4.58}$$

Similarly, for the wavefield $\mathbf{v}_u(z_i^+)$, generating the reflected and transmitted wavefields $\mathbf{v}_d(z_i^+)$ and $\mathbf{v}_u(z_i^-)$, as shown in Figure 4.13b, we obtain two additional reflection and

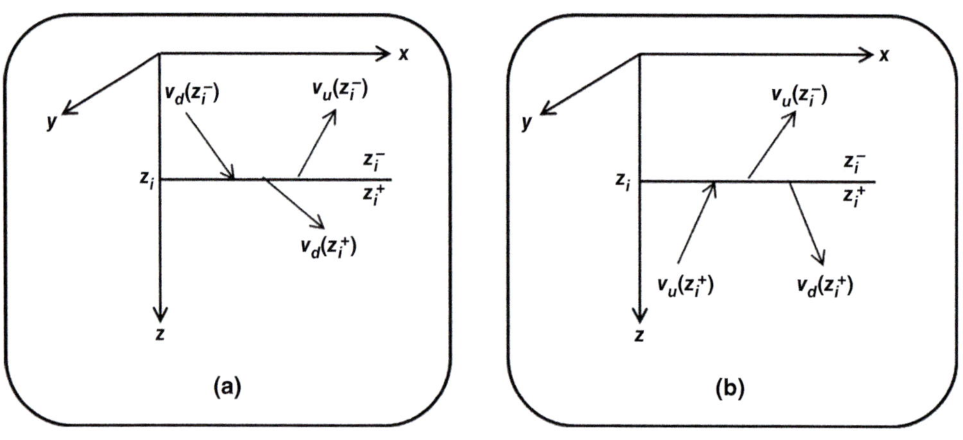

Figure 4.13 The problem of reflection and transmission for an interface separating two media with different elastic properties. (a) Wavefield incident from the top of the interface. (b) Wavefield incident from the bottom of the interface.

transmission coefficient matrices $\mathbf{R}_u$ and $\mathbf{T}_u$ relating them as

$$\left.\begin{aligned}\mathbf{v}_d(z_i^+) &= \mathbf{R}_u\mathbf{v}_u(z_i^+)\\ \mathbf{v}_u(z_i^-) &= \mathbf{T}_d\mathbf{v}_u(z_i^+)\end{aligned}\right\}. \tag{4.59}$$

And, like equations 4.57 and 4.58, we can explicitly write equation 4.59 in component forms. The reflection and transmission coefficient matrices can be obtained from imposing interface boundary conditions to the solution of the wave equation. For a solid–solid welded interface, we must require that the displacement and vertical tractions to be continuous across the interface or

$$\left.\begin{aligned}&u_x(z_i^-) = u_x(z_i^+);\, u_y(z_i^-) = u_y(z_i^+);\\ &u_z(z_i^-) = u_z(z_i^+)\\ &\tau_{xz}(z_i^-) = \tau_{xz}(z_i^+);\, \tau_{yz}(z_i^-) = \tau_{yz}(z_i^+);\\ &\tau_{zz}(z_i^-) = \tau_{zz}(z_i^+)\end{aligned}\right\}. \tag{4.60}$$

When one medium is a fluid, or both media are fluids, we would only require the vertical component of the displacement (u_z) and traction (τ_{zz}) to be continuous across the interface, and the shear stress (τ_{xy}, τ_{xz}, τ_{yz}) must vanish at the interface. Additionally, because fluids are free to move laterally, the horizontal components of the displacement (u_x and u_y) need not be continuous across fluid–fluid, fluid–solid, and solid–fluid interfaces. Imposing these boundary conditions into the solution to the wave equations above and below the interface leads to a set of simultaneous equations from which the downward ($\mathbf{R}_d$, $\mathbf{T}_d$) and upward ($\mathbf{R}_u$, $\mathbf{T}_u$) reflection and transmission matrices can be obtained. A special type of reflection and transmission is at the free surface where all components of the stress tensor vanish, and the dot product between the stress tensor and the normal direction of the free surface should be zero. In the real world, such a free surface condition is approximately valid at the surface of the earth, both for land and ocean environments.

The reflection and transmission matrices are, in general, 3×3 matrices. We have, however, seen from the solution of the Christoffel equation in Chapter 3 that for isotropic and VTI medium, the P-SV system is independent of the SH (see equations 3.104 and 3.105). Therefore, for isotropic and VTI cases, the reflection and transmission matrices decouple into a 2×2 P-SV and transmission matrices and scalars representing SH reflection and transmission coefficients. For example, the downward reflection coefficient matrix $\mathbf{R}_d$ for isotropic and VTI interfaces becomes

$$\mathbf{R}_d = \begin{pmatrix} R_d^{PP} & R_d^{PSV} & 0\\ R_d^{SVP} & R_d^{SVSV} & 0\\ 0 & 0 & R_d^{SHSH}\end{pmatrix}, \tag{4.61}$$

and thus can be written as a 2×2 matrix

$$\mathbf{R}_d^{P-SV} = \begin{pmatrix} R_d^{PP} & R_d^{PSV}\\ R_d^{SVP} & R_d^{SVSV}\end{pmatrix} \tag{4.62}$$

and a scalar R_d^{SHSH}, representing downward SH-to-SH wave reflection coefficient. $\mathbf{T}_d$, $\mathbf{R}_u$, and $\mathbf{T}_u$ can be similarly written. Explicit reflection coefficients for isotropic solid–solid interfaces were obtained by Zoeppritz (1919) and are known as the *Zoeppritz equations*. Here, we follow Aki and Richards (2002) and write down these reflection and transmission coefficients. In Chapter 5, we revisit the problem of reflection and transmission and derive these reflection and transmission coefficients for interfaces separating both solid and fluid layers.

Assume an interface separating two isotropic solids with the P-wave velocity, S-wave velocity, and density for the medium above the interface being α_1, β_1, and ρ_1, and those below the interface being α_2, β_2, and ρ_2 (Figure 4.14). Now, if P wave of unit amplitude is incident from above the interface angle θ_i, it will then generate reflected P and SV waves with magnitudes R_d^{PP} and R_d^{PSV} above the interface and transmitted P and SV waves of magnitudes T_d^{PP} and T_d^{PSV} below the interface (Figure 4.14a). The angles of reflection (θ_r, ϑ_r) and transmission (ξ_t, ζ_t) for P and SV waves are related to θ_i via Snell's law (equation 4.1). Similarly, if SV wave of unit amplitude is incident from the top of the interface, it will generate reflected and transmitted

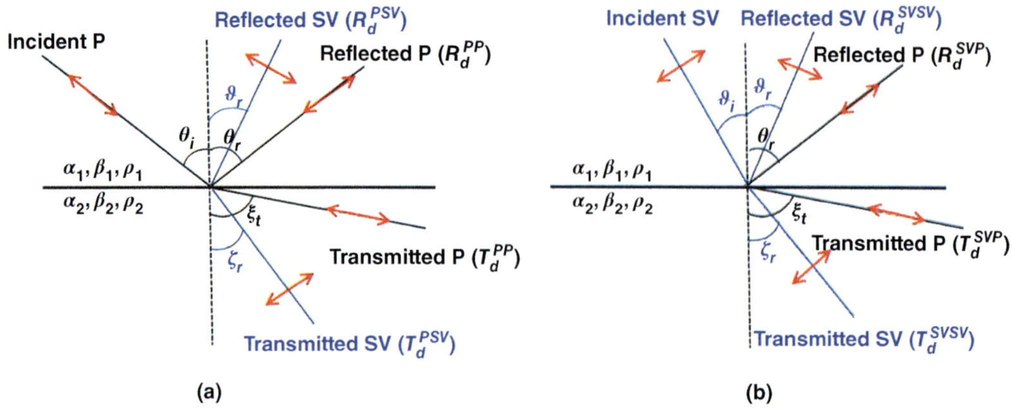

Figure 4.14 P-SV wave reflection and transmission problem for an interface separating two isotropic media. The P-wave velocity, S-wave velocity, and density for the medium above the interface are given respectively by α_1, β_1, and ρ_1 and those for the medium below the interface are given respectively by α_2, β_2, and ρ_2. (a) P wave, incident from the top of the interface. (b) SV wave, incident from the top of the interface. Polarization directions for particle displacements are indicated with red arrows.

waves of magnitudes $R_d^{SVP}, R_d^{SVSV}, T_d^{SVP}$, and T_d^{SVSV} (Figure 4.14b). In a similar fashion, we can obtain the upward reflection and transmission coefficients for the P-SV system (Figure 4.15) and the SH system (Figure 4.16).

It is convenient to express the reflection and transmission coefficients as a function of the ray-parameter p instead of the incidence angle because p remains constant irrespective of the type of reflection or transmission coefficient ($R_d^{PP}, R_u^{SVP}, T_d^{SVSV}$, etc.). Additionally, if these coefficients are expressed as functions of p, we can always use Snell's law to express them as functions of the incidence angle.

To conveniently express the reflection and transmission coefficient as functions of p, which we previously shown to be the horizontal slowness, we first define the P- and S-wave vertical slowness values $q_{\alpha1}$, $q_{\beta1}$, $q_{\alpha2}$, and $q_{\beta2}$ for the medium above and below the interface

$$\left.\begin{aligned} q_{\alpha1} &= \sqrt{\frac{1}{\alpha_1^2 - p^2}};\ q_{\beta1} = \sqrt{\frac{1}{\beta_1^2} - p^2} \\ q_{\alpha2} &= \sqrt{\frac{1}{\alpha_2^2 - p^2}};\ q_{\beta2} = \sqrt{\frac{1}{\beta_2^2} - p^2} \end{aligned}\right\}. \tag{4.63}$$

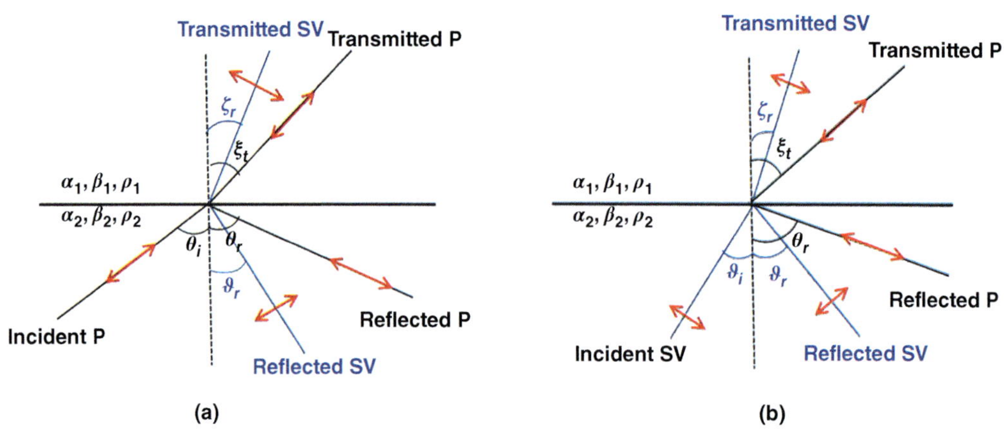

Figure 4.15 Same as Figure 4.14, but when the P and SV waves are incident from the bottom (a and b) of the interface.

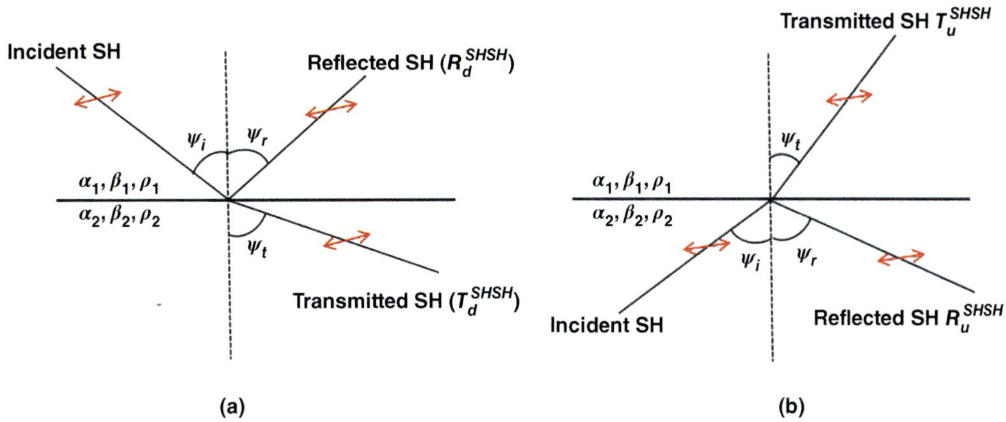

Figure 4.16 SH to SH reflection and transmission problem for the wavefield incident from (a) top and (b) bottom of the interface. The polarization directions indicated with red arrows are directed perpendicular to the plain of the paper.

We also define

$$\left.\begin{aligned} &\Gamma_1 = \rho_1(1 - 2\beta_1^2 p^2); \Gamma_2 = \rho_2(1 - 2\beta_2^2 p^2) \\ &a = \Gamma_2 - \Gamma_1; b = a + \rho_1; c = \rho_2 - a; \\ &d = \rho_2 - \rho_1 - a; E = bq_{\alpha 1} + cq_{\alpha 2}; \\ &F = bq_{\beta 1} + cq_{\beta 2}; G = a - dq_{\alpha 1}q_{\beta 2}; \\ &H = a - dq_{\alpha 2}q_{\beta 1}; D = EF + GHp^2 \end{aligned}\right\}. \tag{4.64}$$

Having defined the above parameters, the exact reflection and transmission coefficients for an isotropic solid–solid interface are given in Aki and Richards (2002) as

$$R_d^{PP} = \frac{\{(bq_{\alpha 1} - cq_{\alpha 2})F - (a + dq_{\alpha 1}q_{\beta 2})Hp^2\}}{D}, \tag{4.65}$$

$$R_d^{PSV} = -2\frac{q_{\alpha 1}(ab + cdq_{\alpha 2}q_{\beta 2})\alpha_1 p}{\beta_1 D}, \tag{4.66}$$

$$R_d^{SVP} = -2\frac{q_{\beta 1}(ab + cdq_{\alpha 2}q_{\beta 2})\beta_1 p}{\alpha_1 D}, \tag{4.67}$$

$$R_d^{SVSV} = \frac{\{(bq_{\beta 1} - cq_{\beta 2})E + (a + dq_{\alpha 1}q_{\beta 2})Hp^2\}}{D}, \tag{4.68}$$

$$T_d^{PP} = 2\frac{\rho_1 q_{\alpha 1} F \alpha_1}{\alpha_2 D}, \tag{4.69}$$

$$T_d^{PSV} = 2\frac{\rho_1 q_{\alpha 1} H \alpha_1 p}{\beta_2 D}, \tag{4.70}$$

$$T_d^{SVP} = -2\frac{\rho_1 q_{\beta 1} G \beta_1 p}{\alpha_2 D}, \tag{4.71}$$

$$T_d^{SVSV} = 2\frac{\rho_1 q_{\beta 1} E \beta_1}{\beta_2 D}, \tag{4.72}$$

$$R_u^{PP} = -\frac{\{(bq_{\alpha 1} - cq_{\alpha 2})F + (a + dq_{\alpha 2}q_{\beta 1})Gp^2\}}{D}, \tag{4.73}$$

$$R_u^{PSV} = 2\frac{q_{\alpha 2}(ac + bdq_{\alpha 1}q_{\beta 1})\alpha_2 p}{\beta_2 D}, \tag{4.74}$$

$$R_u^{SVP} = 2\frac{q_{\beta 2}(ac + bdq_{\alpha 1}q_{\beta 1})\beta_2 p}{\alpha_2 D}, \tag{4.75}$$

$$R_u^{SVSV} = -\frac{\{(bq_{\beta 1} - cq_{\beta 2})E - (a + dq_{\alpha 2}q_{\beta 1})Gp^2\}}{D}, \tag{4.76}$$

$$T_u^{PP} = 2\frac{\rho_2 q_{\alpha 2} F \alpha_2}{\alpha_1 D}, \tag{4.77}$$

$$T_u^{PSV} = -2\frac{\rho_2 q_{\alpha 2} G \alpha_2 p}{\beta_1 D}, \tag{4.78}$$

$$T_u^{SVP} = 2\frac{\rho_2 q_{\beta 2} H \beta_2 p}{\alpha_1 D}, \tag{4.79}$$

$$T_u^{SVSV} = 2\frac{\rho_2 q_{\beta 2} E \beta_2}{\beta_1 D}, \tag{4.80}$$

$$R_d^{SHSH} = \frac{\rho_2 q_{\beta 1} - \rho_1 q_{\beta 2}}{\rho_2 q_{\beta 1} + \rho_1 q_{\beta 2}}, \tag{4.81}$$

$$T_d^{SHSH} = 1 - R_d^{SHSH}, \tag{4.82}$$

$$R_u^{SHSH} = -R_d^{SHSH}, \tag{4.83}$$

and

$$T_u^{SHSH} = 1 + R_d^{SHSH}. \tag{4.84}$$

Equations 4.65–4.84 are the exact reflection and transmission coefficients for an interface between two isotropic solid media.

Before proceeding further, it is important to discuss the concept of *critical angle*. Assume P-wave propagating through an interface separating two elastic media as shown in Figure 4.17. Also assume that $\alpha_2 > \alpha_1$. Then, per Snell's law, if a P-wave is incident at the interface with an incident angle θ_i, it will generate a reflected P wavefield with the same reflection angle and a transmitted P wavefield with the transmission angle ξ_t such that $\xi_t > \theta_i$. This is shown as black rays in Figure 4.17.

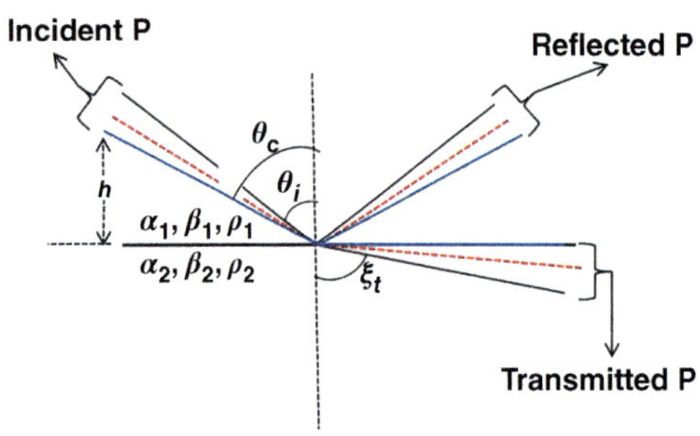

Figure 4.17 Concept of the critical angle and the head wave.

Thus, if we increase the incidence angle further (see dashed red rays in Figure 4.17), the reflection and transmission angles will also increase, until we reach an incidence angle θ_c (shown in blue in Figure 4.17), where the transmission angle will become 90° and the transmitted waves will graze through the interface.

The angle θ_c at which the transmission angle becomes 90° is called the critical angle. From Snell's law, it is obvious that

$$\theta_c = \sin^{-1}\left(\frac{\alpha_1}{\alpha_2}\right); \quad \alpha_2 > \alpha_1. \tag{4.85}$$

The transmitted wavefield at the grazing incidence is called the *head waves*. On the seismic records, these head waves are identifiable as first-arrival events with linear moveout with slopes proportional to the velocity of the medium below the interface. Identifying these first-arrival waves and analyzing them for the subsurface velocity structures is the fundamental basis for refraction seismology (Dobrin and Savit, 1988; Telford et al., 1990). In Figure 4.18, we show a simple layer over a half-space model (Figure 4.18a) and the computed seismograms, where the head-wave arrivals, direct arrivals, and the reflection arrivals are identified.

Now that we discussed the concepts of the critical angle and head waves, we go back to

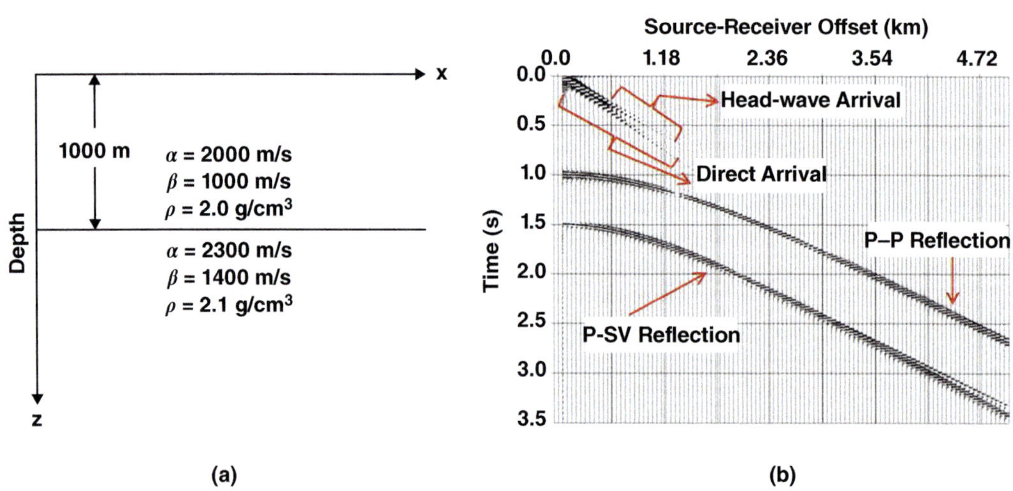

Figure 4.18 (a) Layer over a half-space model. (b) Synthetic seismograms showing the vertical component of the particle displacement, computed using the model shown in (a).

the problem of reflection and transmission. The exact reflection coefficients, given by equations 4.65–4.84, although not difficult to compute, they appear complicated, especially when we want to analyze which parameters are responsible for their variations as functions of the ray parameter or the angle of incidence. This, in turn, led to analyzing these coefficients and deriving their linearized approximations. These approximations were derived by Bortfeld (1961) and Richards and Frasier (1976). Approximations for the P–P reflection coefficient in relation with amplitude-variation-with-angle (AVA) applications were given by Shuey (1985), Mallick (1993), among others. Rüger (1998) among others extended these approximations for interfaces between two anisotropic solids. Finally, Shaw and Sen (2004) showed that all these approximations can be obtained directly from the stationary phase approximation to the Born integral.

As shown in Figure 4.19, if P wave of unit amplitude is incident from above an interface between two isotropic solids at an incidence angle θ_1, then according to Snell's law, it will generate the reflected P wave of amplitude R_d^{PP} with the angle-of-reflection θ_1. Additionally, it will also generate a reflected SV wave of magnitude R_d^{PSV} and reflection angle η_1, transmitted P-wave of magnitude T_d^{PP} and

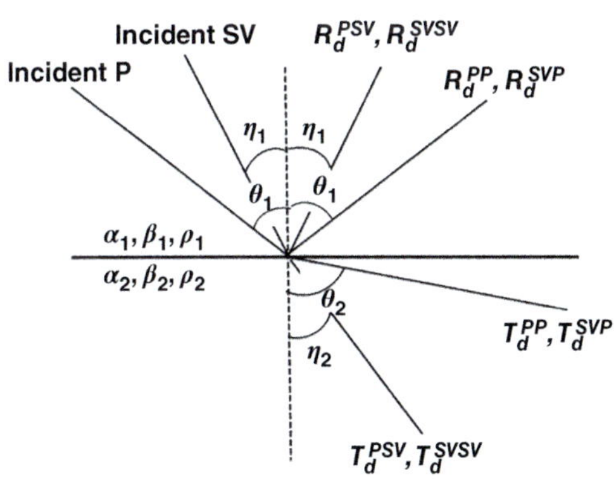

Figure 4.19 Illustration for deriving the approximate (linearized) reflection and transmission coefficients for an interface separating two isotropic solids.

transmission angle θ_2, and a transmitted SV wave of magnitude T_d^{PSV} with transmission angle η_2 such that θ_2, η_1, and η_2 are related to θ_1 via the relation

$$\frac{\sin\theta_1}{\alpha_1} = \frac{\sin\theta_2}{\alpha_2} = \frac{\sin\eta_1}{\beta_1} = \frac{\sin\eta_2}{\beta_2}. \tag{4.86}$$

Therefore, if an SV wave of unit amplitude is incident downward from the top of the interface with an incidence angle η_1 (see Figure 4.19), it will then generate a reflected P wave of magnitude R_d^{PSV} and reflection angle θ_1, a reflected SV wave of magnitude R_d^{SVSV} and reflection angle η_1, a transmitted P wave of magnitude T_d^{SVP} and transmission angle θ_2, and a transmitted SV wave of amplitude T_d^{SVSV} and transmission angle η_2. In deriving the linearized approximations to the reflection and transmission coefficients, we assume that $\alpha_1 \approx \alpha_2$, $\beta_1 \approx \beta_2$, and $\rho_1 \approx \rho_2$. This assumption leads to $\theta_1 \approx \theta_2$ and $\eta_1 \approx \eta_2$. We now define the average properties

$$\left.\begin{aligned} \alpha &= \frac{\alpha_1 + \alpha_2}{2} \\ \beta &= \frac{\beta_1 + \alpha_2}{2} \\ \rho &= \frac{\rho_1 + \rho_2}{2} \end{aligned}\right\}, \tag{4.87}$$

contrast properties

$$\left.\begin{aligned} \Delta\alpha &= \alpha_2 - \alpha_1 \\ \Delta\beta &= \beta_2 - \beta_1 \\ \Delta\rho &= \rho_2 - \rho_1 \end{aligned}\right\}, \tag{4.88}$$

average angles

$$\left.\begin{aligned} \theta &= \frac{\theta_1 + \theta_2}{2} \\ \eta &= \frac{\eta_1 + \eta_2}{2} \end{aligned}\right\}, \tag{4.89}$$

and the angle contrasts

$$\left.\begin{aligned} \Delta\theta &= \theta_2 - \theta_1 \\ \Delta\eta &= \eta_2 - \eta_1 \end{aligned}\right\}. \tag{4.90}$$

Having defined the averages and the contrasts as above (equations 4.87–4.90), the linearized (approximate) equivalents to the exact downward reflection and transmission coefficients given in equations 4.65–4.72 become

$$R_d^{PP} \approx \frac{1}{2}\left(\frac{\Delta\alpha}{\alpha}+\frac{\Delta\rho}{\rho}\right) + \left\{\frac{1}{2}\frac{\Delta\alpha}{\alpha}-2\left(\frac{\beta}{\alpha}\right)^2\left(\frac{\Delta\rho}{\rho}+2\frac{\Delta\beta}{\beta}\right)\right\} \times \sin^2\theta + \frac{1}{2}\frac{\Delta\alpha}{\alpha}\sin^2\theta\tan^2\theta, \tag{4.91}$$

$$R_d^{PSV} \approx -\frac{1}{2}\left(\frac{\alpha}{\beta}\right)\left\{\frac{\Delta\rho}{\rho}-2\left(\frac{\Delta\rho}{\rho}+2\frac{\Delta\beta}{\beta}\right) \times \left(\sin^2\eta - 2\frac{\beta}{\alpha}\cos\theta\cos\eta\right)\right\}\tan\eta, \tag{4.92}$$

$$R_d^{SVP} \approx \frac{\beta\cos\eta}{\alpha\cos\theta}R_d^{PSV}, \tag{4.93}$$

$$R_d^{SVSV} \approx -\frac{1}{2}\left(\frac{\Delta\beta}{\beta}+\frac{\Delta\rho}{\rho}\right)+\left(2\frac{\Delta\rho}{\rho}-\frac{7\Delta\beta}{2\beta}\right) \times \sin^2\eta - \frac{1}{2}\sin^2\eta\tan^2\eta, \tag{4.94}$$

$$T_d^{PP} \approx 1+\frac{1}{2}\left(\frac{\Delta\alpha}{\alpha}-\frac{\Delta\rho}{\rho}\right)+\frac{1}{2}\frac{\Delta\alpha}{\alpha}\tan^2\theta, \tag{4.95}$$

$$T_d^{PSV} \approx \frac{1}{2}\left(\frac{\alpha}{\beta}\right)\left\{\frac{\Delta\rho}{\rho}-2\left(\frac{\Delta\rho}{\rho}-2\frac{\Delta\beta}{\beta}\right) \times \left(\sin^2\eta + 2\frac{\beta}{\alpha}\cos\theta\cos\eta\right)\right\}\tan\eta, \tag{4.96}$$

$$T_d^{SVP} \approx -\frac{\beta\cos\eta}{\alpha\cos\theta}T_d^{PSV}, \tag{4.97}$$

and

$$T_d^{SVSV} \approx 1-\frac{1}{2}\left(\frac{\Delta\rho}{\rho}-\frac{\Delta\beta}{\beta}\tan^2\eta\right). \tag{4.98}$$

By defining the contrasts $\Delta\alpha$, $\Delta\beta$, and $\Delta\rho$ respectively as $\alpha_1-\alpha_2$, $\beta_1-\beta_2$, and $\rho_1-\rho_2$, or by negating their definitions given by equation 4.88, the same set of equations 4.91–4.98 provides the approximate upward reflection and transmission coefficients. We must note that these approximate coefficients are valid not only when the property contrasts $\Delta\alpha$, $\Delta\beta$, and $\Delta\rho$ are small but also for the incidence angles below the critical angle and not near 90°.

Table 4.1 P-wave velocity (α_2), S-wave velocity (β_2), and density (ρ_2) for the layer below the interface for four models with the exact and approximate P-wave reflection coefficients are shown in Figure 4.20.

Model	α_2 (m/s)	β_2 (m/s)	ρ_2 (g/cm^3)
1	2100	1050	2.10
2	1800	1200	1.95
3	2400	1600	2.20
3	3000	2200	2.40

Accuracy of the approximate reflection and transmission coefficients can be verified by comparing them with the exact coefficients. Here, we just verify the accuracy of R_d^{PP} using four different models. We kept the P- and S-wave velocity and density for the layer above the interface (α_1, β_1, and ρ_1) to 2000 m/s, 1000 m/s, and 2 g/cm^3, respectively, for all four models. For the layer below the interface, we varied these properties (α_2, β_2, and ρ_2) to the values shown in Table 4.1.

As can be seen, the properties of the medium below the interface (layer 2) increases from those for layer 2 from model 1 to model 4. In Figure 4.20, we compare the exact and approximate R_d^{PP} for all four models. Note that for small contrasts in the properties (small values of $\Delta\alpha$, $\Delta\beta$, and $\Delta\rho$) like models 1 and 2, the exact and approximate R_d^{PP} are nearly the same. However, when the contrasts are large, like those for models 3 and 4, the approximate formula can duplicate the exact values of R_d^{PP} only for incidence angles less than 5–10°. Another point to note regarding the computation of R_d^{PP} shown in Figure 4.20 is the value of the incidence angle. The exact R_d^{PP}, shown by equation 4.65, is a function of the ray parameter p, which in Figure 4.20 we show as a function of the incidence angle θ_1. To compute the exact R_d^{PP} for $\theta_1 = 20°$, for example, we thus compute $p = \frac{\sin(20°)}{\alpha_1}$, which for $\alpha_1 = 2000$ m/s is about 0.17 s/km and use equation 4.65. The approximate R_d^{PP}, shown by equation 4.91

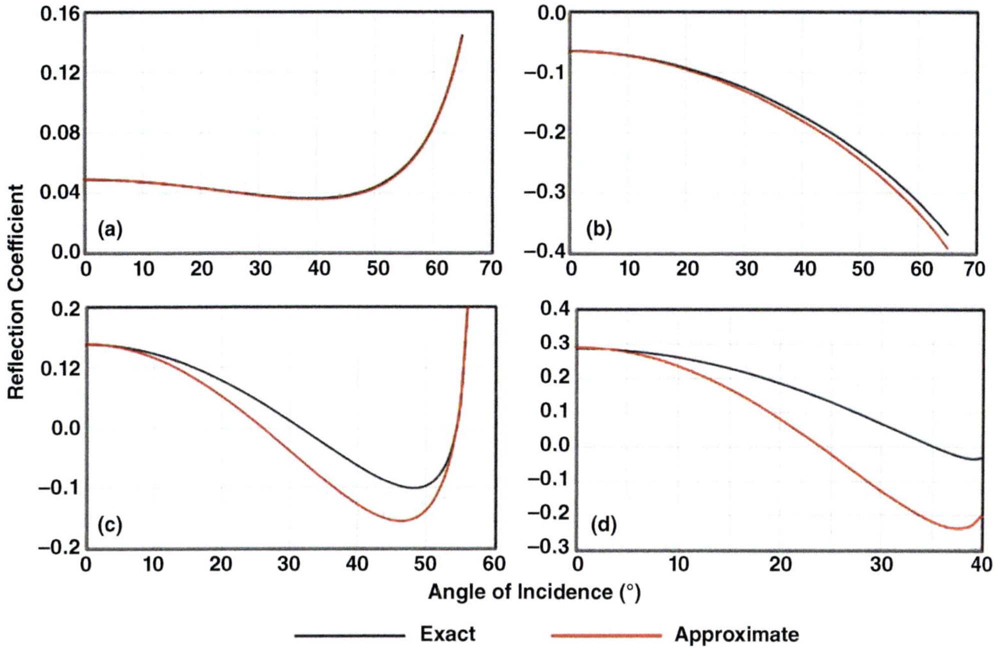

Figure 4.20 Exact (black) and approximate (red) P–P reflection coefficients R_d^{PP} for (a) model 1, (b) model 2, (c) model 3, and (d) model 4.

on the other hand, is a function of the average of the incidence angle θ_1 and the transmitted angle θ_2. So, to compare exact R_d^{PP} computed for $\theta_1 = 20°$ with its approximate counterpart, we must use Snell's law to compute θ_2, compute average angle $\theta = (\theta_1 + \theta_2)/2$, and then use equation 4.9 to compute approximate R_d^{PP}.

4.2.7 Ray-Theoretical Seismogram Computations

In the above section, we discussed how to trace rays in a smoothly varying medium and compute the ray-theoretical amplitudes. We also discussed the reflection and transmission problem for an interface and defined the upward and downward reflection and transmission coefficients. These, in turn, allow us to compute ray-theoretical synthetic seismic responses.

For simplicity, consider a 2D medium with two reflecting interface shown in blue and green (Figure 4.21). To compute P-wave synthetic response for interface 1, we start a seismic wavelet, generated from the source location (shown as an asterisk) and trace rays down until we hit reflector 1. At the reflector, we scale the wavelet amplitude by the downward reflection coefficient from interface 1, $R_d^{PP}(1)$, and continue tracing rays to the receiver location (blue triangle). This ray tracing is shown with red curves in Figure 4.21 for single source–receiver, and when we do this for all receivers, we obtain the synthetic seismogram with the reflection from interface 1. For interface 2 (rays shown in purple), we transmit downward at the interface 1 using $T_d^{PP}(1)$, reflect at the interface 2 with $R_d^{PP}(2)$, transmit at interface 1 with $T_u^{PP}(1)$, and then continue to the receiver location. Computing responses for any other reflections, for example P–S reflections, multiple reflections, etc., can be similarly computed. Note that in ray tracing we must separately compute responses for each event. Therefore, ray-trace-based seismograms are partial responses. Additionally, 2D

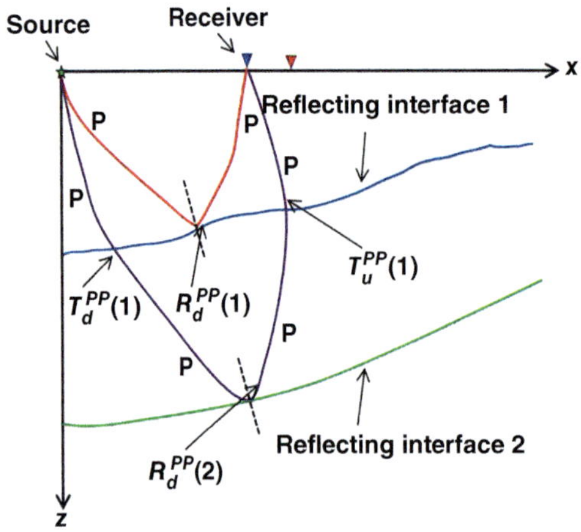

Figure 4.21 Computation of ray-theoretical synthetic seismograms.

ray tracing can easily be extended to 3D ray tracing.

About computing ray-theoretical seismograms, we must note one important thing. Starting with rays with a given takeoff angle, we can trace rays as above and reach a location, but that location may not coincide with the receiver location. In Figure 4.21 for example, we have shown the reflections from interfaces 1 and 2 arriving at a location shown by blue triangle. However, in field seismic data, the source and receiver locations are always fixed. For example, the true receiver location near the blue triangle is the one shown by the red triangle. It is therefore important in ray tracing to find the exact takeoff angle with which we must start for the ray to reach the exact location of the receiver. This, in turn, is an optimization problem and must be iteratively found. We postpone discussion of this problem to Chapter 6 when we discuss optimization of functions. Additionally, Kline and Kay (1966) and Červený et al. (1978) provide excellent mathematical details on ray theory.

4.2.8 Ray Tracing in an Anisotropic Medium

In our discussion of group and phase in an anisotropic medium, we noted that the travel times on the observed seismic data are controlled by the group velocities and group angles, and the reflection and transmission are controlled by the phase angle and slowness (inverse phase velocity). This fact adds an additional complexity for ray tracing in an anisotropic medium. As we compute travel times, we must move along the ray direction and use the group angle and group velocity. However, when we hit an interface for reflection or transmission, it is necessary to find the corresponding phase and phase velocity and use Snell's law. Anisotropic ray tracing must therefore consist of the following steps:

1) Start with a given group angle (φ). Compute the corresponding group velocity
   a) If we recall from Chapter 3, this requires optimization (see Figure 3.18). For the given φ, we must iteratively find the corresponding phase angle θ, phase velocity $V_{ph}(\theta)$, and then group velocity $V_g[\varphi(\theta)]$.
2) Continue tracing rays where travel times are computed from φ and $V_g[\varphi(\theta)]$, but the corresponding θ and $V_{ph}(\theta)$ are used in Snell's law for reflection and transmission.

We will not go into further details on anisotropic ray tracing. Further details on anisotropic ray tracing can be found in Gajewski and Pšenčík (1987) and Pšenčík and Farra (2014).

4.2.9 Other Methods for Computing Partial Seismic Response

Ray tracing is one of the popular methods for computing partial synthetic seismic responses for its ease of use. It is however important to point out some of the drawbacks of the method. First, we pointed out before, ray trace seismic responses are approximate, not exact. Additionally, the method fails in the presence of a caustics where the rays focus, and the area of the ray tube lends to zero. These drawbacks led to other methods for computing partial seismic responses. Two such popular methods are Cagniard de-Hoop method or the generalized ray method (de Hoop, 1960) and multifold path integral method (Frazer, 1987; Sen and Frazer, 1987, 1991).

4.3 Amplitude-Variation-With-Angle Synthetic Seismograms

Besides ray tracing, AVA, also known as the amplitude-variation-with-offset method, is another popular method for computing approximate synthetic seismograms for horizontally stratified models. AVA synthetic seismograms assume that the time and angle-of-incidence domain synthetic seismic responses can be approximately described as the convolution of reflection coefficients with the seismic wavelet.

Computing the AVA synthetic seismograms requires an input elastic earth model in time domain. In Figure 4.22, for example, we show an isotropic elastic model of P- and S-wave velocity (Figure 4.22a) and density (Figure 4.22b). Using this model, we can compute the downward PP reflection coefficients (R_d^{PP}) as functions of incidence angle (θ) and time (t) using either the exact (equation 4.65) or approximate (equation 4.91) formula. Then, for a given source wavelet $w(t)$, the PP AVA synthetic seismic response $S(\theta, t)$ is given by

$$S(\theta, t) = w(t) * R_d^{PP}(\theta, t). \tag{4.99}$$

Figure 4.22c shows the example of PP synthetic seismograms for the model shown in Figure 4.22a,b using the exact R_d^{PP} formula (equation 4.73) and a Hanning window wavelet (equation 2.61). Like R_d^{PP}, we can compute synthetic AVA response for PSV, SVP, and SVSV reflections using the appropriate formula.

Computing AVA synthetic seismic responses are computationally efficient. In addition, unlike ray-theoretical synthetic seismograms in which reflections for each layer interface require separate computation, in AVA we can compute reflections from all layer interfaces in a single computation. However, it must be noted that AVA synthetic

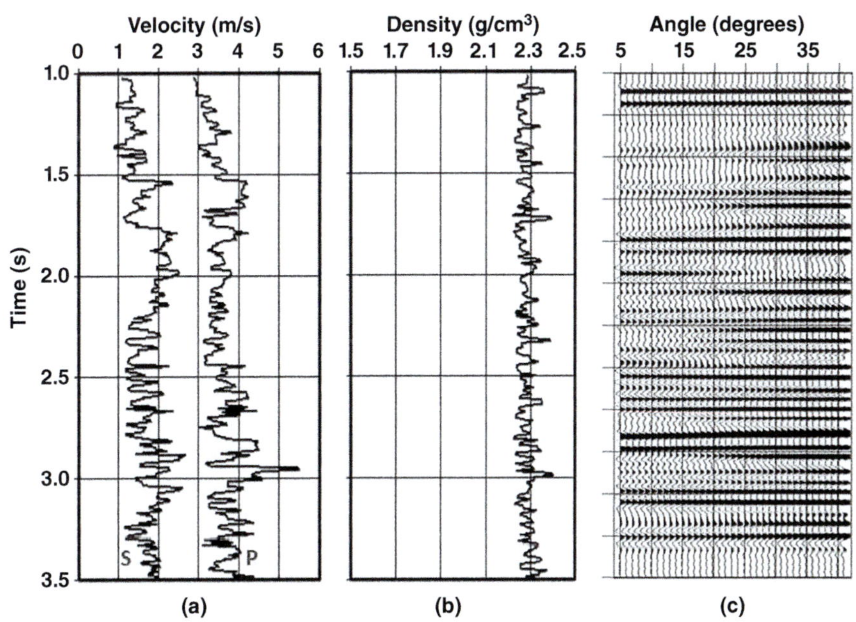

Figure 4.22 Computation of the AVA synthetic seismograms. (a) P- and S-wave velocity model. (b) Density model. (c) AVA synthetic seismograms.

seismograms are approximate. The convolutional assumption inherent to AVA implies primary-only reflection with no transmission loss. In a detailed modeling study, Mallick (2007) showed that such an assumption is approximately valid for relatively small incidence angles $\theta < 30°$. For $\theta > 30°$, complex wave effects, such as the transmission loss, interference with short-period multiple reflections and mode conversions, ray bending, etc., interfere with the primary reflections in such a way that the AVA synthetic seismograms are increasingly inaccurate with increasing incidence angle requiring computation of wave-equation-based synthetic seismograms to model the observed (real) seismic data. We discuss the computation of such wave-equation-based synthetic seismograms in the next chapter (Chapter 5).

4.4 Chapter Summary

In this chapter, we used ray theory to extend the radiation patterns from seismic sources in a homogeneous medium to a medium where the elastic properties smoothly vary with depth. We then discussed the reflection and transmission problem for an interface and showed how to use ray theory to compute partial seismic responses in media with sharp discontinuities. We also briefly outlined other methods for computing partial synthetic seismic responses. Finally, we discussed the computation of approximate synthetic seismograms using AVA.

4.5 Exercises

Problem 4.1 Referring to Figure 4.5a, show that the energy flux along the ray tube with P-wave propagation is given by $E\alpha dA$, where E is defined in equation 4.13, the area dA is defined in Figure 4.5, and α is the P-wave velocity.

Problem 4.2 Starting from the far-field P-wave radiation pattern from a moment-tensor source ($\mathbf{u}_P^{ff}$) given by equation 3.224, use Figure 4.5 and derive equation 4.17.

Problem 4.3 Derive equation 4.19 by taking the limit in equation 4.18.

Problem 4.4 Assuming a homogeneous medium where $\hat{\boldsymbol{\eta}}_0 = \hat{\mathbf{r}}$, show that the expression for $\mathbf{u}_P$ in equation 4.22 is equivalent to $\mathbf{u}_P^{ff}$, given equation 3.224.

Problem 4.5 From equation 4.18 derive equation 4.20.

Problem 4.6 Derive equation 4.33 and then equation 4.35.

Problem 4.7 Derive equation 4.36.

Problem 4.8 Using any programming language of your preference, write a code to compute exact and approximate R_d^{PP} (equations 4.65 and 4.91). Apply your code to four models provided in Table 4.1 and duplicate the computed R_d^{PP} curves shown in Figure 4.20.

Problem 4.9 Repeat problem 4.9 for R_d^{PSV}.

Problem 4.10 Refer to two ray paths from the second reflecting interface shown in Figure 4.23.

Based on the discussion on computing ray-theoretical synthetic seismograms, how would you proceed to compute responses for the two events shown in Figure 4.23?

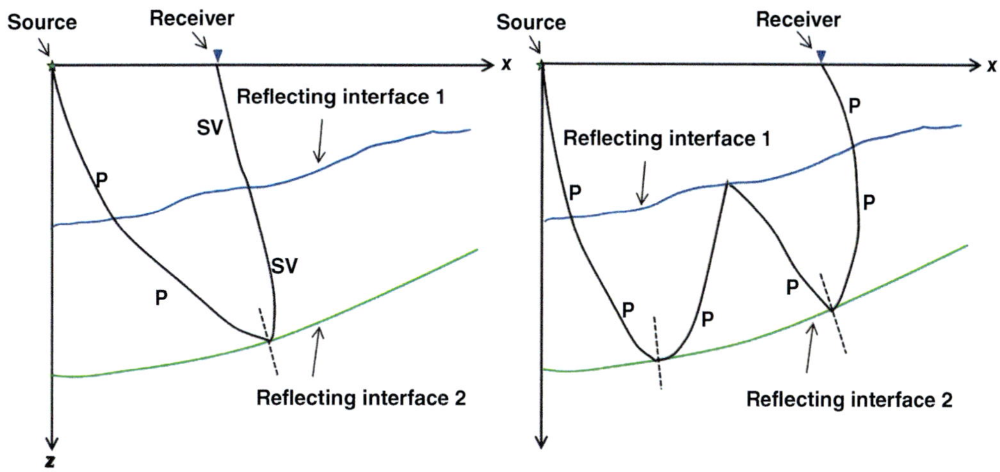

Figure 4.23 Ray paths showing P-SV reflection (left) and multiple P–P reflection from the second interface.

References

Aki, K., and Richards, P.G., 2002, *Quantitative Seismology*, Second Edn, University Science Books, ISBN 0-978-1891389634.

Bortfeld, R., 1961, Approximations to the reflection and transmission coefficients of plane longitudinal and transverse waves, *Geophysical Prospecting*, *9*, 485–502, https://doi.org/10.1111/j.1365-2478.1961.tb01670.x.

Červený, V., Molotkov, I.A., and Pšenčik, I., 1978, *The Ray Method in Seismology*, Charles University Press.

de Hoop, A.T., 1960, Modification of Cagniard's method for solving seismic pulse problems, *Applied Science Research*, *B8*, 349–356, https://doi.org/10.1007/BF02920068.

Dobrin, M.B., and Savit, C.H., 1988, *Introduction to Geophysical Prospecting*, Fourth Edn, McGraw-Hill ISBN 978-0070171961.

Frazer, L.N., 1987, Synthetic seismograms using multifold path integrals-I. Theory, *Geophysical Journal International*, *88*, 621–646, https://doi.org/10.1111/j.1365-246X.1987.tb01649.x.

Gajewski, D., and Pšenčik, I., 1987, Computation of seismic wavefields in 3-D laterally inhomogeneous anisotropic media, *Geophysical Journal International*, *91*, 383–411, https://doi.org/10.1111/j.1365-246X.1987.tb05234.x.

Jenkins, F., and White, H., 2001, *Fundamentals of Optic*, Fourth Edition, 768 pages Edn, McGraw Hill.

Kline, M., and Kay, I.W., 1966, *Electromagnetic Theory and Geometrical Optics*, Interscience, ISBN 0470491191, 9780470491195.

Mallick, S., 1993, A simple approximation to the P-wave reflection coefficient and its implication in the inversion of amplitude variation with offset data, *Geophysics*, *58*, 544–552, https://doi.org/10.1190/1.1443437.

Mallick, S., 2007, Amplitude-variation-with-offset, elastic impedance, and prestack waveform inversion, a modeling study, *Geophysics*, *72*, C1–C7, https://doi.org/10.1190/1.2387108.

Mallick, S., and Frazer, L.N., 1987, Practical aspects of reflectivity modeling, *Geophysics*, *52*, 1355–1364, https://doi.org/10.1190/1.1442248.

Mallick, S., and Frazer, L.N., 1988, Rapid computation of multioffset vertical seismic profile synthetic seismograms for layered media, *Geophysics*, *53*, 479–491, https://doi.org/10.1190/1.1442479.

Mallick, S., and Frazer, L.N., 1990, Computation of synthetic seismograms for stratified azimuthally anisotropic media, *Journal of Geophysical Research*, *95*, 8513–8526, https://doi.org/10.1029/JB095iB06p08513.

Mallick, S., and Frazer, L.N., 1991, Reflection/transmission coefficients and azimuthal anisotropy in marine seismic studies, *Geophysical Journal International*, *105*, 241–252, https://doi.org/10.1111/j.1365-246X.1991.tb03459.x.

Pšenčik, I., and Farra, V., 2014, First-order P-wave ray synthetic seismograms in inhomogeneous, weakly anisotropic, layered media, *Geophysical Journal International*, *198*, 298–307, https://doi.org/10.1093/gji/ggu135.

Richards, P.G., and Frasier, C.W., 1976, Scattering of elastic waves from depth-dependent inhomogeneities, *Geophysics*, *41*, 441–458, https://doi.org/10.1190/1.1440625.

Rüger, A., 1998, Variation of P-wave reflectivity with offset and azimuth in anisotropic media, *Geophysics*, *63*, 935–947, https://doi.org/10.1190/1.1444405.

Sen, M.K., and Frazer, L.N., 1987, Synthetic seismograms using multifold path integrals – II. Computations, *Geophysical Journal International*, *88*, 647–671, https://doi.org/10.1111/j.1365-246X.1987.tb01650.x.

Sen, M.K., and Frazer, L.N., 1991, Multifold phase space path integral synthetic seismogram, *Geophysical Journal International*, *104*, 479–487, https://doi.org/10.1111/j.1365-246X.1991.tb05695.x.

Telford, W.M., Geldart, M.P., and Sheriff, R.E., 1990, *Applied Geophysics*, Second Edn, Cambridge University Press, ISBN: ISBN 978-0521339384.

Zoeppritz, K., 1919, Erdbebenwellen VIII B, Über Reflexion ad Durchgang seismischer wellen durch Unstetigkeitsflachen, *Göttinger Nachr*, *1*, 66–84.

5

Computation of Synthetic Seismograms in Inhomogeneous Medium: Exact Solutions

5.1 Introduction

In Chapter 4, we first discussed propagation of elastic waves in a heterogeneous medium with smooth variations of elastic properties and then moved onto media with sharp boundaries and discussed the computation of partial and approximate synthetic seismograms. In this chapter, we go beyond these partial and approximate solutions and discuss how to compute complete synthetic seismic responses for a given subsurface earth model. First, we focus on a horizontally stratified earth and show how exact seismic responses for such a model can be analytically computed. Next, we move onto numerical computation of seismic responses for a three-dimensional (3D) inhomogeneous medium. Primary motivation of this chapter is to build the background for solving the forward problem, an essential element to all model-based seismic inversion or optimization methods.

5.2 Motivations Behind Computing a Complete Synthetic Seismic Response

We discussed the computation of partial seismic responses using ray theory in Chapter 4. Although fast and compute-efficient, ray theory allows computing response of single event at a time. In Figure 4.21, for example, computing P–P reflection responses from two interfaces are shown. If we want to compute the responses for the mode-converted (P-SV) and multiple reflections, they must be separately computed and added. In subsurface characterization, we often require seismic responses for a geologically complex medium with multiple layers. Using ray theory to compute response from such a medium is cumbersome, and therefore, not practical. Additionally, as we noted in Chapter 4, ray-theoretical seismograms are approximate, not exact. The amplitude-variation-with-angle (AVA) synthetic seismograms based on convolutional assumption, which we also discussed in Chapter 4, allow computing responses from multiple layer boundaries. The convolution assumption assumes that the synthetic response from a stack of layers is composed of primary reflections only, and they are not affected by interferences caused from wave propagation. In a detailed modeling study, Mallick (2007) demonstrated that the interference-free, primary-only assumption is valid for simple models, and as the models complicate, they become progressively inaccurate, especially for large incidence angles (source–receiver offsets). In Figures 5.1–5.3, we use progressively complex models to demonstrate the difference between the synthetic seismic responses computed using the convolutional (AVA) method and the wave-equation-based method, which we discuss in this chapter. In all these figures, we compute the AVA seismograms using the exact

Computation, Optimization, and Machine Learning in Seismology, Advanced Textbook 8, First Edition. Subhashis Mallick.

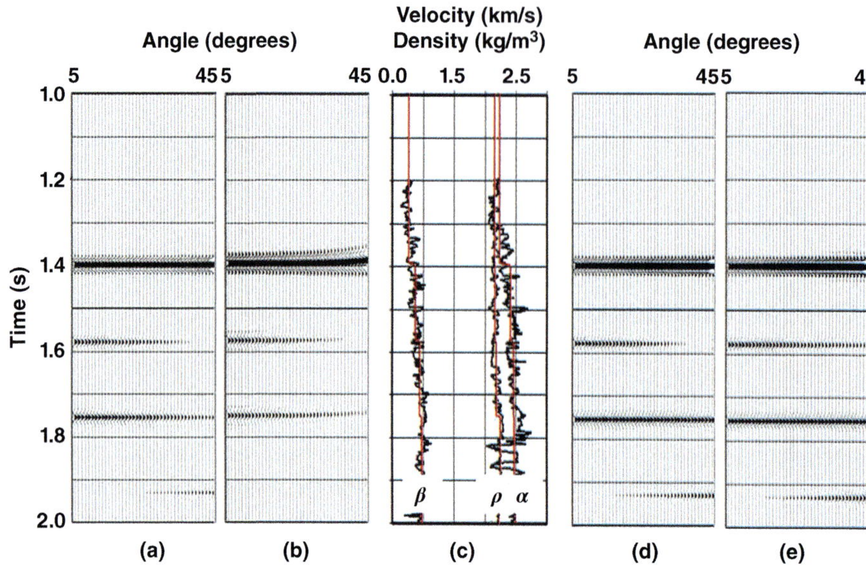

Figure 5.1 Comparison between the AVA and wave-equation-based synthetic seismic response for a well-log model, blocked at 10 wavelengths of the dominant seismic frequency of 30 Hz. (a) AVA response using the exact P–P reflection coefficient formula. (b) Wave-equation-based response. (c) Input model (black: original well-log model; red: blocked model). (d) AVA response using the linearized, two-term reflection coefficient formula. (e) AVA response using the linearized, three-term reflection coefficient formula. Adapted from Mallick (2007).

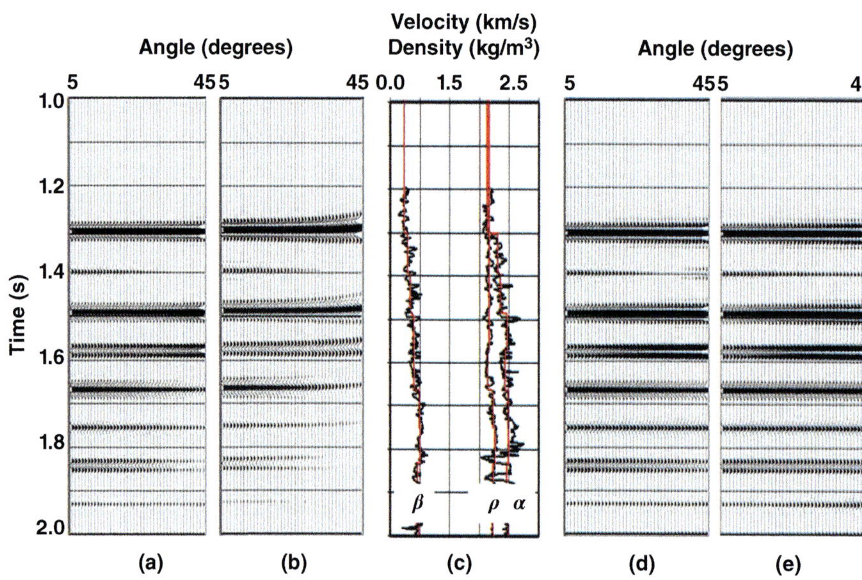

Figure 5.2 Same as Figure 5.1 except for the model being blocked at five wavelengths of the dominant seismic frequency. Adapted from Mallick (2007).

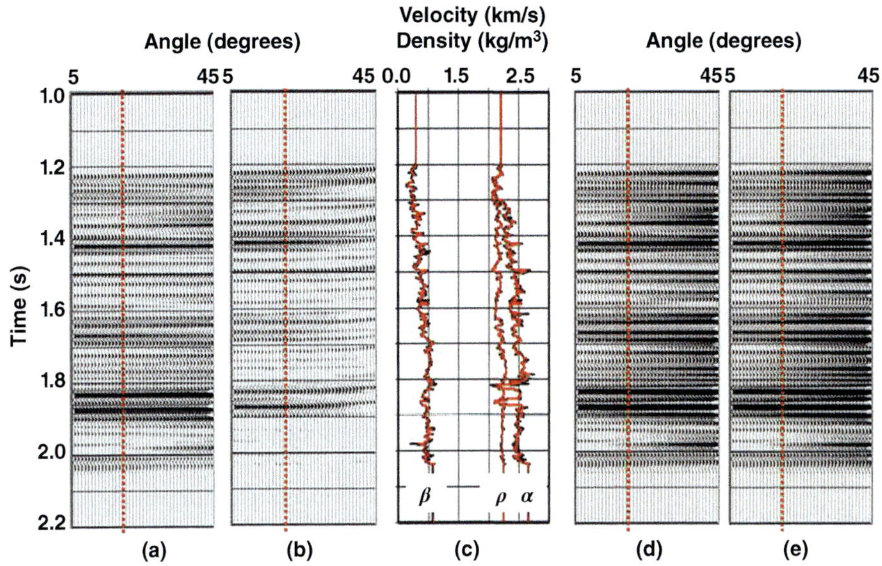

Figure 5.3 Same as Figure 5.1 except for the model being blocked at quarter (0.25) wavelengths of the dominant seismic frequency. The dashed line in red, shown in (a), (b), (d), and (e), separates the large and small incidence angles. Adapted from Mallick (2007).

P–P reflection coefficient formula (equation 4.65) and call them "AVA, Exact Zoeppritz"; a linearized P–P reflection coefficient formula (equation 4.91) and call them "AVA, Three-term"; and the linearized P–P reflection coefficient formula of equation 4.91, but ignoring the third term on the right-hand side of the equation (that is, by ignoring $\frac{1}{2}\frac{\Delta\alpha}{\alpha}\sin^2\theta\tan^2\theta$, and using only the first two terms on the right-hand side of equation 4.91) and call them "AVA, Two-term." We use the same well-log data with measurements of P- and S-wave velocities (α and β) and density (ρ) and show them in black in Figures 5.1c, 5.2c, and 5.3c. Assuming a dominant seismic frequency of 30 Hz, we block these well logs at blocking intervals of 10, 5, and quarter (0.25) wavelengths and compute the corresponding AVA and wave-equation-based synthetic seismic responses, which we show respectively in Figures 5.1–5.3. We also show the blocked well logs with different blocking lengths in these figures (red curves in Figures 5.1c, 5.2c, and 5.3c). Note that in wave-equation-based responses (Figures 5.1b, 5.2b, and 5.3b), the reflections are slightly bent upward as compared with the AVA responses. Reason for this is because the wave-equation synthetics are computed in time and offset and we used the offset-to-angle transformation method, outlined by Todd and Backus (1985) and Resnick (1993) to obtain the time and angle domain responses; see Mallick (2007) for details.

When the model is blocked at sufficiently large blocking interval of 10 wavelengths (Figure 5.1), the thickness of the individual layers is sufficiently large (red curves in Figure 5.1c), and the primary (P–P) reflections are not contaminated by other wave modes. Consequently, convolution-based AVA and wave-equation responses are comparable. However, as the models complicate by blocking the input well logs at smaller intervals (5 and 0.25 wavelengths), the wave-equation synthetics become substantially different from the AVA synthetics. In seismic inverse or optimization problems, we typically intend to resolve subsurface layer properties at least at quarter wavelengths of the dominant seismic frequency. And, we can see from Figure 5.3 that at quarter-wavelength blocking, the wave-equation synthetic responses (Figure 5.3b) are quite different from the AVA responses (Figure 5.3a,d,e), except for the small incidence angles on the left-hand side of the of these figures. Besides the difference in reflection amplitudes, also note that the wave-equation synthetics (Figure 5.3b) does not show any visible response for time $t \geq 2$ s. The AVA synthetics (Figure 5.3a,d,e), on the other hand, show visible reflections at $t \geq 2$ s. For the input model (the red curve in Figure 5.3c), the P waves reach a turning point just below 2 s. The wave-equation method correctly models this turning point, and therefore, no reflections are visible below 2 s in Figure 5.3b. Because there is no way to model a turning point in convolutional modeling, the AVA synthetics (Figure 5.3a,c,e) computes strong reflections for $t \geq 2$ s, as predicted by the angle-dependent reflection coefficients for the model. These reflections, however, should not be present in the computed response, which AVA fails to model.

Subsurface characterization using an optimization method with convolutional AVA-based synthetic computation as forward modeling engine is quite common, especially in the oil and gas industry (Russell, 1988; Hampson, 1991). From the modeling study shown in Figures 5.1–5.3, it is, however, evident that the convolutional modeling assumption of AVA is not strictly valid for large incidence angles, and to correctly characterize the subsurface, it is necessary to use wave-equation-based modeling.

5.3 Analytical Computation of Exact Synthetic Seismograms for a Horizontally Stratified Earth Model

In a horizontally stratified medium, the subsurface elastic properties are functions of depth only and are laterally invariant. Because of this reason, such media are also called one-dimensional (1D) media. Although 1D medium is an oversimplification of earth's subsurface, computing 1D synthetic seismograms is useful for the following reasons:

1) Compared to other methods, computing 1D seismograms is quicker; therefore, it gives us quick insight into how the seismic waves propagate in an inhomogeneous earth with smooth lateral variations and sharp vertical discontinuities.
2) In many areas, earth's subsurface is nearly horizontally stratified where computing 1D synthetic seismograms is adequate. Even for moderately simple geology, applying prestack time or depth migration (Sava and Hill, 2009) approximately corrects prestack seismic data such that 1D assumption is adequate to analyze data.
3) While finite-difference, finite-element, and other similar methods allow computing synthetics for media with any arbitrary inhomogeneity, they are extremely expensive to compute. To circumvent this computational challenge, elastic wave equations are often approximated as acoustic wave equations. Such an approximation is adequate for solving imaging problems but not for subsurface characterization (Mallick and Adhikari, 2015; Pafeng et al., 2017). A much more practical approach is prestack migration, followed by using 1D method for forward modeling in an iterative optimization scheme for subsurface characterization.

5.3.1 Conventions and Notations

Unless otherwise specified, we restrict to a right-handed Cartesian coordinate system as our frame of reference with x (or x_1) and y (or x_2) being the horizontal axes and z (or x_3) being the vertical axis positive downward (Figure 4.3). In our equations and discussions, we therefore freely switch between x and x_1, y and x_2, and z and x_3 to represent the coordinate axes. Additionally, we use the following notations for variables:

$\boldsymbol{\tau} \rightarrow$ Stress tensor, with its components represented as τ_{ij}, where $i, j = 1, 2, 3$ or x, y, z.

$\boldsymbol{\varepsilon} \rightarrow$ Strain tensor, with its components represented as ε_{ij}, where $i, j = 1, 2, 3$ or x, y, z.

$\mathbf{c} \rightarrow$ Stiffness tensor, with its components represented as c_{ijkl}, where $i, j, k, l = 1, 2, 3$ or x, y, z.

$\mathbf{u} \rightarrow$ Particle displacement vector, with its components represented by u_i, with $i = 1, 2, 3$ or x, y, z.

$\mathbf{p} \rightarrow$ Slowness vector, with its components represented by p_i, where $i = 1, 2, 3$ or x, y, z.

$\mathbf{k} \rightarrow$ Wave number vector, with its components represented by k_i, where $i = 1, 2, 3$ or x, y, z.

$\omega \rightarrow$ Temporal frequency.

$t \rightarrow$ Time.

$u_{i,j} \rightarrow \dfrac{\partial u_i}{\partial x_j}; i, j = 1, 2, 3 \text{ or } x, y, z.$

$\tau_{ij,k} \rightarrow \dfrac{\partial \tau_{ij}}{\partial x_k}; i, j, k = 1, 2, 3 \text{ or } x, y, z.$

$\rho \rightarrow$ Density.

Unless otherwise specified, we also assume that repeated indices always mean summing over them (for example, $u_i v_i = u_1 v_1 + u_2 v_2 + u_3 v_3 = u_x v_x + u_y v_y + u_z v_z$). Lastly, we follow the sign conventions for Fourier transforms, previously discussed in Chapter 4, equations 4.8 and 4.9.

5.3.2 The Elastic System in 1D

The 1D elastic system was derived by Woodhouse (1974), and below we briefly outline this derivation.

In the constitutive relation $\tau_{ij} = c_{ijkl}\varepsilon_{kl}$ (equation 3.17), note that the strain tensor, defined by equation 3.13, can be expressed as

$$\varepsilon_{kl} = \frac{1}{2}(u_{k,l} + u_{l,k}). \tag{5.1}$$

Using equation 5.1, and remembering the symmetry properties of the elastic stiffness tensor c_{ijkl}, the constitutive relation (equation 3.17) can be written as

$$\tau_{ij} = c_{ijkl} u_{k,l}. \tag{5.2}$$

Now, assume that the Latin indices (i, j, k, l, etc.) take values 1, 2, and 3 (or x, y, or z), while the Greek indices ($\alpha, \beta, \gamma, \delta$, etc.) only take values 1 and 2 (or x and y). Using these assumptions, we may then rewrite equation 5.2 as

$$\tau_{ij} = c_{ijkl} u_{k,l} = c_{ijk\upsilon} u_{k,\upsilon} + c_{ijk3} u_{k,3}. \tag{5.3}$$

Equation 5.3 is equivalent to the following two equations:

$$\left.\begin{aligned} \tau_{i3} &= c_{i3k\upsilon} u_{k,\upsilon} + c_{i3k3} u_{k,3} \\ \tau_{i\sigma} &= c_{i\sigma k\upsilon} u_{k,\upsilon} + c_{i\sigma k3} u_{k,3} \end{aligned}\right\}. \tag{5.4}$$

We now define the following matrices and vectors:

$$\left.\begin{aligned} (\mathbf{E}_{jl})_{ik} &= c_{ijkl} \\ (\mathbf{t})_i &= \tau_{i3} \\ (\mathbf{t}_\sigma)_i &= \tau_{\sigma i} \\ (\mathbf{u})_i &= u_i \end{aligned}\right\}. \tag{5.5}$$

Note that all $\mathbf{E}_{jl}$ matrices are 3×3 matrices. Equation 5.5 leads equation 5.4 to

$$\left.\begin{aligned} (\mathbf{t})_i &= (\mathbf{E}_{3\upsilon})_{ik} u_{k,\upsilon} + (\mathbf{E}_{33})_{ik} u_{k,3} \\ (\mathbf{t}_\sigma)_i &= (\mathbf{E}_{\sigma\upsilon})_{ik} u_{k,\upsilon} + (\mathbf{E}_{\sigma 3})_{ik} u_{k,3} \end{aligned}\right\}. \tag{5.6}$$

Equation 5.6 is equivalent to the following two equations:

$$\mathbf{t} = \mathbf{E}_{3\upsilon}\frac{\partial \mathbf{u}}{\partial x_\upsilon} + \mathbf{E}_{33}\frac{\partial \mathbf{u}}{\partial x_3} = \mathbf{E}_{3\upsilon}\frac{\partial \mathbf{u}}{\partial x_\upsilon} + \mathbf{E}_{33}\frac{\partial \mathbf{u}}{\partial z} \tag{5.7}$$

and

$$\mathbf{t}_\sigma = \mathbf{E}_{\sigma\upsilon}\frac{\partial \mathbf{u}}{\partial x_\upsilon} + \mathbf{E}_{\sigma 3}\frac{\partial \mathbf{u}}{\partial x_3} = \mathbf{E}_{\sigma\upsilon}\frac{\partial \mathbf{u}}{\partial x_\upsilon} + \mathbf{E}_{\sigma 3}\frac{\partial \mathbf{u}}{\partial z}. \tag{5.8}$$

From equation 5.7, we write

$$\frac{\partial \mathbf{u}}{\partial z} = (\mathbf{E}_{33})^{-1}\left(\mathbf{t} - \mathbf{E}_{3\upsilon}\frac{\partial \mathbf{u}}{\partial x_\upsilon}\right). \tag{5.9}$$

And, substituting equation 5.9 for $\partial \mathbf{u}/\partial z$ in the right-hand side of equation 5.8, we obtain, after some algebra,

$$\mathbf{t}_\sigma = \mathbf{Q}_{\sigma\upsilon}\frac{\partial \mathbf{u}}{\partial x_\upsilon} + \mathbf{E}_{\sigma 3}(\mathbf{E}_{33})^{-1}\mathbf{t}, \tag{5.10}$$

in which

$$\mathbf{Q}_{\sigma\upsilon} = \mathbf{E}_{\sigma\upsilon} - \mathbf{E}_{\sigma 3}(\mathbf{E}_{33})^{-1}\mathbf{E}_{3\upsilon}. \tag{5.11}$$

Now, using the notations we introduced in Section 5.3.1, and assuming the conventions for the Latin and Greek indices, we write the momentum equation (equation 3.69) as

$$\tau_{i\sigma,\sigma} + \tau_{i3,3} = \rho\partial_t^2 u_i - f_i. \tag{5.12}$$

And, from the definitions provided equation 5.5, we can further rewrite equation 5.12 as

$$\partial_z \mathbf{t} = -\partial_\sigma \mathbf{t}_\sigma + \rho\partial_t^2 \mathbf{u} - \mathbf{f}. \tag{5.13}$$

Now, from equation 5.10, and assuming a laterally invariant medium such that the derivatives ∂_σ of the elastic property $\mathbf{Q}_{\sigma v}$ vanish, we write equation 5.13 as

$$\partial_z \mathbf{t} = (\rho\partial_t^2 - \mathbf{Q}_{\sigma\upsilon}\partial_\sigma\partial_\upsilon)\mathbf{u} + \mathbf{E}_{\sigma 3}\mathbf{E}_{33}^{-1}\partial_\sigma \mathbf{t} - \mathbf{f}. \tag{5.14}$$

We now apply triple Fourier transforms to equations 5.9 and 5.14 with the sign conventions introduced by equation 4.9. This replaces ∂_σ with $i\omega p_\sigma$ and ∂_t with $-i\omega$. This reduces equations 5.9 and 5.14 to

$$\left.\begin{aligned} \partial_z \mathbf{u} &= -i\omega p_\sigma (\mathbf{E}_{33})^{-1}\mathbf{E}_{3\sigma}\mathbf{u} + i\omega(\mathbf{E}_{33})^{-1}\left(\frac{1}{i\omega}\mathbf{t}\right) \\ \partial_z\left(\frac{1}{i\omega}\mathbf{t}\right) &= i\omega(\rho\mathbf{I} - p_\sigma p_\upsilon \mathbf{Q}_{\sigma\upsilon})\mathbf{u} - i\omega p_\sigma \\ &\quad \times \mathbf{E}_{\sigma 3}\mathbf{E}_{33}^{-1}\left(\frac{1}{i\omega}\mathbf{t}\right) + \frac{i}{\omega}\mathbf{f} \end{aligned}\right\}. \tag{5.15}$$

Two equations shown in equation 5.15 can be combined as a matrix equation

$$\partial_z \mathbf{b} = i\omega\mathbf{A}\mathbf{b} + \mathbf{F}, \tag{5.16}$$

with

$$\left.\begin{aligned} \mathbf{b} &= \left[\mathbf{u}, \frac{1}{i\omega}\mathbf{t}\right]^T \\ &= \left[u_x, u_y, u_z, \frac{1}{i\omega}\tau_{xz}, \frac{1}{i\omega}\tau_{yz}, \frac{1}{i\omega}\tau_{zz}\right]^T, \\ \mathbf{F} &= \left[\mathbf{0}, \frac{i}{\omega}\mathbf{f}\right]^T \\ &= \left[0, 0, 0, \frac{i}{\omega}f_x, \frac{i}{\omega}f_y, \frac{i}{\omega}f_z\right]^T, \\ \mathbf{A} &= \begin{pmatrix} \mathbf{T} & \mathbf{C} \\ \mathbf{S} & \mathbf{V} \end{pmatrix} \\ \mathbf{T} &= -p_\sigma \mathbf{E}_{33}^{-1}\mathbf{E}_{3\sigma}; \mathbf{C} = \mathbf{E}_{33}^{-1}; \\ \mathbf{S} &= \rho\mathbf{I} - p_\sigma p_\upsilon \mathbf{Q}_{\sigma\upsilon}; \mathbf{V} = -p_\sigma \mathbf{E}_{\sigma 3}\mathbf{E}_{33}^{-1} \end{aligned}\right\}. \tag{5.17}$$

Equation 5.16 is called the *displacement–stress* relation for *1D elastic system* and the 6×6 matrix **A** defined in equation 5.17 is called the *elastic system matrix*. It can be verified that the 3×3 submatrices **S** and **C** are symmetric matrices and $\mathbf{V} = \mathbf{T}^T$, and therefore we can write the elastic system matrix as

$$\mathbf{A} = \begin{pmatrix} \mathbf{T} & \mathbf{C} \\ \mathbf{S} & \mathbf{T}^T \end{pmatrix}. \tag{5.18}$$

In isotropic and VTI medium, the 6×6 system shown by equation 5.17 decouples into 4×4 P-SV and 2×2 SH systems, which is left as an exercise to verify (Problem 5.6).

There are many approaches to solve the 1D elastic system given by equation 5.16. Here, we will follow Kennett's propagator matrix approach (Kennett, 1983), extended for an arbitrary anisotropic medium by Fryer and Frazer (1984).

5.3.3 Solution of the Elastic System: A Homogeneous Region

Assume that **D** is the eigenvector matrix of the elastic system matrix **A** with Λ being the diagonal matrix of eigenvalues of **A**, i.e.,

$$\mathbf{AD} = \mathbf{D}\Lambda. \tag{5.19}$$

We also define the vector **v** as

$$\left.\begin{aligned} \mathbf{v} &= \mathbf{D}^{-1}\mathbf{b} \\ \mathbf{b} &= \mathbf{D}\mathbf{v} \end{aligned}\right\}. \tag{5.20}$$

We can then write the displacement–stress relation (equation 5.16) as

$$\partial_z \mathbf{v} = i\omega\Lambda\mathbf{v}. \tag{5.21}$$

Equation 5.21 is clearly the wave equation, and therefore the vector **v** is called the *wave vector*. For a homogeneous region, if $\mathbf{v}(z_0)$ is the wave vector at depth $z = z_0$, then the solution to equation 5.21 at any arbitrary depth z is

$$\mathbf{v}(z) = e^{i\omega\Lambda(z - z_0)}\mathbf{v}(z_0), \tag{5.22}$$

where

$$e^{i\omega\Lambda(z-z_0)} = \begin{pmatrix} e^{i\omega\lambda_1(z-z_0)} & 0 & 0 & 0 & 0 & 0 \\ 0 & e^{i\omega\lambda_2(z-z_0)} & 0 & 0 & 0 & 0 \\ 0 & 0 & e^{i\omega\lambda_3(z-z_0)} & 0 & 0 & 0 \\ 0 & 0 & 0 & e^{i\omega\lambda_4(z-z_0)} & 0 & 0 \\ 0 & 0 & 0 & 0 & e^{i\omega\lambda_5(z-z_0)} & 0 \\ 0 & 0 & 0 & 0 & 0 & e^{i\omega\lambda_6(z-z_0)} \end{pmatrix} \quad (5.23)$$

is the *wave propagator* that propagates the wavefield from depth z_0 to depth z. Clearly, $\lambda_i, i = 1, 2, \ldots, 6$, are the eigenvalues of the elastic system matrix **A**. The wave propagator is denoted as $\mathbf{Q}(z, z_0)$, i.e.,

$$\mathbf{Q}(z, z_0) = e^{i\omega\Lambda(z-z_0)}. \quad (5.24)$$

From the definition of the wave propagator (equation 5.20), it is also straightforward to show that

$$\mathbf{b}(z) = \mathbf{D}e^{i\omega\Lambda(z-z_0)}\mathbf{D}^{-1}\mathbf{b}(z_0). \quad (5.25)$$

It is therefore evident from equation 5.25 that the matrix

$$\mathbf{P}(z, z_0) = \mathbf{D}e^{i\omega\Lambda(z-z_0)}\mathbf{D}^{-1} \quad (5.26)$$

propagates the displacement–stress vector **b** from depth z_0 to z in a homogeneous region and is therefore called the *propagator matrix*.

Thus, if we consider a homogeneous region with three depth points as shown in Figure 5.4, we can readily verify the following properties of the wave propagator and propagator matrices:

$$\left.\begin{aligned} &\mathbf{Q}(z_n, z_1) = \mathbf{Q}(z_n, z_k)\mathbf{Q}(z_k, z_1) \\ &\mathbf{P}(z_n, z_1) = \mathbf{P}(z_n, z_k)\mathbf{P}(z_k, z_1) \\ &z_n > z_k > z_1 \end{aligned}\right\}. \quad (5.27)$$

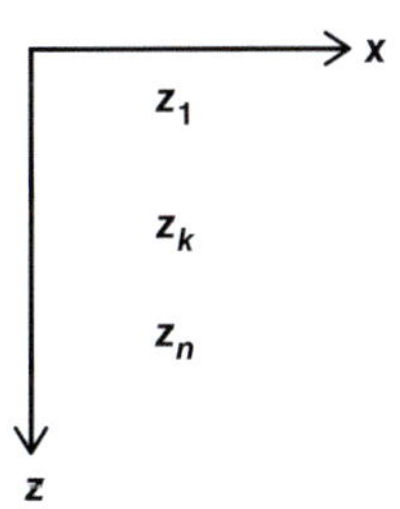

Figure 5.4 A homogeneous elastic medium with three depth points $z_1 < z_k < z_n$.

We can extend equation 5.27 to an arbitrary set of n depth points as

$$\left.\begin{aligned} &\mathbf{Q}(z_n, z_1) = \mathbf{Q}(z_n, z_{n-1})\mathbf{Q}(z_{n-1}, z_{n-2}) \\ &\qquad \ldots \mathbf{Q}(z_3, z_2)\mathbf{Q}(z_2, z_1) \\ &\mathbf{P}(z_n, z_1) = \mathbf{P}(z_n, z_{n-1})\mathbf{P}(z_{n-1}, z_{n-2}) \\ &\qquad \ldots \mathbf{P}(z_3, z_2)\mathbf{P}(z_2, z_1) \\ &z_n > z_{n-1} > z_{n-2} > \ldots z_3 > z_2 > z_1 \end{aligned}\right\}. \quad (5.28)$$

Before proceeding further, it is important to clarify important characteristics of the eigenvector matrix **D**, the eigenvalue matrix **Λ**, wave propagator matrix **Q**, wave vector **v**, and the propagator matrix **P**. The eigenvector matrix is a 6×6 matrix of form

$$\mathbf{D} = \begin{pmatrix} d_{11} & d_{12} & d_{13} & d_{14} & d_{15} & d_{16} \\ d_{21} & d_{22} & d_{23} & d_{24} & d_{25} & d_{26} \\ d_{31} & d_{32} & d_{33} & d_{34} & d_{35} & d_{36} \\ d_{41} & d_{42} & d_{43} & d_{44} & d_{45} & d_{46} \\ d_{51} & d_{52} & d_{53} & d_{54} & d_{55} & d_{56} \\ d_{61} & d_{62} & d_{63} & d_{64} & d_{65} & d_{66} \end{pmatrix}. \quad (5.29)$$

First three columns of **D** are the eigensolutions for the upgoing qP, qS1, and qS2 wavefields, and the next three columns are the same for the downgoing qP, qS1, and qS2 wavefields. In each column of **D**, the top three elements represent the displacement field, and the lower three elements represent the vertical traction. The eigenvalue matrix is of form

$$\Lambda = \begin{pmatrix} \lambda_1 & 0 & 0 & 0 & 0 & 0 \\ 0 & \lambda_2 & 0 & 0 & 0 & 0 \\ 0 & 0 & \lambda_3 & 0 & 0 & 0 \\ 0 & 0 & 0 & \lambda_4 & 0 & 0 \\ 0 & 0 & 0 & 0 & \lambda_5 & 0 \\ 0 & 0 & 0 & 0 & 0 & \lambda_6 \end{pmatrix}, \quad (5.30)$$

in which λ_1, λ_2, λ_3 are the vertical slowness of the upgoing qP, qS1, and qS2 waves and λ_4, λ_5, λ_6 are those for the downgoing qP, qS1, and qS2 waves. In a homogenous region between depths z and $z+\Delta z$, the wave propagator $\mathbf{Q}(z, z+\Delta z)$ is given as

$$\mathbf{Q}(z, z+\Delta z) = e^{i\omega\Lambda\Delta z}$$
$$= \begin{pmatrix} e^{i\omega\lambda_1\Delta z} & 0 & 0 & 0 & 0 & 0 \\ 0 & e^{i\omega\lambda_2\Delta z} & 0 & 0 & 0 & 0 \\ 0 & 0 & e^{i\omega\lambda_3\Delta z} & 0 & 0 & 0 \\ 0 & 0 & 0 & e^{i\omega\lambda_4\Delta z} & 0 & 0 \\ 0 & 0 & 0 & 0 & e^{i\omega\lambda_5\Delta z} & 0 \\ 0 & 0 & 0 & 0 & 0 & e^{i\omega\lambda_6\Delta z} \end{pmatrix}$$
$$= \begin{pmatrix} \mathbf{E}_u & \mathbf{0} \\ \mathbf{0} & \mathbf{E}_d \end{pmatrix}. \tag{5.31}$$

The matrices $\mathbf{E}_u$ and $\mathbf{E}_d$ in equation 5.31 are respectively the upward and downward wavefield propagators,

$$\mathbf{E}_u = \begin{pmatrix} e^{i\omega\lambda_1\Delta z} & 0 & 0 \\ 0 & e^{i\omega\lambda_2\Delta z} & 0 \\ 0 & 0 & e^{i\omega\lambda_3\Delta z} \end{pmatrix}, \tag{5.32}$$

and

$$\mathbf{E}_d = \begin{pmatrix} e^{i\omega\lambda_4\Delta z} & 0 & 0 \\ 0 & e^{i\omega\lambda_5\Delta z} & 0 \\ 0 & 0 & e^{i\omega\lambda_6\Delta z} \end{pmatrix}, \tag{5.33}$$

which propagate the wave vector $\mathbf{v}$ from z to $z+\Delta z$. The wave vector is given as

$$\mathbf{v} = \mathbf{D}^{-1}\mathbf{b} = \begin{pmatrix} \mathbf{v}_u \\ \mathbf{v}_d \end{pmatrix}, \tag{5.34}$$

in which $\mathbf{v}_u$ and $\mathbf{v}_d$ are three-component vectors whose elements are the upgoing and downgoing qP, qS1, and qS2 wavefield amplitudes. Finally,

$$\mathbf{P}(z+\Delta z, z) = \mathbf{D}(z+\Delta z)\mathbf{Q}(z+\Delta z, z)\mathbf{D}^{-1}(z) \tag{5.35}$$

is the propagator matrix, propagating the displacement–stress vector $\mathbf{b}$ from z to $z+\Delta z$.

5.3.4 Reflection and Transmission

After discussing the solution in a homogeneous medium, we now consider an interface at depth $z=z_i$, between two media with different properties (Figure 5.5). As shown in Figure 5.5, we use minus (−) and plus (+) signs to respectively represent the medium just above and just below the interface. The problem of reflection and transmission for an interface between two anisotropic solid layers is given by Fryer and Frazer (1984), and those between fluid–fluid, fluid–solid, and solid–fluid layers is given by Mallick and Frazer (1991). Below, we consider the reflection and transmission problem for solid–solid, fluid–fluid, fluid–solid, and solid–fluid interfaces separately.

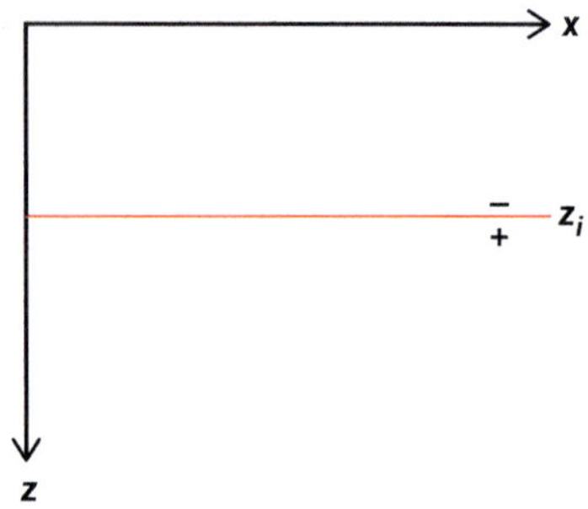

Figure 5.5 An interface at depth $z=z_i$ separating two media with different properties. The minus (−) and plus (+) signs respectively represent the medium infinitesimally above and below the interface.

Solid–Solid Interface: If both media above and below the interface at z_i are solid, then we must have the stress–displacement vector $\mathbf{b}$ (equation 5.17) to be continuous across the interface. Or, in other words, we have

$$\mathbf{b}(z_i^-) = \mathbf{b}(z_i^+). \tag{5.36}$$

From the definition of the wave vector $\mathbf{v}$ (equation 5.20), we can then write equation 5.36 as

$$\mathbf{v}(z_i^+) = \mathbf{D}^{-1}(z_i^+)\mathbf{D}(z_i^-)\mathbf{v}(z_i^-). \tag{5.37}$$

Therefore, equation 5.37 implies that the matrix given as $\mathbf{D}^{-1}(z_i^+)\mathbf{D}(z_i^-)$ is the solid–solid interface wave propagator because it propagates the wave vector (wavefield) across the interface located at depth z_i. This, in turn, leads to the derivation of the reflection and transmission coefficients for an interface between two anisotropic solid layers.

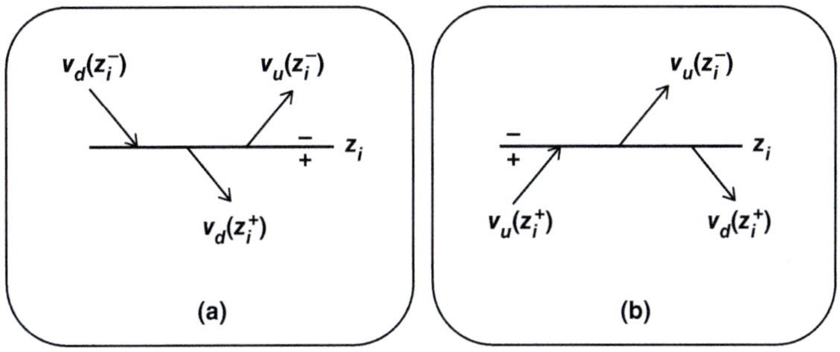

Figure 5.6 Reflection and transmission problems at an interface. (a) Wavefield incident from the top of the interface. (b) Wavefield incident from the bottom of the interface.

We use Figure 5.6 to derive the reflection and transmission coefficients for an interface at depth z_i separating two anisotropic solid layers with different elastic properties. For the wavefield incident from the top of interface (Figure 5.6a), equation 5.37 leads to

$$\begin{pmatrix} \mathbf{0} \\ \mathbf{v}_d(z^+) \end{pmatrix} = \mathbf{D}^{-1}(z^+)\mathbf{D}(z^-)\begin{pmatrix} \mathbf{v}_u(z^-) \\ \mathbf{v}_d(z^-) \end{pmatrix} = \begin{pmatrix} \mathbf{Q}_{11} & \mathbf{Q}_{12} \\ \mathbf{Q}_{21} & \mathbf{Q}_{22} \end{pmatrix}\begin{pmatrix} \mathbf{v}_u(z^-) \\ \mathbf{v}_d(z^-) \end{pmatrix}, \tag{5.38}$$

in which $\mathbf{Q}_{11}$, $\mathbf{Q}_{12}$, $\mathbf{Q}_{21}$, $\mathbf{Q}_{22}$ are 3×3 submatrices of the 6×6 interface wave propagator matrix $\mathbf{Q} = \mathbf{D}^{-1}(z_i^+)\mathbf{D}(z_i^-)$. Now, from the definition of downward reflection and transmission coefficient matrices $\mathbf{R}_d$ and $\mathbf{T}_d$, given in Chapter 4, equation 4.56, we can readily derive $\mathbf{R}_d$ and $\mathbf{T}_d$from equation 5.38 as

$$\left.\begin{aligned} \mathbf{R}_d &= -\mathbf{Q}_{11}^{-1}\mathbf{Q}_{12} \\ \mathbf{T}_d &= \mathbf{Q}_{22} - \mathbf{Q}_{21}\mathbf{Q}_{11}^{-1}\mathbf{Q}_{12} \end{aligned}\right\}. \tag{5.39}$$

For the wavefield incident from below (Figure 5.6b), we can similarly write

$$\begin{pmatrix} \mathbf{v}_u(z^+) \\ \mathbf{v}_d(z^+) \end{pmatrix} = \mathbf{D}^{-1}(z^+)\mathbf{D}(z^-)\begin{pmatrix} \mathbf{v}_u(z^-) \\ \mathbf{0} \end{pmatrix} = \begin{pmatrix} \mathbf{Q}_{11} & \mathbf{Q}_{12} \\ \mathbf{Q}_{21} & \mathbf{Q}_{22} \end{pmatrix}\begin{pmatrix} \mathbf{v}_u(z^-) \\ \mathbf{0} \end{pmatrix}. \tag{5.40}$$

And, from the definitions of upward reflection and transmission coefficients $\mathbf{R}_u$ and $\mathbf{T}_u$ (Chapter 4, equation 4.59), we obtain from equation 5.40

$$\left.\begin{aligned} \mathbf{T}_u &= \mathbf{Q}_{11}^{-1} \\ \mathbf{R}_u &= \mathbf{Q}_{21}\mathbf{Q}_{11}^{-1} \end{aligned}\right\}. \tag{5.41}$$

In deriving equations 5.39 and 5.41, we did not make any assumption for the elastic symmetries for the media above and below the interface. These equations, therefore, represent exact 3×3 reflection and transmission coefficient matrices for an interface between two anisotropic solids. As we had previously shown in Chapter 4 (for example, equations 4.57 and 4.58 showing the expressions for $\mathbf{R}_d$ and $\mathbf{T}_d$), diagonal elements of these matrices are primary (qP–qP, qS1–qS2, qS1–qS2) reflection and transmission coefficients and their off-diagonal elements are mode-converted (qP–qS1, qP–qS2, qS1–qP, qS1–qS2, QS2–qP, and qS2–qS1) reflection and transmission coefficients.

We now combine equations 5.39 and 5.41 into a 6×6 reflection matrix $\mathbf{R}$ as

$$\mathbf{R} = \begin{pmatrix} \mathbf{T}_u & \mathbf{R}_d \\ \mathbf{R}_u & \mathbf{T}_d \end{pmatrix} = \begin{pmatrix} \mathbf{Q}_{11}^{-1} & -\mathbf{Q}_{11}^{-1}\mathbf{Q}_{12} \\ \mathbf{Q}_{21}\mathbf{Q}_{11}^{-1} & \mathbf{Q}_{22} - \mathbf{Q}_{21}\mathbf{Q}_{11}^{-1}\mathbf{Q}_{12} \end{pmatrix} = \mathfrak{F}_d(\mathbf{Q}). \tag{5.42}$$

The operator $\mathfrak{F}_d$ defined by equation 5.42 is called the *downward scatterer* operator (Fryer and Frazer, 1984). Note that $\mathfrak{F}_d$ can be defined for any square ($N\times N$) matrix; if the matrix is the wave propagator for a zone, applying this operator provides the reflection matrix $\mathbf{R}$ for the zone.

Note that in equations 5.38 and 5.40 we related the wavefield below the interface on left to that above the interface to right via the interface wave propagator $\mathbf{Q} = \mathbf{D}^{-1}(z_i^+)\mathbf{D}(z_i^-)$ propagating the wavefield downward from the top to the bottom of the interface. We may similarly define the upward wave propagator

$\mathbf{Q}^u = \mathbf{D}^{-1}(z_i^-)\mathbf{D}(z_i^+)$ and write

$$\begin{pmatrix} \mathbf{v}_u(z^-) \\ \mathbf{v}_d(z^-) \end{pmatrix} = \mathbf{D}^{-1}(z^-)\mathbf{D}(z^+)\begin{pmatrix} \mathbf{0} \\ \mathbf{v}_d(z^+) \end{pmatrix} = \begin{pmatrix} \mathbf{Q}_{11}^{\mathbf{u}} & \mathbf{Q}_{12}^{\mathbf{u}} \\ \mathbf{Q}_{21}^{\mathbf{u}} & \mathbf{Q}_{22}^{\mathbf{u}} \end{pmatrix}\begin{pmatrix} \mathbf{0} \\ \mathbf{v}_d(z^+) \end{pmatrix} \tag{5.43}$$

and

$$\begin{pmatrix} \mathbf{v}_u(z^-) \\ \mathbf{0} \end{pmatrix} = \mathbf{D}^{-1}(z^-)\mathbf{D}(z^+)\begin{pmatrix} \mathbf{v}_\mathbf{u}(z^+) \\ \mathbf{v}_\mathbf{d}(z^+) \end{pmatrix} = \begin{pmatrix} \mathbf{Q}_{11}^{\mathbf{u}} & \mathbf{Q}_{12}^{\mathbf{u}} \\ \mathbf{Q}_{21}^{\mathbf{u}} & \mathbf{Q}_{22}^{\mathbf{u}} \end{pmatrix}\begin{pmatrix} \mathbf{v}_\mathbf{u}(z^+) \\ \mathbf{v}_\mathbf{d}(z^+) \end{pmatrix}, \tag{5.44}$$

and obtain

$$\mathbf{R} = \begin{pmatrix} \mathbf{T}_u & \mathbf{R}_d \\ \mathbf{R}_u & \mathbf{T}_d \end{pmatrix} = \begin{pmatrix} \mathbf{Q}_{11}^{\mathbf{u}} - \mathbf{Q}_{12}^{\mathbf{u}}(\mathbf{Q}_{22}^{\mathbf{u}})^{-1}\mathbf{Q}_{21}^{\mathbf{u}} & \mathbf{Q}_{12}^{\mathbf{u}}(\mathbf{Q}_{22}^{\mathbf{u}})^{-1} \\ -(\mathbf{Q}_{22}^{\mathbf{u}})^{-1}\mathbf{Q}_{21}^{\mathbf{u}} & (\mathbf{Q}_{22}^{\mathbf{u}})^{-1} \end{pmatrix} = \mathfrak{I}_u(\mathbf{Q}^{\mathbf{u}}). \tag{5.45}$$

Like $\mathbf{Q}_{11}, \mathbf{Q}_{12}, \mathbf{Q}_{21}, \mathbf{Q}_{22}, \mathbf{Q}_{11}^u, \mathbf{Q}_{12}^u, \mathbf{Q}_{21}^u, \mathbf{Q}_{22}^u$ in equations 5.43–5.45 are the 3×3 submatrices of the 6×6 upward wave propagator matrix $\mathbf{Q}^u = \mathbf{D}^{-1}(z_i^-)\mathbf{D}(z_i^+)$. Also, the operator $\mathfrak{I}_u$ in equation 5.45 is the *upward scatterer* operator, and like $\mathfrak{I}_d$, $\mathfrak{I}_u$ is defined for a square matrix, and if the matrix is the wave propagator for a zone, it gives the reflection matrix $\mathbf{R}$.

Thus, for any square matrix $\mathbf{M}$, we define the upward and downward scatterer operators as

$$\mathfrak{I}_u(\mathbf{M}) = \begin{pmatrix} \mathbf{M}_{11} - \mathbf{M}_{12}\mathbf{M}_{22}^{-1}\mathbf{M}_{21} & \mathbf{M}_{12}\mathbf{M}_{22}^{-1} \\ -\mathbf{M}_{22}^{-1}\mathbf{M}_{21} & \mathbf{M}_{22}^{-1} \end{pmatrix} \tag{5.46}$$

and

$$\mathfrak{I}_d(\mathbf{M}) = \begin{pmatrix} \mathbf{M}_{11}^{-1} & -\mathbf{M}_{11}^{-1}\mathbf{M}_{12} \\ \mathbf{M}_{21}\mathbf{M}_{11}^{-1} & \mathbf{M}_{22} - \mathbf{M}_{21}\mathbf{M}_{11}^{-1}\mathbf{M}_{12} \end{pmatrix}, \tag{5.47}$$

and note down the following properties of $\mathfrak{I}_u$ and $\mathfrak{I}_d$:

$$\left.\begin{aligned} \mathfrak{I}_d(\mathfrak{I}_d(\mathbf{M})) &= \mathbf{M} \\ \mathfrak{I}_u(\mathfrak{I}_u(\mathbf{M})) &= \mathbf{M} \\ \mathfrak{I}_d(\mathbf{M})\mathfrak{I}_u(\mathbf{M}) &= \mathbf{I} \\ \mathfrak{I}_u(\mathbf{M})\mathfrak{I}_d(\mathbf{M}) &= \mathbf{I} \end{aligned}\right\}. \tag{5.48}$$

The matrix $\mathbf{I}$ in equation 4.48 is an identity matrix of the same size as that of the matrix $\mathbf{M}$. The validity of equation 5.58 can be verified by straightforward matrix algebra and left as an exercise. From here on, we work with the downward scatterer only and omit the subscript "d," or, in other words, we assume $\mathfrak{I} \equiv \mathfrak{I}_d$. Note that from equation 5.42, we have $\mathbf{R} = \mathfrak{I}(\mathbf{Q})$. Therefore, from equation 5.48, we can write

$$\mathfrak{I}(\mathbf{R}) = \mathfrak{I}[\mathfrak{I}(\mathbf{Q})] = \mathbf{Q}. \tag{5.49}$$

We use equation 5.48 later when we describe the reflection and transmission coefficients for a stack of layers.

Fluid–Fluid Interface

When both layers above and below the interface (Figure 5.5) are fluids, only the vertical components of the displacement and traction are continuous, and the shear stresses must vanish. In other words, a fluid–fluid interface must satisfy the boundary condition, given as

$$\left.\begin{aligned} u_z(z_i^+) &= u_z(z_i^-) \\ \tau_{zz}(z_i^+) &= \tau_{zz}(z_i^-) \\ \tau_{zx}(z_i^+) &= \tau_{zx}(z_i^-) = 0 \\ \tau_{zy}(z_i^+) &= \tau_{zy}(z_i^-) = 0 \end{aligned}\right\}. \tag{5.50}$$

Because shear stresses vanish fluids, the displacement–stress vector $\mathbf{b}$ (equation 5.17) for fluids is of the form

$$\mathbf{b} = \left[u_x, u_y, u_z, 0, 0, \frac{1}{i\omega}\tau_{zz}\right]^T. \tag{5.51}$$

Consequently, the eigenvector matrix $\mathbf{D}$ for fluids is of the form

$$\mathbf{D} = \begin{pmatrix} D_{11} & 0 & 0 & D_{14} & 0 & 0 \\ D_{21} & 0 & 0 & D_{24} & 0 & 0 \\ D_{31} & 0 & 0 & D_{34} & 0 & 0 \\ 0 & 0 & 0 & 0 & 0 & 0 \\ 0 & 0 & 0 & 0 & 0 & 0 \\ D_{61} & 0 & 0 & D_{54} & 0 & 0 \end{pmatrix}. \tag{5.52}$$

Because only P waves exist in fluids, the second, third, fifth, and sixth columns of $\mathbf{D}$ are zero (null) vectors. Additionally, because shear stresses τ_{xz} and τ_{yz} are zero for fluids, the fourth and fifth elements of the first and fourth columns of **D** are zero.

By rewriting the fluid–fluid boundary conditions (equation 5.50) in terms of **D** and the wave vector **v** relating the upgoing and downgoing wavefields, the reflection and transmission coefficient matrices for a fluid–fluid interface can be derived as (Mallick and Frazer, 1991)

$$\left.\begin{aligned}
\mathbf{R}_d &= \begin{pmatrix} r_d^{PP} & 0 & 0 \\ 0 & 0 & 0 \\ 0 & 0 & 0 \end{pmatrix} \\
\mathbf{T}_d &= \begin{pmatrix} t_d^{PP} & 0 & 0 \\ 0 & 0 & 0 \\ 0 & 0 & 0 \end{pmatrix} \\
\mathbf{R}_u &= \begin{pmatrix} r_u^{PP} & 0 & 0 \\ 0 & 0 & 0 \\ 0 & 0 & 0 \end{pmatrix} \\
\mathbf{T}_u &= \begin{pmatrix} t_u^{PP} & 0 & 0 \\ 0 & 0 & 0 \\ 0 & 0 & 0 \end{pmatrix}
\end{aligned}\right\}, \tag{5.53}$$

with

$$\left.\begin{aligned}
r_d^{PP} &= \frac{D_{64}(z_i^-)D_{34}(z_i^+) - D_{34}(z_i^-)D_{64}(z_i^+)}{D_{31}(z_i^-)D_{64}(z_i^+) - D_{61}(z_i^-)D_{34}(z_i^+)}, \\
t_d^{PP} &= \frac{D_{31}(z_i^-)D_{64}(z_i^-) - D_{61}(z_i^-)D_{34}(z_i^-)}{D_{31}(z_i^-)D_{64}(z_i^+) - D_{61}(z_i^-)D_{34}(z_i^+)}, \\
r_u^{PP} &= \frac{D_{61}(z_i^+)D_{31}(z_i^-) - D_{31}(z_i^+)D_{61}(z_i^-)}{D_{34}(z_i^+)D_{61}(z_i^-) - D_{64}(z_i^+)D_{31}(z_i^1)}, \\
t_u^{PP} &= \frac{D_{34}(z_i^+)D_{61}(z_i^+) - D_{64}(z_i^+)D_{31}(z_i^+)}{D_{34}(z_i^+)D_{61}(z_i^-) - D_{64}(z_i^+)D_{31}(z_i^-)}
\end{aligned}\right\}. \tag{5.54}$$

Fluid–Solid Interface

The boundary conditions for the case where the medium above the interface is fluid and the one below is solid in Figure 5.5 are given as

$$\begin{pmatrix} u_z(z_i^-) \\ 0 \\ 0 \\ \tau_{zz}(z_i^-) \end{pmatrix} = \begin{pmatrix} u_z(z_i^+) \\ \tau_{zx}(z_i^+) \\ \tau_{zy}(z_i^+) \\ \tau_{zz}(z_i^+) \end{pmatrix}. \tag{5.55}$$

The eigenvector matrix **D** for the fluid layer above the interface is of the type shown in equation 5.52, and we represent $\mathbf{D}(z_i^+)$ for the solid layer below the interface as

$$\mathbf{D}(z_i^+) = \begin{pmatrix} \mathbf{M}_u(z_i^+) & \mathbf{M}_d(z_i^+) \\ \mathbf{N}_u(z_i^+) & \mathbf{N}_d(z_i^+) \end{pmatrix}, \tag{5.56}$$

with $\mathbf{M}_u$, $\mathbf{M}_d$, $\mathbf{N}_u$, and $\mathbf{N}_d$ being 3×3 submatrices of **D**. The reflection and transmission coefficient matrices for a fluid–solid interface is then derived by Mallick and Frazer (1991) as

$$\left.\begin{aligned}
\mathbf{R_d} &= \begin{pmatrix} r_d^{PP} & 0 & 0 \\ 0 & 0 & 0 \\ 0 & 0 & 0 \end{pmatrix} \\
\mathbf{T_d} &= \begin{pmatrix} t_d^{PP} & 0 & 0 \\ t_d^{S1P} & 0 & 0 \\ t_d^{S2P} & 0 & 0 \end{pmatrix} \\
\mathbf{R_u} &= \begin{pmatrix} r_u^{PP} & r_u^{PS1} & r_u^{PS2} \\ r_u^{S1P} & r_u^{S1S1} & r_u^{S1S2} \\ r_u^{S2P} & r_u^{S2S1} & r_u^{S2S2} \end{pmatrix} \\
\mathbf{T_u} &= \begin{pmatrix} t_u^{PP} & t_u^{PS1} & t_u^{PS2} \\ 0 & 0 & 0 \\ 0 & 0 & 0 \end{pmatrix}
\end{aligned}\right\}, \tag{5.57}$$

where

$$r_d^{PP} = \frac{D_{34}(z_i^-) - (\mathbf{d}_1^T\mathbf{x})D_{64}(z_i^-)}{(\mathbf{d}_1^T\mathbf{x})D_{61}(z_i^-) - D_{31}(z_i^-)}, \tag{5.58}$$

$$[t_d^{PP}, t_d^{S1P}, t_d^{S2P}]^T = m\mathbf{x}, \tag{5.59}$$

$$[t_u^{PP}, t_u^{PS1}, t_u^{PS2}] = \frac{1}{n}(\mathbf{d}_1^T\mathbf{B} - \mathbf{d}_2^T), \tag{5.60}$$

and

$$\mathbf{R}_u = D_{61}(z_i^-)\mathbf{x}[t_u^{PP}, t_u^{PS1}, t_u^{PS2}] - \mathbf{B}. \tag{5.61}$$

In equations 5.58, $D_{IJ}(z_i^-)$ stand for the IJ component of the eigenvector matrix **D** for the fluid layer above the interface. The vectors $\mathbf{d}_1$, $\mathbf{d}_2$, **x**, matrix **B**, and the scalars m and n in above are defined as

$$\mathbf{d}_1 = [D_{34}(z_i^+), D_{35}(z_i^+), D_{36}(z_i^+)]^T, \tag{5.62}$$

$$\mathbf{d}_2 = [D_{31}(z_i^+), D_{32}(z_i^+), D_{33}(z_i^+)], \tag{5.63}$$

$$\mathbf{x} = [\{\mathbf{N}_d^{-1}(z_i^+)\}_{13}, \{\mathbf{N}_d^{-1}(z_i^+)\}_{23}, \{\mathbf{N}_d^{-1}(z_i^+)\}_{33}]^T, \tag{5.64}$$

$$\mathbf{B} = \mathbf{N}_d^{-1}(z_i^+)\mathbf{N}_u(z_i^+), \tag{5.65}$$

$$m = \frac{D_{61}(z_i^-)D_{34}(z_i^-) - D_{64}(z_i^-)D_{31}(z_i^-)}{\mathbf{d}_1^T\mathbf{x}D_{61}(z_i^-) - D_{31}(z_i^-)}, \tag{5.66}$$

and

$$n = \mathbf{d}_1^T\mathbf{x}D_{61}(z_i^-) - D_{31}(z_i^-). \tag{5.67}$$

Solid–Fluid Interface

For a solid over a fluid interface, the boundary conditions are just the reverse of the fluid over solid interface and are given by

$$\begin{pmatrix} u_z(z_i^-) \\ \tau_{zx}(z_i^-) \\ \tau_{zy}(z_i^-) \\ \tau_{zz}(z_i^-) \end{pmatrix} = \begin{pmatrix} u_z(z_i^+) \\ 0 \\ 0 \\ \tau_{zz}(z_i^+) \end{pmatrix}. \tag{5.68}$$

By expressing the eigenvector matrices $\mathbf{D}(z_i^-)$ as solid (equation 5.56) and $\mathbf{D}(z_i^+)$ as fluid (equation 5.52), Mallick and Frazer (1991) derived the reflection and transmission matrices for the solid–fluid interface as

$$\left.\begin{aligned} \mathbf{R}_d &= \begin{pmatrix} r_d^{PP} & r_d^{PS1} & r_d^{PS2} \\ r_d^{S1P} & r_d^{S1S1} & r_d^{S1S2} \\ r_d^{S2P} & r_d^{S2S1} & r_d^{S2S2} \end{pmatrix} \\ \mathbf{T}_d &= \begin{pmatrix} t_d^{PP} & t_d^{PS1} & t_d^{PS2} \\ 0 & 0 & 0 \\ 0 & 0 & 0 \end{pmatrix} \\ \mathbf{R}_u &= \begin{pmatrix} R_u^{PP} & 0 & 0 \\ 0 & 0 & 0 \\ 0 & 0 & 0 \end{pmatrix} \\ \mathbf{T}_u &= \begin{pmatrix} t_u^{PP} & 0 & 0 \\ t_u^{S1P} & 0 & 0 \\ t_u^{S2P} & 0 & 0 \end{pmatrix} \end{aligned}\right\}, \tag{5.69}$$

with their individual elements given as

$$\mathbf{R}_d = \begin{pmatrix} r_d^{PP} & r_d^{PS1} & r_d^{PS2} \\ r_d^{S1P} & r_d^{S1S1} & r_d^{S1S2} \\ r_d^{S2P} & r_d^{S2S1} & r_d^{S2S2} \end{pmatrix} = D_{64}(z_i^+)\mathbf{y}[t_d^{PP}, t_d^{PS1}, t_d^{PS2}] - \mathbf{F}, \tag{5.70}$$

$$[t_d^{PP}, t_d^{PS1}, t_d^{PS2}] = \frac{1}{n}(\mathbf{d}_4^T\mathbf{F} - \mathbf{d}_3^T), \tag{5.71}$$

$$r_u^{PP} = \frac{D_{32}(z_i^+) - \mathbf{d}_4^T\mathbf{y}D_{61}(z_i^+)}{\mathbf{d}_4^T\mathbf{y}D_{64}(z_i^+) - D_{34}(z_i^+)}, \tag{5.72}$$

and

$$[t_u^{PP}, t_u^{S1P}, t_u^{S2P}]^T = m\mathbf{y}. \tag{5.73}$$

The vectors $\mathbf{y}$, $\mathbf{d}_3$, $\mathbf{d}_4$, the matrix $\mathbf{F}$, and the scalars m, n in equations 5.70–5.73 are given as

$$\mathbf{y} = [\{\mathbf{N}_u^{-1}(z_i^-)\}_{13}, \{\mathbf{N}_u^{-1}(z_i^-)\}_{23}, \{\mathbf{N}_u^{-1}(z_i^-)\}_{33}]^T, \tag{5.74}$$

$$\mathbf{d}_3 = [D_{34}(z_i^-), D_{35}(z_i^-), D_{36}(z_i^-)]^T, \tag{5.75}$$

$$\mathbf{d}_4 = [D_{31}(z_i^-), D_{32}(z_i^-), D_{33}(z_i^-)]^T, \tag{5.76}$$

$$\mathbf{F} = \mathbf{N}_u^{-1}(z_i^-)\mathbf{N}_d(z_i^-), \tag{5.77}$$

$$m = \frac{D_{64}(z_i^+)D_{31}(z_i^+) - D_{61}(z_i^+)D_{34}(z_i^+)}{\mathbf{d}_4^T\mathbf{y}D_{64}(z_i^+) - D_{34}(z_i^+)}, \tag{5.78}$$

and

$$n = \mathbf{d}_4^T\mathbf{y}D_{64}(z_i^+) - D_{34}(z_i^+). \tag{5.79}$$

5.3.5 The Eigenvalue and Eigenvector Matrices and the Inverse of the Eigenvector Matrix

In the above section, we obtained exact reflection and transmission coefficients for solid–solid, solid–fluid, fluid–solid, and fluid–fluid interfaces. Before going further into finding a solution for a layered system using these coefficients, it is first important to clarify a few concepts we have so far developed.

In Chapter 3, we derived the Christoffel equation, given as $p^2\mathbf{\Gamma} = \rho\mathbf{u}$ (equation 3.77). And in equation 5.16, we derived the 1D elastic system, which, in a source-free region, is given as $\partial_z\mathbf{b} = i\omega\mathbf{A}\mathbf{b}$. To solve the Christoffel equation, we needed the eigenvalues and eigenvectors of the Christoffel matrix $\mathbf{\Gamma}$. To solve the 1D elastic system also, we need the eigenvalues and eigenvectors of the elastic system matrix $\mathbf{A}$. So, what is the difference between these eigenvalues and eigenvectors? In Chapter 3, we also noted that the eigenvalues and eigenvectors of $\mathbf{\Gamma}$ can be analytically computed for isotropic and transversely isotropic symmetries. For other symmetries, however, they must be numerically computed except for some specific propagation directions. This is also true for **A**, and so we may ask what is an efficient numerical method to compute the eigenvalues and eigenvectors of the elastic system matrix **A**? Additionally, computing reflection and transmission coefficients requires computing $\mathbf{D}^{-1}$, the inverse of the eigenvector matrix **D** of **A**. Although efficient algorithms to compute the inverse of the 6×6 matrices are available, we must note that to compute the seismic response, this inverse must be computed for every set of the ray-parameter pair p_x and p_y and for every frequency ω, which could easily result in millions of such inverse computations and be computationally demanding. We may therefore ask: given that each column of **D** are $[\mathbf{u}, \mathbf{t}]^T$ where **u** is the displacement and **t** is the vertical traction, is there a simple physics to compute $\mathbf{D}^{-1}$without computing the actual inverse? In the following, we address each of these issues.

The Christoffel Matrix and the 1D Elastic System Matrix

From our discussions in Chapter 3, following are the properties of the Christoffel matrix $\mathbf{\Gamma}$:

1) It is a 3×3 matrix.
2) The determinant of the matrix $\mathbf{\Gamma} - \rho\mathbf{I}$ is cubic in p^2.
   a) Eigenvalues are the slowness (p) for the upgoing and downgoing qP, qS1, and qS2 wavefields, and the eigenvectors are the corresponding displacement fields. To solve Christoffel equation, we solve for the vertical slowness p_z for a given set of the horizontal slowness values p_x and p_y.

The 1D elastic system matrix **A**, on the contrary, is a 6×6 matrix whose elements are functions of the elastic stiff coefficients (elements of the 6×6 matrix **C**), density (ρ), and the horizontal slowness p_x and p_y.

1) Eigenvalues of **A** are the vertical slowness (p_z) of the upgoing and downgoing qP, qS1, and qS2 wavefields.
2) Eigenvectors of **A** are the solutions of the displacement–stress $[\mathbf{u}, \mathbf{t}]^T$ vectors for the upgoing and downgoing qP, qS1, and qS2 wavefields.

Therefore, the Christoffel equation provides the eigensolutions for the particle displacement field while the 1D elastic system provides the eigensolutions for both particle displacement and vertical traction.

Compute-Efficient Implementation of the Eigenvalues and Eigenvectors of the 1D Elastic System Matrix

Eigenvalue and the eigenvector matrices (the matrices $\mathbf{\Lambda}$ and **D** in equation 5.19) of the elastic system matrix (**A**) for a general anisotropic (triclinic symmetry) system require numerical computation. However, for monoclinic, orthorhombic, hexagonal, and isotropic symmetries, Fryer and Frazer (1987) derived analytical expressions for both $\mathbf{\Lambda}$ and **D**. For practical implementation of computing synthetic seismograms, Mallick and Frazer (1990) implemented a compute-efficient numerical method to compute $\mathbf{\Lambda}$ and **D** and observed the following:

Analytic computation of $\mathbf{\Lambda}$ and **D** is

1) slightly faster than the numerical computation for an isotropic symmetry,
2) takes about the same time as the numerical computation for hexagonal symmetry, and
3) slower than the numerical computation for orthorhombic and monoclinic systems.

In view of the above, it is therefore important to discuss some details of numerical computations of $\mathbf{\Lambda}$ and **D**.

Frazer and Fryer (1989) looked at two different methods to numerically compute $\mathbf{\Lambda}$ and **D**: (1) perturbation theory and (2) Jacobi iteration. Backgrounds for both these methods can be found in Frazer and Fryer (1989), Pease (1965), and Golub and Van Loan (2013). In high-performance parallel computing platforms, the Jacobi iteration method can be efficiently parallelized, and Mallick and Frazer (1990) concluded that this method is superior to the perturbation theory. Below, we briefly outline the fundamentals of computing $\mathbf{\Lambda}$ and $\mathbf{D}$ using Jacobi iteration. For further details of this method, see Pease (1965), Golub and Van Loan (2013), Press et al. (1989), and Frazer and Fryer (1989).

The Jacobi method to iteratively compute the eigenvalues and eigenvectors for complex symmetric matrices is given in Golub and Van Loan (2013). Our elastic system matrix $\mathbf{A}$, on the other hand, is of the form given in equation 5.18, which is complex but not symmetric. Therefore, to use the Jacobi method for complex symmetric matrices, given in Golub and Van Loan (2013), we must compute a 6×6 matrix $\mathbf{A}'$

$$\mathbf{A}' = \mathbf{S}^{-1}\mathbf{A}\mathbf{S}, \tag{5.80}$$

such that $\mathbf{A}'$ is a complex symmetric matrix. Using the properties of the elastic system matrix $\mathbf{A}$, Frazer and Fryer (1989) showed that the matrix $\mathbf{A}'$ is symmetric for

$$\mathbf{S} = \frac{e^{-\frac{i\pi}{4}}}{\sqrt{2}} \begin{pmatrix} \mathbf{I} & i\mathbf{I} \\ i\mathbf{I} & \mathbf{I} \end{pmatrix}, \tag{5.81}$$

with its inverse given as

$$\mathbf{S}^{-1} = \frac{e^{\frac{i\pi}{4}}}{\sqrt{2}} \begin{pmatrix} \mathbf{I} & -i\mathbf{I} \\ -i\mathbf{I} & \mathbf{I} \end{pmatrix}. \tag{5.82}$$

In equations 5.81 and 5.82, $i = \sqrt{-1}$ and $\mathbf{I}$ is a 3×3 identity matrix. Once we compute $\mathbf{A}'$ using equations 5.80–5.82, obtaining the eigenvalues and eigenvectors of $\mathbf{A}'$ using the Jacobi iteration is straightforward. However, how are these eigenvalues and eigenvectors related to those of our original elastic system matrix $\mathbf{A}$, defined in equation 5.18? Frazer and Fryer (1989) had shown that the eigenvalues of $\mathbf{A}'$ are same as those for $\mathbf{A}$. In addition, Frazer and Fryer (1989) had also shown that if $\mathbf{D}'$ is the eigenvector matrix of $\mathbf{A}'$, then the eigenvector matrix $\mathbf{D}$ of the elastic system matrix $\mathbf{A}$ is given as

$$\mathbf{D} = \mathbf{S}^{-1}\mathbf{D}'. \tag{5.83}$$

It is therefore possible to use equation 5.83 and compute $\mathbf{D}$ once we compute $\mathbf{D}'$ via the Jacobi method.

Inverse of the Eigenvector Matrix of the Elastic System Matrix

To find out how to efficiently compute $\mathbf{D}^{-1}$, we first define the inner product for two sets of vectors $\mathbf{b}_1 = [\mathbf{u}_1, \mathbf{t}_1]^T$ and $\mathbf{b}_2 = [\mathbf{u}_2, \mathbf{t}_2]^T$ as

$$\langle \mathbf{b}_1, \mathbf{b}_2 \rangle = \mathbf{u}_1^T \mathbf{t}_2 + \mathbf{u}_2^T \mathbf{t}_1, \tag{5.84}$$

and show that if $\mathbf{b}_1$ and $\mathbf{b}_2$ are the eigenvectors of the elastic system matrix $\mathbf{A}$ defined by equation 5.18, then $\langle \mathbf{b}_1, \mathbf{b}_2 \rangle = 0$. So, let $\mathbf{b}_1 = [\mathbf{u}_1, \mathbf{t}_1]^T$ and $\mathbf{b}_2 = [\mathbf{u}_2, \mathbf{t}_2]^T$ are two eigenvectors of the elastic system matrix $\mathbf{A}$ with eigenvalues λ_1 and λ_2. We can then write

$$\lambda_1 \langle \mathbf{b}_1, \mathbf{b}_2 \rangle = \langle \lambda_1 \mathbf{b}_1, \mathbf{b}_2 \rangle = \langle \mathbf{A}\mathbf{b}_1, \mathbf{b}_2 \rangle. \tag{5.85}$$

Because the elastic system matrix $\mathbf{A}$ is given by equation 5.18, we can write equation 5.85 as

$$\begin{aligned} \lambda_1 \langle \mathbf{b}_1, \mathbf{b}_2 \rangle &= \langle \mathbf{A}\mathbf{b}_1, \mathbf{b}_2 \rangle \\ &= \left\langle \begin{pmatrix} \mathbf{T} \cdot \mathbf{u}_1 + \mathbf{C} \cdot \mathbf{t}_1 \\ \mathbf{S} \cdot \mathbf{u}_1 + \mathbf{T}^T \cdot \mathbf{t}_1 \end{pmatrix}, \begin{pmatrix} \mathbf{u}_2 \\ \mathbf{t}_2 \end{pmatrix} \right\rangle \\ &= (\mathbf{T} \cdot \mathbf{u}_1 + \mathbf{C} \cdot \mathbf{t}_1)^T \cdot \mathbf{t}_2 + \mathbf{u}_2^T \\ &\quad \cdot (\mathbf{S} \cdot \mathbf{u}_1 + \mathbf{T}^T \cdot \mathbf{t}_1) \\ &= (\mathbf{t}_1^T \cdot \mathbf{C}^T + \mathbf{u}_1^T \cdot \mathbf{T}^T) \cdot \mathbf{t}_2 + \mathbf{u}_2^T \\ &\quad \cdot (\mathbf{S} \cdot \mathbf{u}_1 + \mathbf{T}^T \cdot \mathbf{t}_1). \end{aligned} \tag{5.86}$$

We can similarly write

$$\begin{aligned} \lambda_2 \langle \mathbf{b}_1, \mathbf{b}_2 \rangle &= \langle \mathbf{b}_1, \lambda_2 \mathbf{b}_2 \rangle = \langle \mathbf{b}_1, \mathbf{A}\mathbf{b}_2 \rangle \\ &= \left\langle \begin{pmatrix} \mathbf{u}_1 \\ \mathbf{t}_1 \end{pmatrix}, \begin{pmatrix} \mathbf{T} \cdot \mathbf{u}_2 + \mathbf{C} \cdot \mathbf{t}_2 \\ \mathbf{S} \cdot \mathbf{u}_2 + \mathbf{T}^T \cdot \mathbf{t}_2 \end{pmatrix} \right\rangle \\ &= \mathbf{u}_1^T \cdot (\mathbf{S} \cdot \mathbf{u}_2 + \mathbf{T}^T \cdot \mathbf{t}_2) \\ &\quad + (\mathbf{T} \cdot \mathbf{u}_2 + \mathbf{C} \cdot \mathbf{t}_2)^T \cdot \mathbf{t}_1 \\ &= \mathbf{u}_1^T \cdot (\mathbf{S} \cdot \mathbf{u}_2 + \mathbf{T}^T \cdot \mathbf{t}_2) \\ &\quad + (\mathbf{u}_2^T \cdot \mathbf{T}^T + \mathbf{t}_2^T \cdot \mathbf{C}^T) \cdot \mathbf{t}_1. \end{aligned} \tag{5.87}$$

From equations 5.86 and 5.87, we obtain

$$(\lambda_1 - \lambda_2) \langle \mathbf{b}_1, \mathbf{b}_2 \rangle = 0. \tag{5.88}$$

Therefore, if the eigenvalues λ_1 and λ_2 are different, then we must have

$$\langle \mathbf{b}_1, \mathbf{b}_2 \rangle = 0. \tag{5.89}$$

Equation 5.89 implies that the eigenvectors of the elastic system matrix **A** are orthogonal. It also implies that if all eigenvalues of **A** are distinct, and if we normalize each column **b** of the eigenvector matrix **D** such that

$$\langle \mathbf{b}, \mathbf{b} \rangle = 1, \tag{5.90}$$

then

$$(\mathbf{D}^{-1})^T = \mathbf{JD}, \tag{5.91}$$

in which

$$\mathbf{J} = \begin{pmatrix} \mathbf{0} & \mathbf{I} \\ \mathbf{I} & \mathbf{0} \end{pmatrix}. \tag{5.92}$$

As before, **I** in equation 5.92 is a 3×3 identity matrix. The matrix **J** is called the *inner product matrix*.

Equations 5.90–5.92 provide an easy way to compute $\mathbf{D}^{-1}$ using the inner product matrix, without computing the actual inverse. However, what happens if the eigenvalues are not distinct? This happens in the isotropic medium where the two shear waves propagate with the same wave speed, and thus have the same eigenvalues. For hexagonally symmetric and orthorhombic media also, the eigenvalues are the same along certain directions of propagation. For all these cases, however, the eigenvalues will at most have a multiplicity of two. Assume that $\mathbf{b}_1$ and $\mathbf{b}_2$ are such two eigenvectors of **A** with the same eigenvalue. If, in the eigenvector matrix **D**, we replace $\mathbf{b}_1$ by

$$\mathbf{b}_1^{new} = \mathbf{b}_1 - \frac{\langle \mathbf{b}_1, \mathbf{b}_2 \rangle}{\langle \mathbf{b}_2, \mathbf{b}_1 \rangle} \mathbf{b}_2, \tag{5.93}$$

then the resulting $\mathbf{b}_1^{new}$ will be orthogonal to $\mathbf{b}_2$. The method of normalization shown by equation 5.93 is *Gram–Schmidt* normalization (Pease, 1965).

In view of the above discussion, below is the method to compute $\mathbf{D}^{-1}$:

1) Do Gram–Schmidt normalization (equation 5.93) to each pair if eigenvector of **D** with same eigenvalue.
2) Normalize each column **b** of the resulting eigenvector matrix **D** after Gram–Schmidt normalization such that $\langle \mathbf{b}, \mathbf{b} \rangle = 1$. We can do this simply by dividing each component of **b** by $\langle \mathbf{b}, \mathbf{b} \rangle^{1/2}$.
3) Compute $(\mathbf{D}^{-1})^T$ using the inner product matrix (equation 5.91).
4) Compute $\mathbf{D}^{-1}$ by taking the transpose of $(\mathbf{D}^{-1})^T$.

5.3.6 Reflection and Transmission in a Homogeneous Medium

Having discussed how to compute the eigenvalue and eigenvector matrices $\mathbf{\Lambda}$ and **D** and $\mathbf{D}^{-1}$, we are now ready to compute the reflection and transmission response for a stack of layers.

We first consider a homogeneous region. As shown in Figure 5.7, we consider two interfaces at depths z_{i-1} and z_i bounding a homogeneous layer of thickness Δz. As before, we use z_{i-1}^+ to denote immediately beneath the interface at z_{i-1} and z_i^- to denote immediately above the interface at z_i (see Figure 5.7). From our definition of the wave propagator for a homogeneous region (equations 5.23, 5.24, 5.31–5.33) and the reflection matrix **R** using the downward scatterer (equation 5.42), we obtain

$$\mathbf{R}(z_i^-, z_{i-1}^+) = \Im \begin{pmatrix} \mathbf{E}_u & \mathbf{0} \\ \mathbf{0} & \mathbf{E}_d \end{pmatrix} = \begin{pmatrix} \mathbf{E}_u^{-1} & \mathbf{0} \\ \mathbf{0} & \mathbf{E}_d \end{pmatrix}$$
$$= \begin{pmatrix} e^{-i\omega q_P^u \Delta z} & 0 & 0 & 0 & 0 & 0 \\ 0 & e^{-i\omega q_{S1}^u \Delta z} & 0 & 0 & 0 & 0 \\ 0 & 0 & e^{-i\omega q_{S2}^u \Delta z} & 0 & 0 & 0 \\ 0 & 0 & 0 & e^{i\omega q_P^d \Delta z} & 0 & 0 \\ 0 & 0 & 0 & 0 & e^{i\omega q_{S1}^d \Delta z} & 0 \\ 0 & 0 & 0 & 0 & 0 & e^{i\omega q_{S2}^d \Delta z} \end{pmatrix}. \tag{5.94}$$

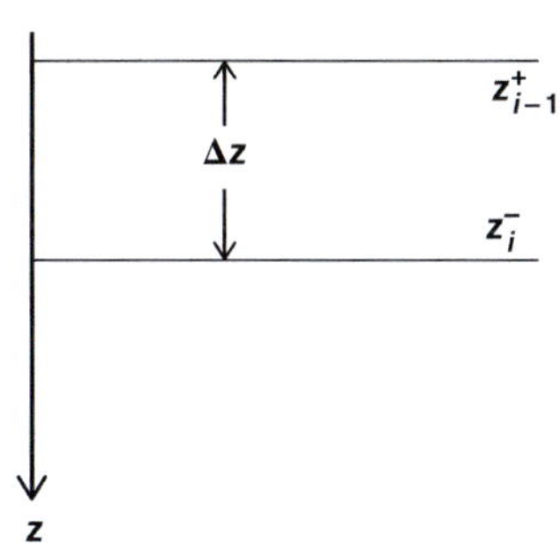

Figure 5.7 Reflection and transmission in a homogeneous region between depths z_{i-1} and z_i.

In equation 5.94, we used the downward scatterer operator shown in equation 5.42 but omitted the subscript "*d*."

5.3.7 Reflection and Transmission in a Stack of Layers

Moving from a homogeneous region, we now consider a stack of layers, shown in Figure 5.8, which form the basic building blocks for computing synthetic seismograms for 1D media.

From our definition of the reflection matrix **R** (equation 5.42), we write

$$\mathbf{R}(z_n^+, z_0^-) = \mathfrak{F}\mathbf{Q}(z_n^+, z_0^-). \tag{5.95}$$

But, using the property of the wave propagator **Q** shown in equation 5.27, we may write equation 5.95 as

$$\mathbf{R}(z_n^+, z_0^-) = \mathfrak{F}[\mathbf{Q}(z_n^+, z_k^+)\mathbf{Q}(z_k^+, z_0^-)]. \tag{5.96}$$

As before, in equations 5.95 we used the downward scatterer and omitted the subscript "*d*." Finally, from equation 4.49, we write equation 5.96 as

$$\mathbf{R}(z_n^+, z_0^-) = \mathfrak{F}[\mathfrak{F}\{\mathbf{R}(z_n^+, z_k^+)\}\mathfrak{F}\{\mathbf{R}(z_k^+, z_0^-)\}]. \tag{5.97}$$

Expanding the right-hand side of equation 5.97 using the definition of the downward scatterer operator (equation 5.42), equation 5.97 defines the components $\mathbf{T}_u$, $\mathbf{R}_d$, $\mathbf{R}_u$, and $\mathbf{T}_d$ of $\mathbf{R}(z_n^+, z_0^-)$, defined by equation 5.42 in terms those of $\mathbf{R}(z_n^+, z_k^+)$ and $\mathbf{R}(z_k^+, z_0^-)$ as follows:

$$\left.\begin{aligned}
\mathbf{T}_u^{n0} &= \mathbf{T}_u^{k0}(\mathbf{I} - \mathbf{R}_d^{nk}\mathbf{R}_u^{k0})^{-1}\mathbf{T}_u^{nk} \\
\mathbf{R}_d^{n0} &= \mathbf{R}_d^{k0} + \mathbf{T}_u^{k0}\mathbf{R}_d^{nk}(\mathbf{I} - \mathbf{R}_u^{k0}\mathbf{R}_d^{nk})^{-1}\mathbf{T}_d^{k0} \\
\mathbf{R}_u^{n0} &= \mathbf{R}_u^{nk} + \mathbf{T}_d^{nk}\mathbf{R}_u^{k0}(\mathbf{I} - \mathbf{R}_d^{nk}\mathbf{R}_u^{k0})^{-1}\mathbf{T}_u^{nk} \\
\mathbf{T}_d^{n0} &= \mathbf{T}_d^{nk}(\mathbf{I} - \mathbf{R}_u^{k0}\mathbf{R}_d^{nk})^{-1}\mathbf{T}_d^{k0}
\end{aligned}\right\}. \tag{5.98}$$

Equation 5.98 is a very useful equation because it allow building the reflection and transmission coefficient matrices over a zone from those given for its parts. In general, consider two separate zones, (1) A to B and (2) B to C, as shown in Figure 5.9. Then, the reflection and transmission coefficients $\mathbf{T}_u^{AC}, \mathbf{R}_d^{AC}, \mathbf{R}_u^{AC}, \mathbf{T}_d^{AC}$ for zones A to C can be expressed in terms of the reflection and transmission coefficients $\mathbf{T}_u^{AB}, \mathbf{R}_d^{AB}, \mathbf{R}_u^{AB}, \mathbf{T}_d^{AB}$ for zones A to B and $\mathbf{T}_u^{BC}, \mathbf{R}_d^{BC}, \mathbf{R}_u^{BC}, \mathbf{T}_d^{BC}$ for zones B to C as

$$\left.\begin{aligned}
\mathbf{R}_d^{AC} &= \mathbf{R}_d^{AB} + \mathbf{T}_u^{AB}\mathbf{R}_d^{BC}(\mathbf{I} - \mathbf{R}_u^{AB}\mathbf{R}_d^{BC})^{-1}\mathbf{T}_d^{AB} \\
\mathbf{T}_d^{AC} &= \mathbf{T}_d^{BC}(\mathbf{I} - \mathbf{R}_u^{AB}\mathbf{R}_d^{BC})^{-1}\mathbf{T}_d^{AB} \\
\mathbf{R}_u^{AC} &= \mathbf{R}_u^{BC} + \mathbf{T}_d^{BC}\mathbf{R}_u^{AB}(\mathbf{I} - \mathbf{R}_d^{BC}\mathbf{R}_u^{AB})^{-1}\mathbf{T}_u^{BC} \\
\mathbf{T}_u^{AC} &= \mathbf{T}_u^{AB}(\mathbf{I} - \mathbf{R}_d^{BC}\mathbf{R}_u^{AB})^{-1}\mathbf{T}_u^{BC}
\end{aligned}\right\}. \tag{5.99}$$

Additionally, if $\mathbf{T}_u^{AC}, \mathbf{R}_d^{AC}, \mathbf{R}_u^{AC}, \mathbf{T}_d^{AC}$ and $\mathbf{T}_u^{AB}$, $\mathbf{R}_d^{AB}, \mathbf{R}_u^{AB}, \mathbf{T}_d^{AB}$ are given, we can easily invert

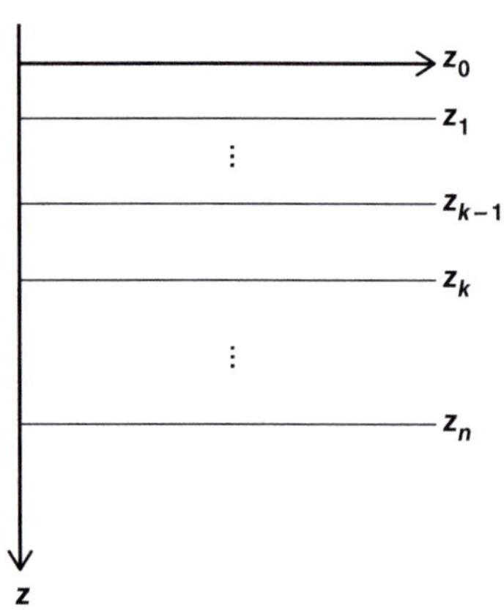

Figure 5.8 Reflection and transmission for a stack of layers.

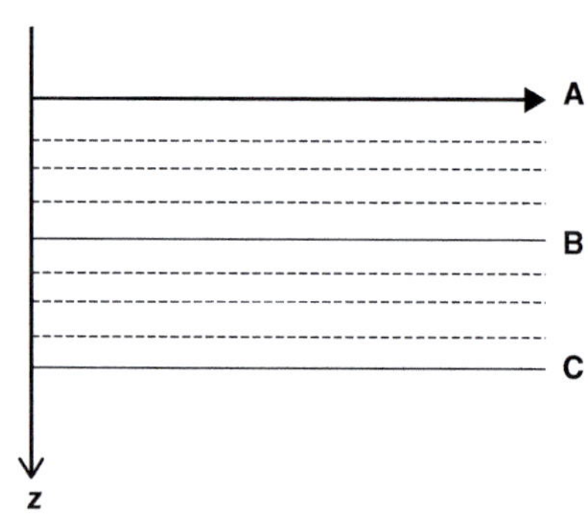

Figure 5.9 Computation of the reflection and transmission coefficients for zones A to B in terms of those for zones A to B and B to C.

equation 5.99 and compute $\mathbf{T}_u^{BC}, \mathbf{R}_d^{BC}, \mathbf{R}_u^{BC}, \mathbf{T}_d^{BC}$ as

$$\left.\begin{aligned}\mathbf{R}_d^{BC} &= (\mathbf{R}_d^{AC} - \mathbf{R}_d^{AB})(\mathbf{T}_d^{AB})^{-1}\\ &\quad \times [\mathbf{T}_u^{AB} + (\mathbf{R}_d^{AC} - \mathbf{R}_d^{AB})(\mathbf{T}_d^{AB})^{-1}\mathbf{R}_u^{AB}]^{-1}\\ \mathbf{T}_d^{BC} &= \mathbf{T}_d^{AC}(\mathbf{T}_d^{AB})^{-1}(\mathbf{I} - \mathbf{R}_u^{AB}\mathbf{R}_d^{BC})\\ \mathbf{R}_u^{BC} &= \mathbf{R}_u^{AC} - \mathbf{T}_d^{AB}\mathbf{R}_u^{AB}(\mathbf{I} - \mathbf{R}_d^{BC}\mathbf{R}_u^{AB})^{-1}\mathbf{T}_u^{BC}\\ \mathbf{T}_u^{BC} &= (\mathbf{I} - \mathbf{R}_d^{BC}\mathbf{R}_u^{AB})(\mathbf{T}_u^{AB})^{-1}\mathbf{T}_u^{AC}\end{aligned}\right\}. \tag{5.100}$$

Mallick and Frazer (1988) used equation 5.100 for computing synthetic seismic response for vertical seismic profile geometries. Sen and Roy (2003) used equation 5.100 to analytically compute the gradient of the objective function, a vital element to any local optimization method, which we discussed in Chapter 1. We revisit equation 5.100 in Chapter 7 when we discuss local optimization methods.

5.3.8 Reflection and Transmission in a Homogeneous Layer and an Interface

We now consider a homogeneous region from depth z_{i-1}^+ and the entire layer and then go past the interface at depth z_i to derive the reflection matrix $\mathbf{R}(z_i^+, z_{i-1}^+)$, as shown in Figure 5.10.

By following the same procedure as before, we can write

$$\begin{aligned}\mathbf{R}(z_i^+, z_{i-1}^+) &= \mathfrak{I}\{\mathbf{Q}(z_i^+, z_{i-1}^+)\}\\ &= \mathfrak{I}[\mathbf{Q}(z_i^+, z_i^-)\mathbf{Q}(z_i^-, z_{i-1}^+)]\\ &= \mathfrak{I}\{\mathfrak{I}(\mathbf{R}(z_i^+, z_i^-)(\mathfrak{I}(\mathbf{R}(z_i^-, z_{i-1}^+))\}.\end{aligned} \tag{5.101}$$

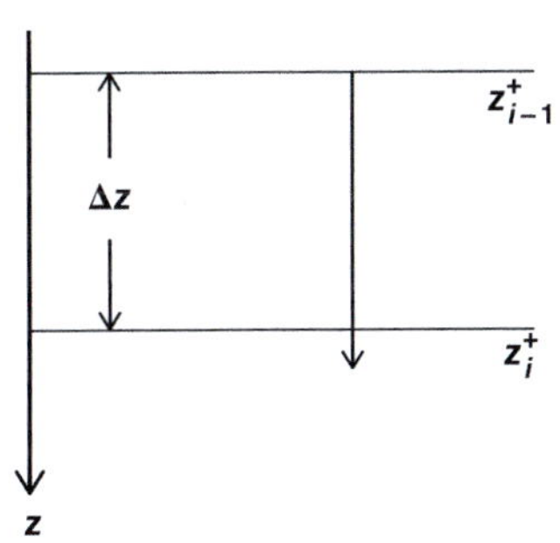

Figure 5.10 Reflection and transmission coefficients for a homogeneous region and an interface between depths z_{i-1}^+ and z_i^+.

On the right-hand side of equation 5.101, the first term is reflection matrix for the interface at z_i (equation 5.42 for a solid–solid interface and equivalent expressions for fluid–fluid, fluid–solid, and solid–fluid interfaces). The second term on the right-hand side of equation 5.101 on the other hand is the reflection matrix for a homogeneous region between depths z_{i-1}^+ and z_i^- given by equation 5.94. We thus explicitly write equation 5.101 as

$$\mathbf{R}(z_i^+, z_{i-1}^+) = \mathfrak{I}\left[\mathfrak{I}\begin{pmatrix}\mathbf{t}_u & \mathbf{r}_d\\ \mathbf{r}_u & \mathbf{t}_d\end{pmatrix}\mathfrak{I}\begin{pmatrix}\mathbf{E}_u^{-1} & \mathbf{0}\\ \mathbf{0} & \mathbf{E}_d\end{pmatrix}\right]. \tag{5.102}$$

For reflection and transmission coefficient matrices, in equation 5.102, we used the lowercase letters $\mathbf{t}_u$, $\mathbf{r}_d$, $\mathbf{r}_u\mathbf{t}_d$ instead of uppercase letters $\mathbf{T}_u$, $\mathbf{R}_d$, $\mathbf{R}_u$, $\mathbf{T}_d$ which we previously used in equation 5.42. The reason for this choice will be obvious in the next subsection when we will derive the iteration equations. Finally, using the downward scatterer operator defined in equation 5.42, we obtain

$$\mathbf{R}(z_i^+, z_{i-1}^+) = \begin{pmatrix}\mathbf{E}_u^{-1}\mathbf{t}_u & \mathbf{E}_u^{-1}\mathbf{r}_d\mathbf{E}_d\\ \mathbf{r}_u & \mathbf{t}_d\mathbf{E}_d\end{pmatrix}. \tag{5.103}$$

5.3.9 Iteration Equations

We now consider a stack of n, bounded above either by a half-space or a free surface and below by a half-space (Figure 5.11). We also assume that we already computed the reflection matrix $\mathbf{R}$ from the top and up to the interface at z_{i-1}^+ (shown by blue arrow in Figure 5.11), which is given as

$$\mathbf{R}(z_{i-1}^+, z_0^-) = \begin{pmatrix}\mathbf{T}_\mathbf{u} & \mathbf{R}_\mathbf{d}\\ \mathbf{R}_\mathbf{u} & \mathbf{T}_\mathbf{d}\end{pmatrix}. \tag{5.104}$$

To these reflection and transmission coefficient matrices $\mathbf{T}_u$, $\mathbf{R}_d$, $\mathbf{R}_u$, $\mathbf{T}_d$, we want to add an additional layer and an interface, shown by the red arrow in Figure 5.11, and build the new reflection and transmission coefficient matrices $\mathbf{T}_u^{new}, \mathbf{R}_d^{new}, \mathbf{R}_u^{new}, \mathbf{T}_d^{new}$ through z_i^+. We have already shown that the reflection matrix for the layer and the interface we are adding is given by equation 5.103. Using the same argument as

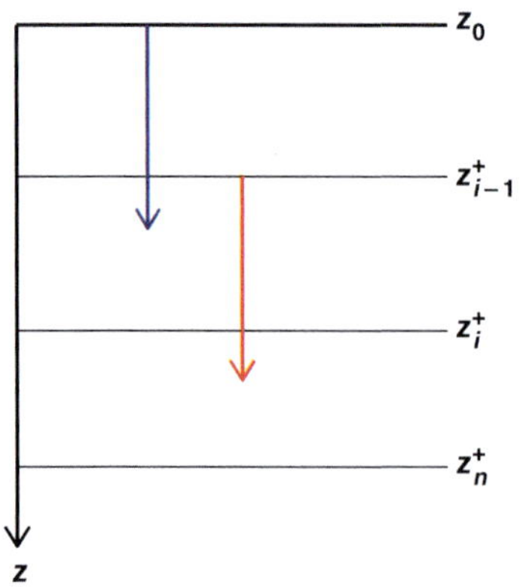

Figure 5.11 Building of the iteration equation by successively adding a layer and an interface.

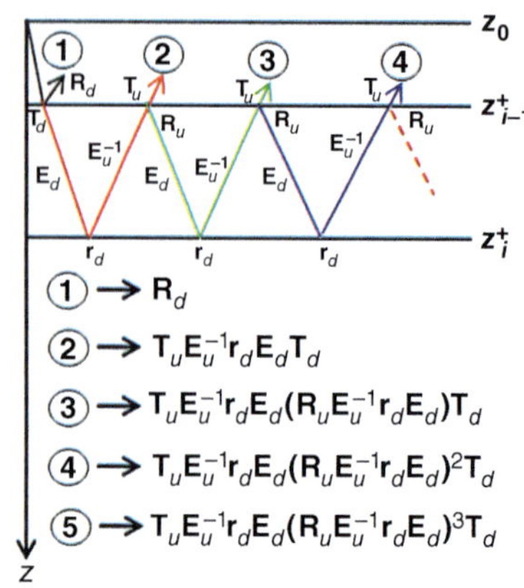

Figure 5.12 A physical explanation for $\mathbf{R}_d^{new}$, shown in equation 5.106.

before, we therefore write

$$\begin{aligned}\mathbf{R}(z_i^+, z_0^-) &= \Im\{\mathbf{Q}(z_i^+, z_0)\}\\ &= \Im\{\mathbf{Q}(z_i^+, z_{i-1}^+)\mathbf{Q}\left(z_{i-1}^+, z_0^-\right)\}\\ &= \Im[\Im\{\mathbf{R}(z_i^+, z_{i-1}^+)\}\Im\{\mathbf{R}(z_{i-1}^+, z_0^-)\}]\\ &= \Im\left[\Im\begin{pmatrix}\mathbf{E}_u^{-1}\mathbf{t}_u & \mathbf{E}_u^{-1}\mathbf{r}_d\mathbf{E}_d\\ \mathbf{r}_u & \mathbf{t}_d\mathbf{E}_d\end{pmatrix}\Im\begin{pmatrix}\mathbf{R}_u & \mathbf{T}_d\\ \mathbf{T}_u & \mathbf{R}_d\end{pmatrix}\right].\end{aligned} \tag{5.105}$$

Expanding right-hand side of equation 5.105 using the definition of the downward scatterer operator (equation 5.42) provides $\mathbf{T}_u^{new}$, $\mathbf{R}_d^{new}$, $\mathbf{R}_u^{new}$, $\mathbf{T}_d^{new}$ as

$$\left.\begin{aligned}\mathbf{T}_u^{new} &= \mathbf{T}_u(\mathbf{I} - \mathbf{E}_u^{-1}\mathbf{r}_d\mathbf{E}_d\mathbf{R}_u)^{-1}\mathbf{E}_u^{-1}\mathbf{t}_u\\ \mathbf{R}_d^{new} &= \mathbf{R}_d + \mathbf{T}_u\mathbf{E}_u^{-1}\mathbf{r}_d\mathbf{E}_d(\mathbf{I} - \mathbf{R}_u\mathbf{E}_u^{-1}\mathbf{r}_d\mathbf{E}_d)^{-1}\mathbf{T}_d\\ \mathbf{R}_u^{new} &= \mathbf{r}_u + \mathbf{t}_d\mathbf{E}_d\mathbf{R}_u(\mathbf{I} - \mathbf{E}_u^{-1}\mathbf{r}_d\mathbf{E}_d\mathbf{R}_u)^{-1}\mathbf{E}_u^{-1}\mathbf{t}_u\\ \mathbf{T}_d^{new} &= \mathbf{t}_d\mathbf{E}_d(\mathbf{I} - \mathbf{R}_u\mathbf{E}_u^{-1}\mathbf{r}_d\mathbf{E}_d)^{-1}\mathbf{T}_d\end{aligned}\right\}. \tag{5.106}$$

There are physical explanations for each term of equation 5.106. We use Figure 5.12 to provide a physical explanation of $\mathbf{R}_d^{new}$ (the second equation in the equation set 5.105). Explanations for $\mathbf{T}_u^{new}$, $\mathbf{R}_u^{new}$, and $\mathbf{T}_d^{new}$ can be similarly given.

For convenience, we used simple ray diagrams with different colors and numbers within circles to physically interpret each event within $\mathbf{R}_d^{new}$ in Figure 5.12 and explain each of them as below:

1) Downward propagating wavefield from the top of the layer will be reflected from the interface z_{i-1} with the downward reflection coefficient $\mathbf{R}_d$ for the stack of layers and propagate to the top. This is the first term in $\mathbf{R}_d^{new}$, and shown as 1 and as a black ray.
2) Besides reflecting at interface z_{i-1}, the wavefield will be transmitted downward with the downward transmission coefficient matrix $\mathbf{T}_d$ for the stack of layers to depth z_{i-1}, downward propagate through layer i with transmission coefficient $\mathbf{E}_d$, reflected from interface z_i with the downward reflection coefficient $\mathbf{r}_d$, upward propagate through layer i with the transmission coefficient $\mathbf{E}_u^{-1}$, and be transmitted through the stack of layers to the top with the upward transmission coefficient $\mathbf{T}_u$ for the stack. Because we are looking at the event from top, the order of multiplication of these matrices must be reversed, and thus the first (primary) reflection from the interface at z_i, shown in red ray and as event 2 in Figure 5.12, is $\mathbf{T}_u\mathbf{E}_u^{-1}\mathbf{r}_d\mathbf{E}_d\mathbf{T}_d$.
3) Using the same argument as above, we can readily see different multiple reflections from layer i as below:
   a) First-order multiple (shown in green ray and as event 3 in Figure 5.12):
      i) $\mathbf{T}_u\mathbf{E}_u^{-1}\mathbf{r}_d\mathbf{E}_d(\mathbf{R}_u\mathbf{E}_u^{-1}\mathbf{r}_d\mathbf{E}_d)\mathbf{T}_d$.
   b) Second-order multiple (blue ray and event 4 in Figure 5.12):
      i) $\mathbf{T}_u\mathbf{E}_u^{-1}\mathbf{r}_d\mathbf{E}_d(\mathbf{R}_u\mathbf{E}_u^{-1}\mathbf{r}_d\mathbf{E}_d)^2\mathbf{T}_d$.

c) Third-order multiple (not shown as a ray but shown as event 5 in Figure 5.12):
   i) $\mathbf{T}_u\mathbf{E}_u^{-1}\mathbf{r}_d\mathbf{E}_d(\mathbf{R}_u\mathbf{E}_u^{-1}\mathbf{r}_d\mathbf{E}_d)^3\mathbf{T}_d$.

d) We can similarly obtain the expressions for the multiples of other (higher than three) orders.

From the above arguments, we can thus write $\mathbf{R}_d^{new}$ as an infinite series

$$\mathbf{R}_d^{new} = \mathbf{R}_d + \mathbf{T}_u\mathbf{E}_u^{-1}\mathbf{r}_d[\mathbf{I} + \mathbf{R}_u\mathbf{E}_u^{-1}\mathbf{r}_d\mathbf{E}_d + (\mathbf{R}_u\mathbf{E}_u^{-1}\mathbf{r}_d\mathbf{E}_d)^2 + (\mathbf{R}_u\mathbf{E}_u^{-1}\mathbf{r}_d\mathbf{E}_d)^3 + \cdots]\mathbf{T}_d. \quad (5.107)$$

The term inside the square brackets on the right-hand side of equation 5.107 is an infinite series of type $\mathbf{I} + \mathbf{A} + \mathbf{A}^2 + \mathbf{A}^3 + \cdots$, where $\mathbf{A}$ is a 3×3 matrix. From the fundamentals of matrix algebra (for example, Pease, 1965), if the determinant of $\mathbf{A}$ is less than 1, this infinite series is convergent and can be expressed as $(\mathbf{I} - \mathbf{A})^{-1}$, For our case, $\mathbf{A} = \mathbf{R}_u\mathbf{E}_u^{-1}\mathbf{r}_d\mathbf{E}_d$, and which are reflection and transmission coefficients. Determinant of $\mathbf{A}$ is therefore less than 1, and we can write equation 5.107 as

$$\mathbf{R}_d^{new} = \mathbf{R}_d + \mathbf{T}_u\mathbf{E}_u^{-1}\mathbf{r}_d(\mathbf{I} - \mathbf{R}_u\mathbf{E}_u^{-1}\mathbf{r}_d\mathbf{E}_d)^{-1}\mathbf{T}_d, \quad (5.108)$$

which is the same as the expression for $\mathbf{R}_d^{new}$ in equation 5.106. Besides showing the equivalence of $\mathbf{R}_d^{new}$ between equation 5.108 and previously shown in equation 5.106, which we derived from the scatterer operator, our physical explanation also points out the following important facts:

1) The terms of the form $(\mathbf{I} - \mathbf{A})^{-1}$ in equation 5.106 can be considered as the *reverberation operator*, which computes the primary and all multiple reflections. Equation 5.106 can thus be considered the equation providing the complete reflection and transmission coefficients, which includes the primaries, mode conversions, and multiple reflections.
2) The infinite series $(\mathbf{I} - \mathbf{A})^{-1} = \mathbf{I} + \mathbf{A} + \mathbf{A}^2 + \mathbf{A}^3 + \cdots$ in equation 5.106 can be truncated to construct compute partial responses. For example, replacing $(\mathbf{I} - \mathbf{A})^{-1}$ by $\mathbf{I}$ would give primary reflection only. If we replace $(\mathbf{I} - \mathbf{A})^{-1}$ by $\mathbf{I} + \mathbf{A}$, we obtain the primary and multiple reflections of first order, and so on.

Equations shown in equation 5.105 are called the iteration equations and provide the fundamental building blocks for computing the reflection and transmission coefficients for a stack of layers. In Figure 5.13, we provide the pseudo-code showing how to use equation 5.106 and build the reflection and transmission coefficient matrices $\mathbf{T}_u$, $\mathbf{R}_d$, $\mathbf{R}_u$, $\mathbf{T}_d$ from scratch. We consider a stack of n layers and recursively use equation 5.106 to build $\mathbf{T}_u$, $\mathbf{R}_d$, $\mathbf{R}_u$, $\mathbf{T}_d$. At the beginning, we set the downward reflection coefficient matrix $\mathbf{R}_d$ to a zero (null) matrix and the upward and downward transmission coefficient matrices $\mathbf{T}_u$ and $\mathbf{T}_d$ to an identity matrix. If we consider the stack of layers is bounded on the top by a half-space, we set the upward reflection coefficient matrix $\mathbf{R}_u$ to a null matrix. For many practical applications, we may, however, like the stack of layers to be bounded on top by a free surface, where the stresses vanish,

```
Start
    Set R_d = 0; T_u = T_d = I; and R_u = 0 or R_u^F (the free surface reflection coefficient).
    Layer Loop (1 to n in steps of 1)
        Use iteration equations (equation 5.105) to add layer i and the interface i to existing
        T_u, R_d, R_u, T_d from the previous step.
    End Layer loop
    Output the reflection and transmission coefficient matrices end exit
End
```

Figure 5.13 Pseudo-code to build the reflection and transmission coefficient matrices for a stack of layers.

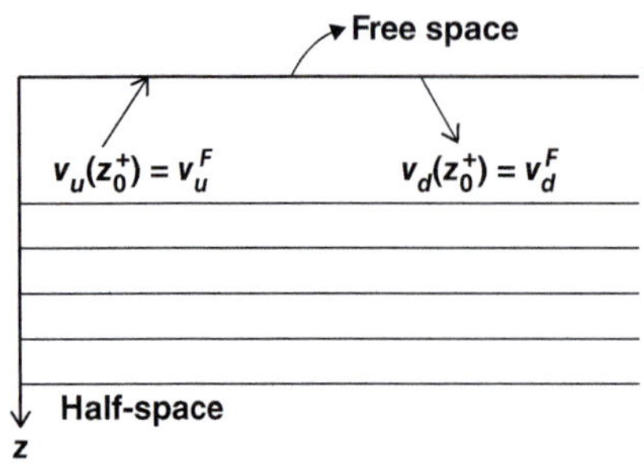

Figure 5.14 Computation of the free surface reflection coefficient.

in which case we set $\mathbf{R}_u$ to the free surface reflection coefficient matrix $\mathbf{R}_u^F$. We then use equation 5.106 and iteratively add one layer and an interface in a layer loop. Once we come to the end of the stack of layers, we output the reflection and transmission coefficient matrices and exit.

In the above discussion, we mentioned that at starting we may either set $\mathbf{R}_u$ to a null matrix or the free surface reflection coefficient matrix $\mathbf{R}_u^F$. But how do we compute $\mathbf{R}_u^F$? Strictly speaking, the free surface boundary condition is valid for an interface when one of the media is vacuum and the stresses vanish. The surface of the earth is characterized by an interface with air on the top. Because the elastic properties of the air are much smaller than that of the earth (solid in land and water in ocean), for all practical purposes, the surface of the earth can be considered as a free surface. We use Figure 5.14 to illustrate how to compute the free surface reflection coefficient. As shown in Figure 5.14, we consider the surface of the earth to be a free surface, located at z_0. If an upgoing wavefield $\mathbf{v}_u(z_0^+) = \mathbf{v}_u^F$ is incident at the free surface and generates a downgoing reflected wavefield $\mathbf{v}_d(z_0^+) = \mathbf{v}_d^F$, we define the free surface reflection coefficient $\mathbf{R}_u^F$ to be 3×3 matrix, which relates these upgoing and downgoing wavefields as

$$\mathbf{v}_d^F = \mathbf{R}_u^F \mathbf{v}_u^F. \tag{5.109}$$

Now recall the relation between the wave vector and the stress–displacement vector (equation 5.20). Because the stress vanishes at the free surface, we have

$$\begin{bmatrix} \mathbf{u}(z_0^+) \\ \mathbf{0} \end{bmatrix} = \mathbf{D}(z_0^+) \begin{bmatrix} \mathbf{v}_u(z_0^+) \\ \mathbf{v}_d(z_0^+) \end{bmatrix}. \tag{5.110}$$

We replace $\mathbf{u}(z_0^+)$ by $\mathbf{u}^F$, $\mathbf{v}_u(z_0^+)$ by $\mathbf{v}_u^F$, and $\mathbf{v}_d(z_0^+)$ by $\mathbf{v}_d^F$ to write equation 5.110 as

$$\begin{bmatrix} \mathbf{u}^F \\ \mathbf{0} \end{bmatrix} = \begin{pmatrix} \mathbf{M}_u^F & \mathbf{M}_d^F \\ \mathbf{N}_u^F & \mathbf{N}_d^F \end{pmatrix} \begin{bmatrix} \mathbf{v}_u^F \\ \mathbf{v}_d^F \end{bmatrix}. \tag{5.111}$$

The matrices $\mathbf{M}_u^F, \mathbf{M}_d^F, \mathbf{N}_u^F, \mathbf{N}_d^F$ in equation 1.110 are the 3×3 submatrices of the eigenvector matrix $\mathbf{D}(z_0^+) = \mathbf{D}^F$ for the top layer immediately below the free surface at z_0. Equation 5.111 is equivalent to following two equations:

$$\mathbf{M}_u^F \mathbf{v}_u^F + \mathbf{M}_d^F \mathbf{v}_d^F = \mathbf{u}^F \tag{5.112}$$

and

$$\mathbf{N}_u^F \mathbf{v}_u^F + \mathbf{N}_d^F \mathbf{v}_d^F = \mathbf{0}. \tag{5.113}$$

From equation 5.113, we have

$$\mathbf{v}_d^F = -(\mathbf{N}_d^F)^{-1} \mathbf{N}_u^F \mathbf{v}_u^F. \tag{5.114}$$

By comparing equation 5.114 with equation 5.109, we thus have the reflection coefficient matrix for the free surface to be

$$\mathbf{R}_u^F = -(\mathbf{N}_d^F)^{-1} \mathbf{N}_u^F. \tag{5.115}$$

Having obtained $\mathbf{R}_u^F$, we now have the complete plan to build the reflection and transmission coefficient matrices $\mathbf{T}_u$, $\mathbf{R}_d$, $\mathbf{R}_u$, $\mathbf{T}_d$ for a stack of layers.

5.3.10 The Source Term

So far, we solved the homogeneous (source-free) part of equation 5.16 by defining the wave vector (equation 5.20) and building the reflection and transmission coefficient matrices $\mathbf{T}_u$, $\mathbf{R}_d$, $\mathbf{R}_u$, $\mathbf{T}_d$ for a stack of layers. Our recorded seismograms are, however, not these reflection and transmission coefficients. In land, we record particle displacement, velocity, or acceleration, and in water, we record pressure. We compute our responses in the frequency (ω) and ray-parameter (p_x, p_y) domain. Therefore, if we compute particle displacement, the particle velocity and acceleration are respectively the first- and second-time derivatives of the displacement, which we can obtain by multiplying the displacement by $i\omega$ and $-\omega^2$. Here, we thus concentrate on computing the displacement field for the land environments and the pressure field for the marine environments. To do that, we first need to introduce the source term, which, in equation 5.17, we derived as

$$\mathbf{F} = \frac{i}{\omega}[0,0,0,f_x,f_y,f_z]^T. \quad (5.116)$$

The components f_x, f_y, and f_z on the right-hand side of equation 5.116 are respectively the x, y, and z components of the external body force that the source generates. In Chapter 3, we discussed the radiation patterns from point sources, where we considered three types of sources: (1) point explosion (uniform), (2) point force, and (3) point double couple (moment tensor). Here, we combine them all and consider a point source located at $\mathbf{r}_s = [0, 0, z_s]$ with the body force equivalent given as

$$\mathbf{f} = \mathbf{h}\delta(\mathbf{r}-\mathbf{r}_s) - \mathbf{M}\nabla\delta(\mathbf{r}-\mathbf{r}_s). \quad (5.117)$$

Note that equation 5.117 combines all three source types we previously considered in Chapter 3. For example, if we set $\mathbf{M} = \mathbf{0}$, then equation 5.117 is a point force. If we set $\mathbf{h} = \mathbf{0}$, then equation 5.117 is a point double couple when $\mathbf{M}$ is a moment tensor and an explosion if $\mathbf{M}$ is an identity tensor (matrix).

Assuming the source located at $\mathbf{r}_s = [0, 0, z_s]$, we have

$$\left.\begin{aligned}&\delta(\mathbf{r}-\mathbf{r}_s) = \delta(x)\delta(y)\delta(z-z_s)\\ &\nabla\delta(\mathbf{r}-r_s) = [\delta'(x)\delta(y)\delta(z-z_s),\\ &\qquad \delta(x)\delta'(y)\delta(z-z_s), \delta(x)\delta(y)\delta'(z-z_s)]\end{aligned}\right\}. \quad (5.118)$$

We also recall from our discussions in Chapter 3 that $\mathbf{M}$ in equation 5.117 is the symmetric second-order tensor, which can be written as

$$\mathbf{M} = \begin{pmatrix} M_{xx} & M_{xy} & M_{xz}\\ M_{xy} & M_{yy} & M_{yz}\\ M_{xz} & M_{yz} & M_{zz}\end{pmatrix}. \quad (5.119)$$

Because we work in $\omega - p_x - p_y$ domain, we first forward Fourier transform equation 5.117 and convert the variables x and y to p_x and p_y using the convention we introduced in Chapter 4 (equation 4.9). In this domain and for $\mathbf{r}_s = [0, 0, z_s]$, equation 5.117 becomes

$$\begin{aligned}\mathbf{f} = {}&\mathbf{h}\delta(z-z_s) - i\omega p_x\mathbf{M}_x\delta(z-z_s)\\ &- i\omega p_y\mathbf{M}_y\delta(z-z_s) - \mathbf{M}_z\delta'(z-z_s),\end{aligned} \quad (5.120)$$

with

$$\left.\begin{aligned}\mathbf{M}_x &= [M_{xx}, M_{xy}, M_{xz}]^T\\ \mathbf{M}_y &= [M_{xy}, M_{yy}, M_{yz}]^T\\ \mathbf{M}_z &= [M_{xz}, M_{yz}, M_{zz}]^T\end{aligned}\right\}. \quad (5.121)$$

Deriving equation 5.120 with the vectors $\mathbf{M}_x$, $\mathbf{M}_y$, $\mathbf{M}_z$ defined by equation 5.121 is not difficult and left as an exercise. From equations 5.120 and 5.121, we can write the source term $\mathbf{F}$ in equation 5.116 as

$$\mathbf{F} = \mathbf{F}_1\delta(z-z_s) + \mathbf{F}_2\delta'(z-z_s), \quad (5.122)$$

where

$$\mathbf{F}_1 = \begin{pmatrix} 0\\ 0\\ 0\\ \frac{i}{\omega}h_x + p_xM_{xx} + p_yM_{xy}\\ \frac{i}{\omega}h_y + p_xM_{xy} + p_yM_{yy}\\ \frac{i}{\omega}h_z + p_xM_{xz} + p_yM_{yz}\end{pmatrix} \quad (5.123)$$

and

$$\mathbf{F}_2 = \begin{pmatrix} 0\\ 0\\ 0\\ -\frac{i}{\omega}M_{zx}\\ -\frac{i}{\omega}M_{zy}\\ -\frac{i}{\omega}M_{zz}\end{pmatrix}. \quad (5.124)$$

Deriving equation 5.122 is also left as an exercise at the end of the chapter.

In our original inhomogeneous equation $\partial_z\mathbf{b} = i\omega\mathbf{A}\mathbf{b} + \mathbf{F}$ (equation 5.16), we now have the complete expression for the source term $\mathbf{F}$ in terms of the seismic source parameters given in equation 5.117, and we are ready to solve the equation. In doing so, we first recall that the solution of the homogeneous equation $\partial_z \mathbf{b} = i\omega \mathbf{A}\mathbf{b}$ is the propagator matrix $\mathbf{P}(z, z_0)$, which we previously derived (equation 5.26), and propagates the stress–displacement vector $\mathbf{b}$ (equation 5.17) from depth z_0 to any arbitrary depth z. If $\mathbf{P}(z, z_0)$ is the solution to the homogeneous system $\partial_z \mathbf{b} = i\omega \mathbf{A}\mathbf{b}$, then the corresponding solution for the inhomogeneous system $\partial_z \mathbf{b} = i\omega \mathbf{A}\mathbf{b} + \mathbf{F}$ can be written as

$$\mathbf{b}(z) = \mathbf{P}(z, z_0)\mathbf{b}(z_0) + \int_{z_0}^{z} \mathbf{P}(z, \xi)\mathbf{F}(\xi)d\xi. \quad (5.125)$$

Validity of equation 5.125 can be verified by direct substitution and left as an exercise.

To use the solution provided by equation 5.125, we first need to solve the integral on the right-hand side of equation 5.125. Assuming a point source, located at depth $z = z_s$ with minus (–) and plus (+) signs representing the zones immediately above and below the source (Figure 5.15), we can combine equation 5.125 with equation 5.122 and write

$$\mathbf{b}(z_s^+) = \mathbf{P}(z_s^+, z_s^-)\mathbf{b}(z_s^-) + \int_{z_s^-}^{z_s^+} \mathbf{P}(z_s^+, \xi) \times \{\mathbf{F}_1 \delta(\xi - z_s) + \mathbf{F}_2 \delta'(\xi - z_s)\} d\xi. \quad (5.126)$$

Note that z_s is the source location, not an interface boundary. Thus, for the definition of the propagator matrix (equation 5.26), $\mathbf{P}(z_s^+, z_s^-)$ is an identity matrix, and therefore $\mathbf{P}(z_s^+, z_s^-)\mathbf{b}(z_s^-) = \mathbf{b}(z_s^-)$. Therefore, we write equation 5.126 as

$$\mathbf{b}(z_s^+) = \mathbf{b}(z_s^-) + \mathbf{F}_1 \int_{z_s^-}^{z_s^+} \mathbf{P}(z_s^+, \xi)\delta(\xi - z_s)d\xi + \mathbf{F}_2 \int_{z_s^-}^{z_s^+} \mathbf{P}(z_s^+, \xi)\delta'(\xi - z_s)d\xi. \quad (5.127)$$

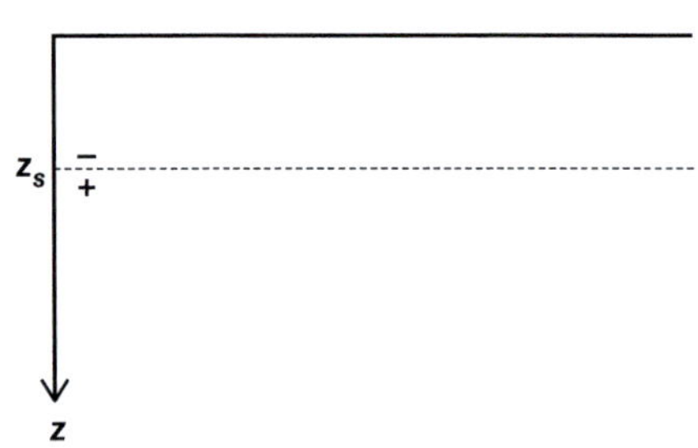

Figure 5.15 A point source located at depth z_s. The minus (–) and plus (+) signs respectively represent the zones immediately above and immediately below the source location.

In the second term on the right-hand side of equation 5.127 $\mathbf{P}(z_s^+, \xi)$ within the region of integration from z_s^- to z_s^+ is always an identity matrix, and thus from the definition of the delta function, we evaluate this term as

$$\mathbf{F}_1 \int_{z_s^-}^{z_s^+} \mathbf{P}(z_s^+, \xi)\delta(\xi - z_s) = \mathbf{F}_1. \quad (5.128)$$

As for the third term on the right-hand side of equation 5.127, we use integration by parts and write this term as

$$\mathbf{F}_2 \int_{z_s^-}^{z_s^+} \mathbf{P}(z_s^+, \xi)\delta'(\xi - z_s)d\xi = \mathbf{F}_2 \{\mathbf{P}(z_s^+, \xi)\delta(\xi - z_s)\}\Big|_{z_s^-}^{z_s^+} - \mathbf{F}_2 \int_{z_s^-}^{z_s^+} \delta(\xi - z_s)\partial_\xi \mathbf{P}(z_s^+, \xi)d\xi. \quad (5.129)$$

Clearly, the first term on the right-hand side of equation 5.129 is zero and

$$\int_{z_s^-}^{z_s^+} \delta(\xi - z_s)\partial_\xi \mathbf{P}(z_s^+, \xi)d\xi = [\partial_\xi \mathbf{P}(z_s, \xi)]_{\xi = z_s}. \quad (5.130)$$

Using equations 5.128–5.130, we thus write equation 5.127 as

$$\mathbf{b}(z_s^+) = \mathbf{b}(z_s^-) + \mathbf{F}_1 - \mathbf{F}_2[\partial_\xi \mathbf{P}(z_s, \xi)]_{\xi = z_s}. \quad (5.131)$$

Using straightforward matrix algebra and the properties of the propagator matrix, given as an exercise at the end of the chapter, it can be shown that

$$\partial_\xi [\mathbf{P}(z, \xi)] = -i\omega \mathbf{P}(z, \xi)\mathbf{A}(\xi). \quad (5.132)$$

And, using equation 5.132 leads equation 5.131 to

$$\mathbf{b}(z_s^+) = \mathbf{b}(z_s^-) + \mathbf{F}_1 + i\omega\mathbf{A}(z_s)\mathbf{F}_2. \quad (5.133)$$

Deriving equation 5.133 is also straightforward and left as an exercise. From equation 5.133, we therefore conclude that having a point source at $z = z_s$ leads to a jump in the displacement–stress vector and given as

$$\mathbf{s}(z_s) = \mathbf{b}(z_s^+) - \mathbf{b}(z_s^-) = \mathbf{F}_1 + i\omega\mathbf{A}(z_s)\mathbf{F}_2, \quad (5.134)$$

in which $\mathbf{A}(z_s)$ is the elastic system matrix (equation 5.18) at the source location, and $\mathbf{F}_1$ and $\mathbf{F}_2$ are respectively defined in equations 5.123 and 5.124. Because the wave vector $\mathbf{v}$ is defined in equation 5.20 as $\mathbf{v} = \mathbf{D}^{-1}\mathbf{b}$, where $\mathbf{D}$ is the eigenvector of the elastic system matrix $\mathbf{A}$, we can also write the corresponding jump in the wave vector due to the source as

$$\begin{aligned}\mathbf{\Sigma} &= [-\Sigma_u, \Sigma_d]^T = \mathbf{D}^{-1}(z_s)\mathbf{s}(z_s)\\ &= \mathbf{D}^{-1}(z_s)[\mathbf{F}_1 + i\omega\mathbf{A}(z_s)\mathbf{F}_2]. \quad (5.135)\end{aligned}$$

We have thus introduced the source term as a jump in the displacement–stress vector $\mathbf{b}$ and the wave vector $\mathbf{v}$.

5.3.11 Computation of the Source Wavefield

Having introduced the source term, we are now ready to compute the synthetic seismic response. Here, we compute the source wavefield, both above and below the source, and next we show how to compute the seismic response at the receiver location.

Referring to Figure 5.16, we consider a stack of n layers or media with varying elastic properties. On top, we assume that the stack of layers is bounded either by a half-space or a free surface, and at the bottom, it is bounded by a half-space. In addition, we use the following notations for the reflection matrix $\mathbf{R}$ (equation 5.42):

$$\left.\begin{aligned}\mathbf{R}(z_0^-, z_s^-) &= \mathbf{R}^{0s}\\ \mathbf{R}(z_s^+, z_n) &= \mathbf{R}^{sn}\end{aligned}\right\}. \quad (5.136)$$

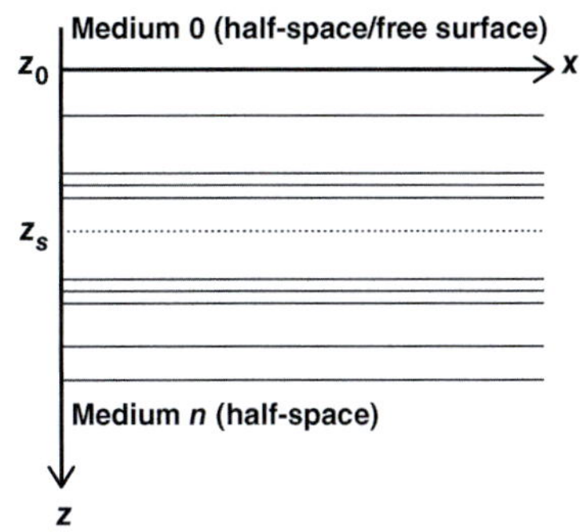

Figure 5.16 Illustration of the computation of the source wavefield for a stack of n layers, bounded above either by a half-space or a free surface and below by a half-space. The source is assumed to be at depth $z = z_s$, shown by a dashed line.

In the above discussion (equation 5.135), we have shown that the source at z_s results in a jump of $\mathbf{\Sigma}$ in the wave vector field. In other words, this means that

$$\mathbf{v}(z_s^-) = \mathbf{v}(z_s^+) - \mathbf{\Sigma}. \quad (5.137)$$

From our definition of the wave propagator, we can also write

$$\mathbf{v}(z_s^+) = \mathbf{Q}(z_s^+, z_n^+)\mathbf{v}(z_n^+), \quad (5.138)$$

in which $\mathbf{Q}(z_s^+, z_n^+)$ is the wave propagator we previously defined (equation 5.24). We also recall the downward and upward scatterer operators $\mathfrak{I}_d$ and $\mathfrak{I}_u$, respectively defined in equations 5.42 and 5.45. In developing the reflection matrix, we so far used the downward scatterer only and omitted the subscript "d." Here, instead we use the properties of the upward scatterer $\mathfrak{I}_u$ to write

$$\mathbf{Q}(z_s^+, z_n^+) = \mathfrak{I}_u(\mathbf{R}^{sn}). \quad (5.139)$$

And, from equations 5.138 and 5.139, we can thus write equation 5.137 as

$$\mathbf{v}(z_s^-) = \mathfrak{I}_u(\mathbf{R}^{sn})\mathbf{v}(z_n^+) - \mathbf{\Sigma}. \quad (5.140)$$

From our definition of $\mathfrak{I}_u$, given in equation 5.45 and using equation 5.135, we can expand equation 5.140 as

$$\begin{pmatrix}\mathbf{v}_u(z_s^-)\\ \mathbf{v}_d(z_s^-)\end{pmatrix} = \begin{bmatrix}\mathbf{T}_u^{sn} - \mathbf{R}_d^{sn}(\mathbf{T}_d^{sn})^{-1}\mathbf{R}_u^{sn} & \mathbf{R}_d^{sn}(\mathbf{T}_d^{sn})^{-1}\\ -(\mathbf{T}_d^{sn})^{-1}\mathbf{R}_u^{sn} & (\mathbf{T}_d^{sn})^{-1}\end{bmatrix} \times \begin{pmatrix}\mathbf{v}_u(z_n^+)\\ \mathbf{v}_d(z_n^+)\end{pmatrix} - \begin{pmatrix}-\mathbf{\Sigma}_u\\ \mathbf{\Sigma}_d\end{pmatrix}$$

$$= \begin{bmatrix} \mathbf{T}_u^{sn} - \mathbf{R}_d^{sn}(\mathbf{T}_d^{sn})^{-1}\mathbf{R}_u^{sn} & \mathbf{R}_d^{sn}(\mathbf{T}_d^{sn})^{-1} \\ -(\mathbf{T}_d^{sn})^{-1}\mathbf{R}_u^{sn} & (\mathbf{T}_d^{sn})^{-1} \end{bmatrix} \times \begin{pmatrix} \mathbf{v}_u(z_n^+) \\ \mathbf{0} \end{pmatrix} - \begin{pmatrix} -\mathbf{\Sigma}_u \\ \mathbf{\Sigma}_d \end{pmatrix}. \quad (5.141)$$

On the right-hand side of the equation 5.141, we imposed the half-space radiation condition at the end of the stack of layers at z_n^+ such that $\mathbf{v}_d(z_n^+)$ vanishes. Equation 5.141, after some algebra, leads to

$$\left.\begin{aligned} \mathbf{v}_u(z_s^-) &= (\mathbf{I} - \mathbf{R}_d^{ns}\mathbf{R}_u^{0s})^{-1}(\mathbf{R}_d^{ns}\mathbf{\Sigma}_d + \mathbf{\Sigma}_u) \\ \mathbf{v}_d(z_s^-) &= \mathbf{R}_u^{0s}\mathbf{v}_u(z_s^-) \\ \mathbf{v}_u(z_s^+) &= \mathbf{R}_d^{ns}\mathbf{v}_d(z_s^+) \\ \mathbf{v}_d(z_s^+) &= (\mathbf{I} - \mathbf{R}_u^{0s}\mathbf{R}_d^{ns})^{-1}(\mathbf{\Sigma}_d + \mathbf{R}_u^{0s}\mathbf{\Sigma}_u) \end{aligned}\right\}. \quad (5.142)$$

The equation set given in equation 5.142 is the complete solution to the source wavefield. We have previously provided the details of how to compute the reflection coefficient matrices $\mathbf{R}_d^{ns}$ and $\mathbf{R}_u^{0s}$ using the iteration equations (equation 5.106). In equation 5.135, we also provided how to compute $\mathbf{\Sigma}_d$ and $\mathbf{\Sigma}_u$ in terms of $\mathbf{D}^{-1}(z_s)$ and the seismic source parameters $\mathbf{F}_1$ and $\mathbf{F}_2$. Therefore, $\mathbf{v}_u(z_s^-)$ and $\mathbf{v}_d(z_s^+)$ in equation 5.142 are completely given in terms of these reflection coefficient matrices and the source terms. And, once $\mathbf{v}_u(z_s^-)$ and $\mathbf{v}_d(z_s^+)$ are computed, $\mathbf{v}_d(z_s^-)$ and $\mathbf{v}_u(z_s^+)$ can also be computed from them.

5.3.12 Computation of the Receiver Wavefield

Once the source wavefields $\mathbf{v}_u(z_s^-), \mathbf{v}_d(z_s^-), \mathbf{v}_u(z_s^+), \mathbf{v}_d(z_s^+)$ are computed (equation 5.142), we can now compute the receiver wavefields and compute the actual responses we record in our data using seismic receivers. In computing the receiver wavefields, it is, however, necessary to identify two different scenarios: (1) receiver located above the source and (2) receiver located below the source because they are different from one another. We therefore separately consider these two cases. There are, however, other scenarios like source and receiver at the same depth, receiver at the surface, buried source, source at the surface, buried receiver, and so on. However, we do not discuss them here because from the two cases we discuss we can easily compute responses for other cases.

Case 1: Receiver Above Source

The situation of receiver above source is illustrated in Figure 5.17a. As before, we have a stack of n layers, bounded on the top by a half-space or a free surface, and by a half-space at the bottom. The source and receiver are respectively located at depths z_s and z_r with $z_r < z_s$.

The receiver wavefield $\mathbf{v}(z_r)$ can then be computed from the source wavefield $\mathbf{v}(z_s^-)$ using the

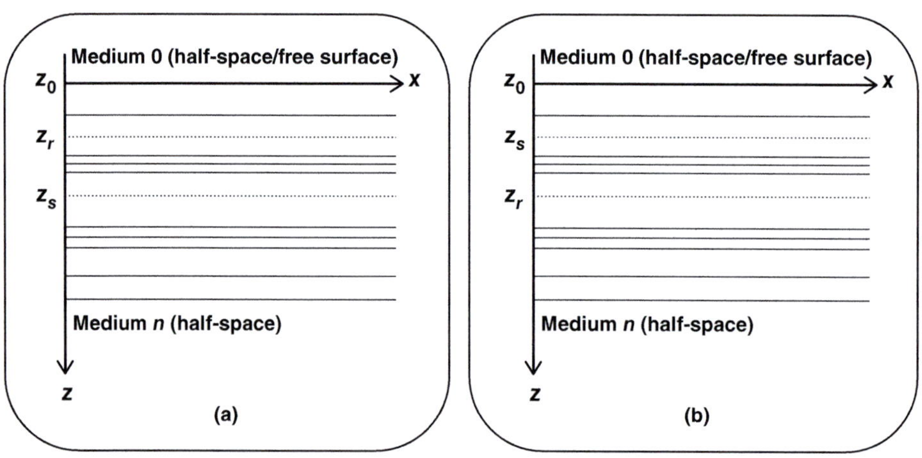

Figure 5.17 Computation of the receiver wavefield response for (a) receiver above source and (b) receiver below source.

wave propagator $\mathbf{Q}(z_r, z_s^-)$ as

$$\mathbf{v}(z_r) = \mathbf{Q}(z_r, z_s^-)\mathbf{v}(z_s^-) = \mathfrak{J}_u(\mathbf{R}^{rs})\mathbf{v}(z_s^-). \quad (5.143)$$

Using our same notations, $\mathbf{R}^{rs}$ is the 6×6 reflection matrix (equations 5.43 and 5.45) between the source and the receiver locations. By carrying out the upward scatterer operation $\mathfrak{J}_u$ (equation 5.147) on $\mathbf{R}^{rs}$, we can then write

$$\begin{pmatrix}\mathbf{v}_u(z_r)\\ \mathbf{v}_d(z_r)\end{pmatrix} = \begin{bmatrix}\mathbf{T}_u^{rs} - \mathbf{R}_d^{rs}(\mathbf{T}_d^{rs})^{-1}\mathbf{R}_u^{rs} & \mathbf{R}_d^{rs}(\mathbf{T}_d^{rs})^{-1}\\ -(\mathbf{T}_d^{rs})^{-1}\mathbf{R}_u^{rs} & (\mathbf{T}_d^{rs})^{-1}\end{bmatrix} \times \begin{pmatrix}\mathbf{v}_u(z_s^-)\\ \mathbf{v}_d(z_s^-)\end{pmatrix}, \quad (5.144)$$

which, after some straightforward algebra and from equation 5.142, gives the upgoing and downgoing receiver wavefields as

$$\left.\begin{aligned}\mathbf{v}_u(z_r) &= (\mathbf{I} - \mathbf{R}_d^{rs}\mathbf{R}_u^{0r})^{-1}\mathbf{T}_u^{rs}(\mathbf{I} - \mathbf{R}_d^{ns}\mathbf{R}_u^{0s})^{-1}\\ &\quad \times(\mathbf{R}_d^{ns}\boldsymbol{\Sigma}_d + \boldsymbol{\Sigma}_u)\\ \mathbf{v}_d(z_r) &= \mathbf{R}_u^{0r}\mathbf{v}_u(z_r)\end{aligned}\right\}. \quad (5.145)$$

Equation 5.145 gives the wave vector. To compute the displacement – stress vector $\mathbf{b} = \left[\mathbf{u}, \frac{i}{\omega}\mathbf{t}\right]^T$, defined in equation 5.17, we use the relation between $\mathbf{v}$ and $\mathbf{b}$, $\mathbf{b} = \mathbf{D}\mathbf{v}$ (see equation 5.20), and write

$$\mathbf{b}(z_r) = \mathbf{D}(z_r)\mathbf{v}(z_r). \quad (5.146)$$

Or, explicitly, as

$$\begin{pmatrix}\mathbf{u}(z_r)\\ \frac{i}{\omega}\mathbf{t}(z_r)\end{pmatrix} = \begin{pmatrix}\mathbf{M}_u^r & \mathbf{M}_d^r\\ \mathbf{N}_u^r & \mathbf{N}_d^r\end{pmatrix}\begin{pmatrix}\mathbf{v}_u(z_r)\\ \mathbf{v}_d(z_r)\end{pmatrix}. \quad (5.147)$$

The matrices $\mathbf{M}_u^r, \mathbf{M}_d^r, \mathbf{N}_u^r, \mathbf{N}_d^r$ in equation 5.147 are the 3×3 submatrices of the eigenvector matrix $\mathbf{D}(z_r)$ at the receiver location. Finally, expanding equation 5.147 and using equations 5.145 and 5.142, we obtain the particle displacement response at the receiver location as

$$\begin{aligned}\mathbf{u}(z_r) &= (\mathbf{M}_u^r + \mathbf{M}_d^r\mathbf{R}_u^{0r})(\mathbf{I} - \mathbf{R}_d^{rs}\mathbf{R}_u^{0r})^{-1}\\ &\quad \times \mathbf{T}_u^{rs}(\mathbf{I} - \mathbf{R}_d^{ns}\mathbf{R}_u^{0s})^{-1}(\mathbf{R}_d^{ns}\boldsymbol{\Sigma}_d + \boldsymbol{\Sigma}_u)\end{aligned} \quad (5.148)$$

and the traction response as

$$\begin{aligned}\frac{i}{\omega}\mathbf{t}(z_r) &= (\mathbf{N}_u^r + \mathbf{N}_d^r\mathbf{R}_u^{0r})(\mathbf{I} - \mathbf{R}_d^{rs}\mathbf{R}_u^{0r})^{-1}\\ &\quad \times \mathbf{T}_u^{rs}(\mathbf{I} - \mathbf{R}_d^{ns}\mathbf{R}_u^{0s})^{-1}(\mathbf{R}_d^{ns}\boldsymbol{\Sigma}_d + \boldsymbol{\Sigma}_u).\end{aligned} \quad (5.149)$$

Equation 5.148 is the particle displacement response, and in marine environments the third (τ_{zz}) component of equation 5.149 is the pressure.

Case 2: Receiver Below Source

The receiver below source is illustrated in Figure 5.17b. Using the same procedures, we used from case 1, we can derive the receiver wavefields as

$$\left.\begin{aligned}\mathbf{v}_d(z_r) &= (\mathbf{I} - \mathbf{R}_u^{rs}\mathbf{R}_d^{nr})^{-1}\mathbf{T}_d^{rs}(\mathbf{I} - \mathbf{R}_u^{0s}\mathbf{R}_d^{ns})^{-1}\\ &\quad \times(\boldsymbol{\Sigma}_d + \mathbf{R}_u^{0s}\boldsymbol{\Sigma}_u)\\ \mathbf{v}_u(z_r) &= \mathbf{R}_d^{nr}\mathbf{v}_d(z_r)\end{aligned}\right\}, \quad (5.150)$$

and particle displacement and traction response as

$$\left.\begin{aligned}\mathbf{u}(z_r) &= (\mathbf{M}_u^r\mathbf{R}_d^{nr} + \mathbf{M}_d^r)(\mathbf{I} - \mathbf{R}_u^{rs}\mathbf{R}_d^{nr})^{-1}\mathbf{T}_d^{rs}\\ &\quad \times(\mathbf{I} - \mathbf{R}_u^{0s}\mathbf{R}_d^{ns})^{-1}(\boldsymbol{\Sigma}_d + \mathbf{R}_u^{0s}\boldsymbol{\Sigma}_u)\\ \frac{i}{\omega}\mathbf{t}(z_r) &= (\mathbf{N}_u^r\mathbf{R}_d^{nr} + \mathbf{N}_d^r)(\mathbf{I} - \mathbf{R}_u^{rs}\mathbf{R}_d^{nr})^{-1}\\ &\quad \times\mathbf{T}_d^{rs}(\mathbf{I} - \mathbf{R}_u^{0s}\mathbf{R}_d^{ns})^{-1}(\boldsymbol{\Sigma}_d + \mathbf{R}_u^{0s}\boldsymbol{\Sigma}_u)\end{aligned}\right\} \quad (5.151)$$

5.3.13 Response Computation in Different Domains

The above method computes seismic response in frequency (ω) and ray-parameter (p_p, p_y) domain. Therefore, any component of the receiver response in equations 5.148, 5.149, and 5.151 can be considered as $R(\omega, p_x, p_y, z)$. Using our Fourier transform convention (equation 4.9), the time and offset domain responses can be computed as

$$R(x, y, z, t) = \iiint R(\omega, p_x, p_y, z)e^{i\omega(t - p_x x - p_y y)} \times \omega^2 dp_x dp_y d\omega. \quad (5.152)$$

Instead, if we just transform frequency ω to time, we obtain the $\tau - p_x - p_y$ response

$$R(p_x, p_y, z, \tau) = \int R(\omega, p_x, p_y, z)e^{i\omega\tau}d\omega. \quad (5.153)$$

For azimuthal isotropy such as isotropic and VTI type of medium, it is not necessary to distinguish between p_x and p_y and compute responses as functions of the horizontal slowness $p = \sqrt{p_x^2 + p_y^2}$ and ω. In such cases our computed response is $R(\omega, p, z)$, and equation 5.152 can be replaced by the Sommerfeld integral using the Bessel functions (See Aki and Richards, 2002 for details).

5.3.14 Inelastic Attenuation

So far, we discussed perfectly elastic medium in which a plane wave $\delta(t - x/c)$ propagating with wave speed c retains its shape exactly, and all frequency components travel coherently with the same wave speed c. It is, however, in contrast to such idealized behavior, commonly observed that as the seismic wave propagates through real materials wave amplitudes attenuate. There are many reasons for this attenuation, and in seismology, they all are macroscopically described as *internal friction*. A convenient way to incorporate such attenuation is through the introduction of an attenuation parameter Q. In seismology, Q is defined as if a volume of material is cycled in stress at a frequency ω, then a dimensionless measure of the internal friction or anelasticity is given by

$$\frac{1}{Q(\omega)} = -\frac{\Delta E}{2\pi E}, \tag{5.154}$$

in which E is the peak strain energy stored in the volume and ΔE is the energy lost in each cycle due to imperfections in the elasticity of the material.

Although equation 5.154 is the correct definition of Q (Aki and Richards, 2002), it is difficult to use. The reason for this is the fact that it is impossible to drive a material element in the field with stress waves of unchanging amplitude and period, and we must seek a more convenient way to define Q so that we can use it in practice, and we can commonly observe the following:

1) Temporal decay of amplitude in a standing wave at a fixed wave number.
2) Spatial decay in a propagating wave at fixed frequency.

In seismology, we are specifically interested in the second because we can decompose the recorded seismic data into individual plane wave components and then analyze the decay of the propagating wavefield for different frequencies with the distance of propagation.

Assuming a linear stress–strain relation (Hooke's law), if A is the wavefield amplitude then

$$E \propto A^2. \tag{5.155}$$

Therefore,

$$\Delta E \propto 2A\Delta A. \tag{5.156}$$

And thus equation 5.154 can be written as

$$\frac{1}{Q(\omega)} = -\frac{1}{\pi}\frac{\Delta A}{A}. \tag{5.157}$$

If we want to study the attenuation of the amplitude at a fixed frequency in the from $A = A(x)$, we can then write ΔA as

$$\Delta A = \left(\frac{dA}{dx}\right)\lambda, \tag{5.158}$$

in which the wavelength λ can be expressed in terms of the frequency ω and phase velocity c as

$$\lambda = \frac{2\pi c}{\omega}. \tag{5.159}$$

We therefore obtain

$$\frac{dA}{dx} = -\frac{\omega}{2cQ}A(x). \tag{5.160}$$

With the initial condition that the amplitude at $x = 0$ is A_0, the solution to the differential equation 5.160 is

$$A(x) = A_0 \exp\left[-\frac{\omega x}{2cQ}\right]. \tag{5.161}$$

We now consider a plane wave of the form $\delta(t - x/c)$ propagating in the positive x direction with the phase velocity c. We also assume that each Fourier component of this impulse

$$\int_{-\infty}^{\infty} \delta\left(t - \frac{x}{c}\right)\exp(i\omega t)dt = \exp\left[\frac{i\omega x}{c}\right] \tag{5.162}$$

to be attenuated by a factor $\exp[-\alpha(\omega)/x]$ such that the attenuation rate is given by

$$\alpha(\omega) = \frac{\omega}{2cQ}. \tag{5.163}$$

To compute the pulse shape $p(x, t)$ of such an attenuating wave at a distance x and at a given time t, we first make the phase velocity c complex but frequency independent and given by

$$c = c_0 \exp\left[-\frac{i}{2Q}\right]. \tag{5.164}$$

Therefore,

$$\begin{aligned}\frac{x}{c} &= \frac{x}{c_0}\exp\left[\frac{i}{2Q}\right] \\ &= \frac{x}{c_0}\left[\cos\left(\frac{1}{2Q}\right) + i\sin\left(\frac{1}{2Q}\right)\right].\end{aligned} \tag{5.165}$$

For $Q \gg 1$, where $\cos(1/2Q) \approx 1$ and $\sin(1/2Q) \approx 1/2Q$, we may also approximate equation 5.165 as

$$\frac{x}{c} \approx \frac{x}{c_0}\left(1 + i\frac{1}{2Q}\right). \tag{5.166}$$

Recalling the original propagating wavefield with its Fourier components given by equation 5.162, the pulse shape of the attenuated wavefield thus becomes

$$p(x,t) = \frac{1}{2\pi}\int_{-\infty}^{\infty} \exp\left[i\omega\left(\frac{x}{c} - t\right)\right] d\omega, \tag{5.167}$$

which we can also write as

$$p(x,t) = \frac{1}{2\pi}\int_{-\infty}^{\infty} \exp\left[-\frac{|\omega|x}{2c_0 Q}\right] \exp\left[i\omega\left(\frac{x}{c} - t\right)\right] d\omega. \tag{5.168}$$

For a constant Q, the integral on the right-hand side of equation 5.168 can be integrated by parts, which gives

$$p(x,t) = \frac{1}{\pi}\left[\frac{\frac{x}{2c_0 Q}}{\left(\frac{x}{2c_0 Q}\right)^2 + \left(\frac{x}{c_0} - t\right)^2}\right]. \tag{5.169}$$

Note that the fundamental notion of causality of a propagating wavefield implies that if the wave is propagating with velocity c_0 in the positive x direction, then we should not expect any wave arrival at distance x before time $t = x/c_0$. However, the pulse shape given by equation 5.169 does not satisfy this causality condition because it has a finite value for $t < x/c_0$. We therefore conclude that making the velocity complex but frequency independent using equation 5.164 violates the fundamental notion of causality for the propagating wavefield. This, in turn, implies that to model attenuating wavefield, velocities (or the elastic stiffness constants for anisotropy) not only need to be complex, but they must also be frequency dependent.

This requirement of causality, on the other hand, leads to a very important issue for practical implementation of computing the seismic responses. In our discussions on computing the reflection and transmission coefficient matrices for an interface and for a stack of layers, note that having the velocities and therefore elastic stiffness constants complex and frequency dependent implies that these matrices must be computed not only for each ray parameter (p_x and p_y) but also for every frequency (ω), which is a big computational burden. Fortunately, for relatively short propagation distances (less than 10 km), typically used in exploration seismology, using a constant Q like the one shown in equation 5.163 does not severely affect the computed seismic responses, and such an assumption is fine to use (Mallick and Frazer 1990). For very large propagation distances (more than 1000 km), we typically deal with, in earthquake seismology, the constant Q assumption of equation 5.163, however, which can severely affect the computed seismic responses.

Causality requires that the velocity c to be a function of frequency ω such that the pulse shape $p(x, t)$ is zero for all $t < x/c_\infty$, where c_∞ is the limit of $c(\omega)$ as $\omega \to \infty$. Such a causality relation is satisfied when

$$\frac{\omega}{c_\infty} + \hbar[\alpha(\omega)] = 2Q\alpha(\omega). \tag{5.170}$$

In equation 5.170, $\hbar$ stands for the Hilbert transform. Equation 5.170 is called the Kramers–Krönig relation (Aki and Richards, 2002), and to satisfy this relation, Q must be frequency dependent (for details, see Aki and

Richards, 2002). So, to satisfy causality, the velocities (elastic stiffness constants) must not only be complex, but they must also be frequency dependent.

There are many attenuation laws satisfying causality, and Aki and Richards (2002) provide detailed in-depth discussion on this topic. Here, we just list two popular methods that are easy to implement.

Absorption-Band Attenuation Law: This law is given in Liu et al. (1976) and given as

$$c(\omega) = c_\infty \left\{ 1 + \frac{2}{\pi Q_m} \ln \left(\frac{\omega_2(\omega_1 - i\omega)}{\omega_1(\omega_2 - i\omega)} \right) \right\}^{\frac{1}{2}}. \tag{5.171}$$

Liu et al. (1976) verified that the attenuation $Q(\omega)$ provided by equation 5.171 is nearly constant with a value slightly less than Q_m within the band $\omega_1 \le \omega \le \omega_2$ and rapidly goes to infinity outside this band (Figure 5.18). If $\omega_1 \ll \omega \ll \omega_2$, equation 5.171 can be approximated as

$$c(\omega) \approx c_\infty \left[1 + \frac{1}{\pi Q_m} \ln \left(\frac{\omega}{\omega_1} \right) \right] \exp \left(-\frac{i}{2Q_m} \right). \tag{5.172}$$

Power Attenuation Law: Power attenuation law is given by Strick (1967, 1970) as

$$\frac{1}{c(\omega)} = \frac{1}{c_\infty} + \frac{k_0}{\sin\left(\frac{\sigma\pi}{2}\right)} \frac{1}{(\varepsilon - i\omega)^\sigma};$$

$$-\frac{\pi}{2} < \arg(\omega + i'') < \frac{3\pi}{2}. \tag{5.173}$$

As we note from equation 5.173, there are three attenuation parameters σ, k_0, and ε in the power attenuation law which require additional explanation. In implementing power attenuation law, Mallick and Frazer (1987) recommend the following:

1) The first attenuation parameter σ should be chosen as $0 < \sigma < 1$ and for most geological materials $\sigma \approx 0.1$.
2) If Q is defined at the reference frequency ω_0, then second attenuation parameter k_0 should be chosen to be

$$k_0 \approx \frac{\omega_0^\sigma}{2c_\infty Q(\omega_0)}. \tag{5.174}$$

3) The third attenuation parameter (ε) is used to ensure that that $1/c(\omega)$ is finite as $\omega \to 0$ and a small value of 0.001 for this parameter provides good results.

In the modeling of the oceanic P and S waves (Po/So) from a Japan earthquake, Mallick and Frazer (1990) implemented the power attenuation law using the above-recommended parameters and obtained very good results up to propagation distance of 2000 km.

Before ending, we would like to discuss the Hilbert transform $\hbar$ shown as a part of the Kramers–Krönig relation (equation 5.170). In seismology, Hilbert transform is used to obtain the analytical signal and compute the amplitude envelop. Aki and Richards (2002) provide the theoretical details of the Hilbert transform and its use. Here, we just outline how to compute this transform.

Assume that our recorded seismic data in time domain are given as $S(t)$. Then, to compute the Hilbert transform $\hbar[S(t)](t)$, we first Fourier transform $S(t)$ and compute $S(\omega)$, which, from our Fourier transform sign convention, is

$$S(\omega) = \int_{-\infty}^{+\infty} S(t) e^{-i\omega t} dt. \tag{5.175}$$

After computing $S(\omega)$, we compute a new frequency-domain attribute, say $S'(\omega)$, just by phase advancing each component of $S(\omega)$ by $\pi/2$. Because such $\pi/2$ phase advance is equivalent to multiplication by $e^{i\pi/2} = i$, this process means that if a given component of $S(\omega)$ is represented as $a + ib$ then the corresponding component of $S'(\omega)$ is $-b + ia$.

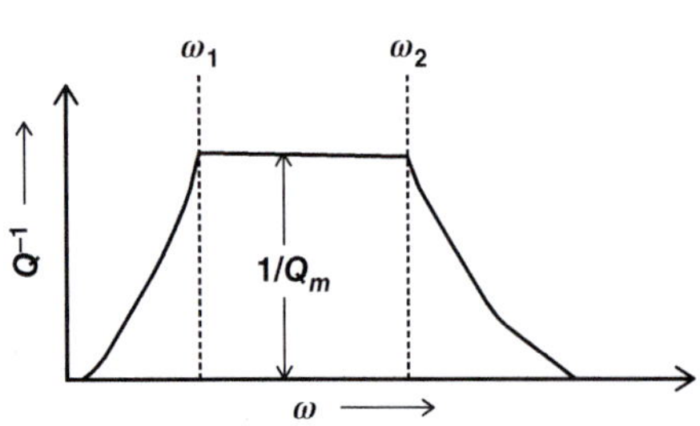

Figure 5.18 Absorption-band attenuation law.

After computing $S'(\omega)$, we compute $S'(t)$ by inverse Fourier transforming

$$S'(t) = \frac{1}{2\pi} \int_{-\infty}^{+\infty} S'(\omega) e^{i\omega t} d\omega. \tag{5.176}$$

The Hilbert transform of $S(t)$ is the real part of $S'(t)$ or

$$\hbar[S(t)](t) = Re[S'(t)]. \tag{5.177}$$

5.4 Synthetic Seismograms for Vertically and Laterally Varying Media

In the above section, we discussed elastic wave propagation in horizontally stratified (1D) medium and computing synthetic seismic response. We also discussed how to include attenuation in these computations. Although 1D is an oversimplification of the true earth, it is still very important to consider for many practical reasons. First, 1D method is superior to convolutional AVA type of modeling, which is still a routine practice, especially in the seismic industry. By honoring complex wave effects, the 1D method, we model real seismic data much greater accurately than AVA, which we already demonstrated using Figures 5.1–5.3 at the beginning of this chapter. Second, because of its analytical (or strictly, semi-analytical because the eigenvector matrix **D** needs numerical computation for a general anisotropic medium) nature, computing 1D synthetic seismic responses is much faster than the other numerical methods that we discuss in this section. As we briefly discussed in Chapter 1, and discuss in more detail in later chapters, inverse or optimization problems require multiple iterations of forward computation of synthetic responses, and thus the compute efficiency to solve such problems is of vital importance. Although 1D assumption is not strictly correct, there are many situations where after prestack time or depth migration, a 1D assumption for every common midpoint or common conversion point location is generally adequate, and therefore, using 1D modeling for inversion/optimization is more practical than using other approaches that assume a fully inhomogeneous subsurface. Full waveform inversion (FWI) by iteratively using complete numerical modeling approaches has gained popularity over the past decade. The computational expense for these methods is, however, still prohibitive, and these methods are mostly used to obtain low-resolution subsurface properties (primarily velocities), suitable for imaging. For high-resolution subsurface characterization for lithological and fluid properties, a 1D assumption is still a practical compromise for routine applications.

The primary stumbling block for high-resolution FWI for subsurface characterization is that the model space is so overwhelmingly large that it is prohibitively expensive to compute. Recent advances in the deep-learning (DL) methods, however, have shown a lot of promise for high-resolution FWI. Biswas et al. (2019), for example, developed a physics-guided convolutional neural network approach to poststack and prestack AVA inversion. Zhang and Alkhalifah (2020) have shown the DL-aided FWI can resolve high-resolution subsurface elastic properties. Therefore, even though current use of high-resolution FWI is limited, the method has a tremendous future potential, and therefore it is important to discuss the numerical modeling methods used in FWI.

Numerically computing synthetic seismic response for an inhomogeneous model is a subject by itself discussing which is beyond the scope of this book. Here, we briefly discuss the fundamental principles. There are many excellent textbooks on this subject. Fichtner (2011) for example gives an excellent overview of FWI with detailed discussions of different numerical modeling methods, including their numerical implementations.

5.4.1 Governing Equations

The elastodynamic equation, which we derived in Chapter 3, equation 3.72, is the primary governing equation for any numerical modeling. In equation 3.72, we ignored the source (body-force density) term (**f**) and here we include it in the equation. In addition, we denote **r** to be our position vector and assume that the density (ρ) and elastic stiffness tensor (**c**) are functions of **r**, and the particle displacement field (**u**) and the body-force density (**f**) are functions of **r** and time (t). Or, in other words,

$$\left.\begin{aligned} \rho &= \rho(\mathbf{r}) \\ \mathbf{c} &= \mathbf{c}(\mathbf{r}) \\ \mathbf{u} &= \mathbf{u}(\mathbf{r}, t) \\ \mathbf{f} &= \mathbf{f}(\mathbf{r}, t) \end{aligned}\right\}. \tag{5.178}$$

Using these conventions, we may then write the time-domain version of the elastodynamic equation as

$$\rho(\mathbf{r})\ddot{\mathbf{u}}(\mathbf{r}, t) - \nabla \cdot [\mathbf{c}(\mathbf{r}) : \nabla\mathbf{u}(\mathbf{r}, t)] = \mathbf{f}(\mathbf{r}, t). \tag{5.179}$$

Note the time-domain version of the elastodynamic equation after including the source term is equivalent to equation 5.179 and is left as an exercise to verify. Equation 5.179, given in terms of the displacement field **u**, is the *displacement formulation* of the elastic wave equation.

In defining the 1D elastic system, we derived our constitutive relation (Hooke's law) as $\tau_{ij} = c_{ijkl}u_{k,l}$ (see equation 5.2), which in the notation of equation 5.178 is equivalent to

$$\boldsymbol{\tau}(\mathbf{r}, t) = \mathbf{c}(\mathbf{r}) : \nabla\mathbf{u}(\mathbf{r}, t). \tag{5.180}$$

From equation 5.180, it is therefore evident that the second term on the left-hand side of equation 5.179 is $-\nabla \cdot \boldsymbol{\tau}(\mathbf{r}, t)$. By replacing the second time derivative $\ddot{\mathbf{u}}(\mathbf{r}, t)$ by the first time derivative $\dot{\mathbf{v}}(\mathbf{r}, t)$ where $\mathbf{v}(\mathbf{r}, t)$ is the particle velocity in equation 5.179, and taking the time derivative of equation 5.180, we obtain an equivalent, *velocity–stress formulation* of the elastic wave equation and given as

$$\left.\begin{aligned} \rho(\mathbf{r})\dot{\mathbf{v}}(\mathbf{r}, t) - \nabla \cdot \boldsymbol{\tau}(\mathbf{r}, t) &= \mathbf{f}(\mathbf{r}, t) \\ \dot{\boldsymbol{\tau}}(\mathbf{r}, t) - \mathbf{c}(\mathbf{r}) : \nabla\mathbf{v}(\mathbf{r}, t) &= \mathbf{0} \end{aligned}\right\}. \tag{5.181}$$

We must note that **c** in equations 5.178–5.181 is the fourth-order elastic stiffness tensor, not the 6×6 elastic stiffness matrix **C**, which we previously used in defining the 1D elastic system. Although we had shown in Chapter 3 that the stiffness matrix **C** is derived from the stiffness tensor **c** by arranging its 21 independent components in the form of a 6×6 symmetric matrix (equations 3.18–3.22), to avoid confusion, it is still important to make this distinction here.

For a horizontally stratified subsurface, it can be easily verified that the velocity–stress formulation of equation 5.181 is equivalent to the 1D elastic system, shown in equations 5.16 and 5.17, for which we already discussed how to obtain analytical/semi-analytical solutions in this chapter. In obtaining the radiation patterns from seismic sources in a homogeneous medium, we solved a variation of the same set of equations in Chapter 3. In developing approximate ray-theoretical solutions also, we used the same equations in Chapter 4. Equations 5.179 and 5.181 are the most general form of the elastic wave equation, which we may consider as the *elastic system for a heterogenous subsurface*, a solution for which must be numerically obtained.

Although there are many methods to obtain such numerical solutions, they all require (1) spatial discretization of the subsurface model (elastic stiffness and density) and the physical variables (particle displacement/velocity and stress), solutions to which we are seeking, and (2) temporal discretization of the physical variables. This process of discretization allows to forward propagate these governing equations (equations 5.179 and 5.181), either in time or in frequency domain, and obtain solutions.

Before discussing discretization, it is important to briefly consider the *acoustic wave equation*, which is a simplified version of the elastic wave equation (equation 5.179 or 5.181). Although strictly valid for fluids, an acoustic wave equation is still commonly used in exploration seismology (Pratt, 1999; Virieux and Operto, 2009; Warner et al., 2013, for example). Primary reasons for this are (1) an acoustic wave equation is computationally efficient and easier to implement than elastic wave equation, and (2) majority of FWI applications in exploration seismology focus on velocity estimation for depth imaging of single (vertical) component seismic data for which an acoustic assumption is adequate.

Because the rigidity modulus $\mu = 0$ in fluids, the momentum equation for an isotropic medium (equation 3.148) becomes

$$\rho\ddot{\mathbf{u}} - \lambda\nabla(\nabla \cdot \mathbf{u}) = \mathbf{f}. \tag{5.182}$$

From the definition of pressure (P) in fluids, given as $P = -\lambda\nabla . \mathbf{u}$ (Aki and Richards, 2002), we write equation 5.182 as

$$\rho\ddot{\mathbf{u}} + \nabla P = \mathbf{f}. \tag{5.183}$$

If we now divide both sides of equation 5.183 by ρ and take the divergence on both sides, we obtain

$$\nabla \cdot \ddot{\mathbf{u}} + \nabla \cdot \left(\frac{1}{\rho}\nabla P\right) = \nabla \cdot \left(\frac{1}{\rho}\mathbf{f}\right). \tag{5.184}$$

Because $P = -\lambda\nabla . \mathbf{u}$, we eliminate the displacement $\mathbf{u}$ and write equation 5.184 as

$$\frac{\ddot{P}}{\lambda} - \nabla \cdot \left(\frac{\nabla P}{\rho}\right) = -\nabla \cdot \left(\frac{\mathbf{f}}{\rho}\right). \tag{5.185}$$

If the variation of density ρ is much slower than that of the pressure P and the source term $\mathbf{f}$, we finally simplify equation 5.185 and obtain the scalar partial differential equation, known as the *acoustic wave equation*, as

$$\ddot{P} - V^2\Delta P = -V^2\nabla \cdot \mathbf{f}, \tag{5.186}$$

in which

$$V = \sqrt{\frac{\lambda}{\rho}} \tag{5.187}$$

is the wave speed. It can be readily seen from equations 5.186 and 5.187 that in an acoustic (fluid) media the wave motion is described by the pressure P, whose properties depend on the source term $\mathbf{f}$ and the spatial distribution of the wave speed V. Below, we discuss elastic wave equation only. However, the implementation of these methods to acoustic wave equation (equation 5.186) is simple, straightforward, and obvious from what we discuss below.

5.4.2 Spatial Discretization

Meaning of spatial discretization is approximating the continuous wavefield, say $\mathbf{u}(\mathbf{r}, t)$ of equation 5.179 as discrete values $\tilde{\mathbf{u}}(t)$. This, in turn, means that $\tilde{\mathbf{u}}(t)$ is an n-dimensional vector, given as $[\tilde{u}_1(t), \tilde{u}_2(t), \cdots, \tilde{u}_n(t)]^T$, approximated from $\mathbf{u}(\mathbf{r}, t)$. We may do such discretization from sampling $\mathbf{u}(\mathbf{r}, t)$ at discrete points in the spatial domain. Discretizing $\mathbf{u}(\mathbf{r}, t)$ as an n-dimensional vector $\tilde{\mathbf{u}}(t)$ would also require discretizing the subsurface model (density and stiffness tensor) as $n \times n$ matrices and the body-force density $\mathbf{f}(\mathbf{r}, t)$ as an n-dimensional vector, which would lead equation 5.179 to the form

$$\mathbf{M} \cdot \ddot{\tilde{\mathbf{u}}}(t) + \mathbf{S} \cdot \tilde{\mathbf{u}}(t) = \tilde{\mathbf{f}}(t). \tag{5.188}$$

The $n \times n$ matrices $\mathbf{M}$ and $\mathbf{S}$ in equation 5.188 are commonly referred as the mass matrix and the stiffness matrix, and $\tilde{\mathbf{f}}(t)$ in equation 5.188 is the discretized version of the body-force density $\mathbf{f}(\mathbf{r}, t)$. In practice, the matrices $\mathbf{M}$ and $\mathbf{S}$ are generally sparse matrices. They do not require explicit computations because computing the vector–matrix products is only required in the numerical implementations.

Depending upon the actual numerical method being used, either displacement formulation or velocity–stress formulation may be more advantageous to use over the other. Therefore, we provide below the discretized versions for both formulations.

Spatially Discrete Form of the Displacement Formulation: To obtain the complete set of equations for the spatially discrete form of the displacement formulation, we use the momentum equation (equation 3.69), using the notations given by equation 5.188, and write as

$$\rho(\mathbf{r})\ddot{\mathbf{u}}(\mathbf{r}, t) - \nabla.\boldsymbol{\tau}(\mathbf{r}, t) = \mathbf{f}(\mathbf{r}, t), \tag{5.189}$$

and combine with the constitutive relation (equation 5.180) and obtain the spatially discrete version of the displacement formulation as

$$\left.\begin{array}{l}\mathbf{M}\cdot\ddot{\tilde{\mathbf{u}}}(t)+\mathbf{D}\cdot\tilde{\mathbf{t}}(t)=\tilde{\mathbf{f}}(t)\\ \tilde{\mathbf{t}}(t)=\mathbf{S}\cdot\tilde{\mathbf{u}}(t)\end{array}\right\}. \quad (5.190)$$

In equation 5.190, $\mathbf{D}$ is the $n\times n$ matrix, spatially discretizing the divergence operator in the momentum equation, and $\tilde{\mathbf{t}}(t)$ is the spatially discrete version of the stress tensor $\tau(\mathbf{r},t)$.

Spatially Discrete Form of the Velocity–Stress Formulation: Following the same procedure as before where we derived the continuous version of the velocity–stress formulation (equation 5.181), we write its spatially discrete form as

$$\left.\begin{array}{l}\mathbf{M}\cdot\dot{\tilde{\mathbf{v}}}(t)+\mathbf{D}\cdot\tilde{\mathbf{t}}(t)=\tilde{\mathbf{f}}(t)\\ \dot{\tilde{\mathbf{t}}}-\mathbf{S}\cdot\tilde{\mathbf{v}}(t)=\mathbf{0}\end{array}\right\}. \quad (5.191)$$

It is evident that $\tilde{\mathbf{v}}(t)$ in equation 5.191 is the discrete version of the continuous velocity field $\mathbf{v}(\mathbf{r},t)$.

Note that by transforming either of the formulations from continuous to a discrete system reduces the forward computation to the solution of a large linear system of equations and an ordinary differential equation in time. In our analytical solutions in a homogeneous medium (Chapter 3), approximate solutions using ray theory (Chapter 4), and for the analytical/semi-analytical solutions for 1D media (early part of this chapter), we started from two fundamental equations: (1) constitutive relation linearly relating the stress and tensors and (2) the momentum equation relating the rate of change momentum to the forces. In deriving the displacement or velocity–stress formulations too, we used the same two equations. In elastic wave theory, these two equations are therefore the fundamental building blocks.

5.4.3 Temporal Discretization

Via spatial discretization, we converted the original partial differential equations (equations 5.179 and 5.181) to ordinary differential equations in time (equations 5.190 and 5.191). These equations can be solved in the time domain. Alternatively, we can also Fourier transform them to frequency domain and solve in the frequency domain. Both these approaches have their respective advantages and disadvantages, and the choice depends on the specific type of problem that we are attempting to solve and the available computational resources. Here, we briefly describe both these approaches.

Time-Domain Computations

From equation 5.188, we can write

$$\ddot{\tilde{\mathbf{u}}}=\mathbf{M}^{-1}[\tilde{\mathbf{f}}(t)-\mathbf{S}\cdot\tilde{\mathbf{u}}(t)]. \quad (5.192)$$

Now, if we approximate the second-order derivative $\ddot{\mathbf{u}}(t)$ via second-order finite difference (FD) as

$$\ddot{\tilde{\mathbf{u}}}\approx\frac{1}{(\Delta t)^2}[\tilde{\mathbf{u}}(t+\Delta t)-2\tilde{\mathbf{u}}(t)+\tilde{\mathbf{u}}(t-\Delta t)], \quad (5.193)$$

where Δt is the time step, and plug into equation 5.192, we obtain

$$\tilde{\mathbf{u}}(t+\Delta t)\approx 2\tilde{\mathbf{u}}(t)-\tilde{\mathbf{u}}(t-\Delta t) + (\Delta t)^2\mathbf{M}^{-1}[\tilde{\mathbf{f}}(t)-\mathbf{S}\cdot\tilde{\mathbf{u}}(t)]. \quad (5.194)$$

Equation 5.194 is thus an explicit scheme to compute the displacement field $\tilde{\mathbf{u}}(t+\Delta t)$ for time step $t+\Delta t$ from the displacement fields for the previous time steps t and $t-\Delta t$. The scheme given by equation 5.194 is the simplest way for time marching. There are, however, more elegant methods, which, including their stabilities, can be found in Fichtner (2011). Although we used the displacement formulation of equation 5.188 for time marching, the one for equation 5.190 and for the velocity–stress formulation (equation 5.191) can be similarly obtained.

Frequency-Domain Computations

For frequency-domain computations, the governing equations are Fourier transformed to frequency domain. Frequency-domain representation of equation 5.188, for example, can be written as

$$-\omega^2\mathbf{M}\cdot\tilde{\mathbf{u}}(\omega)+\mathbf{S}\cdot\tilde{\mathbf{u}}(\omega)=\tilde{\mathbf{f}}(\omega). \quad (5.195)$$

Defining the matrix

$$\mathbf{L}(\omega) = -\omega^2\mathbf{M} + \mathbf{K}, \quad (5.196)$$

often referred to as the impedance matrix, equation 5.195 can be written as a linear system

$$\mathbf{L}(\omega) \cdot \widetilde{\mathbf{u}}(\omega) = \widetilde{\mathbf{f}}(\omega), \quad (5.197)$$

which can be solved for the wavefield for frequency ω. The frequency-domain method is elegant because there are many compute-efficient matrix diagonalization methods to solve the linear system like equation 5.197 (for example, see Press et al., 1989). Additionally, once the system is solved, it can be reused for any new source. Another big advantage of the frequency-domain approach is that it is easy to incorporate attenuation into the linear system using the methods we outlined above. For large 3D problems, however, the memory requirements to solve the linear system are enormous and difficult to apply. For FWI, Pratt (1999) used frequency-domain computations at a few discrete frequencies and demonstrated the method to be compute-efficient.

5.4.4 Overview of Different Numerical Methods

There are many numerical methods for computing synthetic seismic response in 3D heterogeneous media. They all have their relative merits and demerits. The main difference between these methods lies in the spatial discretization process that transforms the spatial derivatives into a system of equations, which can be directly solved. Below is a brief overview of some of these methods. For details, refer to the references included in the descriptions of these methods.

Finite-Difference Method

Finite difference is the oldest, and yet, a very popular, method. Some early examples of the FD method in seismology can be found in Alterman and Karal (1968), Boore (1970, 1972), Alford et al. (1974), Kelly et al. (1976), among others. The fundamental concept behind FD is the following.

Let $f(x)$ be a function representing any dynamic field, such as the displacement, velocity, stress, etc. In FD, this function is defined in a finite number of evenly spaced grid points $x_i, i = 1, 2, \dots, n$, and we replace the derivative $\partial_x f(x_i)$ at the grid point i by its finite-difference approximation using the value of this function at the neighboring grid points. There are many ways to make such an approximation, out of which, the one known to be most numerically stable is the second-order central finite difference (Aki and Richards, 2002) and given as

$$\partial_x[f(x_i)] \approx \frac{1}{2\Delta x}[f(x_i + \Delta x) - f(x_i - \Delta x)], \quad (5.198)$$

in which Δx is the grid spacing or the regular interval at which $f(x)$ is discretized. The error in the above approximation of the derivative is $O(\Delta x)^2$, therefore, finer the grid spacing, more accurate is the approximation of the derivative.

Although initially developed to solve two-dimensional (2D) acoustic problems, the FD methods have been extended to solve 3D fully elastic (and anisotropic) problems. Feasibility of this extension lies in the concept of the *staggered grid* where the dynamic fields such as $f(x)$ are defined at different grid positions in contrast with the *conventional grid* where they are coincident. Details of this concept can be found in Madariaga (1976); Virieux (1984, 1986), Igel et al. (1995), Moczo et al. (2000, 2002), among others. Result of a staggered grid is a reduced grid spacing, thereby reducing grid dispersion due to discretization.

Pseudo-spectral Method

Pseudo-spectral method also discretizes the spatial derivatives like FD. However, this discretization follows the following three steps:

1) Wavefield, sampled at a finite number of grid points, is transformed to the wave number ($k = \omega p$ in 2D or $k_x = \omega p_x$; $k_y = \omega p_y$ in 3D) domain using spatial Fourier transformations (see equations 4.8 and 4.9).

2) The Fourier transformed wavefield is then multiplied by ik (ik_x or ik_y in 3D). Note that this multiplication in the wave number domain is equivalent to the derivative in the space (x in 2D or x and y in 3D).
3) Using the inverse Fourier transform (equations 4.8 and 4.9), the wavefield is transformed into the space (x for 2D or x and y for 3D) domain.

Note that computing the derivative by multiplying by ik in 2D or by ik_x or ik_y in 3D is exact up to the Nyquist wave number. Consequently, only two grid points per wavelength are adequate for sampling the wavefield.

Pseudo-spectral method is more compute-efficient than the FD. However because this method approximates the derivative using Fourier transforms and it works well for relatively smooth models as compared with FD. For further details on the pseudo-spectral method, see Kosloff and Baysal (1982), Kosloff et al. (1990), Tessmer and Kosloff (1994), Furumura et al. (1998), Igel (1999), and Mizutani et al. (2000).

Finite-Element Method

In contrast with the FD or pseudo-spectral method, the finite-element (FE) method is based on the weak form of the wave equation (Li and Mallick, 2015). In FE, the computation domain is decomposed (discretized) into different disjoint subdomains, which are called *elements*, and within each of these elements, the dynamic fields (displacement, velocity, stress, etc.) are approximated as low-order polynomials, where the continuity between these elements is explicitly imposed (Bao et al., 1998; Li and Mallick, 2015). The resulting elastic wave equation then reduces to a system of a sparse-discrete system of polynomial coefficients that can be numerically solved.

Major advantage of FE is its efficiency in handling irregular geometries and the free surface (Li and Mallick, 2015). Despite this advantage, use of FE in seismology so far has been limited (Lysmer and Drake, 1972; Toshinawa and Ohmachi, 1992; Bao et al., 1998). The primary reason for this is the numerical dispersion from the low-order polynomial approximations used by FE. This led to combine FE with FD with some successful compute-efficient hybrid implementations (Moczo et al., 1997).

Li and Mallick (2015) thoroughly investigated the drawbacks of the FE method and developed a compute-efficient anisotropic FE method that does not have the numerical issues as above, and yet capable of handling irregular geometries in an efficient manner. Furthermore, by thoroughly inspecting the sparse-discrete system of the polynomial coefficients, Li and Mallick (2015) rearranged them for parallel computing so that it can efficiently compute 3D synthetic seismic response.

In view of this, although FE is still not the preferred choice to compute the synthetic seismic response for 3D heterogenous earth models, it still has the potential to be the preferred method in future, especially for using in an iterative way in high-resolution FWI.

Other Numerical Methods

Besides the methods discussed above, the following methods are also popular, included as below along with references.

1) *Spectral Element Method*:
   Patera (1984), Priolo et al. (1994), Seriani et al. (1995), Seriani (1998), Faccioli et al. (1997), Komatitsch and Vilotte (1998), Komatitsch and Tromp (2002a, 2002b), Chaljub et al. (2003), Chaljub and Valette, (2004), Nissen-Meyer et al. (2007, 2008).
2) *Direct Solution Method*:
   Geller and Ohminato (1994), Cummins et al. (1994a, 1994b), Geller and Takeuchi (1995), Kawai et al. (2008).
3) *Discontinuous Galerlkin Method*:
   Dumbser and Käser (2006), Käser et al. (2007), de la Puente et al. (2007, 2008, 2009).

5.4.5 Boundary Conditions

For 1D elastic wave propagation, we previously observed how to impose either a free surface or a half-space radiation condition at the top and half-space radiation condition at the bottom. Because of the inherent assumption of lateral invariance, imposing only these two radiation conditions are adequate for computing responses without any reflections from the edges. This is, however, not true for heterogeneous earth models. Imposing the free surface condition of vanishing stress fields (equation 5.111) is not difficult to impose for such cases. However, if the free surface reflections are not desired, these methods would still produce undesired reflections from the top of the model and interfere with the signal. In addition, they also produce reflections from all edges of the model and interfere with the signal. One way to avoid these undesired reflections is to continue the model indefinitely to all sides so that these edge reflections have late arrival times and do not interfere. This, however, would require unnecessary computational burden, which we want to avoid. To reduce such undesired computational burden, it is thus necessary to find ways to minimize these undesired events. There are two common ways to minimize these undesired reflections

1) *Absorbing boundary conditions.*
2) *Absorbing boundary layers.*

Absorbing boundary conditions provide conditions to absorb the wavefield and its derivatives along an artificial boundary around all sides of the model. Original methods developed by Engquist and Majda (1977) and Clayton and Engquist (1977) were of low order and valid only for small incident angles. High-order methods were later developed by Keys (1985), Higdon (1991), among others to overcome these issues.

Absorbing boundary layers introduce a thin region along all sides and modify the original wave equation such that amplitudes of the incident wavefield rapidly decay. Some of the popular methods include the Gaussian taper method (Cerjan et al., 1985) and the perfectly match layer method (Bérenger, 1994).

Further details on these boundary conditions, including their numerical implementations, can be found in Fichtner (2011).

5.4.6 Summary of Different Methods for Computing Synthetic Seismic Responses for Heterogeneous Media

Various method for computing synthetic seismic responses for 2D and 3D heterogeneous earth models have been proposed to date. They all require discretizing the wave equation, converting it to an algebraic set of equations and solve the set. The major numerical issue of these methods lies in the process of discretization. These methods have their own advantages and disadvantages and are suited to solve specific sets of problems.

5.5 Chapter Summary

In this chapter, we obtained complete solutions to the elastic wave equation. First, we obtained analytical/semi-analytical solution to the 1D elastic wave equation and showed how to include anelastic attenuation in the computations. Then, we discussed how to extend the 1D method to heterogeneous earth models and obtain numerical solutions.

5.6 Exercises

Problem 5.1 Explicitly write down all $\mathbf{E}_{jl}$ matrices ($\mathbf{E}_{11}$, $\mathbf{E}_{12}$, $\mathbf{E}_{13}$, $\mathbf{E}_{21}$, $\mathbf{E}_{22}$, $\mathbf{E}_{23}$, $\mathbf{E}_{31}$, $\mathbf{E}_{31}$, $\mathbf{E}_{33}$) defined by equation 5.5 in terms of the components of the 6×6 elastic stiffness matrix defined by equation 3.22. Verify that $\mathbf{E}_{ji} = (\mathbf{E}_{ij})^T$ for all $i, j = 1, 2, 3$.

Problem 5.2 Verify that equations 3.69 and 5.12 are equivalent.

Problem 5.3 From equations 5.12 and 5.5, obtain equation 5.13.

Problem 5.4 Starting from equation 5.13, assume lateral invariance and use equation 5.10 to obtain equation 5.14.

Problem 5.5 Derive the 1D elastic system (equation 5.17). Also verify that the matrices **S** and **C** are symmetric matrices and $\mathbf{V}=\mathbf{T}^T$.

Problem 5.6 Assuming a VTI medium with the elastic stiffness matrix given by equation 3.51 and show that for such a medium the 1D elastic system decouples into a 4×4 P-SV system

$$\left.\begin{aligned} \partial_z \mathbf{b}_{PSV} &= i\omega \mathbf{A}_{PSV} + \mathbf{F}_{PSV} \\ \mathbf{b}_{PSV} &= \left[u_x, u_z, \frac{1}{i\omega}\tau_{xz}, \frac{1}{i\omega}\tau_{zz}\right]^T \\ \mathbf{F}_{PSV} &= \left[0, 0, \frac{i}{\omega}f_x, \frac{i}{\omega}f_z\right]^T \end{aligned}\right\} \tag{5.199}$$

and a 2×2 SH system

$$\left.\begin{aligned} \partial_z \mathbf{b}_{SH} &= i\omega \mathbf{A}_{SH} + \mathbf{F}_{SH} \\ \mathbf{b}_{SH} &= \left[u_y, \frac{1}{i\omega}\tau_{yz}\right]^T \\ \mathbf{F}_{SH} &= \left[0, \frac{i}{\omega}f_y\right]^T \end{aligned}\right\}. \tag{5.200}$$

Also write down the explicit expressions for the matrices $\mathbf{A}_{PSV}$ and $\mathbf{A}_{SH}$.

Problem 5.7 Verify the properties of the wave propagator and propagator matrices given by equation 5.27.

Problem 5.8 Start from equation 5.38 and use equation 4.56 to derive the expressions for the downward reflection and transmission coefficients $\mathbf{R}_d$ and $\mathbf{T}_d$, shown in equation 5.39.

Problem 5.9 Start from equation 5.40 and use equation 4.59 to derive the expressions for the upward reflection and transmission coefficients $\mathbf{R}_u$ and $\mathbf{T}_u$, shown in equation 5.41.

Problem 5.10 Start from equations 5.43 and 5.44, use equations 4.56 and 4.59, and derive equation 5.45.

Problem 5.11 Verify the properties of the upward and downward scatterer operators $\mathfrak{F}_u$ and $\mathfrak{F}_d$, given in equation 5.48.

Problem 5.12 Starting from equation 5.96, use equation 5.42 and derive equation 5.98.

Problem 5.13 Derive equation 5.100 from equation 5.99.

Problem 5.14 Derive equation 5.103 from equation 5.102.

Problem 5.15 Expand equation 5.105 to obtain equation 5.106.

Problem 5.16 Start from the source term represented as $\mathbf{f}=\mathbf{h}\delta(\mathbf{r}-\mathbf{r}_s)-\mathbf{M}\,\nabla\,\delta(\mathbf{r}-\mathbf{r}_s)$, shown in equation 5.117. Assume that the source location is $\mathbf{r}_s=[0,\ 0,\ z_s]$. Then, Fourier transform the spatial variables x and y to the ray parameters p_x and p_y and show that in this domain the source term is given as $\mathbf{f}=\mathbf{h}\delta(z-z_s)-i\omega p_x\mathbf{M}_x\delta(z-z_s)-i\omega p_y\mathbf{M}_y\delta(z-z_s)-\mathbf{M}_z\delta'(z-z_s)$ with $\mathbf{M}_x=[M_{xx}, M_{xy}, M_{xz}]^T$, $\mathbf{M}_y=[M_{xy}, M_{yy}, M_{yz}]^T$, and $\mathbf{M}_z=[M_{xz}, M_{yz}, M_{zz}]^T$.

Problem 5.17 Using the results from the previous problem (Problem 5.16), verify that the equations 5.122–5.124 are correct.

Problem 5.18 Prove the validity of equation 5.125 via direct substitution into the equation $\partial_z\mathbf{b}=i\omega\mathbf{A}\mathbf{b}+\mathbf{F}$.

Problem 5.19 Use the properties of the propagator matrix and derive equation 5.132.

Problem 5.20 Derive equation 5.133.

Problem 5.21 Start from equation 5.141 and obtain the source wavefield responses (equation 5.142).

Problem 5.22 Derive equations 5.145, 5.148–5.151.

Problem 5.23 Carry out the integral on the right-hand side of equation 5.168 and derive the pulse shape $p(x, t)$ given in equation 5.169.

Problem 5.24 The time-domain version of the elastodynamic equation (equation 3.172) with the source term included becomes $\rho\ddot{\mathbf{u}} - \nabla_S \mathbf{C}(\nabla_S)^T\mathbf{u} = \mathbf{f}$, where the operator ∇_S is given by equation 3.173. Show that this equation and equation 5.179 are equivalent.

Problem 5.25 Obtain the velocity–stress formulation, shown in equation 5.180.

Problem 5.26 From equations 5.192 and 5.193 derive equation 5.194.

References

Aki, K. and Richards, P.G. (2002). *Quantitative Seismology*, Seconde. University Science Books ISBN 0-978-1891389634.

Alford, R.M., Kelly, K.R., and Boore, D.M. (1974). Accuracy of finite difference modeling of the acoustic wave equation. *Geophysics 39*: 834–842. https://doi.org/10.1190/1.1440470.

Alterman, Z. and Karal, F.C. (1968). Propagation of elastic waves in layered media by finite difference methods. *Bulletin of the Seismological Society of America 58*: 367–398. https://doi.org/10.1785/BSSA0580010367.

Bao, H., Bielak, J., Ghattas, O. et al. (1998). Large-scale simulation of elastic wave propagation in heterogeneous media on parallel computers. *Computer Methods in Applied Mechanics and Engineering 152*: 85–102. https://doi.org/10.1016/S0045-7825(97)00183-7.

Bérenger, J.-P. (1994). A perfectly match layer for the absorption of electromagnetic waves. *Journal of Computational Physics 114*: 185–200. https://doi.org/10.1006/jcph.1994.1159.

Biswas, R., Sen, M.K., Das, V., and Mukerji, T. (2019). Prestack and poststack inversion using a physics-guided convolutional neural network. *Interpretation 7*: SE161–SE174. https://doi.org/10.1190/INT-2018-0236.1.

Boore, D.M. (1970). Love waves in nonuniform waveguides: finite difference calculations. *Journal of Geophysical Research 75*: 1512–1527. https://doi.org/10.1029/JB075i008p01512.

Boore, D.M. (1972). Finite-difference methods for seismic wave propagation in heterogeneous materials. In: *Methods in Computational Physics*, vol. *11*. Academic Press https://doi.org/10.1016/B978-0-12-460811-5.50006-4.

Cerjan, C., Kosloff, D., Kosloff, R., and Reshef, M. (1985). A nonreflecting boundary condition for discrete acoustic and elastic wave equations. *Geophysics 50*: 705–708. https://doi.org/10.1190/1.1441945.

Chaljub, E., Capdeville, Y., and Vilotte, J.P. (2003). Solving elastodynamics in a fluid–solid heterogeneous sphere: a parallel spectral-element approximation on non-conforming grids. *Journal of Computational Physics 187*: 457–491. https://doi.org/10.1016/S0021-9991(03)00119-0.

Chaljub, E. and Valette, B. (2004). Spectral element modelling of three-dimensional wave propagation in a self-gravitating earth with and arbitrarily stratified outer core. *Geophysical Journal International 158*: 131–141. https://doi.org/10.1111/j.1365-246X.2004.02267.x.

Clayton, R.W. and Engquist, B. (1977). Absorbing boundary conditions for acoustic and elastic wave equations. *Bulletin of the Seismological Society of America 67*: 1529–1540. https://doi.org/10.1785/BSSA0670061529.

Cummins, P., Geller, R., Hatori, T., and Takeuchi, N. (1994a). DSM synthetic seismograms: SH, spherically symmetric case. *Geophysical Research Letters 21*: 533–536. https://doi.org/10.1029/GL021i007p00533.

Cummins, P., Geller, R., and Takeuchi, N. (1994b). DSM complete synthetic seismograms: P-SV, spherically symmetric case. *Geophysical Research Letters 21*: 1663–1666. https://doi.org/10.1029/94GL01281.

de la Puente, J., Ampuero, J.-P., and Käser, M. (2009). Dynamic rupture modeling on unstructured meshes using a discontinuous Galerkin method. *Journal of Geophysical Research 114*: 1–17. https://doi.org/10.1029/2008JB006271.

de la Puente, J., Dumbser, M., Käser, M., and Igel, I. (2008). Discontinuous Galerkin methods for wave propagation in poroelastic media. *Geophysics 73*: T77–T97. https://doi.org/10.1190/1.2965027.

de la Puente, J., Käser, M., Dumbser, M., and Igel, I. (2007). An arbitrary high-order discontinuous Galerkin method for elastic waves on unstructured meshes-IV. Anisotropy. *Geophysical Journal International 169*: 1210–1228. https://doi.org/10.1111/j.1365-246X.2007.03381.x.

Dumbser, M. and Käser, M. (2006). An arbitrary high-order discontinuous Gelerkin method for elastic waved on unstructured meshes-II. The three-dimensional isotropic case. *Geophysical Journal International 167*: 319–336. https://doi.org/10.1111/j.1365-246X.2006.03120.x.

Engquist, B. and Majda, A. (1977). Absorbing boundary conditions for the numerical solutions of waves. *Mathematics of Computation 31*: 629–651. https://doi.org/10.2307/2005997.

Faccioli, E., Maggio, F., Paolucci, R., and Quarteroni, A. (1997). 2D and 3D elastic wave propagation by a pseudo-spectral domain decomposition method. *Journal of Seismology 1*: 237–251. https://doi.org/10.1023/A:1009758820546.

Fichtner, A. (2011). *Full Seismic Waveform Modeling and Inversion*. Springer ISBN: 978-3-642-15806-3, https://doi.org/10.1007/978-3-642-15807-0.

Frazer, L.N. and Fryer, G.J. (1989). Useful properties of the system matrix for a homogeneous anisotropic viscoelastic medium. *Geophysical Journal International 97*: 173–177. https://doi.org/10.1111/j.1365-246X.1989.tb00493.x.

Fryer, G.J. and Frazer, L.N. (1984). Seismic waves in stratified anisotropic media. *Geophysical Journal International 78*: 691–710. https://doi.org/10.1111/j.1365-246X.1984.tb05065.x.

Fryer, G.J. and Frazer, L.N. (1987). Seismic waves in stratified anisotropic media- 11. Elastodynamic eigensolutions for some anisotropic systems. *Geophysical Journal International 91*: 73–101. https://doi.org/10.1111/j.1365-246X.1987.tb05214.x.

Furumura, T., Kennett, B.L.N., and Furumura, M. (1998). Seismic wavefield calculation for laterally heterogeneous whole earth models using the pseudospectral method. *Geophysical Journal International 135*: 845–860. https://doi.org/10.1046/j.1365-246X.1998.00682.x.

Geller, R.J. and Ohminato, T. (1994). Computation of synthetic seismograms and their partial derivatives for heterogeneous media with arbitrary boundary conditions using the direct solution method. *Geophysical Journal International 116*: 421–446. https://doi.org/10.1111/j.1365-246X.1994.tb01807.x.

Geller, R.J. and Takeuchi, N. (1995). A new method for computing highly accurate DSM synthetic seismograms. *Geophysical Journal International 123*: 449–470. https://doi.org/10.1111/j.1365-246X.1995.tb06865.x.

Golub, G.H. and Van Loan, C.F. (2013). *Matrix Computations*. Johns Hopkins University Press.

Hampson, D. (1991). AVO inversion, theory and practice. *Leading Edge 10*: 39–42. https://doi.org/10.1190/1.1436820.

Higdon, R.L. (1991). Absorbing boundary conditions for elastic waves. *Geophysics 56*: 231–241. https://doi.org/10.1190/1.1443035.

Igel, H. (1999). Wave propagation in three-dimensional spherical sections by the Chebyshev spectral method. *Geophysical Journal International 136*: 559–566. https://doi.org/10.1046/j.1365-246x.1999.00758.x.

Igel, H., Mora, P., and Riollet, B. (1995). Anisotropic wave propagation through finite-difference grids. *Geophysics 60*: 1203–1216. https://doi.org/10.1190/1.1443849.

Käser, M., Dumbser, M., de la Puente, J., and Igel, H. (2007). An arbitrary high-order discontinuous Gelerkin method for elastic waves on unstructured meshes-III.

Viscoelastic attenuation. *Geophysical Journal International 168*: 224–242. https://doi.org/10.1111/j.1365-246X.2006.03193.x.

Kawai, K., Takeuchi, N., and Geller, R.J. (2008). Complete synthetic seismograms up to 2 Hz for transversely isotropic spherically symmetric media. *Geophysical Journal International 164*: 411–424. https://doi.org/10.1111/j.1365-246X.2005.02829.x.

Kelly, K.R., Ward, R.W., Treitel, S., and Alford, R.M. (1976). Synthetic seismograms: a finite-difference approach. *Geophysics 41*: 2–27. https://doi.org/10.1190/1.1440605.

Kennett, B.L.N. (1983). *Seismic Wave Propagation in Stratified Media*. Cambridge University Press.

Keys, R.G. (1985). Absorbing boundary conditions for acoustic media. *Geophysics 50*: 892–902. https://doi.org/10.1190/1.1441969.

Komatitsch, D. and Tromp, J. (2002a). Spectral-element simulations of global seismic wave propagation-I, validation. *Geophysical Journal International 149*: 390–412. https://doi.org/10.1046/j.1365-246X.2002.01653.x.

Komatitsch, D. and Tromp, J. (2002b). Spectral-element simulations of global seismic wave propagation-II, 3D models, oceans, rotation and self-gravitation. *Geophysical Journal International 150*: 303–318. https://doi.org/10.1046/j.1365-246X.2002.01716.x.

Komatitsch, D. and Vilotte, J.P. (1998). The spectral element method: an efficient tool to simulate the seismic response of 2D and 3D geological structures. *Bulletin of the Seismological Society of America 88*: 368–392. https://doi.org/10.1785/BSSA0880020368.

Kosloff, D.D. and Baysal, E. (1982). Forward modeling by a Fourier method. *Geophysics 47*: 1402–1412. https://doi.org/10.1190/1.1441288.

Kosloff, D.D., Kessler, D., Filho, A.Q. et al. (1990). Solution of the equations of dynamic elasticity by a Chebyshev spectral method. *Geophysics 55*: 734–748. https://doi.org/10.1190/1.1442885.

Li, T. and Mallick, S., 2015, Anisotropic seismic forward modeling using finite-element method and its implication towards developing an anisotropic full waveform inversion, SEG Technical Program Expanded Abstracts, https://doi.org/10.1190/segam2015-5884573.1.

Liu, H.P., Anderson, D.L., and Kanamori, H. (1976). Velocity dispersion due to anelasticity: implications for seismology and mantle composition. *Geophysical Journal International 47*: 41–58. https://doi.org/10.1111/j.1365-246X.1976.tb01261.x.

Lysmer, J. and Drake, L.A. (1972). A finite-element method for seismology. *Methods in Computational Physics 11*: 181–216. https://doi.org/10.1016/B978-0-12-460811-5.50009-X.

Madariaga, R. (1976). Dynamics of an expanding circular fault. *Bulletin of the Seismological Society of America 67*: 163–182. https://doi.org/10.1785/BSSA0660030639.

Mallick, S. (2007). Amplitude-variation-with-offset, elastic impedance, and prestack waveform inversion, a modeling study. *Geophysics 72*: C1–C7. https://doi.org/10.1190/1.2387108.

Mallick, S. and Adhikari, S. (2015). Amplitude-variation-with-offset and prestack waveform inversion: a direct comparison using a real data example from the Rock Springs uplift, Wyoming, USA. *Geophysics 80*: B45–B59. https://doi.org/10.1190/GEO2014-0233.1.

Mallick, S. and Frazer, L.N. (1987). Practical aspects of reflectivity modeling. *Geophysics 52*: 1355–1364. https://doi.org/10.1190/1.1442248.

Mallick, S. and Frazer, L.N. (1988). Rapid computation of multioffset vertical seismic profile synthetic seismograms for layered media. *Geophysics 53*: 479–491. https://doi.org/10.1190/1.1442479.

Mallick, S. and Frazer, L.N. (1990). Computation of synthetic seismograms for stratified azimuthally anisotropic media. *Journal of Geophysical Research 95*: 8513–8526. https://doi.org/10.1029/JB095iB06p08513.

Mallick, S. and Frazer, L.N. (1991). Reflection/transmission coefficients and azimuthal anisotropy in marine seismic studies. *Geophysical Journal International 105*: 241–252. https://doi.org/10.1111/j.1365-246X.1991.tb03459.x.

Mizutani, H., Geller, R.J., and Takeuchi, N. (2000). Comparison of accuracy and efficiency of time-domain schemes for calculating synthetic seismograms. *Physics of the Earth and Planetary Interiors 119*: 75–97. https://doi.org/10.1016/S0031-9201(99)00154-5.

Moczo, P., Bystrický, E., Kristek, J. et al. (1997). Hybrid modeling of P-SV seismic motion at inhomogeneous viscoelastic topographic structures. *Bulletin of the Seismological Society of America 87*: 1305–1323. https://doi.org/10.1785/BSSA0870051305.

Moczo, P., Kristek, J., and Halada, L. (2000). 3D fourth order staggered-grid finite-difference schemes: stability and grid dispersion. *Bulletin of the Seismological Society of America 90*: 587–603. https://doi.org/10.1785/0119990119.

Moczo, P., Kristek, J., Vavrycuk, V. et al. (2002). 3D heterogeneous staggered-grid finite-difference modeling of seismic motion with volume harmonic and arithmetic averaging of elastic moduli. *Bulletin of the Seismological Society of America 92*: 3042–3066. https://doi.org/10.1785/0120010167.

Nissen-Meyer, T., Fournier, A., and Dahlen, F.A. (2007). A two-dimensional spectral-element method for computing spherical-earth seismograms-I. Moment-tensor source. *Geophysical Journal International 168*: 1067–1092. https://doi.org/10.1111/j.1365-246X.2006.03121.x.

Nissen-Meyer, T., Fournier, A., and Dahlen, F.A. (2008). A 2-D spectral-element method for computing spherical-earth seismograms-II. Waves in solid–fluid media. *Geophysical Journal International 174*: 873–888. https://doi.org/10.1111/j.1365-246X.2008.03813.x.

Pafeng, J., Mallick, S., and Sharma, H. (2017). Prestack waveform inversion of three-dimensional seismic data – an example from the Rock Springs uplift, Wyoming, USA. *Geophysics 82*: B1–B12. https://doi.org/10.1190/GEO2016-0079.1.

Patera, T. (1984). A spectral element method for fluid dynamics: laminar flow in a channel expansion. *Journal of Computational Physics 53*: 468–488. https://doi.org/10.1016/0021-9991(84)90128-1.

Pease, M.P. (1965). *Methods of matrix algebra*. Academic Press.

Pratt, R.G. (1999). Seismic waveform inversion in the frequency domain. Part 1: theory and verification in a physical scale model. *Geophysics 64*: 888–901. https://doi.org/10.1190/1.1444597.

Press, W.H., Flannery, B.P., Teulolsky, S.A., and Vetterling, W. (1989). *Numerical Recipes- The Art of Scientific Computing (FORTRAN Version)*. Cambridge University Press ISBN 0-521-38330-7.

Priolo, E., Carcione, J.M., and Seriani, G. (1994). Numerical simulation of interface waves by high-order spectral modeling techniques. *The Journal of the Acoustical Society of America 95*: 681–693. https://doi.org/10.1121/1.408428.

Resnick, J. (1993). Seismic data processing for AVO and AVA analysis. In: *Offset Dependent Reflectivity: Theory and Practice of AVO and AVA Analysis* (ed. J.P. Castagna and M.M. Backus), 175–189. SEG Publications https://doi.org/10.1190/1.9781560802624.

Russell, B.H. (1988). *Introduction to Seismic Inversion Methods*, SEG Course Notes, Series 2. Society of Exploration Geophysicists.

Sava, P. and Hill, S.J. (2009). Overview and classification of wavefield seismic imaging methods. *Leading Edge 28*: 170–183. https://doi.org/10.1190/1.3086052.

Sen, M.K. and Roy, I.G. (2003). Computation of differential seismograms and iteration adaptive regularization in prestack waveform inversion. *Geophysics 68*: 2026–2039. https://doi.org/10.1190/1.1635056.

Seriani, G. (1998). 3-D large scale wave propagation modeling by a spectral element method on a Cray T3E multiprocessor. *Computer Methods in Applied Mechanics and Engineering 164*: 235–247. https://doi.org/10.1016/S0045-7825(98)00057-7.

Seriani, G., Priolo, E., and Prefarz, A., 1995, Modelling waves in anisotropic media by a spectral element method, Proc. of the 3rd Int. Conf. On Math. And Numer. Aspects of Wave Prop., 289–298.

Strick, E. (1967). The determination of Q, dynamic viscosity, and transient creep curves from wave propagation measurements. *Geophysical Journal International 13*: 197–218. https://doi.org/10.1111/j.1365-246X.1967.tb02154.x.

Strick, E. (1970). A predicted pedestal effect for pulse propagation in constant-Q solids. *Geophysics 35*: 387–403. https://doi.org/10.1190/1.1440101.

Tessmer, E. and Kosloff, D.D. (1994). 3D elastic modeling with surface topography by a Chebyshev spectral method. *Geophysics 59*: 464–473. https://doi.org/10.1190/1.1443608.

Todd, C.P., and Backus, M.M., 1985, Offset-dependent reflectivity in a structural context, SEG Technical Program Expanded Abstracts, 586-588, https://doi.org/10.1190/1.1892805.

Toshinawa, T. and Omachi, T. (1992). Love-wave propagation in a three-dimensional sedimentary basin. *Bulletin of the Seismological Society of America 82*: 1661–1677. https://doi.org/10.1785/BSSA0820041661.

Virieux, J. (1984). SH-wave propagation in heterogeneous media: velocity–stress finite-difference method. *Geophysics 49*: 1933–1957. https://doi.org/10.1190/1.1441605.

Virieux, J. (1986). P-SV wave propagation in heterogeneous media: velocity–stress finite-difference method. *Geophysics 51*: 889–901. https://doi.org/10.1190/1.1442147.

Virieux, J. and Operto, S. (2009). An overview of full-waveform inversion in exploration geophysics. *Geophysics 74*: WCC1–WCC26. https://doi.org/10.1190/1.3238367.

Warner, M., Ratcliffe, A., Nangoo, T. et al. (2013). Anisotropic 3D full-waveform inversion. *Geophysics 78*: R59–R80. https://doi.org/10.1190/geo2012-0338.1.

Woodhouse, J.H. (1974). Surface waves in a laterally varying layered structure. *Geophysical Journal International 37*: 461–490. https://doi.org/10.1111/j.1365-246X.1974.tb04098.x.

Zhang, Z. and Alkhalifah, T. (2020). High-resolution reservoir characterization using deep learning-aided elastic full-waveform inversion: the North Sea field data example. *Geophysics 85*: WA137–WA146. https://doi.org/10.1190/geo2019-0340.1.

6

Optimization of Functions

6.1 Introduction

After outlining the mathematical foundations, followed by discussions of the fundamentals of different aspects of computational seismology in Chapters 2–5, we now begin our discussion on the optimization methods. As we have seen in Chapter 1, optimization methods require either minimization or maximization of the objective. In short, such a problem of minimization or maximization is the following:

1) We are given a function f that depends on one or more independent variables.
2) We want to find the value of those variables where f takes on a minimum or a maximum value.
3) We may also wish to calculate what value of f is achieved at the maximum or minimum.

Solving such a problem is fundamental to all optimization methods. This is because if f is what we defined earlier as the objective in Chapter 1, which, in turn, is a function of the model. Minimizing or maximizing f would thus provide the model estimate.

Before proceeding further, we note that the tasks of minimization and maximization are related to each other because if f is defined such that we want to find its minimum, it is equivalent to finding the maximum of $-f$. So, in this chapter, we first discuss finding the minimum or maximum of single-dimensional functions, and then move onto finding the minima or maxima in multiple dimensions. The methods we discuss in this chapter are fundamental to all local optimization methods. Additionally, they are the optimization frameworks for most of the machine-learning methods. Therefore, the methods and concepts we develop in this chapter are vital to Chapter 7 where we discuss local inverse problems in geophysics and to Chapter 9 where we discuss machine-learning methods.

Most geophysical inverse problems, including seismic, are nonunique, meaning that there could be multiple solutions where the function f attains the optimum (minimum or maximum). Practical optimization problems, therefore, not only require estimating the models (the set of variables) for which f attains the optima and evaluating f, but also they require estimating the uncertainties associated with these estimations. We postpone the discussion of this uncertainty estimations to Chapter 8 after discussing the global optimization methods.

6.2 One-Dimensional Optimization

To optimize one-dimensional (1D) functions, such as $f(x)$, there are one set of methods which require explicit computation of the derivative of the function, i.e., computing $d[f(x)]/dx$, and another set which do not need the derivative. In the below section, we discuss them both.

Computation, Optimization, and Machine Learning in Seismology, Advanced Textbook 8, First Edition. Subhashis Mallick.

6.2.1 Golden Section Search in One Dimension

Golden section search method is one of the classical and oldest methods for finding the minimum of the function $f(x)$, and the procedure to find the maximum of $f(x)$ is identical. This method is, in theory, like the "bisection method" of bracketing the roots of a polynomial in an interval (a, b) and slowly reducing the interval to a level that the bracketing interval is small enough to say that the solution is found (Press et al., 2007).

Consider a polynomial in one dimension, which is given as

$$f(x) = \sum_{i=0}^{n} a_i x^i. \tag{6.1}$$

The roots of $f(x) = 0$ is known to be bracketed within the interval (a, b) with $a < b$ when $f(x)$ has opposite signs at those two points. A minimum on the other hand is known to be bracketed only when there is a triplet of points $a < b < c$ such that $f(b)$ is less than both $f(a)$ and $f(c)$. In such a case, we know that if $f(x)$ is non-singular, then it has a minimum point in the interval (a, c). The golden section search is based on this concept and tries to find the minimum of the function by trial and error.

We use Figure 6.1 to illustrate the way the golden section search method works. We first find the three points a, b, c, where $a < b < c$, and $f(b)$ is less than both $f(a)$ and $f(c)$. In Figure 6.1, the point a is shown as point 1, the point b is shown as point 3, and the point c is shown as point 2. After finding this initial triplet, we next choose a new point x, either between a and b or between b and c. Referring to Figure 6.1, this means that our new point x is either between points 1 and 3 or between points 2 and 3. Suppose, we make the latter choice, i.e., we choose the new point to be between the points b and c (points 3 and 2 in Figure 6.1). We then evaluate $f(x)$. Now, if $f(b) < f(x)$, the new triplet of points would then be $a < b < x$. On the contrary, if $f(b) > f(x)$, then the new bracketing triplet becomes $a < x < c$. In Figure 6.1, for example, the function is evaluated at point 4 (i.e., x is point 4) and $f(b) > f(x)$. Thus, the new bracketing triplets are points 1, 3, and 4. We continue this process in such a way that the middle point of the new triplet is the abscissa whose ordinate is the best minimum achieved so far, and we continue this process of bracketing until the distance between the two outer points of the triplet is tolerably small (see the illustration shown in Figure 6.1).

Before proceeding to further details of the method, let us first clarify the practical meaning of *tolerably small*, which we stated above. For a minimum of $f(x)$ located at $x = b$, we may think that if we can bracket our solution

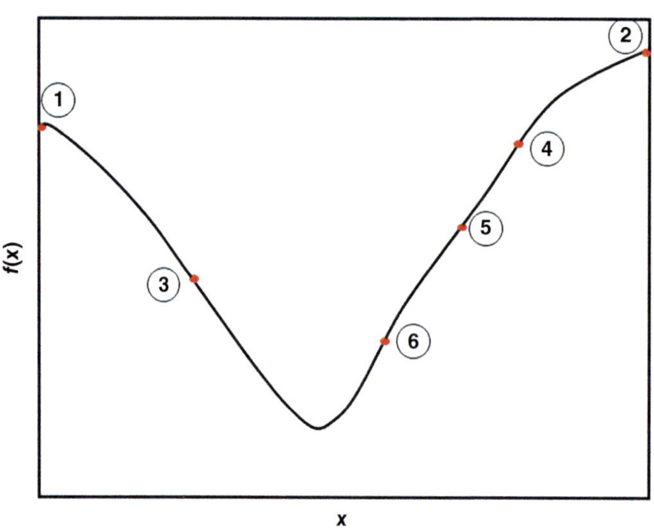

Figure 6.1 Successive bracketing of the minimum used by the golden section search. The minimum is originally bracketed by points 1, 3, and 2. The function is then evaluated at 4, which replaces 2. It is then evaluated at 5 which replaces 1, then at 6 which replaces 4, and so on. The rule at each step is to keep a center point that is lower than the two outside points. After the steps shown here, the minimum is bracketed by points 3, 6, and 5.

in the range $(1-\varepsilon)b < b < (1+\varepsilon)b$, with ε being the computer's floating-point precision, we have solved the problem. Unfortunately, attempting to so may lead to unnecessary computations. In general, the shape of $f(x)$ near $x = b$ is given by the Taylor's theorem as

$$f(x) \approx f(b) + \frac{1}{2}\frac{\partial^2 f}{\partial x^2}\bigg|_{x=b}(x-b)^2$$
$$= f(b) + \frac{1}{2}f''(b)(x-b)^2. \tag{6.2}$$

The second term on the right-hand side of equation 6.2 is negligible compared to the first. To be precise, it will be a factor of ε smaller, and it will just act like zero when added to it whenever. Or, in other words (for details, see Press et al., 2007),

$$|x-b| < \sqrt{\varepsilon}b\sqrt{\frac{2|f(b)|}{b^2 f''(b)}}. \tag{6.3}$$

The bottom line is that as a rule of thumb it is useless to ask for a bracketing interval of width $\sqrt{\varepsilon}$ times the central value, where ε is the machine precision. Knowing this fact can save unnecessary bracketing. In this chapter, we provide flowcharts for different methods for finding the minimum of a function, which helps with the subsequent exercises. In all these flowcharts, there is a convergence check, which must be set to the square root of the machine's floating-point precision. Doing so, in turn, would avoid useless computations with no additional gain.

Now, starting from an initial bracketing interval $(a, b.\ c)$, it remains to define the strategy to find the new point x. Suppose that the point b is a fraction W of the way between the points a and c. In other words,

$$\left.\begin{aligned} \frac{b-a}{c-a} &= W \\ \frac{c-b}{c-a} &= 1-W \end{aligned}\right\}. \tag{6.4}$$

Also assume that our new trial point x is an additional fraction Z beyond point b, or

$$\frac{x \quad b}{c-a} = Z. \tag{6.5}$$

Depending upon the location of the new point, the next bracketing segment would then be either of length $W + Z$ relative to the current one or else of length $1 - W$. If we want to minimize the worst-case possibility, then we should choose Z such that $W + Z = 1 - W$, or

$$Z = 1 - 2W. \tag{6.6}$$

We see at once that the new point is the symmetric point to b in the interval, namely, with $|b-a|$ equal to $|x-c|$. This, in turn, implies that the new point x lies in the larger of the two segments (Z is positive only if $W < 1/2W$). But where is the larger segment? Where did the valuc W itself come from? Presumably, it came from the previous stage of applying our same strategy. Therefore, if Z is chosen to be optimal, then so was W before it. This "scale similarity" implies that x should the same fraction of the way from b to c (if that is the bigger segment) as was b from a to c. In other words,

$$\frac{Z}{1-W} = W. \tag{6.7}$$

Combining equations 6.6 and 6.7 leads to the quadratic equation

$$W^2 - 3W + 1 = 0, \tag{6.8}$$

with solution $W \approx 0.38197$. In other words, the optimum bracketing interval $a < b < c$ has its middle point b a fractional distance 0.38197 away from one end (say, a) and 0.61803 from the other end (say, c). These fractions are those of so-called *golden mean* or *golden section*. This optimal method of function minimization, which is an analog of finding the zeros (roots) of polynomials, is thus called the *golden section search*. A summary of the method is the following:

> Given at each stage, a bracketing triplet of points, the next point to be tried is that which is a fraction 0.38197 into the larger of the two intervals, measuring from the center of the triplet. If we start from a bracketing triplet whose segments are not in the golden ratios, the procedure of choosing the successive points at the golden mean point of the larger segment quickly converges to the proper, self-replicating ratios.

The golden section search guarantees that each new function evaluation will (after self-replicating ratios have been achieved) bracket the minimum to an interval just 0.61803 times the size of the preceding interval.

In the following, we provide two flowcharts for the golden section search method in one dimension. The first flowchart is algorithm for bracketing (*Minimum-Bracket*). Note that in this flowchart (and in all subsequent flowcharts for other methods) there is a function evaluation $f(x)$. To use these flowcharts, one needs write an independent code for the function (objective) evaluation in one dimension and call in from the Minimum-Bracket code. Essentially, Minimum-Bracket does the following:

> Given two initial points $X_A < X_C$, and the evaluation of the single-dimensional objective function at those points given as $f(X_A) = F_A$ and $f(X_C) = F_C$, it uses the golden search method to find the new point X_B between X_A and X_C such that $f(X_B) = F_B$ is less than both F_A and F_C. It returns the triplet of points $X_A < X_B < X_C$ and the evaluated values of the objective functions F_A, F_B, and F_C.

The second flowchart *Golden* is the one for the actual golden search. Given the objective evaluations F_A, F_B, and F_C, and the bracketing triplet of abscissas $X_A < X_B < X_C$ in the golden ratios coming from Minimum-Bracket, this flowchart does the golden search and returns the minimum. To use these flowcharts to find the minimum for 1D optimization problems, it is necessary to write the actual codes from the outlines provided in these flowcharts. In addition, it is also necessary to write a main code with the following functionalities:

1) Choose any three points $X_A < X_B < X_C$ within which we know that the minimum is located. This means that if the evaluated objectives at these three points are respectively F_A, F_B, and F_C, then F_B must be less than both F_A and F_C.
2) Call the code Minimum-Bracket to bracket the triplet of points within the golden ratio.
3) Call the code Golden to find the minimum.
4) Estimate the minimum point.

In both flowcharts given in Figures 6.2 and 6.3, there is a convergence check, where it is necessary to use the square root of the machine's floating-point precession ε as previously discussed.

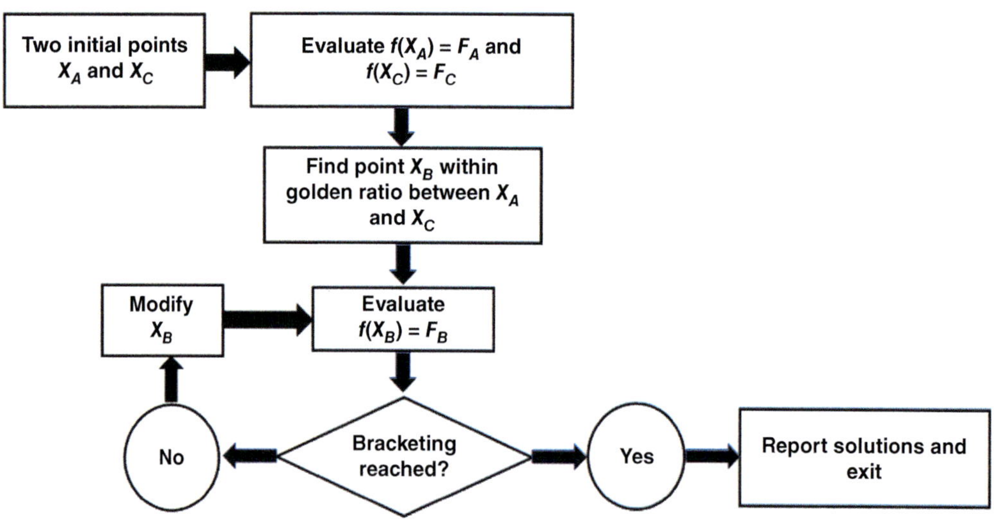

Figure 6.2 Flowchart Minimum-Bracket for bracketing solution.

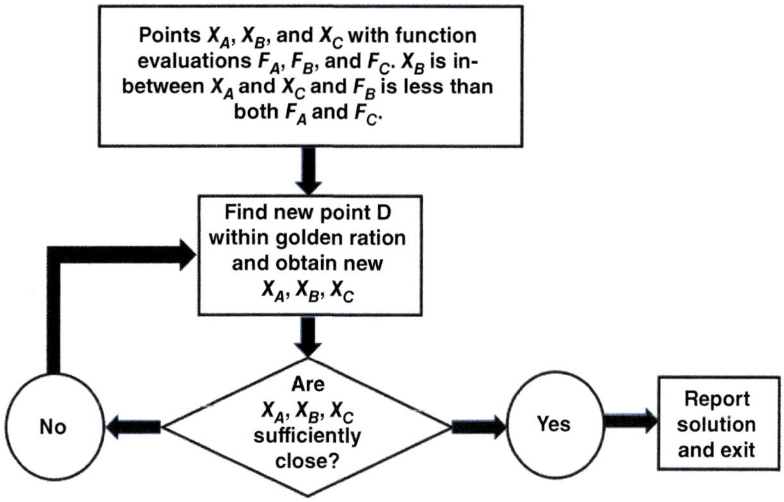

Figure 6.3 Flowchart Golden for obtaining the minimum of a function using golden section search method.

6.2.2 Inverse Parabolic Interpolation and Brent's Method in One Dimension

The golden search method is designed to handle the worse possible case for 1D function optimization. In many situations, the situation is not as bad, and the function is nearly parabolic near to the minimum, then a parabola, fitted through any three points, should take us to the minimum or at least very close to the minimum. Because we want to find the abscissa rather than the ordinate, this procedure is called the *inverse parabolic interpolation*. The concept of this is illustrated in Figure 6.4. As shown in Figure 6.4, starting from three initial points, say 1, 2, 3, a parabola is fit and is shown as the blue curve. The function is then evaluated at the minimum point of this fitted parabola, shown as point 4 in Figure 6.4. Another parabola is then fit through points 1, 4, 2, which is shown in red in Figure 6.4. The minimum point of this new parabola (point 5 in Figure 6.4) is again found, and this process is repeated until the minimum of the original function is obtained, as defined by the machine's floating-point precision. As shown in Figure 6.4, point 5 is, in fact, very close to the true minimum of the original function.

Now, the formula for the abscissa x which is the minimum of a parabola fitted through three points a, b, c is given by (Press et al., 2007)

$$x = b - \frac{1}{2}\frac{(b-a)^2[f(b)-f(c)] - (b-c)^2[f(b)-f(a)]}{(b-a)[f(b)-f(c)] - (b-c)[f(b)-f(a)]}. \tag{6.9}$$

The above formula fails only if the three points (a, b, c) are collinear, in which case the denominator is zero, implying that the minimum of the parabola in infinitely far away.

Brent's method (Brent, 1973) is an inverse parabolic interpolation method for finding the minimum or maximum of functions in one dimension and works as follows:

1) At each stage, it keeps track of six function points, not necessarily all distinct, a, b, u, v, w, x:
   a) The minimum is bracketed between a and b.
   b) The value of x is the point with the least function value found so far or the most recent one in case of a tie.
   c) The point that corresponds to the second least function value is denoted as w.
   d) The point v is the value of w in the previous step of iteration.
   e) The point at which the function was most recently evaluated is u.

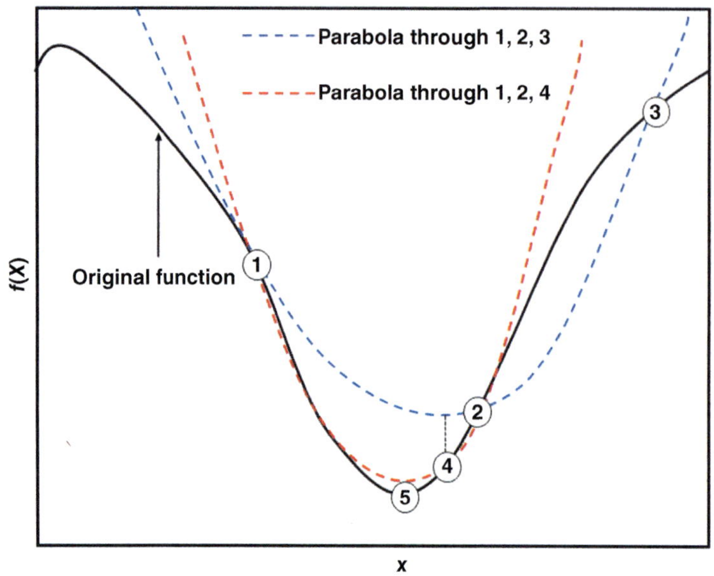

Figure 6.4 Illustration of the inverse parabolic interpolation method for finding the minimum of a single-dimensional function.

   f) In addition to the above, the point x_m is the midpoint between the points a and b, but the function is not evaluated there.
2) At each step, parabolic interpolation is attempted (equation 6.9), fitting through the points x, v, w. To be acceptable, the parabolic step must
   a) fall within the bounding interval (a, b) and
   b) imply a movement from the best current value x that is less than half the movement of the step before the last. This second criterion ensures that the parabolic steps are converging rather than bouncing around a non-convergent limit cycle.
3) It must also be noted here that the method should not evaluate the function less than a distance $\sqrt{\varepsilon}$ away from a point already evaluated, else the evaluation will be within the limits of the roundoff error.

A typical ending configuration of the Brent's method is that a and b are $2 \times \sqrt{\varepsilon}$ apart with x_{min} (the estimated minimum point) at the midpoint of a and b, and therefore fractionally accurate to $\pm\sqrt{\varepsilon}$. Figure 6.5 shows the flowchart for the Brent's inverse parabolic interpolation method. Note that in Figure 6.5 and for other 1D optimization methods, we describe next, we start from the triplet of points within which the solution is bracketed. This initial triplet can be obtained from the Minimum-Bracket (Figure 6.2) or any similar method.

6.2.3 Van Wijngaarden–Dekker–Brent Method

Strictly speaking, this method is not an optimization but a root-finding methodology. There are, however, many practical applications of this methodology.

The Van Wijngaarden–Dekker–Brent method was originally developed in the 1960s by Van Wijngaarden, Dekker, and others, and was later improved by Brent (1973). Here we refer to the final form of the algorithm and call it Brent's method. The Brent's method discussed above in Section 6.2.2 uses inverse hyperbolic interpolation between three points. In contrast with this, the Van Wijngaarden–Dekker–Brent method uses inverse quadratic interpolation between these points. To distinguish between these two methods, we refer the method described in Section 6.2.2 as the Brent's inverse hyperbolic method and the method we describe in this section as the Brent's inverse quadratic method. Brent's inverse quadratic

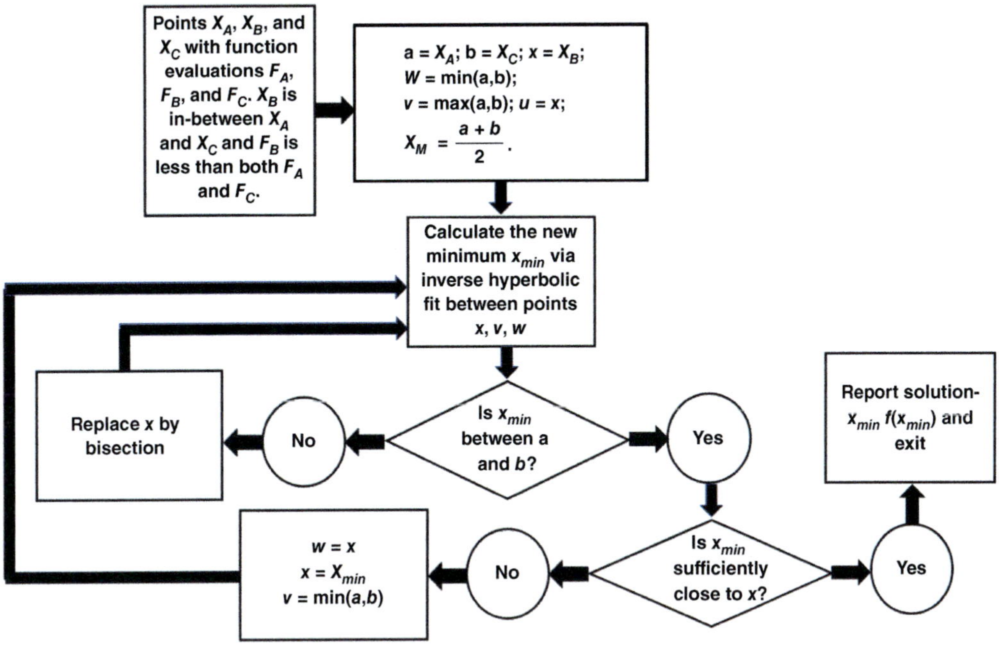

Figure 6.5 Flowchart for the Brent's method of inverse parabolic interpolation.

method is guaranteed to converge, so long as the function, evaluated within the initial interval, is known to contain a root.

Brent's inverse quadratic method combines root bracketing (finding an interval within which the root lies), bisection (cutting the interval), and the inverse quadratic interpolation to converge from the neighborhood of a zero crossing. The inverse quadratic interpolation uses three prior points to fit an inverse quadratic function (i.e., x as a function of y) whose value at $y = 0$ is taken as the next estimate of the root of x. The method can additionally handle the situations when the root falls outside the bracketing interval.

If the three points are $[a, f(a)]$, $[b, f(b)]$, and $[c, f(c)]$, then the quadratic interpolation formula is given by (Press et al., 2007)

$$x = \frac{[y-f(a)][y-f(b)]c}{[f(c)-f(a)][f(c)-f(b)]} + \frac{[y-f(b)][y-f(c)]a}{[f(a)-f(b)][f(a)-f(c)]} + \frac{[y-f(c)][y-f(a)]b}{[f(b)-f(c)][f(b)-f(a)]}. \quad (6.10)$$

Unlike inverse hyperbolic interpolation (equation 6.9), inverse quadratic interpolation, shown in equation 6.10, is derived from the Taylor series expansion of $y = f(x)$, truncated after the second derivative term. If we set $y = 0$ in equation 6.10, we readily find the result for the next root estimate to be (see Press et al., 2007 for details)

$$x = b + \frac{P}{Q}, \quad (6.11)$$

where

$$P = S[T(R-T)(c-b) - (1-R)(b-a)] \quad (6.12)$$

and

$$Q = (T-1)(R-1)(S-1), \quad (6.13)$$

with

$$\left.\begin{aligned} R &= \frac{f(b)}{f(c)} \\ S &= \frac{f(b)}{f(a)} \\ T &= \frac{f(a)}{f(c)} \end{aligned}\right\}. \quad (6.14)$$

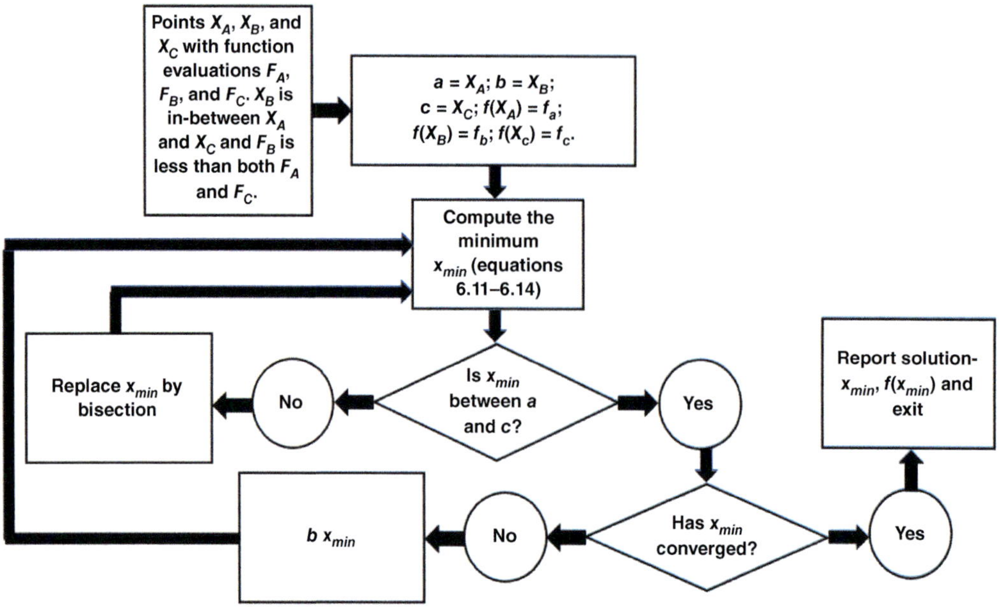

Figure 6.6 Flowchart for the Van Wijngaarden–Dekker–Brent method of the Brent's inverse quadratic interpolation method.

In practice, b is the current best estimate of the root and P/Q is a small correction. As a general rule of thumb, quadratic methods work well when the function behaves smoothly, and they run into serious problems of giving very bad estimate of the next root or causing division-by-zero error when $Q \approx 0$. Brent's method guards against this problem by maintaining the brackets on the root and checking where the interpolation would land before carrying out the division. When the method finds that the correction P/Q would not land within the bounds, or the bounds are not collapsing rapidly enough, it takes a bisection step (Press et al., 2007).

A sample flowchart for the method is provided in Figure 6.6.

6.2.4 One-Dimensional Optimization Using First Derivatives

After showing some 1D optimization methods used in Geophysics in the next section, we then move onto multidimensional optimizations. But, prior to looking into the multidimensional problems, it is important to look at another class of 1D optimization, which will be useful in multiple dimensions.

So far, we considered 1D optimization methods that require function evaluation, i.e., solving the forward problem, evaluating the objective, and optimizing the objective without any need for computing the derivative of the objective. However, it will be obvious very soon that having the ability to compute the derivative of the objective is very powerful for optimization, specifically in multiple dimensions.

So, here we want to accomplish precisely the same goal, i.e., optimization in one dimension by isolating the functional minimum (or maximum) that is bracketed by the triplet of abscissas $a < b < c$ but utilizing an additional capability to compute the first derivative of the function in addition to its value. In principle, we might simply search for a zero of the derivative, ignoring the function value information, using a root finder such as either of the two Brent's methods we just discussed. But it does not take too long to reject that idea because by doing so we cannot distinguish the maxima from minima. If we cast our optimization as

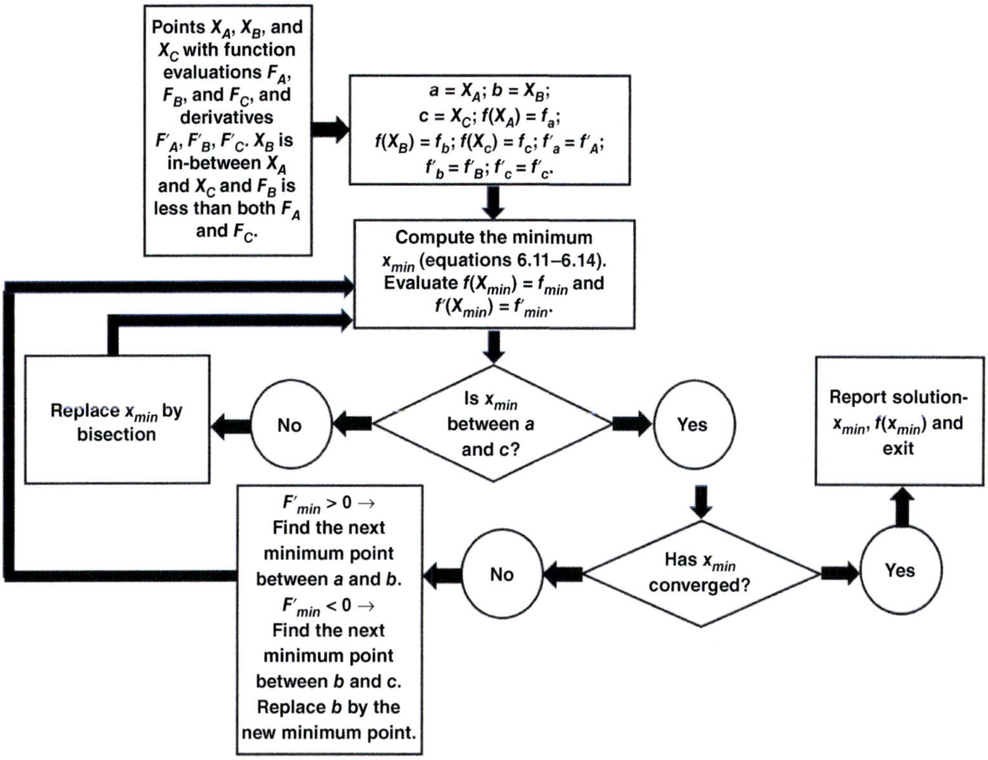

Figure 6.7 Flowchart for the Brent's method for minimizing function with derivative information.

minimization, for example, where do we go from the initial conditions where the derivatives on one or both outer bracketing points indicate that "downhill" is in the direction out of the "bracketed" interval? We do not always want to give up our strategy of maintaining a rigorous bracket on the minimum. And the only way to keep such a bracket is to update it using the function (not derivative) information, with the center point in the bracketing triplet having the lowest function value. Thus, the role of the derivatives can only be to help us choose the new trial points within the bracket.

This leads us to use derivative information in the following way for a minimization problem:

1) Find the sign of the derivative at the center point of the bracketing triplet $a < b < c$.
2) If the sign is positive, then the next test point should be taken in the interval (a, b).
3) If the sign is negative, then the test point should be in the interval (b, c).

Figure 6.7 shows the flowchart for 1D optimization using the derivative information. It is like the Brent's method discussed before. We use this concept and elaborate it more when we move onto the multidimensional optimization problems.

6.2.5 Practical Examples of One-Dimensional Optimization

In the above section, we discussed 1D optimization problems. Most optimization problems we deal with in geophysics, inclusive of seismic optimization, are multidimensional problems. However, before going into multidimensional optimization, it is important to point out and discuss the following two single-dimensional optimization problems of practical interest that are often encountered in seismology:

1) The problem of phase and group velocities and angles in an anisotropic medium.
2) The ray tracing problem.

We briefly discussed the phase/group problem in Chapter 3 and the ray tracing in Chapter 4, and here we show how to use a single-dimensional optimization method to solve them. Besides the above two, single-dimensional optimization problems are also encountered elsewhere in seismology, and we postpone their discussions for later.

Phase and Group Velocities and Angles in an Anisotropic Medium

Recall our discussion of the Christoffel equation (equation 3.77) in Chapter 3, Section 3.8:

1) The equation in cubic in p^2, in which p is the slowness. Therefore, there are three solutions in p^2 which correspond to the upgoing and downgoing qP, qS_1, and qS_2 wavefields.
2) These solutions are the plane waves with slowness $p(\theta, \varphi)$ for the corresponding wavefields, propagating at different phase angles θ and azimuth φ.
3) Slowness solution $p(\theta, \varphi)$ for different values of θ and φ is a surface in three dimensions, known as the slowness surface. The directed object from the origin to the slowness surface is the slowness vector $\mathbf{p}(\theta, \varphi)$, whose magnitude is $p(\theta, \varphi)$ (see Figure 3.13).
4) The inverse of $p(\theta, \varphi)$ is the phase velocity $V_{ph}(\theta, \varphi)$, and the phase velocity vector $\mathbf{V}_{ph}(\theta, \varphi)$ is directed parallel to $\mathbf{p}(\theta, \varphi)$ with magnitude $V_{ph}(\theta, \varphi)$.
5) Although the solution to the Christoffel equation in a general (triclinic) anisotropic medium must be numerically obtained, analytical solutions exist for many other forms of anisotropic symmetry.

Also recall the discussion of the ray (group) velocity and ray (group) angle in Chapter 3, Section 3.9:

1) While the plane waves with constant phase propagate along (θ, φ), the energy propagates along the outward normal direction to the slowness surface.
2) The ray or group direction is along this outward normal, and the group velocity vector is related to the slowness vector and phase velocity vector via equation 3.111 using which both its magnitude and direction (i.e., the group angle) can be obtained. For a general anisotropic medium, it must be numerically obtained as shown in Figure 3.12 and equations 3.112 and 3.113. There are, however, analytical solutions for specific symmetry types.
   a) For the P-SV propagation in a transversely isotropic medium with vertical symmetry axis (VTI) medium for example, equations 3.121–3.126 and 3.130 provide such an analytical solution.

In an isotropic elastic medium where the slowness surface is a sphere, the outward normal to the slowness surface and the direction of $\mathbf{p}(\theta, \varphi)$ are same, and therefore, there is no difference between the phase and group directions and the phase and group velocities (see Figure 3.13). For other anisotropic systems, however, this is not true. Travel times on the seismic data recorded in offset and time provide the group velocities and group angles (Figures 3.15–3.17). The elastic properties, on the other hand, are directly related to the phase velocity as a function of phase angle. Additionally, the reflection coefficients controlling the amplitude behavior on the recorded seismic reflection data are also functions of slowness or phase velocity and phase angle. To obtain elastic properties from seismic reflection records, it is therefore necessary to map group angles and group velocities to their corresponding phase angles and phase velocities. As will be seen later that a plane-wave decomposition of time and offset $(x-t)$ domain data into intercept time and slowness $(\tau-p)$ domain allows this mapping of group to phase. However, $x-t$ to $\tau-p$ transformation is sometimes difficult for limited offset data (Mukhopadhyay and Mallick, 2011). It is therefore convenient to find a way to directly map the group angles and group velocities onto phase angles and phase velocities.

From the relationship between group and phase (equation 3.111 for general numerical solution, or equations 3.121–3.126 and equation 3.130 for analytical solution in a VTI medium), we discussed before in Chapter 3, Section 3.9.3, we found that while it is straightforward to compute the group velocity and group angle from the phase velocity and angle, either numerically or analytically (for specific symmetry types), the reverse is not true and must be iteratively estimated via an optimization. We also gave an outline of such an optimization in Chapter 3 (Figure 3.18). In this outline, we first set the observed angle α and the observed velocity $V_{obs}(\alpha)$ respectively to group angle ϕ and group velocity $V_g(\phi)$. Then, by initially setting the group angles and phase angles to be equal ($\theta = \phi$), we compute the group velocity and angle from the phase angle θ and iteratively modify θ until computed group angle and group velocity ϕ and $V_g(\phi)$ converge to the observed α and $V_{obs}(\alpha)$. In this outline, however, we did not mention the process of modifying θ so that θ and $V_{ph}(\theta)$ can be estimated from α and $V_{obs}(\alpha)$.

Figure 6.8 shows a flow diagram for solving the group/phase problem for the P-SV system in a VTI medium using any 1D optimization method we discussed above. The procedure to obtain solution for any other anisotropic system is similar. Note that to use any of the 1D optimization methods such as the golden section search, Brent's inverse hyperbolic, Brent's inverse quadratic, etc., it is first necessary to bracket the solution. For the group/phase problem, the observed velocity V_{obs} as a function of the observed angle α is given from the data, and in Chapter 3 we have seen that α is the group angle ϕ and V_{obs} is the group velocity $V_g(\phi)$. If we want to estimate the corresponding phase angle θ and the associated phase velocity $V_{ph}(\theta)$, then we must first find two phase angles θ_1 and θ_2 between which the actual value lies. In other words, we must find the bracketing interval $\theta_1 < \theta < \theta_2$ in which θ is the desired solution to the phase angle corresponding to group angle $\phi = \alpha$. In Figure 6.8, we first find this bracketing interval $\theta_1 < \theta < \theta_2$ as follows. For convenience, we use different colors to highlight different branches of this computation.

1) For any observed angle and velocity pair α, $V_{obs}(\alpha)$, we first assume $\theta = \phi$ and calculate the phase velocity $V_{ph}(\theta)$. For P-SV waves in a VTI medium, for example, we can

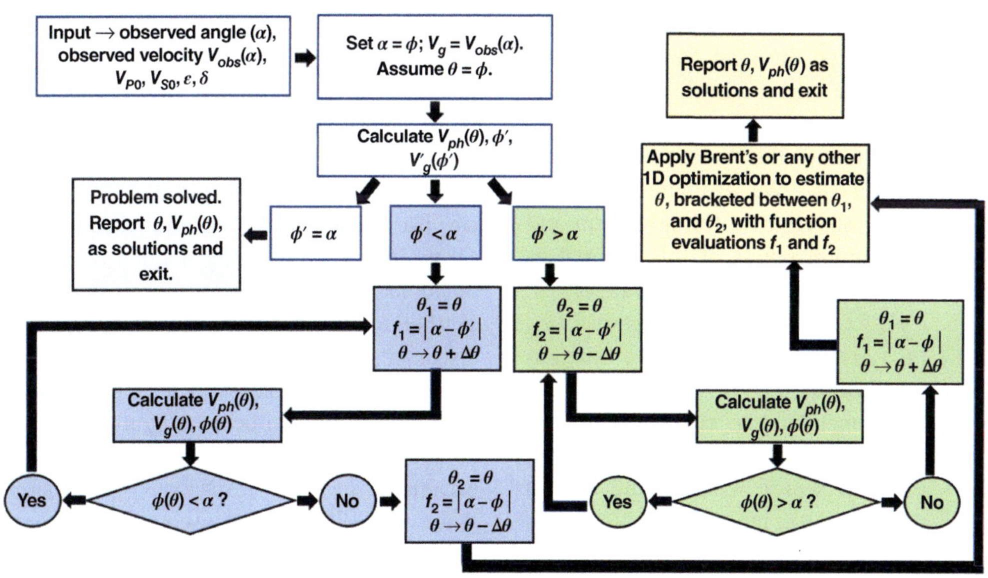

Figure 6.8 Flow diagram for implementing one-dimensional optimization to estimate the phase angle, phase velocity, and group velocity from the observed velocity V_{obs} as a function of the observed incidence angle α (see Figure 3.16b), and the known VTI model of V_{P0}, V_{S0}, ε, and δ.

compute this from equations 3.96 and 3.97. Once we compute the phase velocity, we can compute the group velocity and group angle. Restricting to P-SV waves in a VTI medium, we can analytically compute them from equations 3.121–3.126 and 3.130. In general, this computed group angle and group velocity will not be equal to $\alpha = \phi$ and $V_{obs}(\alpha) = V_g(\phi)$, which we obtain from seismic data. In Figure 6.8, therefore, we respectively denote them as ϕ' and $V_g(\phi')$.

2) We now compare ϕ' obtained from step 1 with α. This comparison leads to three different possibilities:
   a) $\phi' = \alpha$ (the region highlighted in gray in Figure 6.8). This will always happen for an isotropic medium and for some specific directions of propagation in a VTI medium. Under this scenario, the problem is solved. We therefore report θ and $V_{ph}(\theta)$ to be the solutions and exit.
   b) $\phi' < \alpha$ (highlighted in blue in Figure 6.8). Under this scenario, we are on the left-hand side of the bracketing region. Therefore, we must find another point on the right-hand side of the bracketing region. Thus, we set $\theta = \theta_1$ and iteratively increase θ to $\theta + \Delta\theta$ for a conveniently chosen value of $\Delta\theta$ until we reach on the right-hand side of the bracketing interval. In that case, we set the value of θ to θ_2.
      i) We found the bracketing interval $\theta_1 < \theta < \theta_2$. We evaluate our function (objective) at the minimum and maximum points of this interval as $f_1 = \lceil \alpha - \phi(\theta_1) \rceil$ and $f_2 = |\alpha - \phi(\theta_2)|$. We feed this bracketing interval (θ_1, θ_2) and the corresponding function evaluations (f_1,f_2) to an 1D optimization method like either one of the two Brent's methods or the golden section search method and find θ and $V_{ph}(\theta)$, report these solutions and exit. The 1D optimization part is highlighted in yellow in Figure 6.8.
   c) $\phi' > \alpha$ (highlighted in green in Figure 6.8). In contrast to the above step (step b), here we are on the right-hand side of the bracketing interval. Therefore, the procedure here is opposite of step b. That is, we keep decreasing θ by $\Delta\theta$ until we find the left-hand side of the bracketing interval. Once found, we feed the interval (θ_1, θ_2) and the corresponding objective evaluations to the same (f_1, f_2) into 1D optimization part highlighted in yellow to estimate θ and $V_{ph}(\theta)$, and report solutions and exit.

The procedure outlined above is left as an 1D optimization exercise at the end of the chapter.

Ray Tracing Problem

Ray tracing is an essential element in the computation of ray-theoretical synthetic seismograms, which we discussed in Chapter 4, Section 4.2.7. Because we acquire seismic data that at fixed receiver locations away from the source location, finding the exact takeoff angle of the ray, generated from the source location, downward propagate, reflect from an interface, and upward propagate at a given receiver location cannot be analytically obtained. In Chapter 4, we mentioned that this is an optimization problem and here we show how any of the 1D optimization methods introduced above can solve this problem.

For simplicity, we assume a horizontally stratified earth model and consider primary (P–P) reflection from an interface located at depth h from the surface (Figure 6.9). It will be however clear that the optimization method we discuss below can be used for nonhorizontal layers. Ray tracing frequently used in seismology not only to compute ray-theoretical synthetic seismograms which we already discussed, but also in ray-based imaging (migration) methods, such as the Kirchhoff migration, Gaussian-beam migration, etc., to downward propagate the source and receiver wavefields to their respective *image points* (Sava and Hill, 2009). Consequently, the optimization needed in ray tracing is of great practical importance.

As shown in Figure 6.9, the source are receiver locations are fixed such that the

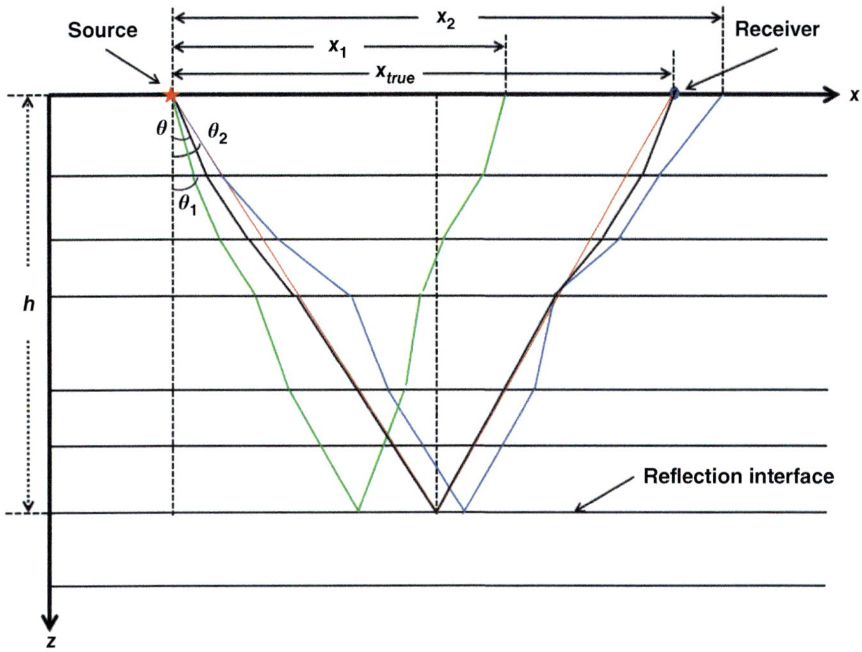

Figure 6.9 Illustration of the ray tracing problem between fixed source (red asterisk) and receiver (blue dot) locations for reflection from an interface at depth *h*. Different colors were used to represent different ray paths: red represents the straight ray path between the source and receiver, black represents the true ray path, and green and blue represent the ray paths bracketing the true ray path on either side.

receiver is at a distance x_{true} away from the source. In ray tracing, we need to find the exact ray that takes off at an angle, say θ, and reaches the specific receiver location. This ray is shown as solid black line in Figure 6.9. But how do we find this ray? The only way to find it is via trial and error through an optimization. Like the group/phase velocity/angle problem, we first bracket the solution by finding two ray paths with takeoff angles, say θ_1 and θ_2, within which the actual ray lies. For example, if we connect the source to the reflection point, and the reflection point to the receiver by straight lines (solid red lines in Figure 6.9), then the angle θ_{try} these straight lines make with the vertical is given as

$$\theta_{try} = \sin^{-1}\left(\frac{x_{true}/2}{\sqrt{\left(\frac{x_{true}}{2}\right)^2 + h^2}}\right). \tag{6.15}$$

Using θ_{try} as the takeoff angle, we trace ray through the model as shown by the blue lines in Figure 6.9. Of course, this ray tracing will take us to an incorrect receiver location to the surface. Based on the actual model, this incorrect location could be either further away or closer to the source than the true receiver location. So, like the group/phase velocity problem illustrated in Figure 6.8, we keep increasing or decreasing the takeoff angle until we find two angles θ_1 and θ_2 with corresponding receiver locations x_1 and x_2 between which the true receiver location (x_{true}) lies as denoted by the green and blue ray paths in Figure 6.9. Once the solution is bracketed between θ_1 and θ_2, any 1D optimization method can be used to find the actual ray. Figure 6.10 shows an outline of such a procedure. Note that the steps shown in Figure 6.10 are not much different from those in Figure 6.8, and therefore, we do not explain them in detail.

Although we demonstrated this optimization method for ray tracing for horizontally stratified medium and for primary (P–P) reflections, this method would still work

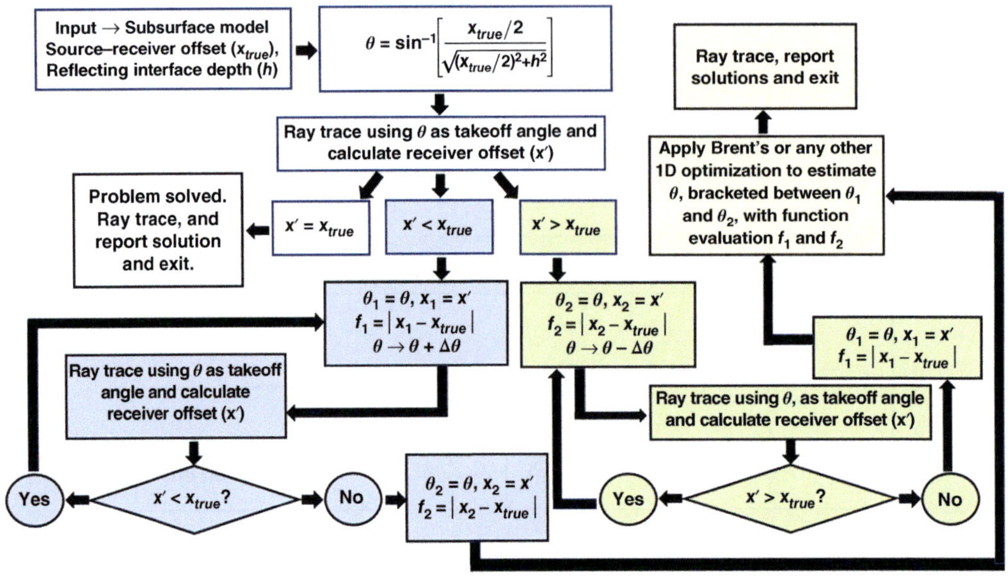

Figure 6.10 1D optimization for the ray tracing problem.

when the layers are not horizontal and for mode-converted reflections. It will also work for three-dimensional models. For primary reflections, when the layers are horizontal, the reflection point is always at the midpoint between the source and receiver locations. But this is not true for the mode-converted reflections even when the layers are horizontal, and for nonhorizontal layers, the reflection point is not the midpoint even for primary reflections. It is, however, straightforward to bracket the solution and use the method outlined in Figure 6.10 for all such cases.

Finally, going from an isotropic to an anisotropic medium, the same method works. But for anisotropic media, one must be careful about using the phase angle and phase velocity for reflection/refraction and group angles and group velocities for travel times, which we previously discussed in Chapter 4.

6.3 Multidimensional Optimization

Now that we discussed 1D optimization in detail, showed how to solve practical geophysical optimization problems, we now move onto multidimensional optimization. Again, we look at multidimensional from the minimization point of view, keeping in mind that casting the problem as maximization instead of minimization is trivial.

6.3.1 Fundamental Concepts

Consider a multidimensional function $f(\mathbf{x})$, where $\mathbf{x} = [x_1, x_2, \dots, x_N]^T$. If we start at a point $\mathbf{a}$ in N-dimensional space and proceed from there along some vector direction $\mathbf{n}$, then the function $f(\mathbf{a})$ can be minimized along the line by N one-dimensional methods previously discussed by changing the direction $\mathbf{n}$ at each step. We can thus view various multidimensional methods to be composed of sequences of such single-dimensional or line minimizations. Different methods such as optimization differ only by how, at each stage, they try the next direction $\mathbf{n}$ to be. In fact, all such methods presume the existence of a *black box* sub-algorithm (say, LINMIN), whose definition we take from Press et al., (2007) as

LINMIN: Given as input the vectors **a** and **n**, and the function $f(\mathbf{a})$, find the scalar λ that minimizes $f(\mathbf{a} + \lambda \mathbf{n})$. Replace **a** by $\mathbf{a} + \lambda\mathbf{n}$ and **n** by $\lambda\mathbf{n}$.

In our discussions, we cast all multidimensional methods under the general scheme of successive line minimization using an algorithm LINMIN defined as above. We first consider a set of methods whose choice of successive directions does not involve explicit computation of the function's gradient. Then, we consider the set that require calculating the gradient. In fact, the choice of whether use the gradient depends upon the problem in hand. While using the gradient information can drastically reduce the number of steps needed to reach the minimum, calculating the gradient could be a computational challenge for some problems. We specifically discuss these issues, including line minimization in Chapter 7 where we consider seismic inverse problems.

Let us first consider optimizing without using the gradients. So, to start, we can think of a very simple method:

1) In the N-dimensional space, assume the unit vectors $\hat{\mathbf{x}}_1, \hat{\mathbf{x}}_2, \ldots, \hat{\mathbf{x}}_N$ as a set of directions.
2) Using LINMIN, move along the first direction to its minimum, then from there along the second direction to its minimum, and so on, cycling through the entire set of directions as many times as necessary.
3) Stop when the function does not decrease any more.

A simple method outlined above may work fine for many functions but could be inefficient for many other functions.

For example, consider a function in two dimensions whose contour map define a long, narrow valley, oriented at some angle to the coordinate basis vectors (Figure 6.11). It can be seen from this figure that the only way to go down the length of the valley by going along the coordinate basis vectors at each stage is by a series of many tiny steps. More generally, in N dimensions, if the function's second derivatives are much larger in magnitude in some directions than the others, then many cycles through all N basis vectors will be necessary to obtain anywhere close to the minimum (optimum) point.

To summarize, we need a better set of directions than the basis vectors $\hat{\mathbf{x}}_i$ and the one illustrated in Figure 6.11. All multidimensional optimizations therefore consist of procedures for updating the set of directions as

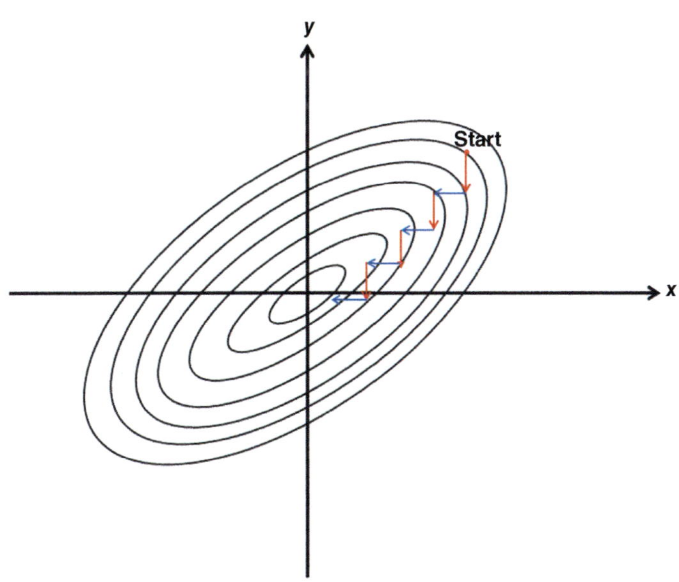

Figure 6.11 Successive minimizations along the coordinate directions along a long, narrow valley. Unless the valley is optimally oriented, this method is very inefficient.

they progress and attempt to come up with a set which either (1) includes very good directions that will take us far along the narrow valleys or (2) includes some number of *non-interfering* directions with the special property that the minimization along one is not spoiled or affected by subsequent minimization along the other so that intermediate cycling through the set of directions can be avoided.

6.3.2 Conjugate Directions

The concept of non-interfering directions mentioned above is known as the *conjugate directions*. Here, we develop the fundamental mathematical concept of the conjugate directions, which we use later to develop the conjugate gradient method of multidimensional function optimization.

First, note that if we minimize a function along a direction **u**, then the gradient of the function must be perpendicular to **u** at the line minimum; else there would still be a non-zero directional derivative along **u**. If we assume some point **a** to be the origin of the coordinate system, then any function $f(x_1, x_2, \ldots, x_N)$ in the neighborhood of **a** can be written by its Taylor series as

$$f(\mathbf{x}) = f(\mathbf{a}) + \sum_{i=1}^{N} \left.\frac{\partial f}{\partial x_i}\right|_{\mathbf{a}} x_i + \frac{1}{2}\sum_{i=1}^{N}\sum_{j=1}^{N} \left.\frac{\partial^2 f}{\partial x_i \partial x_j}\right|_{\mathbf{a}} x_i x_j + \cdots \approx f(\mathbf{a}) - \mathbf{b}\cdot\mathbf{x} + \frac{1}{2}\mathbf{x}^T\cdot\mathbf{H}\cdot\mathbf{x}. \quad (6.16)$$

where

$$\mathbf{b} = -\nabla f(\mathbf{a}); \quad |\mathbf{H}|_{ij} = \left.\frac{\partial^2 f}{\partial x_i \partial x_j}\right|_{\mathbf{a}}. \quad (6.17)$$

Note that on the right-hand side of equation 6.16 we approximated $f(\mathbf{a})$ in a quadratic form by ignoring the derivatives greater than two. All multidimensional optimization methods we discuss are based on this quadratic approximation of the function f. In equation 6.17 **H** is a symmetric $N \times N$ matrix whose components are second-order partial derivatives of f and is called the *Hessian matrix* of $f(\mathbf{a})$, which can be explicitly written as

$$\mathbf{H} = \begin{pmatrix} \frac{\partial^2 f}{\partial x_1^2} & \frac{\partial^2 f}{\partial x_1 \partial x_2} & \cdots & \frac{\partial^2 f}{\partial x_1 \partial x_N} \\ \frac{\partial^2 f}{\partial x_2 \partial x_1} & \frac{\partial^2 f}{\partial x_2^2} & \cdots & \frac{\partial^2 f}{\partial x_2 \partial x_N} \\ \cdot & \cdot & \cdots & \cdot \\ \cdot & \cdot & \cdots & \cdot \\ \cdot & \cdot & \cdots & \cdot \\ \frac{\partial^2 f}{\partial x_N \partial x_1} & \frac{\partial^2 f}{\partial x_N \partial x_2} & \cdots & \frac{\partial^2 f}{\partial x_N^2} \end{pmatrix}_{\mathbf{a}}. \quad (6.18)$$

From the truncated Taylor series approximation of equation 6.16, we can also readily verify that

$$\nabla f(\mathbf{x}) = \mathbf{H}\cdot\mathbf{x} - \mathbf{b}. \quad (6.19)$$

Note that equation 6.19 implies that the gradient will vanish, and the function will be at an extremum at a value **x** near the point **a** can be obtained by solving the linear set of equations given as $\mathbf{H}\cdot\mathbf{x} = \mathbf{b}$. This, in turn, implies that if we can estimate either the Hessian matrix or its inverse, we solve the optimization problem assuming that the function f can be accurately represented in a quadratic form expressed via equations 6.16–6.18. We return to this idea in the later part of this chapter and more explicitly in Chapter 7 when we discuss specific geophysical optimization (inverse) problems.

Now, how does $\nabla f(\mathbf{x})$ change as we move along some direction? We obtain from equation 6.19

$$\delta[\nabla f(\mathbf{x})] = \mathbf{H}\cdot(\delta\mathbf{x}). \quad (6.20)$$

Suppose that we have moved along some direction **u** to a minimum and now propose to move along a new direction **v**. The condition that the motion along **v** will not spoil our previous minimization along **u** is just that the gradient stays perpendicular to **u**, i.e., the change in the gradient be perpendicular to **u**. From equation 6.20, we express this explicitly as

$$0 = \mathbf{u}\cdot\delta[\nabla f(\mathbf{x})] = \mathbf{u}\cdot\mathbf{H}\cdot(\delta\mathbf{x}) = \mathbf{u}\cdot\mathbf{H}\cdot\mathbf{v}. \quad (6.21)$$

When equation 6.21 holds for two vectors **u** and **v**, they are said to be *conjugate*. When this relation holds pairwise for all members of a set of vectors, they are called a *conjugate set*. If we do successive line minimizations of a function along a conjugate set of directions, then we do not need to redo any of those directions.

The success of the direction set methods lies in coming up with a set of N linearly independent, mutually conjugate directions. Then, a single pass of N line minimizations will take us exactly at thc minimum of a quadratic form given by equation 6.16. For functions that are not exactly quadratic forms, it will not be exactly at the minimum, but repeated cycles of N line minimizations will, in due course, converge to the minimum.

6.3.3 Steepest Descent (Gradient Descent) Method

We start with the quadratic form of the Taylor series approximation of a multidimensional function, given by equations 6.16–6.18, and consider the classes of optimization methods where not only the values of the N-dimensional function $f(\mathbf{x})$ but also its gradient ∇f (i.e., the vector of first partial derivatives) are available. Note from equations 6.16–6.18, the number of unknown parameters in f is equal to the number of free parameters in **b** and **H**, which is $(1/2)N(N+1)$, or of the order of N^2. Because finding the location of the minimum requires all these parameters, we should not expect to find the minimum until we have an equivalent information content of the order of N^2 numbers.

If the gradient information is not used, it is therefore necessary to perform N^2 separate line minimizations (or maximizations), each requiring function evaluations. From our discussions in Chapter 5, for seismic inverse problems these function evaluations mean computing synthetic seismic responses. For local optimization, we must start from an initial model, say $\mathbf{x}_0$, and iteratively modify the model via N^2 forward computations before we are anywhere close to the solution. This, in turn, is a computational challenge, even in high-performance parallel computing environments.

Now, if we use the gradient information, it will bring us N new components of information. And, if we use them wisely, we should need only N line minimizations instead of N^2, which, in turn, could be a big saving of computational resources for large-scale optimization problems frequently encountered in geophysics. Here, we discuss how to efficiently (and wisely) use the gradient information to solve multidimensional optimization problems.

Before proceeding further, we must note here that a factor of N improvement is not necessarily implied in these methods we discuss. As a rough estimate, we might imagine that the calculation of each component of the gradient takes about as long as evaluating the function itself. In that case, there will be of the order of N^2 function evaluations, both with and without gradient information. Even if the advantage is not of order N, it could be nevertheless substantial because of the following reasons.

1) Each calculated component of the gradient will typically save not just one function evaluation, but many, which, in turn, is equivalent to a whole line minimization.
2) In many practical problems, there is often a high degree of redundancy in the formulas of the various components of the function's gradient. If this is so, especially when there is also redundancy with the calculation of the function being optimized, then the calculation of the gradient may cost significantly less than N function evaluations. We discuss this issue further in Chapter 7 in discussing practical geophysical optimization problems where we see how to efficiently compute the gradient.

One of the early applications of the gradient information for optimizing functions in multiple dimensions is the *steepest descent* or *gradient descent* method, which can be summarized as follows:

> *Steepest Descent/Gradient Descent Method*: Start at a point $\mathbf{x}_0$. Iteratively move from point $\mathbf{x}_i$ to the point $\mathbf{x}_{i+1}$ by minimizing in the direction of the local downhill gradient $-\nabla f(\mathbf{x}_i)$.

This method has been successfully used to solve geophysical inversions, which we discuss further in Chapter 7. However, the problem with this method is like what we previously demonstrated in successive minimization along the coordinate axes (Figure 6.11). The gradient descent method tends to make many small steps in going down a long, narrow valley, even if the valley is a perfect quadratic form (equation 6.16). In two dimensions, for example, we may expect that the first step will take the point to the valley floor the second step directly down the long axis. But remember that the new gradient at the minimum point of any minimization is perpendicular to the direction just traversed. Therefore, the steepest descent method will make right angle turns in successive steps, which, in general, does not go to the minimum.

A variant of the gradient descent method, known as the Adam optimizer, has been recently introduced in connection with deep-learning applications in machine learning (Kingma and Ba, 2015). Adam optimizer has been shown to be computationally efficient for large-scale data science related applications.

6.3.4 Conjugate Gradient Method

We now refer to equation 6.21, where $\mathbf{H}$ is the Hessian matrix (equation 6.18) and the vectors $\mathbf{u}$ and $\mathbf{v}$ represent the conjugate directions. We now want to find a way of proceeding not down the new gradient like the steepest descent method, but in a direction that is somehow constructed to be the conjugate of the old gradient, and insofar possible, to all previous directions traversed. The methods accomplishing this construction are called the *conjugate gradient* methods.

Mostly used conjugate gradient methods are (1) *Fletcher–Reeves* method and (2) *Polak–Ribiere* method (Press et al., 2007). Both these methods are closely related and are based on the following theorem:

Let $\mathbf{H}$ be a symmetric, positive-definite, $N \times N$ matrix, and $\mathbf{g}_0$ be an arbitrary N-dimensional vector and let $\mathbf{h}_0 = \mathbf{g}_0$. Now, for $i = 1, 2, \ldots, N$, we define the two sequences of vectors

$$\left.\begin{aligned} \mathbf{g}_{i+1} &= \mathbf{g}_i - \lambda_i \mathbf{H} \cdot \mathbf{h}_i \\ \mathbf{h}_{i+1} &= \mathbf{g}_{i+1} + \gamma_i \mathbf{h}_i \end{aligned}\right\}, \tag{6.22}$$

in which the scalars λ_i and γ_i are chosen such that $\mathbf{g}_{i+1} \cdot \mathbf{g}_i = 0$ and $\mathbf{h}_{i+1} \cdot \mathbf{H} \cdot \mathbf{h}_i = 0$. This, in turn, means that

$$\left.\begin{aligned} \lambda_i &= \frac{\mathbf{g}_i \cdot \mathbf{g}_i}{\mathbf{g}_i \cdot \mathbf{H} \cdot \mathbf{h}_i} \\ \gamma_i &= \frac{\mathbf{g}_{i+1} \cdot \mathbf{H} \cdot \mathbf{h}_i}{\mathbf{h}_i \cdot \mathbf{H} \cdot \mathbf{h}_i} \end{aligned}\right\}. \tag{6.23}$$

Unless the denominators in equation 6.23 are zero, then for all $i \neq j$

$$\left.\begin{aligned} \mathbf{g}_i \cdot \mathbf{g}_j &= 0 \\ \mathbf{h}_i \cdot \mathbf{H} \cdot \mathbf{h}_j &= 0 \end{aligned}\right\}. \tag{6.24}$$

The above means that the procedure defined by equations 6.22 and 6.23 makes each $\mathbf{g}_i$ orthogonal to its immediate predecessor and each $\mathbf{h}_i$ conjugate to its immediate predecessor. In fact, this procedure generates a sequence of $\mathbf{g}$ vectors that are mutually orthogonal and a sequence of $\mathbf{h}$ vectors that are mutually conjugate. We can prove this theorem straightforward induction.

Knowing that equation 6.24 holds, it is also possible to start from equation 6.22 and derive the following equation that is equivalent to equation 6.24

$$\left.\begin{aligned} \gamma_i &= \frac{\mathbf{g}_{i+1} \cdot \mathbf{g}_{i+1}}{\mathbf{g}_i \cdot \mathbf{g}_i} = \frac{(\mathbf{g}_{i+1} - \mathbf{g}_i) \cdot \mathbf{g}_{i+1}}{\mathbf{g}_i \cdot \mathbf{g}_i} \\ \lambda_i &= \frac{\mathbf{g}_i \cdot \mathbf{h}_i}{\mathbf{h}_i \cdot \mathbf{H} \cdot \mathbf{h}_i} \end{aligned}\right\}. \tag{6.25}$$

We are now set to develop a formalism to minimize (or maximize) a function, approximated by a quadratic form via Taylor series (equation 6.16). Let us first assume that we know the Hessian matrix $\mathbf{H}$. We can then use the construction outlined via equations 6.22–6.25 to find the successive conjugate directions $\mathbf{h}_i$ along which to line minimize. After N such line minimizations, we should efficiently arrive at the minimum of the quadratic form. But the problem is that we do not know the Hessian matrix $\mathbf{H}$ unless we explicitly compute it. Computing $\mathbf{H}$ is, however, a challenge. And, even if we assume that we have efficient means to compute it on large supercomputers, it is just not computing but the entire matrix must be stored in the computer memory, which could even be a bigger challenge that computing. For example, if we just restrict to a simple seismic inversion problem at a single location assuming isotropy, the model parameters are the P- and S-wave velocities (V_P, V_S) and density (ρ) for each depth point. If our deepest target is 5000 m and we want to discretize the model at 5 m depth intervals, i.e., a total of 1000 depth points, then $\mathbf{H}$ would be a matrix of size 3000×3000 elements. If we go from isotropy to anisotropy, the number of the model parameters and therefore the size of $\mathbf{H}$ will further increase and storing these large matrices on the computer memory is difficult.

To save the burden of computing and storing the Hessian matrix, below is another theorem along with the proof.

Let $\mathbf{g}_i$ and $\mathbf{h}_i$ be the same sequence of vectors as previously defined. Suppose that we happen to have $\mathbf{g}_i = -\nabla f(\mathbf{p}_i)$ for some point $\mathbf{p}_i$, where f is of the quadratic form given by equation 6.16. Suppose that we proceed from $\mathbf{p}_i$ along the direction $\mathbf{h}_i$ to a local minimum of f at some point $\mathbf{p}_{i+1}$ and then set $\mathbf{g}_{i+1} = -\nabla f(\mathbf{p}_{i+1})$, then $\mathbf{g}_{i+1}$ is the same vector as would have been constructed from equation 6.22, and we have constructed it from the gradient without any knowledge of the Hessian matrix $\mathbf{H}$.

Proof: Setting $\mathbf{x} = \mathbf{p}_i$ in equation 6.19, we obtain

$$\nabla f(\mathbf{p}_i) = \mathbf{H} \cdot \mathbf{p}_i \quad \mathbf{b}. \tag{6.26}$$

Therefore,

$$\nabla f(\mathbf{p}_i) = \mathbf{H} \cdot \mathbf{p}_i - \mathbf{b}. \tag{6.27}$$

Because $\mathbf{g}_i = -\nabla f(\mathbf{p}_i)$, we have $\mathbf{g}_{i+1} = -\nabla f(\mathbf{p}_{i+1})$, and per the theorem above, we write from equation 6.27

$$\mathbf{g}_i = -\nabla f(\mathbf{p}_i) = -\mathbf{H} \cdot \mathbf{p}_i + \mathbf{b} \tag{6.28}$$

and

$$\begin{aligned} \mathbf{g}_{i+1} &= -\mathbf{H} \cdot (\mathbf{p}_i + \lambda \mathbf{h}_i) + \mathbf{b} \\ &= -\mathbf{H} \cdot \mathbf{p}_i + \mathbf{b} - \lambda \mathbf{H} \cdot \mathbf{h}_i \\ &= \mathbf{g}_i - \lambda \mathbf{H} \cdot \mathbf{h}_i, \end{aligned} \tag{6.29}$$

with λ suitably chosen to take us to the line minimum. But at the line minimum we have

$$\mathbf{h}_i \cdot \nabla f = -\mathbf{h}_i \cdot \mathbf{g}_{i+1} = 0. \tag{6.30}$$

Combining equations 6.29 and 6.30, we obtain

$$\begin{aligned} 0 &= \mathbf{h}_i \cdot \mathbf{g}_{i+1} \\ &= \mathbf{h}_i \cdot (\mathbf{g}_i - \lambda \mathbf{H} \cdot \mathbf{h}_i) \\ &= \mathbf{h}_i \cdot \mathbf{g}_i - \lambda \mathbf{h}_i \cdot \mathbf{H} \cdot \mathbf{h}_i. \end{aligned} \tag{6.31}$$

Or

$$\lambda = \frac{\mathbf{g}_i \cdot \mathbf{h}_i}{\mathbf{h}_i \cdot \mathbf{H} \cdot \mathbf{h}_i}. \tag{6.32}$$

Thus, λ in equation 6.32 is equivalent to λ_i in equation 6.23. Thus, the definition of $\mathbf{g}_{i+1}$ in equation 6.31 is equivalent to that in equation 6.22, and we proved the theorem.

Based on the above, we now have the fundamental basis for the conjugate gradient algorithm that neither requires the evaluation of the Hessian matrix nor the memory necessary to store this matrix. In our algorithm, a sequence of directions $\mathbf{h}_i$ is constructed using only line minimizations, evaluations of the gradient vector, and an auxiliary vector to store the latest sequence of $\mathbf{g}$ vectors. So far, everything above is applicable to both Fletcher–Reeves and Polak–Ribiere methods. But there is one difference between the two methods. Fletcher–Reeves uses the first expression of γ_i in equation 6.25, while Polak–Ribiere uses the second expression of γ_i in equation 6.25. But are not they identical? Yes, they are for exact quadratic forms (equation 6.16). In real world, however, the function being optimized is not exactly quadratic. Thus, after arriving at the supposed minimum of the quadratic

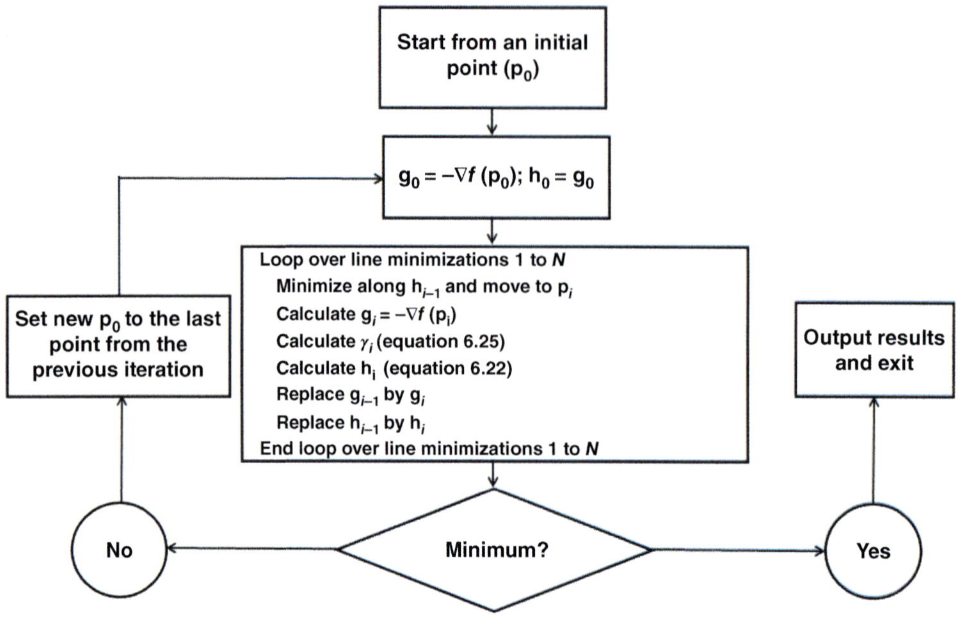

Figure 6.12 Basic flow diagram for implementing a conjugate gradient optimization. Note that for calculating λ_i, if the first expression of equation 6.25 is used, it is the Fletcher–Reeves method, and if the second expression is used, it is then the Polak–Ribiere method. In addition, also note that starting from an initial point $\mathbf{p}_0$, the line minimization moves to point $\mathbf{p}_i$ from point $\mathbf{p}_{i-1}$ in each step.

form, it is necessary to proceed for another set of iterations, and there is evidence that the Polak–Ribiere method accomplishes transition to further iterations more gracefully than the Fletcher–Reeves method (Press et al., 2007).

Irrespective of whether the Polak–Ribiere or Fletcher–Reeves method is used, Figure 6.12 shows the basic flowchart for the conjugate gradient method of optimization. Starting from an initial point $\mathbf{p}_0$ (the initial model in the model or decision space, which we previously discussed in Chapter 1), it first finds $\mathbf{g}_0 = -\nabla f(\mathbf{p}_0)$ and sets $\mathbf{h}_0 = \mathbf{g}_0$ as we described above. This is followed by successive line minimizations along $\mathbf{h}_{i-1}$ to move to the new point (model) $\mathbf{p}_i$ to find $\mathbf{g}_i$ and $\mathbf{h}_i$ for the next iteration (see Figure 6.12). The method for line minimization is directed by the actual problem in hand, which we discuss in Chapter 7.

6.3.5 Variable Metric Method

Variable metric method, which is also known as the *quasi-Newton* method, is not much different from the conjugate gradient methods of accumulating information from successive line minimizations so that N such minimizations lead to the minimum of a quadratic form in N dimensions. Both variable metric and conjugate gradient methods require the ability to compute the gradient or the first partial derivative of the function to be optimized at arbitrary points. The variable metric approach differs from the conjugate gradient in the way it stores and updates the information that is accumulated in successive minimizations. So, instead of requiring an intermediate storage of the order N, the variable metric method requires storing a matrix of size $N \times N$, and consequently, it could be a disadvantage for moderate to large-sized optimization problems.

In general, there is not any overwhelming advantage that the variable metric methods hold over the conjugate gradient methods. Developed somewhat earlier, and more widely propagated, the variable metric methods, however, are more widely used than the conjugate gradient methods in most of the science and engineering applications. Consequently, many implementations of the variable metric methods to efficiently handle practical issues such as the minimization of the roundoff errors, special conditions, etc., are available within the variable metric method.

Two most frequently implemented variable metric methods are the Davidson–Fletcher–Powell or DFP algorithm and the Boyden–Fletcher–Goldfarb–Shanno or BFGS algorithm. It has been demonstrated that these two schemes differ only in details of their roundoff error, convergence tolerance, etc., which are not discussed here. It is however empirically demonstrated that the BFGS scheme is somewhat superior to the DFP scheme (Press et al., 2007).

As discussed before, we assume that our function can be locally approximated by a quadratic form (equations 6.16–6.18). The basic idea behind the variable metric methods is to iteratively build a good approximation of the inverse of the Hessian matrix ($\mathbf{H}^{-1}$) or, in other words, construct a sequence of matrices $\mathbf{A}_i$ with the property

$$\lim_{i\to\infty} \mathbf{A}_i = \mathbf{H}^{-1}. \tag{6.33}$$

In practice of course, we would like to achieve this limit in N iterations instead of ∞. The advantage of knowing $\mathbf{H}$ or $\mathbf{H}^{-1}$ can be readily seen from equation 6.19—$\nabla f(\mathbf{x}) = \mathbf{H}\cdot\mathbf{x} - \mathbf{b}$. Thus, if the minimum is achieved at $\mathbf{x}=\mathbf{x}_m$, the gradient must vanish, i.e., $\nabla f(\mathbf{x}_m) = \mathbf{0}$, which leads equation 6.19 to

$$\mathbf{H}\cdot\mathbf{x}_m = \mathbf{b}. \tag{6.34}$$

Therefore, if we construct $\mathbf{A}=\mathbf{H}^{-1}$, we can then find our desired answer—the minimum point $\mathbf{x}_m$. The variable metric methods attempt to iteratively construct $\mathbf{A}=\mathbf{H}^{-1}$. At any stage of iteration, if we are at a point $\mathbf{x}_i$, which is not minimum, we then obtain from equation 6.19

$$\mathbf{H}\cdot\mathbf{x}_i = \nabla f(\mathbf{x}_i) + \mathbf{b}. \tag{6.35}$$

Subtracting equation 6.35 from equation 6.34 and multiplying both sides of the result by $\mathbf{H}^{-1}$ gives

$$\mathbf{x}_m - \mathbf{x}_i = \mathbf{H}^{-1}[-\nabla f(\mathbf{x}_i)]. \tag{6.36}$$

The left-hand side of equation 6.36 is the finite step we need to obtain to the minimum since the right-hand side of this equation is known once we have estimated an accurate $\mathbf{A}\approx\mathbf{H}^{-1}$. We provide below an outline of how to construct $\mathbf{H}^{-1}$ or, specifically, the matrix $\mathbf{A}_{i+1}$ for step $i+1$ from $\mathbf{A}_i$ estimated at the preceding step i is such a way that this matrix converges to $\mathbf{H}^{-1}$ with the progress of the iteration. We provide these steps without rigorous derivations and refer to Davidson (1959), Fletcher (1987), and Nocedal and Wright (1999) for further details.

From equation 6.35, at iteration $i+1$ we must have

$$\mathbf{H}\cdot\mathbf{x}_{i+1} = \nabla f(\mathbf{x}_{i+1}) + \mathbf{b}. \tag{6.37}$$

Subtracting equation 6.36 from equation 6.37 and multiplying both sides by $\mathbf{H}^{-1}$, we write

$$\mathbf{x}_{i+1} - \mathbf{x}_i = \mathbf{H}^{-1}(\nabla f_{i+1} - \nabla f_i), \tag{6.38}$$

in which $\nabla f_j \equiv \nabla f(\mathbf{x}_j)$. Having made the step from $\mathbf{x}_i$ to $\mathbf{x}_{i+1}$, we would like our new approximation $\mathbf{A}_{i+1}$ to satisfy equation 6.38 as if it were true $\mathbf{H}^{-1}$. Or, in other words, we would like to have

$$\mathbf{x}_{i+1} - \mathbf{x}_i = \mathbf{A}_{i+1}(\nabla f_{i+1} - \nabla f_i). \tag{6.39}$$

In addition, it is also desirable to have our iteration formula from $\mathbf{A}_i$ to $\mathbf{A}_{i+1}$ to be of the form $\mathbf{A}_{i+1} = \mathbf{A}_i +$ correction. At any stage of iteration, we have two vectors, $\mathbf{x}_{i+1} - \mathbf{x}_i$, $\nabla f_{i+1} - \nabla f_i$ (see equation 6.39). Additionally, we also have the estimate of the matrix $\mathbf{A}_i$ from previous iteration. Therefore, this correction term must be a matrix made from $\mathbf{x}_{i+1} - \mathbf{x}_i$, $\nabla f_{i+1} - \nabla f_i$, and $\mathbf{A}_i$ in such a way that equation 6.39 holds. The DFP updating formula is one such way to compute $\mathbf{A}_{i+1}$ and is given as (Press et al., 2007)

$$\mathbf{A}_{i+1} = \mathbf{A}_i + \frac{(\mathbf{x}_{i+1} - \mathbf{x}_i)\otimes(\mathbf{x}_{i+1} - \mathbf{x}_i)}{(\mathbf{x}_{i+1} - \mathbf{x}_i)\cdot(\nabla f_{i+1} - \nabla f_i)} - \frac{[\mathbf{A}_i\cdot(\nabla f_{i+1} - \nabla f_i)]\otimes[\mathbf{A}_i\cdot(\nabla f_{i+1} - \nabla f_i)]}{(\nabla f_{i+1} - \nabla f_i)\cdot\mathbf{A}_i\cdot(\nabla f_{i+1} - \nabla f_i)}. \tag{6.40}$$

The symbol $\otimes$ in equation 6.40 is a matrix, given as the *direct* or *outer* product of two vectors, and is given as

$$\mathbf{u} \otimes \mathbf{v} = \begin{pmatrix} u_1v_1 & u_1v_2 & \cdots & u_1v_N \\ u_2v_1 & u_2v_2 & \cdots & u_2v_N \\ \cdot & \cdot & \cdots & \cdot \\ \cdot & \cdot & \cdots & \cdot \\ \cdot & \cdot & \cdots & \cdot \\ u_Nv_1 & u_Nv_2 & \cdots & u_Nv_N \end{pmatrix}. \tag{6.41}$$

It can be verified by direct substitution that $\mathbf{A}_{i+1}$ given by equation 6.40 satisfies equation 6.39.

The BFGS updating formula is the same as the DFP formula (equation 6.40) with one additional term

$$\cdots + [(\nabla f_{i+1} - \nabla f_i) \cdot \mathbf{A}_i \cdot (\nabla f_{i+1} - \nabla f)]\mathbf{u} \otimes \mathbf{u}, \tag{6.42}$$

where

$$\mathbf{u} = \frac{(\mathbf{x}_{i+1} - \mathbf{x}_i)}{(\mathbf{x}_{i+1} - \mathbf{x}_i) \cdot (\nabla f_{i+1} - \nabla f_i)} - \frac{\mathbf{A}_i \cdot (\nabla f_{i+1} - \nabla f_i)}{(\nabla f_{i+1} - \nabla f_i) \cdot \mathbf{A}_i \cdot (\nabla f_{i+1} - \nabla f_i)}. \tag{6.43}$$

Like DFP, it can be verified by direct substitution that $\mathbf{A}_{i+1}$ given by the BFGS algorithm (equations 6.40–6.42) satisfies equation 6.39. Furthermore, it can also be shown that the DFP updating formula (equation 6.40) and the BFGS updating formula with an additional term (equations 6.42 and 6.43) converge to the inverse of the Hessian matrix $\mathbf{H}^{-1}$ in N steps if the function f is of the quadratic form. However, if f is not quadratic, like the conjugate gradient method, it may require additional steps to converge to $\mathbf{H}^{-1}$.

So far, we provided two algorithms to construct $\mathbf{H}^{-1}$, but we have not yet discussed what is the most optimal way to go from $\mathbf{x}_i$ to $\mathbf{x}_{i+1}$. In addition, we also did not discuss how to choose the matrix $\mathbf{A}$ at the beginning of the iteration. The variable metric methods attempt to minimize (or maximize) the gradient. Thus, at each step i, if the problem is cast as minimization, it is then fair to go from point $\mathbf{x}_i$ along $-\nabla f(\mathbf{x}_i)$ and find the point $\mathbf{x}_{new}$ along the path where $f(\mathbf{x}_{new})$ is minimum and set the new updated point for next iteration as $\mathbf{x}_{i+1} = \mathbf{x}_{new}$. Also, for computing $\mathbf{A} \approx \mathbf{H}^{-1}$, it is fair to set $\mathbf{A}$ to be an identity matrix at the beginning and gradually build it with iteration.

Assuming a minimization problem, the complete plan for implementing the variable metric or quasi-Newton optimization method is provided in the flow diagram shown in Figure 6.13.

Advanced Implementation of the Variable Metric Method

Although rare, but it can conceivably happen that the roundoff errors cause the matrix $\mathbf{A}_i$ to become nearly singular or non-positive-definite. This can be serious because the supposed search directions might then not lead downhill and because nearly singular $\mathbf{A}_i$ matrices tend to give the subsequent $\mathbf{A}_i$ matrices that are also singular. There is however a simple fix to this problem. In case of any doubt, the algorithm should be restarted at the claimed minimum point and see if it goes anywhere. This is a simple, but not an elegant fix. Modern implementations of variable metric methods deal with this problem in a more sophisticated way, which we discuss further in Chapter 7.

Instead of building an approximation to $\mathbf{H}^{-1}$, it is possible to build up an approximation to $\mathbf{H}$ itself. Then, instead of calculating the left-hand side of equation 6.39, we can solve the following set of linear equations:

$$\mathbf{H} \cdot (\mathbf{x}_m - \mathbf{x}_i) = -\nabla f(\mathbf{x}_i). \tag{6.44}$$

Solving equation 6.44 appears to be a process of the order of N^3. Additionally, it does not apparently seem to solve the roundoff error. The primary trick here is not to store $\mathbf{H}$ but a triangular decomposition of $\mathbf{H}$, somewhat like the LU decomposition, specialized for the symmetric, positive-definite matrices, called *Cholesky decomposition*. The updating formula Cholesky decomposition of $\mathbf{H}$ is such as to guarantee that the matrix remains positive-definite and non-singular even in the presence of finite roundoff.

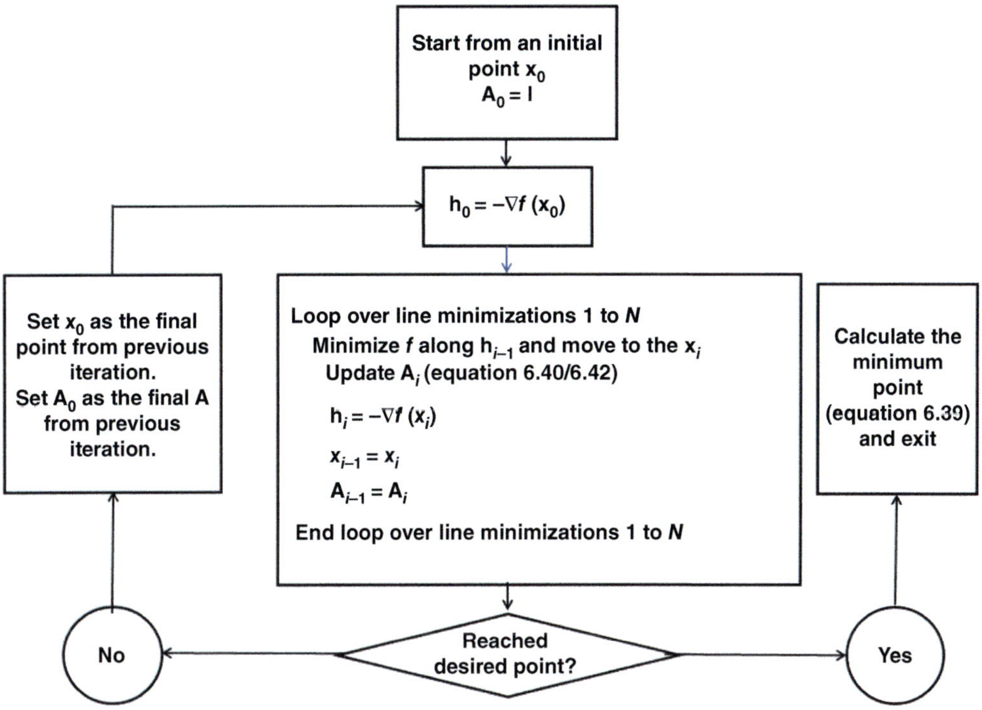

Figure 6.13 Flow diagram for implementing variable metric optimization in multiple dimensions.

Another issue is the question of how exact the line searches need to be so that an overall good convergence of the method is maintained. It may be a complete wastage of computer resources to ensure accurate convergence for each line minimization that may not have any significance to the optimizing an N-dimensional function. One useful fact to know here is that the BFGS updating is more tolerant than the DFP updating. Finally, the implementation of an optimization method is problem dependent. In Chapter 7, we discuss geophysical optimization problems where we describe different flavors of the method in further detail.

6.3.6 Other Popular Methods

With recent advances in machine learning, especially for deep-learning applications, many other optimization methods have become popular. Out of them, the ones worth mentioning are the following:

1) *Gauss–Newton* method: This is a variant of the variable metric or quasi-Newton method, which has also been successfully implemented to solve geophysical inverse problems (for example, see Padhi et al., 2015).
2) *Adam Optimizer*, or simply *Adam*: This is a variant of the gradient descent (steepest descent) method. We briefly mentioned Adam in Section 6.3.3. This method was first introduced by Kingma and Ba (2015), and this method is an extended version of the stochastic gradient descent method, especially designed to efficiently solve large-scale optimization problems encountered in deep-learning applications.
3) *Levenberg-Marquardt (L-M)* algorithm: This method was originally introduced by Levenberg (1944) and later by Marquardt (1963). This method is a generalization of the Gauss–Newton and steepest descent methods.

We discuss the Gauss–Newton method in Chapter 7, and Adam and L-M methods in Chapter 9.

6.3.7 A Final Note to Multidimensional Optimization Problems

The multidimensional gradient-based methods we discussed in this chapter (steepest descent, conjugate gradient, variable metric, etc., and the other methods such as Adam and L-M) are based on the principle of least squares. In all these methods, we first approximate $f(\mathbf{x})$ in quadratic form (equation 6.16). Then, by defining the gradient $\nabla f(\mathbf{x})$ in terms of the Hessian matrix $\mathbf{H}$ (equation 6.19), we cast the problem as a pure least-square minimization by setting $\nabla f(\mathbf{x})=\mathbf{0}$ during each iteration. As we previously discussed in Chapter 1, for practical problems, a pure least-square minimization requires matrix inversion (equation 1.23), which can be numerically unstable. To stabilize matrix inversion, we also mentioned the method of damped least squares (equation 1.24) where a small term is added to the diagonal of the matrix before inversion. In this chapter, we were primarily interested in discussing the principles of different gradient-based multidimensional optimization problems and therefore deliberately omitted this damping term in our derivations and discussions. In practice, however, this damping term is needed, which we discuss in Chapter 7, where we deal with local optimization/inverse problems in seismology.

6.4 Chapter Summary

In this chapter, we discussed different methods for optimizing functions. Starting from function optimization in single dimension, we thoroughly discussed two practical 1D optimizations: (1) group/phase velocity and angle problem encountered in seismic wave propagation in an anisotropic medium and (2) ray tracing problem for estimating the takeoff angle for rays for fixed source-to-receiver offsets. We then moved onto function optimization in multiple dimensions, discussed the concept of the conjugate directions, and introduced the conjugate gradient and variable metric methods for optimizing multidimensional functions. We also introduced the L-M optimization and Adam optimizer, which are popular in machine-learning-related applications. The concept developed in this chapter is the fundamental basis for solving practical optimization problems in seismology, which is discussed in the subsequent chapters.

6.5 Exercises

Problem 6.1 Group-to-phase mapping. In Problem 3.12, we provided the elastic parameters in Thomsen (1986) with notation and density for both Taylor Sandstone and Green River Shale, which are again provided below:

Rock Formation	V_{PO} (m/s)	V_{SO} (m/s)	ε	δ	γ	ρ (g/cm^3)
Taylor Sandston	3368	1829	0.11	−0.035	0.255	2.5
Green River Shale	3392	1768	0.195	−0.22	0.18	2.075

1) Using equations provided in Chapter 3, write a code to compute the exact P-wave phase velocity (V_{ph}), group velocity (V_g), and group angle (ϕ) as functions of the phase angle (θ) and compute them for both Taylor Sandstone and Green River Shale for $\theta = 0°$ to $65°$ in steps of $1°$.
2) Write a code to compute phase angle and phase velocity from group angle and group velocity based on the flowchart given in Figure 6.8.
3) Apply your code from step 2 to compute phase angle and phase velocity from the group angle and group velocity for both Taylor Sandstone and Green River Shale, obtained from step 1 above, and demonstrate that your code estimates correct values.

Problem 6.2 Repeat Problem 6.1 for SV waves.

Problem 6.3 Repeat Problem 6.1 for SH waves.

Problem 6.4 Ray tracing problem. Refer to the horizontally stratified earth model as given below.

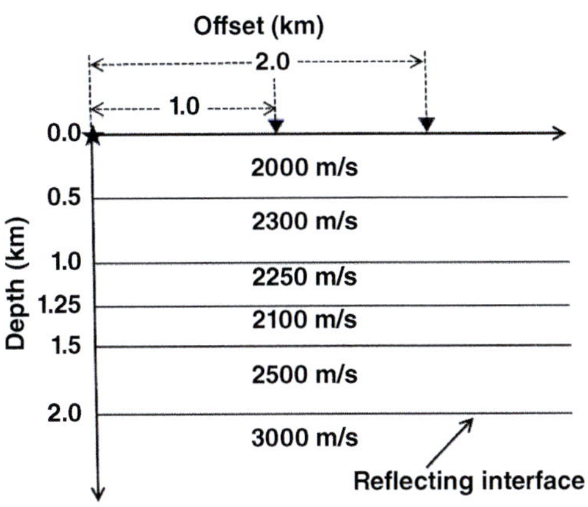

Horizontal layers with different P-wave velocities and layer thicknesses are given in this figure. The source location is shown by an asterisk, and two receiver locations, denoted as inverted triangles, are shown to be located at 1 and 2 km away from the source.

Based on the flowchart provided in Figure 6.9, write a code to trace rays from the source-to-receiver locations, both located at the surface. Use your code and trace rays from the source to the two receiver locations shown in the figure for P–P reflection from the interface at the depth of 2 km from the surface.

Problem 6.5 From equations 6.22, 6.23, and 6.24 show that equations 6.23 and 6.25 are equivalent.

References

Brent, R.P., 1973, *Algorithms for Minimization Without Derivatives*, Dover Publications, ISBN- 978-0-486-41998-5.

Davidson, W.C., 1959, Variable Metric Method for Minimization, AEC Research and Development Report ANL-5990, https://doi.org/10.2172/4252678.

Fletcher, R., 1987, *Practical Methods of Optimization*, Second Edn, Willie, ISBN 978-0-471-91547-8, https://doi.org/10.1002/9781118723203.

Kingma, D.P., and Ba, J.L., 2015, Adam: a method for stochastic optimization, arXiv:1412.6980v9, https://doi.org/10.48550/arXiv.1412.6980.

Levenberg, K., 1944, A method for the solution of certain non-linear problems in least squares, *Quarterly of Applied Mathematics*, *2*, 164–168, https://doi.org/10.1090/qam/10666.

Marquardt, D., 1963, An algorithm for least-squares estimation of nonlinear parameters, *SIAM Journal on Applied Mathematics*, *11*, 431–441, https://doi.org/10.1137/0111030.

Mukhopadhyay, P.K., and Mallick, S., 2011, An accurate ray-based offset-to-angle transform from normal moveout uncorrected multicomponent data in a transversely isotropic medium with vertical symmetry axis, *Geophysics*, *76*, C41–C51, https://doi.org/10.1190/1.3565182.

Nocedal, J., and Wright, S.J., 1999, *Numerical Optimization*, Springer, ISBN 0-387-98793-2.

Padhi, A., Mallick, S., Fortin, W., and Holbrook, W.S., 2015, 2-D ocean temperature and salinity images from pre-stack seismic waveform inversion methods: an example from the South China Sea, *Geophysical*

Journal International, *202*, 800–810, https://doi.org/10.1093/gji/ggv188.

Press, W.H., Teulolsky, S.A., Vetterling, W., and Flannery, B.P., 2007, *Numerical Recipes The Art of Scientific Computing*, Third Edn, Cambridge University Press, ISBN 978-0-521-88068-8.

Sava, P., and Hill, S.J., 2009, Overview and classification of wavefield seismic imaging methods, *The Leading Edge*, *28*(2), 170–183, https://doi.org/10.1190/1.3086052.

Thomsen, L., 1986, Weak elastic anisotropy, *Geophysics*, *51*, 1954–1966, https://doi.org/10.1190/1.1442051.

7

Local Optimization Methods in Seismology

7.1 Introduction

In Chapter 6, we thoroughly discussed optimization of functions. We first discussed optimizing functions in single dimension and considered some problems of practical interest. Following this, we then moved onto multidimensional optimization and discussed the gradient descent (steepest descent), conjugate gradient, and quasi-Newton (also known as the variable metric or Gauss–Newton method). We also briefly mentioned some recent methods such as the Adam optimizer and Levenberg-Marquardt (L-M) methods, which became popular in solving the problems related to machine-learning applications. In this chapter, we use these concepts and develop the fundamentals of local or gradient-based optimization methods in seismology.

Assume that we have a set of observed data in the data space defined as the vector **d**:

$$\mathbf{d} = [d_1, d_2, \dots, d_M]^T, \tag{7.1}$$

where the subscripts $1, 2, \dots, M$ denote different data points, and the superscript denotes a transpose. Also assume that we have the model vector **m** in the model/decision space defined as

$$\mathbf{m} = [m_1, m_2, \dots, m_N]^T. \tag{7.2}$$

We also assume that there is a function $g(\mathbf{m})$ that maps the model vector **m** into the synthetic data vector **s** at the same points as the observed data vector **d**, i.e.,

$$\mathbf{s} = g(\mathbf{m}) = [s_1, s_2, \dots, s_M]^T. \tag{7.3}$$

Note that in Chapter 1, while dealing with the amplitude-variation-with-angle (AVA) inversion, we found the mapping function (operator) $g(\mathbf{m})$ to be a matrix. This is, however, not in general true. For example, for seismic waveform inversion, $g(\mathbf{m})$ is the wave equation solver for a model comprising the subsurface elastic properties. For electromagnetic (EM) problem, on the other hand, $g(\mathbf{m})$ is the EM wave equation solver and so on. In Chapter 4 and 5, we discussed different methods to compute $g(\mathbf{m})$ for both homogeneous and inhomogeneous earth models. At the beginning of this chapter, we treat $g(\mathbf{m})$ to be our *physics of the forward problem*, mapping the model vector onto the synthetic (predicted) data vector (equation 7.3), which can be any of the methods discussed in Chapters 4 and 5. We thus define our objective E as a function of the real and synthetic data vectors **d** and **s** (and other regularization parameters) and develop the theoretical foundations of different local optimization methods. As we discussed in Chapter 6, these local methods not only require computing E but also its gradient ∇E. In developing the theory, we first assume that we know how to compute ∇E. We then revisit the computation of $g(\mathbf{m})$ and discuss how ∇E can be efficiently computed.

Besides the local methods, which we focus on in this chapter, there is also another class

Computation, Optimization, and Machine Learning in Seismology, Advanced Textbook 8, First Edition. Subhashis Mallick.

of optimization methods known as the global optimization methods, which we also introduced in Chapter 1. In contrast to the local methods, these global methods do not require computing ∇E, and we postpone the discussion of these gradient free methods to Chapter 8.

Having defined the data vector **d** (equation 7.1), the model vector **m** (equation 7.2), and the synthetic data vector **s** via the mapping function $g(\mathbf{m})$ (equation 7.3), any optimization method, local or global, requires defining the objective that we want to optimize. Casting our optimization to be a minimization problem, we first define the *error* or *misfit* between the real and predicted (synthetic) data vectors as

$$\mathbf{e} = |\mathbf{d} - \mathbf{s}| \tag{7.4}$$

and find a way to minimize **e**. Local methods approach this minimization in a least-squares sense. So, we define the objective as the sum of the squares of the error as

$$E(\mathbf{m}) = \mathbf{e}^T\mathbf{e}^* = |\mathbf{s} - \mathbf{d}|^T|\mathbf{s} - \mathbf{d}|^* \tag{7.5}$$

and find a way to minimize $E(\mathbf{m})$. Before proceeding further, note that the superscript T on the right-hand side of equation 7.5 denotes a transpose and the superscript * (asterisk) denotes a complex conjugate. For geophysical optimization problems, including seismic problems, the observed and synthetic data are sometimes real and sometimes they both are complex-valued. So, defining the objective by equation 7.5 is valid for both real- and complex-valued quantities.

7.2 Fréchet Derivative (Jacobi) Matrix and the Computation of the Gradient of the Objective Function

Having defined the error or misfit (equation 7.4) and the objective (equation 7.5), we now define the gradient of $E(\mathbf{m})$

$$\nabla E(\mathbf{m}) = \frac{\partial}{\partial \mathbf{m}}E(\mathbf{m}) = \frac{\partial}{\partial \mathbf{m}}|\mathbf{s} - \mathbf{d}|^T|\mathbf{s} - \mathbf{d}|^*. \tag{7.6}$$

But

$$\begin{aligned}|\mathbf{s} - \mathbf{d}| = [&|s_{1R} - d_{1R}| + i|s_{1I} - d_{1I}|,\\ &\times |s_{2R} - d_{2R}| + i|s_{2I} - d_{2I}|, \dots,\\ &\times |s_{MR} - d_{MR}| + i|s_{MI} - d_{MI}|]^T.\end{aligned} \tag{7.7}$$

In equation 7.7, $i = \sqrt{-1}$ and for any $j = 1, 2, \dots, M$, s_{jR} and d_{jR} are the real and s_{jI} and d_{jI} are the imaginary parts of the jth component of **s** and **d**.

We can similarly write

$$\begin{aligned}|\mathbf{s} - \mathbf{d}|^* = [&|s_{1R} - d_{1R}|1i|s_{1I} - d_{1I}|,\\ &\times |s_{2R} - d_{2R}| - i|s_{2I} - d_{2I}|, \dots,\\ &\times |s_{MR} - d_{MR}| - i|s_{MI} - d_{MI}|].\end{aligned} \tag{7.8}$$

Therefore,

$$\begin{aligned}|\mathbf{s} - \mathbf{d}|^T|\mathbf{s} - \mathbf{d}|^* &= (s_{1R} - d_{1R})^2 + (s_{2R} - d_{2R})^2\\ &+ \cdots + (s_{MR} - d_{MR})^2 + (s_{1I} - d_{1I})^2\\ &+ (s_{2I} - d_{2I})^2 + \cdots + (s_{MI} - d_{MI})^2.\end{aligned} \tag{7.9}$$

From equations 7.8 and 7.9, equation 7.6 becomes

$$\begin{aligned}\nabla E(\mathbf{m}) &= \frac{\partial}{\partial \mathbf{m}}E(\mathbf{m}) = \frac{\partial}{\partial \mathbf{m}}|\mathbf{s} - \mathbf{d}|^T|\mathbf{s} - \mathbf{d}|^*\\ &= 2\begin{pmatrix} |s_{1R} - d_{1R}|\frac{\partial s_{1R}}{\partial m_1} + |s_{1I} - d_{1I}|\frac{\partial s_{1I}}{\partial m_1} + \cdots + |s_{MR} - d_{MR}|\frac{\partial s_{MR}}{\partial m_1} + |s_{MI} - d_{MI}|\frac{\partial s_{MI}}{\partial m_1}\\ |s_{1R} - d_{1R}|\frac{\partial s_{1R}}{\partial m_2} + |s_{1I} - d_{1I}|\frac{\partial s_{1I}}{\partial m_2} + \cdots + |s_{MR} - d_{MR}|\frac{\partial s_{MR}}{\partial m_2} + |s_{MI} - d_{MI}|\frac{\partial s_{MI}}{\partial m_2}\\ \cdot\\ \cdot\\ \cdot\\ |s_{1R} - d_{1R}|\frac{\partial s_{1R}}{\partial m_N} + |s_{1I} - d_{1I}|\frac{\partial s_{1I}}{\partial m_N} + \cdots + |s_{MR} - d_{MR}|\frac{\partial s_{MR}}{\partial m_N} + |s_{MI} - d_{MI}|\frac{\partial s_{MI}}{\partial m_M} \end{pmatrix}.\end{aligned} \tag{7.10}$$

We now define the $M \times N$ matrix $\mathbf{J}$ with elements $J_{ij} = \partial s_i/\partial m_j$, i.e., the partial derivative of the ith component of the synthetic data vector with respect to the jth component of the model vector. This matrix $\mathbf{J}$ is known as the *Jacobi* or the *Fréchet derivative* matrix. An explicit and complex-valued form of $\mathbf{J}$ can be written as

$$\mathbf{J} = \begin{pmatrix} \frac{\partial s_{1R}}{\partial m_1} + i\frac{\partial s_{1R}}{\partial m_1} & \frac{\partial s_{1R}}{\partial m_2} + i\frac{\partial s_{1R}}{\partial m_2} & \cdots & \frac{\partial s_{1R}}{\partial m_N} + i\frac{\partial s_{1R}}{\partial m_N} \\ \frac{\partial s_{2R}}{\partial m_1} + i\frac{\partial s_{2R}}{\partial m_1} & \frac{\partial s_{2R}}{\partial m_2} + i\frac{\partial s_{2R}}{\partial m_2} & \cdots & \frac{\partial s_{2R}}{\partial m_N} + i\frac{\partial s_{2R}}{\partial m_N} \\ \cdot & \cdot & \cdots & \cdot \\ \cdot & \cdot & \cdots & \cdot \\ \cdot & \cdot & \cdots & \cdot \\ \frac{\partial s_{MR}}{\partial m_1} + i\frac{\partial s_{MR}}{\partial m_1} & \frac{\partial s_{MR}}{\partial m_2} + i\frac{\partial s_{MR}}{\partial m_2} & \cdots & \frac{\partial s_{MR}}{\partial m_N} + i\frac{\partial s_{MR}}{\partial m_N} \end{pmatrix}. \quad (7.11)$$

Finally, combining equations 7.10 and 7.11, we write the gradient of the objective as

$$\nabla E(\mathbf{m}) = \frac{\partial}{\partial \mathbf{m}} E(\mathbf{m}) = \frac{\partial}{\partial \mathbf{m}} |\mathbf{s} - \mathbf{d}|^T |\mathbf{s} - \mathbf{d}|^* = \mathrm{Re}(\mathbf{J}^T |\mathbf{s} - \mathbf{d}|^*). \quad (7.12)$$

In equation 7.12, "Re" stands for the real part of a complex-valued quantity. Note that equation 7.12 is the most general representation of the gradient of the objective function when both observed and synthetic data are complex-valued. An equivalent formulation for real-valued observed and synthetic data is straightforward—just remove the complex conjugate and the extraction of the real part of a complex quantity from equation 7.12 and thus the solution is obtained. The derivation of equation 7.12 is straightforward and is left as an exercise.

Now that we have defined the objective $E(\mathbf{m})$ and its gradient $\nabla E(\mathbf{m})$ via equations 7.5, 7.11, and 7.12, we can use any of the multidimensional optimization method outlined in Chapter 6, such as the gradient descent method, conjugate-gradient method, quasi-Newton method, etc., for optimization. Because these optimizations are already discussed in Chapter 6, we not repeat them here. Instead, we focus on the practical implementation issues of local optimization problems. They include the following:

1) Regularization of the objective to ensure stability.
2) Overview of the implementation aspects of different optimization methods.
3) Efficient ways to compute the gradient $\nabla E(\mathbf{m})$.
4) Overview of different optimization problems of geophysical interest.

7.3 Regularization of the Objective

Note that in equation 7.5 we have defined the objective to be the sum of the squares of errors (misfit), which we attempt to minimize. However, in most optimization problems, the objective defined by equation 7.5 is rarely used. The primary reason for this is because irrespective of the optimization method being used, a stable solution demands the stability of the Hessian matrix, or to be specific, the inverse of the Hessian matrix, even if the method does not compute it directly. The problem is like the least-squares AVA inversion problem introduced in Chapter 1, where we introduced the concepts of a "pure" and a "damped" least-squares solution (equations 1.23 and 1.24). Note that the term $\varepsilon^2 \mathbf{I}$ on the right-hand side of equation 1.24 is added to ensure the stability of the matrix inversion. Here, we generalize this concept to formulate a numerically stable optimization strategy. However, before proceeding further, it is first necessary to introduce a few fundamental concepts. So, in the following, we first introduce these concepts and then formulate the general strategies for the objective regularization.

7.3.1 Variance of the Model Parameter Estimates

In most practical optimization methods, we deal with noisy data that cause errors in the estimation of the model parameters. Note that all local optimizations, such as the conjugate gradient, quasi-Newton, etc., and estimates of the model parameters in each iteration step are linear functions of data of the form

$$\mathbf{m}^{est} = \mathbf{Md} + \mathbf{v}, \tag{7.13}$$

where $\mathbf{M}$ is some matrix and $\mathbf{v}$ is some vector. In successive line minimizations in any of the methods, such as gradient descent, conjugate gradient, quasi-Newton, etc., discussed in Chapter 6, the general formula for estimated model at each iteration can be represented by equation 7.13, although the nature of the matrix $\mathbf{M}$ and the vector $\mathbf{v}$ varies depending upon the method being used. Therefore, if we assume that the data vector $\mathbf{d} = [d_1, d_2, \ldots, d_M]^T$ has a distribution characterized by the data covariance matrix $[\text{cov}\,\mathbf{d}] = \mathbf{C}_d$, where

$$[\mathbf{C}_d]_{ij} = \frac{1}{M-1}(d_i - \bar{d})(d_j - \bar{d}), \tag{7.14}$$

with $\bar{d}$ being the mean value of the data vector, then the estimates of the model parameters will have a distribution characterized by

$$\mathbf{C}_m = [\text{cov}\,\mathbf{m}] = \mathbf{M}[\text{cov}\,\mathbf{d}]\mathbf{M}^T = \mathbf{MC}_d\mathbf{M}^T. \tag{7.15}$$

The matrix $\mathbf{C}_m$ in equation 7.15 is the model covariance matrix or the covariance of the solution, which can be calculated from the data covariance matrix and the matrix $\mathbf{M}$ in a straightforward fashion. If the data are uncorrelated and are of equal variance, say σ_d^2, i.e., $\mathbf{C}_d = \sigma_d^2\mathbf{I}$, then simple formulas exist for many optimization problems. One such is provided as an exercise (Exercise 7.2) at the end of this chapter.

7.3.2 Variance and Prediction Error of the Least-Squares Solutions

General least-squares solutions involve minimization of the objective given as $E(\mathbf{m}) = \mathbf{e}^T\mathbf{e}^*$ (see equation 7.5). As demonstrated in Figure 7.1, if $E(\mathbf{m})$ has very sharp minimum near the estimated solution $\mathbf{m}^{est}$ (Figure 7.1a), we would then expect that the solution is well determined because small errors (ΔE) in determining the shape of $E(\mathbf{m})$ caused by random fluctuations in the data will lead to only small errors ($\Delta\mathbf{m}$) in the estimated model $\mathbf{m}^{est}$. On the other hand, if $E(\mathbf{m})$ has a broad minimum (Figure 7.1b), then it has a large variance, and even small errors in determining $E(\mathbf{m})$ will lead to large errors in $\mathbf{m}^{est}$.

As can be seen from Figure 7.1, it is the curvature of $E(\mathbf{m})$ near its minimum that predicts the accuracy of $\mathbf{m}^{est}$. The curvature of

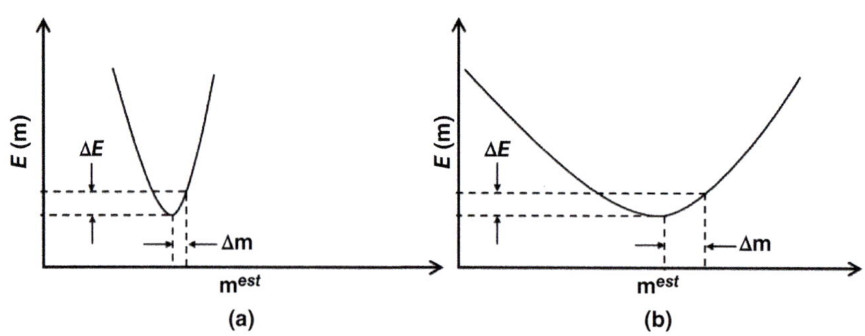

Figure 7.1 Demonstration of the dependence of the shape of $E(\mathbf{m})$ on the accuracy of the predicted model $\mathbf{m}^{est}$.

$E(\mathbf{m})$, in turn, can be measured by its second derivative. We thus define the prediction error $\Delta E = E(\mathbf{m}^{true}) - E(\mathbf{m}^{est})$ in which $\mathbf{m}^{true}$ is the true model, where $E(\mathbf{m})$ is minimum and $\mathbf{m}^{est}$ is the estimated model. We then expand ΔE in a truncated Taylor series around the minimum to write

$$\begin{aligned}\Delta E &= E(\mathbf{m}^{true}) - E(\mathbf{m}^{est}) \\ &= [\mathbf{m}^{true} - \mathbf{m}^{est}]^T \frac{1}{2}\frac{\partial^2 E}{\partial \mathbf{m}^2}\bigg|_{\mathbf{m}=\mathbf{m}^{est}} \\ &\quad \times [\mathbf{m}^{true} - \mathbf{m}^{est}]. \qquad (7.16)\end{aligned}$$

Note that we set the first-order term in the Taylor series expansion of equation 7.16 to zero because the expansion is done around the minimum point and therefore must vanish. Now recall (equation 6.16) that the term $\frac{1}{2}\partial^2 E/\partial \mathbf{m}^2$ on the right-hand side of equation 7.16 is the Hessian matrix $\mathbf{H}$. Therefore,

$$\Delta E = [\mathbf{m}^{true} - \mathbf{m}^{est}]^T \mathbf{H} [\mathbf{m}^{true} - \mathbf{m}^{est}]. \qquad (7.17)$$

Therefore, the Hessian matrix controls the quality of the estimated model as compared with the true model. Now, from equation 7.6

$$\frac{\partial^2 E}{\partial \mathbf{m}^2} = \frac{\partial}{\partial \mathbf{m}}\left(\frac{\partial E}{\partial \mathbf{m}}\right) = \frac{\partial}{\partial \mathbf{m}}[\nabla E(\mathbf{m})]. \qquad (7.18)$$

And, combining equations 7.18 and 7.12, we write

$$\frac{\partial^2 E}{\partial \mathbf{m}^2} = \frac{\partial}{\partial \mathbf{m}}[\mathrm{Re}\{\mathbf{J}^T|\mathbf{s} - \mathbf{d}|^*\}]. \qquad (7.19)$$

7.3.3 Data Resolution Matrix

Although the local optimization methods are sometimes referred as nonlinear least-squares method (for example, see Padhi et al., 2015), they all come to the minimum point by taking successive linear steps. Therefore, in contrast with the true nonlinear optimization methods, such as the genetic algorithm simulated annealing, particle swarm optimization, Markov Chain Monte Carlo, etc., which we discuss in Chapter 8, here we refer all local methods as the linearized optimization methods. In each iteration step in these local or linearized methods, we thus cast the physics of the forward problem which predicts the observed data $\mathbf{d} = g(\mathbf{m})$ as $\mathbf{d} = \mathbf{G}\mathbf{m}$, where $\mathbf{G}$ is a matrix and our estimated model at each iteration is $\mathbf{m}^{est} = \mathbf{G}^{-g}\mathbf{d}$. Note here that $\mathbf{G}^{-g}$ is not the true inverse of $\mathbf{G}$, but a best estimate of its inverse at any stage of iteration and so $\mathbf{G}\mathbf{G}^{-g} \neq \mathbf{I}$ unless $\mathbf{m}^{est} = \mathbf{m}^{true}$. Therefore, unlike operator-based methods, our $\mathbf{G}$ is just our best approximation of $g(\mathbf{m})$ at a specific step and varies with iteration until $\mathbf{G}\mathbf{G}^{-g}$ becomes nearly an identity matrix. Having said so, we may ask: how well the estimated model fits the data, which is given as

$$\begin{aligned}\mathbf{d}^{pre} &= \mathbf{G}\mathbf{m}^{est} = \mathbf{G}(\mathbf{G}^{-g}\mathbf{d}^{obs}) \\ &= \mathbf{G}\mathbf{G}^{-g}\mathbf{d}^{obs} = \mathbf{N}\mathbf{d}^{obs}. \qquad (7.20)\end{aligned}$$

In equation 7.20, we explicitly express the observed data as $\mathbf{d}^{obs}$. Note that for the data vector with M components (equation 7.1), the matrix $\mathbf{N}$ in equation 7.20 is of size $M \times M$ and is called the *data resolution matrix*. This matrix describes how well the predictions match the observed data. Note that if $\mathbf{N}$ is an identity matrix, then $\mathbf{d}^{pre} = \mathbf{d}^{obs}$, and the prediction error is zero. Ideally, we should start from a point where the data resolution matrix is diagonally dominant and make its off-diagonal elements as small as possible as we proceed through iterations.

7.3.4 Model Resolution Matrix

The data resolution matrix (equation 7.20) represents how well the observed data can be predicted or resolved from optimization. So, we can ask the same question about the model parameters. Assume that there is a true but unknown set of model parameters given by the model vector $\mathbf{m}^{true}$ that solves the set of equations given as $\mathbf{G}\mathbf{m}^{true} = \mathbf{d}^{obs}$; we may then ask how closely a particular set of model parameters, given as $\mathbf{m}^{est}$, fits the true model $\mathbf{m}^{true}$? Plugging in the expression for the observed data $\mathbf{G}\mathbf{m}^{true} = \mathbf{d}^{obs}$ into the estimate of the estimated model $\mathbf{m}^{est} = \mathbf{G}^{-g}\mathbf{d}^{obs}$ gives

$$\begin{aligned}\mathbf{m}^{est} &= \mathbf{G}^{-g}\mathbf{d}^{obs} = \mathbf{G}^{-g}[\mathbf{G}\mathbf{m}^{true}] \\ &= \mathbf{G}^{-g}\mathbf{G}\mathbf{m}^{true} = \mathbf{R}\mathbf{m}^{true}. \qquad (7.21)\end{aligned}$$

In equation 7.21 $\mathbf{R}$ is the $N \times N$ *model resolution matrix*. Like the data resolution matrix, if $\mathbf{R} = \mathbf{I}$, then each model parameter is uniquely determined. But, if $\mathbf{R}$ is not an identity matrix, then the estimates of the model parameters are weighted averages of the true model parameters. In linear (local) optimizations, our objective is to obtain $\mathbf{R}$ as close to an identity matrix as possible via successive iterations

7.3.5 Objective Regularization

Now that we have defined the basic terminologies, we are ready to revisit the objective, defined by equation 7.5. As mentioned before, definition of the objective $E(\mathbf{m})$ using equation 7.5 is unstable and must be regularized. There are many ways to regularize $E(\mathbf{m})$, and they all are based on the concept of damped least squares introduced briefly in Chapter 1. Here, we follow the steps adopted by Sen and Roy (2003), which are also implemented by Padhi et al. (2015) and Mukhopadhyay and Mallick (2016). In this scheme, $E(\mathbf{m})$ is defined as

$$E(\mathbf{m}) = [\mathbf{s} - \mathbf{d}]^T \mathbf{W}_D [\mathbf{s} - \mathbf{d}]^* + (\mathbf{m} - \mathbf{m}_{prior}) \mathbf{W}_M (\mathbf{m} - \mathbf{m}_{prior}). \tag{7.22}$$

In equation 7.22, $\mathbf{s}$ and $\mathbf{d}$ are defined as before and $\mathbf{m}$ is the current model vector. The vector $\mathbf{m}_{prior}$ in equation 7.22 is the prior model vector, i.e., the model vector from the previous iteration. The matrices $\mathbf{W}_D$ and $\mathbf{W}_M$ in equation 7.22 are the data and model weighting matrices, and they are diagonal or diagonally dominant. A convenient way to define $\mathbf{W}_D$, for example, is to set it as the inverse of the covariance matrix (equation 7.14) which, in theory, accounts for the noise in the data (Sen and Roy, 2003). So, assuming uncorrelated data, all equal variance, we can set $\mathbf{W}_D = (1/\sigma^2)\mathbf{I}$ or $\mathbf{W}_D = [\sigma]^T \mathbf{I}$, where

$$[\boldsymbol{\sigma}]_i = \frac{1}{M-1} \frac{1}{(d_i - \overline{d})^2}, \tag{7.23}$$

with $\overline{d}$ being the mean data vector. Another, and possibly more elegant, way is to set $\mathbf{W}_D$ to be the data resolution matrix $\mathbf{N}$ (equation 7.20) which will vary with iteration. We revisit the idea of setting $\mathbf{W}_D$ to the data resolution matrix later, and for now let us assume $\mathbf{W}_D$ to be the diagonal with its elements being the inverse of the data covariance matrix as suggested by Sen and Roy (2003). Also, for now, we assume the model weighting matrix ($\mathbf{W}_M$) to be diagonal and we revisit defining $\mathbf{W}_M$ later.

Now, following the same procedure as before, i.e., starting from $E(\mathbf{m})$ defined by equation 7.5 and deriving $\nabla E(\mathbf{m})$, shown in equation 7.12, if we start from equation 7.22, we then obtain

$$\nabla E(\mathbf{m}) = \frac{\partial E}{\partial \mathbf{m}} = \mathrm{Re}[\mathbf{J}^T \mathbf{W}_D |\mathbf{s} - \mathbf{d}|^*] + \mathbf{W}_M (\mathbf{m} - \mathbf{m}_{prior}). \tag{7.24}$$

In equation 7.24, $\mathbf{J}$ is the Jacobi or the Fréchet derivative matrix (equation 7.11). Therefore, once we figure out how to construct $\mathbf{W}_D$ and $\mathbf{W}_M$ and assuming that we know how to compute the Fréchet derivative matrix (discussion of which we postpone until the end of this chapter when we discuss specific optimization problems), we can compute $\nabla E(\mathbf{m})$ from equation 7.24 and use any of the methods discussed in Chapter 6 to find the minimum of $E(\mathbf{m})$.

7.4 Implementation of Local Optimization Methods

Starting from the regularized objective $E(\mathbf{m})$ defined by equation 7.22, we now provide outlines for implementing different local optimization methods for solving practical seismic inverse problems. Here, we outline following three approaches, which we previously discussed in Chapter 6:

1) Steepest descent (gradient descent).
2) Conjugate gradient.
3) Quasi-Newton.

From our previous discussions (Chapter 6), we note that the fundamental concepts of the steepest descent and conjugate-gradient methods are similar because none of them require explicit computation of the Hessian. The only way these two methods differ is in the process they do minimization in each iteration. While the steepest descent method tends to find the minimum along $-\nabla E(\mathbf{m})$ for each iteration, the conjugate-gradient method tends to find the minimum along a direction that is conjugate to the directions it used in the previous iterations. In contrast to this, the quasi-Newton or Gauss–Newton method attempts to estimate the Hessian (or to be specific, the inverse of the Hessian) via successive iterations. In the following, we first discuss the steepest and conjugate-gradient methods together, following which we discuss the Gauss–Newton method.

7.4.1 Steepest Descent (Gradient Descent) and Conjugate-Gradient Methods

The use of the steepest descent method to solve geophysical inverse problems is not new. McAulay (1985), Pan and Phinney (1989), Kormendi and Dietrich (1991), and Pan et al. (1994), for example, provide fundamentals of the use of the method to invert noisy seismic data. Like steepest descent, conjugate-gradient method is also used to solve many large geophysical inverse problems such as the least-squares Radon transforms, travel time tomography, and full waveform inversion (FWI) (Witte et al., 2018).

Here, we discuss both methods together by adopting the approach of Sen and Roy (2003). Although Sen and Roy (2003) used steepest descent method for optimization, it will be seen below that their implementations can easily be adopted into the conjugate-gradient method. In addition, Sen and Roy (2003) also provide an efficient way to compute the Jacobi, which we discuss later in Section 7.5.

We start from the objective $E(\mathbf{m})$ as defined by equation 7.22, where $\mathbf{s}-g(\mathbf{m})$ is the synthetic data vector, $\mathbf{d}$ is the observed data vector, $\mathbf{m}$ is the current model vector, and $\mathbf{m}_{prior}$ is the prior model vector. Considering that the observed and synthetic data vectors have M components (see equations 7.1 and 7.3), $\mathbf{W}_D$ in equation 7.22 is the inverse of the data covariance matrix, i.e., $\mathbf{W}_D = \mathbf{C}_d^{-1}$, and is a matrix of size of size $M \times M$, which can be estimated from the data (equation 7.14). Finally, $\mathbf{m}$ and $\mathbf{m}_{prior}$ are N-component vectors (see equation 7.2) and $\mathbf{W}_M$ is a matrix of size $N \times N$. Sen and Roy (2003) suggested $\mathbf{W}_M = \alpha \mathbf{C}_m^{-1}$, where $\mathbf{C}_m$ is the model covariance matrix (equation 7.15) and α is a scalar that is varied with iteration, which we discuss below. Because the matrix $\mathbf{M}$ for each iteration (equation 7.13) is known and the data covariance matrix $\mathbf{C}_d$ is also known, the model covariance matrix $\mathbf{C}_m$ can be computed from equation 7.15 and so is $\mathbf{W}_M$. Instead of defining the iteration-dependent $\mathbf{W}_M$ as above, Padhi et al. (2015) and Mukhopadhyay and Mallick (2016) used $\mathbf{W}_M = \alpha \mathbf{I}$ where α is defined the way Sen and Roy (2003) defined it and $\mathbf{I}$ is an $N \times N$ identity matrix.

Having $\mathbf{W}_D$ and $\mathbf{W}_M$ computed as above and assuming we know how to compute the Fréchet derivative matrix $\mathbf{J}$, we can use equation 7.24 to compute $\nabla E(\mathbf{m})$. Once $\nabla E(\mathbf{m})$ is computed, we are now ready to implement any optimization method for minimizing $E(\mathbf{m})$. The method used by Sen and Roy (2003) and Mukhopadhyay and Mallick, among others, is one such method based on the steepest descent. So, for each iteration, this method attempts to find the steepest direction and finds the minimum point in that direction. We come back to this steepest descent approach later but let us first investigate how a conjugate-gradient method can be implemented. Recall the required steps outlined in Chapter 6 (Figure 6.13). Here, we repeat these steps in the context of minimizing $E(\mathbf{m})$, given by equation 7.22.

We start from an initial model $\mathbf{m}_0$ and compute the initial synthetic data vector $\mathbf{s}_0 = g(\mathbf{m}_0)$. Next, we use equations 7.22 and 7.24 to compute $E(\mathbf{m}_0)$ and $\nabla E(\mathbf{m}_0)$. We then set the vectors $\mathbf{g}_0 = \nabla E(\mathbf{m}_0)$ and $\mathbf{h}_0 = \mathbf{g}_0$. Having the initial vectors $\mathbf{g}_0$ and $\mathbf{h}_0$, we follow the steps outlined in Figure 6.13 to successively minimize $E(\mathbf{m})$ along the conjugate directions. For any iteration step n, we find the minimum point $\mathbf{m}_n$ in the model space along the direction $\mathbf{h}_{n-1}$. Having found $\mathbf{m}_n$, we compute $E(\mathbf{m}_n)$, Fréchet derivative matrix $\mathbf{J}_n$, and the gradient $\nabla E(\mathbf{m}_n)$ using equations 7.11, 7.22, and 7.24. Next, we set $\mathbf{g}_0 = -\nabla E(\mathbf{m}_n)$ and compute the scalar γ_n either as

$$\gamma_n = \frac{\mathbf{g}_n \cdot \mathbf{g}_n}{\mathbf{g}_{n-1} \cdot \mathbf{g}_{n-1}} \tag{7.25}$$

or as

$$\gamma_n = \frac{(\mathbf{g}_n - \mathbf{g}_{n-1}) \cdot \mathbf{g}_n}{\mathbf{g}_{n-1} \cdot \mathbf{g}_{n-1}}, \tag{7.26}$$

and calculate the vector $\mathbf{h}_n = \mathbf{g}_n + \gamma_n \mathbf{h}_{n-1}$. Note here that if we use equation 7.25 for γ_n, the conjugate-gradient method becomes the Fletcher–Reeves algorithm. Instead, if we use equation 7.26 for γ_n, the method is the Polak–Ribiere algorithm.

In contrast to the conjugate gradient, in steepest descent method, we just find the minimum along the direction of the gradient of the objective, i.e., along $-\nabla E(\mathbf{m})$ for each iteration. We therefore have the necessary steps to implement both steepest descent and conjugate-gradient method, except for a robust method to find the minimum of the objective function along a direction. Once we find such a method, if we set the direction along the steepest gradient, the method is the steepest descent method. Instead, if we set this direction to be the conjugate directions, the method then becomes the conjugate-gradient method.

To find the minimum of the objective along any direction, we adopt the procedure of Sen and Roy (2003), who provided a compact expression for minimization along the steepest direction and modify the method to find the minimum along the conjugate direction such that we can implement both steepest descent and conjugate-gradient methods. Below, we first provide the Sen and Roy (2003) method without rigorous mathematical detail. We then focus on how this steepest descent method can be modified to a conjugate-gradient method. Finally, we provide detailed workflows for both methods.

From the model $\mathbf{m}_n$, estimated at iteration n, the model $\mathbf{m}_{n+1}$ for the next iteration $n+1$ is given by Sen and Roy (2003) as

$$\mathbf{m}_{n+1} = \mathbf{m}_n - \nu_n\{(\mathbf{W}_M + \mathbf{J}_n^T \mathbf{W}_D \mathbf{J}_n^*)^{-1} \nabla E(\mathbf{m})\}. \tag{7.27}$$

And, by explicitly plugging in the explicit expression for $\nabla E(\mathbf{m})$ from equation 7.24 and the synthetic data vector for the model $\mathbf{m}_n$ (equation 7.3), we rewrite equation 7.27 as

$$\begin{aligned}\mathbf{m}_{n+1} = \mathbf{m}_n - \nu_n(\mathbf{W}_M + \mathbf{J}_n^T \mathbf{W}_D \mathbf{J}_n^*)^{-1} \\ \times [\mathrm{Re}\{\mathbf{J}_n^T \mathbf{W}_D |g(\mathbf{m}_n) - \mathbf{d}|^*\} \\ \times \mathbf{W}_M(\mathbf{m}_n - \mathbf{m}_{prior})].\end{aligned} \tag{7.28}$$

In equation 7.28, $\mathbf{W}_D$ and $\mathbf{W}_M$ are the data and model weighting matrices as previously defined, $\mathbf{J}_n$ is the Jacobi (Fréchet derivative) matrix for the current model vector $\mathbf{m}_n$ (equation 7.11), and $\mathbf{m}_{prior}$ is the initial model ($\mathbf{m}_0$). Finally, the scalar υ_n in equation 7.28 is an iteration-adaptive step length defined as

$$\nu_n = \frac{\boldsymbol{\beta}_n^T \mathbf{W}_M \boldsymbol{\theta}_n^*}{\boldsymbol{\theta}_n^T \mathbf{W}_M \boldsymbol{\theta}_n^* + \mathbf{b}_n^T \mathbf{W}_D \mathbf{b}_n^T}, \tag{7.29}$$

with

$$\left.\begin{aligned}\boldsymbol{\beta}_n &= \mathbf{W}_M^{-1} \mathbf{J}_n^T \mathbf{W}_D |g(\mathbf{m}_n) - \mathbf{d}| + (\mathbf{m}_n - \mathbf{m}_{n-1}) \\ \boldsymbol{\theta}_n &= \mathbf{F}_n \boldsymbol{\beta}_n \\ \mathbf{F}_n &= (\mathbf{I} - \mathbf{W}_M^{-1} \mathbf{J}_n^T \mathbf{W}_D \mathbf{J}_n^*)^{-1} \\ \mathbf{b}_n &= \mathbf{J}_n \boldsymbol{\theta}_n\end{aligned}\right\}. \tag{7.30}$$

In above, we modified the original equations of Sen and Roy (2003) to include complex values of the synthetic data $\mathbf{s} = g(\mathbf{m}_n)$ and the observed data $\mathbf{d}$.

Modifying this minimization from the steepest descent direction to the conjugate direction is straightforward. At any iteration step n with the model $\mathbf{m}_{n-1}$ and the direction $\mathbf{h}_{n-1}$ from the previous iteration $n-1$, the model $\mathbf{m}_n$ for the current iteration can be found simply by modifying equation 7.27 to minimize along $\mathbf{h}_{n-1}$ instead of minimizing along the steepest direction $\nabla E(\mathbf{m}_n)$ as

$$\mathbf{m}_n = \mathbf{m}_{n-1} + (\mathbf{W}_{M-1} + \mathbf{J}_{n-1}^T \mathbf{W}_D \mathbf{J}_{n-1}^*)^{-1} \mathbf{h}_{i-1}. \tag{7.31}$$

Note that the iteration-adaptive step length for conjugate-gradient method is already included in the vector $\mathbf{h}_{n-1}$ via the scalar γ_{n-1} computed from equation 7.25 or 7.26 in the previous iteration. Having found $\mathbf{m}_n$, we compute $E(\mathbf{m}_n)$ from equation 7.22, $\mathbf{J}_n$ from equation 7.11, and the gradient $\nabla E(\mathbf{m}_n)$ from equation 7.24. Next, we set $\mathbf{g}_n = -\nabla E(\mathbf{m}_n)$, compute the scalar γ_n using either equation 7.25 or 7.26, and calculate the vector $\mathbf{h}_n = \mathbf{g}_n + \gamma_n \mathbf{h}_{n-1}$. Next, we replace $\mathbf{m}_{n-1}$ by $\mathbf{m}_n$ and $\mathbf{h}_{n-1}$ by $\mathbf{h}_n$ and move to the next iteration.

Therefore, if we know how to compute the Jacobi matrix $\mathbf{J}$, we are all set to implement both conjugate-gradient and steepest descent methods, except for one small thing. We previously noted that the data weighting matrix can be set as $\mathbf{W}_D = \mathbf{C}_d^{-1}$, but what about the model weighting matrix $\mathbf{W}_M$? We previously mentioned that Sen and Roy (2003) defined $\mathbf{W}_m = \alpha \mathbf{C}_m^{-1}$ and Padhi et al. (2015) and Mukhopadhyay and Mallick (2016) defined $\mathbf{W}_M = \alpha \mathbf{I}$. But, from where we obtain the scalar α?

Fundamentally, α is a regularization parameter, which is computed and updated at each iteration. Although there are several ways to compute this parameter, Engl's modified discrepancy principle, given in Engl (1987), is computationally efficient. In fact, Roy (2005) calculated an explicit expression for α as

$$\alpha = \frac{\delta^{\frac{p}{q+2}}}{\left(\|\Delta \mathbf{m}_n\|^2\right)^{\left(\frac{1}{q}+2\right)}}. \tag{7.32}$$

In equation 7.32, δ is the noise level on data and p and q are positive non-zero numbers satisfying the relation

$$1.5\,p - 2 = q > 1, \tag{7.33}$$

and $\Delta \mathbf{m}_n$ is the computed change in the model at the iteration n. Typically, the value of α is chosen to be a large number ($\approx$10) at the beginning. After computing $\Delta \mathbf{m}_n$, α is then updated using equations 7.32 and 7.33 in each subsequent iteration. But how do we obtain the noise level on data (the parameter δ in equation 7.32)? This can be obtained using the L-curve technique, outlined in Hansen (1992)

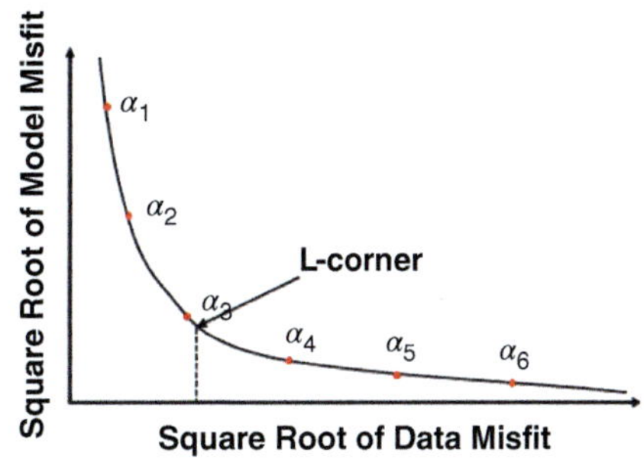

Figure 7.2 Demonstration of the L-curve technique.

and Hansen and O'Leary (1993). Fundamental concept behind the L-curve method is that if we plot data misfit in the horizontal axis versus the model misfit in the vertical axis using a suitable scale for varying δ values (and so for the regularization weight α, see equations 7.32 and 7.33), it produces an L-shaped curve (Figure 7.2). The value of δ (and therefore α) at the corner or knee of the curve plays an important role in that the regularization corresponding to that point is the optimum value for the regularization weight (see Figure 7.2).

Although Figure 7.2 uses the square root of the misfit for both axes, another popular method of scaling is the logarithmic scaling. Padhi et al. (2015) claims that using logarithmic scaling produces better convergence of the algorithm than using the square-root scaling. However, the choice of scaling may be entirely dependent upon the specific problem being solved.

Let us examine the L-curve technique in some detail. The error $E(\mathbf{m})$, defined by equation 7.22 can be written as

$$E(\mathbf{m}) = E_d + \alpha E_m, \tag{7.34}$$

where

$$E_d = |\mathbf{s} - \mathbf{d}|^T \mathbf{W}_D |\mathbf{s} - \mathbf{d}|^*, \tag{7.35}$$

and

$$\begin{aligned} E_m &= [\mathbf{m} - \mathbf{m}_{prior}]^T \mathbf{W}_M [\mathbf{m} - \mathbf{m}_{prior}]^T \\ &= [\Delta \mathbf{m}]^T \mathbf{W}_M [\Delta \mathbf{m}]. \end{aligned} \tag{7.36}$$

In equation 7.36, we used $\Delta \mathbf{m} = \mathbf{m} - \mathbf{m}_{prior}$. Note from equation 7.34, as $\alpha \to 0$, $E(\mathbf{m}) \to E_d$. And this is also true when $\Delta \mathbf{m} \approx \mathbf{0}$, the null vector. The primary philosophy of linear optimization is that the perturbation in the synthetic data vector $\mathbf{s} = g(\mathbf{m})$ due to perturbation in the model parameter stays linear. Assuming this linearity and that the first-order derivatives of $\mathbf{s} = g(\mathbf{m})$ are continuous and the second-order derivatives exist, we write $\mathbf{s}$ in the neighborhood of the model as

$$\mathbf{s}(\mathbf{m} + \Delta\mathbf{m}) \approx \mathbf{s}(\mathbf{m}) + \frac{\partial \mathbf{s}}{\partial \mathbf{m}}\Delta\mathbf{m} = \mathbf{s}(\mathbf{m}) + \mathbf{J}\Delta\mathbf{m}. \tag{7.37}$$

Thus, considering linear variations of $\mathbf{s} = g(\mathbf{m})$ around the neighborhood of $\mathbf{m}$, we use equation 7.37 and write $E(\mathbf{m})$ as

$$E(\mathbf{m}) \approx |\mathbf{u} - \mathbf{J}\Delta\mathbf{m}|^T \mathbf{W}_D |\mathbf{u} - \mathbf{J}\Delta\mathbf{m}|^* + \alpha\{[\Delta\mathbf{m}]^T \mathbf{W}_M \Delta\mathbf{m}\}. \tag{7.38}$$

In equation 7.38, $\mathbf{u}$ is the data residual vector $|\mathbf{s} - \mathbf{d}|$, the difference between the computed (synthetic) and the observed data.

Note that out of the two terms on the right-hand side of equation 7.38, the first term is the data misfit, and the second term is the model misfit. The L-curve method essentially plots these two terms with a suitably chosen scale, such as the square root, logarithmic, etc., for different values of α and uses the value of α at the L-corner point (see Figure 7.2) as the optimum regularization weight. Note that although we use Figure 7.2 to demonstrate the concept of the L-corner point, in practice, it is not necessary to plot the two scaled misfit function to manually find this corner point. It is always analytically obtained from the curvature of the misfit function shown in Figure 7.2. Equations 7.32 and 7.33, for example, provide an analytical expression for α when logarithmic scaling is used.

Besides L-curve method, Sen and Biswas (2015) proposed a regularization weight to be computed as an annealing type of parameter and given as

$$\alpha = 0.9e^{-(2.3\times10^7 I)}, \tag{7.39}$$

where I is the iteration number. Ayani et al. (2020) used this approach to regularize $E(\mathbf{m})$ to solve the optimization problem to invert EM data.

We now have the complete set of steps to implement both steepest descent and conjugate-gradient methods, which we provide in Figures 7.3 and 7.4.

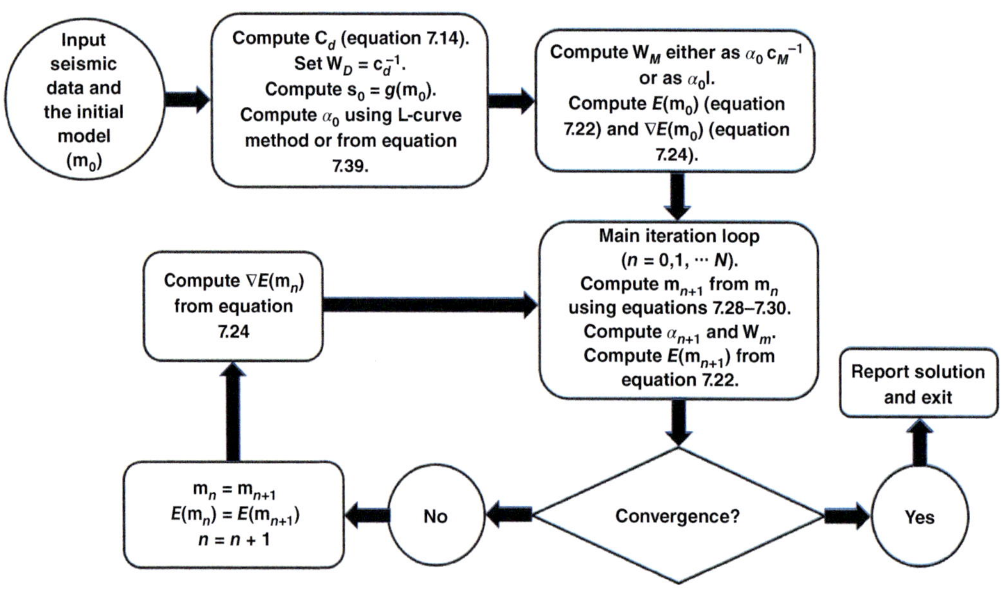

Figure 7.3 Flowchart for the steepest descent method.

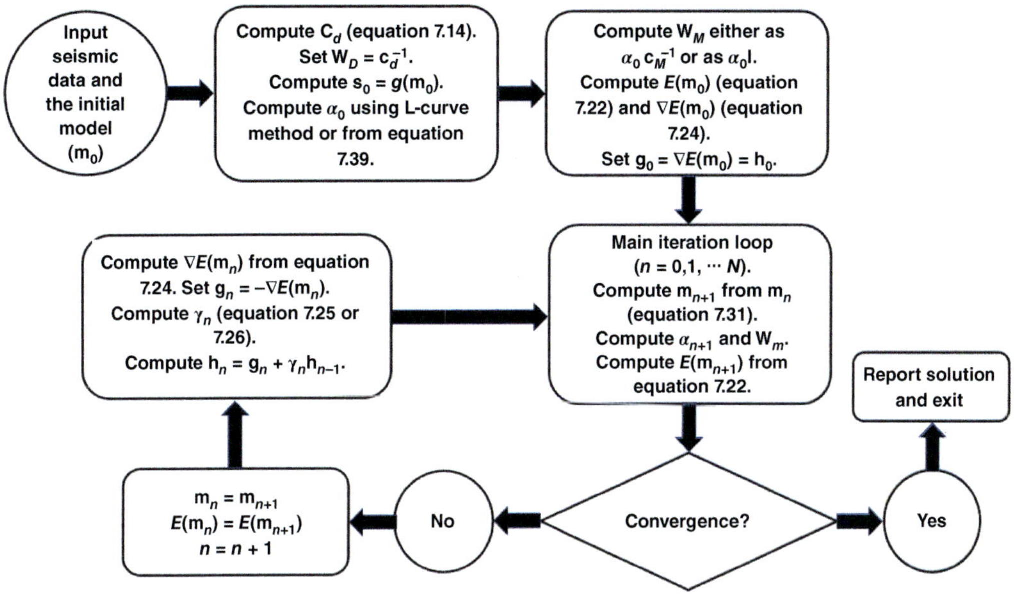

Figure 7.4 Flowchart for the conjugate-gradient method.

7.4.2 Gauss–Newton Method

Like the steepest descent and conjugate-gradient methods, Gauss–Newton method is also popular for solving seismic and other geophysical inverse problems. A good account of the Gauss–Newton method can be found in Loke and Dahlin (2002), Abubakar et al. (2009), Padhi et al. (2015), among others. Below, we provide a brief overview of the method.

In our implementations of the steepest descent and conjugate-gradient methods, we used diagonal matrices for both $\mathbf{W}_D$ and $\mathbf{W}_M$. However, ideally, they both should be diagonally dominant, not exactly diagonal. In fact, the off-diagonal elements of $\mathbf{W}_D$ and $\mathbf{W}_M$ should be wide (but diagonally dominant) at the beginning and become narrow as the iteration progresses. In other words, $\mathbf{W}_D$ should converge to the data resolution matrix $\mathbf{N}$ (equation 7.20) and $\mathbf{W}_M$ should converge to the model resolution matrix $\mathbf{R}$ (equation 7.21) with iteration.

Now recall the fundamentals of different flavors of the quasi-Newton method outlined in Chapter 6. The Gauss–Newton is similar. Over the iteration i, this method iteratively build a matrix $\mathbf{A}$ in such a way that it converges to the inverse of the Hessian matrix $\mathbf{H}$ at the end of the iteration, i.e.,

$$\lim_{i\to\infty} \mathbf{A}_i = \mathbf{H}^{-1}. \tag{7.40}$$

Primary concept of iteratively building the matrix $\mathbf{A}$ is because if $E(\mathbf{m})$ is defined by equation 7.22 in the neighborhood of a given point, say $\mathbf{m}_0$, in the model space, we may then expand it in a Taylor series to write

$$\begin{aligned} E(\mathbf{m}) = E(\mathbf{m}_0) &+ \sum_{i=1}^{N} \left.\frac{\partial E}{\partial m_i}\right|_{m_i} dm_i \\ &+ \frac{1}{2}\sum_{i=1}^{N}\sum_{j=i}^{N} \left.\frac{\partial^2 E}{\partial m_i \partial m_j}\right|_{m_i, m_j} dm_i dm_j \\ &+ \ldots \approx c - \mathbf{b}\cdot\mathbf{m} + \frac{1}{2}\mathbf{m}^T\cdot\mathbf{H}\cdot\mathbf{m}, \end{aligned} \tag{7.41}$$

with

$$\left.\begin{aligned} c &= E(\mathbf{m}_0) \\ \mathbf{b} &= -\nabla E(\mathbf{m})|_{\mathbf{m}_0} \end{aligned}\right\}, \tag{7.42}$$

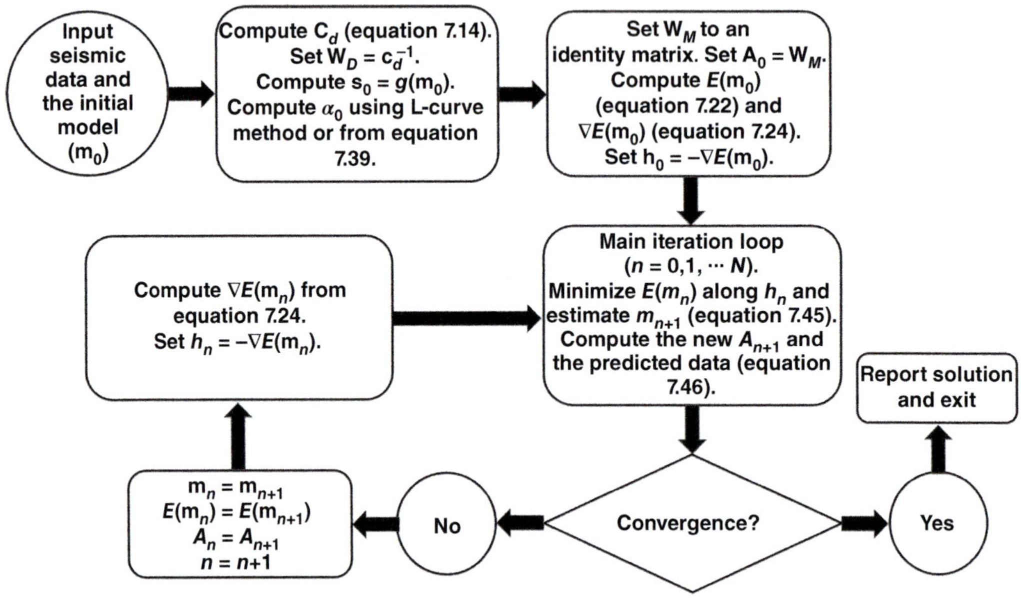

Figure 7.5 Flowchart for the Gauss–Newton method.

and **H** is the Hessian matrix given as

$$\mathbf{H} = \begin{pmatrix} \frac{\partial^2 E}{\partial m_1^2} & \frac{\partial^2 E}{\partial m_1 \partial m_2} & \cdots & \frac{\partial^2 E}{\partial m_1 \partial m_N} \\ \frac{\partial^2 E}{\partial m_2 \partial m_1} & \frac{\partial^2 E}{\partial m_2^2} & \cdots & \frac{\partial^2 E}{\partial m_2 \partial m_N} \\ \cdot & \cdot & \cdots & \cdot \\ \cdot & \cdot & \cdots & \cdot \\ \cdot & \cdot & \cdots & \cdot \\ \frac{\partial^2 E}{\partial m_N \partial m_1} & \frac{\partial^2 E}{\partial m_N \partial m_2} & \cdots & \frac{\partial^2 E}{\partial m_N^2} \end{pmatrix}. \tag{7.43}$$

From equation 7.41 we also have

$$\nabla E(\mathbf{m}) = \mathbf{H} \cdot \mathbf{m} - \mathbf{b}. \tag{7.44}$$

Therefore, for successive iterations if we set $\nabla E(\mathbf{m}) = \mathbf{0}$, then the estimated model at any iteration step i is given as

$$\mathbf{m}_i^{est} = \mathbf{H}^{-1} \cdot \mathbf{b}_i = \mathbf{A}_i \cdot \mathbf{b}_i, \tag{7.45}$$

where $\mathbf{A}_i$ is our best estimate of the inverse of the Hessian matrix at that iteration. Thus, once the model is estimated from equation 7.45, we can attempt to predict **b** in equation 7.45 as

$$\mathbf{b}_i^{pre} = \mathbf{A}_i^{-1} \mathbf{m}_i^{est} = \mathbf{A}_i^{-1} \mathbf{A}_{i-1} \mathbf{b}^{obs}. \tag{7.46}$$

Thus, the product of the inverse of **A** generated at the current step and **A** generated from the previous iteration is like the model resolution matrix, which can be used for the model weighting matrix $\mathbf{W}_M$. A Gauss–Newton implementation for minimizing $E(\mathbf{m})$ following these concepts is shown in Figure 7.5.

7.4.3 Other Methods

Besides the methods discussed above, a few new optimization methods have been developed to efficiently handle large-scale data-science-related problems. These methods, however, can be used to solve seismic and other geophysical inverse problems as well. In Chapter 6, we briefly mentioned about two such methods: (1) Adam and (2) Levenberg-Marquardt (L-M). In essence, these methods are variations of the classical methods we discussed above, specially designed for computational efficiency to handle very large multidimensional optimization problems. We postpone discussion of these methods to Chapter 9 when we discuss different machine-learning-related applications.

7.5 Computation of the Jacobi (Fréchet Derivative) Matrix

In the above section, we outlined how to compute the data and model weighting matrices ($\mathbf{W}_D$ and $\mathbf{W}_M$) and provided details of some implementation aspects of different methods. We have however skipped the details of computing the Jacobi or the Fréchet derivative matrix $\mathbf{J}$ (equation 7.11). In general, for the data vector with M components (equation 7.1) and the model vector with N components (equation 7.2), $\mathbf{J}$ is a matrix of size $M \times N$ and can be numerically computed. For example, in any iteration step n with the model vector $\mathbf{m}_n$, assume that the synthetic data vector is computed as

$$\mathbf{s}_n = g(\mathbf{m}_n). \tag{7.47}$$

Then, if we compute another synthetic data vector with the ith component of the model vector changed from m_i to $m_i + \Delta m_i$ and keep other components of the model vector unchanged as

$$\mathbf{s}'_n = g(\mathbf{m}_n|_{m_i = m_i + \Delta m_i}), \tag{7.48}$$

then the vector $\mathbf{v}$ given as

$$\mathbf{v} = \frac{\mathbf{s}'_n - \mathbf{s}_n}{\Delta m_i} \tag{7.49}$$

represents the ith column of $\mathbf{J}$, and repeating this process for each component of the model vector provides the entire matrix. Therefore, for the model vector with N components, computing $\mathbf{J}$ requires $N+1$ synthetic computations at each iteration. There is nothing wrong about computing $\mathbf{J}$ this way if we choose Δm_i to be sufficiently small so that the finite differencing, represented by equation 7.49, approximates the derivatives with sufficient accuracy and represents the ith column of $\mathbf{J}$. In addition, for complex optimization problems, computing $\mathbf{J}$ via a finite-differencing approach as above may be the only option. However, for many optimization problems of practical geophysical interest, the problem could be simplified such that $\mathbf{J}$ can be efficiently computed. Below, we discuss a few of them.

7.5.1 Amplitude-Variation-With-Angle Inversion

In AVA, we are concerned with estimating the subsurface earth properties using either a linearized or an exact form of the reflection coefficient. The primary assumption in AVA is that when the offset-domain seismic data are transformed into angles, the amplitudes on the data are proportional to the reflection coefficients as a function of the incidence angle. For simplicity, let us assume P-wave (vertical component or pressure) seismic data converted into angles (Figure 1.7). Also assume that we use the linearized form of the P-wave reflection coefficient (Bortfeld, 1961; Richards and Frasier, 1976; Shuey, 1985; Mallick, 1993), given by equations 1.7–1.16.

In AVA, the offset-domain prestack data are first converted into angles (Figure 1.8). We then discretize the reflections on the angle domain data (Figure 1.8) to represent them as the observed data vector and relate to the model vector as $\mathbf{d} = \mathbf{Gm}$. In Chapter 1, we have shown how to cast this problem as an operator-based pure or damped least-squares inverse problem to estimate the model (equations 1.18–1.24). Besides such operator-based inversion, this problem can also be cast as a local model-based inversion by iteratively solving the forward problem $\mathbf{d} = \mathbf{Gm}$, which requires computing the Jacobi. We previously discussed that a model-based method is always a better choice than an operator-based method for estimating the model from noisy data (see the discussions in Chapter 1). Consequently, casting the AVA problem as a model-based local optimization using any of the methods we discussed above is of practical interest. Additionally, as we see below, computing the Jacobi for the AVA problem is analytical and does not require any numerical derivative estimation by finite differencing (equation 7.49).

Consider a single reflection event of the seismic data after offset-to-angle transformation, shown by the red dashed line in Figure 7.6, such that the picked reflection amplitudes can be represented by an M-component data vector

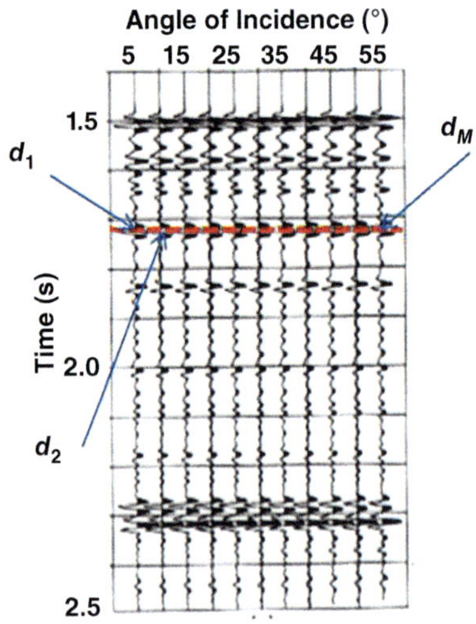

Figure 7.6 Example of an angle gather showing reflection amplitudes as a function of the angle of incidence and time. The dashed red line represents a single reflection event in which the picked amplitudes can be represented as an M-component data vector with individual components represented as $d_1, d_2, \ldots, d_M$.

$\mathbf{d} = [d_1, d_2, \ldots, d_M]^T$ as shown by equation 7.1. From our discussions from Chapter 1, we may then represent this data vector as $\mathbf{d} = \mathbf{Gm}$ in which

$$\mathbf{G} = \begin{pmatrix} 1 & \sin^2\theta_1 & \sin^2\theta_1 \tan^2\theta_1 \\ 1 & \sin^2\theta_2 & \sin^2\theta_2 \tan^2\theta_2 \\ \cdot & \cdot & \cdot \\ \cdot & \cdot & \cdot \\ \cdot & \cdot & \cdot \\ 1 & \sin^2\theta_M & \sin^2\theta_M \tan^2\theta_M \end{pmatrix}, \quad (7.50)$$

and $\mathbf{m}$ is the three-component model vector (that is, N in equation 7.2 is 3) and given as

$$\mathbf{m} = [A, B, C]^T. \quad (7.51)$$

In equation 7.50, $\theta_1, \theta_2, \ldots, \theta_M$ are the individual angles (5°, 10°, 15°, etc.) in the angle domain data (Figure 7.6), and A, B, and C in equation 7.51 are the AVA intercept, AVA gradient, and AVA curvature, respectively, previously defined by equations 1.8–1.10. The data vector $\mathbf{d} = [d_1, d_2, \ldots, d_M]^T$ and the forward problem $\mathbf{d} = \mathbf{Gm}$, with $\mathbf{G}$ and $\mathbf{m}$ defined respectively by equations 7.50 and 7.51, represent the underlying physics for the AVA optimization. Thus, starting with a prior model we can compute the synthetic data using $\mathbf{s} = \mathbf{Gm}$ and keep updating the model until we satisfactorily match the data. Also note that, for simplicity, here we demonstrated the AVA problem by picking the reflection amplitudes along a single event (see Figure 7.6), and if we pick them for all events on the data, we obtain a subsurface model of A, B, and C. Once they are estimated, we may then use the background (low-frequency) model and equations 1.9–1.16 to obtain the P-wave velocity (V_P), S-wave velocity (V_S), and density (ρ).

Now, from the linearized AVA equation (equation 1.7), we immediately see that

$$\left.\begin{aligned} \frac{\partial R_{PP}}{\partial A} &= 1 \\ \frac{\partial R_{PP}}{\partial B} &= \sin^2\theta \\ \frac{\partial R_{PP}}{\partial C} &= \sin^2\theta \tan^2\theta \end{aligned}\right\}. \quad (7.52)$$

From equations 7.50–7.52, and the definition of the Jacobi (equation 7.11), we immediately see that for the AVA inverse problems we have

$$\mathbf{J} = \mathbf{G}. \quad (7.53)$$

Therefore, for AVA inverse problems using the linearized Zoeppritz equations (equation 1.7) as the fundamental physics of the forward problem $\mathbf{s} = \mathbf{Gm}$, the Jacobi matrix can be analytically computed, and we may use any of the local optimization methods (steepest descent, conjugate gradient, Gauss–Newton, etc.) for model estimation.

7.5.2 Full Waveform Inversion

Full waveform inversion uses the full wave equation, $\mathbf{s} = g(\mathbf{m})$, as the fundamental physics of the forward problem. Consequently, computing the Jacobi matrix for FWI problems is not as straightforward as it is for the AVA inversion problems. In Chapter 5, we have discussed different wave equation methods for computing synthetic seismograms. In these discussions, we discussed these computing methods under two separate categories:

1) Computation of synthetic seismic response for a horizontally stratified (one-dimensional or 1D) subsurface model, in which the model is heterogeneous in depth but laterally invariant.
2) Computation of synthetic response for complete three-dimensional (3D) subsurface, in which the model is heterogeneous in all three dimensions, i.e., both vertically and laterally.

Consequently, computing the Jacobi for 1D and 3D (or 2D) inverse problems must be separately considered.

Inversion with Locally 1D Assumption

Although a fully 3D FWI is ideal for every application, they are too expensive, even for the high-performance computers. Additionally, in many situations in practice, the subsurface geology is such that a 3D FWI is unnecessary.

Figure 7.7, for example, shows one inline and one crossline section from a 3D seismic survey from the Rock-Springs Uplift (RSU), Wyoming, USA. These two inline and crossline sections intersect each other approximately at the middle of the survey. It can be seen from Figure 7.7 that RSU subsurface geology is nearly horizontal with gentle dips. Under such a situation, a relative-amplitude-preserved (RAP) processing workflow, including a prestack time or depth migration, would move the reflection image points such that the prestack seismic data, sorted in the common midpoint (CMP) domain, can be approximately considered as locally 1D (Resnick, 1993; Mallick, 1999; Pafeng et al., 2017). Of course, such local 1D structure would change as we move from one CMP location to the next, but each CMP location can be independently inverted using 1D wave equation as the fundamental physics of the forward problem in computing the synthetic data as $\mathbf{s} = g(\mathbf{m})$ and iteratively matching with the observed data $\mathbf{d}$ using any of the methods such as the steepest descent, Gauss–Newton, conjugate gradient, etc., we discussed above. This, however, would require computing the Jacobi. Although computing synthetic data under such local 1D assumption is computationally efficient, computing Jacobi by straightforward finite differencing via equation 7.49 can be computationally inefficient because it would require many forward synthetic seismogram computations, which could be expensive even under such 1D assumption.

To further elaborate the computational expense involved in computing the Jacobi, let us first look at the steps to compute it using a pure finite-difference approach, given by equation 7.49.

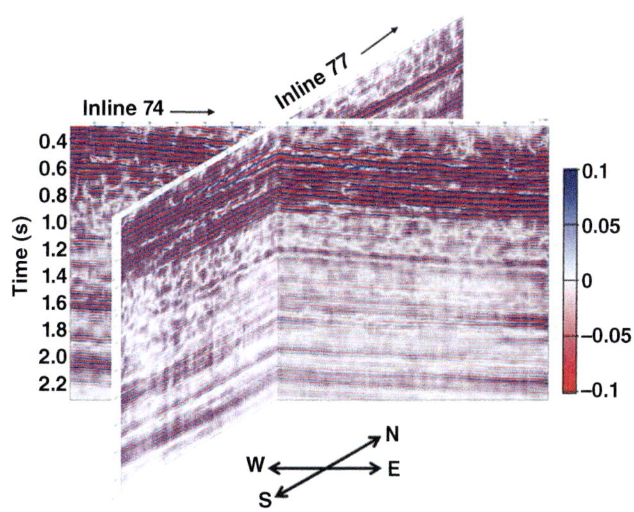

Figure 7.7 Stacked section from the 3D seismic data from the Rock-Springs Uplift, Wyoming, USA. One inline (Inline 74) and one crossline (Xline 77) out of entire 3D data volume are displayed. These two sections cross each other approximately in the middle of the entire survey.

1) At any iteration step, we have the data vector $\mathbf{d} = [d_1, d_2, \dots, d_M]^T$ (equation 7.1), the model vector $\mathbf{m} = [m_1, m_2, \dots, m_N]^T$ (equation 7.2), and the synthetic data vector $\mathbf{s} = g(\mathbf{m}) = [s_1, s_2, \dots, s_M]^T$ (equation 7.3).
2) We perturb each model parameter m_i for each subsurface layer by a small amount, say Δm_i, and compute another synthetic data vector $\mathbf{s}_i + \Delta\mathbf{s}_i = [s_1 + \Delta s_1, s_2 + \Delta s_2, \dots, s_M + \Delta s_M]^T$. Then, $\{(\mathbf{s}_i + \Delta\mathbf{s}_i) - \mathbf{s}\}/\Delta m_i$ is the ith column of the Jacobi matrix $\mathbf{J}$.
3) Repeating step 2 above for all model parameters $i = 1, 2, \dots, N$ would give us the entire $M \times N$ Jacobi matrix.

The approach outlined above, although straightforward to implement, is not very compute-efficient. This procedure will not only require computing the synthetic data for the given model, but it would also require computing additional synthetic data in which each one of the N model parameters must be perturbed leaving the rest constant. As an example, assume that we invert a time window of 2 s under an isotropic elastic assumption where there are three model parameters, namely, the P-wave velocity (V_P), S-wave velocity (V_S), and density (ρ) for each layer to be estimated. To resolve the subsurface at a quarter wavelength of the dominant seismic frequency of 30–40 Hz, a 2 s time window would require the model to be discretized with about 500 layers, meaning there will be 1500 model parameters (three for each layer). Thus, computing $\mathbf{J}$ using the procedure outlined above would require computing 1501 synthetic data at each iteration step, one for the original model, and 1500 additional with each model parameter is perturbed keeping the rest constant. This, in turn, could be a huge computational burden. For an anisotropic inversion where we would like to estimate additional model parameters, the computational burden is even higher.

It is, therefore, worthwhile to investigate the methods to efficiently compute the Jacobi matrix. One such approach was originally proposed by Mallick and Frazer (1988) for computing synthetic vertical seismic profile response and was later adopted by Sen and Roy (2003) to efficiently compute the Jacobi to solve seismic inverse problem using a local optimization scheme. To understand that how the Jacobi could be efficiently computed using the Sen and Roy (2003) approach, it is first necessary to review some of the forward computation methods that we previously discussed in Chapter 5.

In Chapter 5, Section 5.3, we discussed the computation of synthetic seismograms for horizontally stratified or 1D earth models. For convenience so that the readers do not have to switch back and forth between chapters, we repeat here the essential equations with accompanying figures, which are the fundamental building blocks for efficiently computing the Jacobi matrix.

The 1D wave equation method, discussed in Chapter 5, iteratively builds the reflection and transmission (R/T) coefficient matrix for a stack of layers by successively adding one layer and an interface to the previous R/T matrix. As shown in Figure 7.8, we start from the top and add layer 1 and interface 1, then layer 2 and interface 2, and so on, until we reach layer N, the end of the stack of layers where we apply the radiation boundary condition.

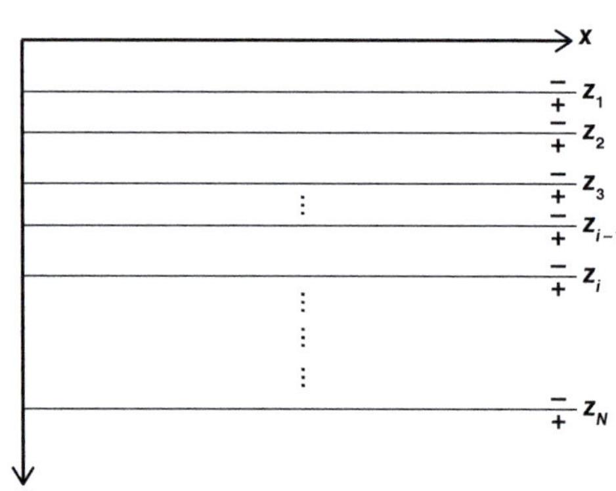

Figure 7.8 Procedure to build the R/T matrices for a stack of N layers. The plus (+) and minus (−) signs represent the media immediately below or immediately above any interface.

Essential equations to remember from Chapter 5 are the following:

R/T Matrices for a Layer and an Interface: Consider an interface at depth $z = z_i$ (see Figure 7.8). Also consider that R/T matrix for the interface at z_i are $\mathbf{t}_u$, $\mathbf{r}_d$, $\mathbf{r}_u$, and $\mathbf{t}_d$. We have discussed how to compute these matrices for solid–solid, solid–fluid, fluid–solid, and fluid–fluid interfaces in Chapter 5. Then, the R/T matrix for the entire layer $i-1$ (i.e., from depth z_{i-1}^+) through the interface i (i.e., to z_i^+) is given as

$$\mathbf{R}(z_i^+, z_{i-1}^+) = \begin{pmatrix} \mathbf{E}_u^{-1}\mathbf{t}_u & \mathbf{E}_u^{-1}\mathbf{r}_d\mathbf{E}_d \\ \mathbf{r}_u & \mathbf{t}_d\mathbf{E}_d \end{pmatrix}. \quad (7.54)$$

The matrices $\mathbf{E}_u^{-1}$ and $\mathbf{E}_d$ in equation 7.54 are given by equation 5.94.

Iteration Equations to Build R/T Matrices for a Stack of Layers: As mentioned above, the R/T matrices for a stack of layers are computed by successively adding a layer and an interface, and the iteration equations are the building blocks for this successive addition. Suppose at any stage, for example through the top of layer $i-1$ at depth z_{i-1}^+ (Figure 7.8), we have built the R/T matrices $\mathbf{T}_u$, $\mathbf{R}_d$, $\mathbf{R}_u$, $\mathbf{T}_d$. Then, if we add the homogeneous layer $i-1$ and the interface at z_i (Figure 6.8), iteration equations to compute the new R/T matrices $\mathbf{T}_u^{new}, \mathbf{R}_d^{new}, \mathbf{R}_u^{new}$, and $\mathbf{T}_d^{new}$ to depth z_i^+ are

$$\left.\begin{aligned}
\mathbf{T}_u^{new} &= \mathbf{T}_u(\mathbf{I} - \mathbf{E}_u^{-1}\mathbf{r}_d\mathbf{E}_d\mathbf{R}_u)^{-1}\mathbf{E}_u^{-1}\mathbf{t}_u \\
\mathbf{R}_d^{new} &= \mathbf{R}_d + \mathbf{T}_u\mathbf{E}_u^{-1}\mathbf{r}_d\mathbf{E}_d(\mathbf{I} - \mathbf{R}_u\mathbf{E}_u^{-1}\mathbf{r}_d\mathbf{E}_d)^{-1}\mathbf{T}_d \\
\mathbf{R}_u^{new} &= \mathbf{r}_u + \mathbf{t}_d\mathbf{E}_d\mathbf{R}_u(\mathbf{I} - \mathbf{E}_u^{-1}\mathbf{r}_d\mathbf{E}_d\mathbf{R}_u)^{-1}\mathbf{E}_u^{-1}\mathbf{t}_u \\
\mathbf{T}_d^{new} &= \mathbf{t}_d\mathbf{E}_d(\mathbf{I} - \mathbf{R}_u\mathbf{E}_u^{-1}\mathbf{r}_d\mathbf{E}_d)^{-1}\mathbf{T}_d
\end{aligned}\right\}. \quad (7.55)$$

Note that we have derived both equations 7.54 and 7.55 in Chapter 5 (see equations 5.102 and 5.105), and here we just show their final expressions for the convenience of the readers.

Two Other Fundamental Equations to Efficiently Compute the Jacobi: Besides equations 7.54 and 7.55, two additional equations are important for computing the Jacobi matrix, which we repeat here for convenience.

We refer to Figure 7.9 and assume that the R/T matrices for zones A to C are $\mathbf{T}_u^{AC}, \mathbf{R}_d^{AC}$,

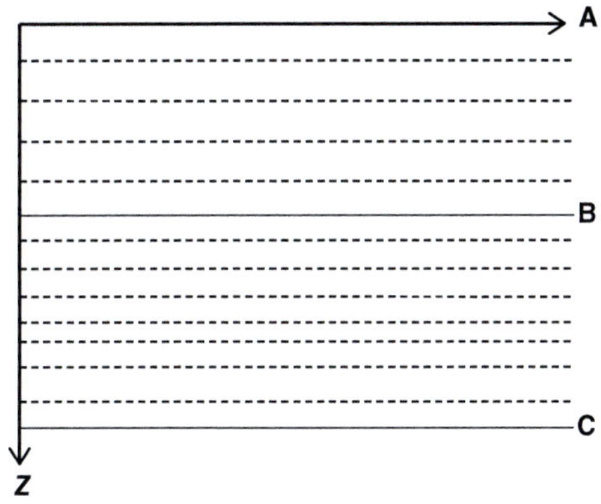

Figure 7.9 Computation of the reflection and transmission coefficients for zones A to B in terms of those for zones A to B and B to C.

$\mathbf{R}_u^{AC}, \mathbf{T}_d^{AC}$, for zones A to B are $\mathbf{T}_u^{AB}, \mathbf{R}_d^{AB}, \mathbf{R}_u^{AB}$, $\mathbf{T}_d^{AB}$, and for zone B to C are $\mathbf{T}_u^{BC}, \mathbf{R}_d^{BC}, \mathbf{R}_u^{BC}, \mathbf{T}_d^{BC}$. Then, using the iteration equations, we can express the R/T matrices for zones A to C in terms of those for zones A to B and B to C as

$$\left.\begin{aligned}
\mathbf{R}_d^{AC} &= \mathbf{R}_d^{AB} + \mathbf{T}_u^{AB}\mathbf{R}_d^{BC}(\mathbf{I} - \mathbf{R}_u^{AB}\mathbf{R}_d^{BC})^{-1}\mathbf{T}_d^{AB} \\
\mathbf{T}_d^{AC} &= \mathbf{T}_d^{BC}(\mathbf{I} - \mathbf{R}_u^{AB}\mathbf{R}_d^{BC})^{-1}\mathbf{T}_d^{AB} \\
\mathbf{R}_u^{AC} &= \mathbf{R}_u^{BC} + \mathbf{T}_d^{BC}\mathbf{R}_u^{AB}(\mathbf{I} - \mathbf{R}_d^{BC}\mathbf{R}_u^{AB})^{-1}\mathbf{T}_u^{BC} \\
\mathbf{T}_u^{AC} &= \mathbf{T}_u^{AB}(\mathbf{I} - \mathbf{R}_d^{BC}\mathbf{R}_u^{AB})^{-1}\mathbf{T}_u^{BC}
\end{aligned}\right\}. \quad (7.56)$$

Now, if $\mathbf{T}_u^{AC}, \mathbf{R}_d^{AC}, \mathbf{R}_u^{AC}, \mathbf{T}_d^{AC}$ and $\mathbf{T}_u^{AB}, \mathbf{R}_d^{AB}$, $\mathbf{R}_u^{AB}, \mathbf{T}_d^{AB}$, i.e., if the R/T matrices for zones A to C and those for zone A to B are given, we can easily invert equation 7.56 and obtain the R/T matrices for zones B to C as

$$\left.\begin{aligned}
\mathbf{R}_d^{BC} &= (\mathbf{R}_d^{AC} - \mathbf{R}_d^{AB})(\mathbf{T}_d^{AB})^{-1} \\
&\quad \times [\mathbf{T}_u^{AB} + (\mathbf{R}_d^{AC} - \mathbf{R}_d^{AB})(\mathbf{T}_d^{AB})^{-1}\mathbf{R}_u^{AB}]^{-1} \\
\mathbf{T}_d^{BC} &= \mathbf{T}_d^{AC}(\mathbf{T}_d^{AB})^{-1}(\mathbf{I} - \mathbf{R}_u^{AB}\mathbf{R}_d^{BC}) \\
\mathbf{R}_u^{BC} &= \mathbf{R}_u^{AC} - \mathbf{T}_d^{AB}\mathbf{R}_u^{AB}(\mathbf{I} - \mathbf{R}_d^{BC}\mathbf{R}_u^{AB})^{-1}\mathbf{T}_u^{BC} \\
\mathbf{T}_u^{BC} &= (\mathbf{I} - \mathbf{R}_d^{BC}\mathbf{R}_u^{AB})(\mathbf{T}_u^{AB})^{-1}\mathbf{T}_u^{AC}
\end{aligned}\right\}. \quad (7.57)$$

Again, both equations 7.56 and 7.57 have been shown in Chapter 5 as equations 5.98 and 5.99.

Equations 7.54, 7.55, 7.56, and 7.57 are the basic building blocks to efficiently compute the Jacobi matrix. As we build the R/T matrices

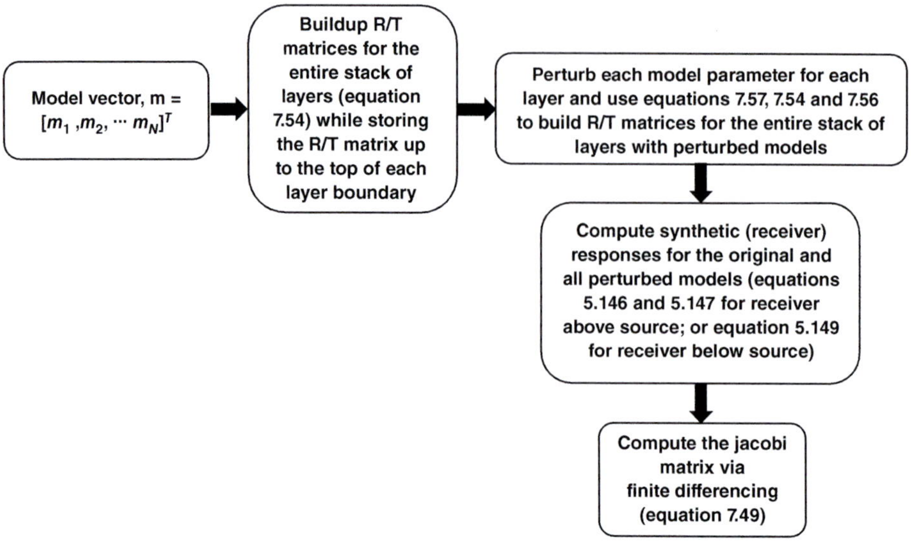

Figure 7.10 Flowchart for efficiently computing the Jacobi matrix. Adapted from Sen and Roy (2003).

from the top using the iteration equations (equation 7.55 or 5.105), we store the computed coefficients for up to each layer. So, for any given layer i, we have the following:

1) R/T matrices from top to layer $i-1$ (we consider these to be equivalent to $\mathbf{T}_u^{AB}, \mathbf{R}_d^{AB}, \mathbf{R}_u^{AB}$, $\mathbf{T}_d^{AB}$ of equations 7.56/5.98 and 7.57/5.99).
2) R/T matrices from top to layer i (we consider these to be equivalent to $\mathbf{T}_u^{AC}, \mathbf{R}_d^{AC}$, $\mathbf{R}_u^{AC}, \mathbf{T}_d^{AC}$ of equations 7.56/5.99 and 7.57/5.99).
3) R/T matrices for the entire stack of N layers. These are the R/T matrices for the entire model, and we refer them as $\mathbf{T}_u^{AN}, \mathbf{R}_d^{AN}$, $\mathbf{R}_u^{AN}, \mathbf{T}_d^{AN}$.

So, for any specific layer (i), with the stored R/T matrices $\mathbf{T}_u^{AB}, \mathbf{R}_d^{AB}, \mathbf{R}_u^{AB}, \mathbf{T}_d^{AB}$ up to layer $i-1$ and those up to the layer i ($\mathbf{T}_u^{AC}, \mathbf{R}_d^{AC}, \mathbf{R}_u^{AC}$, $\mathbf{T}_d^{AC}$), we can use equation 7.57/5.99 to compute the R/T matrices for that layer. Then by perturbing each model parameter for the specific layer, we can compute the new R/T matrices for the given layer using equation 7.54/5.102). Finally, using equation 7.56/5.98, we can build the R/T matrices for the entire stack of layers with layer i perturbed. Once the R/T matrices are built for the original and perturbed models, we can then compute the receiver wavefields for all models (see Chapter 5 for details) and compute the Jacobi by finite differencing (equation 7.49). Figure 7.10 shows the flowchart for computing the Jacobi matrix using this compute-efficient approach of Sen and Roy (2003).

The procedure above would require computing synthetic data only for the number of model parameters for each layer. Thus, referring to the isotropic elastic inversion over a 2 s window with 500 layers, where a pure finite differencing would need 1501 synthetic computations per iteration, this method would require only 3 computations per iteration. For a transversely isotropic medium it would require 6 synthetic computations (five elastic constants and density for each layer) per iteration, for an orthorhombic elastic medium it would require 10 synthetic computations (nine elastic constants and density for each layer) per iteration, and so on. Thus, the Sen and Roy (2003) procedure to compute the Jacobi matrix is compute-efficient.

Full 3D Inversion

For areas with simple to moderate geology, RAP processing with prestack time or depth migration allows applying inversion with local

1D assumption. There are, however, many areas with complex subsurface geology where such local 1D assumption is not possible. Some such examples include

1) Imaging and characterization in the vicinity of a salt dome.
2) Imaging and characterization below the basalts.
3) Mapping and characterizing a subducting plate boundary.

The subsurface cannot be approximated as locally 1D for such complex geology. Therefore, it is not possible to use 1D method for the forward modeling engine under such situations, and we must use a fully 3D forward modeling method in a 3D optimization scheme for inversion.

In geophysical literature, 3D seismic optimizations are generally referred as the FWI. These optimizations use a complete 3D wave equation solver for the forward modeling engine in the optimization. Note that the 1D problem, we discussed above, is also a subset of FWI because it also uses a wave-equation-based forward modeling engine where the subsurface is assumed horizontally stratified.

In Chapter 5, Section 5.4, we have discussed different variations of such 3D forward modeling method, both in time and frequency domains. As compared with the 1D modeling, these 3D modeling methods are much more computationally demanding. Additionally, as we discuss below, computing the Jacobi for such 3D FWI is an additional computational burden. For time-domain forward modeling using finite-difference or finite-element method, the computational burden for FWI is too high. Consequently, most of the FWI applications apply forward modeling in the frequency domain using discrete frequency bands (Pratt, 1999; Warner et al., 2013; among others). A very good overview of the different numerical approaches, their relative merits/demerits, and their implementations in high-performance computing environments is given by Virieux and Operto (2009) and Virieux et al. (2017). Fundamental steps for 3D FWI are the following:

1) *Forward Modeling*: Numerically solve the elastodynamic equation (equation 5.177 for a complete anisotropic system or its subset for other simpler elastic systems, such as acoustic, isotropic elastic, transversely isotropic, orthorhombic, etc.). Note that the elastodynamic equation is a partial differential equation and for a general 3D problem it must be numerically solved using the methods outlined in Chapter 5, Section 5.4.
2) *Computation of the Jacobi (Fréchet Derivative) Matrix*: 3D FWI methods require the Jacobi to be numerically computed. Computing Jacobi for 3D inverse problems is not as computationally efficient as they are for AVA or 1D waveform inverse problems. There are, however, some recent developments, which have made application of 3D FWI feasible, and below we briefly outline some of these methods.
3) *Optimization Method*: Once we have efficient ways for the forward modeling method and computing the Jacobi matrix, any of the optimization methods discussed above could be used to solve 3D FWI problems.

Efficient Computation of the Jacobi for 3D FWI Problems

As we discussed above, once we find an efficient way to compute the Jacobi matrix **J**, we can then use any of the methods such as the steepest descent, conjugate-gradient, Gauss–Newton, etc., in 3D FWI. We can always compute **J** by straightforward finite differencing (equation 7.49). For large 3D problems, however, computing **J** using this approach is not feasible. To approximate equation 7.49 as derivatives, unless Δm_i is small, it may lead truncation errors. And, smaller is the value of Δm_i, more challenging is to compute **J**. Therefore, for large-scale 3D inverse problems, computing **J** columnwise using equation 7.49 is difficult. Consequently, efficiently computing **J** for large-scale 3D FWI problems is a subject of active research.

In contrast to straightforward finite differencing by perturbing each model parameter m_i by Δm_i and using equation 7.49 to compute **J** columnwise, another approach, known as the *adjoint approach*, is a more computationally efficient way to compute the derivatives and therefore the Jacobi matrix **J**. Below, we provide a brief outline of the adjoint method, and further details about this method can be found in Chavent (2009) and Fichtner (2011).

Fundamental Concept of the Adjoint Approach In Chapter 2, Section 2.2.4, we discussed adjoint differential forms (equation 2.143), and in equation 2.150, we also provided the conditions for the differential operators $L(v)$ and $M(u)$ to be self-adjoint or the conditions when $L(u) = M(u)$. The adjoint approach for computing gradient is derived from these concepts. In the following, provide the outline of the general formulation of the adjoint approach.

Irrespective of the method used, our computed synthetic seismic data are functions of the position vector **r**, time t (or frequency ω), and the model vector **m**. In other words,

$$\mathbf{s} = \mathbf{s}(\mathbf{m}; \mathbf{r}, t). \tag{7.58}$$

In equation 7.58, the model vector belongs to the model/decision space and contains all possible parameters (V_P, V_S, ρ, etc.), and the semicolon (;) on the right-hand side of equation 7.58 indicates that while the computed synthetic data **s** evolves in space and time (or frequency), the model vector **m** remains fixed for a given realization of **s**. Here, we restrict to FWI and refer **s** to be the computed seismic wavefield response that is computed via wave equation methodology, previously discussed in Chapter 5. Following the terminologies, we previously used in defining the adjoint forms in Chapter 2, and those we used to define our 3D wave equation in Chapter 5, we symbolically refer this wave equation methodology as

$$\mathbf{L}(\mathbf{s}, \mathbf{m}) = \mathbf{f}, \tag{7.59}$$

where **L** is an operator, **f** is the source term, given as the external body-force density, and the density ρ and the elastic stiffness coefficients **C** (or the elastic stiffness tensor **c**) are embedded within the model vector **m** (see Chapter 5, equation 5.177, or the equivalent velocity–stress formulation, given by equation 5.179).

Generally, it is not **s**, but a scalar objective, which we may represent as $E[\mathbf{s}(\mathbf{m})]$, which we want to optimize by computing its gradient. Restricting to the optimization problems related to 3D FWI, $E[\mathbf{s}(\mathbf{m})]$, for example, is the misfit (sum of the squared errors). For simplicity, we ignore the regularization terms and write this objective as

$$\begin{aligned} E(\mathbf{m}) &= \int_T dt \int_{\mathfrak{R}^3} \{\mathbf{s}(\mathbf{m}; \mathbf{r}, t) - \mathbf{d}(\mathbf{r}, t)\}^2 d^3\mathbf{r} \\ &= \langle E(\mathbf{m}) \rangle. \end{aligned} \tag{7.60}$$

In equation 7.60, $\mathbf{d}(\mathbf{r}, t)$ represents the observed seismic data, T represents time (or frequency), and $\mathfrak{R}^3$ represents a volume in 3D Euclidean space. We also assume that this volume is enclosed by the surface $\partial\mathfrak{R}^3$. The symbol $\langle\cdot\rangle$ on the far right-hand side of equation 7.60 is the abbreviated notation for the integral over time (or frequency) and space.

For local optimization, we are interested in the derivative $\nabla_m E\delta\mathbf{m}$ of $E\{\mathbf{s}(\mathbf{m})\}$ with respect to the model vector **m** in the direction $\delta\mathbf{m}$, which can be written from equation 7.60 using the chain rule of derivatives as

$$\nabla_m E\delta\mathbf{m} = \nabla_s E\delta\mathbf{s} = \langle\nabla_s E\delta\mathbf{s}\rangle, \tag{7.61}$$

where $\delta\mathbf{s}$ is the derivative of the computed synthetic data vector with respect to the model parameters, that is

$$\delta\mathbf{s} = \nabla_m \mathbf{s}\delta\mathbf{m}. \tag{7.62}$$

Note that $\nabla_m\mathbf{s}$ on the right-hand side of equation 7.62 is the Jacobi, previously defined in a discretized matrix form by equation 7.11. A finite-differencing approach (equation 7.49) for computing $\boldsymbol{\delta}\mathbf{s}$ by perturbing the model parameters would provide the derivative $\nabla_m E\delta\mathbf{m}$ from equations 7.61 and 7.62, but it can be too expensive and often impractical, and we seek to find an alternative (and practical) way to compute $\nabla_m E\delta\mathbf{m}$ where we do not need to explicitly compute $\delta\mathbf{s}$.

Applying the chain rule of differentiation to equation to our symbolic representation to the wave equation (equation 7.59), we obtain

$$\nabla_m \mathbf{L}\delta\mathbf{m} + \nabla_s \mathbf{L}\delta\mathbf{s} = \mathbf{0}. \tag{7.63}$$

By multiplying equation 7.63 by an arbitrary vector, $\mathbf{s}^\dagger$, and then applying the integral $\langle\cdot\rangle$, we obtain from equation 7.63

$$\langle \mathbf{s}^\dagger \cdot \nabla_m \mathbf{L}\delta\mathbf{m}\rangle + \langle \mathbf{s}^\dagger \cdot \nabla_s \mathbf{L}\delta\mathbf{s}\rangle = 0. \tag{7.64}$$

We now add equations 7.61 and 7.64 and obtain

$$\nabla_m E\delta\mathbf{m} = \langle \nabla_s E\delta\mathbf{s}\rangle + \langle \mathbf{s}^\dagger \cdot \nabla_m \mathbf{L}\delta\mathbf{m}\rangle + \langle \mathbf{s}^\dagger \cdot \nabla_s \mathbf{L}\delta\mathbf{s}\rangle. \tag{7.65}$$

We now refer to the definition of the adjoint operators $\nabla_s E^\dagger$ and $\nabla_s \mathbf{L}^\dagger$, which, according to equation 2.143, can be defined as

$$\left.\begin{aligned} \langle \nabla_s E\delta\mathbf{s}\rangle &= \langle \delta\mathbf{s} \cdot \nabla_s E^\dagger\rangle \\ \langle \mathbf{s}^\dagger \cdot \nabla_s \mathbf{L}\delta\mathbf{s}\rangle &= \langle \delta\mathbf{s} \cdot \nabla_s \mathbf{L}^\dagger E^\dagger\rangle \end{aligned}\right\}, \tag{7.66}$$

for any $\delta\mathbf{s}$ and $\mathbf{s}^\dagger$. We then obtain from equation 7.65

$$\nabla_m E\delta\mathbf{m} = \langle \delta\mathbf{s} \cdot [\nabla_s E^\dagger + \nabla_s \mathbf{L}^\dagger \mathbf{s}^\dagger]\rangle + \langle \mathbf{s}^\dagger \cdot \nabla_m \mathbf{L}\delta\mathbf{m}\rangle. \tag{7.67}$$

Equation 7.67 shows that we can compute the derivative of the objective $\nabla_m E\delta\mathbf{m}$ without explicitly computing $\delta\mathbf{s}$ via finite differencing if we can determine the vector field $\mathbf{s}^\dagger$ that satisfies

$$\nabla_s \mathbf{L}^\dagger \mathbf{s}^\dagger = -\nabla_s E^\dagger. \tag{7.68}$$

And, from our discussions in Chapter 2, we immediately recognize that equation 7.68 is the adjoint of equation 7.59 with $-\nabla_s E^\dagger$ being the adjoint source. Once the solution $\mathbf{s}^\dagger$ of the adjoint equation (equation 7.68) is found, the derivative of the objective becomes

$$\nabla_m E\delta\mathbf{m} = \langle \mathbf{s}^\dagger \cdot \nabla_m \mathbf{L}\delta\mathbf{m}\rangle. \tag{7.69}$$

And we compute this derivative without explicitly computing $\delta\mathbf{s}$ by finite differencing. To do this, we, however, need to find the adjoint operator $\nabla_s \mathbf{L}^\dagger$ and a solution to the adjoint equation (equation 7.68).

The Fréchet (Jacobi) Kernels Before proceeding further, it is first important to introduce an important parameter, known as the *Jacobi kernels* or the *Fréchet kernels.* Our physical intuition for FWI-related optimization problems is based on the physical interpretation of the sensitivity or the Fréchet kernels, defined as the volume derivatives (or volumetric densities) of the Jacobi (Fréchet derivative) $\nabla_m E$, which is given as

$$K_m(\mathbf{r}) = \frac{\partial}{\partial V}\nabla_m E = \int_T \mathbf{s}^\dagger \cdot \nabla_m \mathbf{L} dt. \tag{7.70}$$

Using K_m defined by equation 7.70, we can therefore express equation 7.69 as

$$\nabla_m E\delta\mathbf{m} = \int_{\Re^3} K_m(\mathbf{r})\delta\mathbf{m}(\mathbf{r})d^3\mathbf{r}. \tag{7.71}$$

Note that $K_m(\mathbf{r})$ represent how the objective $E(\mathbf{m})$ are influenced by the changes by each component m of the model vector $\mathbf{m}$ at a given position in space. Consequently, analyzing $K_m(\mathbf{r})$ for different seismic waves and objectives for each model component m allows designing efficient FWI schemes.

From Continuous to Discrete Model Space In the above section, we restricted our analysis to continuous model space in the enclosed volume $\Re^3$. In practice, we must, however, discretize the model space as discrete entities to define the model vector as

$$\mathbf{m}(\mathbf{r}) = [m_1(\mathbf{r}), m_2(\mathbf{r}), \dots, m_N(\mathbf{r})]^T, \tag{7.72}$$

for each position $\mathbf{r}$ in space. Each $m_i(\mathbf{r})$ in equation 7.72 with $i = 1, 2, \dots, N$ may be represented by a linear combination of $J < \infty$ basis functions, say $b_j(\mathbf{r})$, as (Fichtner, 2011)

$$m_i(\mathbf{r}) = \sum_{j=1}^{J} \kappa_{ij} b_j(\mathbf{r}). \tag{7.73}$$

Thus, from equation 7.73, the model vector $\mathbf{m}$ and the objective $E(\mathbf{m})$ are completely determined by the coefficients or the model parameters κ_{ij}. Thus, to compute the gradient of the objective, we are primarily interested in computing the partial derivatives

$$\begin{aligned}\frac{\partial E}{\partial \kappa_{ij}} &= \lim_{\varepsilon\to 0}\frac{E(\dots,\kappa_{ij}+\varepsilon,\dots)-E(\dots,\kappa_{ij},\dots)}{\varepsilon}\\ &= \lim_{\varepsilon\to 0}\frac{E(\dots,m_i+\varepsilon b_j,\dots)-E(\dots,m_i,\dots)}{\varepsilon}\\ &= \nabla_{m_i}Eb_j = \int_{\mathfrak{R}^3} K_{m_i}(\mathbf{r})b_j(\mathbf{r})d^3\mathbf{r}. \qquad (7.74)\end{aligned}$$

Therefore, from equation 7.74, it follows that the gradient $\partial E/\partial \kappa_{ij}$ is given by the projection of K_{m_i} onto b_j,

Summary of the Adjoint Approach Before proceeding further, below are the key equations for the adjoint method:

1) The forward problem, given by equation 7.59.
2) The objective, given by equation 7.60.
3) The adjoint equation, given by equation 7.68.
4) Derivative of the objective, given by equation 7.71.
5) Fréchet (Jacobi) kernel, given by equation 7.70

Application to the Elastic Wave Equation Having summarized the fundamental concepts of the adjoint approach above, we now move onto the elastic wave equation and the derivation of the corresponding adjoint equations. We use the following notations previously introduced in Chapter 5:

Position vector → $\mathbf{r}$.
Displacement field → $\mathbf{u}(\mathbf{r})$.
Elastic stiffness tensor → $\mathbf{c}(\mathbf{r})$, with elastic stiff matrix given as $\mathbf{C}(\mathbf{r})$
Density → $\rho(\mathbf{r})$.
Stress tensor → $\boldsymbol{\tau}(\mathbf{r}, t)$.
External body-force density → $\mathbf{f}(\mathbf{r}, t)$.

Closed volume in 3D space → $\mathfrak{R}^3$ with the enclosing surface defined as $\partial\mathfrak{R}^3$.

Using the above notations, we may then write the elastodynamic equation (equation 5.177) as

$$\rho(\mathbf{r})\ddot{\mathbf{u}}(\mathbf{r},t) - \nabla\cdot\boldsymbol{\tau}(\mathbf{r},t) = \mathbf{f}(\mathbf{r},t). \qquad (7.75)$$

Additionally, the stress tensor $\boldsymbol{\tau}$ and the gradient of the displacement field $\nabla\mathbf{u}$ are related via the constitutive relation (see equation 5.2) as

$$\boldsymbol{\tau}(r,t) = \int_{\eta=t_0}^{t} \dot{\mathbf{c}}(\mathbf{r}, t-\eta) : \nabla\mathbf{u}(\mathbf{r},\eta)d\eta. \qquad (7.76)$$

Note that the frequency domain representation of the constitutive relation, previously derived in Chapter 5 (equation 5.2), is represented as a convolution in time domain by equation 7.76. We combine equations 7.75 and 7.76 and write our elastodynamic wave equation as

$$\mathbf{L}(\mathbf{u},\rho,\mathbf{c}) = \mathbf{f}, \qquad (7.77)$$

where $\mathbf{L}$ can be considered as the elastic wave operator given as

$$\mathbf{L}(\mathbf{u},\rho,\mathbf{c}) = \rho(\mathbf{r})\ddot{\mathbf{u}}(\mathbf{r},t) - \nabla\cdot\int_{\eta=t_0}^{t} \dot{\mathbf{c}}(\mathbf{r},t-\eta) : \nabla\mathbf{u}(\mathbf{r},\eta)d\eta. \qquad (7.78)$$

The operator $\mathbf{L}$ defined by equation 7.78 is also accompanied by the initial and boundary conditions and given as

$$\left.\begin{aligned}\mathbf{u}|_{t\le t_0} = \dot{\mathbf{u}}|_{t\le t_0} = \mathbf{0}\\ \hat{\mathbf{n}}\cdot\boldsymbol{\tau}|_{\mathbf{r}\in\partial\mathfrak{R}^3} = \mathbf{0}\end{aligned}\right\}. \qquad (7.79)$$

In equation 7.79 $\partial\mathfrak{R}^3$ is the external bounding surface to the closed volume $\mathfrak{R}^3$ under consideration and $\hat{\mathbf{n}}$ is the outward unit normal vector to this surface. The first equation in equation 7.79 is the initial condition of causality of the propagating seismic waves, implying that there is no wavefield prior to a certain arrival time $t = t_0$. The second equation of 7.84 is the traction-free (free surface) boundary condition at the bounding surface $\partial\mathfrak{R}^3$ of the 3D volume. We have previously discussed the free surface boundary condition in Chapter 5 (see equations 5.108–5.114).

Note that the displacement field $\mathbf{u}$ in the elastic wave equation is equivalent to the synthetic data **s**, which we previously used in developing the fundamental concepts of the adjoint approach. To compute the derivative of the objective E, defined for the elastic wave equation with respect to the model parameters (the density ρ and the elastic stiffness tensor **c**), we must, therefore, find the adjoint operator $\nabla_u \mathbf{L}^\dagger$ of $\nabla_u \mathbf{L}$. From the second equation of equation 7.66, we thus write

$$\langle \mathbf{u} \cdot \nabla_u \mathbf{L}^\dagger \mathbf{u}^\dagger \rangle = \langle \mathbf{u}^\dagger \cdot \nabla_u \mathbf{L} \mathbf{u} \rangle. \tag{7.80}$$

For the elastic wave equation (equation 7.77) with the operator **L** defined by equation 7.79, $\langle \mathbf{u}^\dagger \cdot \nabla_u \mathbf{L} \mathbf{u} \rangle$ can be written explicitly as

$$\begin{aligned}\langle \mathbf{u}^\dagger \cdot \nabla_u \mathbf{L}\mathbf{u} \rangle &= \int_T dt \int_{\Re^3} \mathbf{u}^\dagger \cdot \nabla_u \mathbf{L}\mathbf{u}\, d^3\mathbf{r} \\ &= \int_T dt \int_{\Re^3} \rho \mathbf{u}^\dagger \cdot \ddot{\mathbf{u}}\, d^3\mathbf{r} - \int_T dt \int_{\Re^3} \mathbf{u}^\dagger \\ &\quad \cdot \left\{ \nabla \cdot \int_{\eta=t_0}^{t} \dot{\mathbf{c}}(t-\eta) : \nabla \mathbf{u}(\eta)\, d\eta\, d^3\mathbf{r} \right\}.\end{aligned} \tag{7.81}$$

Integrating the first term on the right-hand side of equation 7.81 by parts, we obtain

$$\begin{aligned}\int_T dt \int_{\Re^3} \rho \mathbf{u}^\dagger \cdot \ddot{\mathbf{u}}\, d^3\mathbf{r} &= \int_T dt \int_{\Re^3} \rho \mathbf{u} \cdot \ddot{\mathbf{u}}^\dagger d^3\mathbf{r} \\ &\quad + \int_{\Re^3} \rho \dot{\mathbf{u}} \cdot \mathbf{u}^\dagger d^3\mathbf{r}|_{t=t_1} - \int_{\Re^3} \rho \mathbf{u} \cdot \dot{\mathbf{u}}^\dagger d^3\mathbf{r}|_{t=t_1}.\end{aligned} \tag{7.82}$$

In equation 7.82, the initial conditions (equation 7.79) are already imposed. We now impose the terminal conditions,

$$\mathbf{u}^\dagger|_{t\geq t_1} = \dot{\mathbf{u}}^\dagger|_{t\geq t_1} = \mathbf{0}, \tag{7.83}$$

for $t_1 > t_0$. This terminal condition restricts the seismic wavefield to be within the time interval $t_0 < t < t_1$. This condition leads equation 7.82 to

$$\langle \mathbf{u}^\dagger \cdot \rho \ddot{\mathbf{u}} \rangle = \langle \mathbf{u} \cdot \rho \ddot{\mathbf{u}}^\dagger \rangle. \tag{7.84}$$

Clearly, equation 7.84 provides the first ingredient of the adjoint operator for the elastodynamic wave equation, and we now turn to the second term on the right-hand side of equation 7.81. We use the symbol Υ for $\langle \mathbf{u}^\dagger \cdot (\nabla . \boldsymbol{\tau}) \rangle$, i.e., we define

$$\begin{aligned}\Upsilon = \langle \mathbf{u}^\dagger \cdot (\nabla \cdot \boldsymbol{\tau}) \rangle &= \int_T dt \int_{\Re^3} \mathbf{u}^\dagger \\ &\quad \cdot \left\{ \nabla \cdot \int_{\eta=t_0}^{t} \dot{\mathbf{c}}(t-\eta) : \nabla \mathbf{u}(\eta) d\eta \right\} d^3\mathbf{r}.\end{aligned} \tag{7.85}$$

Now, recalling the symmetries of the stiffness tensor $c_{ijkl} = c_{klij} = c_{jikl} = c_{ijlk} = c_{jilk} = \cdots$, it is straightforward to show that

$$\begin{aligned}\mathbf{u}^\dagger \cdot \{\nabla \cdot (\dot{\mathbf{c}} : \nabla \mathbf{u})\} &= \nabla \cdot (\mathbf{u}^\dagger \cdot \dot{\mathbf{c}} : \nabla \mathbf{u}) \\ &\quad - \nabla \cdot (\mathbf{u} \cdot \dot{\mathbf{c}} : \nabla \mathbf{u}^\dagger) + \mathbf{u} \cdot \{\nabla \cdot (\dot{\mathbf{c}} : \nabla \mathbf{u}^\dagger)\}.\end{aligned} \tag{7.86}$$

Using equation 7.86 and Gauss' divergence theorem (equation 2.286), we can write equation 7.85 as

$$\begin{aligned}\Upsilon &= \int_{\partial\Re^3} d^2\mathbf{r} \int_{t=t_0}^{t_1} \mathbf{u}^\dagger(t) \cdot \left\{ \int_{\eta=t_0}^{t} \dot{\mathbf{c}}(t-\eta) : \nabla \mathbf{u}(\eta) d\eta \right\} \cdot \hat{\mathbf{n}} dt \\ &\quad - \int_{\partial\Re^3} d^2\mathbf{r} \int_{\eta=t_0}^{t_1} \mathbf{u}(\eta) \cdot \left\{ \int_{t=\eta}^{t_1} \dot{\mathbf{c}}(t-\eta) : \nabla \mathbf{u}^\dagger(t) dt \right\} \cdot \hat{\mathbf{n}} d\eta \\ &\quad + \int_{\Re^3} d^3\mathbf{r} \int_{\eta=t_0}^{t_1} \mathbf{u}(\eta) \cdot \left\{ \nabla \cdot \int_{t=\eta}^{t_1} \dot{\mathbf{c}}(t-\eta) : \mathbf{u}^\dagger(t) dt \right\} d\eta.\end{aligned} \tag{7.87}$$

Now, the boundary condition (second equation of equation 7.79) implies that the first term on the right-hand side of equation 7.87 to vanish because the expression within the curly brackets is the stress tensor, which must vanish at the bounding surface $\partial\Re^3$. Because $\mathbf{u}^\dagger$ is constrained by the terminal condition (equation 7.83), imposing the condition

$$\hat{\mathbf{n}} \cdot \boldsymbol{\tau}^\dagger|_{\mathbf{r}\in\partial\Re^3} = \mathbf{0} \tag{7.88}$$

forces the second term on the right-hand side of equation 7.87 to also vanish with the adjoint stress tensor $\boldsymbol{\tau}^\dagger$ being defined as

$$\boldsymbol{\tau}^\dagger = \int_{\eta=t}^{t_1} \dot{\mathbf{c}}(\eta - t) : \nabla \mathbf{u}^\dagger(\eta) d\eta. \tag{7.89}$$

Finally, combining equations 7.81, 7.84, 7.87–7.89 gives

$$\begin{aligned}\langle \mathbf{u} \cdot \nabla_u \mathbf{L}^\dagger \mathbf{u}^\dagger \rangle &= \langle \mathbf{u} \cdot \rho \ddot{\mathbf{u}}^\dagger \rangle - \Upsilon \\ &= \langle \mathbf{u} \cdot (\rho \ddot{\mathbf{u}}^\dagger - \nabla \cdot \boldsymbol{\tau}^\dagger) \rangle.\end{aligned} \tag{7.90}$$

Thus, the adjoint operator $\nabla_u \mathbf{L}^\dagger$ is given by

$$\nabla_u \mathbf{L}^\dagger \mathbf{u}^\dagger = \rho \ddot{\mathbf{u}}^\dagger - \nabla \cdot \boldsymbol{\tau}^\dagger. \tag{7.91}$$

Consequently, from equation 7.77, to compute the derivative of the objective $\nabla_m E$, we just need to solve the adjoint equation, which is given as

$$\rho \ddot{\mathbf{u}}^\dagger - \nabla \cdot \boldsymbol{\tau}^\dagger = -\nabla_u E^\dagger, \tag{7.92}$$

along with the terminal and boundary conditions given by equations 7.83 and 7.88.

Note that the adjoint equation (equation 7.92) is also a wave equation, and therefore, it can be numerically solved using the same forward modeling methods outlined in Chapter 5. Additionally, for perfectly elastic media, the operator $\mathbf{L}$ is self-adjoint, i.e., $\mathbf{L} = \mathbf{L}^\dagger$ and can be solved along with the solution of the forward problem. Once the adjoint equation is solved, computing the derivatives without finite differencing is straightforward.

Primary numerical issue with solving the adjoint equation comes from the terminal conditions (7.83 and 7.88), which require the adjoint fields to vanish at the end of the observation (i.e., at $t = t_1$). For computer implementations, this condition can be met only by solving the adjoint equation backward in time or by reversing the time axis from t_0 to t_1 to t_1 to t_0. By doing so, these terminal conditions act as zero initial conditions for numerical simulations. We must note here that such time reversal is common in many applications, including the reverse-time migration (for example, see Lamat et al., 2006), most of which are closely related to the adjoint method we discussed here. Further details, including the numerical implementation aspects of the adjoint method, can be found in Chavent (2009) and Fichtner (2011).

7.6 Examples

There are many successful applications of local optimization methods to solve seismic inverse problems, ranging from inversion of poststack seismic data to prestack inversion using both AVA and full wave equation methods. Below, we show a few selected examples from this wide range.

7.6.1 Poststack Inversion

Poststack inversion methods are one of the oldest seismic inversion methods. With the introduction of high-performance computers, although these poststack methods are being superseded by prestack seismic inversions, they are still widely used. Primary assumption for poststack inversion is that the amplitudes in the processed stacked seismic data are proportional to zero-offset (normal-incidence) reflection coefficients, given by equation 1.3, from which contrasts or high-frequency variations in acoustic impedance (velocity times density) can be extracted. Using these high-frequency variations, when added to the low-frequency background acoustic impedance, we can estimate the broadband absolute acoustic impedance. A very good account of the theory of poststack seismic inversion methods can be found in Oldenburg et al. (1983). Additionally, practical aspects of implementing these methods to invert noisy seismic data have been addressed by Sandø et al. (1990).

There are many applications of poststack inversion. In Figure 7.11, we show one such example of poststack acoustic impedance inversion from Morozov and Ma (2009) with the stacked seismic data on the left and the inverted acoustic impedance on the right. Note that a low-impedance gas sand and a high-impedance carbonate formation underneath the gas sand could be clearly identified from the inversion.

For poststack inversion, the physics of the forward problem is represented by the convolution of the time-domain seismic wavelet

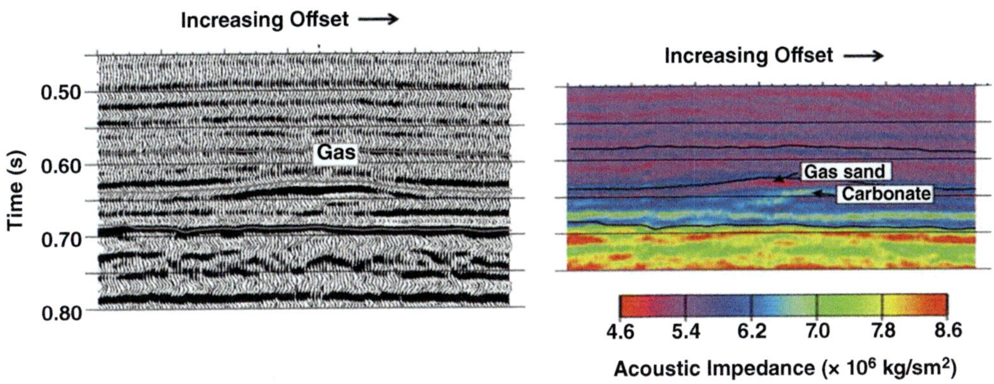

Figure 7.11 Poststack seismic example from Morozov and Ma (2009). On the left is the stacked seismic data in which a gas-bearing sand formation is marked. On the right is the acoustic impedance section from poststack inversion. The low-impedance gas-bearing sand and the high-impedance carbonate formation immediately below the sand are marked. Note that while the stacked data on the left are shown for 0.45–0.80 s time window, the inverted impedance section is shown for 0.5–0.75 s window. Morozov and Ma (2009)/Society of Exploration Geophysicists.

with normal-incidence reflection coefficients (equation 1.3), for which the Jacobi can be analytically computed. Therefore, implementing any of local optimization methods we discussed in this chapter is straightforward.

7.6.2 Prestack Inversion

Poststack seismic inversion methods are still popular, primarily because they are computationally efficient, robust, and easy to use. We must note, however, that these poststack methods provide a single piece of information, namely the acoustic impedance. With the advancements of computational technology, using prestack seismic data for a detailed analysis of subsurface lithological and fluid properties slowly became popular from the mid-1980s. This, in turn, led to the developments of prestack seismic inversion methods.

Prestack inversion methods can be broadly classified into two categories: (1) AVA inversion methods and (2) FWI methods. Below, we show examples from both these methods.

Amplitude-Variation-With-Angle (AVA)-Based Inversion

AVA inversion methods assume that the reflection amplitudes on prestack seismic data after careful RAP processing (Resnick, 1993) are represented by the convolution of the angle-dependent reflection coefficients with the seismic wavelet. Therefore, if the time and offset-domain seismic data are converted to time and incidence-angle domain (Todd and Backus, 1985; Resnick, 1993; Mukhopadhyay and Mallick, 2011), their amplitudes are proportional to the angle-dependent reflection coefficients, convolved with the seismic wavelet.

Restricting to P-wave reflections only, the linearized P–P reflection coefficient formula R_{PP} is previously given as (see equation 1.7)

$$R_{PP} \approx A + B\sin^2\theta + C\sin^2\theta\tan^2\theta. \tag{7.93}$$

where θ is the average incidence angle. The coefficients A, B, and C in equation 7.93 are respectively known as the AVA intercept, gradient, and curvature. For isotropic elastic medium, A, B, and C are functions of the P- and S-wave velocities (V_P, V_S) and density (ρ) via equations 1.8–1.16. In the presence of anisotropy, they are also functions of the other anisotropic parameters besides V_P, V_S, and ρ (Rüger, 1998). The primary concept behind AVA inversion is to cast the problem as $\mathbf{d} = \mathbf{Gm}$ with $\mathbf{d}$ being the observed data and **G** and **m** defined by equations 7.50 and 7.51 in a local optimization scheme to estimate A, B, and C. As we have seen before in equation 7.52, the Jacobi matrix for such a problem can be analytically computed, which leads to a computationally efficient optimization method. From the estimates of A, B, and C, the high-frequency variations ΔV_P, ΔV_S, and $\Delta\rho/\rho$ (and other anisotropic parameters in the presence of anisotropy) can be estimated. A very good account on the background theory and practical implementation of AVA inversion can be found in Hampson (1991).

Primary problem in AVA inversion using a linearized R_{PP} (equation 7.93) is shown in Figure 7.12. While the second term ($\sin^2\theta$) of the AVA equation 7.93 is well behaved (Figure 7.12a), the third term ($\sin^2\theta\ \tan^2\theta$) is not so well behaved as it rapidly goes to infinity as $\theta \rightarrow 90°$ (Figure 7.12b). Because $\sin^2\theta \tan^2\theta$ is nearly negligible for $\theta < 25°$ (see Figure 7.12c), many past applications of AVA inversion restricted to two terms by ignoring the third term on the right-hand side of equation 7.93 and by restricting the angle domain data with θ less than $25-30°$. These two-term AVA methods are robust and provide estimates of the AVA intercept A and the AVA gradient B only. And, from equations 1.8 and 1.9, we immediately see that the estimates of A and B from such a two-term AVA inversion in conjunction with the knowledge of the background S-to-P-wave velocity ratio (V_S/V_P) would provide P- and S-wave impedances $I_P = \rho V_P$ and $I_S = \rho V_S$.

An example of two-term AVA inversion is shown in Figures 7.13 and 7.14. Figure 7.13 shows a stacked seismic section from the deepwater Andaman Sea, India. The reflection indicated with arrows is the bottom-simulating reflector (BSR) marking the base of the gas-hydrate stability zone. Inverted P- and S-wave impedances are shown respectively in Figure 7.14a,b. Note that the BSR is clearly visible in the P-impedance section. On the S-impedance section, on the other hand, BSR is not so clearly visible. Figure 7.14c shows the Poisson's ratio for the zone within the black box shown in Figure 7.14a,b. Note that from the P- and S-wave impedances I_P and I_S, the

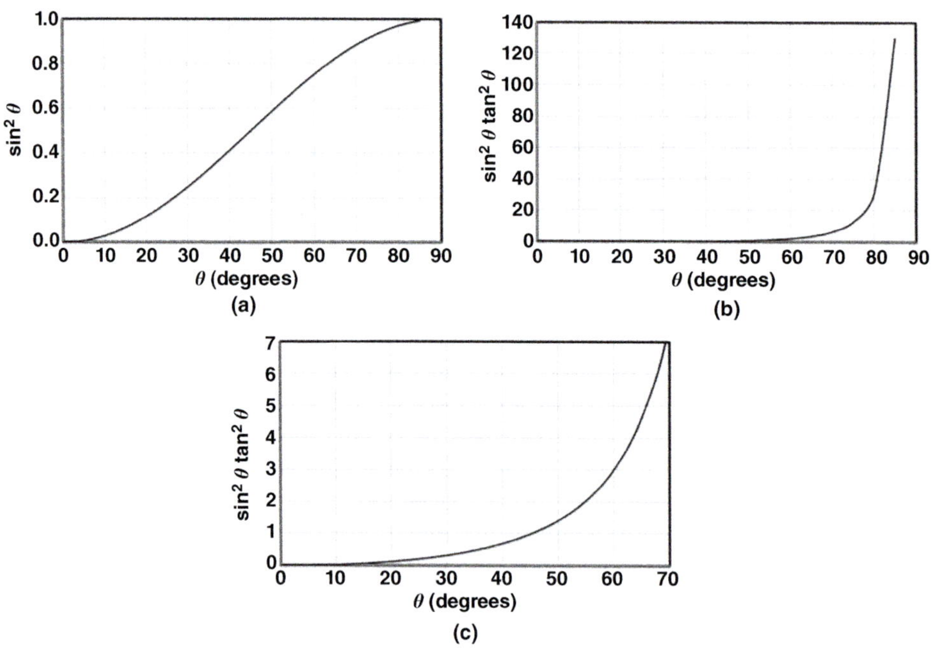

Figure 7.12 Numerical analysis of different terms in the AVA inverse problem using the linearized P–P reflection coefficient formula (equation 7.93). (a) $\sin^2\theta$ and (b) $\sin^2\theta\tan^2\theta$ terms as functions of θ. (c) Same as (b) with maximum angle θ restricted to 70°.

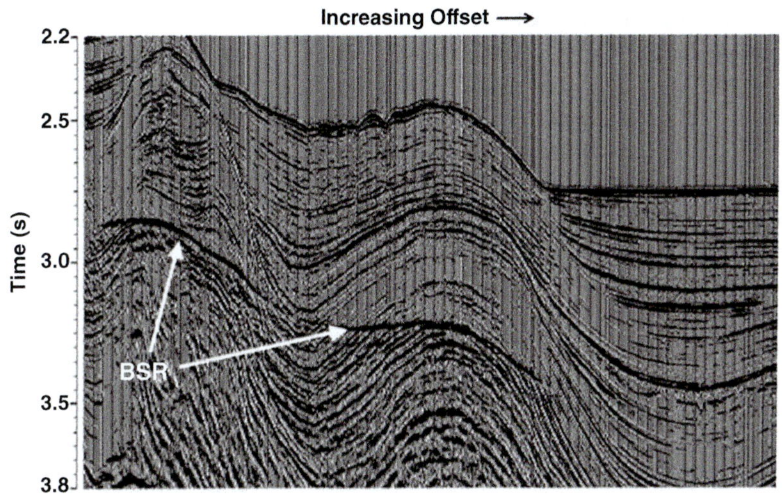

Figure 7.13 Stacked seismic section from the deepwater west Andaman Sea, India, showing a bottom-simulating reflector (BSR) indicated by arrows. Reproduced from Mallick et al. (2000)/Society of Exploration Geophysicists.

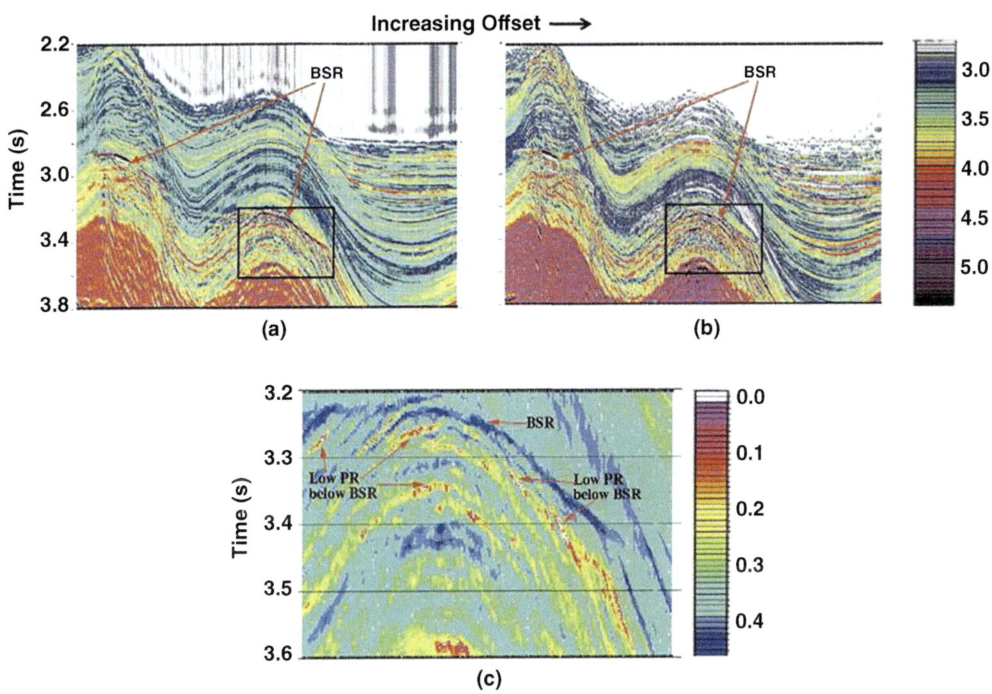

Figure 7.14 Two-term AVA and hybrid inversion result of the Andaman Sea seismic data. (a, b) P- and S-wave impedances (× 10^6 kg/sm^2), and (c) Poisson's ratio for the portion within the box in (a) and (b). The impedance scale on top right is for P wave and that for the S-wave is half of those values. Reproduced from Mallick (2001)/Society of Exploration Geophysicists.

S-to-P-wave velocity ratio γ can be computed as $\gamma = I_S/I_P$. And the Poisson's ratio ν is then given as (Mallick et al., 2000)

$$\nu = \frac{1-2\gamma^2}{2(1-\gamma^2)}. \tag{7.94}$$

Poisson's ratio is known to be a useful lithology discriminator. Note, from Figure 7.14c, that the BSR is indicated by a Poisson's ratio of about 0.4, which is consistent with independent laboratory measurements of the properties of gas hydrates (Sloan, 1998).

Additionally, the pockets of low Poisson's ratio (<0.15) below the BSR are indicative of free gas below the BSR. The two-term AVA results shown in Figure 7.14 were called as hybrid inversion results by Mallick et al. (2000) and Mallick (2001). AVA provides the high-frequency contrasts in the layer properties, and to obtain the absolute values of these properties, their low-frequency background trends must be obtained from other sources. Because Mallick et al. (2000) obtained these background properties from a global FWI at sparse locations, the results shown in Figure 7.14 were called hybrid inversion results. However, they were essentially obtained from a two-term AVA inversion.

A two-term AVA inversion is superior to poststack inversion as it provides I_S in addition to I_S, from which useful properties like Poisson's ratio can be estimated. It is, however, desirable to find a way to obtain individual estimates of V_P, V_S, and ρ instead of I_P and I_S. Using Gassmann fluid substitution methods (Smith et al., 2003), the following properties can be readily verified regarding the sensitivity of V_P, V_S, and ρ to the presence of fluids in the subsurface rocks:

1) V_P is not as sensitive to the presence of fluids in the rocks as ρ.
   a) In the presence of small amount of gas (<10%) in the rocks, V_P drops drastically and stays nearly constant with increasing gas saturation.
   b) The drop in density ρ, on the other hand, is more gradual with increasing gas saturation.
2) The S-wave velocity V_S is not sensitive to the presence of fluids in the rock pores because they do not propagate in fluids.

In view of the above discussion, having the ability to estimate V_P, V_S, and ρ instead of I_P and I_S is useful in identifying subsurface fluids, which, in turn, is useful for both global and exploration seismology.

Recent advances of AVA methods (Kelly et al., 2001; Skidmore et al., 2001) in conjunction with the concepts of elastic impedance (Connolly, 1999) and extended elastic impedance (Whitcombe et al., 2002) have allowed applications of three-term AVA inversions feasible. It is, however, important to note that an estimation of density from such inversion must be treated with caution. Using synthetic data with various levels of noise, Mallick (2001) demonstrated that a two-term AVA is more robust than a three-term AVA. Additionally, Mallick (2001) also showed that even for a three-term AVA inversion, the uncertainty in the estimation of I_P and I_S is much less than the corresponding uncertainties in the estimation of V_P, V_S, and ρ. For inverting 3D seismic data from the Sleipner Field, North Sea, Buland and Omre (2003) also made the same conclusion.

Figures 7.15 and 7.16 show an example of estimating V_P, V_S, and ρ from a three-term AVA inversion. Figure 7.15 shows the stacked seismic section for one inline, and Figure 7.16 shows the estimated V_P, V_S, and ρ.

Like poststack inversion, there are a large number of successful applications of both two- and three-term AVA inversions, which are too many to list. Therefore, here we just show a few examples.

Full Waveform Inversion (FWI)

Like poststack and AVA, successful FWI applications using local optimization methods are too many to list, and here we restrict to a few of them.

Figures 7.17 and 7.18 are FWI examples in which the subsurface is assumed to be locally 1D. Figure 7.17 shows an acoustic example from Padhi et al. (2015) for estimating the oceanwater sound speed. Figure 7.18, on the other hand, is an isotropic elastic example at a single location, obtained by Mukhopadhyay and Mallick (2016).

The acoustic example of Figure 7.17 is from real seismic data from the South-China Sea. From a visual comparison between the initial and inverted sound-speed models shown, it may appear that they are similar. There are, however, subtle differences. One quality check in any model-based inversion method

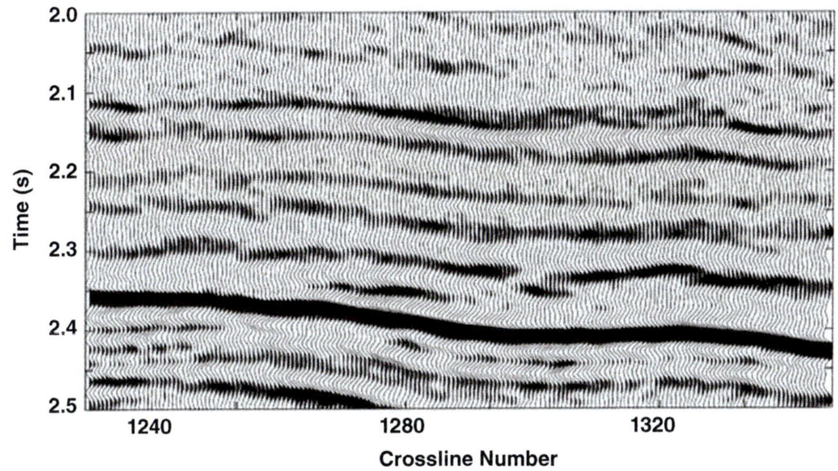

Figure 7.15 Stacked seismic data for a single inline from the 3D seismic data from the Sleipner Field, North Sea. The strong reflection event ranging from 2.35 s on the left to approximately 2.45 s on the right is interpreted as the top Shetland horizon (base reservoir). Reproduced from Buland and Omre (2003)/Society of Exploration Geophysicists.

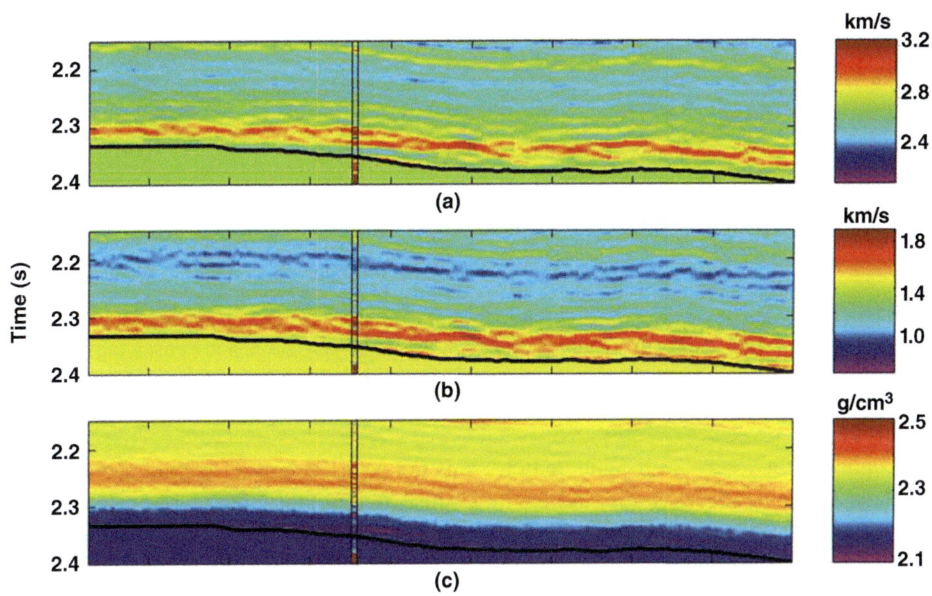

Figure 7.16 Result of a three-term AVA inversion of the data shown in Figure 7.15. (a) V_P, (b) V_S, and (c) ρ. Buland and Omre (2003)/Society of Exploration Geophysicists.

is to compare the observed data with synthetic data predicted by the inverted model. Figure 7.19 compares the synthetic stacked data computed using the initial and inverted models (Figure 7.17) with the real (observed) stacked data. The synthetic stacked data, shown in Figure 7.19a,b, were computed using 1D wave equation synthetic seismograms for the same range of source–receiver offsets as they were in the real data, followed by geometrical spreading compensation, normal moveout correction, and stacking. Note that the synthetic stacks generated from the initial model (Figure 7.19a) do not match the real

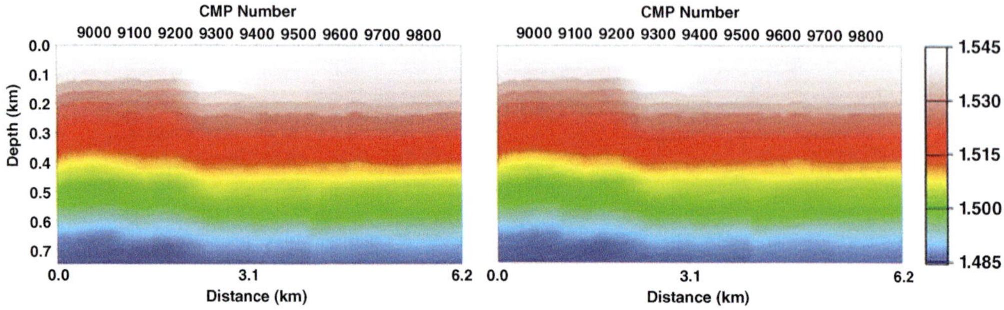

Figure 7.17 Acoustic FWI example with locally 1D assumption for estimating the oceanwater sound speed. The initial model is on the left and the inverted model is on the right. The color scale for the sound speed (in km/s) is on the far right. Adapted from Padhi et al. (2015).

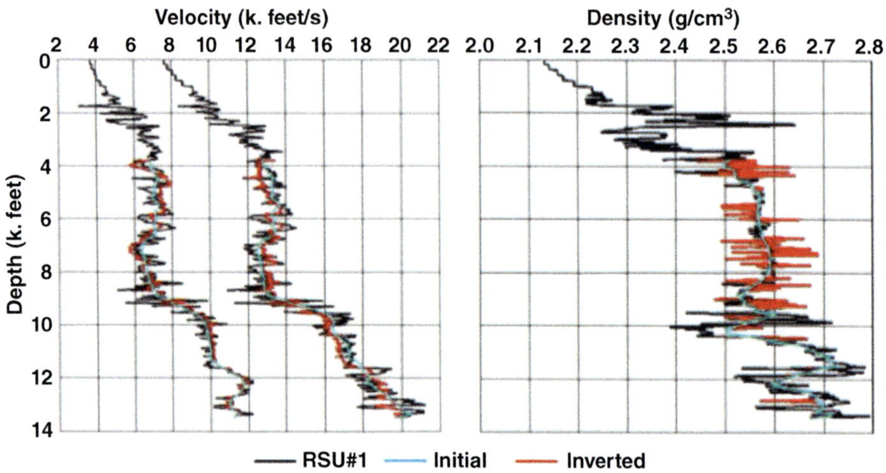

Figure 7.18 Isotropic elastic FWI example with locally 1D assumption for real seismic data at a single location. The location coincided with a well location with P- and S-wave sonic and density logs. The well data are shown in black; the initial model is shown in cyan; and the inverted model is shown in red. Mukhopadhyay and Mallick (2016)/Society of Exploration Geophysicists.

stacks (Figure 7.19c) very well. The match between the synthetic stacks computed from the inverted model (Figure 7.19b) and the real stacked data (Figure 7.19c) on the other hand is very good.

The elastic inversion result (Figure 7.18) is from the RSU at the location where the seismic data coincided with the well data. Note that V_P and V_S estimated from inversion, shown in the left panel of Figure 7.18, match with the well data very well. The match between the inverted and the measured density from the well log (right panel of Figure 7.18) is good for depths below 8000 ft. But in 4000–8000 ft depth interval, the match is not so good. From regional geology of RSU, this interval is known to be the Baxter Shale formation (Pafeng et al., 2017) with a uniform density of about 2.56 g/cm^3. The inverted density for this interval, on the other hand, shows wild fluctuations about this mean value. We revisit this issue later in Chapter 8 when we discuss global inversion methods. Figure 7.20 shows the comparison between the real prestack seismic data with the predicted (synthetic) data computed using the inverted model in intercept-time/ray-parameter ($\tau - p$) domain, showing that the inversion satisfactorily captured major features of the observed seismic data.

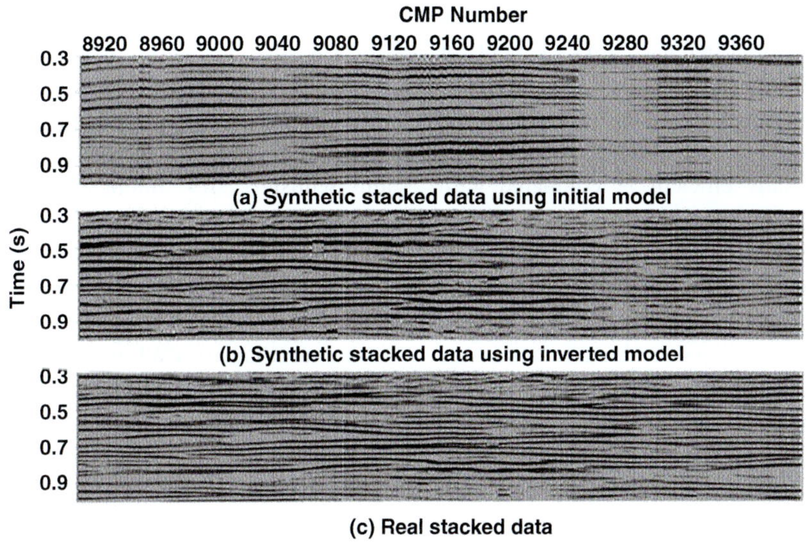

Figure 7.19 Comparison of real stacked data with the synthetic stacked data computed using the initial and inverted models shown in Figure 7.17. Reproduced from Padhi et al. (2015)/Oxford University Press.

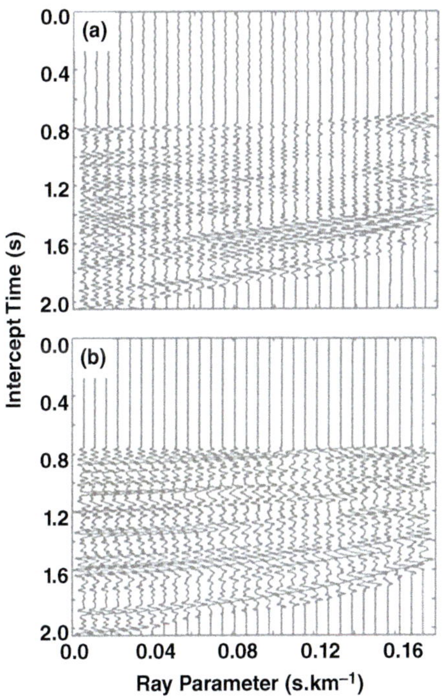

Figure 7.20 Comparison between (a) observed seismic data and (b) synthetic seismic data, computed from the inverted model shown in Figure 7.18. Reproduced from Mukhopadhyay and Mallick (2016)/Society of Exploration Geophysicists.

The 1D FWI methods are generally applied in the plane-wave $(\tau - p)$ domain. From our discussions of 1D wave equation modeling method in Chapter 5, it can be readily seen that computing the response for each ray parameter (p) is independent of one another, implying that the $\tau - p$ domain synthetic computation can be parallelized in high-performance parallel computing platforms. Additionally, the Sen and Roy (2003) method, as discussed above, allows efficient computation of the gradient. Consequently, 1D FWI methods are computationally efficient and allow estimation of the subsurface model properties at a very high resolution, suitable for reservoir characterization applications. Although these 1D methods are subset of FWI, to distinguish them from the other FWI methods where the subsurface model properties are allowed to vary both laterally and vertically, these 1D methods are sometimes referred as prestack waveform inversion or PWI (Mallick and Adhikari, 2015; Pafeng et al., 2017).

Complete FWI methods that assume laterally and vertically heterogeneous subsurface are much more computationally challenging than PWI because they require numerical implementation of the forward modeling using the finite-difference, finite-element, or similar methods we discussed in Chapter 5. The adjoint approach allows efficient gradient computation, but still they are not as computationally efficient as compared with PWI. Consequently, subsurface resolutions of the estimated models from FWI are generally lower than that of PWI. Most successful FWI application uses computation of the forward responses in the frequency domain for discrete frequency bands. Pratt et al. (1998) and Pratt (1999) formulated the fundamentals of such frequency domain acoustic FWI approach and demonstrated that the method can resolve the subsurface with much finer details than the travel-time-based tomographic methods.

Initial FWI applications were limited to obtaining a reliable subsurface model for depth imaging (for example, Vigh and Starr, 2008; Warner et al., 2013). Over the past decade, there are, however, many successful FWI applications. Describing all these new developments is beyond the scope of this book. A good overview of FWI can be found in Virieux and Operto (2009) and Virieux et al. (2017). Below, we give a few selected FWI examples. In geophysical literature, there are many variations of FWI, such as interferometric full waveform inversion, reflection waveform inversion, tomographic full waveform inversion, etc. (Biondi and Almomin; 2014; Guo and Alkhalifah, 2017; Ren et al., 2019; Sinha and Schuster, 2019; among others). Because all these methods use local, gradient-based methods for optimization, they are subject to convergence to a local optimum if an incorrect model is used as the initial model. These different variations of FWI are different ways to choose a good initial model, and here we refer them all as FWI.

Our first example is a synthetic example using the Marmousi model. This model is a two-dimensional (2D) model, based on a complex subsurface structure from the North Sea. The original model was an acoustic model, generated by the Institut Français du Pétrole (Brougois et al., 1990), which was later extended to an elastic Marmousi-2 model by the University of Houston (Martin et al., 2006). Figure 7.21 shows the acoustic FWI result for the Marmousi model, obtained by Sinha and Schuster (2019). Note that starting from a smooth initial model, FWI satisfactorily recovered the velocity structure of the true model. FWI application to the elastic Marmousi model

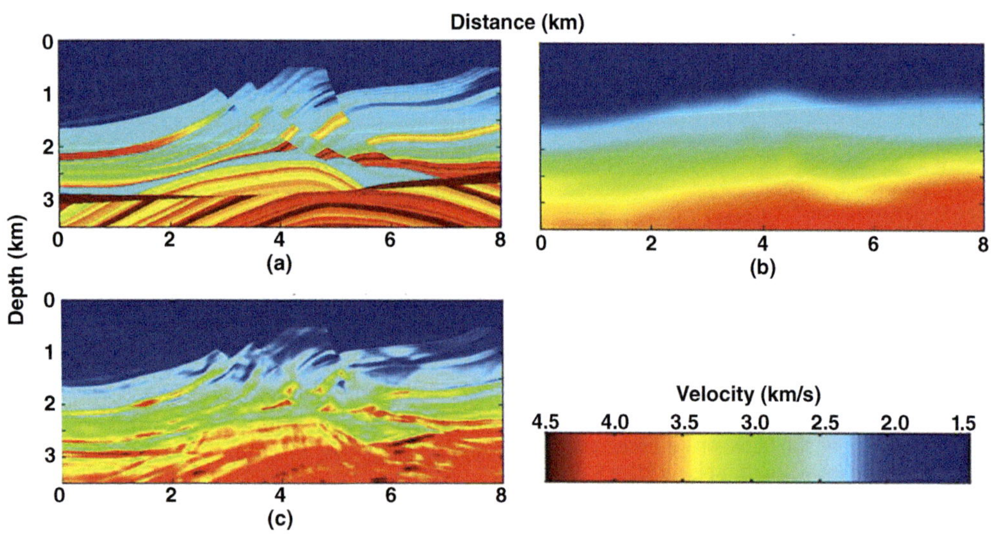

Figure 7.21 Acoustic FWI example. (a) True model, (b) initial model, and (c) inverted model. Sinha and Schuster (2019)/Society of Exploration Geophysicists.

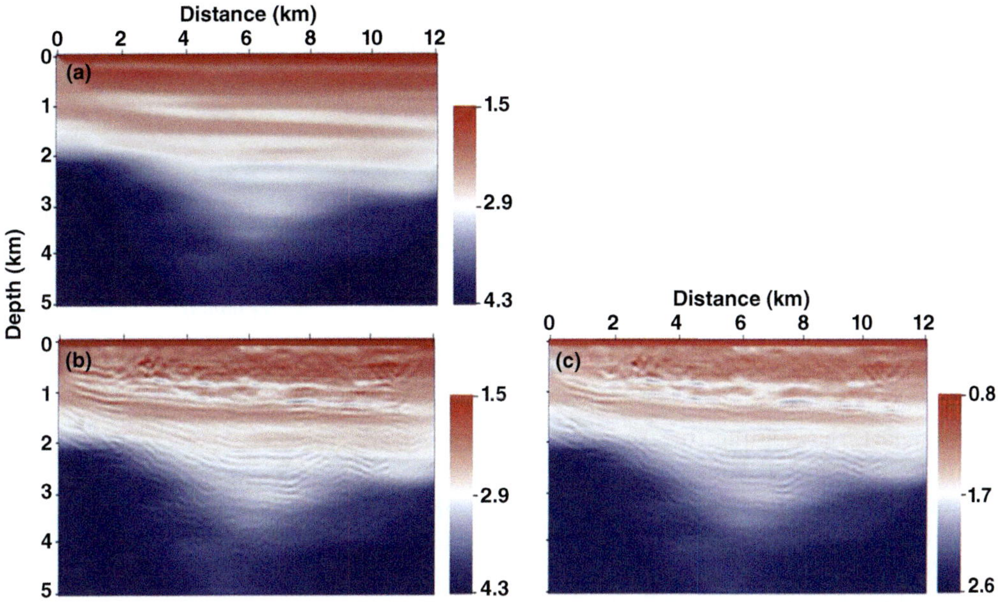

Figure 7.22 Field data example of FWI. (a) Initial V_P. (b, c) Estimated V_P and V_S. The color scale (in km/s) is shown on the right of each panel. Ren et al. (2019)/Society of Exploration Geophysicists.

can be found in Guo and Alkhalifah (2017). Additionally, Guitton and Alkhalifah (2017) provide an FWI example on the anisotropic elastic version of the Marmousi-2 model.

The second example, shown in Figure 7.22, is a field data example from China, obtained by Ren et al. (2019). Starting from a smooth initial V_P model, estimated from data processing (Figure 7.22a) and computing the initial V_S model from V_P by assuming a fixed Poisson's ratio of 0.24, the method obtained reliable estimates of V_P and V_S (Figure 7.22b,c).

Our final example is a 3D synthetic example from Monteiller et al. (2015) demonstrating the feasibility of FWI in global seismological applications. Figure 7.23 shows a section of

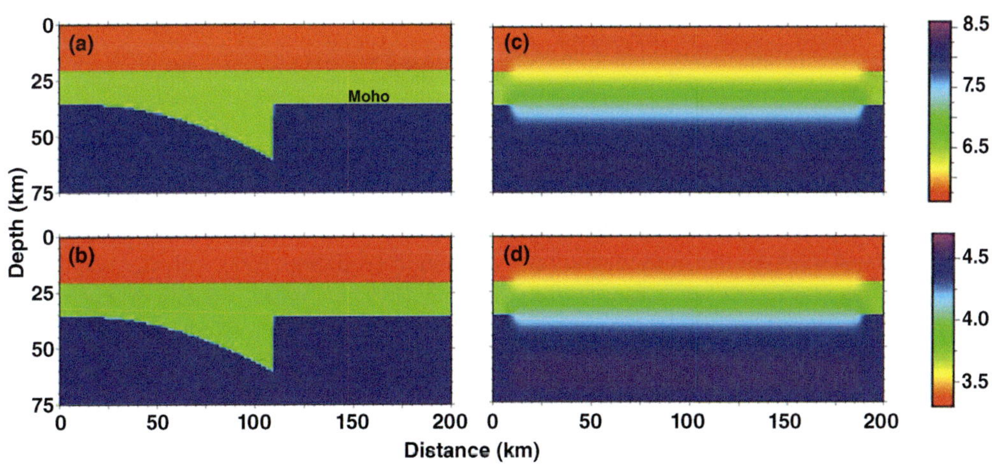

Figure 7.23 Synthetic example for FWI applications related to global seismological application. (a, b) True and (c, d) initial V_P and V_S. The color scale (in km/s) for V_P is on top right and the same for V_S is at the bottom right. Monteiller et. al. (2015)/with permission of Vadim Monteiller.

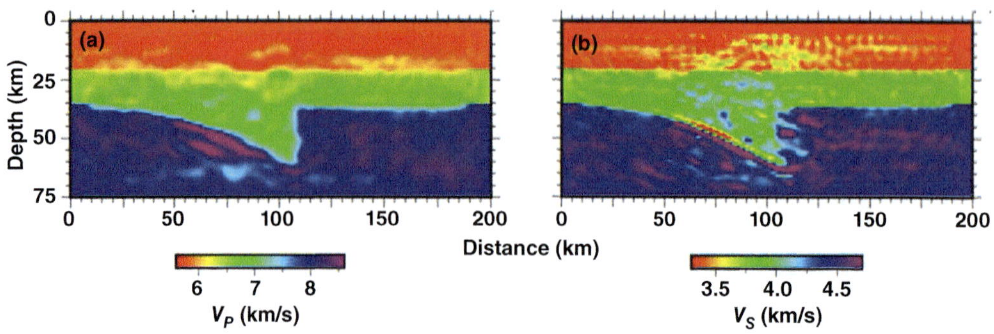

Figure 7.24 FWI result of the synthetic seismic data whose true and initial results are shown in Figure 7.23.

the 3D model, showing the true V_P and V_S (Figure 7.23a,b), and the initial V_P and V_S used in FWI. The true model is an analog of the crust beneath the narrow mountain range (Monteiller et al., 2015). The V_P and V_S estimated from FWI are shown in Figure 7.24.

As a final note, FWI is still in the development stage. Most of the examples shown above were in frequency domain using discrete frequency bands. The optimization methods used in most of them were variations of the Gauss–Newton algorithm, and they used the adjoint-state approach to compute gradients. Further details on implementing these FWI methods, including details on how to use the Fréchet sensitivity kernels for successful applications, can be found in the references cited in the text.

7.7 Chapter Summary

In this chapter, we discussed different local or gradient-based optimization methods to solve seismic inverse problems. We first thoroughly discussed different methods, including their implementation details, assuming the gradient can be explicitly computed. Next, we discussed how the gradient of the objective function (Jacobi) can be efficiently computed for AVA, 1D FWI or PWI, and 3D FWI problems. Finally, we concluded with some selected examples.

7.8 Exercises

Problem 7.1 Verify that the expression given in the extreme right-hand side of equation 7.12, i.e., $\mathrm{Re}(\mathbf{J}^T|\mathbf{s}-\mathbf{d}|^*)\ \mathrm{Re}(\mathbf{J}^T|\mathbf{s}-\mathbf{d}|^*)$, is correct.

Problem 7.2 Assume a pure least-squares AVA problem where the solution is given as (see equation 1.23)

$$\mathbf{m}^{est} = (\mathbf{G}^T\mathbf{G})^{-1}\mathbf{G}^T\mathbf{d}.$$

Also assuming uncorrelated data, the data covariance matrix is given as

$$\mathbf{C}_d = [\mathrm{cov}\ \mathbf{d}] = \sigma_d^2\mathbf{I}.$$

Verify that in such a situation the model covariance is given as

$$\mathbf{C}_m = [\mathrm{cov}\ \mathbf{m}] = \sigma_d^2[\mathbf{G}^T\mathbf{G}]^{-1}.$$

Problem 7.3 Derive equation 7.57 from equation 7.56.

Problem 7.4 Derive equation 7.67.

Problem 7.5 Starting from the elastic wave equation (equation 7.77) with the operator $\mathbf{L}$ defined by equation 7.79, obtain the expression of $\langle \mathbf{u}^\dagger \cdot \nabla_u \mathbf{L}\mathbf{u}\rangle$ as given by equation 7.81.

Problem 7.6 Recalling the symmetries of the elastic stiffness tensor given as $c_{ijkl} = c_{klij} = c_{jikl} = c_{ijlk} = c_{jilk} = \cdots$, verify that the expression given by equation 7.86 is correct.

References

Abubakar, A., Li, M., Liu, J., and Habashy, T.M., 2009, Simultaneous joint inversion of MT and CSEM data using a multiplicative cost function, SEG Technical Program Expanded Abstracts, 719-723, https://doi.org/10.1190/1.3255855.

Ayani, M., MacGregor, L., and Mallick, S., 2020, Inversion of marine controlled source electromagnetic data using a parallel non-dominated sorting genetic algorithm, *Geophysical Journal International*, *220*, 1066–1077, https://doi.org/10.1093/gji/ggz501.

Biondi, B., and Almomin, A., 2014, Simultaneous inversion of full data bandwidth by tomographic full-waveform inversion, *Geophysics*, *79*, WA129–WA140, https://doi.org/10.1190/GEO2013-0340.1.

Bortfeld, R., 1961, Approximation to the reflection and transmission coefficients of plane longitudinal and transverse waves, *Geophysical Prospecting*, *9*, 485–502, https://doi.org/10.1111/j.1365-2478.1961.tb01670.x.

Brougois, A., Bourget, M., Lailly, P., Poulet, M., Ricarte, P., and Versteeg, R., 1990, Marmousi model and data, EAGE Conference and Workshop, 5-16, https://doi.org/10.3997/2214-4609.20141119.

Buland, A., and Omre, H., 2003, Bayesian linearized AVO inversion, *Geophysics*, *68*, 185–198, https://doi.org/10.1190/1.1543206.

Chavent, G., 2009, *Nonlinear Least Squares for Inverse Problems*, Springer, ISBN 978-94-007-3060-1, https://doi.org/10.1007/978-90-481-2785-6.

Connolly, P., 1999, Elastic impedance, *The Leading Edge*, *18*, 438–452, https://doi.org/10.1190/1.1438307.

Engl, H.W., 1987, Discrepancy principles of Tikhonov regularization of ill posed problems leading to optimal convergence rates, *Journal of Optimization Theory and Applications*, *52*, 209–215, https://doi.org/10.1007/BF00941281.

Fichtner, A., 2011, *Full Seismic Waveform Modeling and Inversion*, Springer, ISBN: 978-3-642-15806-3, https://doi.org/10.1007/978-3-642-15807-0.

Guitton, A., and Alkhalifah, T., 2017, A parameterization study for elastic VTI full-waveform inversion of hydrophone components: synthetic and North Sea field data examples, *Geophysics*, *82*, R299–R308, https://doi.org/10.1190/GEO2017-0073.1.

Guo, Q., and Alkhalifah, T., 2017, Elastic reflection-based waveform inversion with a nonlinear approach, *Geophysics*, *82*, R309–R321, https://doi.org/10.1190/GEO2016-0407.1.

Hampson, D., 1991, AVO inversion, theory and practice, *The Leading Edge*, *10*, 39–42, https://doi.org/10.1190/1.1436820.

Hansen, P.C., 1992, Analysis of discrete ill-posed problems by means of L-curve, *SIAM Review*, *34*, 561–580, https://doi.org/10.1137/1034115.

Hansen, P.C., and O'Leary, D.P., 1993, The use of L-curve in the regularization of discrete ill-posed problems, *SIAM Journal on Scientific Computing*, *14*, 1487–1503, https://doi.org/10.1137/0914086.

Kelly, M., Skidmore, C., and Fore, D., 2001, AVO inversion, part 1: isolating rock property contrasts, *The Leading Edge*, *20*, 320–323, https://doi.org/10.1190/1.1438940.

Kormendi, F., and Dietrich, M., 1991, Nonlinear waveform inversion of plane-wave seismograms instratified elastic media, *Geophysics*, *56*, 664–674, https://doi.org/10.1190/1.1443083.

Lamat, C., Montagner, J.P., Fink, M., Capdeville, Y., Tourin, A., and Clévédé, V., 2006, Time-reversal imaging of seismic sources and application to the great Sumatra earthquake, *Geophysical Research Letters*, *33*, https://doi.org/10.1029/2006GL026336.

Loke, M.H., and Dahlin, T., 2002, A comparison of the Gauss–Newton and quasi-Newton methods in resistivity imaging inversion, *Journal of Applied Geophysics*, *49*, 149–162, https://doi.org/10.1016/S0926-9851(01)00106-9.

Mallick, S., 1993, A simple approximation of the P-wave reflection coefficient and its implication in the inversion of amplitude variation with offset data, *Geophysics*, *58*, 544–552, https://doi.org/10.1190/1.1443437.

Mallick, S., 1999, Case history: some practical aspects of prestack waveform inversion using a genetic algorithm: an example from the East Texas woodbine gas sand, *Geophysics*, *64*, 326–336, https://doi.org/10.1190/1.1444538.

Mallick, S., 2001, AVO and elastic impedance, *The Leading Edge*, *20*, 1094–1104, https://doi.org/10.1190/1.1487239.

Mallick, S., and Adhikari, S., 2015, Amplitude-variation-with-offset and prestack waveform inversion: a direct comparison using a real data example from the Rock Springs uplift, Wyoming, USA, *Geophysics*, *80*, B45–B59, https://doi.org/10.1190/GEO2014-0233.1.

Mallick, S., and Frazer, L.N., 1988, Rapid computation of multioffset vertical seismic profile synthetic seismograms for layered media, *Geophysics*, *53*, 479–491, https://doi.org/10.1190/1.1442479.

Mallick, S., Huang, X., Lauve, J., and Ahmad, R., 2000, Hybrid seismic inversion: a reconnaissance tool for deepwater exploration, *The Leading Edge*, *19*, 1230–1237, https://doi.org/10.1190/1.1438512.

Martin, G.S., Wiley, R., and Marfurt, K.J., 2006, Marmousi2: an elastic upgrade for Marmousi, *The Leading Edge*, *25*, 156–166, https://doi.org/10.1190/1.2172306.

McAulay, A.D., 1985, Prestack inversion with plane layer point source modeling, *Geophysics*, *50*, 77–89, https://doi.org/10.1190/1.1441839.

Monteiller, V., Chevrot, S., Komatitsch, D., and Wang, Y., 2015, Three-dimensional full waveform inversion of short-period teleseismic wavefields based upon the SEM–DSM hybrid method, *Geophysical Journal International*, *202*, 811–827, https://doi.org/10.1093/gji/ggv189.

Morozov, I.B., and Ma, J., 2009, Accurate poststack acoustic-impedance inversion by well-log calibration, *Geophysics*, *74*, R59–R67, https://doi.org/10.1190/1.3170687.

Mukhopadhyay, P.K., and Mallick, S., 2011, An accurate ray-based offset-to-angle transform from normal moveout uncorrected multicomponent data in a transversely isotropic medium with vertical symmetry axis, *Geophysics*, *76*, C41–C51, https://doi.org/10.1190/1.3565182.

Mukhopadhyay, P.K., and Mallick, S., 2016, Prestack elastic full-waveform inversion using a nonlinear least-square approach, SEG Technical Program Expanded Abstracts, https://doi.org/10.1190/segam2016-13974384.1.

Oldenburg, D.W., Scheuer, T., and Levy, S., 1983, Recovery of acoustic impedance from reflection seismogram, *Geophysics*, *48*, 1318–1337, https://doi.org/10.1190/1.1441413.

Padhi, A., Mallick, S., Fortin, W., and Holbrook, W.S., 2015, 2-D ocean temperature and salinity images from pre-stack seismic waveform inversion methods: an example from the South China Sea, *Geophysical Journal International*, *202*, 800–810, https://doi.org/10.1093/gji/ggv188.

Pafeng, J., Mallick, S., and Sharma, H., 2017, Prestack waveform inversion of three-dimensional seismic data – an example from the Rock Springs uplift, Wyoming, USA, *Geophysics*, *82*, B1–B12, https://doi.org/10.1190/GEO2016-0079.1.

Pan, G.S., and Phinney, R.A., 1989, Full-waveform inversion of plane wave seismograms in stratified acoustic media: applicability and limitations, *Geophysics*, *54*, 368–380, https://doi.org/10.1190/1.1442662.

Pan, G.S., Young, C.Y., and Castagna, J.P., 1994, An integrated target oriented prestack elastic waveform inversion, *Geophysics*, *59*, 1392–1404, https://doi.org/10.1190/1.1443697.

Pratt, R.G., 1999, Seismic waveform inversion in the frequency domain. Part 1: theory and verification in a physical scale model, *Geophysics*, *64*, 888–901, https://doi.org/10.1190/1.1444597.

Pratt, R.G., Shin, C., and Hicks, G.H., 1998, Gauss–Newton and full Newton methods in frequency-space seismic waveform inversion, *Geophysical Journal International*, *133*, 341–362, https://doi.org/10.1046/j.1365-246X.1998.00498.x.

Ren, Z., Li, Z., and Gu, B., 2019, Elastic reflection waveform inversion based on the decomposition of sensitivity kernels, *Geophysics*, *84*, R235–R250, doi 10.1190/GEO2018-0220.1.

Resnick, J., 1993, Seismic data processing for AVO and AVA analysis, in J.P. Castagna, and M.M. Backus, eds., *Offset Dependent Reflectivity: Theory and Practice of AVO and AVA Analysis*, 175–189, SEG, https://doi.org/10.1190/1.9781560802624.

Richards, P.G., and Frasier, C.W., 1976, Scattering of elastic waves from depth-dependent inhomogeneities, *Geophysics*, *41*, 441–458, https://doi.org/10.1190/1.1440625.

Roy, I.G., 2005, Iteratively adaptive regularization in inverse modeling with Bayesian outlook – application on geophysical data, *Inverse Problems in Science and Engineering*, *13*, 655–670, https://doi.org/10.1080/17415970500171076.

Rüger, A., 1998, Variation of P-wave reflectivity with offset and azimuth in anisotropic media, *Geophysics*, *63*, 935–947, https://doi.org/10.1190/1.1444405.

Sandø, I.A., Skålnes, Å., and Kanestrom, R., 1990, Poststack inversion techniques: Potential and limitations, SEG Technical Program Expanded Abstracts, https://doi.org/10.1190/1.1889933.

Sen, M.K., and Biswas, R., 2015, Choice of regularization weight in basis pursuit reflectivity inversion, *Journal of Geophysics and Engineering*, *12*, 70–79, https://doi.org/10.1088/1742-2132/12/1/70.

Sen, M.K., and Roy, I.G., 2003, Computation of differential seismograms and iteration adaptive regularization in prestack waveform inversion, *Geophysics*, *68*, 2026–2039, https://doi.org/10.1190/1.1635056.

Shuey, R.T., 1985, A simplification of the Zoeppritz equations, *Geophysics*, *50*, 609–614, https://doi.org/10.1190/1.1441936.

Sinha, M., and Schuster, G.T., 2019, Interferometric full-waveform inversion, *Geophysics*, *84*, R45–R60, https://doi.org/10.1190/GEO2018-0047.

Skidmore, C., Kelly, M., and Cotton, R., 2001, AVO inversion, part 2: isolating rock property contrasts, *The Leading Edge*, *20*, 425–428, https://doi.org/10.1190/1.1438966.

Sloan, E.D., 1998, Gas hydrates: review of physical/chemical properties, *Energy & Fuels*, *12*, 191–196, https://doi.org/10.1021/ef970164.

Smith, T.M., Sondergeld, C.H., and Rai, C.S., 2003, Gassmann fluid substitutions: a tutorial, *Geophysics*, *68*, 430–440, https://doi.org/10.1190/1.1567211.

Todd, C.P., and Backus, M.M., 1985, Offset-dependent reflectivity in a structural context, SEG Technical Program Expanded Abstracts, 586–588, https://doi.org/10.1190/1.1892805.

Vigh, D., and Starr, E.W., 2008, 3D prestack plane-wave, full-waveform inversion, *Geophysics*, *73*, VE135–VE144, https://doi.org/10.1190/1.2952623.

Virieux, J., Asnaashari, A., Brossier, R., Métivier, L., Ribodetti, A., and Zhou, W., 2017, An introduction to full waveform inversion, in *Encyclopedia of Exploration Geophysics*, R1–R40 https://doi.org/10.1190/1.9781560803027.entry6.

Virieux, J., and Operto, S., 2009, An overview of full-waveform inversion in exploration geophysics, *Geophysics*, *74*, WCC1–WCC26, https://doi.org/10.1190/1.3238367.

Warner, M., Ratcliffe, A., Nangoo, T., Morgan, J., Umpleby, A., Shah, N., Vinje, V., Štekl, I., Guasch, L., Win, C., Conroy, G., and Bertrand, A., 2013, Anisotropic 3D full-waveform inversion, *Geophysics*, *78*, R59–R80, https://doi.org/10.1190/geo2012-0338.1.

Whitcombe, D.N., Connolly, P.A., Reagan, R.L., and Redshaw, T.C., 2002, Extended elastic impedance for fluid and lithology prediction, *Geophysics*, *67*, 63–67, https://doi.org/10.1190/1.1451337.

Witte, P., Louboutin, M., Lensink, K., Lange, M., Kukreja, N., Luporini, F., Gorman, G., and Herrman, F.J., 2018, Full-waveform inversion, part 3: optimization, *The Leading Edge*, *37*, 142–145, https://doi.org/10.1190/tle37020142.1.

8

Global Optimization Methods in Seismology

8.1 Introduction

In Chapter 7, we dealt with local, gradient-based optimization methods. These methods require an initial model, using which they compute synthetic (predicted) data, match with the observed data, and iteratively modify the model. When the match between the observed and predicted data is satisfactory to a user-defined accuracy, they conclude that the problem is solved, stop iterating, and report the final model as solution. By computing the gradient of the objective, given either as similarity or dissimilarity between the observed data and its corresponding prediction, these methods attempt to reach the optimum point by taking (small) linear steps in the objective space in each iteration. We thoroughly discussed different flavors of these gradient-based methods in Chapter 7 and found them to be powerful optimization tools for solving a variety of complex seismic inverse problems including full waveform inversion (FWI). It is, however, important to note that they may sometimes fail to converge to the global optimum or the true solution if the initial model is far from the true model. For single-parameter optimization cast as maximization, in Chapter 1 we used Figure 1.13 to briefly illustrate this problem of the gradient-based methods. Here, we use some seismic inversion examples and further elaborate this.

Figure 8.1 shows two sound-speed models: Model 1 and Model 2. Both these models were estimated from prestack waveform inversion of the water-column reflections from a real marine seismic dataset from the South China Sea. Model 1 was obtained by Padhi et al. (2015) and Model 2 was obtained by Mallick and Chakraborty (2022). Note that background (low frequency) sound-speed (V) values for both models have approximately similar characters with the following three zones:

1) Zone 1 (red), approximately between 0.3 and 0.5 s time window with $1.545 \geq V \geq 1.515$ km/s.
2) Zone 2 (yellow/green), approximately between 0.5 and 0.85 s time window with $1.515 \geq V \geq 1.5$ km/s.
3) Zone 3 (blue/cyan), for times more than 0.85 s where $V \leq 1.5$ km/s.

While the background trends for both models are alike, their subtle high-frequency features above this trend are substantially different with Model 2 being of much higher resolution than Model 1.

Figure 8.2 compares real stacked data with the synthetic stacked data computed using both models. The synthetic stacked data shown in Figure 8.2c were generated by computing wave-equation-based prestack synthetic seismic data for the same range of offsets as they were in the real data for each common midpoint (CMP) location, followed by geometrical spreading compensation, normal moveout (NMO) correction, and stacking. A visual comparison between the real stacked data and the

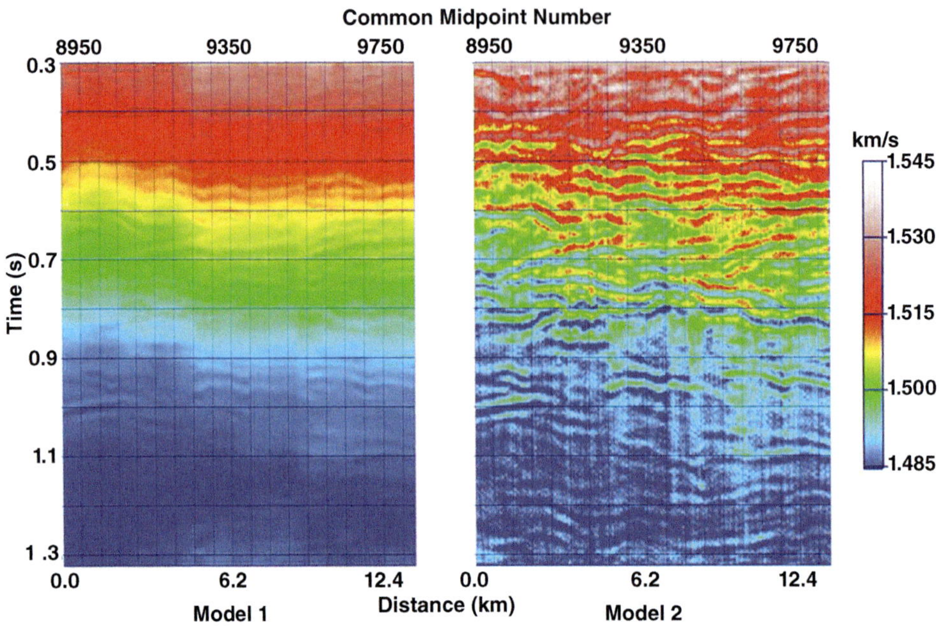

Figure 8.1 Two sound-speed models, Model 1 (left) and Model 2 (right), from two separate prestack waveform inversions of water-column reflections off a real seismic data from the South China Sea. Model 1 was obtained by Padhi et al. (2015) and Model 2 was obtained by Mallick and Chakraborty (2022). Reproduced from Mallick and Chakraborty (2022).

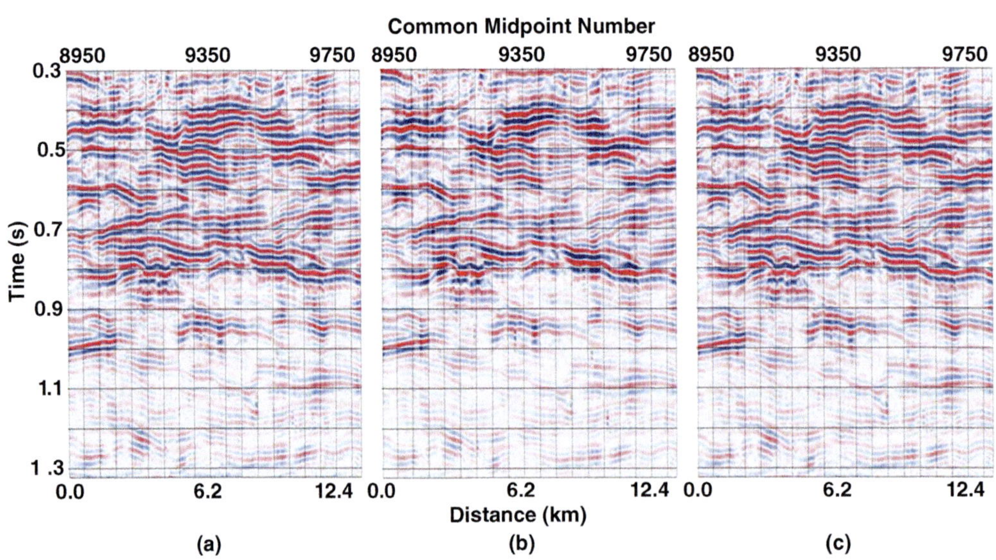

Figure 8.2 Comparison between the synthetic and real stacked data. (a) Synthetic stacked data computed using Model 1. (b) Real stacked data. (c) Synthetic stacked data computed using Model 2. Mallick and Chakraborty (2022)/Society of Exploration Geophysicists.

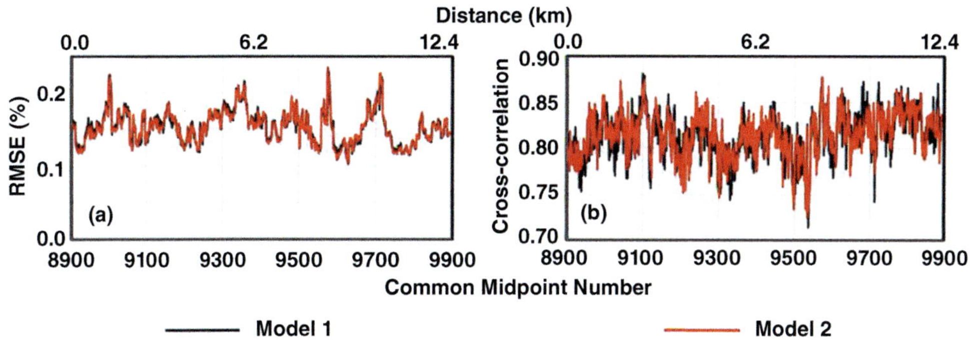

Figure 8.3 Quantitative comparison between Model 1 and Model 2 in predicting observation. (a) RMSE in percent. (b) Normalized cross-correlation. Mallick and Chakraborty (2022)/Society of Exploration Geophysicists.

synthetic stacked data, as shown in Figure 8.2, demonstrates that both models can predict observation with (almost) equal accuracy.

To quantify the visual comparison shown in Figure 8.2, Figure 8.3 shows the root-mean-square error (RMSE) and normalized cross-correlation between the observed and synthetic data, indicating that both Model 1 and Model 2 predict the observation with approximately equal accuracy with an average RMSE of about 0.15% and an average normalized cross-correlation of about 0.83. Therefore, from an inversion point of view, both these models may be considered as valid models.

The reason for the above demonstration of showing two different models predicting observation with almost equal accuracy is to emphasize that seismic inversion is a *nonunique* problem. It is not just two models, Model 1 and Model 2, but there are a range of models in the model space that can fit observation with almost equal accuracy. As illustrated in Chapter 1 (Figure 1.13), if we use a gradient-based optimization to estimate the model, it will converge to any one of this range of models or even to a local optimum depending upon the choice of the initial model.

We may, of course, argue that one out of this wide range of models is the correct model, and we appropriately choose our initial model for the gradient-based optimization such that the method converges to that model. If we have a very good idea about the area from other constraints, such as the regional/local geology, well logs, etc., it is practical to start from such a good initial model. For oil and gas exploration in mature fields, the subsurface model is often known with sufficient accuracy. For example, as the field develops and production begins, there are enough subsurface well-log measurements. From these measurements, an accurate subsurface model can be built. Consequently, if time-lapse seismic data are acquired during production, it is usually possible to build an accurate initial model for a gradient-based method to predict the change in the reservoir fluid properties during different stages of production. In new areas, however, we may not be as fortunate to build an initial model close to the true model, and yet we would like to find the true model with sufficient accuracy. Going past to oil and gas exploration, for deep crustal exploration in land, or for deep marine exploration for finding the structure of the oceanic crust and the upper mantle, we may not have a good initial model. In global seismology also, we may not have such an initial model.

To summarize, gradient-based methods are powerful optimization tools, but they tend to converge to a model, close to the initial model we choose. Although it is possible to build a good initial model for many situations, there are situations where this is difficult. In such situations, where our initial model is far from the true model, gradient-based methods may converge to a local instead of a global optimum. In addition, even if we have a good initial model, we may still want to find the range of models fitting observation with equal accuracy because finding so would allow us to find not only the range of models but also quantify the uncertainty associated with each model parameter estimate.

Using a single-parameter optimization example in Chapter 1, we used Figures 1.14–1.17 to introduce a new class of optimization methods, known as the *global optimization* methods. Instead of starting from an initial model, these methods start from a range of models, randomly generated within a user-specified search window or a prior distribution in the model space (see Figure 1.14). Then, they explore the model space to find the model (or models) that fit the observation (see Figures 1.15–1.17). Unlike the local, gradient-based methods, these global methods do not require computing the gradient of the objective. However, because of the nature of their search, they require many forward computations of synthetic data, and therefore, they are usually much more computationally demanding than their local, gradient-based counterparts. In this chapter, we discuss these global optimization methods.

In our optimization methods, we so far restricted to optimizing a single objective. There are instances, however, where more than one objective must be optimized. Restricting to seismic, for example, we often acquire multicomponent seismic data using three component seismic sensors (geophones), providing seismic response (particle displacement, particle velocity, or particle acceleration) along the vertical and two horizontal (radial and transverse) directions. Consequently, optimizing multicomponent seismic data would require optimizing multiple objectives, one for each data component. Additionally, we sometimes combine seismic with other geophysical data, such as gravity/magnetic, electrical/electromagnetic (EM), etc., in a multi-objective, multi-physics, and multi-scale optimization. One way to handle such problems is to define a scaler objective as a weighted sum of the individual objectives, defined for different data components as

$$\left.\begin{aligned} E &= \sum_{k=1}^{M} w_k E_k \\ \sum_{k=1}^{M} w_k &= 1 \end{aligned}\right\}. \tag{8.1}$$

In equation 8.1, M is the total number of data components with E_k, $k = 1, 2, \ldots, M$ being the objective defined for the data component k, and w_k is the weight defined for the kth objective. Having defined the scalar objective by equation 8.1, we may then use any single-objective method for optimization. However, as we see later, multi-objective optimizations are inherently nonunique with multiple solutions. By defining a scalar objective like equation 8.1, followed by a single-objective optimization, local or global cannot correctly handle this inherent nonuniqueness. We may run optimization many times with different weights and combine them all to handle nonuniqueness. But this is cumbersome and computationally inefficient. To correctly handle multiple objectives, it is thus necessary to use multi-objective optimization methods, which belong to the class of global optimization in which the objective is treated as an objective vector instead of a scalar. We discuss these multi-objective methods at the end of this chapter.

8.2 Bayesian Approach to Optimization Problems

The above example of two different models fitting observation with almost equal accuracy demonstrates the issue of nonuniqueness, associated with seismic inverse problems. However, it is not just seismic inversion, nonuniqueness is inherent to any geophysical inverse problem. Recall our forward problem where we define synthetic or predicted data vector $\mathbf{s}$ from the model vector $\mathbf{m}$ as $\mathbf{s} = g(\mathbf{m})$. Also recall that in local optimization methods, we linearize this forward problem as $\mathbf{s} = \mathbf{G}\mathbf{m}$, where **G** is a matrix. This linearization, in turn, allows us to analyze variance and prediction errors in least-squares solutions and define the data and model resolution matrices **N** and **R** via equations 7.20 and 7.21. These concepts of data and model resolution matrices are the classical approach to mathematically describe nonuniqueness for local, gradient-based optimization methods (Menke, 1989). This classical approach has also been modified to a *maximum likelihood method* where the probabilistic model of error in data and the prior information on the model are included for estimating the solution to an inverse problem. Formulation of the inverse problem from such theoretical and statistical viewpoints can be found in the classic textbook, written by Tarantola (1987).

Tarantola (1987) approach can be formally derived from the *Bayesian* approach to solve the optimization problem, proposed by Backus (1970a, b, c) and Backus and Gilbert (1967, 1968, 1970). This approach is important not only to handle the inherent nonuniqueness associated with the inverse problems but also to quantify the uncertainty in the model parameter estimates. Seismic data, for example, are always noisy. Therefore, even if we assume that out of many solutions, there is a single solution that is geologically meaningful, start from an initial model close to this so-called true solution and use a local optimization method to estimate it; we can never estimate it exactly where the objective function, given as the sum of the least squares (equation 7.5), or its regularized equivalent (equation 7.22) is exactly zero. Goals of any optimization method are, therefore, not only to find the model parameters but also to quantify the uncertainty associated with their estimates. Casting optimization in a Bayesian framework allows estimating the posterior probability density function (PDF) for each model parameter estimate, which is the fundamental building block for uncertainty quantification.

In our introduction to global optimization in Chapter 1 for estimating a single model parameter, the first step was to generate a set of random models within a specified window (see Figure 1.14). In this context, we may consider the model parameters to be random variables, being drawn from a prior distribution. In Figure 1.14, for example, we specified this prior distribution as uniform such that any value of this parameter can be chosen with equal probability within the specified window. There is, however, no harm in specifying this prior distribution to be something else. For example, we may specify an initial model and a Gaussian distribution around this initial model to be our prior and choose our initial suite of random models.

The primary concept of casting an optimization problem in a Bayesian framework follows from the fact that by treating the model parameters as random variables, we start from their prior distribution and combine it with the corresponding physics of the forward problem to compute the corresponding posterior PDFs. To compute the exact PDF, it is necessary to exhaustively search the entire model space, which is difficult in practice; therefore, we aim to obtain a best approximation of the exact PDF. In local optimization methods, for example, we first start from an initial model, say $\mathbf{x}_0$, and compute the objective $f(\mathbf{x}_0)$. Then, by expanding $f(\mathbf{x}_0)$ as a truncated Taylor series around the initial model given in a quartic form shown by equations 6.16–6.18, we compute the gradient of the objective function $\nabla f(\mathbf{x}_0)$ (equation 6.19). We then find a new model $\mathbf{x}_1$ by minimizing $f(\mathbf{x}_0)$ along a preferred direction, and we continue doing so in an iterative manner until a certain iteration N for which we think that $f(\mathbf{x}_N)$ has attained the desired minimum point. Although there are many ways to do these successive minimizations, such as gradient (steepest) descent, conjugate-gradient, Gauss–Newton, etc., they all are based on the method of least squares. This least-squares criterion is, in turn, based on the Gaussian hypothesis (Tarantola, 1987). Local optimization methods therefore assume a Gaussian distribution of model parameters in each iteration step as the prior. So, if we can store all models and their objective evaluations during all iterations of any local optimization, and follow the steps outlined below, we can estimate the posterior PDF. Although this posterior PDF is not the true PDF, it can still be considered its best approximation near the initial model we started.

To understand the Bayesian formulation to the inverse problem, we start from equations 7.1 and 7.2 where we represented the data and model vectors **d** and **m** with M and N components, respectively, and assume that each data and model components d_i, $i = 1, 2, \ldots, M$, and $m_j, j = 1, 1, \ldots, N$, are random variables. We also use the following notations:

1) PDF of **d** for given model **m** is $p(\mathbf{d} \mid \mathbf{m})$.
2) Conditional PDF of **m** for given **d** is $p(\mathbf{m} \mid \mathbf{d})$.
3) The PDF of the data and the model are respectively $p(\mathbf{d})$ and $p(\mathbf{m})$.

Therefore, we can write from the definition of the conditional probability

$$p(\mathbf{m} \mid \mathbf{d})p(\mathbf{d}) = p(\mathbf{d} \mid \mathbf{m})p(\mathbf{m}). \tag{8.2}$$

From equation 8.2, we therefore write the conditional PDF of the model **m**, given the measured data **d** to be

$$p(\mathbf{m} \mid \mathbf{d}) = \frac{p(\mathbf{d} \mid \mathbf{m})p(\mathbf{m})}{p(\mathbf{d})}. \tag{8.3}$$

Equation 8.3, providing the state of information on the model **m**, given the data **d** is formally called the Bayes' rule. Note that $p(\mathbf{d})$ in the denominator of equation 8.3 does not depend upon the model **m** and therefore can be considered as a constant factor in the inverse problem (Duijndam, 1988; Sen and Stoffa, 1996). So, we replace the denominator of equation 8.3 by a constant and write

$$p(\mathbf{m} \mid \mathbf{d}) \propto p(\mathbf{d} \mid \mathbf{m})p(\mathbf{m}). \tag{8.4}$$

Note $p(\mathbf{m} \mid \mathbf{d})$, the conditional PDF of the model given the data, on the left-hand side of equation 8.4, is the posterior PDF that we want to estimate via optimization. The second term $p(\mathbf{m})$ on the right-hand side of equation 8.4 is the PDF of the model, or the prior distribution (uniform/Gaussian, etc., around the estimated initial model), which we provide. Finally, we obtain the first term $p(\mathbf{d} \mid \mathbf{m})$, the PDF of data for a given model on the right-hand side of equation 8.4 via forward computation of the synthetic or predicted data for different models and matching them with observation. The Bayes' rule therefore formally represents how the prior distribution $p(\mathbf{m})$ is modified by $p(\mathbf{d} \mid \mathbf{m})$ through the physics of the forward problem in the estimation of the posterior PDF.

Now, if we denote our measured or observed data as $\mathbf{d}_{obs}$ and substitute $\mathbf{d} = \mathbf{d}_{obs}$ in $p(\mathbf{d} \mid \mathbf{m})$, then the result is interpreted as a function of the model **m** denoted by $l(\mathbf{d}_{obs} \mid \mathbf{m})$, which is known as *likelihood function* (Box and Tiao, 1973; Cary and Chapman, 1988; Sen and Stoffa, 1996, among others). We therefore write equation 8.4 as

$$p(\mathbf{m} \mid \mathbf{d}) \propto l(\mathbf{d}_{obs} \mid \mathbf{m})p(\mathbf{m}) = \pi(\mathbf{m}). \tag{8.5}$$

In equation 8.5, the PDF $p(\mathbf{m})$ is the probability of the model independent of the data. In other words, it contains the information on the model without any knowledge of the data and therefore $p(\mathbf{m})$ is the *prior*. Similarly, the PDF $p(\mathbf{m} \mid \mathbf{d})$ in equation 8.5 can be considered as the state of information on the model, given the data, and it is the *posterior* PDF. The right-hand side of equation 8.5 is not the true posterior PDF but proportional to it, which we refer here as $\pi(\mathbf{m})$. Sen and Stoffa (1996) normalized the right-hand side of equation 8.5 and called it the posterior probability density (PPD). Here, we consider $p(\mathbf{m} \mid \mathbf{d})$ as the posterior PDF instead of PPD and refer $\pi(\mathbf{m}) = l(\mathbf{d} \mid \mathbf{m})p(\mathbf{m})$ to be a distribution, which is not the true posterior PDF but proportional to it, which we aim to find via optimization. The posterior PDF is thus given by the product of the likelihood function and the prior with some normalizing factor in the denominator. Therefore, the prior knowledge on the model is modified by the likelihood function in the posterior PDF. Thus, if we choose a uniform prior like the example shown by Figure 1.14, the shape of the posterior PDF will be dominated by the likelihood function. For multi-parameter optimization problems, with which we are primarily concerned, the situation where the prior PDF dominates the likelihood function is unlikely. For most optimization problems, the prior PDF dominates a subspace of the model parameter space while the likelihood function dominates other, usually much larger subspaces (Sen and Stoffa, 1996).

Fundamentally, the likelihood function depends upon the distribution of noise (or error) in the data. Consequently, estimating the likelihood function requires the knowledge of the error distribution. This error can be due to the measurement errors, caused by the instruments measuring the data. In addition, it can be because of using an inexact theory in the prediction of the data via the forward modeling $\mathbf{s} = g(\mathbf{m})$. If we assume Gaussian errors, the likelihood function becomes

$$l(\mathbf{d}_{obs} \mid \mathbf{m}) \propto \exp[-E(\mathbf{m})]. \tag{8.6}$$

In equation 8.6, $E(\mathbf{m})$ is the objective or error that we want to minimize. We already defined $E(\mathbf{m})$ in equation 7.22 via the data and model weighting matrices $\mathbf{W}_D$ and $\mathbf{W}_M$. Strictly speaking, $\mathbf{W}_D$ should account for the noise in the data or the measurement errors and $\mathbf{W}_M$ should account for the noise generated from using an inexact theory (theoretical error) in the forward modeling $\mathbf{s} = g(\mathbf{m})$. By setting $\mathbf{W}_D$ as the inverse of the data covariance matrix, we handled the measurement errors in Chapter 7 (see equation 7.23). In defining $\mathbf{W}_M$, however, we used an iteration-adaptive scalar α which can be computed using the L-curve technique (Figure 7.2) or from an iteration-adaptive annealing type of equation (equation 7.39). By cross-plotting the data misfit and model misfit terms on the right-hand side of equation 7.38 for different values of α and then choosing the corner point as its optimum value, we may loosely say that the L-curve method handles the theoretical error within $\mathbf{W}_M$. Using an annealing type of equation (equation 7.39) to compute α and then computing $\mathbf{W}_M$ do not, however, include any theoretical error in $\mathbf{W}_M$. Strictly speaking, no theory for forward modeling in $\mathbf{s} = g(\mathbf{m})$ is exact. It is therefore practically impossible to model every feature on the observed data $\mathbf{d}_{obs}$ in our synthetic or predicted data from $\mathbf{s} = g(\mathbf{m})$, no matter how sophisticated our modeling method is. In geophysical inverse problems, we thus consider our theoretical errors as data errors. In other words, the features we cannot model in $\mathbf{d}_{obs}$, because of using an inexact theory, we treat them as noise on data. Under such assumption that the observational features we cannot model are observational errors, we may argue that we do not need to account for the theoretical errors in our optimization and treat all sources of errors as data noise.

If we assume the prior PDF $p(\mathbf{m})$ to be a Gaussian, we can then easily derive the expression for the posterior PDF. Note that for Gaussian prior PDF the posterior PDF is non-Gaussian. This is because of the presence of the nonlinear operator $g(\mathbf{m})$ within the objective function $E(\mathbf{m})$. If the prior PDF is uniform, like our simple example of Figure 1.14, the maximum of the posterior PDF coincides with the minimum of the objective function $E(\mathbf{m})$.

By combining equations 8.5 and 8.6, we write the expression for the normalized posterior PDF as (Sen and Stoffa, 1996)

$$p(\mathbf{m} \mid \mathbf{d}) = \frac{\exp[-E(\mathbf{m})]p(\mathbf{m})}{\int d\mathbf{m} \exp[-E(\mathbf{m})]p(\mathbf{m})}. \tag{8.7}$$

The integral in the denominator of equation 8.7 is the normalization term, spanning over the entire model space. Note that if this posterior PDF can be exactly computed, the optimization problem is completely solved. In real world, however, exactly computing the posterior PDF for multi-dimensional optimization problem is difficult and often impractical. As pointed out by Tarantola and Valette (1982), the exact and most accurate way to compute the posterior PDF is by computing the right-hand side of equation 8.7 at each point in the model space, which requires solving the forward problem $\mathbf{s} = g(\mathbf{m})$ for those points. This is extremely compute-intense and practically impossible for most of the geophysical problems. So, the best we can do is find an approximation to it.

Another challenging task for multi-dimensional optimization problems is to find a convenient way to display the posterior PDF. As a result, several other measures are often used for convenience. For example, the marginal or conditional PDF of any model parameter m_i is given by setting all model parameters except m_i as constants, which can be written as

$$p(m_i \mid \mathbf{d}) = \int dm_1 \int dm_2 \cdots \int dm_{i-1} \int dm_{i+1} \cdots \int dm_N p(\mathbf{m} \mid \mathbf{d}). \tag{8.8}$$

Similarly, the mean model $\langle \mathbf{m} \rangle$ and the posterior model covariance matrix $\mathbf{C}_M$ are given respectively as

$$\langle \mathbf{m} \rangle = \int \mathbf{m} p(\mathbf{m} \mid \mathbf{d}) d\mathbf{m} \tag{8.9}$$

and

$$\mathbf{C}_M = \int [\mathbf{m} - \langle \mathbf{m} \rangle][\mathbf{m} - \langle \mathbf{m} \rangle]^T p(\mathbf{m} \mid \mathbf{d}) d\mathbf{m}. \tag{8.10}$$

In the above equation, $\mathbf{C}_M$ is of specific interest for quantifying uncertainty. Note from equation 8.10, if the model vector is of size N (and so is the mean model), $\mathbf{C}_M$ is a square matrix of size $N \times N$. The diagonal elements C_{ii}, $i = 1, 2, \ldots, N$, of this matrix are the measures of variances of the ith model parameter, with their square roots being the standard deviations (uncertainties) in their estimations. Additionally, the off-diagonal elements C_{ij}, $i \neq j$, of the matrix $\mathbf{C}_M$ represent the inter-dependence between the model parameters i and j (Mallick, 1995).

From the above discussion, casting an optimization problem in a Bayesian framework consists of the following steps:

1) Define the prior distribution $p(\mathbf{m})$. We can define this as a uniform distribution within a specified window of lower and upper bounds. Alternatively, we may choose an initial model and define $p(\mathbf{m})$ to be a Gaussian or any other distribution around this initial model.
2) Define the objective $E(\mathbf{m})$. If cast as minimization, we may define $E(\mathbf{m})$ as given by equation 7.22 or any equivalent form. For many problems, however, it is convenient to cast the optimization as a maximization instead of minimization and define $E(\mathbf{m})$ as normalized cross-correlation (for example, see Sen and Stoffa, 1992). We discuss this in detail later.
3) Store all models and their corresponding objectives for the entire course of the optimization process.
4) Evaluate the posterior PDF (equation 8.7), marginal PDFs (equation 8.8), mean model (equation 8.9), posterior model covariance matrix (equation 8.10), etc.

As we see in the global optimization methods, we discuss below that these methods allow us to estimate the distribution $\pi(\mathbf{m})$ as shown by equation 8.5. This distribution is proportional to the true posterior PDF. Therefore, to estimate the posterior PDF, it is necessary to find the normalizing factor (see equation 8.7) such that its integral over the entire model space is unity. Primary challenge to the Bayesian approach lies in the evaluation of the integrals shown in equations 8.7–8.10. Note that in theory these integrals must be carried out in the entire model space. However, for the regions within this space, where the prior $p(\mathbf{m})$ is zero, we can safely ignore those regions in our integral evaluations. However, if the prior $p(\mathbf{m})$ is nonzero and our optimization method ignores those regions, then the posterior PDF will be underestimated and the estimate of any marginal PDF, mean model, posterior covariance, etc., will also be incorrect. In general, the integrals, shown in equations 8.7–8.10, are of the form $\int f dV$, carried out in a multi-dimensional space, in which the integrand, given by the function f, is randomly sampled. Evaluating such integrals is the topic of the *Monte Carlo integration*, a thorough discussion of which can be found in Press et al. (2007). Here, we briefly summarize the method.

8.2.1 Simple Monte Carlo Integration

Consider that we pick N random points, say $x_1, x_2, \ldots, x_N$, uniformly distributed in a multi-dimensional space V. The basic theorem of Monte Carlo then evaluates $\int f dV$ as (Press et al., 2007)

$$\int f dV \approx V\langle f \rangle \pm V\sqrt{\frac{\langle f^2 \rangle - \langle f \rangle^2}{N}}, \quad (8.11)$$

with

$$\left.\begin{aligned} \langle f \rangle &= \frac{1}{N}\sum_{i=1}^{N} f(x_i) \\ \langle f^2 \rangle &= \frac{1}{N}\sum_{i=1}^{N} [f(x_i)]^2 \end{aligned}\right\}. \quad (8.12)$$

The term with $\pm$ sign on the right-hand side of equation 8.11 is one standard deviation error estimate of the integral. Also, this error is not necessarily distributed as a Gaussian and thus should be taken as a rough estimate of the probable error in evaluating the integral (Press et al., 2007). Comparing equation 8.11 with equation 8.7 for the posterior PDF, we can see that $f \equiv \exp[-E(\mathbf{m})]p(\mathbf{m})$.

The primary challenge in evaluating the integrals in equations 8.7–8.10 is thus the evaluation of the Monte Carlo integrals. For example, assume a two-dimensional case where $f = f(x, y)$ is shown in Figure 8.4. Out of the entire domain, our objective is therefore to find the region (or regions) R within which $f(x, y)$ is nonzero and aim to evaluate the integral as accurately as possible. Note that R may have a complicated shape, and our first step is to find the region V that includes R which can be easily sampled. We may then define our function f to be equal to g for points in R and zero for the points outside of R. We try to make V to enclose R as closely as possible. This is because including the zero values of f will increase the error estimate term of equation 8.11. From these arguments, out of three such choices of V, denoted as V_1, V_2, and V_3 in Figure 8.4, it is obvious that for an accurate integral evaluation V_1 is the worst choice and V_2 is the best choice. Note that the Monte Carlo integration, which we discussed above, not only requires separating the regions within which our multi-dimensional function is nonzero, but also requires those regions to be uniformly sampled. Evaluation of the Monte Carlo integrals this way is called simple Monte Carlo integration. As we see later that, in practice, such a uniform sampling within these regions is often impossible, and the best we can do is to nonuniformly sample them in proportion to the distribution $\pi(\mathbf{m})$, which we estimate via optimization. We postpone discussion of evaluating such integrals with nonuniform sampling for later. To build the mathematical

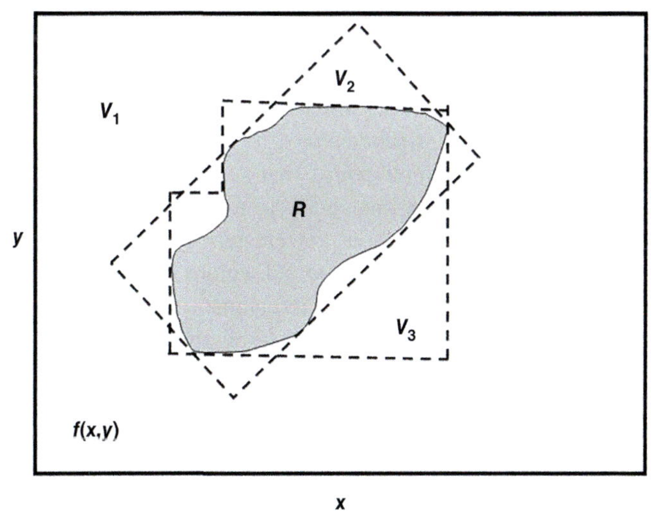

Figure 8.4 Evaluation of the Monte Carlo integrals demonstrated using a two-dimensional function $f(x, y)$. Out of the entire multi-dimensional space V, we must first choose the region or regions R within which $f(x, y)$ is nonzero. Then, by randomly sampling $f(x. y)$ within R, we attempt to evaluate the integral as accurately as possible.

foundations for such nonuniform sampling and accurate computation of the integrals, it is, however, necessary to discuss a few fundamental concepts, which will help developing our understanding of different flavors of the global optimization methods.

Quasi-random Sequences

As we observed in equation 8.11, a uniform and random choice of N points is an n-dimensional space leads to an error term in a simple Monte Carlo integration that decreases in proportion to $\left(\sqrt{N}\right)^{-1}$. This does not, however, mean that this $\left(\sqrt{N}\right)^{-1}$ error is always inevitable. For example, assume that we uniformly choose our sample points in a Cartesian grid such that each point in the grid is sampled only once. Under such situation, the Monte Carlo integration becomes a deterministic quadrature scheme with its fractional error decreasing in proportion to N^{-1}. However, the problem with such grid-based sampling is that we must decide in advance how fine the gird must be, and once we decide this grid size, we must then complete all its sample points. If the evaluation of the function is computationally demanding, such grid-based sampling is expensive, and it is useful to implement a convergence or stopping criterion such that we sample up to the point where we feel that the integral is evaluated with sufficient accuracy and avoid unnecessary computational burden. Using grid-based sampling, it is difficult to implement such criterion.

In view of the above discussion, we may then ask whether there is a scheme in which we randomly pick our sample points that spread out within the region of our choice in a random but uniform way without any clustering or bias of dense points in some subregions and sparse points in some other subregions. If yes, we can then pick these points to evaluate integral until we think the convergence criterion is satisfied. Fortunately, there are many pseudorandom number generators such as the Halton's sequence (Halton, 1960), Sobol' sequence (Sobol', 1967; Antonov and Saleev, 1979; Bartley and Fox, 1988; Lambert, 1988; Neiderreiter, 1988), etc., which can generate quasi-random sequences and achieve this goal. Although these quasi-random sequence generators are not purely random, they tend to generate sequences by maximally avoiding one another and thus avoiding any clustering or bias. By successively moving from coarse to finer scales, these quasi-random sequence generators can be used in which we can easily implement a stopping criterion and thus avoid unnecessary computations; see Press et al. (2007) for further details on how to implement such quasi-random sequence generators for an accurate evaluation of the Monte Carlo integrals.

Importance Sampling

Assume that we can express the function f in equation 8.11 as the product of a function h that is almost constant and another positive function g. We can then write the integral on the right-hand side of equation 8.11 as

$$\int f dV = \int \left(\frac{f}{g}\right) g dV = \int h g dV. \tag{8.13}$$

We can interpret equation 8.13 means that we can integrate f by sampling h with a nonuniform density gdV. Now, consider that the points are chosen within the volume V with a probability density p such that

$$\int p dV = 1. \tag{8.14}$$

Then, from equations 8.11, the integral I of any function f can be estimated from N sample points as

$$I = \int f dV = \int \left(\frac{f}{p}\right) p dV \approx \left\langle \frac{f}{p} \right\rangle \pm \sqrt{\frac{\left\langle \frac{f^2}{p^2} \right\rangle - \left\langle \frac{f}{p} \right\rangle^2}{N}}, \tag{8.15}$$

with $\langle h \rangle$ representing the arithmetic mean of any function h over the N sample points. Like equations 8.11, the $\pm$ in equation 8.15 is one standard deviation error estimate of the integral. Comparing equations 8.15 and 8.11, we can easily verify that equation 8.11 is a special case of equation 8.15 where p is a constant and is equal to $1/V$. Now, how do we optimally choose the sampling density for p? As we observed above in equation 8.13, the primary idea for this optimal choice is to make $h = f/p$ as close to a constant as possible. To go into further details, we first note that the numerator inside the square root on the right-hand side of equation 8.15 is the variance per sample point, which we must aim to make as small as possible. In other words, an accurate evaluation of the Monte Carlo integral requires that

$$S = \left\langle \frac{f^2}{p^2} \right\rangle - \left\langle \frac{f}{p} \right\rangle^2 \tag{8.16}$$

is small. Now from equations 8.15 and 8.16, we can write

$$S = \left\langle \frac{f^2}{p^2} \right\rangle - \left\langle \frac{f}{p} \right\rangle^2 \approx \int \left(\frac{f}{p}\right)^2 p dV - \left[\int \frac{f}{p} p dV\right]^2 = \int \frac{f^2}{p} dV - \left[\int f dV\right]^2. \tag{8.17}$$

We now apply the constraint given by equation 8.14 to find the optimal p from the functional variation

$$\frac{\delta}{\delta p}\left[\int \frac{f^2}{p} dV - \left(\int f dV\right)^2 + \lambda \int p dV\right] = 0, \tag{8.18}$$

in which λ is the Lagrange multiplier. The middle term within the square brackets on the left-hand side of equation 8.18 is independent of p. Therefore, the variation given by equation 8.17 leads to

$$p = \frac{|f|}{\sqrt{\lambda}} = \frac{|f|}{\int |f| dV}. \tag{8.19}$$

Note that in deriving equation 8.19, the Lagrange multiplier λ has been chosen to enforce the constraint given by equation 8.14. In our case, the function f is always positive. And, if f has a single sign in the entire domain of integration, we obtain the obvious result that the optimal choice of p, if we can figure out a practical way of effective sampling, is that it be proportional to $|f|$. In such a case, the variance is reduced to zero. It is also seen to be true that $p \propto |f|$ is also optimal even when the function f takes both signs in the domain of integration. In such a case, from equations 8.17 and 8.19, we obtain the variance per sample point as

$$S = S_{optimal} = \left(\int f dV\right)^2 - \left(\int f dV\right)^2. \tag{8.20}$$

Furthermore, we can always add a constant K to the integrand and make it all a single sign. This changes the integral by KV, and then the optimal choice of p always gives zero variance or a perfect integral evaluation. Thus, a perfect knowledge of p in equation 8.19 requires a perfect knowledge of $\int |f| dV$, which, in turn, is equivalent to saying that we already know the integral that we are trying to compute. If the function f takes on a known constant value for most of the volume V, it is then a good idea to add a constant to make that value to zero. If we do that, the accuracy attainable by importance sampling does not depend on how small the $S_{optimal}$ given by equation 8.20 is, but on how small is equation 8.17 for an implementable p, likely only a crude approximation to the ideal (Press et al., 2007).

In the following, we discuss different global optimization methods where we cast them in Bayesian framework and revisit efficient ways to sample the model space for each method such that the estimations of the Monte Carlo integrals, and therefore the posterior PDF and other related parameters, are sufficiently accurate.

8.3 Global Optimization Methods

Monte Carlo inversion (MCI) is one of the earliest applications of global optimization method in geophysics. Foundation of the modern MCI was developed by Metropolis and Ulam (1949). Pioneering work of Backus and Gilbert (1967, 1968, 1970) formalized the concepts of the geophysical inverse theory and showed that nonuniqueness is inherent to any geophysical inverse problem and how a Bayesian approach allows describing it. This followed the application of MCI method to solve geophysical inverse problems (Keilis-Borok and Yanovskaya, 1967; Press, 1968 1970a, b; Anderson et al., 1972; among others). These early applications were based on the following steps (Sambridge and Mosegaard, 2002):

1) Randomly generate discrete earth models in a uniform fashion between the pairs of lower and upper bounds of each model parameter.
2) Test each model generated in step 1 for its suitability to the available data, and either accept or reject the model.
3) Interpret the final set of accepted models.

With the introduction of digital computers in the mid-1960s, the application of MCI became feasible to solve some simple (yet important) geophysical inverse problems. These applications included inversion of seismic body-wave (P and S waves) travel times, mass moment, and the eigenperiods of the earth's free oscillation variations to estimate the P- and S-wave velocity (V_P, V_S) and density (ρ) of the earth (Press, 1968, 1970a, b; Anderson et al., 1972). Figure 8.5 shows the flowchart of such an approach, modified from Press (1968). Starting from a random V_P model between its upper and lower bounds, the method first tests its validity with the V_P travel time on the observed seismic data. If the test fails, the model is rejected and a new V_P model is drawn. If the test passes, it then draws a random V_S model and carries out the same travel time test. Once a successful V_P and V_S model is obtained, the method then selects the density (ρ) model and tests its validity with the mass moment. Then, after obtaining the model satisfying the P- and S-travel times and the mass moment, the method then tests the eigenfrequencies in a way that the chosen model matches at least n out of the total of m observations (see Figure 8.5). After successfully finding the model, the process is repeated to obtain a total of j models, which are finally output for interpretation and analysis.

The methodology described by Figure 8.5 is considered as the pure MCI. As can be seen, such a method is based on a pure random walk in the model space without any specific direction.

The primary advantage of MCI is that the method avoids all assumptions of linearity between model parameters and the observed data, which the local gradient-based methods assume. We must note, however, that to thoroughly explore the entire model space and compute the posterior PDF and other relevant parameters, MCI requires many forward calculations. For travel-time-based methods of Press (1968), this may be feasible. However, to match the full waveforms in FWI, MCI is not practical, even in modern high-performance parallel computing environments. In addition, one major problem of MCI is that it is never known if sufficiently large number of models has been tested to explore the entire model space. Press (1968), for example, tested five-million earth models out of which only six models passed all tests using the procedure outlined in Figure 8.5. To summarize, although MCI methods to solve geophysical inverse problems were developed in the 1960s and 1970s and showed a lot of advantages over the gradient-based methods, they were not very practical to use. Consequently, gradient-based methods remained to be the most popular optimization tools in those days.

With the advancement of computational technology and the introduction of the vectorizing and eventually massively parallelizing computing environments, there have been a lot of research on the MCI methods, especially for improving their computational efficiency such that they can be implemented in practice. One such development is the directed random walks with nonuniform sampling of the model space. Although there are many such global optimization methods developed and successfully used to this date, the most popular one is the *Markov chain Monte Carlo* (*MCMC*) and its variations such as the *simulated annealing* (*SA*) and *genetic algorithm* (*GA*) optimization. We discuss below these methods in detail.

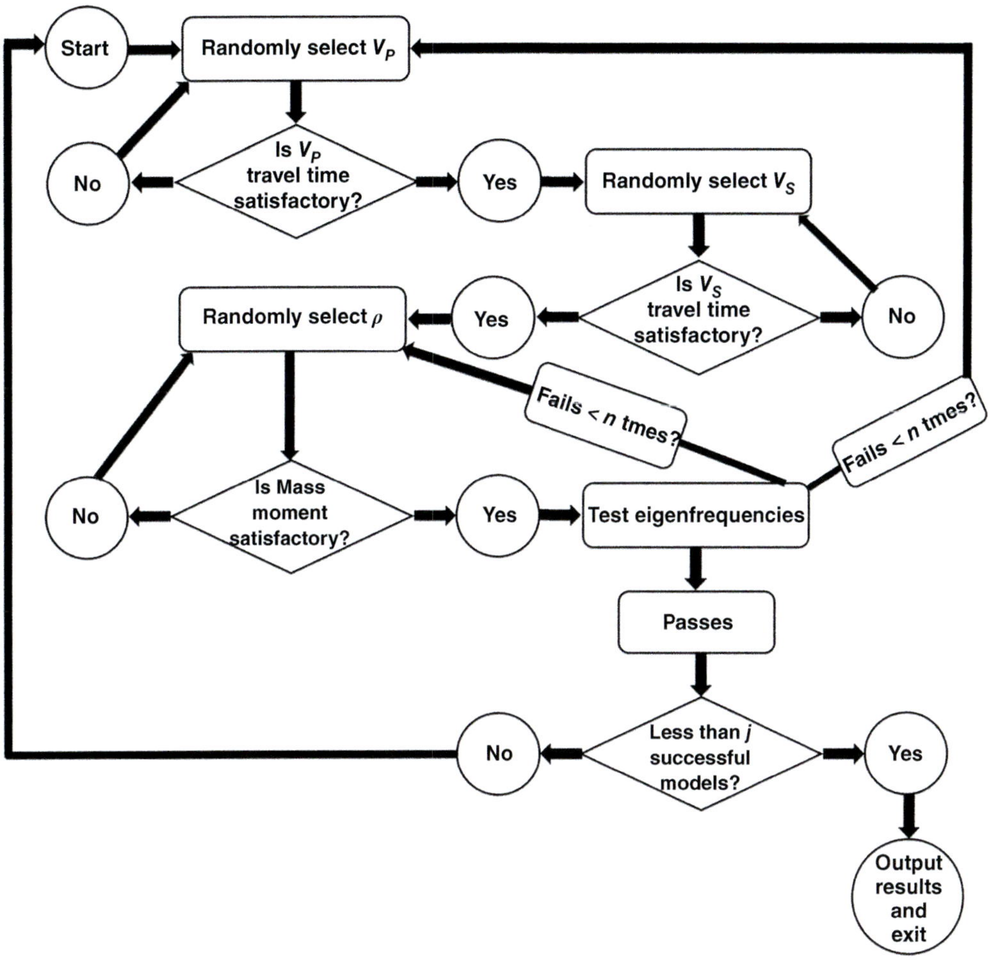

Figure 8.5 Flowchart of the Monte Carlo inversion method. Adapted from Press (1968).

8.3.1 Markov Chain Monte Carlo Optimization

Like Monte Carlo integration, the *MCMC* method is a random sampling method. Monte Carlo integration samples the multidimensional space $\mathcal{M}$ uniformly. The MCMC, on the contrary, attempts to visit the point $\mathbf{m}$ in $\mathcal{M}$ with probability proportional to a given distribution function $\pi(\mathbf{m})$. Additionally, it attempts to generate the samples of $\pi(\mathbf{m})$ in the model space $\mathcal{M}$ where no explicit mathematical expression exists for π and only an algorithm to calculate it at a given point in space is available (Sambridge and Mosegaard, 2002). The distribution function $\pi(\mathbf{m})$ is not necessarily the PDF because it may not be normalized to have a unity integral over the sampled region (equation 8.14). However, $\pi(\mathbf{m})$ is proportional to the probability distribution $p(\mathbf{m})$. This is a typical case in geophysical and other optimization problems we encounter in applied science and engineering where we compute the misfit function $E(\mathbf{m})$ and cast it in a Bayesian framework to define p as (see equations 8.5 and 8.6)

$$p(\mathbf{m}_k) \propto A \exp[-BE(\mathbf{m}_k)], \tag{8.21}$$

in which $\mathbf{m}_k$ is a model at any given point in $\mathcal{M}$ and A and B are constants.

MCMC is a powerful way to cast an optimization in a Bayesian framework to estimate the model parameters $\mathbf{m}$ and the uncertainty associated with their estimates. Given an observed data $\mathbf{d}$, we can calculate $p(\mathbf{d} \mid \mathbf{m})$—the probability of the data, given the model parameters. If we assume the prior as $p(\mathbf{m})$, then the Bayes' rule (see equation 8.4) says that the posterior probability of the model is proportional to $\pi(\mathbf{m}) \equiv p(\mathbf{d} \mid \mathbf{m})p(\mathbf{m})$ but with an unknown normalizing constant. Therefore, although $\pi(\mathbf{m})$ may not be the normalized probability density, if we can find a way to sample from it, we can estimate any quantity of interest such as the mean and variance. We can easily recover the normalized PDF by observing how often we sample a given volume $d\mathbf{m}$, which we already addressed earlier (for example, see equation 8.7), and discuss it further in the later part of this section. In addition, we can also calculate the marginal probability distribution of a given model component m_i by marginalizing (integrating over) all other components (equation 8.8).

MCMC tends to automatically put its sample points preferentially where $\pi(\mathbf{m})$ is large. In a high-dimensional space where computing $\pi(\mathbf{m})$ is expensive, this can be of great advantage over a pure MCI approach of Figure 8.5.

Two fundamental concepts, originally developed by Metropolis et al. (1953), led to the success of the MCMC methods in the modern high-performance computing environments (Sambridge and Mosegaard, 2002; Mosegaard and Sambridge, 2002; Press et al., 2007). The first one is that we should try to sample $\pi(\mathbf{m})$ not as completely unrelated independent random points but by using the property of the *Markov chain*. Markov chain represents a sequence of points $\mathbf{m}_0, \mathbf{m}_1, \mathbf{m}_2, \ldots$ such that each point $\mathbf{m}_i$ is chosen from a distribution that only depends on the value of the immediately preceding point $\mathbf{m}_{i-1}$. In other words, the chain of the visited points has a memory extending to one previous point, and thus, it is completely defined by a transition probability function of two variables $p(\mathbf{m}_i \mid \mathbf{m}_{i-1})$, given as the probability with which the point $\mathbf{m}_i$ is picked given a previous point $\mathbf{m}_{i-1}$.

The second fundamental concept is that if $p(\mathbf{m}_i \mid \mathbf{m}_{i-1})$ is chosen to satisfy the detailed balance equation given as

$$\pi(\mathbf{m}_{i-1})p(\mathbf{m}_i \mid \mathbf{m}_{i-1}) = \pi(\mathbf{m}_i)p(\mathbf{m}_{i-1} \mid \mathbf{m}_i), \tag{8.22}$$

then the Markov chain will ergodically sample $\pi(\mathbf{m})$, meaning that the sequence of the visited points $\mathbf{m}_0, \mathbf{m}_1, \mathbf{m}_2, \ldots$ will visit every point $\mathbf{m}$ in proportion to $\pi(\mathbf{m})$. Equation 8.22, in fact, expresses the concept of the physical equilibrium in the reversible transition

$$\mathbf{m}_{i-1} \leftrightarrow \mathbf{m}_i. \tag{8.23}$$

Equation 8.23 means that if $\mathbf{m}_{i-1}$ and $\mathbf{m}_i$ respectively occur in proportion to $\pi(\mathbf{m}_{i-1})$ and $\pi(\mathbf{m}_i)$, then the overall transition rates in each direction, and each product of a population density, and a transitional probability are the same. This reversibility property, in turn, follows from the fact that the Markov chain itself is ergodic. To see this, we integrate both sides of equation 8.22 with respect to $\mathbf{m}_{i-1}$ to obtain

$$\int \pi(\mathbf{m}_{i-1})p(\mathbf{m}_i \mid \mathbf{m}_{i-1})d\mathbf{m}_{i-1} = \pi(\mathbf{m}_i)\int p(\mathbf{m}_{i-1} \mid \mathbf{m}_i)d\mathbf{m}_i = \pi(\mathbf{m}_i). \tag{8.24}$$

The left-hand side of equation 8.24 is the probability of $\mathbf{m}_i$ that is obtained by integrating over all possible values of the previously visited point $\mathbf{m}_{i-1}$ with their corresponding transition probabilities. The right-hand side of the equation on the other hand is the desired $\pi(\mathbf{m}_i)$. Therefore, equation 8.24 says that if $\mathbf{m}_{i-1}$ is drawn from the distribution π, then so is its successor $\mathbf{m}_i$.

Before proceeding further, we also need to verify that the physical equilibrium in the reversible transition given by equation 8.22 can be rapidly approached from any initial starting point $\mathbf{m}_0$. A formal proof of this is rigorous and beyond the scope of this book. However, note that because of ergodicity even an unlikely value of $\mathbf{m}_0$ will be visited by the equilibrium Markov chain at least once. Because the Markov chain has no past memory, a choice of any starting point $\mathbf{m}_0$ is thus equivalent to picking up the equivalent distribution chain at that point. For practical implementations, we need to recognize that when we start from a very unlikely point, the successor points will also be quite unlikely until we reach a more probable part of the distribution. In practice, there is, therefore, a need to implement a *burn-in* phase to the MCMC chain where we step through the model space and discard a certain number of points before we start sampling. Two sampling strategies, (1) the *Metropolis–Hastings algorithm* and (2) the *Gibbs' Sampler*, provide the way to efficiently sample the model space, and we discuss below both of them.

Metropolis–Hastings Algorithm

Metropolis–Hastings algorithm was originally developed by Hastings (1970) as a generalization of the work of Metropolis et al. (1953). Note that unless we can find the transition probability function $p(\mathbf{m}_i \mid \mathbf{m}_{i-1})$ that satisfies the balance equation (equation 8.22), there is no way to proceed, and the Metropolis–Hastings algorithm provides a plan for it. We explain below the algorithm for how to start from an initial point, say $\mathbf{m}_1$, and move to the next point $\mathbf{m}_2$, and the criteria for moving to other successive points during iteration is the same.

First, we pick a proposal distribution $q(\mathbf{m}_2 \mid \mathbf{m}_1)$. This proposal distribution can be anything if the successive steps of iteration can adequately sample the region of interest. As an example, $q(\mathbf{m}_2 \mid \mathbf{m}_1)$ can be a multivariate normal distribution centered around $\mathbf{m}_1$. Alternatively, $q(\mathbf{m}_2 \mid \mathbf{m}_1)$ could be defined as a uniform distribution between predefined minimum and maximum search limits around $\mathbf{m}_1$.

Now to generate a step, starting from $\mathbf{m}_1$, we first generate a candidate point $\mathbf{m}_{2c}$ by drawing from the proposal distribution. Next, we evaluate an acceptance probability $\alpha(\mathbf{m}_1, \mathbf{m}_{2c})$ as

$$\alpha(\mathbf{m}_1, \mathbf{m}_{2c}) = \min\left\{1, \frac{\pi(\mathbf{m}_{2c})q(\mathbf{m}_1 \mid \mathbf{m}_{2c})}{\pi(\mathbf{m}_1)q(\mathbf{m}_{2c} \mid \mathbf{m}_1)}\right\}. \quad (8.25)$$

Finally, we either accept the candidate point with the probability $\alpha(\mathbf{m}_1, \mathbf{m}_{2c})$ and set $\mathbf{m}_2 = \mathbf{m}_{2c}$, or reject it to leave the point unchanged by setting $\mathbf{m}_2 = \mathbf{m}_1$, that is, revisiting $\mathbf{m}_1$ once again. Thus, the result of this process is a transition probability given as

$$p(\mathbf{m}_2 \mid \mathbf{m}_1) = q(\mathbf{m}_2 \mid \mathbf{m}_1)\alpha(\mathbf{m}_1, \mathbf{m}_2), \quad (8.26)$$

in which $\mathbf{m}_2 \neq \mathbf{m}_1$. We can easily verify that the algorithm satisfies the detailed balance equation. To see this, we multiply both sides of equation 8.25 by the denominator of the second argument on the right-hand side and write

$$\begin{aligned}
&\pi(\mathbf{m}_1)q(\mathbf{m}_2 \mid \mathbf{m}_1)\alpha(\mathbf{m}_1, \mathbf{m}_2)\\
&\quad = \min\{\pi(\mathbf{m}_1)q(\mathbf{m}_2 \mid \mathbf{m}_1), \pi(\mathbf{m}_2)q(\mathbf{m}_1 \mid \mathbf{m}_2)\}\\
&\quad = \min\{\pi(\mathbf{m}_2)q(\mathbf{m}_1 \mid \mathbf{m}_2), \pi(\mathbf{m}_1)q(\mathbf{m}_2 \mid \mathbf{m}_1)\}\\
&\quad = \pi(\mathbf{m}_2)q(\mathbf{m}_1 \mid \mathbf{m}_2)\alpha(\mathbf{m}_2 \mid \mathbf{m}_1), \qquad (8.27)
\end{aligned}$$

which, from equation 8.26, can be recognized as the detailed balance equation. In many instances, it is possible to choose a proposal distribution $q(\mathbf{m}_2 \mid \mathbf{m}_1)$ such that equation 8.25 takes a simple form. For example, if we assume normal distribution with a fixed covariance, then $q(\mathbf{m}_2 \mid \mathbf{m}_1)$ will only depend upon the absolute difference $|\mathbf{m}_1 - \mathbf{m}_2|$. Then, the ratio $q(\mathbf{m}_1 \mid \mathbf{m}_{2c})/q(\mathbf{m}_{2c} \mid \mathbf{m}_1)$ is unity. In many situations, for some components m of $\mathbf{m}$, $q(m_{2c} \mid m_1)$ is lognormally distributed with mode at m_1. Under such situation, the ratio of these components is m_{2c}/m_1 (Press et al., 2007).

Figure 8.6 shows a simplified workflow for the Metropolis–Hastings algorithm, modified from the pseudocode given in Zunino et al. (2015), which they call an *extended Metropolis sampler*. Recalling equation 8.5, given the likelihood function $l(\mathbf{d} \mid \mathbf{m})$ and the prior $p(\mathbf{m})$, our objective is to compute the unnormalized distribution $\pi(\mathbf{m}) = l(\mathbf{d} \mid \mathbf{m})p(\mathbf{m})$. When we believe that a sufficiently large number of samples are drawn from $\pi(\mathbf{m})$, we then normalize $\pi(\mathbf{m})$ as an estimate of the posterior PDF $p(\mathbf{m} \mid \mathbf{d})$. Once we estimate $p(\mathbf{m} \mid \mathbf{d})$, we can also use equations 8.8–8.10 and estimate the conditional PDF $p(m_i \mid \mathbf{d})$ for each model parameter m_i, the mean model $\langle \mathbf{m} \rangle$, and the model covariance $\mathbf{C}_M$.

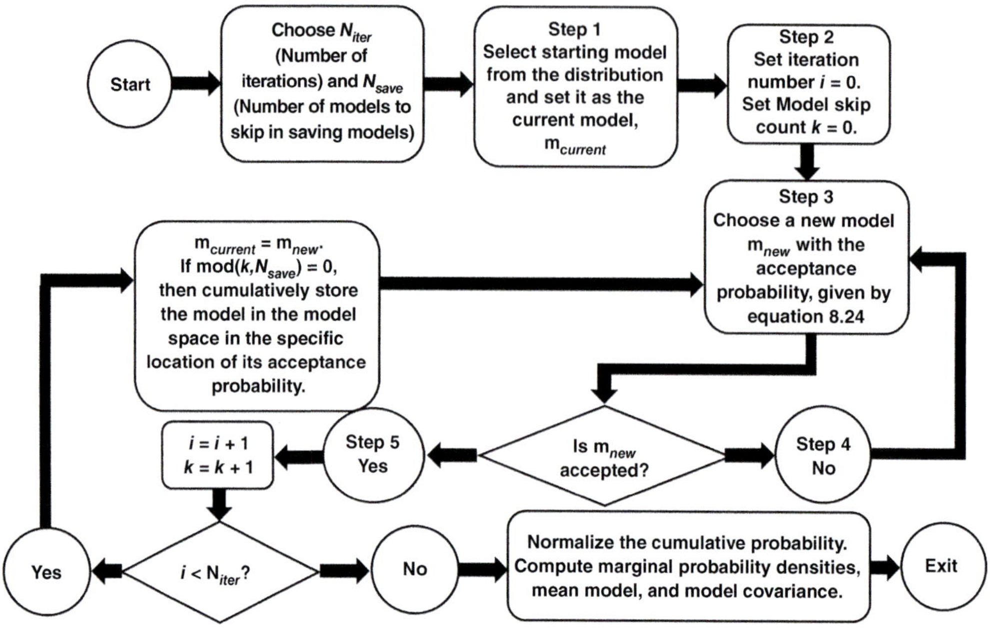

Figure 8.6 A simplified workflow for the Metropolis–Hastings algorithm.

The algorithm of Zunino et al. (2015) uses two user-defined parameters: (1) N_{iter}, the number of iterations, and (2) N_{save}, the number of proposed models to skip before saving the current model in the ensemble of solutions.

The value of N_{iter} must be sufficiently large as to sample the entire model space of interest with sufficient accuracy. Consequently, the choice of this parameter is problem dependent. Additionally, it also depends upon the complexity of the forward problem being solved. Restricting to seismic inverse problems, if we use the convolution as the basis for forward computation of synthetic seismograms such as the poststack or AVA inversion, computing the forward response is fast and we may use a large value for N_{iter}. However, if we use wave-equation-based method for forward modeling like in FWI, computational complexities involved do not allow a very high value for the choice of N_{iter}. As we previously discussed, one major component for the computational efficiency of MCMC is the length of the *burn-in* phase, or the first stage where the MCMC method finds the highly probable regions in the model space where it can begin sampling. If we do not adequately iterate after this burn-in phase, the model space is likely to be under-sampled. For FWI problems, it is therefore necessary to carefully choose our prior $p(\mathbf{m})$ so that we can reduce the length of this burn-in phase. There are new variations of MCMC such as the gradient-based MCMC (Zhao and Sen, 2022), which use the gradient information, and not only reduce the length of the burn-in phase but also improve the overall efficiency in sampling the model space.

The Metropolis–Hastings algorithm produces samples that are correlated. Therefore, to obtain statistically independent models so that the estimated PDF is unbiased, it is important to save only a part of the models, not all of them. The second user-defined parameter N_{save} is used for this purpose. Out of all models generated during iteration, only every N_{save} proposed model is saved in the ensemble of solutions for estimating the PDF and other related parameters. Like N_{iter}, the choice of N_{save} is also problem dependent. Typically, a few short trial runs are usually necessary for finding an optimum value for N_{save} (Zunino et al., 2015).

Following the choice of N_{iter} and N_{save}, the steps shown in Figure 8.6 are the following:

1) Randomly draw a model from the distribution and set as the current model, $\mathbf{m}_{current}$.
   a) Compute the forward synthetic data $\mathbf{s}_{current} = g(\mathbf{m}_{current})$ and match with the observed data $\mathbf{d}$ to compute the objective and the unnormalized distribution $\pi(\mathbf{m}_{current})$.
2) Initialize the iteration counter i and the model saving counter k to zero.
3) Draw a new model $\mathbf{m}_{new}$ from the proposal distribution $q(\mathbf{m}_{new} \mid \mathbf{m}_{current})$.
   a) Compute forward synthetic data $\mathbf{s}_{new} = g(\mathbf{m}_{new})$ and match with the observed data to compute the objective and the unnormalized distribution $\pi(\mathbf{m}_{new})$.
   b) Accept $\mathbf{m}_{new}$ with probability $\alpha(\mathbf{m}_{current} \mid \mathbf{m}_{new})$ using equation 8.25.
4) If $\mathbf{m}_{new}$ is rejected, go back to step 3.
5) If $\mathbf{m}_{new}$ is accepted, increment both iteration counter i and model saving counter k by one and check if the maximum numbers of iterations N_{iter} have been reached.
   a) If not, then set $\mathbf{m}_{current} = \mathbf{m}_{new}$ and check if this iteration step requires saving the model. Note that this checking is done by computing the modulus function mod(k, N_{save}). If this step requires that model must be saved, do so and go to step 4. Otherwise, go directly to step 4 without saving the model.
   b) If the current iteration number has reached N_{iter}, then normalize $\pi(\mathbf{m})$ as an estimate of the posterior PDF $p(\mathbf{m} \mid \mathbf{d})$. Also, compute the mean model $\langle \mathbf{m} \rangle$ and the model covariance $\mathbf{C}_M$ (see equations 8.9 and 8.10) and exit.

Steps 1 through 5 are the general workflow for the Metropolis–Hastings algorithm. In parallel computing environments, major portions of these steps can be parallelly implemented which can improve overall performance of the algorithm. We omit these parallel implementations here and focus on the fundamental workflow and refer to Zunino et al. (2015) for the details on how to modify this algorithm for parallel computing.

Although the above steps provide the workflow of the Metropolis–Hastings, we have not explained step 5b—how to normalize $\pi(\mathbf{m})$ to estimate $p(\mathbf{m} \mid \mathbf{d})$. A convenient way to do this normalization is the binning technique of Frazer and Basu (1990), which was adopted by Sen and Stoffa (1991) and Mallick (1995, 1999). This binning technique works as follows for seismic problems where the model parameters (P- and S-wave velocity, density, etc.) are functions of depth or time:

1) Assign a 2D floating-point array for each model parameter in which the first dimension is the value of the parameter and the second dimension is depth.
   a) For example, assume that the P-wave velocity V_P varies between a lower limit of $V_P^{\min} = 2000$ m/s and an upper limit of $V_P^{\max} = 5000$ m/s. Also, assume that our minimum and maximum depth intervals are $Z_{min} = 0$ m and $Z_{max} = 5000$ m. If we would like to resolve V_P at a resolution of $\Delta V_P = 10$ m/s and our depth resolution $\Delta Z = 5$ m, then the size of the array in the first dimension would be $N_{VP} = \frac{(V_P^{\max} - V_P^{\min})}{\Delta V_P} + 1 = 301$ and the size of the second dimension would be $N_Z = \frac{(Z_{\max} - Z_{\min})}{\Delta Z} + 1 = 1001$.
   b) We can do the same for other model parameters such as the S-wave velocity (V_S), density (ρ), etc. Note that the second dimension (N_Z) for all model parameters would be the same, but the first dimension (N_{VS}, N_ρ, etc.) may vary depending upon their desired resolution, defined by the choice of $V_S^{\min}$, $V_S^{\max}$, ΔV_S, $\rho_{\min}$, $\rho_{\max}$, $\Delta\rho$, etc.
2) Initialize the 2D arrays assigned for each model parameter to zero.
3) Because the number of depth samples is N_Z, for each accepted model $\mathbf{m}_{current}$, the first N_Z samples are $(\mathbf{V}_P)_{current}$, next N_Z samples are $(\mathbf{V}_S)_{current}$, next N_Z samples are $(\rho)_{current}$, and so on. So, for each sample $j = 1, 2, \ldots, N_Z$ for $(\mathbf{V}_P)_{current}$, $(\mathbf{V}_S)_{current}$,

$(\rho)_{current}$, etc., find their location in the first dimension of the assigned 2D array. For example, if the jth sample of $(\mathbf{V}_P)_{current}$ is V_P^j, then its location is $i_P^j = \frac{(V_P^j - V_P^{min})}{\Delta V_P} + 1$. We can similarly calculate i_S^j, i_ρ^j, etc.

4) After calculating i_P^j, i_S^j, i_ρ^j, etc., for each $j = 1, 2, \ldots, N_Z$ from step 3, add the accepted probability $\alpha(\mathbf{m}_{current} \mid \mathbf{m}_{new})$ to its existing value at the location of the arrays at i_P^j, i_S^j, i_ρ^j, etc., for each depth sample j.
5) At the end of the run, the 2D arrays are the cumulative probability for each model parameter. So, if we divide the first dimensions of each array by the total number of stored models, it is then the PDF for each model parameter. From the workflow of Figure 8.6, it is evident that this normalizing constant is simply $k - 1$.

Gibbs' Sampler

The Gibbs sampler is an important special case of the Metropolis–Hastings algorithm whose details can be found in Geman and Geman (1984), Gelfand and Smith (1990), Gelfand et al. (1990), Casella and George (1992), among others. This is based on the fact that any multivariate distribution is uniquely determined by the set of all of its full conditional distributions (Press et al., 2007).

In equation 8.8, we defined the full conditional distribution. Using this definition and following the notations used in the previous section, a full conditional distribution of $\pi(\mathbf{m})$ is obtained by setting all components of $\mathbf{m}$ constant, except one, which we may refer as m_i, and then by sampling as a function of m_i only. This, in other words, means that we sample through the distribution $\pi(\mathbf{m})$ along a single coordinate direction, with fixed values along other directions. We denote this full conditional distribution as $\pi(m_i \mid \mathbf{m}^{i-})$ in which $\mathbf{m}^{i-}$ represents values of all coordinates except i being kept constant.

We now construct the Metropolis–Hastings algorithm in which we allow only one component m_i to vary and keep the rest constant. Under this condition, equation 8.25 becomes

$$\alpha(m_{i1}, m_{i2c} \mid \mathbf{m}^{i-}) = \min \times \left\{ 1, \frac{\pi(m_{i2c} \mid \mathbf{m}^{i-}) q(m_{i1} \mid m_{i2c} \mid \mathbf{m}^{i-})}{\pi(m_{i1} \mid \mathbf{m}^{i-}) q(m_{i2c} \mid m_{i1}, \mathbf{m}^{i-})} \right\}. \tag{8.28}$$

In equation 8.28, m_{i1} is the ith component of $\mathbf{m}_1$ under consideration and m_{i2c} is the ith component of the candidate point $\mathbf{m}_{2c}$ which we are considering, keeping all other components constant. We now pick our proposal distribution as

$$q(m_{i2} \mid m_{i1}, \mathbf{m}^{i-}) = \pi(m_{i2} \mid \mathbf{m}^{i-}). \tag{8.29}$$

By doing so, the second argument in the right-hand side of equation 8.28 becomes 1. Therefore, the acceptance probability α in equation 8.28 is 1. This, in other words, means that if we propose a value m_{i2} from the full conditional probability distribution $\pi(m_i \mid \mathbf{m}^{i-})$, we can then always accept it. The advantage of using the Gibbs sampler is therefore obvious. However, its disadvantage is that the full conditional distribution must be normalized as a true probability distribution. Because unless it is normalized to represent PDF, we cannot use it as a conditional probability distribution. Consequently, to take advantage of the Gibbs sampler, it is necessary to calculate the normalizing constant $\int \pi(m_i \mid \mathbf{m}^{i-}) dm_i$ for every $\mathbf{m}^{i-}$. In addition, we must have a practical and working algorithm to draw m_2 from the normalized distribution. However, we must also emphasize that because one-dimensional normalizing constants are much easier to compute than the multi-dimensional ones, if we can find ways to compute the one-dimensional normalizing constants, such as the binning technique of Frazer and Basu (1990), then the Gibbs sampler would surely have a lot of numerical advantage over the Metropolis–Hastings.

To implement the Gibbs sampler, we can proceed as follows:

1) Cycle through each component m_i of $\mathbf{m}$. We can do so either in a fixed cyclic order or in a random order.

2) For each component i, we hold all other components $j \neq i$ constant and draw a new value of m_i from the full conditional distribution $\pi(m_i \mid \mathbf{m}^{i-})$ of all possible values of the ith component.
   a) To properly draw this new value, it will be necessary to normalize $\pi(m_i \mid \mathbf{m}^{i-})$ by computing $\int \pi(m_i \mid \mathbf{m}^{i-})dm_i$, either analytically or numerically for every $\mathbf{m}^{i-}$ and dividing it to $\pi(m_i \mid \mathbf{m}^{i-})$.
3) Accept m_i and proceed to the next component of $\mathbf{m}$.

We further discuss the implementation aspects of the Gibbs sampler in relation with the SA method.

Importance Sampling and MCMC

In Section "Importance Sampling," we discussed how the error in the Monte Carlo integration could be reduced by importance sampling. We start from equation 8.15 and write using the notations used in the previous section as

$$I = \int f(\mathbf{m})d\mathbf{m} = \int \frac{f(\mathbf{m})}{p(\mathbf{m})} p(\mathbf{m})d\mathbf{m} \quad (8.30)$$

for a carefully chosen $p(\mathbf{m})$. In Section "Importance Sampling," we found that an optimal p would resemble f in functional form and would allow for an efficient method to sample uniformly over $p(\mathbf{m})d\mathbf{m}$. However, it may appear that the MCMC is a general-purpose way to sample over any p, and the importance sampling is quite easy to implement. This is, however, not true. Like the Gibbs sampler, the problem is the normalizing constant. The primary advantage of MCMC is that it allows sampling over a distribution $\pi(\mathbf{m})$ without any need for normalization. The normalized probability distribution $p(\mathbf{m})$ that is being sampled over is obviously

$$p(\mathbf{m}) \frac{\pi(\mathbf{m})}{\int \pi(\mathbf{m})d\mathbf{m}}. \quad (8.31)$$

Using equation 8.31, we can therefore write equation 8.30 as

$$I = \left\{ \int \pi(\mathbf{m})d\mathbf{m} \right\} \left\{ \int \frac{f(\mathbf{m})}{\pi(\mathbf{m})} p(\mathbf{m})d\mathbf{m} \right\}. \quad (8.32)$$

In equation 8.32, the differential $p(\mathbf{m})d\mathbf{m}$ can be sampled over by MCMC if we only know $\pi(\mathbf{m})$. The denominator $\pi(\mathbf{m})$ in equation 8.32 can also be computed. There is, however, no easy way to compute the normalizing constant $\int \pi(\mathbf{m})d\mathbf{m}$, except using the binning technique of Frazer and Basu (1990). Although straightforward to implement, it is important to ensure that the binning technique adequately samples the model space so that it is a close approximation of the normalizing constants for each model parameter under consideration.

8.3.2 Simulated Annealing Optimization

Simulated annealing mimics the annealing process for global optimization. It is well known from crystallography that when a crystalline material is slowly cooled from its melting point, it leads to the formation of highly ordered crystals. The slower the rate of cooling, more perfect is the crystal growth, and lower is the lattice energy. This entire physical process can be used as an optimization procedure if we treat our objective function like the lattice energy and attempt to minimize it by slowly reducing a temperature-like control parameter.

The annealing process can be simulated to solve large optimization problems. In such a simulation, we identify the model parameters as the state space variables and the objective function of the optimization problem with the energy E. In each step or iteration, we then simulate thermal fluctuations by perturbing the model parameters and another controlling parameter T like the temperature in the physical annealing. While we randomly perturb the model parameter in each iteration, we vary (slowly decrease) the controlling parameter T using a specified rate of cooling, which we refer as the *annealing schedule*. Because the procedure simulates the process of physical annealing, the method is called the *SA* optimization.

Simulated annealing optimization method, formulated by Kirkpatrick et al. (1983), has been used to geophysical problems by Rothman (1985, 1986), Basu and Frazer (1989), Frazer and Basu (1990), Sen and Stoffa (1991), among others. A computationally efficient variation of SA, known as the *very fast simulated re-annealing* (*VFSR*), was later developed by Ingber (1989). In geophysics, the application of VFSR to solve seismic inverse problems can be found in Zhao et al. (1996) and Sen and Stoffa (2013).

Kirkpatrick Algorithm

In each iteration step j of the *Kirkpatrick algorithm*, the model $\mathbf{m}_j$ is randomly perturbed and objective function (energy) E is calculated. The model is then accepted if E decreases. However, if E increases, then the model is accepted with the probability

$$p_{accept} = \exp\left(\frac{-\Delta E}{T}\right), \tag{8.33}$$

where ΔE is the change of the objective function and T is the temperature. If the new model is rejected, then another new perturbation is attempted, and this process is repeated until a prespecified number of iterations N_{iter} or the objective function ceases to decrease further.

It can easily be seen that if T is kept constant, the Kirkpatrick algorithm is actually the Metropolis–Hastings algorithm that samples the probability distribution

$$p(\mathbf{m}) = \frac{\exp\left[-\dfrac{E(\mathbf{m})}{T}\right]}{Z(T)}, \tag{8.34}$$

in which $Z(T)$ is the normalizing constant. In statistical physics, the distribution given by equation 8.34 is known as the *Gibbs–Boltzmann distribution*. In SA, however, T is slowly decreased from a high value down to zero. For decreasing T, the Gibbs–Boltzmann distribution converges toward a distribution having all its probability mass in the global minimum for E. Thus, as the temperature gradually approaches zero, the probability that our system is close to the global minimum for its objective function approaches one.

Simulated annealing allows a convenient way for sampling the Gibbs–Boltzmann distribution using the Gibbs sampler. Implementation of the Gibbs sampler for SA was formulated by Rothman (1985, 1986) to solve residual statics problem in seismic data processing. This method is a modified version of the Kirkpatrick algorithm and is known as the *heat-bath algorithm*. Later, Frazer and Basu (1990) and Sen and Stoffa (1991) applied this method to solve seismic waveform inversion problems. We briefly summarize below this method using the notations of Sen and Stoffa (1991).

In heat-bath algorithm, each model parameter is sequentially visited for all its possible values while keeping the other model parameters fixed, and then the Gibbs–Boltzmann distribution is evaluated as

$$p(m_{ij}) = \frac{\exp\left[-\dfrac{E(m_{ij})}{T}\right]}{\sum_{j=1}^{M}\exp\left[-\dfrac{E(m_{ij})}{T}\right]}. \tag{8.35}$$

In equation 8.35, the subscript $i = 1, 2, \ldots, N$ represents a specific model parameter and the subscript $j = 1, 2, \ldots, M$ represents the number of possible values that the model parameter m_i can take. E refers to the energy and T is a controlling parameter analogous to the temperature.

First, a random model is drawn from the distribution given by equation 8.34. Considering a seismic inversion problem under an isotropic elastic assumption, we can assume our model to be of the form

$$\mathbf{m} = \begin{pmatrix} m_1 \\ m_2 \\ \cdot \\ \cdot \\ \cdot \\ m_N \end{pmatrix} = \begin{pmatrix} V_{P1} \\ \vdots \\ V_{PN/3} \\ V_{S1} \\ \vdots \\ V_{SN/3} \\ \rho_1 \\ \vdots \\ \rho_{N/3} \end{pmatrix} = \begin{pmatrix} \mathbf{V}_P \\ \mathbf{V}_S \\ \boldsymbol{\rho} \end{pmatrix}. \tag{8.36}$$

In equation 8.36, V_{Pi}, V_{Si}, and ρ_i respectively represent the P-wave velocity, S-wave velocity, and density for the depth sample $i = 1.2, \ldots, \frac{N}{3}$, which, on the far right-hand side, we also represent as vectors $\mathbf{V}_P$, $\mathbf{V}_S$, and ρ, each with $N_D = N/3$ depth samples. We also assume that each model parameter m_j can take M different values. In our discussion of the Metropolis–Hastings algorithm, we have discussed how to choose M for each model parameter. Note that M can be different for different model parameters. Having the flexibility to set different values of M for different model parameters is important. In seismic inversion, the lower and upper bounds of V_P, V_S, ρ, etc., generally vary with depth. Therefore, having the ability to choose different values of M for different model parameters m_j, $j = 1, 2, \ldots, N$ in equation 8.36 gives flexibility to the optimization method.

After drawing a random model, we now vary m_1 for all its possible M values but keep m_2 through m_N fixed. For all possible values om m_1 (that is, for m_{11} through m_{1M}), we compute the synthetic data $\mathbf{s}_{1j} = g(\mathbf{m}_{1j})$, $j = 1, 2, \ldots, M$, and match with the observed data $\mathbf{d}$ to compute the objective or the energy $E(m_{1j})$ and evaluate the distribution (equation 8.35). A new value of m_1 is then drawn from this distribution, and we repeat the same process by varying m_2 only and keeping m_1 and m_3 through m_N fixed. We keep repeating the same process by varying each model parameter, one at a time keeping the rest of the parameters fixed. Once all the model parameters are perturbed this way, we say that one iteration or sweep is completed. Then, the temperature is lowered, and the same process is repeated until *crystallization*, that is, when we attain a low-energy state and reach the global minimum.

Note that by keeping all model parameters fixed except one and computing the Gibbs–Boltzmann distribution (equation 8.35), the heat-bath approach estimates the full conditional distribution for each model parameter via the Gibbs sampler. We can use the same binning technique of Frazer and Basu (1990) to compute these PDFs and then compute the mean model and the model covariance. Figure 8.7 shows the workflow for the heat-bath algorithm.

We discussed above all the steps shown in Figure 8.7 except how do we choose the annealing schedule? In other words, how do we choose the initial temperature T and reduce it from one iteration (sweep) to the next? In annealing theory, a perfect crystallization occurs when the rate of cooling is infinitely slow starting from a high temperature above the melting point to the critical temperature T_C where the crystallization occurs (Sambridge and Mosegaard, 2002). In practice, however, we must find an optimum cooling or annealing schedule that will take us from a high temperature to T_C in finite steps and yet the energy of the system reaches near the global minimum. For numerical systems like the seismic or other geophysical inverse problems, Nulton and Salamon (1988) developed such a near-optimum annealing schedule that keeps the system close to the thermal equilibrium during all steps of iteration. In this *Nulton–Salamon annealing*, the mean value of E, given as $\langle E \rangle$, is kept at a constant distance

$$v = \frac{\langle E \rangle - \langle E \rangle_{eq}}{\sigma_E(T)} \tag{8.37}$$

from an equilibrium mean value $\langle E \rangle_{eq}$, which we never realize. The denominator $\sigma_E(T)$ on the right-hand side of equation 8.37 is the standard deviation of the fluctuating objective (energy) function, and the distance v is called the thermodynamic speed. Using the equilibrium condition of equation 8.37, Nulton and Salamon (1988) obtained the differential equation for the annealing schedule as

$$\frac{dT}{dt} = -\frac{vT}{\varepsilon(T)\sqrt{C(T)}}, \tag{8.38}$$

in which $\varepsilon(T)$ and $C(T)$ are respectively the relaxation time and the heat capacity. Although the differential equation 8.38 can be solved to obtain a near-optimum annealing schedule for T as a function of time (iteration) t (for details, see Sambridge and Mosegaard, 2002), in practice any annealing schedule works fine as long as the rate of cooling is slow. For residual statics problems, Rothman (1986) proposed an annealing schedule and given as

$$T = T_0(0.99)^k, \tag{8.39}$$

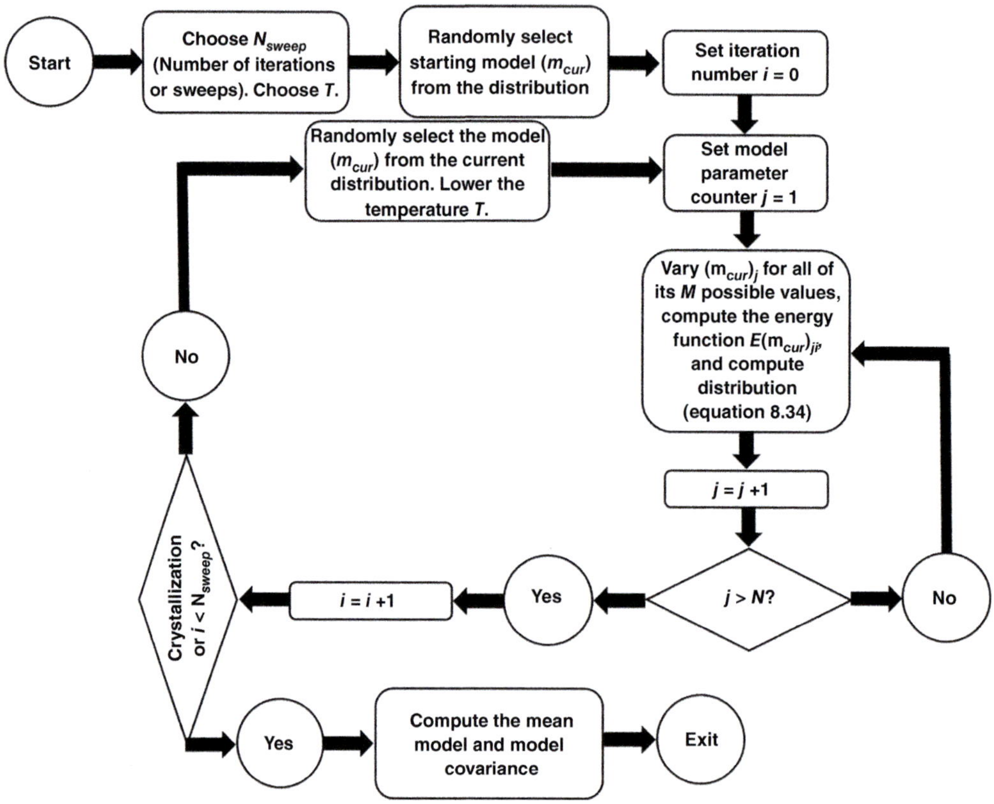

Figure 8.7 Workflow for the heat-bath algorithm.

in which k is the iteration number. For seismic waveform inversion problems, Sen and Stoffa (1991) used this schedule for different values of T_0 and demonstrated that the best results can be obtained for $T_0 = 0.005$ where the low-energy state was achieved in much fewer iteration steps compared to using higher values of T_0 (0.5, 0.1, and 0.05). To summarize, a simple annealing schedule of equation 8.39 works fine, but the choice of the parameter T_0 is problem-dependent and may need a few trial runs before finding an optimum value for this parameter.

VFSR Algorithm

VFSR algorithm, developed by Ingber (1989), is a modification of the SA method that is designed for a rapid convergence of the algorithm to the global optimum and has been found to be suitable for geophysical applications (Zhao et al., 1996; Sen and Stoffa, 2013).

As we discussed above, the SA chooses a new model from the entire model space from the Gibbs–Boltzmann type of distribution and compares with the current model. VFSR, on the contrary, chooses new models from a Cauchy distribution, which depends only on the current model and the temperature parameter. By lowering the temperature in successive iterations, it can be shown that VFSR attains the low-energy state (global optimum) in much fewer iteration steps than SA. Because algorithmic aspects of VFSR is not different from SA, we do not repeat them here and refer to Zhao et al. (1996) for details.

8.3.3 Genetic Algorithm Optimization

Genetic algorithm is a global optimization method based on the natural analogy from genetics and evolution theory (Holland, 1975; Goldberg, 1989). In geophysics, GA has been used in a variety of inverse problems (Stoffa and Sen, 1991; Sen and Stoffa, 1992, 1996; Mallick, 1995, 1999; Du and MacGregor, 2010; Mallick and Adhikari, 2015; Mazzotti et al., 2016; Pafeng et al., 2017; Ayani et al., 2020, among others).

Here, we discuss the GA methodology with reference to the seismic inversion problem. However, applying this method to solve other optimization problems will be clear from these discussions. For simplicity, we restrict to isotropic elastic inverse problem where our model vector $\mathbf{m}$ consists of N distinct components containing the P- and S-wave velocities ($\mathbf{V}_P$ and $\mathbf{V}_S$) and density (ρ) for each depth sample with $N/3$ being the total number of depth samples (equation 8.36). In gradient-based methods we discussed in Chapter 7, it is first necessary to define an initial or starting model of $\mathbf{V}_P$, $\mathbf{V}_S$, and ρ. Following this, these methods use the gradient of the objective function to search around this initial model and obtain the estimated model. In global methods, on the other hand, we randomly select the models from a distribution. Although sampling methods such as the Metropolis–Hastings and Gibbs sampler are computationally efficient, global methods such as MCMC, SA, and GA require a lot of sampling of the prior $p(\mathbf{m})$ and combine them with the likelihood function $l(\mathbf{d} \mid \mathbf{m})$ to compute the unnormalized posterior PDF $\pi(\mathbf{m})$ and its normalized version $p(\mathbf{m} \mid \mathbf{d})$ as shown in equations 8.5 and 8.7. Computing the likelihood function $l(\mathbf{d} \mid \mathbf{m})$, in turn, requires computing the objective (or energy in SA optimization) $E(\mathbf{m})$. And, to compute $E(\mathbf{m})$, we must compute forward synthetic seismic response and match with observation. We can do this matching as a regularized error or misfit (for example, equation 7.22) and attempt to minimize $E(\mathbf{m})$. Most global optimization problems, specifically SA and GA, however, can be better formulated if cast as maximization rather than minimization. To do so, we set $E(\mathbf{m})$ as the normalized cross-correlation as a measure of the similarity between the observed and synthetic data and attempt to maximize it. Such a cross-correlation can be written as (Sen and Stoffa, 1991)

$$E(\mathbf{m}) = \frac{1}{M}\mathrm{Re}\left(\frac{\mathbf{d}^{\dagger}\mathbf{s} + \mathbf{s}^{\dagger}\mathbf{d}}{\mathbf{d}^{\dagger}\mathbf{d} + \mathbf{s}^{\dagger}\mathbf{s}}\right). \tag{8.40}$$

In equation 8.40, M is the total number of samples in both observed and synthetic data vectors ($\mathbf{d}$ and $\mathbf{s}$), the superscript $\dagger$ denotes a complex conjugate transpose, and $\mathrm{Re}(\cdot)$ stands for the real part of its argument. Irrespective of whether we cast optimization as minimization or maximization, computing $E(\mathbf{m})$ requires expensive computation of the forward synthetic data vector $\mathbf{s} = g(\mathbf{m})$. Our goal is therefore to limit the number of forward synthetic computations and yet sufficiently sample from the distribution such that our estimation of the posterior PDF is sufficiently accurate. For any inverse problem, it is therefore important to obtain an estimate of the initial model. For local gradient-based methods, we can use this initial model as our starting point and use the gradient information to guide optimization. For global methods, on the other hand, we can provide a distribution around this initial model as our prior $p(\mathbf{m})$. By randomly sampling from this prior, computing synthetic data, and matching with observation, we may then estimate the posterior PDF. Before discussing the GA method, we first discuss how an initial model can be estimated for seismic inverse problems because estimating the initial model is fundamental to any optimization, local or global.

Estimation of the Initial Model for Seismic Inverse Problems

Figure 8.8 shows a real vertical-component seismic dataset with estimated P-wave velocity field ($\mathbf{V}_P$) as a function of time. Figure 8.8a shows the prestack gather in the CMP domain from the real 3D seismic data from the Rock-Springs Uplift (RSU), Wyoming, USA. The time axis is increasing vertically downward, and the source-to-receiver offset is horizontal, increasing from left to right. Note the hyperbolic moveout of different reflections in Figure 8.8a as functions of time (t) and

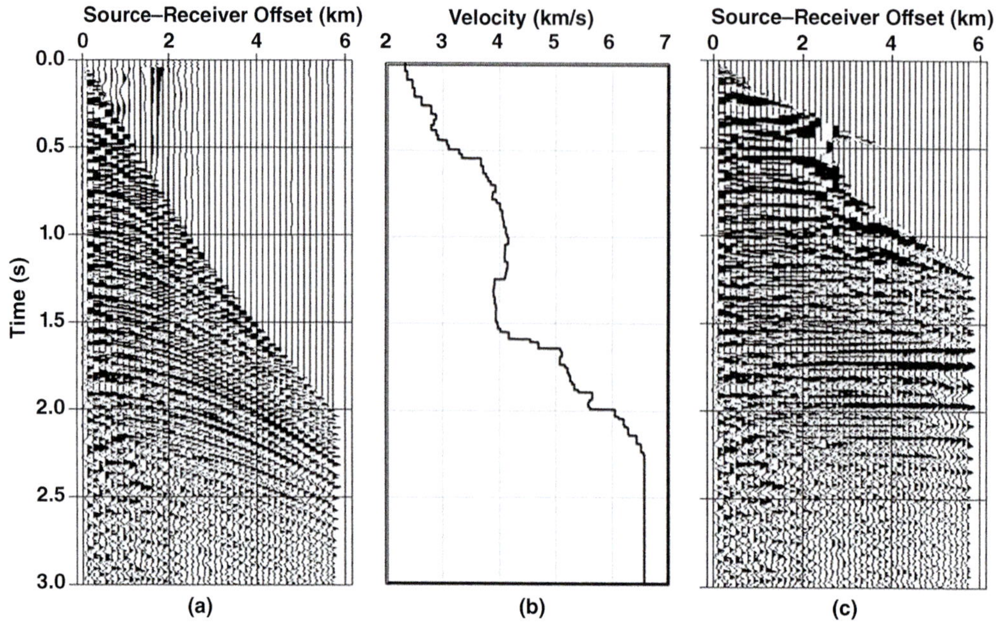

Figure 8.8 Illustration of finding the P-wave velocity from prestack seismic data. (a) Input seismic gather in CMP domain, (b) picked P-wave velocity, and (c) same as (a) after NMO correction using the velocity field shown in (b).

source-to-receiver offset (x). A standard step in seismic analysis is velocity analysis, in which a time-dependent velocity field, as shown in Figure 8.8b, is chosen which optimally corrects the input prestack gather of Figure 8.8a for the NMO, which is shown in Figure 8.8c. The velocity analysis procedure is standard in seismic data processing, details of which can be found in any seismic data processing textbook such as Yilmaz (1987).

The estimated $\mathbf{V}_P$ shown in Figure 8.8b is a function of zero-offset two-way time t which can be estimated directly from the seismic data. For time domain inversion, this $\mathbf{V}_P(t)$ can be use as the initial P-wave velocity model. In waveform inversion, the initial model must be converted to depth domain, which is also straightforward.

It can be seen from Figure 8.8b that the resolution of the estimated $\mathbf{V}_P$ from velocity analysis is low. Such a low-resolution $\mathbf{V}_P$ model is generally adequate as a starting model for geologically simple to moderate areas. In many circumstances, however, subsurface geology can be complex. In oil and gas exploration such a situation is often encountered in the vicinity of the salt domes. In solid earth geophysics, subsurface geology can be complex near a subducting plate boundary. Under such circumstances, a better estimate of an initial model than the one from the velocity analysis is necessary. In seismic processing, a sophisticated velocity estimation tool such as travel time tomography is now a standard practice, which can provide a good initial velocity model suitable for such geologically complex areas. Details of the travel time tomography can be found in Yilmaz (2001).

We discussed above how to obtain an initial $\mathbf{V}_P$ model. For seismic inversion, however, we require initial estimates of other model parameters besides $\mathbf{V}_P$. For isotropic elastic inversion, for example, we need the initial S-wave velocity ($\mathbf{V}_S$) and density (ρ) models. If an inversion is needed for estimating the anisotropic subsurface properties, then we also need initial estimates of other anisotropic parameters. We postpone the anisotropic inversion to the next section on multi-objective optimization. Here, we restrict to isotropic elastic problems only.

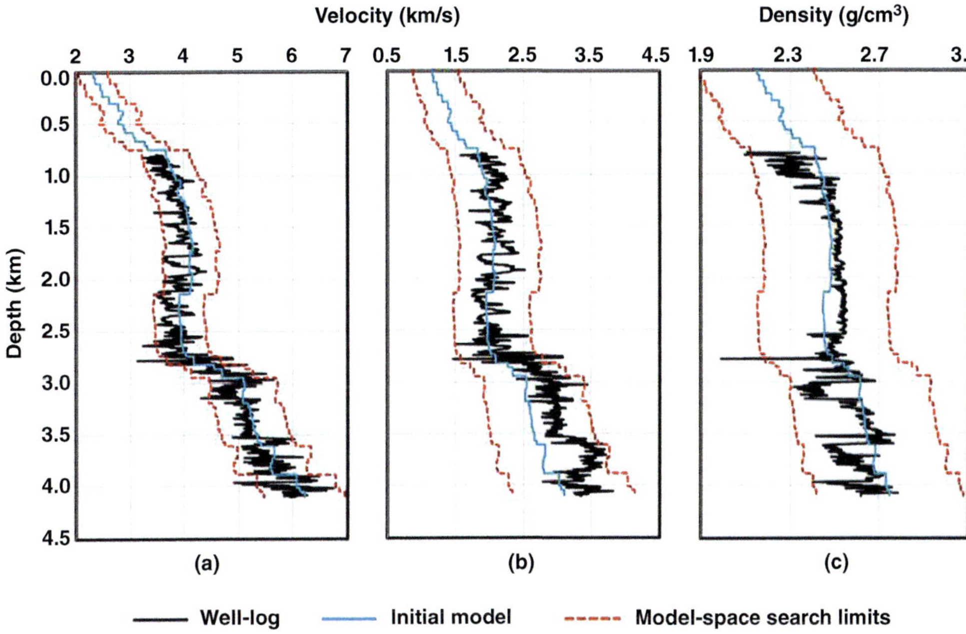

Figure 8.9 Estimation of the initial model for (a) P-wave velocity, (b) S-wave velocity, and (c) density, shown in cyan and displayed with the well-log data in black. The red dashed lines are the search windows assigned for each model parameter within which a global method can randomly generate models.

In Figure 8.9, we discuss how to obtain the initial model for isotropic elastic seismic inversion. The cyan curve, shown in Figure 8.9a, is the same $\mathbf{V}_P$, shown in Figure 8.8b, after time-to-depth conversion. The RSU example shown coincides with a well location with P- and S-wave sonic and density log measurements. The black curve shown in Figure 8.9a is the P-wave velocity from the sonic log at this well location. Note that although of much lower resolution, the $\mathbf{V}_P$ from the velocity analysis and that from the well log compare closely with one another. Once we obtain $\mathbf{V}_P$, initial $\mathbf{V}_S$ and ρ models can be estimated from the $\mathbf{V}_P$ using established V_P versus V_S and V_P versus ρ relations. If well logs are present, these relations can be directly established. The presence of such well logs are common in shallow crustal explorations, such as in oil and gas exploration problems, but not for deep seismic exploration studies. It is therefore worthwhile to establish a method to obtain $\mathbf{V}_S$ and ρ models when well logs are unavailable.

One good way to estimate $\mathbf{V}_S$ field is by using the Poisson's ratio υ, which can be expressed in terms of the squared S-to-P-wave velocity ratio $\gamma = (V_S/V_P)^2$ as

$$\upsilon = \frac{1 - 2\gamma}{2(1 - \gamma)}. \tag{8.41}$$

In shallow explorations used by the oil and gas industry, primary interest is to delineate the sedimentary sequences such as the sands, shales, and carbonates. These sedimentary sequences are mostly composed of shales with occasional presence of sandstones and limestones within which hydrocarbons may be present. The primary objective for these exploration methods is to distinguish these sandstone and carbonate lithologies in the subsurface from the background, and they are mostly composed of the shale-type lithology. Poisson's ratio of shales is about 1/3. Using equation 8.41, it can be readily verified that $\nu = 1/3$ corresponds to $V_S = V_P/2$. The cyan curve in Figure 8.9b shows the $\mathbf{V}_S$ field as a function of depth computed using this relationship and displayed along with the S-wave sonic log at the well location (the black curve in Figure 8.9b). Although not as accurate as $\mathbf{V}_P$, the estimated $\mathbf{V}_S$ using $V_S = V_P/2$ follows the general trend of the well data up to a depth of 2.75 km and a slight departure from this trend below 2.75 km. Setting $V_S = V_P/2$ is generally used as an estimate of the initial S-wave velocity in AVA and elastic impedance inversion (Hilterman, 1990; Connolly, 1999; Hampson et al., 2005). For deep seismic exploration problems, major rocks have Poisson's ratio of 0.25, where setting $V_S = V_P/\sqrt{3}$ is a good way to estimate the initial $\mathbf{V}_S$ from $\mathbf{V}_P$.

For sedimentary sequences, Gardner et al. (1974) provided a relation of type $\rho = \alpha(V_P)^\beta$ for computing the density from P-wave velocity. The cyan curve in Figure 8.9c shows the density field ρ computed using this relation along with the well-log density. Note that the estimated ρ is not as accurate as that of the estimated $\mathbf{V}_S$, which, in turn, means the uncertainty in the estimation of density from single (vertical) component seismic data is expected to be higher than that for the estimation of the P- and S-wave velocities. However, if multicomponent seismic data are used, the density estimation is accurate, which we discuss in the next section. For Gulf of Mexico, Gardner et al. (1974) suggested $\beta = 0.25$ and $\alpha = 0.23$ or 0.31 for V_P in ft/s or m/s. Although a relation like $\rho = \alpha(V_P)^\beta$ is adequate for sedimentary sequences, it is not true for deep seismic exploration. Details of how to provide an initial density for such situations can be found in Barton (1986).

The initial model, shown in Figure 8.9, may be used as the starting model for any gradient-based local optimization method, which we discussed in Chapter 7. For global methods, such as MCMC, SA, or GA, we can define a distribution around these estimations and randomly draw models. In Figure 8.9, for example, we show lower and upper search limits for $\mathbf{V}_P$, $\mathbf{V}_S$, and ρ as dashed red curves. Randomly drawing models within these limits is equivalent to defining a uniform distribution around each model parameter such that any value within these limits may be drawn with equal probability. As an alternative to this uniform distribution, we may define any other distribution around these estimated models in our optimization. It is also important to point out that, in practice, we do not assign a distribution, uniform, or something else, for each depth sample but for a range of samples, dictated by dominant seismic wavelength and the desired resolution of the estimated models. We revisit this issue later in the chapter.

Fundamental Steps for GA Optimization Workflow

Fundamental steps in GA optimization are the following:

1) Generate an initial random suite of N_{pop} models from the distribution defined for each model parameter. In GA, this initial suite of models is called the initial population of models, P_{init}, and we set it as the current model population $P_{current}$.
2) Compute synthetic (predicted) data for each member of $P_{current}$ and match with the observed data to compute their objectives, E_i, $i = 1, 2, \ldots, N_{pop}$.
3) Scale the computed objectives using a suitably chosen scaling function to compute the scaled objectives $(E_{scaled})_i$, $i = 1, 2, \ldots, N_{pop}$, for each member of $P_{current}$.
4) Choose an intermediate model population P_{inter} of size N_{pop} from the current population $P_{current}$ via the GA step of *reproduction* or *selection*.
5) Store P_{inter} in the model space using the scaled objectives of their individual members and the Frazer and Basu (1990) binning technique, we discussed earlier.
6) From the intermediate population P_{inter}, choose another new population of models P_{new}, also of size N_{pop} via GA steps *crossover*, *mutation*, and *elitism*. This step requires setting P_{inter} as the parent population (P_{parent}) and generating the child population P_{child} via crossover and mutation, computing synthetic data for each member of the child population and matching with

observation to compute their objectives, scaling the objectives, and comparing the scaled objectives of the parent population with their child population to choose a new population P_{new} model of size N_{pop}.

7) Check if any of the convergence criterion is reached (we discuss these criteria later).
   a) If no, then set P_{new} from step 6 as $P_{current}$ and go back to step 4. In GA, this step is called advancing from one generation to the next.
   b) If yes, normalize the cumulative probability (step 4) to compute the posterior PDF and compute other parameters such as the covariance $\mathbf{C}_M$ and exit.

The GA workflow with steps 1 through 7 is shown in Figure 8.10. Two user-defined parameters used in this workflow are G_{max} and C_{max}. These two parameters are used for setting up the convergence criteria in step 7, with G_{max} being the maximum number of generations or iterations and C_{max} being the maximum correlation between the synthetic and observed data to be attained during optimization. If either of these criteria is reached, the method exits (see Figure 8.10).

The workflow shown in Figure 8.10 is the fundamental GA workflow, in which steps 1 through 6 require further discussions, as provided below.

Step 1: Generation of the Initial Model Population

The step 1 in Figure 8.10 involves randomly generating an initial population of N_{pop} models. This is equivalent to randomly sampling the prior $p(\mathbf{m})$ for each model parameter in $\mathbf{m}$. We restrict to an isotropic elastic seismic inverse problem where $\mathbf{m}$ consists of $\mathbf{V}_P$, $\mathbf{V}_S$, and ρ, each of which are functions of depth (or time) as shown in Figure 8.9. This random sampling can be performed in two different ways: (1) *lateral sampling* and (2) *vertical sampling*. We explain below both sampling methods by generating five random $\mathbf{V}_P$ models in which a uniform distribution between a minimum and a maximum search limit for each depth (or time) sample is specified as the prior (see Figure 8.9a). Extending this

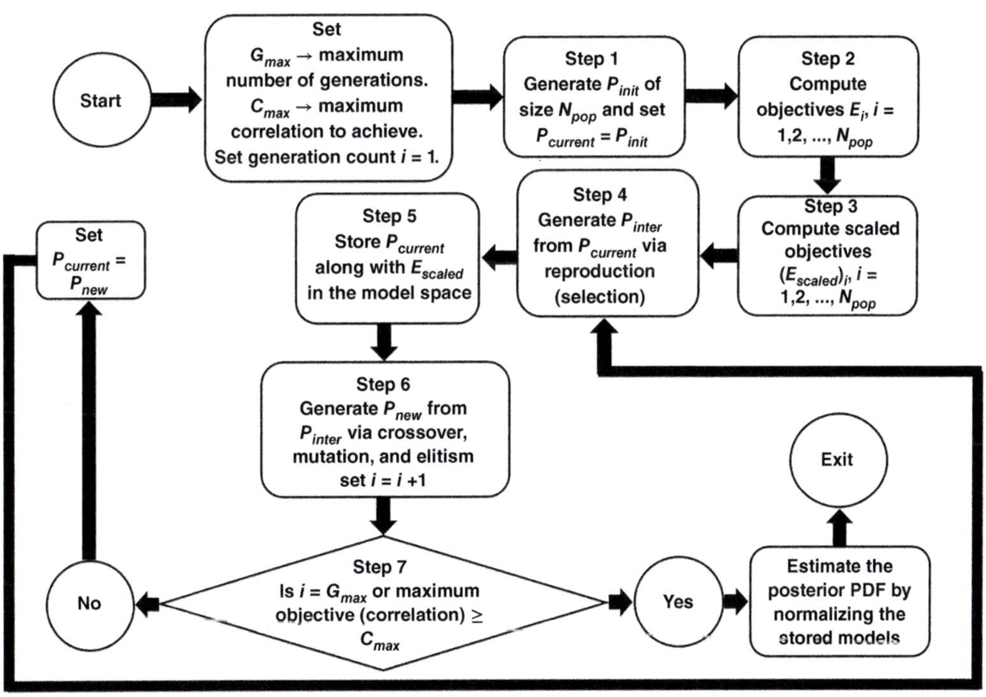

Figure 8.10 An overview of the GA workflow.

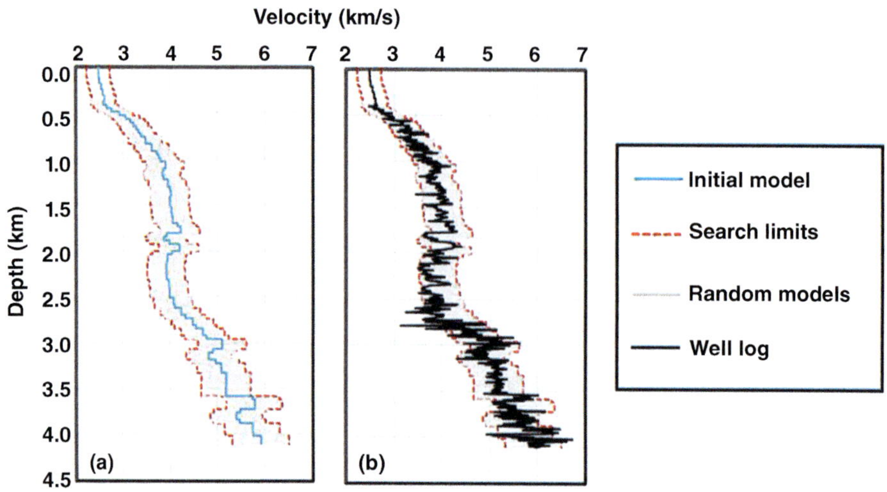

Figure 8.11 Illustration of lateral sampling for the P-wave velocity field. The well-log model, initial model, and the search limits are same as they are in Figure 8.9a. (a) Initial model is compared with the randomly generated models. (b) Well-log model is compared with the randomly generated models.

method for other distributions and for other model parameters ($\mathbf{V}_S$ and ρ) is obvious and straightforward. It must also be noted that although we use only five random models, we do so only for visual representation. For an accurate sampling, it is necessary to use many (typically hundreds or even higher) models.

In *lateral sampling* (Figure 8.11), a value for each depth sample (or, strictly a range of depth samples, as we discussed above) is randomly drawn from the distribution, provided for that sample (or the range samples). Sampling method shown in Figure 8.11 is standard in GA and other global optimization methods such as SA, MCMC, etc. And comparing the random models with the initial models (Figure 8.11a), we see that the random models (gray curves) are bounded by the lower and upper limits (dashed red curves), and they wildly fluctuate around the initial model (cyan curve). From these wildly fluctuating models around the estimated initial model, global methods search the model space and attain the global optimum as an estimate of the true model. We can never exactly estimate this true model, and the best we can do is to find a model that is close to it. Because the example shown here coincides with the location of the well data, we may assume that if our optimization method is able to estimate a model close to the well-log model, we have estimated the true model to the best of our ability. Now, if we compare the random models with the well-log model (Figure 8.11b), we find that by fluctuating around the well-log model, these combined set of random models sample the entire well-log model besides a few spurious points where the well-log model exceeds the specified search interval (for example, few points between 2.5 and 3 km, around 4 km, etc., which are caused from the measurement errors and were explicitly excluded).

Starting from a suite of random models, generated from the prior via lateral sampling shown in Figure 8.11, the objective of any global optimization method is to exhaustively search the model space and reach the global optima. By storing all models in the model space, they also estimate the posterior PDF to obtain the maximum likelihood model, mean model, model covariance, etc. In theory, these methods can estimate the global optima and compute all these parameters. However, because of wild fluctuations of the initially generated models from lateral sampling, they generally require many forward synthetic calculations, which can be computationally challenging, especially for seismic waveform inversion problems. Consequently, an alternative sampling strategy that can lead to a faster convergence of the algorithm than the lateral sampling shown in Figure 8.11 is desirable.

The primary issue with the sampling strategy of Figure 8.11 is the unsmooth nature of the initially generated models because of their wild fluctuations. Due to their unsmooth nature, they tend to take a long time to converge to the global optima. We may therefore ask, is there a different strategy to generate these random models which takes advantage of the smooth nature of the initial model and yet samples the entire model space? Such a sampling is illustrated using Figure 8.12. Using the same lower and upper bounds as in Figure 8.11, we generate a range of smooth models that are parallel/subparallel to the initial model and vary within the specified bounds (Figure 8.12a). Note that although smooth, these random models sample the entire range of variations of the well-log model as shown in Figure 8.12b. In contrast with lateral sampling, the sampling strategy shown in Figure 8.12 may be considered as the *vertical sampling*. Although it may appear that the lateral sampling shown in Figure 8.11 encompasses the entire model space more densely than the vertical sampling of Figure 8.12, this is not true. We demonstrated above both sampling methods using five random models. In practice, where the total number of models is on the order of hundreds or higher, both lateral and vertical sampling methods comparably sample the model space. Because of the additional smooth constraint imposed by vertical sampling, it tends to converge faster than lateral sampling.

Step 2: Objective Computation

Computing the objective (step 2 in Figure 8.10) requires computing the synthetic data via $\mathbf{s} = g(\mathbf{m})$, followed by matching it with the observation $\mathbf{d}$. For seismic inverse problems, synthetic data can be computed using the methods previously discussed in Chapter 5. GA uses the natural analogy of the survival of the fittest of the biological evolution. To match synthetic data with observation and compute the objective $E(\mathbf{m})$, it is therefore convenient to use the normalized cross-correlation as the representation for $E(\mathbf{m})$ so that the optimization is cast as maximization rather than minimization. However, as previously noted in Chapter 6, minimization and maximization are trivially related to one another. Therefore, defining $E(\mathbf{m})$ as cross-correlation using equation 8.40 and then attempting to minimize $-E(\mathbf{m})$ could also be implemented in GA. Alternatively, defining $E(\mathbf{m})$ in terms of the

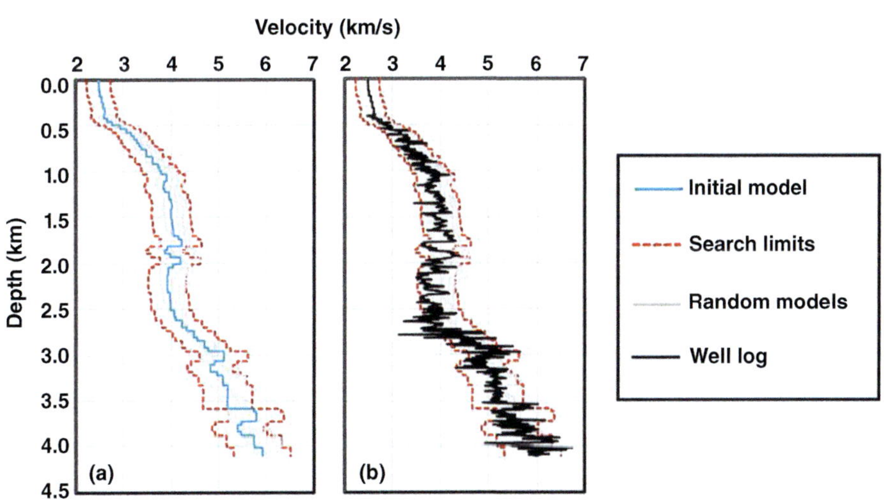

Figure 8.12 Same as Figure 8.11 but with vertical sampling.

regularized misfit (for example, equation 7.22) and then attempting to either minimize $E(\mathbf{m})$ or maximize $-E(\mathbf{m})$ could also be implemented in GA optimization. In summary, computing the objective is completely a user preference. In this book, however, we cast GA as a maximization problem and use cross-correlation (equation 8.40 or any of its equivalent forms) as objective.

Step 3: Objective Scaling

For local optimization problems (Chapter 7), we first defined the objective $E(\mathbf{m})$ as the sum of the squared errors or misfit $|\mathbf{d}-\mathbf{s}|$ via equation 7.5. We then argued that a pure least-squares approach of defining $E(\mathbf{m})$ via equation 7.5 and attempting to minimize it by computing its gradient $\nabla E(\mathbf{m})$ is unstable and discussed the strategies to regularize $E(\mathbf{m})$ via equation 7.22.

Like local, gradient-based optimizations, defining the objective as normalized cross-correlation and then attempting to maximize is not stable. In GA, for example, if we define the objective using equation 8.40, the method would often cluster around a local optimum, which may be far from the true (global) optimum where we wish to reach. The problem of avoiding such convergence to local optimum is avoided by implementing three specific steps: (1) *objective scaling*, (2) *elitism*, and (3) *diversity preservation*. In this section, we discuss objective scaling. And we discuss the elitism and diversity preservation later in Sections "Step 6: Crossover, Mutation, and Elitism" and "Diversity Preservation in GA Optimization."

Objective scaling is equivalent to objective regularization for local, gradient-based optimization methods, which we previously discussed in Chapter 7. The concept of objective scaling is to scale up the low objective values (low-performing members in the population) and scale down the high objective values (high-performing members of the population) in such a way that the models with the high objective values do not dominate the solutions that may eventually lead GA to cluster around a local optimum to which these solutions may belong to.

Goldberg (1989) provided a simple linear method for objective scaling. In this method, objective values $E(\mathbf{m})$ computed as normalized cross-correlation are referred as the *raw objectives* and are linearly scaled to $E^S(\mathbf{m})$, which are referred as the *scaled objectives*. Figure 8.13 illustrates this linear objective scaling of E to E^S as

$$E^S = a + bE. \tag{8.42}$$

The constants a and b in equation 8.42 are obtained from the constraints

$$E^S_{\text{max}} = S_C E_{\text{avg}} \tag{8.43}$$

and

$$E^S_{\text{avg}} = E_{\text{avg}}. \tag{8.44}$$

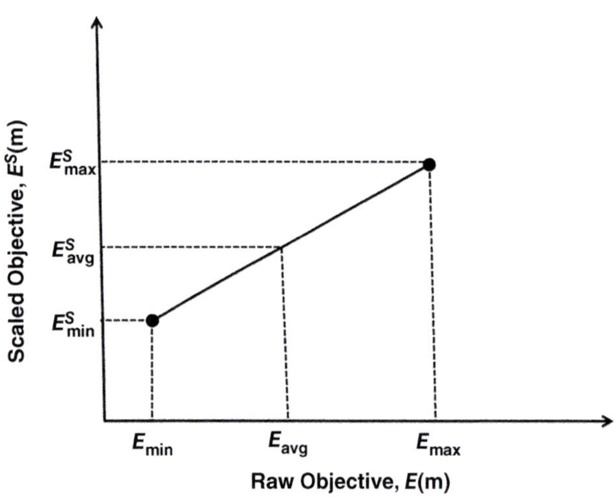

Figure 8.13 Concept of the linear objective scaling.

In equation 8.43, $E^S_{\max}$ and $E_{\max}$ are respectively the maximum values of the raw and scaled objectives and S_C is a user-defined, positive scaling parameter, usually less than 1. The parameters E^S_{avg} and E_{avg} in equation 8.44 are the average values of the scaled and raw objectives, respectively. From equations 8.42–8.44, we can find the constants a and b as

$$\left.\begin{aligned} a &= E_{avg}(1-b) \\ b &= \frac{(S_C-1)E_{avg}}{E_{\max}-E_{avg}} \end{aligned}\right\}. \tag{8.45}$$

Based on the above discussion, steps for linear objective scaling are the following:

1) For all computed objectives E_i, $i = 1, 2, \ldots, N_{pop}$, in each generation with N_{pop} being the population size, find the maximum and average objective value $E_{\max}$ and E_{avg}.
2) Calculate the constants a and b from equation 8.45.
3) Compute the scaled objective as $E^S_i = a + bE_i$ for each member $i = 1, 2, \ldots, N_{pop}$.

The scaling parameter S_C in equation 8.43 represents the degree to which the $E^S_{\max}$ is scaled compared to $E_{\max}$. Thus, using a low value for S_C would scale down $E^S_{\max}$ from $E_{\max}$ more than using a high value for this parameter. In early generations of GA optimization, none of the models are expected to be anywhere close to the global optimum. To avoid better-performing models to dominate and eventually drive the solution to a local instead of the global optimum, we must therefore choose a low value for S_C in early generations of the GA optimization. At the same time, as the generation progresses and we think that we are near the global optima, we must emphasize the better-performing models and therefore use a high value of S_C. To summarize, S_C must be set low at the beginning and gradually increased to a high value at the end. For AVA and seismic waveform inversion problems, Mallick (1995, 1999) found that a choice of S_C between 0.7 and 0.8 at the beginning and linearly increasing it to about 1.8 at the end is a good choice for a generation adaptive S_C.

In contrast to linear scaling, Stoffa and Sen (1991) and Sen and Stoffa (1992) proposed an exponential scaling. This exponential scaling uses a temperature-like control parameter T like the one used in SA optimization to scale the raw objective E_i to the scaled objective E^S_i as

$$E^S_i = \exp(E_i/T). \tag{8.46}$$

As illustrated in Figure 8.14, the exponential scaling effectively stretches the objective function. Higher the value of T, more is the stretching of the scaled objective E^S from the raw objective E. In early generations, we do not expect the models close to the global optima and we want minimal stretching of the objective so that all models can be selected with nearly equal probability during reproduction or selection phase, as discussed in the next section. This is achieved using a high value of T (Figure 8.14a,b). As the generation progresses and we expect to reach near the global optima, we want the high-performing models (models with high raw objective E) to be selected more often in reproduction or selection than the low-performing models. This is achieved by using a low value of T (Figure 8.14c,d). Like SA, Stoffa and Sen (1991) suggest using a generation-dependent T, of type $T = T_0A^{g-1}$, where T_0 is the initial value of T, A is a constant, and g is the generation number. Running GA with different values, Stoffa and Sen found that setting $T_0 = 1.5$ and $A = 0.99$ gives satisfactory results. Choice of these values is, however, problem dependent. In practice, it is a good idea to make a few short trial runs and set optimum values of these parameters. Further analysis of exponential scaling based on the *schemata theorem* (Holland, 1975) can be found in Sen and Stoffa (1992).

Although linear and exponential scaling are both designed to stretch the raw objective to avoid clustering to local optima and expedite convergence, they both can run into numerical issues. In linear scaling, for example, the choice of a high value of the scaling parameter S_C can sometimes lead to negative values for some scaled objectives E^S. In the next section, we discuss the process of selection (reproduction)

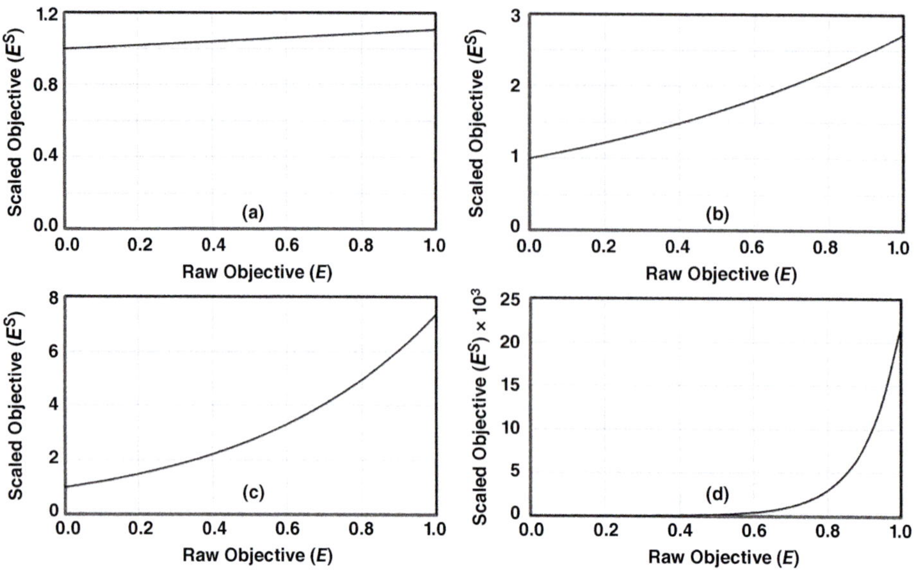

Figure 8.14 Illustration of the exponential scaling for different values of the temperature parameter T: (a) $T = 10.0$, (b) $T = 1.0$, (c) $T = 0.5$, and (d) $T = 0.1$.

where we assign each model with a selection probability, which is a direct function of the scaled objectives. Consequently, having negative E_i^S for a model $\mathbf{m}_i$ in the population would lead to its negative selection probability, which is nonphysical. To handle this problem of negative E^S, it is important to check their values and set them to zero if they are negative. For details, see Goldberg (1989), pp. 77–78.

For a very low value of T, for example $T = 0.001$, exponential scaling runs into floating-point overflow problems. Again, a careful choice for generation-dependent T is important to avoid such numerical issue.

Step 4: Reproduction (Selection)

Following objective scaling, the next step in GA optimization is *reproduction* or *selection* in which a given model $\mathbf{m}_i$ is reproduced or selected with the probability

$$p(\mathbf{m}_i) = \frac{E^S(\mathbf{m}_i)}{\sum_{j=1}^{N_{pop}} E^S(\mathbf{m}_j)}. \quad (8.47)$$

In equation 8.47, E^S is the scaled objective and N_{pop} is the population size or the total number of models. By performing this process N_{pop} times and selecting a model from the current population $P_{current}$, this step generates an intermediate population P_{inter}, which is also of the same size N_{pop}. Based on the probability of selection $p(\mathbf{m}_i)$, for the model $\mathbf{m}_i$ in $P_{current}$, this model thus may be selected more than once in P_{inter} or may not be selected at all.

The simplest way to implement selection or reproduction is roulette wheel selection illustrated in Figure 8.15. We define a roulette wheel in which we assign areas for each model $\mathbf{m}_i$, $i = 1, 2, \ldots, N_{pop}$, in proportion to their respective probability of selection $p(\mathbf{m}_i)$ as defined by equation 8.46. In Figure 8.15, we illustrate these areas shaded in different colors. We now spin the roulette wheel N_{pop} times. When the spinning stops, we select the model coinciding with the location of the blue arrow (see Figure 8.15). It is obvious that the models assigned with higher areas in the roulette wheel are likely to be selected more often than the ones assigned with lower areas.

Although simple and straightforward, roulette wheel selection can lead to serious stochastic errors in dealing with complex problems. Details of these errors can be found in

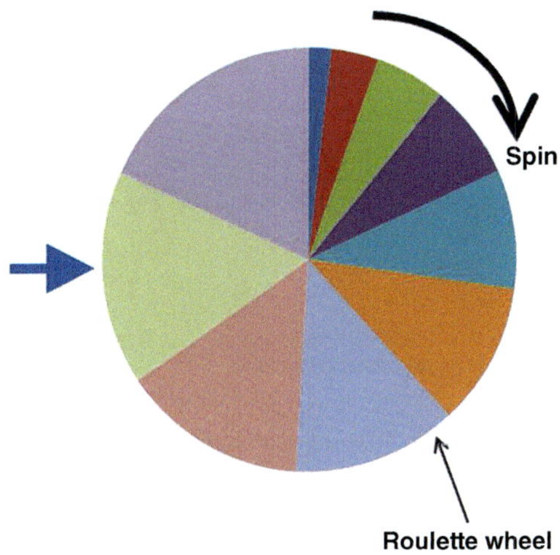

Figure 8.15 Illustration of roulette wheel selection.

Goldberg (1989), which we do not repeat here. To minimize stochastic errors, Goldberg (1989) also listed many other selection alternatives. Mallick (1995, 1999) experimented with these different alternative selection methods and found the *remainder stochastic sampling without replacement* to be a very good method for seismic inverse problems. We briefly describe below this selection method.

In stochastic remainder selection without replacement, the mathematically expected value e_i for each member i of the current population $P_{current}$ is first calculated as

$$e_i = N_{pop}p(\mathbf{m}_i), \tag{8.48}$$

in which $p(\mathbf{m}_i)$ is defined by equation 8.47. After calculating e_i for all members of the current population $P_{current}$, the integer part of e_i is used as the number of copies of the member i to be selected into the intermediate population P_{inter}. For example, if for a given member j, $e_j = 3.67$, then three copies of the jth member of $P_{current}$ would be selected into P_{inter}. Repeating this process for all members $j, j = 1, 2, \ldots, N_{pop}$, would then partially fill up P_{inter}, in which some of the original members would be present once, more than once, or not present at all, depending upon whether the integer part of their mathematically expected values are 1, greater than 1, or less than 1.

Next, the rest of the new population is filled up as follows:

1) A member is randomly selected out of the entire original population $P_{current}$ and a coin is tossed with the probability of heads set to the fractional part of its expected value.
2) Using the same example again, if the member j is randomly selected where $e_j = 3.67$, then the coin would be tossed with the probability of heads set to 0.67.
3) If the outcome of the coin toss is heads, then the one copy of the member j would be selected into P_{inter}.

These steps are repeated until the entire population is filled up, i.e., the selected number of members in the P_{inter} is N_{pop}.

Step 5: Storing of the Models in the Model Space

After selection, step 5 in GA is storing the models. We already discussed this using the binning technique and therefore do not repeat.

Step 6: Crossover, Mutation, and Elitism

Via reproduction or selection, GA obtains the intermediate population P_{inter} from the current population $P_{current}$. However, P_{inter} is just a copy of $P_{current}$ with some members being repeated and some other members being omitted. We just selected members of $P_{current}$ in proportion to their selection probability but not explored the model space to find new models. Crossover, mutation, and elitism are the processes by which the model space is explored to find new models and create a new population P_{new} and advance to the next generation. To explain these processes, it is however important to first discuss parameter coding and decoding.

To mimic the evolution process, GA works on coded parameters instead of the parameters themselves, which, in turn, requires *parameter coding*. Parameter coding can be explained in a simple way using *binary coding*. Restricting to isotropic elastic seismic waveform inversion problems, our models are given as $\mathbf{V}_P$, $\mathbf{V}_S$, and $\boldsymbol{\rho}$ as functions of depth (see equation 8.36). In binary coding, we code these model parameters for each depth sample as binary digits (bits), as illustrated in Figure 8.16. In Figure 8.16, 16 bits are assigned for $(V_P)_i$ and 8 bits are assigned for $(V_S)_i$ and ρ_i for each depth sample i of $\mathbf{V}_P$, $\mathbf{V}_S$, and $\boldsymbol{\rho}$. For each depth sample, we also define the minimum and maximum possible values of the model parameters as $(V_{Pmin})_i$, $(V_{Pmax})_i$, $(V_{Smin})_i$, $(V_{Smax})_i$, $(\rho_{min})_i$, and $(\rho_{max})_i$, which, in general, can vary with depth. We also define that if we set their assigned bits to zeros, it is equivalent to a value of $(V_{Pmin})_i$, $(V_{Smin})_i$, or $(\rho_{min})_i$ for the depth sample i. Similarly, if we set all their assigned bits to ones, it is then equivalent to $(V_{Pmax})_i$, $(V_{Smax})_i$, or $(\rho_{max})_i$ for that depth sample. We now draw a random model from the prior. For each depth sample for the model, we code their actual values as bits using user-defined values of $(V_{Pmin})_i$, $(V_{Pmax})_i$, $(V_{Smin})_i$, $(V_{Smax})_i$, $(\rho_{min})_i$, and $(\rho_{max})_i$ for each depth sample, and concatenate them to form a large string of bits, which becomes a single member of the population, and performing the same N_{pop} times provides a population of size N_{pop} as illustrated in Figure 8.16.

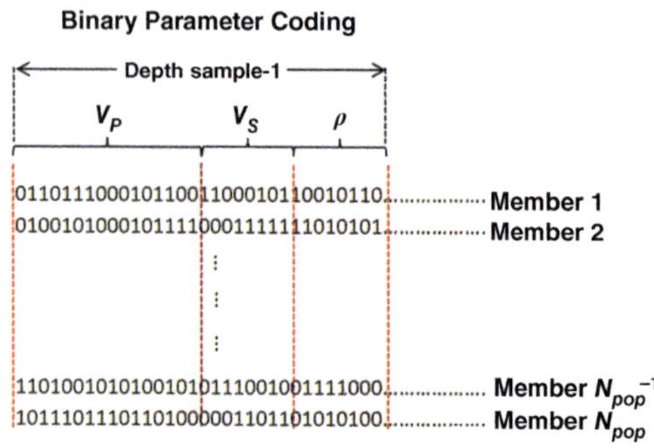

Figure 8.16 Illustration of binary coding. Only the first sample is shown, and the coded strings for each sample are concatenated to from a large string of bits to represent a single member of the population. Performing this operation N_{pop} times produces N_{pop} such strings. The entire set of this N_{pop} strings is called the population. Each string, representing a single member of the population, is called a chromosome and each bit within each chromosome is called a gene. In this figure, 16 bits are used to represent V_P and 8 bits are used to represent V_S and ρ. For illustration, each of these model parameters are separated from one another using dashed red lines.

Note that the choice of using 16 bits for V_P and 8 bits for V_S and ρ is arbitrary and used here for illustration purpose only. Assuming N_D to be the number of depth samples and N_{VP}, N_{VS}, and N_ρ are the number of bits being used to code $(V_P)_i$, $(V_S)_i$, and ρ_i for each depth sample $i = 1, 2, \ldots, N_D$, then the string length or the number of bits (N_{bits}) for each member of the population shown in Figure 8.16 is given as $N_{bits} = N_D N_{VP} N_{VS} N_\rho$. Each of these binary strings of length N_{bits} represents a single member of the population and called a *chromosome*. Within each chromosome, individual bits with either one or zero as their values are called *genes*.

Inverse of parameter coding is *parameter decoding*, which also we must elaborate before proceeding further. Parameter decoding stands for decoding a binary string (chromosome) for their actual values, which is necessary as we progress over the generations. To explain parameter decoding, we stay simple and assume that we use eight bits to code each of the model parameters. Now assume that the binary string for a depth sample i is given as 100101110011100111001001, and the values of $(V_{Pmin})_i$, $(V_{Pmax})_i$, $(V_{Smin})_i$, $(V_{Smax})_i$, $(\rho_{min})_i$, and $(\rho_{max})_i$ are respectively 2500 m/s, 5500 m/s, 1250 m/s, 2250 m/s, 2.0 g/cm^3, and 2.5 g/cm^3. From the parameter coding of Figure 8.16, we know that the first eight bits of the string, i.e., 10010111 is the binary-coded $(V_P)_i$, next eight bits, i.e., 00111001 is the binary-coded $(V_S)_i$, and the last eight bits, i.e., 11001001 is the binary-coded ρ_i. Now the equivalent decimal integer d_{val} of the binary string 10010111 that $(V_P)_i$ represents is $d_{val} = 1\times 2^0 + 0\times 2^1 + 0\times 2^2 + 1\times 2^3 + 0\times 2^4 + 1\times 2^5 + 1\times 2^6 + 1\times 2^7 = 1 + 8 + 32 + 64 + 128 = 233$. Assuming when all eight bits are zero (00000000) is equivalent to $(V_{Pmin})_i = 2500$ m/s and when all of them are one (11111111) is equivalent to $(V_{Pmax})_i = 5500$ m/s, we obtain $(V_P)_i = (V_{Pmin})_i + (d_{val} - 1)\frac{(V_{Pmax})_i - (V_{Pmin})_i}{2^8 - 1} = 2500 + 232\frac{5500-2500}{255} = 5229$ m/s. Values for $(V_S)_i$ and ρ_i can be obtained in a similar way.

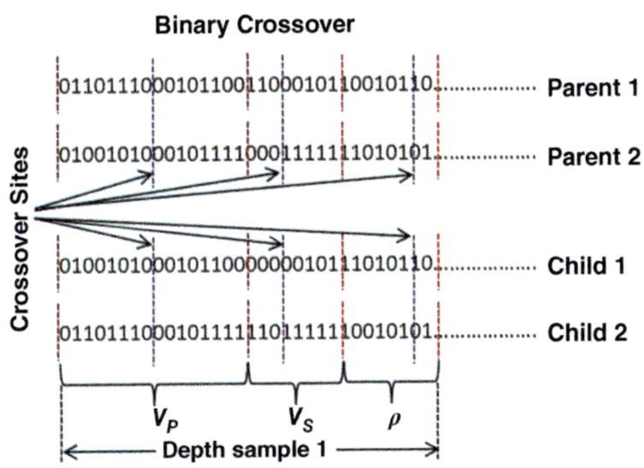

Figure 8.17 Illustration of crossover for binary-coded strings.

Now that we understand the concept of parameter coding and decoding, we are ready to discuss crossover, mutation, and elitism (update). To explore the model space and create a new generation of models from the old generation, the processes that GA uses are crossover, mutation, and elitism (update). Typically, crossover and mutation are combined into a single process, which is then followed by elitism to advance to the new generation.

Figure 8.17 illustrates the binary-coded crossover. First, two members from the reproduced population P_{inter} (see Figure 8.10, step 4) are randomly selected and treated as parents. In Figure 8.17, they are called *Parent* 1 and *Parent* 2. Next, a crossover site within each parameter space, $(V_P)_i$, $(V_S)_i$, and ρ_i, for each depth sample $i = 1, 2, \ldots, N_D$ is randomly selected (dashed purple lines in Figure 8.17). Finally, the bits (genes) on the left-hand side of the crossover sites of each parameter are swapped between the parents to produce two children, denoted as *Child* 1 and *Child* 2 in Figure 8.17. Crossover is performed with a crossover probability P_C. For each depth sample and each parameter, a coin is tossed with chances of heads set to P_C. If the outcome of the coin toss is heads, then the crossover for the given sample and model parameter is performed, else the bits for the parents for the parameter are simply copied into the children. Thus, crossover produces two new (child) members from the two original (parent) members. For seismic inversions, setting $P_C \approx 0.8$ at early generations and slowly reducing it to about 0.65 at the end produces good results (Mallick, 1995, 1999). In mutation, each bit in both child members (*Child* 1 and *Child* 2) is sequentially visited, and a coin is tossed with the chances of heads set to the probability of mutation P_M. If the outcome of the coin toss is heads, then the bit is changed, i.e., if it is 0 it is changed to 1, and if it is 1 it is changed to 0. A starting value of 0.15 and reducing it down to about 0.01 at the end is a good choice for P_M (Mallick, 1995, 1999).

After producing each pair of children from their parents via crossover and mutation, both child members are decoded for their actual model parameters, and objectives for both children are computed and scaled. Next, in elitism the scaled objectives of the children and their corresponding parents are compared, and two members with the highest objective values are allowed to advance to the next generation. Like crossover and update, elitism is also performed with a probability of elitism P_E, and a value of 0.5 at early generations and increasing it to about 0.9 at the end are good choices for P_E. Crossover, mutation, and elitism produces two members of the new population P_{new} from the two parent members from the reproduced population P_{inter}, and performing $N_{pop}/2$ such operations of crossover, mutation, and elitism would produce N_{Pop} members of the new population P_{new} (see Figure 8.10). We then check the convergence criteria (step 7 in Figure 8.10). If neither of the convergence criteria is not reached, we advance to the next generation by setting $P_{current} = P_{new}$ and repeat the process. However, if the convergence is reached, we then normalize the cumulative probabilities to compute the posterior PDF and other parameters and exit (see Figure 8.10).

In view of the above discussion, we used binary coding and decoding to explain crossover and mutation. Although such binary coding (and decoding) allows an easy way to describe crossover and mutation, such a coding discreetly samples the model space. For any model parameter A with minimum and maximum values set as A_{min} and A_{max}, the resolution of this discrete sampling is given as $\frac{A_{\max}-A_{\min}}{2^{N_B}-1}$, where N_B is the number of bits used to code the parameter. Considering that the model space is continuous not discrete, modern GA implementations are *real-coded*, which do not use any coding and deal directly with real numbers. Binary crossover and mutation have been extended for real-coded GA by Deb and Agarwal (1995, 1999) using simulated binary crossover (SBX) and real-parameter mutation (RPM). For seismic inverse problems, they have been used by Padhi and Mallick (2013, 2014), Li and Mallick (2015), Li et al. (2016), and Ayani et al. (2020). Padhi and Mallick (2014) described the procedures of implementing both SBX and RPM, which we describe below in detail.

Assume that for a given model parameter i which, for seismic inversion, is $(V_P)_i$, $(V_S)_i$, or ρ_i for any depth sample i, two parent members in a given generation t are $x_i^{1,t}$ and $x_i^{2,t}$ with $x_i^{2,t} > x_i^{1,t}$. Then, in SBX, two parameters β and α are first computed as

$$\beta = 1 + \frac{2}{x_i^{2,t} - x_i^{1,t}} \min(x_i^{1,t} - x_i^{L,t}, x_i^{U,t} - x_i^{2,t}) \tag{8.49}$$

and

$$\alpha = 2 - \frac{1}{\beta^{\eta+1}}. \tag{8.50}$$

In equation 8.49, $x_i^{L,t}$ and $x_i^{U,t}$ are respectively the lower and upper bounds for the model parameter x_i, and in equation 8.50 η is the crossover index, given as a positive real number, which we explain below. Following the definition of β and α as above, we next define another parameter β_{qi} as

$$\beta_{qi} = \begin{cases} (u_i\alpha)^{\frac{1}{\eta+1}}; u_i \leq \dfrac{1}{\alpha} \\ \left(\dfrac{1}{2 - u_i\alpha}\right)^{\frac{1}{\eta+1}}, \text{ otherwise} \end{cases}. \tag{8.51}$$

In equation 8.51, u_i is a random number between 0 and 1. Finally, the model parameter i for the two child members $x_i^{1,t+1}$ and $x_i^{2,t+1}$ is computed as

$$x_i^{1,t+1} = 0.5\{(x_i^{1,t} + x_i^{2,t}) - \beta_{qi}(x_i^{2,t} - x_i^{1,t})\} \tag{8.52}$$

and

$$x_i^{2,t+1} = 0.5\{(x_i^{1,t} + x_i^{2,t}) + \beta_{qi}(x_i^{2,t} - x_i^{1,t})\}. \tag{8.53}$$

And, repeating the above for all model parameters and for all depths would produce two child members from two parents.

Note that the denominator of equation 8.49 contains $x_i^{2,t} - x_i^{1,t}$. Therefore, when the values of the model parameter i for the two parent members are close to one another, computing β and hence obtaining the parameter values for the two child members is unstable. However, note from equation 8.48 that when $(x_i^{2,t} - x_i^{1,t}) \to 0, \beta \to \infty$. And from equation 8.50, when $\beta \to \infty$, $\alpha \to 2$. Therefore, for computer implementation of SBX, a small number is predefined. When $x_i^{2,t} - x_i^{1,t}$ is greater than this small number, SBX is then performed using equations 8.49 through 8.53. However, when $x_i^{2,t} - x_i^{1,t}$ is less than this predefined small number, then the value of α is set to 2 and then SBX is performed using equations 8.51–8.53. Like binary crossover, SBX is also performed with the crossover probability of P_C. However, besides P_C, SBX requires an additional parameter η—the crossover index (equations 8.50 and 8.51). Also note from equations 8.50–8.53 that smaller is the value of η, higher is the probability of getting the parameter values of the child members far away from their parents. Thus, starting from a small value of η and gradually increasing it as the generation progresses would provide another control to maintain a diverse population for real-coded GA and avoid immature convergence to a local optimum.

To mutate the child parameter value x_i^{child} after crossover into x_i^{mute}, first δ is defined as

$$\delta = \frac{\min\left[\left(x_i^{child} - x_i^L\right), \left(x_i^U - x_i^{child}\right)\right]}{x_i^U - x_i^L}, \tag{8.54}$$

where x_i^U and x_i^L respectively denote the upper and lower bounds of x_i, i.e., they are the same as the $x_i^{U,t}$ and $x_i^{L,t}$ previously defined in equation 8.49. Next, $\overline{\delta}_i$ is defined as

$$\overline{\delta}_i = \begin{cases} [2r_i + (1-2r_i)(1-\delta)^{\kappa+1}]^{\frac{1}{\kappa+1}} - 1; \\ \quad r_i \leq 0.5 \\ 1 - [2(1-r_i) - (1-2r_i)(1-\delta)^{\kappa+1}]^{\frac{1}{\kappa+1}}, \\ \quad \text{otherwise} \end{cases}, \tag{8.55}$$

where r_i is a random number between 0 and 1 and κ is the mutation distribution index. Finally, the mutated child member is computed as

$$x_i^{mute} = x_i^{child} + (x_i^U - x_i^L)\overline{\delta}_i. \tag{8.56}$$

Like SBX, RPM also requires an additional parameter κ, the mutation distribution index, in addition to P_M. This is also a positive number, a small value of which produces the parameter value of the mutated solution far away and a large value produces it close to the original solution. Like crossover index η, varying κ with generation would thus be another control for avoiding premature convergence. We discuss optimum choices for both η and κ later in Section 8.4.

Diversity Preservation in GA Optimization

In the above section, we discussed objective scaling to improve stability of GA optimization and avoid premature convergence. In addition, for real-coded GA, varying the crossover index (η) and the mutation distribution index (κ) with generation allows additional robustness to GA to avoid premature convergence. Despite objective scaling and varying the crossover and mutation indices, GA may still fail to converge. This is because of *genetic drift*—a tendency for GA to cluster toward a single region in the model space when finite population size is used (Goldberg, 1989). One way to avoid genetic drift is to use a very large population size and propagate it to a large number of generations. However, being a nonlinear method, the run-time for GA becomes prohibitively expensive with increasing population size and must be avoided. So, instead of using a large population, running several independent GA optimizations with small population sizes and later combining them is one good way to handle genetic drift (Sen and Stoffa, 1992; Mallick, 1995; Ayani et al., 2020). However, making several such GA runs is still computationally demanding, and it is advisable to find ways to avoid genetic drift in a single rather than several GA runs. For multi-objective optimizations, one way to avoid genetic drift in a single run is to compute the *population diversity*, given as the normalized Euclidean distance of a member in the population from its nearest neighbors measured along all objective axes (Deb et al., 2002; Padhi and Mallick, 2013, 2014; Li and Mallick, 2015; Ayani et al., 2020). Also for single-objective optimization problems, this diversity can be calculated using the pseudocode shown in Figure 8.18 in which the population size is given as N_{Pop} and the measured diversity is stored as a floating-point array "*dist*" of size N_{Pop}. Note that the diversity value calculated

```
maximum_distance = 0.0;
loop over N_pop (ipop=0; ipop < N_pop; ipop++)
     dist[ipop] = 0.0;
     loop over N_pop (jpop=0; jpop < N_pop; jpop++)
          If (ipop ≠ jpop)
               dist[ipop] = dist[ipop] + |objective[ipop]-objective[jpop]|
          endif
     end N_pop loop (jpop=0; jpop < N_pop; jpop++)
     if (dist[ipop] > maximum_distance) dist[ipop] = maximum_distance;
end N_pop loop (ipop=0; ipop< N_pop; ipop++)
loop over N_pop (ipop=0; ipop < N_pop; ipop++)
     dist[ipop] = dist[ipop]/ maximum_distance;
end N_pop loop (ipop=0; ipop< N_pop; ipop++)
end
```

Figure 8.18 Pseudocode to compute the population diversity as normalized Euclidean distance of a member from others, measured along the objective axes.

by the pseudocode of Figure 8.18 is normalized, which can have a value varying between 0 and 1. In addition, higher is the diversity value, more diverse (i.e., more isolated from its neighbors) is the model. Thus, multiplying the raw objective values of each model by their respective diversity values prior to scaling would prefer more diverse models over the ones that are less diverse and clustered in a single region. This, in turn, would add an additional control to avoid premature convergence of GA. Thus, in *diversity-preserved* GA, diversity can be used as an additional regularization parameter for the objective besides scaling.

Implementation of GA in High-Performance Parallel Computing Environments

In the above section, we thoroughly discussed the GA optimization method to solve seismic inverse problems. In these discussions, for example at the beginning of Figure 8.10, we introduced the following three user-defined parameters:

1) Population size (N_{pop}).
2) Maximum number of generations (G_{max}).
3) Maximum correlation to achieve (C_{max}).

We have not, however, discussed the optimum choices for these parameters in relation to seismic inverse problems. Performance of GA and the accuracy of the estimated subsurface models depend upon the optimal choices of N_{pop}, G_{max}, and C_{max}. It is thus important to discuss about these parameters in detail.

The choice of C_{max} depends upon the input data quality. If the input seismic data are relatively noise free (typical for most modern marine seismic data), a correlation as high as 0.9 (or sometimes even higher) can be reached. Land seismic data, on the other hand, are relatively noisy and a correlation above 0.85 is hardly achievable. To avoid unnecessary computations, it is thus important to analyze the signal-to-noise ratio and make an optimum choice of C_{max}. Generally, few (short) trial runs are always helpful for choosing an optimum value of C_{max} and avoid unnecessary computations.

Being a stochastic method, the higher are the values of both N_{pop} and G_{max}, more exhaustive would be the sampling of the model space. However, higher are the choices of these parameters, more computationally demanding is the GA optimization. Additionally, although Figure 8.10 shows the GA optimization with a population of size N_{pop} being progressed to G_{max} generations, it does not, in general, work for real seismic data. Irrespective of how big the choices of these parameters are, and every attempt is made to preserve population diversity, after about $4N_{pop}$ generations GA fails to further explore the model space. So, after about $4N_{pop}$ generations, it is usually necessary to obtain a fresh set of initial model population P_{init} by resetting the prior. And, in practice, this restart is necessary several times.

To develop GA or any other global method that would work on a wide variety of real data, the fundamental issue that must be addressed is that the real data are always noisy, which must be efficiently handled. Any model-based optimization, local or global, is an iterative process that must have ways to stop iteration and exit such that the method does not suffer from unnecessary computational burden. Thus, after each iteration, these algorithms check if the method converged to a reasonable solution. If not, the iteration is continued, else the method reports solutions and exits. This convergence check is shown in local optimization methods in Chapter 7 (Figures 7.3–7.5) and in the global methods we discussed so far in this chapter (see Figure 8.5–8.7 and 8.10). In dealing with noisy real data using a global optimization method, deciding whether to continue with iteration or to stop iterating is not, however, one simple step as shown in Figures 8.5–8.7, or 8.10. Noise levels on real seismic data are known to widely vary, which are controlled by the geological factors, environmental factors, and many other factors directly related to how the data were acquired in the field and later processed in a computational facility. Even within a single area, there are often different noise levels in different parts. To handle noisy data, the stopping criteria must be implemented such that there is convergence check at different points within the algorithm so that the data that are less noisy may converge early and exit, and at the same time, more iterations are allowed for noisy data. Such multipoint convergence check, if correctly implemented, would not only allow additional computations when needed, but it would also avoid unnecessary computations on the data that are relatively less noisy. Such multipoint convergence check, and repeating iteration, is computationally demanding. For computational efficiency, it is thus important to find ways to parallelize the methodology for parallel computing environments.

Considering the above issues, Figure 8.19 shows the GA workflow for seismic waveform inverse problems in parallel computing environments. Workflow for other global methods such as MCMC or SA can be similarly implemented. The workflow shown in Figure 8.19 is complex with different parts or modules, shown in different colors. The first module (Module 1), color coded in light green, is the basic GA optimization step. The second module (Module 2), color coded in yellow, is the first convergence checkpoint. The third module (Module 3), color coded in gray, is the second convergence checkpoint. Finally, the fourth module (Module 4), color coded in light blue, is the output module where the method reports solutions and exits. The main user-defined controlling parameters are (1) N_{pop}, population size; (2) G_{max}, the number of generations; (3) R_{max}, the maximum number of repeats; (4) N_{iter}, the maximum number of iterations; and (5) C_{max}, maximum correlation, i.e., the maximum value of raw objective to be achieved in the optimization process.

At the beginning of Module 1 (top right-hand corner of Figure 8.19), an initial random population P_{init} of N_{pop} models is generated from the prior $p(\mathbf{m})$ and distributed to different processors. Different processors compute the synthetic data and objectives for the subset of the model population they each receive. The master node then collects results from all processors, computes diversity, and performs objective scaling and reproduction (selection) to generate the intermediate

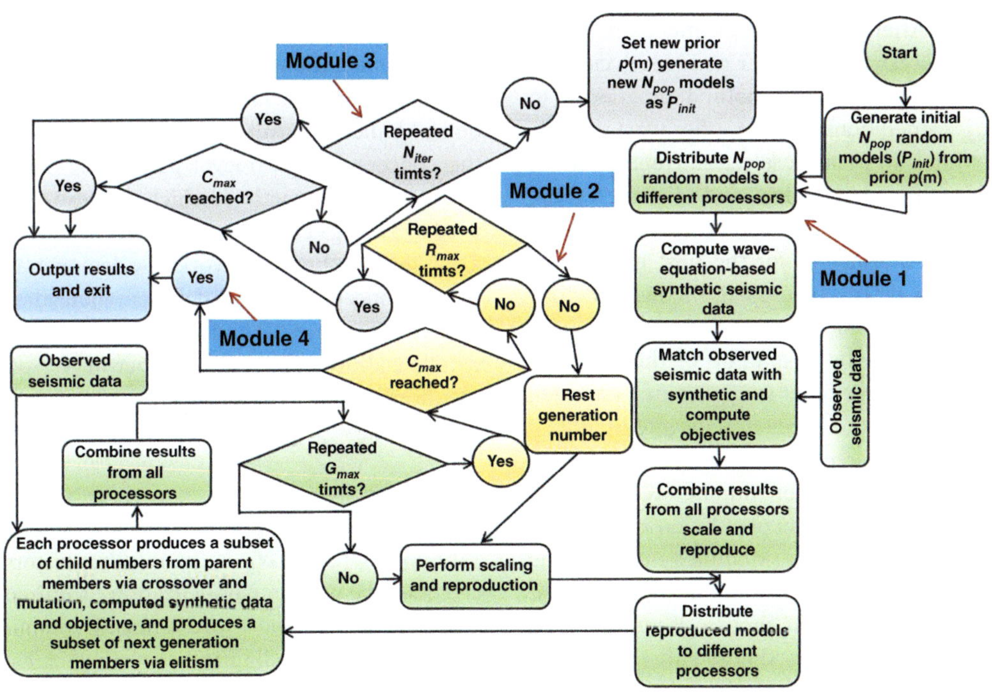

Figure 8.19 Parallel implementation of the GA workflow with multipoint convergence check.

selected population P_{inter}. Next, the selected models are again distributed to different processors such that they each perform crossover, mutation, and elitism, and produce new members of the next generation on a subset of the entire population. The size of the population subset each processor receives and produces new members is decided by the number of processors being used. For example, if $N_{pop} = 100$ and 50 processors are used, then each processor would receive two members from the selected population as parents, produce two child members via crossover and mutation, and then two new members of the new generation via elitism. When the number of processors is less than 50, they each would then receive more than two parents and thus produce more than two new next-generation members. If the number of assigned processors are more than 50, then only the first 50 processors would do the operation and the rest would sit idle. Thus, for a population size of N_{pop}, the maximum number of processors to be used by the method should be $\frac{N_{pop}}{2}$, else the additional processors would not be used. In addition, N_{pop} should be always an even number so that $\frac{N_{pop}}{2}$ is an integer. After generating new members within each processor, the master node receives them all and checks if the method progressed to G_{max} generations. If not, the master node continues to perform selection (reproduction) using the new models and distributes the selected models to different processors for creating the models for the next generation. However, if G_{max} is reached, the methodology then goes into Module 2 (yellow). In this module, whether the best value of the objective achieved at this point is C_{max} is first checked. If yes, the method goes straight into Module 4 (light blue) to report solutions and exits. Otherwise, the generation number is reset to 1 and the current model population is sent back to the green module (Module 1) to continue with another set of GA optimization for G_{max} generations, and this process is repeated for a maximum of R_{max} times with the chance of reporting solutions and exiting if C_{max} is reached during any stage. After such R_{max} repeats, the method goes to Module 3 (gray). In this module, the convergence criterion of reaching the objective to C_{max} is first checked, and if not, the prior $p(\mathbf{m})$ is reset to the posterior PDF from the previous pass, a fresh new set of random models (initial population P_{init}) is generated within the model space, and sent to the top of Module 1 to start a fresh set of GA optimization with R_{max} repeats. Module 3 is repeated for a maximum of N_{iter} times with chances of being sent to Module 4 of reporting solutions and exiting at multiple points (see Figure 8.19).

Although not explicitly mentioned, the workflow of Figure 8.19 was originally developed by Mallick (1999) and was later used by Mallick and Adhikari (2015) and Pafeng et al. (2017). More recently, Jia et al. (2021) successfully demonstrated the workflow of Figure 8.19 using two sets of real seismic data from two different regions. In addition, by making a few short initial trial runs, Jia et al. (2021) also found the optimum choices for different parameters as $N_{pop} = 80$, $G_{max} = 320$, $R_{max} = 5$, $N_{iter} = 7$, and $C_{max} = 0.95$.

The workflow of Figure 8.19 is, however, valid for running GA optimization at a single location out of the entire range of seismic data. To implement this method at multiple locations, Mallick and Adhikari (2015) introduced the concept of a two-level parallelization of GA optimization method, which was also later used by Pafeng et al. (2017) and Jia et al. (2021).

Figure 8.20 shows the workflow demonstrating the concept of the two-level parallelization of GA for inverting 2D and 3D seismic data. For parallelization level 1, the number of N_S sectors from the entire set of available number of processors (N_P) is first calculated as

$$N_S = \frac{N_P}{N_C}. \tag{8.57}$$

In equation 8.57 N_C is the number of processors to be used in parallelization level 2, as explained below. The entire 2D or 3D seismic data and corresponding initial models of V_P, V_S, and ρ (and their priors) are then divided into N_S sectors in which each of these sectors is

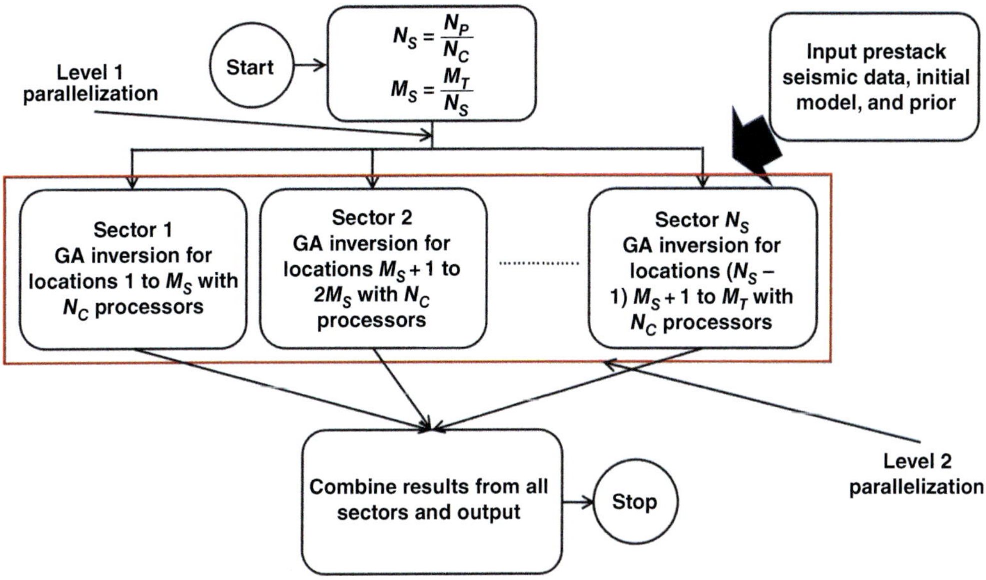

Figure 8.20 GA optimization with two-level parallelization for seismic inversion at multiple locations. Adapted from Jia et al. (2021).

assigned with M_S CMP locations to be inverted and given as

$$M_S = \frac{M_T}{N_S}, \tag{8.58}$$

where M_T is the total number of locations on the input seismic data to be inverted. For parallelization level 2, each set of N_C processors, assigned for sectors i, $i = 1, 2, \ldots, N_S$, receives only a subset of input seismic data, initial model, and their priors for M_S CMP locations assigned for that sector. Following this, M_S locations in each sector are simultaneously inverted, in which each sector performs GA optimization in parallel using N_C processors and the parallel workflow of Figure 8.19. Finally, each sector sends the computed results to the master node to combine them all into a single output.

To explain the concept of this two-level parallelization, we use the example of Jia et al. (2021) in which the total number of CMP locations were 214 (i.e., $M_T = 214$). So, if we use a total of 256 processors (i.e., we set $N_P = 256$) and want to assign 16 processors to be used for each sector (i.e., we set $N_C = 16$), then from equation 8.57, the total number of sectors the input data will be divided into is $N_S = \frac{N_P}{N_C} = \frac{256}{16} = 16$. After computing 16 sectors, the entire set of 256 processors is then divided into 16 sectors and a set of 16 processors is assigned for each sector. Thus, the processors with ranks 0–15 are assigned for sector 1, the ones with ranks 16–31 are assigned for sector 2, and so on. Next, according to equation 8.58, the number of CMP locations to be inverted in each sector is calculated as $M_S = \frac{M_T}{N_S} = \frac{214}{16} = 13.375$, with the nearest whole (integer) number being 13. Now if each sector receives 13 CMP locations to invert, then the total number of inverted CMP locations becomes $N_S M_S = 16 \times 13 = 208$. Because the total number of CMP locations (M_T) is 214, the first six sectors will invert 14 CMP locations in each (i.e., a total of $6 \times 14 = 84$ locations), and the rest of the 10 sectors (sectors 7–16) will invert 13 locations in each (i.e., a total of 130 locations) and thus all 214 locations are inverted.

The Choice of the Number of Depth Samples in the Initial and Inverted Models

In the discussions above, we assumed our models $\mathbf{V}_P$, $\mathbf{V}_S$, and ρ as vectors, each with

N_D depth samples. But we have not yet discussed how do we choose N_D. The choice of N_D is directly related to the resolution of the estimated models and requires further discussion.

Resolution of models obtainable from seismic inversions is dictated by the dominant wavelength. Thus, if f_{dom} is the dominant frequency in our seismic data, then the dominant wavelength λ_{dom} is given by

$$\lambda_{dom} = \frac{V}{f_{dom}}, \tag{8.59}$$

in which V is the velocity. Because P-wave velocity is always higher than the S-wave velocity, it can be readily seen from equation 8.59 that the dominant wavelength of P waves is higher than that of the S waves. For inverting single (vertical or pressure) component seismic data, which we are discussing, it is the P waves that control seismic resolution, and therefore, V in equation 8.59 should be set to the P-wave velocity. As a rule of thumb, maximum resolution obtainable from seismic inversion is typically quarter wavelength or $0.25\lambda_{dom}$. In waveform-based seismic inversions, Mallick and Adhikari (2015) and Pafeng et al. (2017) went beyond quarter wavelength and obtained reasonable results at a resolution of about $0.1\lambda_{dom}$. Irrespective of whether we desire to resolve our model at $0.25\lambda_{dom}$ or go beyond to $0.1\lambda_{dom}$, there is a maximum resolution achievable, which is dictated by λ_{dom} in the observed seismic data.

In seismic waveform inversions using any global optimization method, attempting to obtain the highest desirable resolution ($0.25\lambda_{dom}$ or $0.1\lambda_{dom}$) in a single inversion pass is however difficult because it usually takes a very long time to converge. A more practical approach is to obtain the desired resolution in steps. In GA, for example, Mallick and Adhikari (2015) and Pafeng et al. (2017) suggest that we start from a low resolution of about $0.5\lambda_{dom}$ and slowly increase the resolution until we achieve the resolution we desire.

Recently, *transdimensional seismic inversion* using different variations of the MCMC method was introduced by Sen and Biswas (2017), Biswas and Sen (2017), Zhu and Gibson (2018), Jian et al. (2018), Visser et al. (2019), Dhara et al. (2020), among others. This transdimensional method allows the number of layers (number of depth samples) in the model to dynamically vary with iteration, which, in principle, obtains the desired model resolution in successive steps.

An equivalent of the transdimensional approach in MCMC for GA optimization was developed by Hong and Sen (2009), which they call the *multi-scale GA* optimization method, which we provide as workflow in Figure 8.21.

As shown in Figure 8.21, after setting the user-defined parameters G_{max} and C_{max}, previously defined in Figures 8.10 and 8.19, the multi-scale GA has the following steps:

For generation number 1, defined as $t = 1$, we randomly select a total of N models for R different scales (resolutions, such as λ_{dom}, $0.5\lambda_{dom}$, $0.25\lambda_{dom}$, etc.). This leads that each scale $i = 1, 2, \ldots, R$ will have N/R models.

1) For each scale, we then compute the scaled objective using either linear or exponential scaling methods previously defined.
2) Perform reproduction or selection.
    a) After reproduction, we do the convergence check.
        i) If convergence is accomplished, we report solutions and exit.
        ii) If convergence is not accomplished, we then go to step 4.
3) Perform crossover, mutation, and elitism to advance to next generation.
4) Advance generation number by 1. Compute generation-dependent scaling parameters for each scale (we already discussed how to compute the generation-dependent scaling parameters). Scale the objectives and go to step 3 to continue iteration.

The multi-scale GA approach of Figure 8.21 is designed to simultaneously estimate models at multiple resolutions. In practice, however, applying such multi-scale method is difficult. The first step in the workflow of Figure 8.21 is generating random models with different scales. This must be generated by

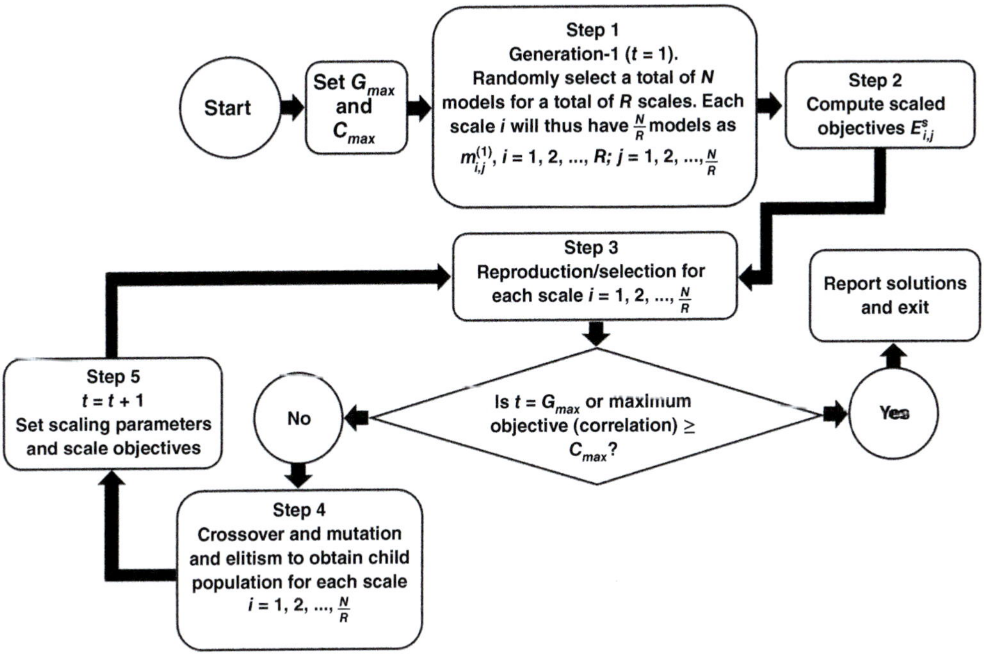

Figure 8.21 Workflow for multi-scale GA optimization. Adapted from Hong and Sen (2009).

randomly sampling the prior $p(\mathbf{m})$ at different scales. After doing so, these initial suite of N generation-1 models $\mathbf{m}_{i,j}^{(1)}, i = 1,, 2, \ldots, R$; $j = 1, 2, \ldots, \frac{N}{R}$, are advanced through generations until either one of the two stopping criteria ($t = G_{max}$ or maximum value of the objective $E_{max} \geq C_{max}$) is reached (see Figure 8.21). Thus, the workflow of Figure 8.21, in principle, is not different from our original seven-step GA workflow of Figure 8.10, except for the fact that in Figure 8.21 we simultaneously apply this workflow to the models that are resolved at different scales. In Section "Implementation of GA in High-Performance Parallel Computing Environments," we discussed that such a simple process is not likely to work in practice for noise-contaminated data, and we must use more complex workflow of Figure 8.19 in which there is convergence check at multiple points such that the method can exit from the workflow early when the data are relatively noise free and it can continue iterating further when the data are noisier. It is therefore necessary to include the workflow of Figure 8.19 for GA optimization at multiple resolutions.

Mallick (1999) obtained estimates of the subsurface model by manually running the GA workflow of Figure 8.19 with successively higher and higher resolutions. In this approach, an initial model is first discretized at a resolution of λ_{dom}, a prior distribution $p(\mathbf{m})$ around this model is defined as a uniform distribution, and GA optimization using the workflow of Figure 8.19 is run. The maximum likelihood model from this run is then used as the initial model and discretized at a resolution of $0.5\lambda_{dom}$, the same prior distribution is set around this initial model, and another GA optimization is run using the workflow of Figure 8.19. This process of setting up the initial model and the corresponding prior $p(\mathbf{m})$ and GA optimization using the workflow of Figure 8.19 is repeated until the desired resolution is reached.

The manual approach of running GA with successively higher resolutions from one pass to the next worked very well in characterizing the patchy pockets of commercial and non-commercial gas accumulations within the deltaic East Texas Woodbine sand deposit and the surrounding formations (see Mallick, 1999 for details). However, the posterior PDFs shown by Mallick (1999) were not true PDFs. To estimate the posterior PDFs, Mallick (1999) used the results from the final inversion run with the desired resolution and ignored other models generated from the previous passes. In each of these runs, the maximum likelihood model from the previous GA run was extracted, the prior $p(\mathbf{m})$ was defined as a uniform distribution around the model, and GA was run with successively higher resolution. The posterior PDFs shown in Mallick (1999) were extracted from the last run with the highest model resolution.

In view of the above discussion, Figure 8.22 shows the workflow of the multi-scale GA methodology, which uses the workflow of Figure 8.19 at multiple scales. Besides setting the user-defined parameters G_{max}, C_{max}, R_{max}, and N_{iter}, needed for the workflow of Figure 8.19, the multi-scale approach of Figure 8.22 needed three additional user-defined parameters: (1) beginning resolution (R_{beg}), (2) ending resolution (R_{end}), and (3) increment in resolution (R_{inc}). Following these definitions, the four-step workflow of Figure 8.22 is the following:

1) Set the current resolution $R_{current} = R_{beg}$. Also, set the prior distribution $p(\mathbf{m})$. This prior distribution can be set by defining an initial model and then setting any distribution (uniform, Gaussian, etc.) around the model.
2) Run GA optimization with $R_{current}$ resolution using the workflow of Figure 8.19. Following this, we check if $R_{current}$ is equal to our desired resolution R_{end}.
   i) If $R_{current} = R_{end}$, we report solutions and exit.
   ii) If $R_{current}$ has not reached the desired resolution R_{end}, we then go to step 3.
3) We increase the current resolution $R_{current}$ by R_{inc}.
4) We set the prior as the posterior PDF estimated from the previous pass (step 2) and go to step 2 to run GA with increased resolution.

In the workflow of Figure 8.22, note that every pass of GA optimization uses the posterior PDF from the previous GA run as the prior

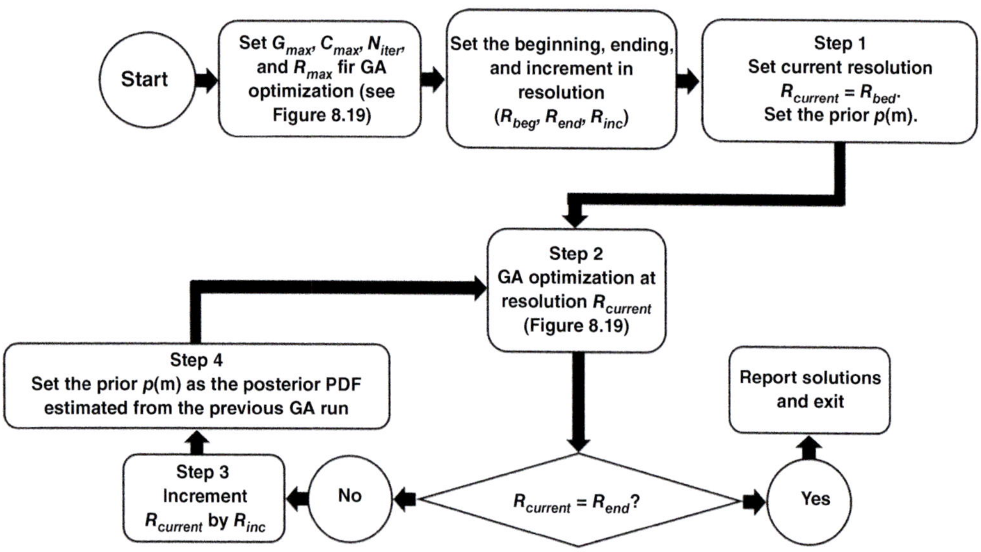

Figure 8.22 An automated multi-scale GA workflow.

to draw random models for GA inversion with successively increased model resolution. In view of the transdimensional MCMC method, the workflow of Figure 8.22 can be considered as a *transdimensional GA* optimization method.

8.4 Multi-Objective Optimization

So far, we have discussed the optimization problems, local and global, in which we attempted to optimize a single objective. Many problems of practical interest, however, require interpreting multiple sets instead of a single set of data. Taking seismic, for example, if we want to acquire multicomponent seismic data, inverting them will require multiple objectives, one for each data component. These multicomponent seismic data may be explained by single physics, namely, the elastic wave theory or the physics of the seismic wave propagation. However, features on the vertical-component seismic data are dominated by the P waves while their horizontal components are dominated by the S waves. And, because dominant wavelengths of the P and S waves are different, these components tend to resolve the subsurface with different scales or resolutions. Multicomponent seismic inversion is therefore a single-physics and multi-scale inverse problem. In recent past, attempts have been made to combine seismic with other geophysical data. Du and MacGregor (2010), for example, attempted to jointly invert controlled source electromagnetic data with seismic data for subsurface reservoir characterization. Karaoulis et al. (2012) made a numerical investigation to jointly invert time-lapse cross-well DC resistivity and seismic data. These inverse problems are not only multi-scale but also multi-physics. Over the past years, these multi-physics, multi-scale, and multi-objective problems have been treated like single-objective optimization, meaning that a single objective is defined as a weighted sum of each objective, which is then followed by a single-objective optimization method. Multi-objective problems are, in general, nonunique with multiple sets of solutions known as the *Pareto-optimal* solutions, none of which could be considered better than the other in terms of satisfying all objectives (Deb et al., 2002). In multi-objective problems, it is important to estimate these Pareto-optimal set of solutions because doing so not only allows finding the correct solution, but it also provides a way to calculate the uncertainty associated with estimating each model parameter. Note that if we cast a multi-objective into a single-objective problem, it gives only one out of the complete set of the Pareto-optimal solutions that may be biased by the weights we choose to define the objective. Of course, by changing weights and rerunning the inversion multiple times can in theory allow estimating the entire Pareto-optimal set, but such a procedure is too cumbersome to apply in practice. Although some techniques for automatically choosing and modifying the weights attached to each objective have been suggested (Hajela and Lin, 1992; Ritzel et al., 1994; Miettinen, 1999), they all give one solution or at most a small fraction of the Pareto-optimal set that may or may not be the most desirable. To handle multi-objective problems, it is therefore advisable to use an optimization method capable of treating all objectives as the components of an *objective vector* and simultaneously minimize (or maximize) them all via an iterative process to estimate these Pareto-optimal set of solutions in a single run.

8.4.1 The Concepts of Pareto-Optimality, Pareto-Optimal Solution Sets, and Dominance

Before discussing multi-objective optimization, we must first define the concepts of *Pareto-optimality*, *Pareto-optimal solution sets*, and *Dominance*, which are the fundamental building blocks for multi-objective optimization methods. We develop these concepts using a two-objective optimization problem. In geophysics, such a two-objective optimization can be encountered if we simultaneously optimize vertical and horizontal (radial) components of seismic data, or we jointly invert single-component seismic with electromagnetic or gravity data. Extending these concepts for more than two objectives is intuitive.

The concepts of Pareto-optimality and the Pareto-optimal solution sets for a two-objective optimization (minimization) problem are illustrated in Figure 8.23. In Figure 8.23a, two solutions to the optimization problem, namely Solution A and Solution B, are plotted as black dots in the objective space with Objective 1 as the horizontal axis and Objective 2 as the vertical axis. Note that the Solution A optimizes (minimizes) Objective 1 very well but does not optimize Objective 2 so well. The Solution B, on the other hand, optimizes Objective 2 but not Objective 1. Because our goal is to simultaneously optimize both objectives to satisfaction, none out of both solutions shown in Figure 8.23a can be considered better than the other. In fact, it is not just these two specific solutions, but there is an entire set of solutions, as shown in Figure 8.23b, in which none of the solutions can be considered better than the other in satisfying both objectives. Note that the solutions A and B of Figure 8.23a are just two extreme corner points of this entire set, as shown in Figure 8.23b. In multi-objective optimization, if there is a set of solutions like Figure 8.23b, none of which are better than the other in satisfying all objectives, we then say that the condition of the Pareto-optimality is satisfied, and the entire set is called the Pareto-optimal solution set. In the objective space, with a total number of N solutions, there could be a multiple such Pareto-optimal solution sets $i = 1, 2, 3, \dots, M$, as shown in different colors in Figure 8.23c. Clearly, if we denote m_i to be the number of solutions for the Pareto-optimal set i, then

$$\sum_{i=1}^{M} m_i = N. \tag{8.60}$$

In multi-objective optimization, all N solutions at any iteration are sorted into M Pareto-optimal sets in which the number of

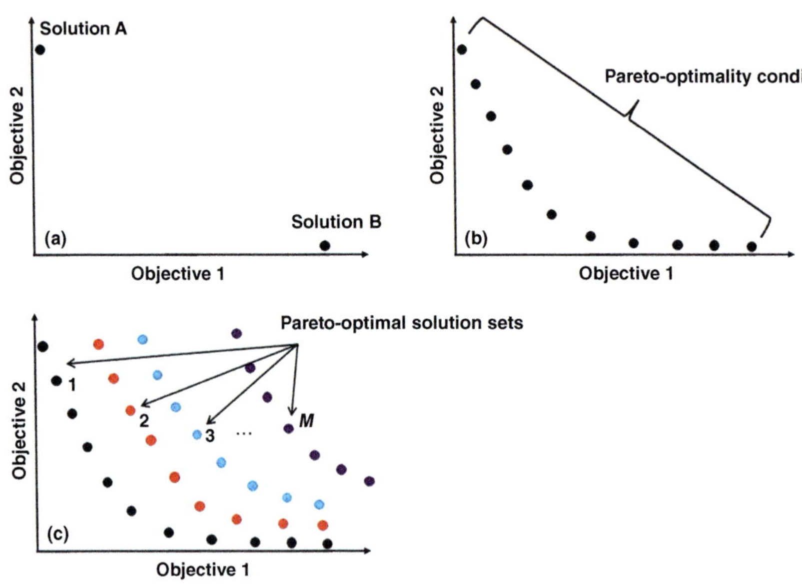

Figure 8.23 The concepts of Pareto-optimality and Pareto-optimal solution sets. (a) Two solutions (solution A and solution B) in the objective space, (b) the condition of the Pareto-optimality in a set of solutions, and (c) the Pareto-optimal solution sets.

solutions m_i in each set sums to the total number of Nsolutions (equation 8.60). This sorting is done using the concept of *dominance*. We mathematically describe dominance later in Section 8.4.3, and here we intuitively develop this concept using M Pareto-optimal solution sets shown in Figure 8.23c.

Restricting to two-objective optimization shown in Figure 8.23, and assuming that we cast optimization as minimization of the objectives, the Pareto-optimal set 1, shown as black dots in Figure 8.23c, can be considered to be better than sets $2, 3, \ldots, M$ shown in red, cyan, purple, etc. For the solutions belonging to set 1, we thus say that they do not dominate each other, but they dominate all solutions belonging to sets $2, 3, \ldots, M$. Similarly, for set 2 (red dots in Figure 8.23c), their individual members do not dominate each other, but they dominate the solutions belonging to sets $3, 4, \ldots, M$ and are dominated by the solutions belonging to set 1. We can continue the same analysis for all M sets and determine which sets they dominate and which sets they are dominated by. Clearly, the last set M (purple dots in Figure 8.23c) is dominated by the rest of the sets $1, 2, 3, \ldots, M-1$.

8.4.2 Why Multi-Objective Methods Are Necessary?

Seismic data have been the most effective tool for delineating and characterizing the subsurface. In exploration geophysics, these seismic data have traditionally been single component. In solid earth geophysics and earthquake seismology, although using both vertical and horizontal components are common, they are typically interpreted using traditional single-objective analysis where objectives of each component are combined as a weighted sum into a single objective and analyzed. The primary focus of the oil and gas industry is now shifting from the conventional to unconventional and naturally fractured reservoirs. To mitigate global warming from fossil-fuel burning, there is also a growing need for carbon dioxide (CO_2) sequestration in deep saline reservoirs. Fractured geothermal reservoirs are also a focus of attention as an alternate source of energy in recent years. To use fuel cells as the future energy source, there is now a growing need for *hydrogen hubs* where hydrogen is produced and stored underground. In most of these recent interests, the subsurface reservoirs are anisotropic, and to correctly handle the anisotropic effects of wave propagation, acquisition, processing, and inversion of the full wavefield, seismic data is required. Such full wavefield data are multicomponent (generally three- and sometimes nine-component), and the corresponding inversion or optimization is single-physics, multi-scale, and multi-objective as we previously discussed. An accurate reservoir description and monitoring requires seismic data to be combined with other geophysical data (electromagnetic, gravity, etc.) and engineering (reservoir flow simulation, geomechanical analysis, etc.) data. For example, while the traditional vertical-component seismic data are effective in delineating different subsurface rock formations (lithology), they are not very sensitive to the different fluids contained within their porosity (oil, gas, water, CO_2, hydrogen, etc.). Electromagnetic data, on the other hand, are not very sensitive to the subsurface lithology but are very sensitive to the formation fluids. Consequently, combining seismic with EM is useful for accurately delineating the subsurface rock and fluid properties (Lang and Grana, 2015). Additionally, reservoir description is not only delineating the properties once but also monitoring them over time. As hydrocarbon is produced from conventional, unconventional, or naturally fractured reservoirs or energy is produced from the hydrothermal reservoirs, for example, geophysical data must be reacquired and combined with the production data to monitor the subsurface fluid movements and optimize subsequent production. All CO_2 sequestration and hydrogen storage require post-injection monitoring to ensure that the injected fluid is in place and does not leak out into the atmosphere. Such monitoring requires combining geophysical data with engineering data such as the reservoir fluid-flow simulation and geomechanical modeling, etc., into a multi-objective optimization that is not only multi-scale but also multi-physics.

To summarize, multi-objective optimization is specifically necessary for future energy needs. Not only do we want to estimate subsurface lithology and fluid properties, but also we want to quantify uncertainties associated with their estimates. And, multi-objective method is the correct choice to meet these needs.

8.4.3 Multi-Objective Optimization: A General Overview

Multi-objective optimization attempts to make a best estimate of the Pareto-optimal solution sets and is relatively a new field of research, especially in geophysics. An exhaustive model-space search would, in theory, allow estimating the Pareto sets. In most cases of practical importance, such an exhaustive search is computationally expensive, and for the complexity of the underlying physics, it is also often unfeasible. Therefore, several stochastic methods, such as evolutionary algorithms, particle swarm optimization, scatter search, SA, etc., have been proposed and developed to handle multi-objective optimization, which attempt to make a reasonable approximation of the Pareto set. A very good overview of the current state of the art for multi-objective optimization can be found in Konak et al. (2006), Tang and Wang (2013), and Giagkiozis et al. (2015).

Different Pareto-based multi-objective optimizations differ from one another in terms of their specific implementation details. However, they all are based on evaluating multiple objectives and then sorting the individual members in terms of their dominance. Without any loss of generality, we treat our multi-objective optimization as minimization, where we evaluate each component y_i of the objective vector as $1 - E_i$ where E_i is computed as the normalized cross-correlation between the corresponding observed and synthetic data components (equation 8.40).

We follow the terminology used in multi-objective optimization and first start with defining the model vector $\mathbf{x}$ in the model or decision space $\boldsymbol{X}$ given as

$$\mathbf{x} = [x_1, x_2, \ldots, x_N]^T, \tag{8.61}$$

where N is the number of model components with x_i, $i = 1, 2, \ldots, N$, being the individual model parameters. Note that $\mathbf{x}$ in equation 8.61 is the same as the model vector $\mathbf{m}$ that we previously defined. For multicomponent seismic inverse problem, individual components of $\mathbf{x}$ are thus depth domain V_P, V_S, and ρ (and other anisotropic parameters for an anisotropic inversion), discretized at the desired resolution, as we previously discussed. If we jointly invert seismic and EM, these components are the depth-domain elastic and electrical properties and so on. We then define a mapping function $\boldsymbol{f}: \mathbf{X} \rightarrow \mathbf{Y}$, which evaluates the quality of a specific solution by assigning an objective vector $\mathbf{y}$ in the objective space $\mathbf{Y}$ as

$$\mathbf{y} = [y_1, y_2, \ldots, y_K]^T. \tag{8.62}$$

Cleary, the parameter K in equation 8.62 is the number of objectives in our multi-objective optimization problem. Here, we evaluate each component of the objective vector $\mathbf{y}$ as

$$y_i = 1 - E_i; 1 = 1, 2, \ldots, K, \tag{8.63}$$

with E_i being the normalized cross-correlation (equation 8.40) between the data component $\mathbf{d}_i$ and its corresponding synthetic (prediction) $\mathbf{s}_i$.

Having described the model and objective vectors via equations 8.61–8.63, we now mathematically formulate the concept of *dominance*, which we intuitively provided in Section 8.4.1. Assume that $\mathbf{x}^{(1)}$ and $\mathbf{x}^{(2)}$ are two sets of solutions (models) in the model (decision) space with their corresponding objectives evaluated as $\mathbf{y}^{(1)}$ and $\mathbf{y}^{(2)}$, which we write using equation 8.62 as

$$\left.\begin{aligned} \mathbf{y}^{(1)} &= [y_1^{(1)}, y_2^{(1)}, \cdots, y_K^{(1)}]^T \\ \mathbf{y}^{(2)} &= [y_1^{(2)}, y_2^{(2)}, \cdots, y_K^{(2)}]^T \end{aligned}\right\}. \tag{8.64}$$

Having defined $\mathbf{x}^{(1)}$, $\mathbf{x}^{(2)}$, $\mathbf{y}^{(1)}$, and $\mathbf{y}^{(2)}$ as above, we say that the set of solutions $\mathbf{x}^{(1)}$ dominates the other set $\mathbf{x}^{(2)}$ (denoted as $\mathbf{x}^{(1)} \succ \mathbf{x}^{(2)}$) when the following two conditions satisfy:

Condition 1

$$y_i^{(1)} \leq y_i^{(2)}, \tag{8.65}$$

for all $i = 1, 2, \dots, K$.

Condition 2

$$y_j^{(1)} < y_j^{(2)}, \tag{8.66}$$

at least for one value of j where $1 \leq j \leq K$.

The conditions for $\mathbf{x}^{(1)} \succ \mathbf{x}^{(2)}$, given by equations 8.65 and 8.66, assume that our multi-objective optimization is cast as minimization of the objectives. Having mathematically formulated dominance, we can now readily associate the dominance levels of different Pareto-optimal solution sets $1, 2, \dots, M$ shown in different colors in Figure 8.23c.

As the first step, all multi-objective optimizations randomly generate a set of N_{pop} solutions (models) $\mathbf{x}^{(i)}$, $i = 1, 2, \dots, N_{pop}$, in the decision (model) space in proportion to their prior distributions and evaluate the objective vectors $\mathbf{y}^{(j)}$, $j = 1, 2, \dots, K$. In multi-objective evolutionary algorithms (MOEAs), which is our focus here, each of these random solutions is called the members, or the individuals, and the entire set is called the population.

After generating the random population, in the next step the solutions are sorted, based on their dominance. Although there are many ways to do this sorting, two popular approaches to sorting are (1) rank-based sorting and (2) domination-count-based or domination-strength-based sorting (Konak et al., 2006). In rank-based sorting, the members are grouped into different ranks. The members belonging to rank 1 are those which do not dominate one another, but they dominate the rest that are not yet ranked. Similarly, the members belonging to rank 2 are dominated by the rank 1 members, but they dominate the rest. This process of such ranking is continued until the entire set of the members of the population is exhausted. Thus, in Figure 8.23, the black dots would represent rank 1, the red dots would represent rank 2, the cyan dots would represent rank 3, and the purple dots would represent rank M. In domination-count-based sorting, each member $\mathbf{x}^{(i)}$, $i = 1, 2, \dots, N_{pop}$, is assigned a strength $S(\mathbf{x}^{(1)})$ which is simply a measure of the number of solutions $\mathbf{x}^{(i)}$ dominates, which is then followed by assigning each member $\mathbf{x}^{(i)}$ a raw fitness value, given as a measure of the number of solutions $\mathbf{x}^{(i)}$ is dominated by. Therefore, if a solution is not dominated by any other solutions, its raw fitness is 0, and if it is dominated by many individuals, its raw fitness is high. Referring to Figure 8.23, we immediately see that the raw fitness of the black dots would be 0 because they are not dominated by any. Similarly, the raw fitness of the red dots would be the number of solutions represented by the black dots, the raw fitness of the cyan dots would be the total number of the black and red dots, and so on.

Sorting the population in terms of ranks or raw fitness provides the niching mechanism based on the Pareto dominance. This process may however fail when most members do not dominate each other. This situation is again due to genetic drift (Goldberg, 1989), which we discussed earlier. Like single-objective GA, the clustering of solutions caused by genetic drift is desirable when the solutions are near the global optimum, but it must be avoided at all other times to avoid premature convergence to a local optimum. MOEA methods avoid clustering by maintaining population diversity using two different methods:

1) Computing the crowding distance (Deb et al., 2002).
2) Computing the distance of a member from its kth nearest neighbor, for a suitably chosen value of k (Silverman, 1986).

We discussed how to compute the normalized crowding distance for single-objective GA (Figure 8.18). In MOEA, normalized crowding distance for all members belonging to the same rank is computed along all objective axes. Then, while choosing between the members of the same rank or raw fitness, the more diverse solution (solution with higher crowding distance) is preferred over a solution that is less diverse. Also in the kth nearest neighbor method, the distance is measured along all objective axes and the inverse of this distance is treated as its measure of population density and added to the raw fitness value as a measure of its fitness. Thus, out of any two individuals with the same rank or raw fitness, the one that is less crowded or less densely populated is preferred over the other. Besides diversity calculation, Li and Mallick (2015) found that an iteration-adaptive linear objective scaling is also a good method to maintain population diversity. Li and Mallick (2015) also showed that using a real-parameter coding with iteration-adaptive choices of the crossover index η and mutation distribution index κ (equations 8.50 and 8.55) provides additional means for diversity preservation for MOEA.

After non-dominated sorting and diversity computations, a population of new members or solutions is created. In MOEA, creating the new population of members from the old one is called advancing from one generation into the next, which can be performed using the GA operations of reproduction (selection), crossover, mutation, and elitism as described earlier. There are two popular approaches to create such a new set: (1) without using any external archive and (2) using an external archive. In the first approach, the reproduction generates the new set of members from the old set, in proportion to their levels of non-dominance and diversity. In crossover, the members after reproduction are treated as parents, and an entirely new set of members, called the children or offsprings, is produced. In mutation, parameters of the children are partially changed using a given probability of mutation. After computing the objectives of the mutated children, in elitism the parents and their children are combined into a single set, sorted according to their dominance levels and diversity, and are compared to select the best set of members to advance to the next generation, and this process is repeated until a predefined stopping criterion is reached. In the second approach, an external set of archive members are maintained along with the current population. The current population and the archive population are sorted according to their dominance and diversity, and the best members are copied into the archive. This is followed by the reproduction from the archive, and crossover and mutation to create the children. Once the children are created, the entire population is replaced by the children, which is again combined with the archive to repeat the same process until some predefined stopping criterion is reached.

Even though all multi-objective optimizations are based on the general ideas described above, their specific implementations are algorithm dependent, and out of many multi-objective methods proposed to date including some hybrid implementations, the two most promising ones are the fast non-dominated sorting GA or NSGA II of Deb et al. (2002) and the improved version of the strength Pareto evolutionary algorithm or SPEA2 of Zitzler et al. (2001). NSGA II, shown in Figure 8.24, uses ranks for sorting, crowding distance for diversity preservation, and does not use any external archive in generating the new members. SPEA2, shown in Figure 8.25, uses domination count for sorting, kth nearest neighbor for diversity preservation, and uses an external archive.

8.4.4 Geophysical Applications of Multi-Objective Optimization

Application of multi-objective optimization to solve geophysical inverse problems is relatively new. Estimation of an anisotropic model of mantle lithosphere using the splitting parameters of teleseismic shear waves and P-wave residual spheres using a multi-objective optimization method was reported by Kozlovskaya et al. (2007). For wave-equation migration velocity inversion, Singh et al. (2008) jointly minimized the semblance and differential semblance using MOEA. To estimate earthquake focal mechanisms, Heyburn and Fox (2010) jointly inverted the teleseismic body wave data and regional surface wave amplitude spectra using a multi-objective optimization method.

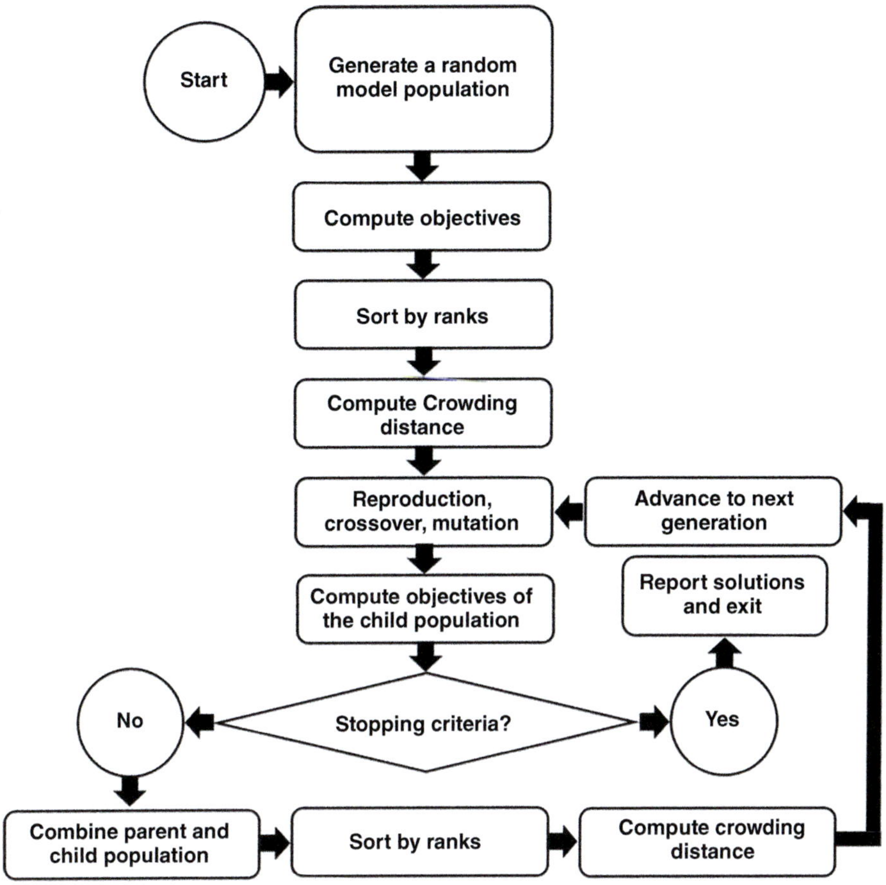

Figure 8.24 Workflow of the fast non-dominated sorting genetic algorithm (NSGA II). Adapted from Deb et al. (2002).

The first application of multi-objective optimization to invert the multicomponent (vertical and horizontal radial) seismic waveform data was carried out by Padhi and Mallick (2013, 2014). They implemented the original version of NSGA II, shown in Figure 8.24, and demonstrated that the method can extract the elastic properties and density from multicomponent seismic data, under both isotropic and anisotropic (VTI) subsurface conditions. In addition, Padhi and Mallick (2014) also developed a practical way to approximately calculate the posterior PDFs and uncertainties associated with estimating each model parameter.

Li and Mallick (2015) extended the work of Padhi and Mallick (2013, 2014) to develop a multi-objective inversion method for orthorhombic (ORT) media properties using three-component seismic data along two source-to-receiver azimuthal directions (i.e., for six objectives). To handle ORT properties, the original NSGA II workflow of Figure 8.24 was modified to the workflow of Figure 8.26.

Major highlights of the modifications shown in Figure 8.26 are the following:

1) Each aspect of NSGA II was carefully analyzed, and the entire method was parallelized. As shown in Figure 8.26, this parallel NSGA II distributes the parent population into different nodes, evaluates the objectives and performs the dominance-based sorting, and sends them back to the master node for subsequent processing.

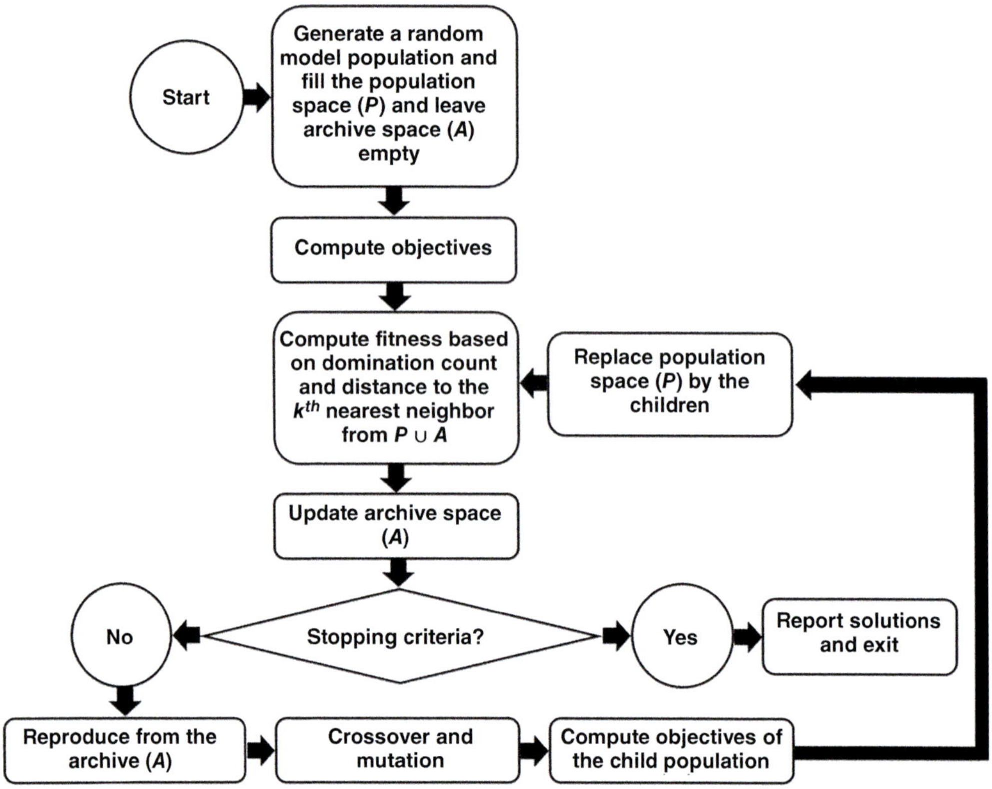

Figure 8.25 Workflow for the strength Pareto evolutionary algorithm (SPEA2). Adapted from Zitzler et al. (2001).

2) The raw objectives were first linearly scaled (see Figure 8.13), and these scaled objectives were used for sorting the population into their respective ranks.
3) The crowding distance to preserve population diversity was defined not only as the function of the scaled objectives, but also as a function of how much a given model is diverse from the others in the model space. Note that, in this respect, this modified NSGA II utilized the SPA2 concept of the kth nearest neighbor as a measure of population diversity, where k is provided from the model space. For the multi-objective optimization problems where the total number of parameters to be estimated is much higher than the number of objectives as is the case for the ORT inverse problem, a modification of the NSGA II to adapt to this concept of SPEA2 was found necessary by Li and Mallick (2015).
4) To achieve continuous sampling of the model space and maintain population diversity, Li and Mallick (2015) used real-coded GA, in which the crossover and the mutation probabilities (P_c and P_m) and their respective indices (η and κ) were varied over the generations.
5) Computing the approximate posterior PDFs as implemented by Padhi and Mallick (2014) was also implemented to approximately quantify the uncertainty associated for each model parameter estimate.

Figure 8.27 shows an example of the Pareto-optimal solution sets for an isotropic inversion using both vertical and radial components reported by Li and Mallick (2015).

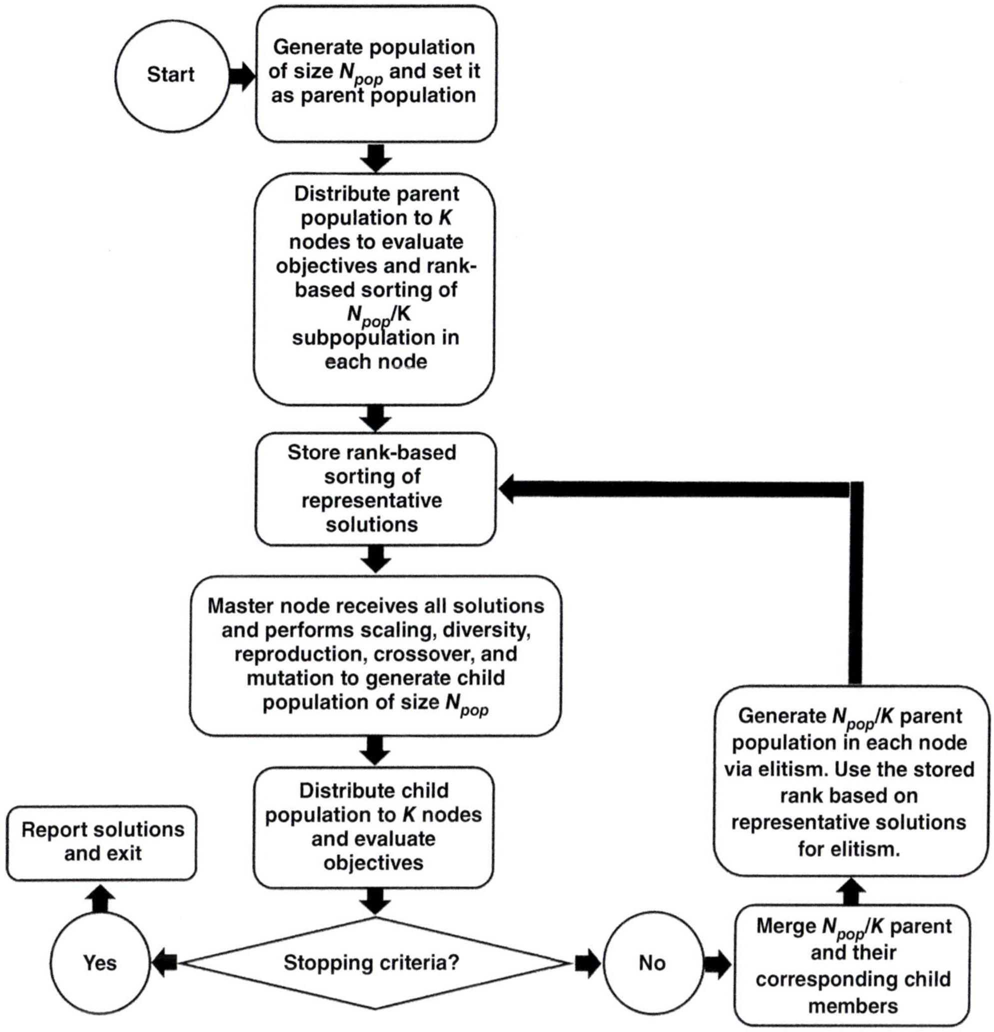

Figure 8.26 Workflow of Li and Mallick (2015) for inverting three-component seismic data along two source-to-receiver azimuths for estimating subsurface anisotropic (orthorhombic) elastic properties and density. Adapted from Li and Mallick (2015).

Note how the random models of the initial population (generation 1, shown as black dots) approach different Pareto solution sets with the progress of generations and attain the shapes like the ones shown in Figure 8.23.

8.5 Examples

There are many examples of successful applications of single-component global inversion methods in seismology and other fields of geophysics. A very good overview of these numerous examples can be found in Sen and Stoffa (2013), which we do not repeat here. Inversion of multicomponent synthetic seismic data examples for VTI and ORT subsurface properties and density can be found in Padhi and Mallick (2013, 2014) and Li and Mallick (2015). Here, we restrict only to some selected real data examples and one synthetic example with a complex subsurface structure.

The first example, shown in Figure 8.28, is a real data example at a single location from

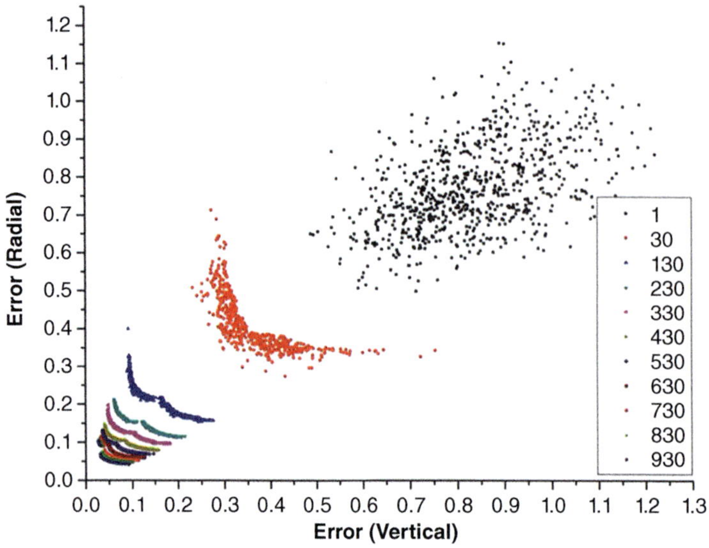

Figure 8.27 Development of the Pareto-optimal solution sets with the progress of generations for a two-objective isotropic seismic inverse problem. Li and Mallick (2015)/with permission of Oxford University Press. Different colors denote different generations as marked.

the Rock-Springs Uplift (RSU), Wyoming, USA, using the multi-scale (transdimensional) GA optimization (Figures 8.19 and 8.22). This example was specifically chosen because the location of the prestack CMP domain seismic data coincided with the RSU#1 well with P- and S-wave sonic and density log measurements such that the inversion results could be directly compared with the well-log data. The model was parameterized as the P-wave velocity (V_P), density (ρ), and Poisson's ratio (υ). As discussed above, the initial V_P model was estimated from velocity analysis. To ensure the stability, the velocity analysis was performed at regular time intervals of 100 ms. As shown in Figure 8.28a, the initial V_P model from velocity analysis is a smoothed version of the well-log V_P measured by the P-wave sonic. Gardner's relation $\rho = \alpha V_P^{\beta}$ was used to compute the initial density model, which is shown and compared with the well-log density in Figure 8.28b. Finally, the initial S-wave velocity was set as $V_S = V_P/2$, which, according to equation 8.41, is equivalent to a constant $\nu = 1/3$, shown in Figure 8.28c. To these initial models, their respective lower and upper bounds (dashed gray curves in Figure 8.28a–c) were set as follows:

1) $\pm 20\%$ for V_P.
2) $\pm 10\%$ for ρ.
3) From -75% to $+20\%$ for ν.

Using a uniform prior distribution between the above search limits, the transdimensional GA optimization (Figures 8.19 and 8.22) was run with the following parameters:

1) $N_{pop} = 80$, $G_{max} = 400$, $R_{max} = 5$, $N_{iter} = 7$, and $C_{max} = 0.99$ (see Figure 8.19).
2) $R_{beg} = 0.5\lambda_{dom}$, $R_{end} = 0.25\lambda_{dom}$, and $R_{inc} = 0.125\lambda_{dom}$ (see Figure 8.22), where λ_{dom} is the dominant seismic wavelength.

Figure 8.28d–f shows the estimated posterior PDFs from GA optimization, and Figure 8.28g–i compares the inverted (mean) models computed from these posterior PDFs with the well logs. Note that although the initial models were far away, especially for ρ and ν, the ones estimated from GA reasonably match with the well logs. Additionally, the posterior PDFs shown in Figure 8.28d–f provide a way to estimate the uncertainty associated with each model parameter estimate.

Our second example is the prestack waveform inversion of the 3D RSU seismic data, a single location inversion of which is shown in Figure 8.28. This example is from Pafeng et al. (2017) with 245 inline and 247 crossline locations and with 214 usable CMP locations for

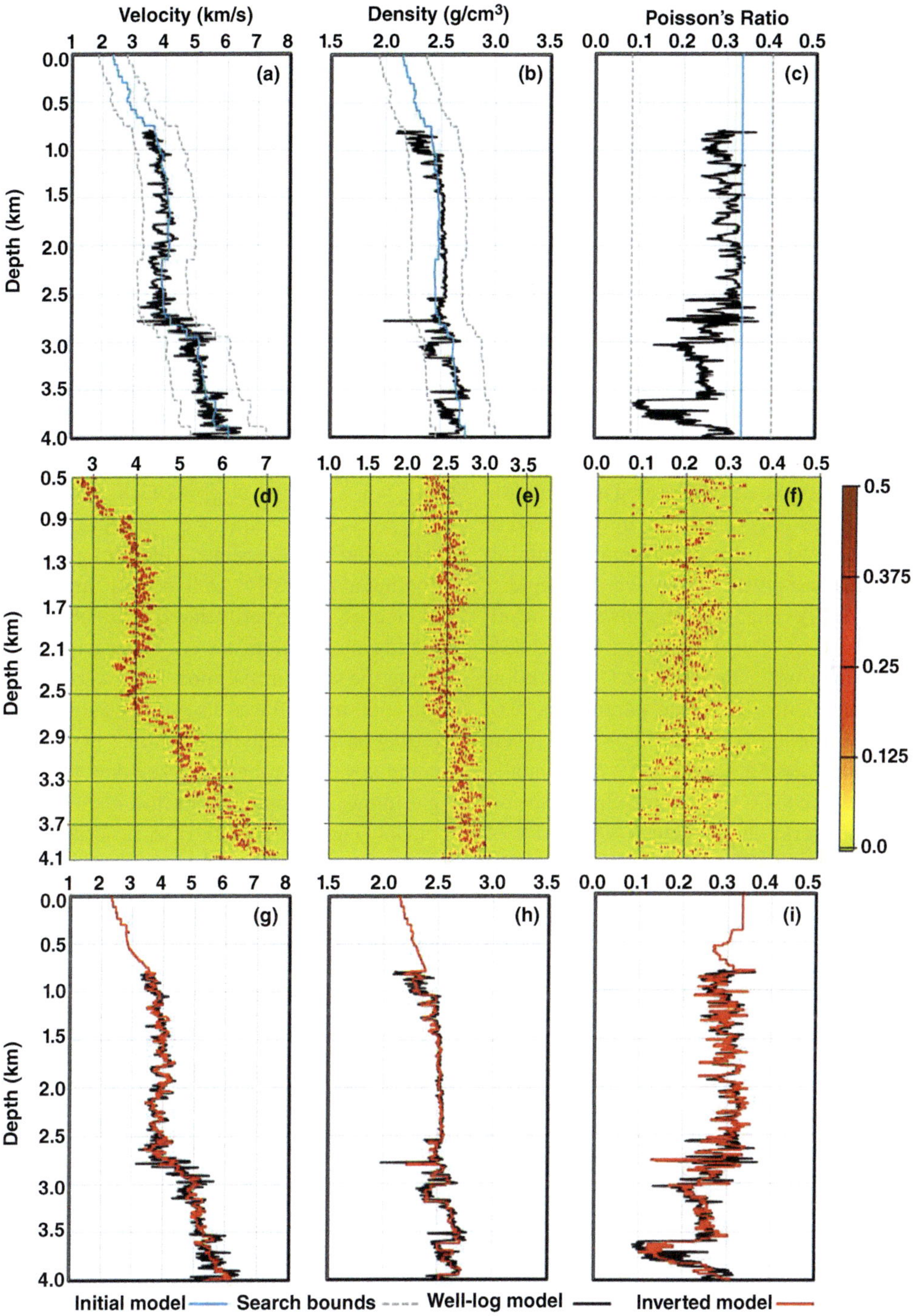

Figure 8.28 Multi-scale (transdimensional) GA inversion of real seismic data. (a–c) Initial model (cyan) and the lower and upper search bounds (dashed gray), compared with the well-log data (black). (d–f) Estimated posterior PDFs for each model parameter. (g–i) Inverted model (red) compared with the well logs (red).

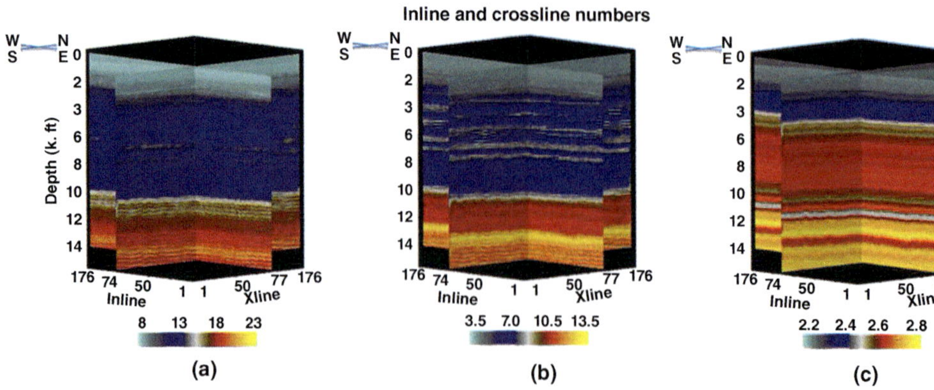

Figure 8.29 Prestack waveform inversion of the RSU seismic data using GA optimization. (a) P-wave velocity, (b) S-wave velocity, and (c) density. Pafeng et al. (2017)/Society of Exploration Geophysicists.

inversion with the RSU#1 located in the middle and was used to generate the initial model of the P- and S-wave velocity and density. This initial model was previously shown in Chapter 1 (Figure 1.19). Figure 8.29 shows the result of prestack waveform inversion of the RSU seismic data using the parallel GA workflow shown in Figure 8.20. The parameters used for this inversion run (see Figure 8.20) were $M_T = 214$, $N_P = 256$, and $N_C = 16$. Additionally, for simultaneous GA runs in different sectors using the workflow of Figures 8.19 and 8.22, the parameters were $N_{pop} = 80$, $G_{max} = 320$, $R_{max} = 5$, $N_{iter} = 0.7$, $C_{max} = 0.95$, $R_{beg} = 0.5\lambda_{dom}$, $R_{end} = 0.125\lambda_{dom}$, and $R_{inc} = 0.125\lambda_{dom}$. One inline and one crossline, which intersect one another at the RSU#1 well location, are shown in Figure 8.29. To verify the accuracy of GA predicting observation, an AVA-based inversion using the same initial model and a local optimization method was also run by Pafeng et al. (2017) on the same data volume. Figure 8.30 compares the synthetic stacked data for a single inline, computed using the AVA inverted model (Figure 8.30a) and GA inverted model (Figure 8.30b) with the observed stacked data

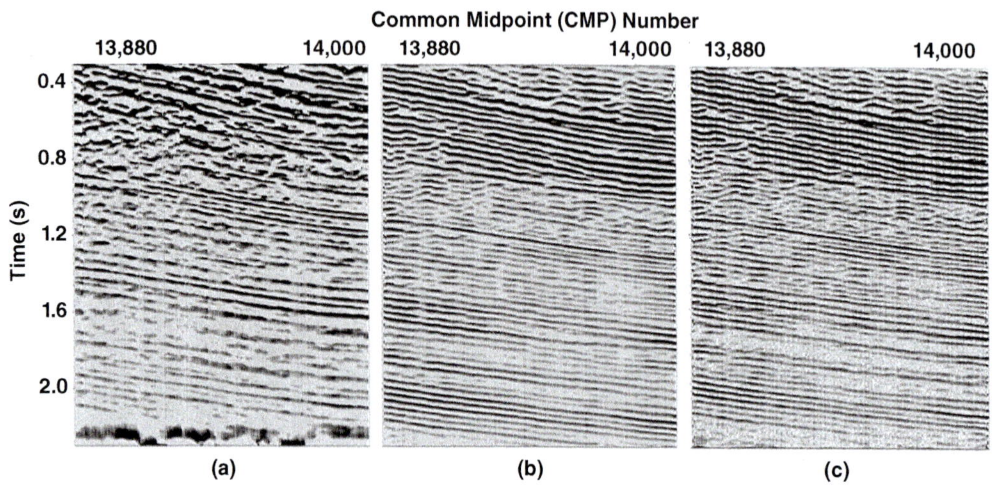

Figure 8.30 (a) Synthetic stacked data computed using the AVA inverted model. (b) Same as (a) but computed using the prestack GA inverted model. (c) Real stacked data. Reproduced from Pafeng et al. (2017)/Society of Exploration Geophysicists.

(Figure 8.30c). Clearly, the GA inverted model predicted the observation with much better accuracy than the AVA inverted model, providing a level of confidence in the accuracy of the method.

The third example is from Zhao and Sen (2022) on the Marmousi synthetic data using the gradient-based MCMC (GMCMC) sampling method. Marmousi is a complex 2D model developed by the Institut Français du Petrol (IFP) with the primary purpose of comparing different depth imaging and velocity estimation methods (Versteeg, 1994). The true and the initial models are shown in Figure 8.31. Marmousi synthetic data are acoustic and were computed using a time-domain finite-difference modeling and a Ricker wavelet with a peak frequency of 10 Hz. For forward computation in GMCMC, Zhao and Sen (2022) used a frequency domain finite-difference method for six evenly spaced frequencies between 6 and 7.5 Hz. Using a uniform prior distribution between prespecified upper and lower limits, Zhao and Sen ran five different tests of the GMCMC method whose results are shown in Figure 8.32. Details of the GMCMC method and these tests can be found in Zhao and Sen (2022). However, note from Figures 8.31 and 8.32 that even if the initial model is quite far away from the true model (Figure 8.31), all tests of the inversion method found the true model to a very good accuracy (Figure 8.32). Although of low frequency, the results shown in Figures 8.31 and 8.32 indicate GMCMC to be a powerful global optimization method. The results, however, are preliminary and require further investigations.

The fourth and final example is a multiobjective optimization of 3D multicomponent and wide-azimuth vertical seismic profile (VSP) data from Li et al. (2016). The objective was to characterize the Niobrara shale oil reservoir from the Wattenberg field of the Denver basin, Northeastern Colorado,

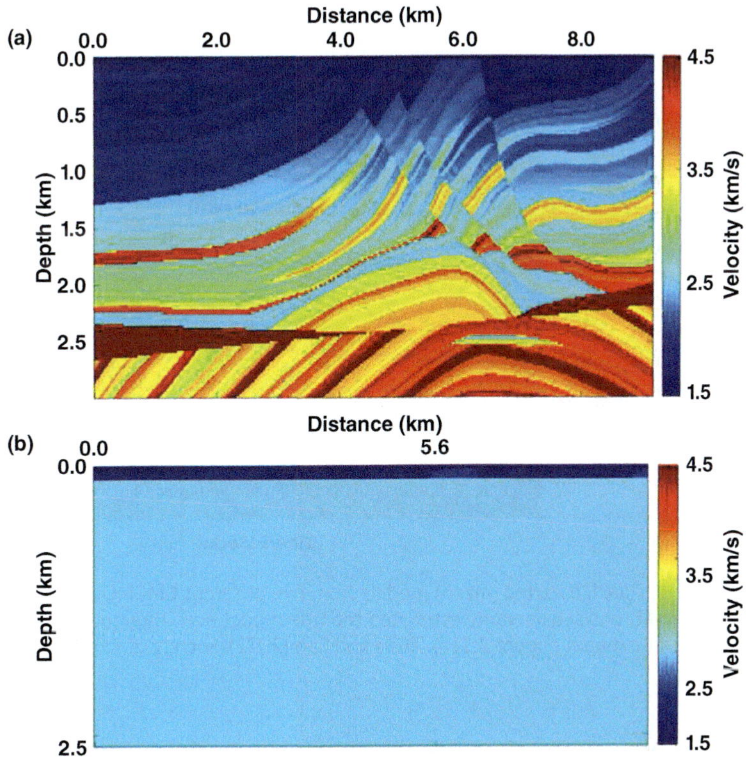

Figure 8.31 (a) True and (b) initial model used by Zhao and Sen (2022) for GMCMC. Zhao and Sen (2022)/Society of Exploration Geophysicists.

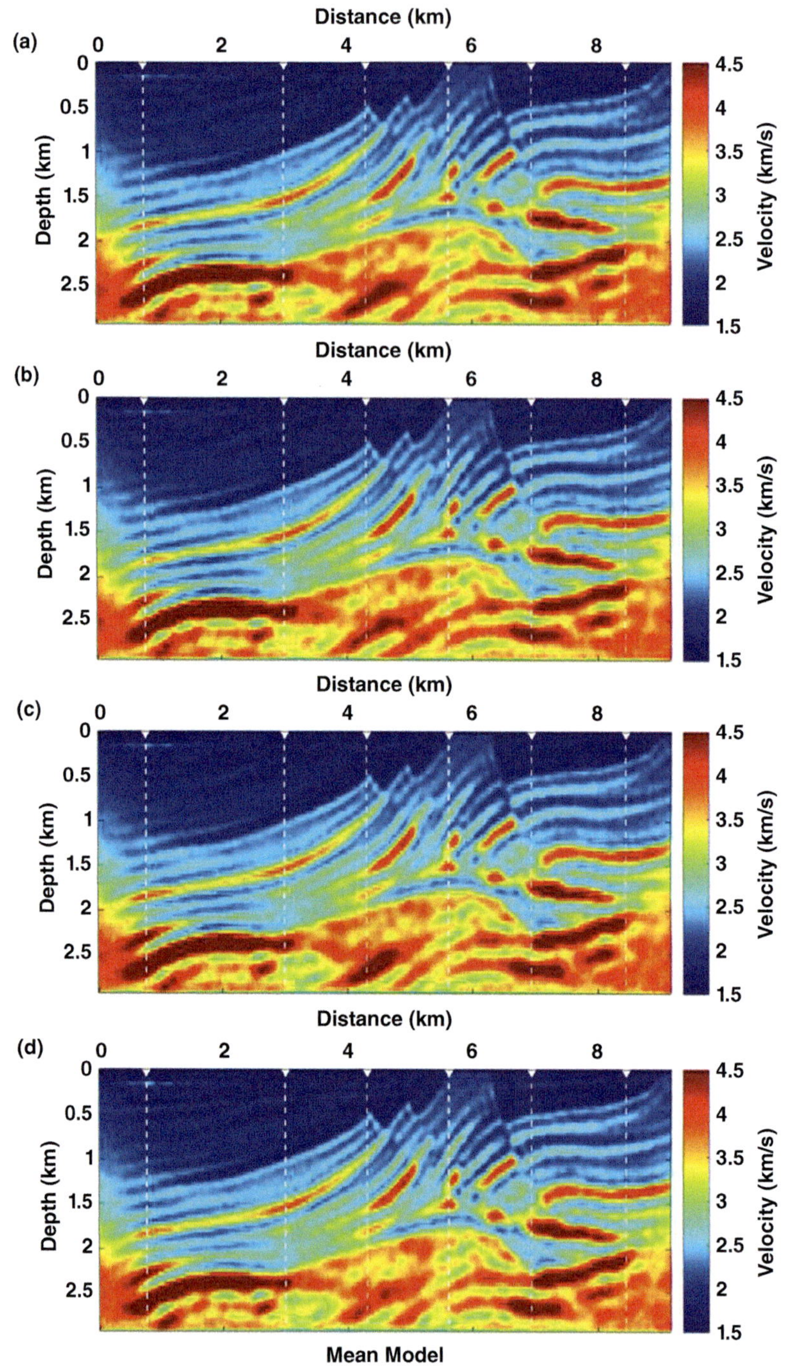

Figure 8.32 (a–d) Inverted (mean) models from five different GMCMC test runs of Zhao and Sen (2022). Note that all runs satisfactorily estimated the true model even though the initial model (Figure 8.31b) is far from the true model (Figure 8.31a). Zhao and Sen (2022)/Society of Exploration Geophysicists.

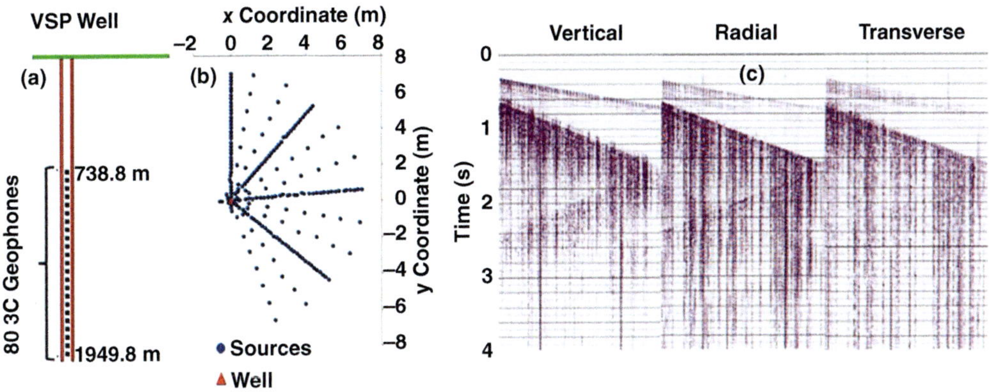

Figure 8.33 Acquisition geometry and example of the three-component (3C) VSP data from the Wattenberg field of the Denver basin. (a) Receiver locations. Eighty 3C geophones were placed in the wellbore between 738.8 and 1949.8 m depths. (b) Source locations at the surface with the *x* and *y* coordinates shown as the horizontal and vertical axes and the well location as the origin. (c) Three-component VSP data from one of the source locations, acquired at the wellbore. Horizontal axis is depth increasing from left to right. Li et al. (2016)/Society of Exploration Geophysicists.

USA (Matuszczak, 1973), for azimuthally anisotropic properties.

Figure 8.33 shows the VSP acquisition geometry and a sample VSP data. Eighty 3C geophones were placed between 738.8 and 1949.8 m depth intervals (Figure 8.33a). Wide-azimuth sources were placed on the surface as shown by blue circles in Figure 8.33b, with well location (red triangle) being the origin of the coordinate system and the positive *x*- and *y*-axes along the north and east. Finally, Figure 8.33c shows an example of the acquired VSP data from a single location.

As can be seen from Figure 8.33b, although the acquisition covered a wide range of azimuths and offsets, the offset coverage was dense along four azimuths: (1) N0°E, (2) N40°E, (3) N85°E, and (4) N130°E. Out of these four, the data along N40°E were of poor quality and therefore 3C data along the other three azimuths, i.e., a total of nine objectives, were inverted using the parallel version of the multi-objective optimization method, developed by Li and Mallick (2015).

For optimization, the Li and Mallick (2015) procedure shown in Figure 8.26 was run using a population size $N_{pop} = 4000$ for a maximum number of generations $G_{max} = 5000$. Using the real-coded GA parameterization, Li et al. (2016) varied the crossover probability (P_C) and the crossover index (η) as

$$\left.\begin{aligned} P_C &= 0.7 - \frac{t}{10t_{\max}} \\ \eta &= 1 + 19\frac{t}{t_{\max}} \end{aligned}\right\}, \tag{8.67}$$

in which t and $t_{\max}$ represent the current generation and the maximum number of generations. Similarly, the mutation probability (P_M) and the mutation index (κ) were varied with generation as

$$\left.\begin{aligned} P_M &= \tfrac{1}{N} + \tfrac{t}{t_{\max}}\left(1 - \frac{1}{N}\right) \\ \kappa &= 100 + t \end{aligned}\right\}. \tag{8.68}$$

In equation 8.68, t and $t_{\max}$ have the same meaning as in equation 8.67 and N is the total number of model parameters. The generation-dependent P_c and η were deliberately chosen such that P_c decreases and η increases with generation such that the model space is widely sampled at the beginning and is slowly reduced over generations as the algorithm approaches the true solution. The choices of generation-dependent P_M and κ (equation 8.68) were also motivated by maintaining diversity in the population at early stage and slowly reducing it over time as the

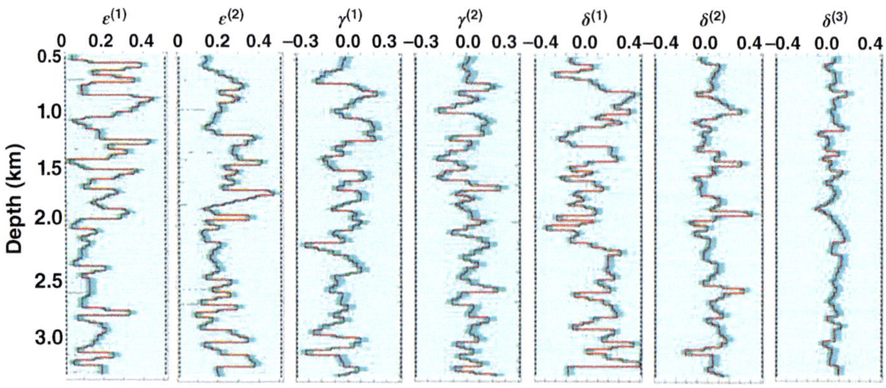

Figure 8.34 3C VSP inversion result for estimating the anisotropic model parameters. The red curves are the estimated model parameters. The dashed curves are the minimum and maximum search limits for each model parameter representing their uniform prior distributions. Cyan background shows the estimated posterior PDFs.

solutions near the global optimum. Additionally, in their implementation, Li et al. (2016) followed the work of Li and Mallick (2015) and computed the crowding distance as a combination of the normalized distance measured in the scaled objective space with that measured in the model space.

Because well-log data with P- and S-wave sonic and density logs were available at the well location, Li et al. (2016) used these logs to constrain the vertical P- and S-wave velocities (V_{P0}, V_{S0}) and density and focused only on the estimation of the other parameters for orthorhombic anisotropy, namely $\varepsilon^{(1)}$, $\varepsilon^{(2)}$, $\gamma^{(1)}$, $\gamma^{(2)}$, $\gamma^{(3)}$, $\delta^{(1)}$, $\delta^{(2)}$, and $\delta^{(3)}$, defined by Tsvankin (1997). These inversion results are shown in Figure 8.34. Note although the priors for each model parameter were uniform and wide, they all were estimated quite well, represented by their estimated posterior PDFs. These results were also found to be consistent (but with much higher resolution) with the independent results obtained from the slowness polarization analysis method of Grechka and Mateeva (2007).

8.6 Chapter Summary

In this chapter, we started with an example of two different models predicting the observed seismic data with almost equal accuracy as an argument for a global, instead of a local optimization, and formally introduced single- and multi-objective global methods for optimization. We then defined the Bayesian approach to solve optimization problems and discussed simple Monte Carlo integration method for random (but uniform) sampling of the model space. Following this, we started our discussion on the global optimization, the concept of a pure MCI, the problems of MCI in solving complex problems, and the need for an exhaustive and nonuniform sampling strategies such as Metropolis–Hastings and Gibbs' Sampler. We then thoroughly discussed MCMC, SA, and GA optimization methods of single-objective optimization. From single objective, we then moved onto discussing the multi-objective optimization methods. Finally, we ended the chapter with a few selected examples.

8.7 Exercises

Problem 8.1 Derive the ergodicity property of the Markov chain (equation 8.24).

Problem 8.2 Verify that the sampling strategy of the Metropolis–Hastings algorithm satisfies the detailed balance equation (equation 8.22).

Problem 8.3 In Metropolis–Hastings algorithm, assume that the proposal distribution $q(\mathbf{m}_2 \mid \mathbf{m}_1)$ to be normal distribution with a fixed covariance. Then, verify that the ratio $q(\mathbf{m}_1 \mid \mathbf{m}_{2c})/q(\mathbf{m}_{2c} \mid \mathbf{m}_1)$ is one.

Problem 8.4 A simplified workflow of the Metropolis–Hastings algorithm, shown in Figure 8.6, is used for serial computing. How would you extend this workflow for parallel computing environments?

Problem 8.5 Assume that you are inverting seismic data only for the estimation of the P-wave velocity (V_P) as a function of depth. Also, define a uniform prior $p(\mathbf{m})$ between the lower and upper bounds of 1800 and 3000 km/s. Starting from an initial set of models, sampled randomly from this $p(\mathbf{m})$, you used any global optimization method (MCMC/SA/GA) and the binning technique and produced the cumulative probabilities as shown in the following table:

In the above table, the first column is depth in meters and the first row shows different V_P values in meters per second between the lower and upper bounds in steps of 100 m/s. Finally, the other rows represent the cumulative probabilities for each depth for different V_P values within the range.

If a total of 1000 models were generated in course of the optimization, then compute and display the posterior PDFs for each depth sample using equation 8.7.

Problem 8.6 From the PDFs computed in Problem 8.5, estimate and display the depth-domain maximum-likelihood and mean models. Hint: The maximum likelihood model is given by joining the peaks of the computed PDFs for each depth sample, and the mean model $\langle \mathbf{m} \rangle$ is given by equation 8.9.

Problem 8.7 From the posterior PDF $p(\mathbf{m} \mid \mathbf{d})$, computed in Problem 8.5, and $\langle \mathbf{m} \rangle$, computed in Problem 8.6, use equation 8.10 and compute the model covariance matrix $\mathbf{C}_M$. Compute the diagonal elements of $\mathbf{C}_M$ and display them as a function of depth. What does this display signify?

V_P (km/s) Depth (m)	1800	1900	2000	2100	2200	2300	2400	2500	2600	2700	2800	2900	3000
0	10	20	570	20	10	0	0	0	0	0	0	0	0
100	10	20	570	20	10	0	0	0	0	0	0	0	0
200	5	15	25	500	100	50	30	20	10	5	0	0	0
300	0	10	30	60	100	300	200	60	20	10	5	0	0
400	0	15	25	50	80	100	450	100	80	40	25	10	0
500	0	10	30	70	110	150	400	100	80	30	10	0	0
600	5	30	50	150	400	150	75	25	15	5	0	0	0
700	5	10	30	50	80	100	120	350	100	50	40	20	10
800	0	10	15	25	50	75	80	120	300	100	75	50	20
900	0	20	30	15	5	30	50	80	90	100	300	150	50
1000	0	5	25	50	200	300	150	100	80	20	10	5	0
1100	0	50	100	150	160	300	100	80	30	10	5	0	0
1200	5	25	75	90	120	130	320	110	50	25	10	0	0
1300	0	0	50	75	90	100	120	280	100	70	50	20	5
1400	0	0	10	30	50	120	130	140	300	100	50	10	5
1500	0	0	10	20	40	60	80	100	120	130	300	50	20

Problem 8.8 Derive equation 8.28 from equation 8.25.

Problem 8.9 Recall the example of the binary-coded GA (Section "Step 6: Crossover, Mutation, and Elitism"):

1) Using eight bits for each model parameter (V_P, V_S, ρ), the binary string for a given depth sample i is given as 100101110011100111001001, and the values of $(V_{Pmin})_i$, $(V_{Pmax})_i$, $(V_{Smin})_i$, $(V_{Smax})_i$, $(\rho_{min})_i$, and $(\rho_{max})_i$ are respectively 2500 m/s, 5500 m/s, 1250 m/s, 2250 m/s, 2.0 g/cm^3, and 2.5 g/cm^3.
2) Using the same procedure that we used in Section "Step 6: Crossover, Mutation, and Elitism" for decoding $(V_P)_i$, decode $(V_S)_i$ and ρ_i.

Problem 8.10 Reverse of parameter decoding is parameter coding, as discussed in Section "Step 6: Crossover, Mutation, and Elitism" and illustrated in Figure 8.16. For binary-coded GA (see Figure 8.19), the initial model population (P_{init}) of size N_{pop} is generated in proportion to the prior distribution $p(\mathbf{m})$. These models are real numbers, which for the isotropic elastic inverse problem are the values of V_P, V_S, and ρ for each depth (or time) sample. After generating P_{init}, they are reproduced to create the intermediate population (P_{inter}). Then, to produce the child population, each member of P_{inter} must be coded as binary strings for crossover (Figure 8.17) and mutation. Write a computer program (in your preferred language) for parameter coding to perform the following:

1) Code the values of $(V_P)_i$, $(V_S)_i$, and ρ_i for any depth sample i as a binary string with specified minimum and maximum values $(V_{Pmin})_i$, $(V_{Pmax})_i$, $(V_{Smin})_i$, $(V_{Smax})_i$, $(\rho_{min})_i$, and $(\rho_{max})_i$.
2) Allow the flexibility to vary the number of bits as $(N_{bits})_{V_P}$, $(N_{bits})_{V_S}$, and $(N_{bits})_\rho$ for each model parameter (V_P, V_S, and ρ) in your code.

Use the decoding example of Section "Step 6: Crossover, Mutation, and Elitism," where $(V_{Pmin})_i$, $(V_{Pmax})_i$, $(V_{Smin})_i$, $(V_{Smax})_i$, $(\rho_{min})_i$, and $(\rho_{max})_i$ are respectively 2500 m/s, 5500 m/s, 1250 m/s, 2250 m/s, 2.0 g/cm^3, and 2.5 g/cm^3, and assume that the model randomly drawn from $p(\mathbf{m})$ has the values of $(V_P)_i = 5229$ m/s and $(V_S)_i$ and ρ_i values, obtained from Problem 8.9, and test your code to verify that for $(N_{bits})_{V_P} = (N_{bits})_{V_S} = (N_{bits})_\rho = 8$, your $(V_P)_i$, $(V_S)_i$, and ρ_i values are used to code the binary string 100101110011100111001001. Repeat the same to code $(V_P)_i$, $(V_S)_i$, and ρ_i with $(N_{bits})_{V_P} =$ 16 and $(N_{bits})_{V_S} = (N_{bits})_\rho = 8$.

References

Anderson, R.S., Worthington, M.H., and Cleary, J.R., 1972, Density modeling by Monte Carlo inversion-I methodology, *Geophysical Journal International*, *29*, 433–444, https://doi.org/10.1111/j.1365-246X.1972.tb06169.x.

Antonov, I.A., and Saleev, V.M., 1979, An economic method of computing LPτ sequences, *USSR Computational Mathematics and Mathematical Physics*, *7*, 86–112, https://doi.org/10.1016/0041-5553(79)90085-5.

Ayani, M., MacGregor, L., and Mallick, S., 2020, Inversion of marine controlled source electromagnetic data using a parallel non-dominated sorting genetic algorithm, *Geophysical Journal International*, *220*, 1066–1077, https://doi.org/10.1093/gji/ggz501.

Backus, G., 1970a, Inference from inadequate and inaccurate data: I, *Proceedings of the National Academy of Sciences*, *65*(1), 1–7, https://doi.org/10.1073/pnas.65.1.1.

Backus, G., 1970b, Inference from inadequate and inaccurate data: II, *Proceedings of the National Academy of Sciences*, *65*(2), 281–287, https://doi.org/10.1073/pnas.65.2.281.

Backus, G., 1970c, Inference from inadequate and inaccurate data: III, *Proceedings of the National Academy of Sciences*, *67*(1), 282–289, https://doi.org/10.1073/pnas.67.1.282.

Backus, G., and Gilbert, F., 1967, Numerical applications of a formalism for geophysical inverse problems, *Geophysical Journal International*, *13*, 247–276, https://doi.org/10.1111/j.1365-246X.1967.tb02159.x.

Backus, G., and Gilbert, F., 1968, Resolving the power of gross earth data, *Geophysical Journal International*, *16*, 169–205, https://doi.org/10.1111/j.1365-246X.1968.tb00216.x.

Backus, G., and Gilbert, F., 1970, Uniqueness in the inversion of inaccurate gross earth data, *Philosophical Transactions of the Royal Society of London*, *266*, 1230192, https://doi.org/10.1098/rsta.1970.0005.

Bartley, P., and Fox, B.L., 1988, Implementing Sobol's quasirandom sequence generators, *ACM Transactions on Mathematical Software*, *14*, 88–100, https://doi.org/10.1145/42288.214372.

Barton, P.J., 1986, The relationship between seismic velocity and density in the continental crust – a useful constraint? *Geophysical Journal International*, *87*, 195–208, https://doi.org/10.1111/j.1365-246X.1986.tb04553.x.

Basu, A., and Frazer, L.N., 1989, Rapid determination of the critical temperature in simulated annealing inversion, *Science*, *249*, 1409–1412, https://doi.org/10.1126/science.249.4975.1409.

Biswas, R., and Sen, M.K., 2017, 2D full waveform inversion and uncertainty estimation using the reversible jump Hamiltonian Monte Carlo, SEG Technical Program Expanded Abstracts, 1280–1285, https://doi.org/10.1190/segam2017-17680416.1.

Box, G.P., and Tiao, G.C., 1973, *Bayesian Inference in Statistical Analysis*, Addison-Wesley, ISBN: 0-471-57428-7.

Cary, P.W., and Chapman, C.H., 1988, Automatic 1-D waveform inversion of marine seismic reflection data, *Geophysical Journal International*, *93*, 527–546, https://doi.org/10.1111/j.1365-246X.1988.tb03879.x.

Casella, G., and George, E.I., 1992, Explaining the Gibbs sampler, *American Statistician*, *46*, 167–174, https://doi.org/10.2307/2685208.

Connolly, P., 1999, Elastic impedance, *The Leading Edge*, *18*, 438–452, https://doi.org/10.1190/1.1438307.

Deb, K., and Agrawal, R.B., 1995, Simulated binary crossover for continuous search space, *Complex Systems*, *9*, 115–148.

Deb, K., and Agrawal, S., 1999, A niched-penalty approach for constraint handling in genetic algorithms, Proceedings of ICANNGA-99, Portoroz, Slovenia, https://doi.org/10.1007/978‐3‐7091‐6384‐9_40.

Deb, K., Pratap, A., Agarwal, S., and Meyarivan, T., 2002, A fast and elitist multi-objective genetic algorithm, NSGA-II, *IEEE Transactions on Evolutionary Computation*, *6*(2), 181–197, https://doi.org/10.1109/4235.996017.

Dhara, A., Sen, M.K., Yang, D., Schmedes, J., Routh, P., and Sain, R., 2020, Facies and reservoir properties estimation by a transdimensional seismic inversion, SEG Technical Program Expanded Abstracts, 255-259, https://doi.org/10.1190/segam2020-3427560.1.

Du, Z., and MacGregor, L.M., 2010, Reservoir characterization from joint inversion of marine CSEM and seismic AVA data using genetic algorithms: a case study based on the Luva gas field, SEG Technical Program Expanded Abstracts, 737-741, https://doi.org/10.1190/1.3513888.

Duijndam, A.J.W., 1988, Bayesian estimation in seismic inversion, part I: principles, *Geophysical Prospecting*, *36*, 878–898, https://doi.org/10.1111/j.1365-2478.1988.tb02198.x.

Frazer, L.N., and Basu, A., 1990, Freeze bath inversion, SEG Technical Program Expanded Abstracts, 1123-1125, https://doi.org/10.1190/1.1889927.

Gardner, G.H.F., Gardner, L.W., and Gregory, A.R., 1974, Formation velocity and density – the diagnostic basics for stratigraphic traps, *Geophysics*, *39*, 770–780, https://doi.org/10.1190/1.1440465.

Gelfand, A.E., Hills, S.E., Racine-Poon, A., and Smith, A.F.M., 1990, Illustration of Bayesian inference in normal data models using Gibbs sampling, *Journal of the American Statistical Association*, *85*, 972–985, https://doi.org/10.2307/2289594.

Gelfand, A.E., and Smith, A.F.M., 1990, Sampling-based approaches to calculating marginal densities, *Journal of the American Statistical Association*, *85*, 398–409, https://doi.org/10.2307/2289776.

Geman, S., and Geman, D., 1984, Stochastic relaxation, Gibbs distributions, and the Bayesian restoration of images, *IEEE Transactions on Pattern Analysis and Machine Intelligence*, *6*, 721–741, https://doi.org/10.1109/TPAMI.1984.4767596.

Giagkiozis, I., Purshouse, R.C., and Fleming, P.J., 2015, An overview of population-based algorithms for multi-objective optimization, *International Journal of Systems Science*, *46*(9), 1572–1599, https://doi.org/10.1080/00207721.2013.823526.

Goldberg, D.E., 1989, *Genetic Algorithms in Search, Optimization and Machine Learning*, Addison Wesley Publishing Company, ISBN: 978-0-201-15767-3.

Grechka, V., and Mateeva, A., 2007, Inversion of P-wave VSP data for local anisotropy: theory and case study, *Geophysics*, *72*, 69–79, doi 10.1190/1.2742970.

Hajela, P., and Lin, C.-Y., 1992, Genetic search strategies in multicriterion optimal design, *Structural Optimisation*, *4*, 99–107, https://doi.org/10.1007/BF01759923.

Halton, J.H., 1960, On the efficiency of certain quasi-random sequences of points in evaluation multi-dimensional integrals, *Numerische Mathematik*, *2*, 84–90, https://doi.org/10.1007/BF01386213.

Hampson, D. P., Russell, B.H., and Bankhead, B., 2005, Simultaneous inversion of pre-stack seismic data, SEG Technical Program Expanded Abstracts, 1633–1637, https://doi.org/10.1190/1.2148008.

Hastings, W.K., 1970, Monte Carlo sampling methods using Markov chains and their applications, *Biometrika*, *57*, 97–109, https://doi.org/10.2307/2334940.

Heyburn, R., and Fox, B., 2010, Multi-objective analysis of body and surface waves from the Market Rasen (UK) earthquake, *Geophysical Journal International*, *181*, 532–544, https://doi.org/10.1111/j.1365-246X.2010.04523.x.

Hilterman, F., 1990, Is AVO the seismic signature of lithology? A case study of Ship Shoal-South Addition, *The Leading Edge*, *10*, 39–42, https://doi.org/10.1190/1.1439744.

Holland, J.H., 1975, *Adaptation in Natural and Artificial System*, University of Michigan Press, ISBN: 978-0262581110.

Hong, T., and Sen, M.K., 2009, A new MCMC algorithm for seismic waveform inversion and corresponding uncertainty analysis, *Geophysical Journal International*, *177*(1), 14–32, https://doi.org/10.1111/j.1365-246X.2008.04052.x.

Ingber, L., 1989, Very fast simulated re-annealing, *Mathematical and Computer Modelling*, *12*, 967–973, https://doi.org/10.1016/0895-7177(89)90202-1.

Jia, L., Mallick, S., and Wang, C., 2021, Data-driven prestack-waveform inversion using genetic algorithm: methodology and examples, *Interpretation*, *9*, T1065–T1084, https://doi.org/10.1190/INT-2020-0201.1.

Jian, W., Lei, Z., Hao, C., and Xiu-ming, W., 2018, Trans-dimensional Bayesian inversion for directional resistivity logging while drilling data, SEG Technical Program Expanded Abstracts, 849-852, https://doi.org/10.1190/segam2018-2998428.1.

Karaoulis, M., Revil, A., Zhang, J., and Werkema, D.D., 2012, Time-lapse joint inversion of crosswell DC resistivity and seismic data: a numerical investigation, *Geophysics*, *77*, D141–D157, https://doi.org/10.1190/geo2012-0011.1.

Keilis-Borok, V.I., and Yanovskaya, T.B., 1967, Inverse problems of seismology, *Geophysical Journal International*, *13*, 223–234, https://doi.org/10.1111/j.1365-246X.1967.tb02156.x.

Kirkpatrick, S.C., Gelatt, D., and Vecchi, M.P., 1983, Optimization by simulated annealing, *Science*, *220*, 671–680, https://doi.org/10.1126/science.220.4598.671.

Konak, W., David, C., and Alice, E.S., 2006, Multi-objective optimization using genetic algorithms: a tutorial, *Reliability Engineering and System Safety*, *91*, 992–1007, https://doi.org/10.1016/j.ress.2005.11.018.

Kozlovskaya, E., Vecsey, L., Plomerova, J., and Raita, T., 2007, Joint inversion of multiple data types with the use of multiobjective optimization: problem formulation and application to the seismic anisotropy investigations, *Geophysical Journal International*, *171*(2), 761–779, https://doi.org/10.1111/j.1365-246X.2007.03540.x.

Lambert, J.P., 1988, Quasi-random sequences in numerical practice, in R.P. Agarwal, Y.M. Chow, and S.J. Wilson, eds., *Numerical Mathematics*, Vol. *86*, 273–284, ISNM, Singapore https://doi.org/10.1007/978‐3‐0348‐6303‐2_22.

Lang, X, and Grana, D., 2015, Bayesian rock physics inversion of acoustic and electrical properties for rock-fluid property estimation, SEG Technical Program Expanded Abstracts, 2931-2935, https://doi.org/10.1190/segam2015-5833999.1.

Li, T., and Mallick, S., 2015, Multicomponent, multi-azimuth pre-stack seismic waveform inversion for azimuthally anisotropic media using a parallel and computationally efficient non-dominated sorting genetic algorithm, *Geophysical Journal International*, *200*, 1136–1154, https://doi.org/10.1093/gji/ggu445.

Li, T., Mallick, S., Tamimi, N., and Davis, T., 2016, Inversion of wide-azimuth multicomponent vertical seismic profile data for anisotropic subsurface properties, SEG Technical Program Expanded Abstracts, 1252-1257, https://doi.org/10.1190/segam2016-13875028.1.

Mallick, S., 1995, Model-based inversion of amplitude-variations-with-offset data using a genetic algorithm, *Geophysics*, *60*, 939–954, https://doi.org/10.1190/1.1443860.

Mallick, S., 1999, Some practical aspects of prestack waveform inversion using a genetic algorithm: an example from the East Texas Woodbine gas sand, *Geophysics*, *64*, 326–336, https://doi.org/10.1190/1.1444538.

Mallick, S., and Adhikari, S., 2015, Amplitude-variation-with-offset and prestack waveform inversion: a direct comparison using a real data example from the Rock Springs uplift, Wyoming, USA, *Geophysics*, *80*, B45–B59, https://doi.org/10.1190/geo2014-0233.1.

Mallick, S., and Chakraborty, D., 2022, Prediction of the ocean water sound speeds via attribute-guided seismic waveform inversion, *Geophysics*, *87*, U67–U79, https://doi.org/10.1190/geo2021-0442.1.

Matuszczak, R.A., 1973, Wattenberg field, Denver basin, Colorado: abstract, *AAPG Bulletin*, *57*, 959, https://doi.org/10.1306/83D90DE5‐16C7‐11D7‐8645000102C1865D.

Mazzotti, A., Bienati, N., Stucchi, E., Tognarelli, A., Aleardi, M., and Sajeva, A., 2016, Two-grid genetic algorithm full-waveform inversion, *The Leading Edge*, *35*, 1068–1075, https://doi.org/10.1190/tle35121068.1.

Menke, W., 1989, *Geophysical Data Analysis: Discrete Inverse Theory*, Academic Press, ISBN: 978-0-12-490921-2.

Metropolis, N., Rosenbluth, A.W., Rosenbluth, M.N., and Teller, A.H., 1953, Equation of state calculations by fast computing machines, *The Journal of Chemical Physics*, *21*, 1087–1092, https://doi.org/10.1063/1.1699114.

Metropolis, N., and Ulam, S., 1949, The Monte Carlo method, *Journal of the American Statistical Association*, *44*, 335–341, https://doi.org/10.1080/01621459.1949.10483310.

Miettinen, K., 1999, *Nonlinear Multiobjective Optimization*, Kluwer Academic Publishers, ISBN: 978-1-4615-5563-6.

Mosegaard, K., and Sambridge, M., 2002, Monte Carlo analysis of inverse problems, *Inverse Problems*, *18*, R29–R54, https://doi.org/10.1088/0266-5611/18/3/201.

Neiderreiter, H., 1988, in H. Brass, and G. Hämmerlin, eds., *Quasi-Monte Carlo methods for multidimensional numerical integration III*, International Series of Numerical Mathematics, Vol. *85*, 157–171, Birkhäuser, Basel https://doi.org/10.1007/978‐3‐0348‐6398‐8_15.

Nulton, J.D., and Salamon, P., 1988, Statistical mechanics of combinatorial optimization, *Physical Review*, *A37*, 1351–1356, https://doi.org/10.1103/physreva.37.1351.

Padhi, A., and Mallick, S., 2013, Accurate estimation of density from the inversion of multicomponent prestack seismic waveform data using a non-dominated sorting genetic algorithm, *The Leading Edge*, *32*, 94–98, https://doi.org/10.1190/tle32010094.1.

Padhi, A., and Mallick, S., 2014, Multicomponent prestack seismic waveform inversion in transversely isotropic media using a non-dominated sorting genetic algorithm, *Geophysical Journal International*, *196*, 1600–1618, https://doi.org/10.1093/gji/ggt460.

Padhi, A., Mallick, S., Fortin, W., and Holbrook, W.S., 2015, 2-D ocean temperature and salinity images from pre-stack seismic waveform inversion methods: an example from the South China Sea, *Geophysical Journal International*, *202*, 800–810, https://doi.org/10.1093/gji/ggv188.

Pafeng, J., Mallick, S., and Sharma, H., 2017, Prestack waveform inversion of three-dimensional seismic data – an example from the Rock Springs uplift, Wyoming, USA, *Geophysics*, *82*(1), B1–B12, https://doi.org/10.1190/geo2016-0079.1.

Press, F., 1968, Earth models obtained by Monte-Carlo inversion, *Journal of Geophysical Research*, *73*, 5223–5234, https://doi.org/10.1029/JB073i016p05223.

Press, F., 1970a, Earth models consistent with geophysical data, *Physics of the Earth and Planetary Interiors*, *3*, 3–22, https://doi.org/10.1016/0031-9201(70)90039-7.

Press, F., 1970b, Regionalized earth models, *Journal of Geophysical Research*, *75*, 6575–6581, https://doi.org/10.1029/JB075i032p06575.

Press, W.H., Teulolsky, S.A., Vetterling, W., and Flannery, B.P., 2007, *Numerical Recipes The Art of Scientific Computing*, Third Edn, Cambridge University Press, ISBN 978-0-521-88068-8.

Ritzel, B.J., Eheart, J.W., and Ranjithan, S., 1994, Using genetic algorithms to solve a multiple objective groundwater pollution containment problem, *Water Resources Research*, *30*(5), 1589–1603, https://doi.org/10.1029/93WR03511.

Rothman, D.H., 1985, Nonlinear inversion statistical mechanics, and residual statics corrections, *Geophysics*, *50*, 2784–2796, https://doi.org/10.1190/1.1441899.

Rothman, D.H., 1986, Automatic estimation of large residual statics corrections, *Geophysics*, *51*, 332–346, https://doi.org/10.1190/1.1442092.

Sambridge, M., and Mosegaard, K., 2002, Monte Carlo methods in geophysical inverse problems, *Reviews of Geophysics*, *40*, 3-1–3-29, https://doi.org/10.1029/2000RG000089.

Sen, M.K., and Biswas, R., 2017, Trans-dimensional seismic inversion using the reversible jump Hamiltonian Monte Carlo algorithm, *Geophysics*, *82*, R119–R134, https://doi.org/10.1190/geo2016-0010.1.

Sen, M.K., and Stoffa, P.L., 1991, Nonlinear one-dimensional seismic waveform inversion using simulated annealing, *Geophysics*, *56*, 1624–1638, https://doi.org/10.1190/1.1442973.

Sen, M.K., and Stoffa, P.L., 1992, Rapid sampling of model space using genetic algorithms: examples from seismic waveform inversion, *Geophysical Journal International*, *108*, 281–292, https://doi.org/10.1111/j.1365-246X.1992.tb00857.x.

Sen, M.K., and Stoffa, P.L., 1996, Bayesian inference, Gibbs' sampler and uncertainty estimation in geophysical inversion, *Geophysical Prospecting*, *44*, 313–350, https://doi.org/10.1111/j.1365-2478.1996.tb00152.x.

Sen, M.K., and Stoffa, P.L., 2013, *Global Optimization Methods in Geophysical*

Inversion, Second Edn, Cambridge University Press, ISBN 978-1-107-01190-8.

Silverman, B.W., 1986, *Density Estimation for Statistics and Data Analysis*, Chapman and Hall, ISBN 978-0412246203.

Singh, V.P., Duquet, B., Léger, M., and Schoenauer, M., 2008, Automatic wave-equation migration velocity inversion using multiobjective evolutionary algorithm, *Geophysics*, *73*, VE61–VE73, https://doi.org/10.1190/1.2966008.

Sobol', I.M., 1967, On the distribution of points in a cube and the approximate evaluation of integrals, *USSR Computational Mathematics and Mathematical Physics*, *7*, 86–112, https://doi.org/10.1016/0041-5553(67)90144-9.

Stoffa, P.L., and Sen, M.K., 1991, Nonlinear multiparameter optimization using genetic algorithms: inversion of plane-wave seismograms, *Geophysics*, *56*, 1794–1810, https://doi.org/10.1190/1.1442992.

Tang, L., and Wang, X., 2013, A hybrid multiobjective evolutionary algorithm for multiobjective optimization problems, *IEEE Transactions on Evolutionary Computation*, *17*(1), 20–46, https://doi.org/10.1109/TEVC.2012.2185702.

Tarantola, A., 1987, *Inverse Problem Theory, Methods for Data Fitting and Model Parameter Estimation*, Elsevier, ISBN: 0-444-42765-1.

Tarantola, A., and Valette, B., 1982, Generalized nonlinear inverse problems solved using the least squares criterion, *Reviews of Geophysics*, *20*(1), 219–232, https://doi.org/10.1029/RG020i002p00219.

Tsvankin, I., 1997, Anisotropic parameters and P-wave velocity for orthorhombic media, *Geophysics*, *62*, 1292–1309, https://doi.org/10.1190/1.1444231.

Versteeg, R., 1994, The Marmousi experience: velocity model determination on a synthetic complex data set, *The Leading Edge*, *13*, 927–936, https://doi.org/10.1190/1.1437051.

Visser, G., Guo, P., and Saygin, E., 2019, Bayesian transdimensional seismic full-waveform inversion with a dipping layer parameterization, *Geophysics*, *84*, R845–R858, https://doi.org/10.1190/geo2018-0785.1.

Yilmaz, O., 1987, *Seismic Data Processing*, Society of Exploration Geophysicists, ISBN: 978-0931830402.

Yilmaz, O., 2001, *Seismic Data Analysis*, Society of Exploration Geophysicists, ISBN: 978-1560800941.

Zhao, L.-S., Sen, M.K., Stoffa, P., and Frohlich, C., 1996, Application of very fast simulated annealing to the determination of the crustal structure beneath Tibet, *Geophysical Journal International*, *125*, 355–370, https://doi.org/10.1111/j.1365-246X.1996.tb00004.x.

Zhao, Z., and Sen, M.K., 2022, Uncertainty quantification for full waveform inversion with a mini-batch gradient based sampling method, Second International meeting of Applied Geoscience & Energy (Expanded Abstracts), https://doi.org/10.1190/image2022-3751538.1.

Zhu, D., and Gibson, R., 2018, Seismic inversion and uncertainty quantification using transdimensional Markov chain Monte Carlo method, *Geophysics*, *83*, R321–R334, https://doi.org/10.1190/geo2016-0594.1.

Zitzler, E., Laumanns, M., and Thiele, L., 2001, SPEA2: Improving the strength Pareto evolutionary algorithm, Comput. Eng. Networks Lab., Swiss Fed. Inst. Technol., Zurich, Switzerland, Tech. Rep. 103, https://doi.org/10.3929/ethz‐a‐004284029.

Zunino, A., Mosegaard, K., Lange, K., Melnikova, Y., and Hansen, T.M., 2015, Monte Carlo reservoir analysis combining seismic reflection data and informed priors, *Geophysics*, *80*, 831–R41, https://doi.org/10.1190/GEO2014-0052.1.

9

Artificial Intelligence for Seismic Inverse Problems

9.1 Introduction

So far, we discussed optimization methods where we solve the physics of the forward problem to compute synthetic data and match with observation in an iterative way until we reach a point when we conclude that the match between the synthetic and observed data is satisfactory and examine results. Our iterative approach can be local or global and single- or multi-objective function optimization. We also discussed how to cast such optimization in a Bayesian framework by combining the prior model distribution $p(\mathbf{m})$ with the physics of the forward problem to approximate the posterior probability density function, $p(\mathbf{m} \mid \mathbf{d})$, which, in turn, allows estimating the uncertainties associated with each model parameter estimate. Irrespective of local or global, these optimizations of iteratively matching synthetics with observation and deriving a model and uncertainties are classical optimization methods.

With the current advances of computing technology, applications of artificial intelligence (AI) made huge impacts in science, engineering, sociology, economics, finance, and many other fields. In geophysics, AI has been used almost from the beginning when it was first introduced. However, AI has been demonstrated to be useful in solving complex seismic and other geophysical inverse problems only recently.

The scope of AI is so broad that it can be the subject of an entire book. And, in fact, there are such excellent books that are solely dedicated to different intelligent methods, including their geophysical applications (for example, Hajian and Styles, 2018). Here, we restrict to a subset of AI, called machine learning (ML). ML methods are dynamic systems, which adapt themselves via learning process and then make informed decisions or predictions. We first start our discussion using the subset of ML that are based on different variations of the artificial neural networks (ANNs), which have been found useful in solving seismic inverse problems. We then move on to other variations of ML that have also been used in geophysics.

9.2 Artificial Neural Network

ANN is one of the most popular ML methods, widely used in geophysics. ANN, in principle, attempts to mimic the function of the brain to solve a problem. For example, a newborn child has no understanding of the complexities of the world. But as the child grows to adulthood, she/he learns these complexities to make informed decisions. This learning process is achieved through *neurons*, the central component of the nervous tissue of the brains of human and other species of the animal kingdom. Because all ANNs tend to design an artificial neuron on the computer based

Computation, Optimization, and Machine Learning in Seismology, Advanced Textbook 8, First Edition. Subhashis Mallick.

on the functions of a biological neuron, it is first important to describe the anatomy of a biological neuron.

9.2.1 Anatomy of a Biological Neuron

Biological neurons are cells that can fire electric signals, called action potentials. They communicate with the other cells via *synapses*, which are specialized connections that generally use chemical neurotransmitters to pass electric signals from the presynaptic neuron to the target cells through the synaptic gaps. These biological neurons are classified into three basic types, based on their operations. The first type are the sensory neurons. These neurons respond to touch, smell, sound, light, and taste, which affect the cells belonging to the sensory organs. By sending signals to the brain, these neurons activate the respective sensory organs. The second type of neurons are the motor neurons, which receive signals from the brain and control muscle movements. Finally, the third type of neurons are the interneurons, which connect to other neurons within the same regions of the brain. To perform complex brain functions, multiple neurons are connected to form a neural circuit.

Figure 9.1 represents a simplified version of the neuron anatomy with the cell body (soma and the nucleus), dendrites, axon, and axon terminal. The cell body (soma and nucleus) is a compact structure, and the dendrites and axon extrude from the soma. Finally, the branched-out tips at the end of the axon are

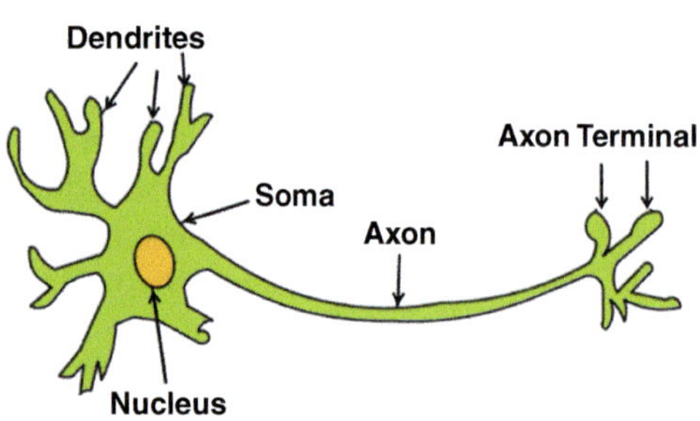

Figure 9.1 A simplified version of the neuron anatomy. Islam and Noor (2020)/Kazi Mejbaul Islam.

the axon terminals. Typically, the neurons receive electrical signals from the dendrites, process them, and send out (transmit) signals to the other relevant cells from the axon terminals across the synapses when the processed electrical signal exceeds a threshold value.

9.2.2 An Equivalent Artificial Neuron

An artificial neuron simulates the process of a biological neurons in the computer as illustrated in Figure 9.2. As can be seen from Figure 9.2, an artificial neuron (soma) receives signals, shown as an N-dimensional vector $[x_1, x_2, x_3, \ldots, x_{N-1}, x_N]^T$ from the input nodes through the dendrites. Soma processes the received signals as a weighted sum,

$$h = \sum_{j=1}^{N} x_j w_j, \tag{9.1}$$

where w_j is the weight assigned to each input x_j, $j = 1, 2, \ldots, N$. The processed signal h, typically referred as the *net*, is then transferred to the axon terminal via the axon. The axon terminal then adds the activation a and bias b, and outputs result as $y = f(h, a, b)$. This activation and bias are equivalent to mimicking the process of the biological neurons firing the electric signals via the axon terminals to the relevant cells when the received signals exceed certain threshold values. These processes are explained as below.

9.2.3 The Activation and Bias

The activation and bias shown in Figure 9.2 are used to mimic the process of different biological neurons to perform different tasks, and consequently, their choices in an artificial neuron must vary with the specific task it is assigned to. In seismic, for example, assume that we want to design a system to predict (interpret) the salt boundaries from the stacked three-dimensional (3D) seismic data. This prediction, in turn, is equivalent to classifying the entire 3D data volume into a binary system—salt and no-salt. Such binary classification is also necessary for many other

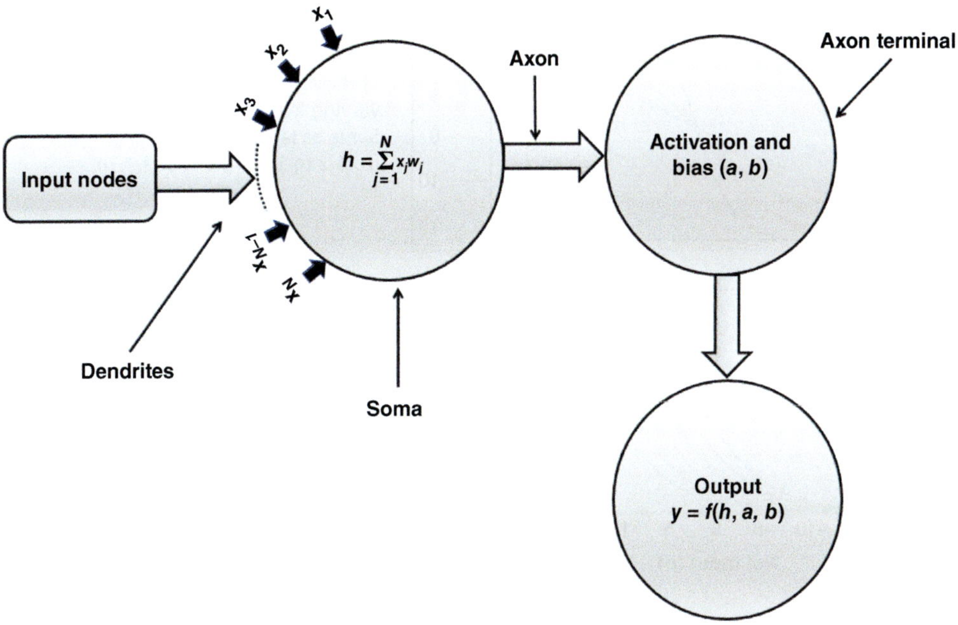

Figure 9.2 Function of an artificial neuron based on that of a biological neuron. Equivalence between the dendrites, soma, axon, and axon terminal is marked.

cases, such as interpreting faults from seismic data, picking of first breaks for refraction seismic analysis, classifying seismic signals from the earthquakes as hazardous or safe conditions, etc. For such binary classifications, a step function (hard limit) like

$$a = \begin{cases} 0; x < 0 \\ 1; x \geq 0 \end{cases} \tag{9.2}$$

can be an appropriate choice for the activation function, and then add a bias b to the step function to move it along the vertical, horizontal, or both axes. A step function a like equation 9.2 in conjunction with the appropriate choice of the bias b can, in theory, classify the entire data into a binary system (salt or no-salt; fault or no-fault; first-break-arrival or not; hazardous or safe; etc.). The process of course is not as simple and would require multiple neurons to be interconnected as a network for such classification.

For seismic inverse problems, the goal is to predict a subsurface model (V_P, V_S, ρ, etc.) from the input seismic data. For such problems, a step function like equation 9.2 in association with a bias is not appropriate and a smooth function is desirable. One popular smooth activation and bias function, typically used in seismic inverse problems, is a *log-sigmoid* function that directly computes the output y as

$$y = \frac{1}{1 + e^{-\frac{h+b}{d}}}, \tag{9.3}$$

where b is the bias, d is the sigmoid shaping factor, and h is the net output processed at the neuron (equation 9.1). Later, we also introduce two other activation functions (1) *tanh* or *sigmoid* function and (2) *rectified linear unit* or *ReLU*; see Hajian and Styles (2018), p. 13, for a complete list of the popular activation functions used in ANNs.

Figure 9.3 shows how the shape of the log-sigmoid function is influenced by the choices of the bias b and the sigmoid shaping factor d. Figure 9.3a shows log-sigmoid function for a fixed value of $b = 2$ and different values of d. Figure 9.3b, on the other hand, shows the same for a fixed value of $d = 4$ for

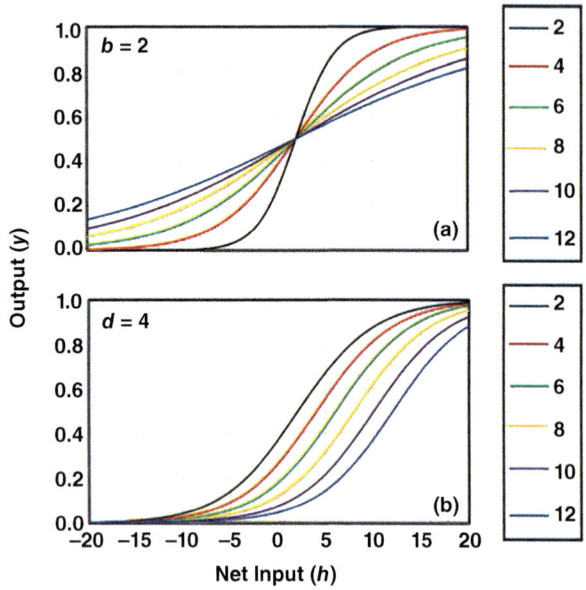

Figure 9.3 Influence of the log-sigmoid activation function on the choices of the shaping factor d and the bias b. (a) Shapes of the activation function for $b = 2$ and d varying from 2 to 12, in steps of 2. (b) Same as (a) but for $d = 4$ and b varying from 2 to 12, in steps of 2.

different values of b. From this figure, we may conclude the following:

1) The shaping factor d controls how gently or steeply varying is the activation function. The higher is the value of d, the more gently varying is the activation function (Figure 9.3a).
2) The bias b controls offset or bias of the activation function toward the positive horizontal (h or the net input) axis. The higher is the value of b, more biased is the activation toward the positive values of h (Figure 9.3b).

As can be seen from Figure 9.3, the choices of b and d must be problem dependent, which we revisit later.

A biological neuron works in two distinct phases: (1) learning or training phase and (2) prediction phase. Out of these two phases, the learning phase is of particular interest where the neuron learns the action from a specific set of inputs. After learning completes, in the prediction phase the neuron attempts to predict the action from other input sets that were not used in training.

Before going onto the neural networks, it is first necessary to discuss the learning phase for a single artificial neuron. For simplicity, we ignore the activation and bias and consider how a single neuron receiving a set of inputs and processing them like equation 9.1 learns a particular action. This is an optimization process in which a set of P input patterns, $[x_{p1}, x_{p2}, \dots, x_{pN}]^T$, $p = 1, 2, \dots, P$, with their corresponding outputs y_p, $p = 1, 2, \dots, P$, are provided to the neuron. Starting from an initial set of weights w_i, $i = 1, 2, \dots, N$, the objective of the training or learning phase is to iteratively adjust all weights such that all training patterns are accurately predicted from the adjusted weights. Thus, once the optimum weights are estimated, we assume that the neuron has learnt how to predict from a similar sets of other input and predict the corresponding actions.

As illustrated in Figure 9.4, assume a single training pattern, which can be given as

$$\mathbf{X}_p = [x_{p1}, x_{p2}, x_{p3}, \dots, x_{pN}]^T, \tag{9.4}$$

producing the output pattern y_p in the neuron as

$$y_p = \mathbf{X}_p \cdot \mathbf{W} = x_{p1}w_1 + x_{p2}w_2 + \dots + x_{pN}w_N. \tag{9.5}$$

Clearly, $\mathbf{W} = [w_1, w_2, \dots, w_N]^T$ in equation 9.5 is the N-component weighting vector that we

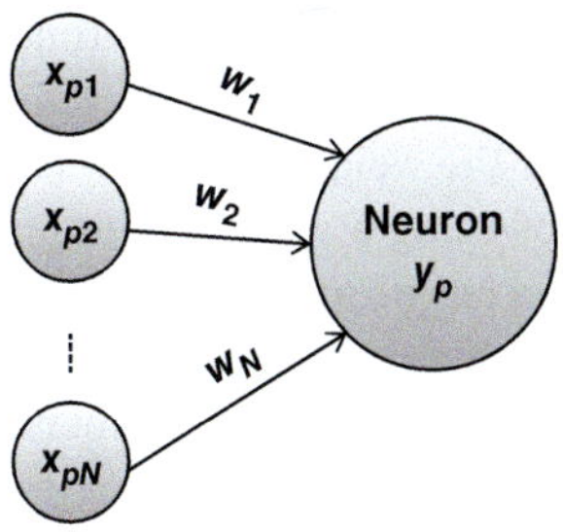

Figure 9.4 Learning phase for a single artificial neuron.

want to estimate during learning. While y_p is the output from the neuron, we also provide d_p, the true (desired) output expected from the input pattern $[x_{p1}, x_{p2}, \ldots, x_{pN}]^T$, which is provided by the teacher. Therefore, if there are P such input patterns, the summed squared error from all of them is given as

$$E = \sum_{p=1}^{P} E_p = \frac{1}{2}\sum_{p=1}^{P}(d_p - y_p)^2. \tag{9.6}$$

Note that E in equation 9.6 is the same as the objective (without any regularization term), which we previously discussed. One simple way to estimate the weighting vector $\mathbf{W}$ is therefore the method of least squares using the steepest descent (gradient descent) method. In this method, we calculate the gradient of the objective ∇E and set it to zero. In other words, we compute the derivatives $\partial E/\partial w_i$, $i = 1, 2, \ldots, N$, and set them to zero. This will lead to N independent equations using which w_i, $i = 1, 2, \ldots, N$, can be solved. We then recompute the output y_p from each input pattern $\mathbf{X}_p$ (equation 9.4) and the newly adjusted weights (equation 9.5), recompute E (equation 9.6), and recompute the weights again by setting $\nabla E = \mathbf{0}$. By iteratively repeating this process until the point when the error or the objective E does not reduce anymore, we may say that the learning is complete via the least-squares function optimization.

Although the process outlined above is simple and straightforward, note that the method is pure least-squares minimization. And, in Chapter 7, we discussed that such pure least-squares method is not stable, and we must use additional regularization (damping) terms for stability. Therefore, as discussed in Chapter 7, for stability we must change the weight in proportion to the negative of the derivative of the error as measured on the current pattern with respect to each weight. Or, in other words, we must have

$$\Delta_p w_j = -\gamma \frac{\partial E_p}{\partial w_j}. \tag{9.7}$$

In equation 9.7, E_p is the error for the training pattern p, $\Delta_p w_j$ is the change in the weight w_j, $j = 1, 2, \ldots, N$, and γ is the constant of proportionality, which is given as a positive number. But

$$\frac{\partial E_p}{\partial w_j} = \frac{\partial E_p}{\partial y_p}\frac{\partial y_p}{\partial w_j}. \tag{9.8}$$

Also from equation 9.6, we have

$$E_p = \frac{1}{2}(d_p - y_p)^2. \tag{9.9}$$

And, from equation 9.5, we can write

$$\frac{\partial y_p}{\partial w_j} = x_{pj}. \tag{9.10}$$

From equations 9.8 and 9.9, we write equation 9.7 as

$$\Delta_p w_j = -\gamma \frac{\partial E_p}{\partial y_p} x_{pj}. \tag{9.11}$$

But from equation 9.9

$$\frac{\partial E_P}{\partial y_p} = -(d_p - y_p) = \delta_p. \tag{9.12}$$

Therefore,

$$\Delta_p w_j = -\gamma \delta_p x_{pj}, \tag{9.13}$$

in which δ_p is the difference between the target output and the network output for pattern p and x_{pj} is the jth element of the input pattern p.

Equation 9.13 establishes the general *delta rule* or Widrow–Hoff rule for the change in connection weights as the basis for all learning algorithms. For sets of N input nodes and M neurons (Figure 9.5), the delta rule for the change Δw_{ij} in connection weight w_{ij} between input i and neuron j is therefore

$$\Delta w_{ij} = -\gamma x_i |d_j - y_j|. \tag{9.14}$$

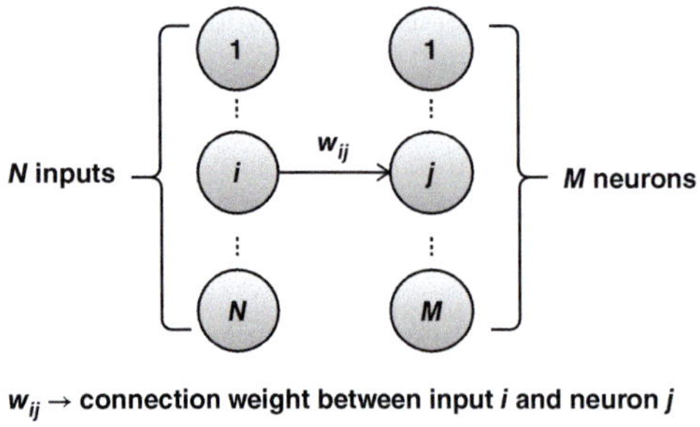

Figure 9.5 The delta rule for the connection weight.

In equation 9.14, x_i is the ith input component, y_j is the computed output at neuron j, and d_j is the desired output at neuron j provided by the teacher. Note that we are interested in the absolute value $|d_j - y_j|$ in the computation of Δw_{ij}. Finally, the positive constant of proportionality γ in equation 9.14 is called the *learning rate*.

In standard gradient descent method, learning rate γ must be small and constant. There is nothing wrong in doing so, but the problem is that this will require many small steps to reach the desired point where the error is minimum, which, in turn, could be computationally demanding. To save computational expense, we would, therefore, like to reach this minimum point with a few steps if possible. In other words, we would like to use a high learning rate γ and reach the minimum point in a compute-efficient fashion. The problem in doing so is that the method becomes extremely oscillatory, which may not even converge and reach the desired point. A practical way to use a large γ and reach the desired point in a few steps is to make the weight change for step $t + 1$ to be dependent upon the weight change at the previous step t by adding a small *momentum* term:

$$\Delta w_{ij}(t+1) = -\gamma x_i |d_j - y_j| + \alpha \Delta w_{ij}(t). \quad (9.15)$$

The parameter α in equation 9.15 is a constant (momentum) that makes the weight change at step $t + 1$ dependent upon the weight change at the previous step (step t). By introducing the time discrepancy, equation 9.15 introduces a differential equation for the weights. Figure 9.6 compares the effect of using a small learning rate with those using a large learning rate with and without using the momentum term. The choices of the learning rate γ and the momentum term α are problem dependent, which we discuss later.

9.2.4 From a Single Neuron to Multiple Neurons: A Simple Neural Network

Like biological neurons, where multiple neurons connect to form a neural circuit, the artificial neurons are also connected to form a neural network (or, specifically an ANN) to solve specific tasks. Figure 9.7 shows one simple example of a neural network, known as the *feed forward back propagation* (*FFBP*) neural network. FFBP is the fundamental building block for all ANNs. To develop the concepts of the ANN architecture, we therefore discuss FFBP first.

FFBP neural network is composed of the three fundamental units or layers, shown inside the red box of Figure 9.7:

Unit 1: The *input layer*. This unit contains N input nodes with each node containing one component of the input data vector $\mathbf{X}$ previously defined.

Unit 2: The *hidden layer*. This unit contains M hidden neurons. From any node i in the input layer, any neuron j in the hidden layer receives the input data component x_i along with the weight w_{ij}, activation a_1, and bias b_1 (see Figure 9.7).

Unit 3: The *output layer*. This unit contains K output nodes in which each node contains a single component of the output vector $\mathbf{Y}$ as the network output. From each hidden layer neuron j, any output node k receives the computed data from the neuron along with the weight w_{jk}, activation a_2, and bias b_2 for computing the network output for the kth output node (see Figure 9.7).

For forward propagation, FFBP receives a total of P input patterns $\mathbf{X}_p = [x_{p1}, x_{p2}, \ldots,$

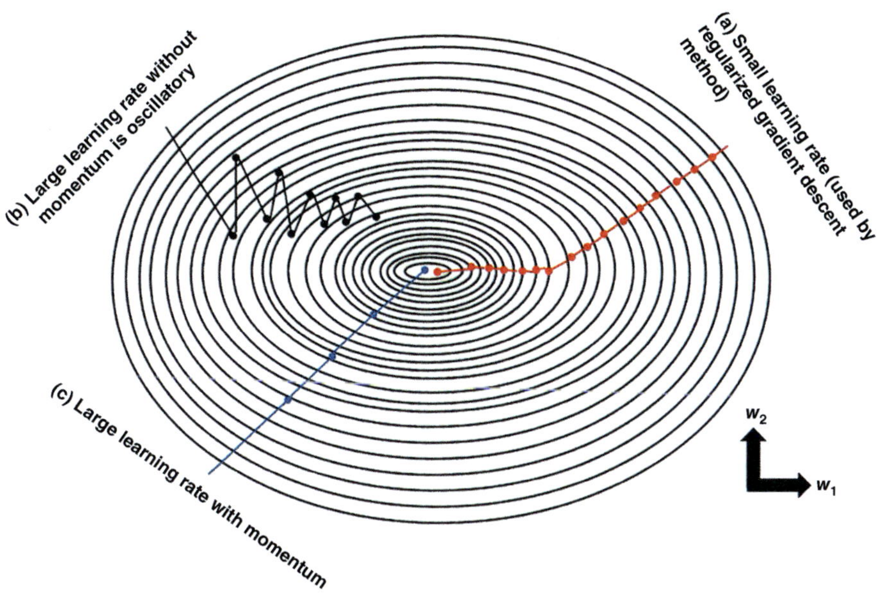

Figure 9.6 Illustration of the estimation of the optimum weights with and without the momentum. The ovals represent the contours of the objective (error) E (equation 9.6) in which only two weights (w_1 and w_2) are to be estimated. The contour values decrease inward with the expected minimum at the center, which we want to find because the values of w_1 and w_2 at the center are the optimum weights that minimize E. Starting from initial values of w_1 and w_2, gradient descent method can, in theory, reach this minimum point to estimate the weights. This, however, requires the learning rate γ to be set to a small constant value. Therefore, the method may take a large number of steps (iterations) to reach the minimum point, which can be computationally demanding. This is shown as scenario (a) with successive minimizations as red lines. On the other hand, if a large learning rate without the momentum term is used for minimization, shown as scenario (b), the method tends to oscillate (see the black lines), which may not even converge to the minimum point. Finally, using a large learning rate and the momentum term, shown as scenario (c), the method can converge very quickly (see the blue lines).

$x_{pN}]^T$, $p = 1, 2, \ldots, P$, as training data. These input data from each node are then fed to each of the M hidden neurons along with the weights w_{ij}, $i = 1, 2, \ldots, N$; $j = 1.2, \ldots, M$, activation a_1, and bias b_1 to compute the signal h_{pj} for the neuron j for each training pattern p. Therefore, the computed signal at any hidden neuron j and for any training pattern p is given as

$$h_{pj} = a_1 \sum_{i=1}^{N} w_{ij} x_{pi} + b_1. \tag{9.16}$$

We must note that the way the activation and bias affect the computation of h_{pj} depends upon their choices. For example, if we choose log-sigmoid with bias b_1 and sigmoid shaping factor d_1, equation 9.16 can then be written as

$$h_{pj} = \frac{1}{1 + e^{-\frac{\sum_{i=1}^{N} w_{ij} x_{pi} + b_1}{d_1}}}. \tag{9.17}$$

Once each h_{pj} are computed, the output y_{pk} at the kth output node for training pattern p becomes

$$y_{pk} = a_2 \sum_{j=1}^{M} w_{jk} h_{pj} + b_2. \tag{9.18}$$

Once again, if we use a log-sigmoid with bias b_2 and shaping factor d_2, we write this output as

$$y_{pk} = \frac{1}{1 + e^{-\frac{\sum_{j=1}^{M} w_{jk} h_{pj} + b_2}{d_2}}}. \tag{9.19}$$

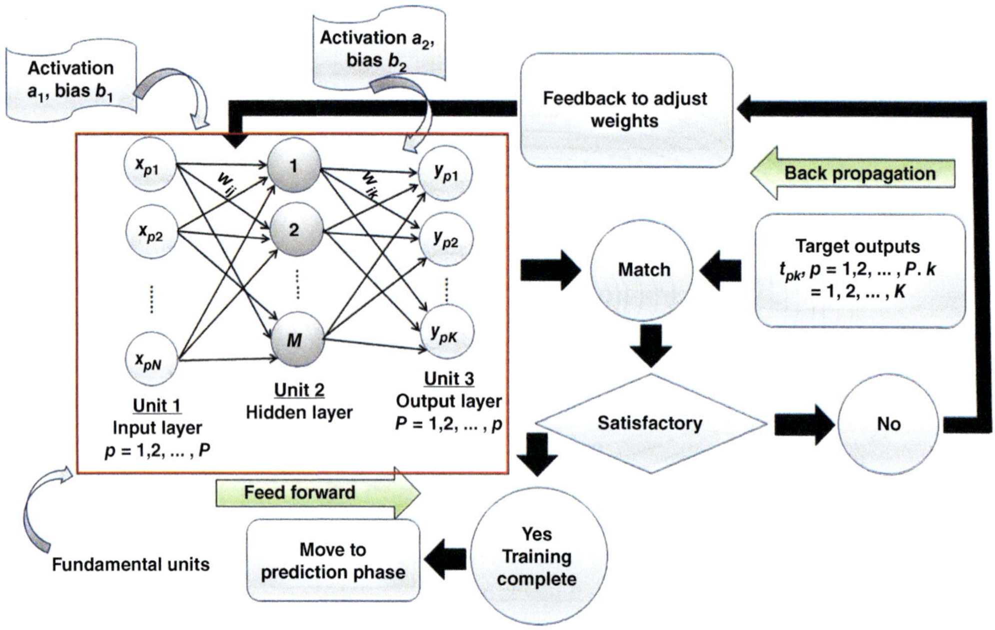

Figure 9.7 Feed forward back propagation neural network.

The forward propagation in FFBP thus consists of computing P output patterns from their corresponding input patterns using the steps outlined above.

Now, during the training or learning phase, the teacher provides the target outputs t_{pk} that corresponds to each of the computed network output y_{pk} (see Figure 9.7). The objective of FFBP is to match y_{pk} with t_{pk} as close as possible. So, to back propagate, we compute the mean squared network error,

$$E = \frac{1}{2P} \sum_{p=1}^{P} \sum_{k=1}^{K} (t_{pk} - y_{pk})^2, \tag{9.20}$$

and use an optimization method to minimize E and compute new network weights w_{ij} and w_{jk}. Following back propagation, we again forward propagate to recompute h_{pj}, y_{pk}, and E. We continue doing this forward and back propagation until convergence or when E does not reduce anymore. At this point, we assume that the learning is complete and move onto the next (prediction) phase (see Figure 9.7).

Optimization Methods for Estimating Weights

Each step of the forward and back propagation discussed above is called a *learning epoch*. The process of back propagation and estimating weights can be performed using any of the local methods, such as the gradient descent (with or without the momentum term), conjugate gradient, and Gauss–Newton, which we previously discussed in Chapter 7. Additionally, any global methods, such as MCMC, SA, GA, etc., which we discussed in Chapter 8, can also be used for back propagation. Hajian and Styles (2018) provide an extensive list of the local learning algorithms that have been successfully used in the past. Out of this extensive list, two learning algorithms, (1) Levenberg–Marquardt (L-M) algorithm and (2) adaptive moment estimation (ADAM) algorithm, have been popular in many geophysical inverse problems. Application of global optimization methods for ANNs are rare in geophysical inverse problems to this date. Few examples include the use of SA using the Boltzmann machines (Phan and Sen, 2019) and GA in seismic attribute selection for well-log prediction (Dorrington and Link, 2004). Here, we discuss L-M and ADAM only. To see how GA or SA can be used instead, see Phan and Sen (2019) and Dorrington and Link (2004).

L-M Learning Algorithm L-M algorithm (Levenberg, 1944; Marquardt, 1963) minimizes E while trying to keep the step between the old and updated weights small at each iteration using the following formula (see Günaydun and Günaydun, 2008):

$$\mathbf{w}_{new} = \mathbf{w}_{old} - [\mathbf{J}^T\mathbf{J} + \gamma\mathbf{I}]^{-1}E(\mathbf{w}_{old}). \tag{9.21}$$

In equation 9.21, $\mathbf{w}_{old}$ and $\mathbf{w}_{new}$ are the old (that is from the previous learning epoch) and the updated weights for the current epoch, $\mathbf{J} = \partial E/\partial \mathbf{w}$ is the Jacobi matrix of E as defined in Chapter 7, and $\mathbf{I}$ is the identity matrix. The parameter γ in equation 9.21 is the step length. It must be noted here that as $\gamma \to 0$, L-M converges to a Gauss–Newton method. On the other hand, as γ become very large, the method tends to become a gradient descent method. This parameter is used to define the adaptive learning rate and is set to a value between 0 and 1, and an additional parameter $0 < \beta < 1$, called the decay rate, is used to dynamically vary γ in each iteration (learning epoch). Thus, for each learning epoch, if E decreases, then the learning rate is increased using the decay rate. And, if E increases, then the learning rate is decreased using the decay factor. From equation 9.21, we can write the change in the weight $\Delta\mathbf{w} = \mathbf{w}_{new} - \mathbf{w}_{old}$ as

$$\Delta\mathbf{w} = -[\mathbf{J}^T\mathbf{J} + \gamma\mathbf{I}]^{-1}E(\mathbf{w}_{old}). \tag{9.22}$$

To dynamically change the learning rate, γ is multiplied by β if $E(\mathbf{w})$ decreases and is divided by β if $E(\mathbf{w})$ increases.

Figure 9.8 shows a simplified workflow for the L-M algorithm. As illustrated in this figure,

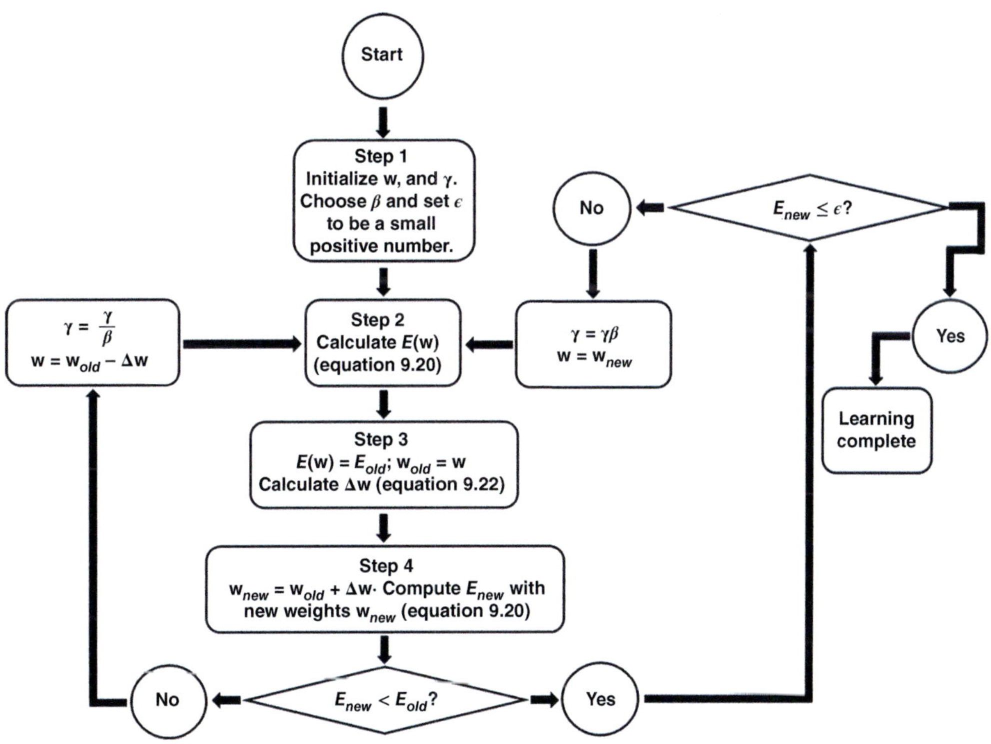

Figure 9.8 A simplified workflow of the Levenberg–Marquardt (L-M) learning algorithm.

the steps for the algorithm are the following:

1) Initialize the weights $\mathbf{w}$, the learning rate γ, and the decay rate β. Typically choosing a random value between 0 and 1 can be used to initialize all of them. We also set the exit criterion by setting another parameter ε to a small number.
2) Calculate the objective $E(\mathbf{w})$ (equation 9.20).
3) Set $E_{old} = E(\mathbf{w})$ and $\mathbf{w}_{old} = \mathbf{w}$ and use equation 9.22 to compute $\Delta\mathbf{w}$.
4) Set the new weights as $\mathbf{w}_{new} = \mathbf{w}_{old} + \Delta\mathbf{w}$ and then use equation 9.20 and compute the new objective E_{new}. After computing E_{new}, compare it with the old objective E_{old} from the previous step.
   a) If $E_{new} > E_{old}$, then set the new learning rate as γ/β, weights as $\mathbf{w}_{old} - \Delta\mathbf{w}$, and go back to step 2 for the next learning epoch.
   b) If $E_{new} < E_{old}$, then first check if desired minimum, given as $E_{new} \leq \varepsilon$, is reached.
      i) If not, then set the new learning rate as $\gamma\beta$, set the updated weights to $\mathbf{w}_{new}$, and go to step 2 for the next learning epoch.
      ii) If yes, then the learning is complete and move to the prediction phase using the estimated weights.

ADAM Learning Algorithm The original version of the *ADAM* was introduced by Kingma and Ba (2014) as an extension of the stochastic gradient descent method. Later, ADAM has been further improved for computational efficiency by Zhang (2018) and Pomerat et al. (2019).

ADAM assumes $E(\mathbf{w})$ to be a stochastic objective function with realizations $E_1(\mathbf{w})$, $E_2(\mathbf{w}), \ldots, E_T(\mathbf{w})$ for each iteration (time step or learning epoch) $t = 1, 2, \ldots, T$. And the goal of the algorithm is to minimize E in these successive steps until a point when E does not reduce any further. To update the weights $\mathbf{w}$, the method uses both first and second moments $\mathbf{m}_t = f(\mathbf{g}_t) = f[\nabla E(\mathbf{w})]$ and $\mathbf{v}_t = f(\mathbf{g}_t \cdot \mathbf{g}_t) = f(\mathbf{g}_t^2)$ in each time step t. As pointed out by Kingma and Ba (2014), for sparse data points, using both $\mathbf{m}_t$ and $\mathbf{v}_t$ stabilizes the optimization. Here, we do not go to these details and only describe the implementation aspects of the ADAM; refer to Kingma and Ba (2014), Zhang (2018), and Pomerat et al. (2019) for further details on ADAM.

Figure 9.9 shows the five-step workflow of the ADAM optimization, modified from Kingma and Ba (2014). In step 1, we start from an initial set of the weights $\mathbf{w}_0$ and set it as the current weight vector $\mathbf{w}$. We also initialize the first and second moments $\mathbf{m}_0$ and $\mathbf{v}_0$ to zero for time step $t = 0$. In addition, we choose the step size α, exponential decay rates β_1 and β_2 for the first and the second moments, and a small number ε as the damping parameter for the least squares. Kingma and Ba (2014) suggested that the well-tested values for these parameters are $\alpha = 0.001$, $\beta_1 = 0.9$, $\beta_2 = 0.999$, and $\varepsilon = 10^{-8}$. In step 2, we use equation 9.20 to compute the objective $E(\mathbf{w})$ for the current weight vector $\mathbf{w}$ and check for convergence. The convergence check can be setup in many ways. For example, we can check if $E(\mathbf{w})$ is less than or equal to a predefined small number and assume that the method converged to the desired accuracy. Alternatively, we can compare $E(\mathbf{w})$ for the current time step t with that of the previous step $t-1$ and decide if the method is converging or diverging and stop iteration. We can also setup a set of criteria like the GA workflow shown in Figure 8.19 and conclude that the method converged if any one of this set of criteria is met. Irrespective of the way we set up this convergence check, if the convergence is achieved, we assume that the learning is complete and output the weight vector to go to the prediction phase. However, if the method has not yet converged to the desired accuracy, we go to steps 3, 4, and 5 to compute new weights and go back to step 2 (see Figure 9.9). In step 3, we first increment the time step by one ($t = t + 1$) and calculate the gradient of the objective for the current time step as $\mathbf{g}_t = \nabla E(\mathbf{w})$. We then compute the biased first and second moments for the

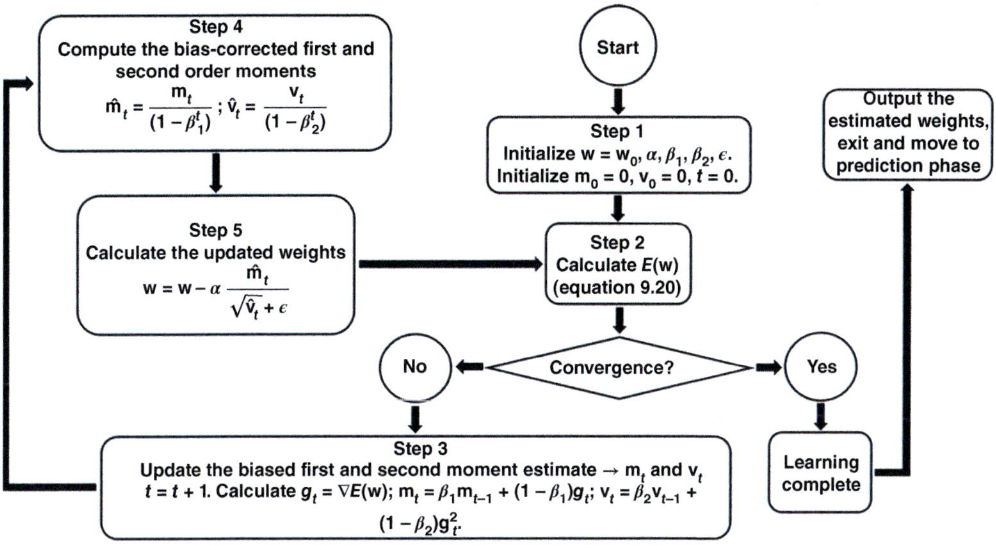

Figure 9.9 The basic ADAM workflow. Adapted from Kingma and Ba (2014).

current time step as

$$\left.\begin{aligned}\mathbf{m}_t &= \beta_1 \mathbf{m}_{t-1} + (1-\beta_1)\mathbf{g}_t \\ \mathbf{v}_t &= \beta_2 \mathbf{v}_{t-1} + (1-\beta_2)\mathbf{g}_t^2\end{aligned}\right\}. \tag{9.23}$$

In equation 9.23, $\mathbf{g}_t^2$ denotes the dot product $\mathbf{g}_t \cdot \mathbf{g}_t$. Following step 3, step 4 computes the bias-corrected first and second moments as

$$\left.\begin{aligned}\hat{\mathbf{m}}_t &= \frac{\mathbf{m}_t}{(1-\beta_1^t)} \\ \hat{\mathbf{v}}_t &= \frac{\mathbf{v}_t}{(1-\beta_2^t)}\end{aligned}\right\}. \tag{9.24}$$

Finally, the updated weights are computed in step 5 as

$$\mathbf{w} = \mathbf{w} - \alpha \frac{\hat{\mathbf{m}}_t}{\sqrt{\hat{\mathbf{v}}_t} + \epsilon}. \tag{9.25}$$

Note that $\mathbf{w}$ on the left-hand side of equation 9.25 is the new (updated) weight vector and that on the right-hand side is the weight vector from the previous step.

The Universal Approximation Theorem Referring to the FFBP neural network, shown in Figure 9.7, we combine equations 9.16 and 9.18 and write the output y_{pk} for the training pattern p at the kth output node as

$$y_{pk} = a_2 \sum_{j=1}^{M} w_{jk} \left(a_1 \sum_{i=1}^{N} w_{ij} x_{pi} + b_1 \right) + b_2. \tag{9.26}$$

For log-sigmoid activation, we can similarly combine equations 9.17 and 9.19 and write

$$y_{pk} = \frac{1}{1 + e^{-\frac{\left(\sum_{j=1}^{M} w_{jk} \left(\frac{1}{1+e^{-\frac{\sum_{i=1}^{N} w_{ij}x_{pi}+b_1}{d_1}}}\right)+b_2\right)}{d_2}}}. \tag{9.27}$$

For each y_{pk}, $p = 1, 2, \ldots, P$; $k = 1, 2, \ldots, K$, we also provide the target (true) output t_{pk}, and the objective of the network is to find the set of optimum weights w_{ij} and w_{jk} (see Figure 9.7) such that each network output y_{pk} is close to its corresponding target output t_{pk}. Restricting to seismic inverse problems, each x_{pi}; $p = 1, 2, \ldots, P$; $i = 1, 2, \ldots, N$, is the set of P input seismic data, each with N samples, and t_{pk} is the kth model sample (P-wave velocity, V_P; S-wave velocity, V_S; density, ρ; etc.), for each input seismic data (*labeled data*), being provided for training. In Chapters 7 and 8, we started from an initial model (or models) of

$\mathbf{V}_P$, $\mathbf{V}_S$, $\boldsymbol{\rho}$, solved the physics of the forward problem to compute synthetic seismic data, matched with the observed seismic data, and iteratively updated model until a satisfactory match between the observed and computed synthetic data is obtained. In such classical approaches as we discussed in Chapters 6–8, we used different local and global optimization methods to iteratively solve the physics of the forward problem. Fundamental assumption in these methods is that the input seismic data can be adequately described as

$$\mathbf{d} = g(\mathbf{m}), \tag{9.28}$$

in which g is the physics of the forward problem (wave equation, convolutional, etc.) and $\mathbf{m}$ the model vector, which for isotropic elastic problems is $[\mathbf{V}_P, \mathbf{V}_S, \boldsymbol{\rho}]^T$.

In contrast with these classical methods, here we provide a set of labeled data (input seismic data and their expected models). By forward propagation using equations 9.26, 9.27, or something equivalent, we then compute the network output (y_{pk}). Then by computing the network error or objective E, given as a sum of the squares of the difference between y_{pk} and their corresponding target or true output t_{pk} (equation 9.20), we attempt to minimize E using an optimization method. For the method to work, the forward propagation, given by equations 9.26 or 9.27, must therefore approximate the model $\mathbf{m} = [\mathbf{V}_P, \mathbf{V}_S, \boldsymbol{\rho}]^T$ as a function of the input seismic data $\mathbf{d}$. Or, in other words, these equations must approximate

$$\mathbf{m} = f(\mathbf{d}). \tag{9.29}$$

The fact that the functional forms given by equations 9.26 or 9.27 can adequately approximate any function that comes from a famous theorem known as the *universal approximation theorem*. Here, we just describe this theorem without proof. Further details including proof of the universal approximation theorem can be found in Funahashi (1989), Cybenko (1989), Hornik et al. (1989), Hornik (1991), and Pinkus (1999).

Universal Approximation Theorem: Let $\varphi(\cdot)$ be a non-constant, bounded, and continuous function. Also let I_m denote an m-dimensional unit hypercube $[0, 1]^m$ with the space of continuous function on I_m being denoted by $C(I_m)$. Then, given any $\varepsilon > 0$ and function $f(\mathbf{x}) = f(x_1, x_2, \ldots, x_m) \in C(I_m)$, there exists an integer N, real constants $v_i, b_i \in \mathfrak{R}$, and real vectors $\mathbf{w}_i \in \mathfrak{R}^m$, where $i = 1, 2, \ldots, N$, such that we may define

$$F(\mathbf{x}) = \sum_{i=1}^{N} v_i \varphi(\mathbf{w}_i^T \cdot \mathbf{x} + b_i) \tag{9.30}$$

as an approximate realization of the function f where f is independent of φ, that is,

$$|F(\mathbf{x}) - f(\mathbf{x})| < \varepsilon \tag{9.31}$$

for all $\mathbf{x} \in I_m$. In other words, functions of the form $F(\mathbf{x})$ are dense in $C(I_m)$. This theorem still holds when I_m is replaced by any compact subset of $\mathfrak{R}^m$.

Comparing equation 9.30 with the network output (equation 9.26), we can immediately see that they are equivalent. Proof of the universal approximation using log-sigmoid activation (equation 9.27) can be found in Cybenko (1989).

As a simple illustration of the universal approximation theorem, Figure 9.10 provides approximating the function $y = x^2$. Figure 9.10a shows the network architecture with a single hidden layer with four neurons that maps input x from the input layer onto the output y in the output layer, and Figure 9.10b shows the result with the true function ($y = x^2$), shown in black, and its universal approximation, shown in red. The *ReLU* (discussed below) was used for the activation function of the network. Note that the function approximation is almost exact and indistinguishable from the true function. By iteratively updating the weights for a given activation and bias, the neural networks thus attempt to use the universal approximation theorem to approximate any function for the optimum set of weights. Approximating $y = x^2$ is simple and therefore can be achieved using a single hidden layer as shown in Figure 9.10. To approximate complex functions, however, a cascade of hidden layers are usually needed, which we discuss next.

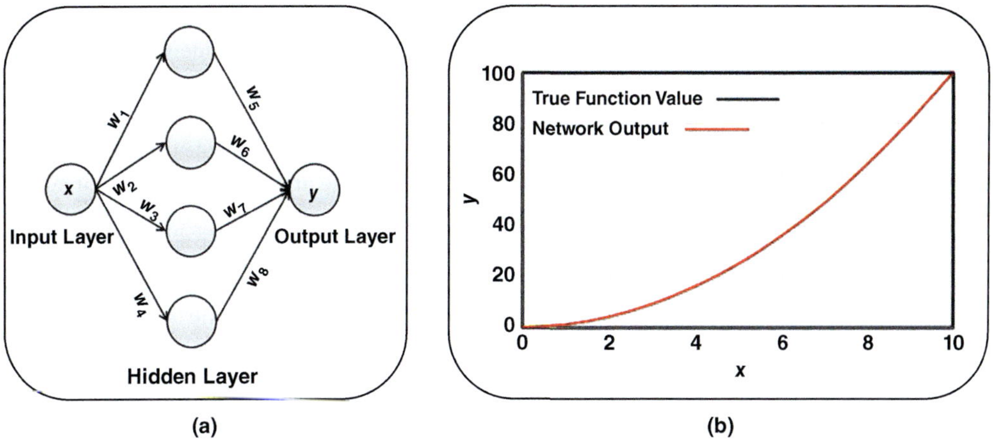

Figure 9.10 Illustration of the universal approximation theorem in approximating the function $y = x^2$. (a) The network architecture. (b) Approximation result.

9.3 Deep Neural Networks

The FFBP neural network architecture (Figure 9.7) with a single hidden layer can be extended to an architecture with a cascade of hidden layers, which forms the fundamental framework for *deep neural network (DNN)* designs. As shown in Figure 9.11, the input layer with N nodes connects to a cascade of L hidden layers with each hidden layer with M neurons. The interconnection between the hidden layers can vary, but they ultimately produce the output in the K output nodes in the output layer.

DNN methods can be broadly divided into two classes: (1) supervised and (2) unsupervised. Supervised learning methods are the ones we have been discussing so far. They require a set of learning patterns and their corresponding target outputs to be provided by the teacher. In addition to supervised learning, there are DNN and other AI methods that use unsupervised learning. These unsupervised methods also require training data, but they do not require any target output for these training data to be provided by the teacher, and learning is performed in an unsupervised fashion. In addition to these two broad classes, a new

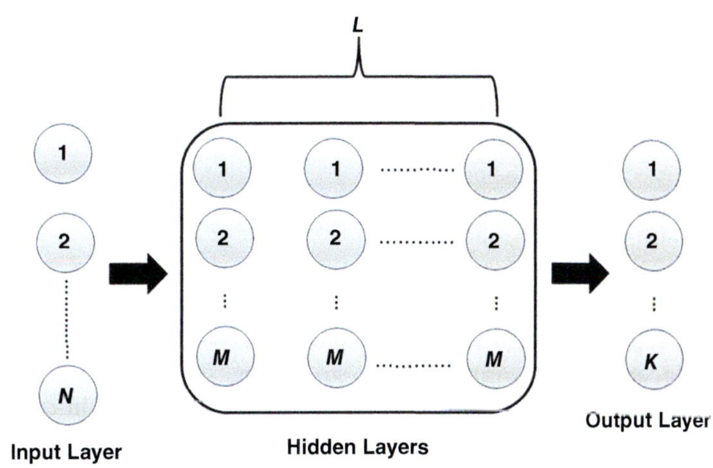

Figure 9.11 Basic architecture of a deep neural network.

class, known as semi-supervised learning, has recently emerged in DNN, where a part of the training is performed in a supervised fashion and a part is performed in an unsupervised fashion. Whether supervised, unsupervised, or semi-supervised, addressing the following five questions is fundamental to DNN:

1) How to optimize to obtain the weight vector **w**?
2) How to choose the number of hidden layers (the parameter L in Figure 9.11)?
3) How to choose the number of hidden neurons per layer (the parameter M in Figure 9.11)?
4) What is the criterion or criteria to decide when to stop training?
5) How to choose the optimum number of training data (the parameter P)?

We already addressed question 1 in our discussion of the FFBP network and extending the process to multiple hidden layers is straightforward. We address below questions 2 through 5. The simplest form of DNN is the multilayer perception (MLP) network where each neuron in each hidden layer is connected to all neurons of the next hidden layer. We first address questions 2–5 with reference to MLP network and then discuss different flavors of DNNs.

9.3.1 Number of Hidden Layers

Determining the number of hidden layers is a critical step in any DNN design, and it entirely depends upon the complexity of the problem. For example, if the relationship between the input and output is linear, then one hidden layer should be sufficient to solve the problem. However, if the relation is complex, it is necessary to use multiple layers. The nonlinearity of the problem is handled not only by using many hidden layers but also by the choice of the activation function, which also attempts to understand the inherent nonlinearity of the problem via learning.

Restricting to seismic inverse problems, assume that we restrict to small incidence angles (<25–30°) so that we can convert time-offset domain seismic data onto time and average angle of incidence (Figure 1.8), and assume that the reflection amplitudes are proportional to the two-term reflection coefficient formula $R_{PP} \approx A + B \sin^2\theta$, where θ is the average angle of incidence. For such a case, if we design the network with the time and angle domain seismic data in the input node and the amplitude-variation-with-angle (AVA) intercept A and the AVA gradient B, both as functions of time in the output node, then the relation between the input and output is linear in $\sin^2 \theta$; therefore, single hidden layer is sufficient to solve the problem. On the other extreme, if we go beyond AVA and use seismic data in time and offset as input (Figure 8.8a) and depth domain elastic earth model as output (Figure 8.28), then the network must approximate the wave equation and need multiple hidden layers.

In general, for complex problems a multilayer system is better than a single layer because it can provide training with less generalization errors. However, there is no clear rule as to how many layers are needed. The choice is problem dependent and must be empirically determined. During learning, supervised or unsupervised, all neural networks generalize a mapping function from the input to output using the universal approximation theorem via the hidden layers. This generalization depends on the complexity of the problem the network is trying to solve. Therefore, a good rule of thumb for any neural network design is to start with a few hidden layers and add more layers as necessary.

9.3.2 Number of Neurons in Each Hidden Layer

Like the number of hidden layers, setting the number of neurons in each hidden layer is also problem dependent. Hajian and Styles (2018) suggest that a good rule of thumb is to set the number of neurons in each hidden layer to half the number of input nodes and then add or delete neurons as necessary. Referring to Figure 9.11, this means that for N input nodes, we first set $M = N/2$. Note the number of input and output nodes (the parameters N and K in Figure 9.11) are fixed and the number of neurons per layer (M) and the number of layers (L) are the free parameters that need to be set in the network design. Therefore, a good way to start is to set $M = N/2$. If the training error (discussed next) does not reduce below the user-specified tolerance, then add more neurons. On the other hand, if the training error quickly reduces below the tolerance, then reduce the number of neurons.

The main issue here is related to any classical optimization problem of whether the problem is overdetermined or underdetermined. If we use too many neurons in each hidden layer, it can lead to overfitting where it will perform extremely well on the training data but poorly on the those which were not seen during training. On the contrary, if we use too few, it will lead to underfitting. The primary issue is therefore to find the right balance such that it is neither over-fitted nor under-fitted. This issue will become clear as we move forward as below where we revisit both the number of hidden layers and the number of neurons per layer after discussing the stopping criteria for training and the optimal choice of the number of training data.

9.3.3 Stopping Criteria for Training

Per universal approximation theorem, the approximation of the MLP is not perfect. As shown by equation 9.31, there is always a finite error $\varepsilon > 0$ between the original function $f(\mathbf{x})$ and its universal approximation $F(\mathbf{x})$ given by equation 9.30. The goal of the neural network design is to make the approximation error ε as small as possible. This approximation error is influenced by the following factors:

1) Learning algorithm and the number of iterations determine how well the error on the training set is minimized.
2) The number of training data determines how well the training samples represent the actual function.
3) The number of hidden units (the number of hidden layers and the number of neurons per layer) determines the "expressive power" of the network. As an example, only a few hidden layers are needed to approximate a smooth function. On the other hand, if the function is widely fluctuating, more hidden layers may be needed to approximate it.

So, to determine how good is our network design, it is first necessary to quantify the appropriate measure of the error. We have previously defined the error using equation 9.20, and here we elaborate further in terms of the optimum number of training data (or training/learning patterns).

From equation 9.20, we first define the error E_p for individual training patterns as

$$E_p = \frac{1}{2} \sum_{k=1}^{K} (t_{pk} - y_{pk})^2. \tag{9.32}$$

As previously defined for supervised learning, y_{pk} is the network output at the kth output node for the pth training pattern and t_{pk} is the corresponding target output provided by the teacher. For unsupervised learning, the target output t_{pk} is not used, and the error E_p for each training pattern is computed differently, which we discuss later. We now use the term $P_{learning}$ for the total number of training patterns to be used for learning in place of P we previously used. The reason for doing this is explained below. Having defined $P_{learning}$, we define *learning error rate* $E_{learning}$ as

$$E_{learning} = \frac{1}{P_{learning}} \sum_{p=1}^{P_{learning}} E_p. \tag{9.33}$$

A common mistake in the neural network design is data overfitting. If the entire set of P learning data is used, the network tends to overfit the data. By overfitting, the network will perform very well on the learning data. It may, however, perform poorly on other data that were not used for training. The primary objective of an efficient network design is to generalize the function approximation as universal approximator, not memorize it. And overfitting forces the network to memorize and perform extremely well on the learning data, but it would not generalize the function approximation and therefore perform poorly on other data. One good way to avoid overfitting is to divide the entire set of P learning data into two sets: (1) a set of $P_{learning}$ data to be used for learning and (2) another set of P_{test} data to be used for testing or validating the network. Clearly, $P = P_{learning} + P_{test}$. For each iteration or learning epoch, we only use the subset $P_{learnig}$ and compute the learning error rate $E_{learning}$ to adjust weights via optimization. Also in each learning epoch, we use the P_{test} test/validation data and compute the validation error rate as

$$E_{test} = \frac{1}{P_{test}} \sum_{p=1}^{P_{test}} E_p. \tag{9.34}$$

Figure 9.12 illustrates a typical variation of $E_{learning}$ and E_{test} with iteration/learning epoch. Note that $E_{learning}$ decreases with iteration. However, E_{test} decreases until it reaches a minimum point and then increases. Clearly, the region on the left-hand side of this minimum point is the underfit region. The region on the right-hand side of the minimum point is, on the other hand, the overfit region where the error on the training data continues to decrease but that on the test/validation set increases because the network fails to generalize and memorize the training data. We should therefore stop at the minimum point and output the estimated weights as the best-fit model.

9.3.4 Optimum Number of Training (and Validation) Data

Now that we have discussed the overfitting and underfitting issue for supervised networks, one last question yet to be answered is the choice of the optimum number of training and validation data.

Figure 9.13 illustrates the learning and validation errors ($E_{learning}$ and E_{test}) as functions of the number of training and validation patterns ($P_{learning}$ and P_{test}). As can be seen from this figure, when $P_{learning}$ and P_{test} are small, $E_{learning}$ can be small. But it does not mean that the network has generalized the function approximation if E_{test} is large. As both $P_{learning}$ and P_{test} are increased, and if the function approximation is adequate, then both $E_{learning}$ and E_{test} should converge to the same value, which should be used as optimal choices for $P_{learning}$ and P_{test} (see Figure 9.13). However, if $E_{learning}$ and E_{test} do not converge, the learning procedure has not converged to find the global minimum of the objective function and it is necessary to revisit the network design, as discussed below.

9.3.5 Revisiting the Network Design

As demonstrated in Figure 9.13, by increasing $P_{learning}$ and P_{test}, if $E_{learning}$ and E_{test} do not tend

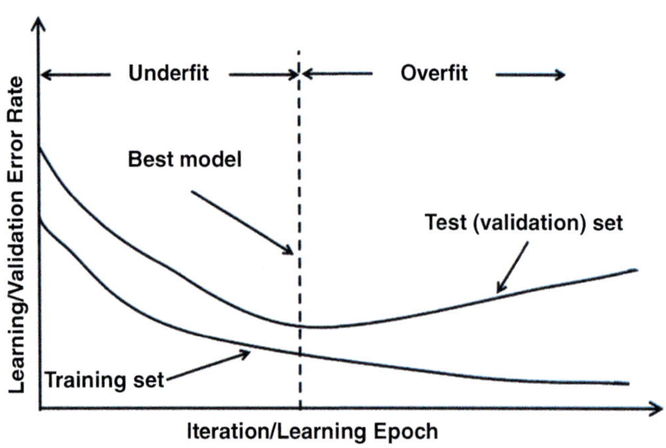

Figure 9.12 Variations of the learning and validation error rates with iteration. Hajian and Styles (2018)/Springer Nature.

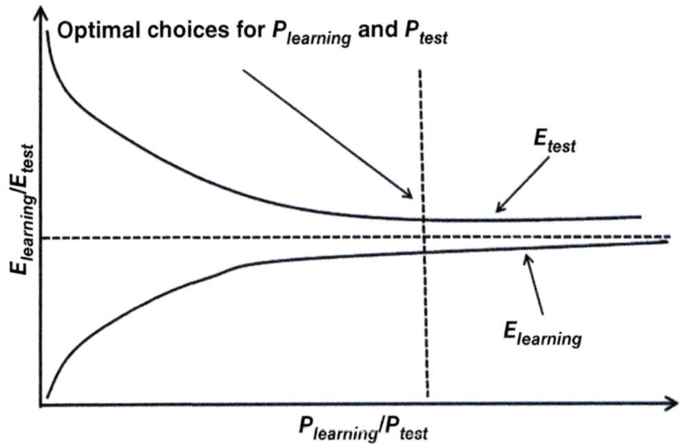

Figure 9.13 Learning and validation error as a function of the number of training and validation patterns. Hajian and Styles (2018)/Springer Nature.

to converge toward each other, then we may conclude that the original network design is not adequate in approximating the functional mapping of the input nodes onto the output nodes via the hidden layers. We must, therefore, revisit our design of the hidden layers. There are only two parameters that control the hidden layers: (1) the number of neurons per layer (M in Figure 9.11) and (2) the number of hidden layers (L in Figure 9.11).

The expressive power of any DNN is therefore controlled by H, the total number of the hidden neurons, given as $H = ML$. As shown in Figure 9.14, as we increase H, the network would keep fitting the learning pattern set $P_{learning}$ with increasing accuracy and thus $E_{learning}$ would keep decreasing. However, this does not necessarily mean that the functional approximation of the network is more accurate as we increase H. In reality, the set of $P_{learning}$ patterns are noise-contaminated seismic data. Thus, increasing H would eventually tend to model the noise instead of making a smooth function approximation. As a result, we will see what is known as the peaking effect, E_{test} will decrease up to a certain point and then increase (see Figure 9.14).

Unfortunately, there is no clear theory as to decide optimum values for M and L. For a single-layer system, we can run the network for different values of M (Figure 9.7), calculate the mean-squared error (MSE) for training for each, plot MSE versus M, and choose the optimum value for M where MSE is the lowest for both training and test data. Hajian et al. (2012) introduced a *multi-start approach* to find the

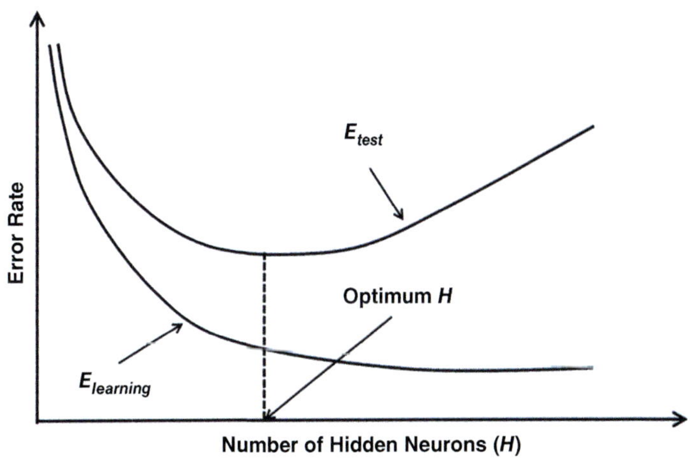

Figure 9.14 Variation of $E_{learning}$ and E_{test} with the number of hidden neurons (H). Hajian and Styles (2018)/Springer Nature.

optimal number of hidden neurons. The main concept behind this multi-start approach is that for a given set of say M neurons, if the network is run J times for the same training and validation data, the MSE errors for both (i.e., $E_{learning}$ and E_{test}) will be different. This is because the initial weights w_{ij} and w_{jk} (see Figure 9.7) in each of these J runs are randomly chosen, and thus, the errors are likely to be different for these runs. Therefore, it is better to run the training process for different values of these initial weights and find the optimum values for the final weights that have the least training and validation errors. So, in multi-start approach, the network is run with different values of M, each with different values of the randomly chosen initial weights. Then, by comparing these values, the optimum number of neurons (M) is chosen. Further details of this multi-start approach can be found in Hajian et al. (2012) and Hajian and Styles (2018).

To move from a single-layer system to a multilayer system of a DNN design, it is not only necessary to find the optimum value of M but we must also find L, the number of hidden layers (see Figure 9.11). The primary concept in any DNN design is deep learning, where the task for each hidden layer is to learn different features of the entire learning process. Taking image recognition as an example, the task of the first few hidden layer may be to recognize the low-order features like the outline/border, and as we go to the deeper layers, they may be designed to recognize more complex parts of the image. Such an image recognition problem is common in seismic exploration in interpreting the faults or salt boundaries from 3D seismic depth images. Properly designing the deep learning networks in conjunction with a proper choice of the training and validation samples is the key to success of these image recognition problems (for example, see Shi et al., 2019).

Although, in the above discussion, we restricted to a fixed number of M neurons in each hidden layer (Figure 9.11), it is not true, and deciding the number of neurons per layer depends upon the complexity of the problem. For FWI problems using physics-informed neural networks, Huang et al. (2021), for example, used eight hidden layers with 64, 64, 32, 32, 16, 16, 8, and 8 neurons in layers 1–8, respectively.

Finally, we must also note that all DNN methods are based on the universal function approximation. Consequently, a proper network design and correctly choosing both training and validation data is the key to their success. We elaborate these aspects in more detail below.

Mathematical Foundation of the Deep Neural Network

Although computationally efficient packages, different DNNs, and other ML methods are now available using which designing any complex network is straightforward, it is still important to formulate the fundamental mathematical framework of these designs and understand how they work in practice.

We develop our mathematical concepts using the MLP network illustrated in Figure 9.15, with N input nodes in the input layer, L hidden layers each with M_l, $l = 1, 2, \dots, L$ neurons, and K output nodes in the output layer. Clearly, there are two distinct components in the network architecture of Figure 9.15: (1) forward propagation where the network output is computed and (2) back propagation where the network output is compared with the target output and fed backward. We discuss below each of them separately.

Forward Propagation To develop mathematical formulation for forward propagation, we use the following conventions:

1) Input layer is defined as layer 0, hidden layer 1 as layer 1, hidden layer 2 as layer 2, and so on. If there are L hidden layers, the last hidden layer would thus be layer L and the output layer would be layer $L + 1$.
2) The general form of the activation and bias between any two consecutive layers i and $i + 1$ is defined respectively as a_i

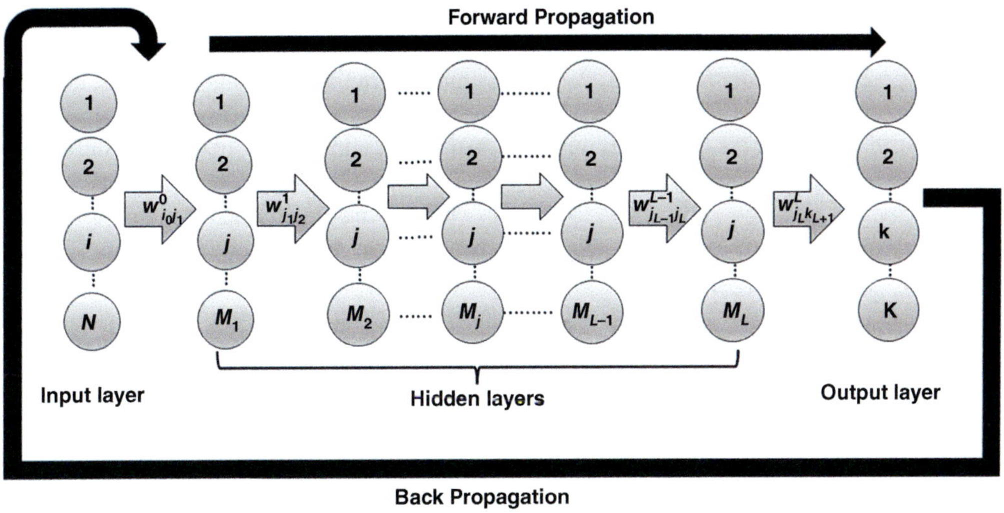

Figure 9.15 A generalized MLP architecture with forward and back propagation. Arrows shaded in gray indicate that each node on the left are connected to each node on the right with respective weights, activation, and bias.

and b_i. Thus, between the input and first hidden layers, they are a_0 and b_0. Similarly, between the last hidden layer (layer L) and the output layer (layer $L+1$), they are a_L and b_L. For log-sigmoid activation (equation 9.3), the bias and sigmoid shaping between layers i and $i+1$ are defined respectively as b_i and d_i.

3) The number of hidden neurons for any hidden layer i, $i = 1, 2, \ldots, L$, is M_i. Thus, for hidden layer 1 it is M_1, for hidden layer 2 it is M_2, and so on.
4) The connection weight between the node j of layer i and node k of the next layer $i+1$ is $w^i_{j_i k_{i+1}}$. Thus, the connection weight between the ith node of the input layer (layer 0) and the jth neuron of the first hidden layer (layer 1) would be $w^0_{i_0 j_1}$. Similarly, the connection weight between the jth nodes of layers 1 and 2 would be $w^1_{j_1 j_2}$, etc. (see Figure 9.15).

From the above conventions, we use equation 9.16 and write the output for training pth pattern at hidden layer 1 at node j, $j = 1, 2, \ldots, M_1$, as

$$h^1_{pj} = a_0 \sum_{i_0=1}^{N} w^0_{i_0 j_1} x^0_{p i_0} + b_0, \tag{9.35}$$

with $x^0_{pi_0}$ being the input data at the ith input node (layer 0) for the pth training pattern. The equivalent expression for a log-sigmoid activation can be written from equation 9.17 as

$$h^1_{pj} = \frac{1}{1 + e^{-\frac{\sum_{i_0=1}^{N} w^0_{i_0 j_1} x^0_{p i_0} + b_0}{d_0}}}. \tag{9.36}$$

We can similarly extend equation 9.35 to write the output at the jth node of hidden layer 2 as

$$\begin{aligned} h^2_{pj} &= a_1 \sum_{j_1=1}^{M_1} w^1_{j_1 j_2} h^1_{p j_1} + b_1 \\ &= a_1 \sum_{j_1=1}^{M_1} w^1_{j_1 j_2} \left(a_0 \sum_{i_0=1}^{N} w^0_{i_0 j_1} x^0_{p i_0} + b_0 \right) + b_1. \end{aligned} \tag{9.37}$$

For log-sigmoid activation, the equivalent of equation 9.37 is

$$\begin{aligned} h^2_{pj} &= \frac{1}{1 + e^{-\frac{\sum_{j_1=1}^{M_1} w^1_{j_1 j_2} h^1_{p j_1} + b_1}{d_1}}} \\ &= \frac{1}{1 + e^{-\frac{\sum_{j_1=1}^{M_1} w^1_{j_1 j_2} \frac{1}{1+e^{-\frac{\sum_{i_0=1}^{N} w^0_{i_0 j_1} x^0_{p i_0} + b_0}{d_0}}} + b_1}{d_1}}}. \end{aligned} \tag{9.38}$$

Equations 9.37 and 9.38 lead to the recursive relation for computing the outputs h^i_{pj} for any hidden layer $2 \le i \le L$ at neurons $j = 1, 2, \ldots, M_i$ from the outputs $h^{i-1}_{pj}, j = 1, 2, \ldots, M_{i-1}$, of the previous hidden layer $i-1$ for any training pattern p as

$$h^i_{pj} = a_{i-1} \sum_{j_{i-1}=1}^{M_{i-1}} w^{i-1}_{j_{i-1}j_i} h^{i-1}_{pj_{i-1}} + b_{i-1}, \quad (9.39)$$

which for the log-sigmoid becomes

$$h^i_{pj} = \frac{1}{1 + e^{\frac{\sum_{j_{i-1}=1}^{M_{i-1}} w^{i-1}_{j_{i-1}j_i} h^{i-1}_{pj-1} + b_{i-1}}{d_{i-1}}}}. \quad (9.40)$$

Finally, from the outputs $h^L_{pj}, j = 1, 2, \ldots, M_L$, of the last hidden layer, we can write the network output y_{pk} for each output node $k = 1, 2, \ldots, K$ as

$$y_{pk} = a_L \sum_{j_L=1}^{M_L} w^L_{j_L j_{L+1}} h^L_{pj_L} + b_L, \quad (9.41)$$

for any general activation and bias, which for log-sigmoid specializes to

$$y_{pk} = \frac{1}{1 + e^{\frac{\sum_{j_L=1}^{M_L} w^L_{j_L j_{L+1}} h^L_{pj_L} + b_L}{d_L}}}. \quad (9.42)$$

Computing y_{pk} using the above steps is the forward propagation component of the network (see Figure 9.15). Here, we formulated this in terms of a general activation and bias and then specialized for a log-sigmoid function. Formulating for other activation functions can be similarly done.

Back Propagation Given training patterns $p = 1, 2, \ldots, P$ with target outputs t_{pk}, $k - 1, 2, \ldots, K$, we now define the network error E by equation 9.20 and attempt to minimize E by back propagation. As previously noted, given a set of P training patterns, we do not use all of them to back propagate. Instead, we divide the entire set into two subsets, (1) $P_{learning}$ and (2) P_{test}, where $P_{learning} + P_{test} = P$, and use the subset $P_{learning}$ for back propagation and the subset P_{test} for validation. To retain simplicity in the governing equations, here we use the set of P patterns for back propagation, keeping in mind that in practice we only use a part only.

Minimizing E requires that its derivatives with respect to the network weights be set to zero. We thus define the weight vector $\mathbf{w}$, which we write from Figure 9.15 as

$$\mathbf{w} = [w^0_{11}, \ldots, w^0_{1M_1}, \ldots, \ldots, w^L_{M_L 1}, \ldots, w^L_{M_L K}]^T. \quad (9.43)$$

Clearly, with N input and K output nodes, and L hidden layers, each with $M_l, l = 1, 2, \ldots, L$ neurons, as shown in Figure 9.14, the number of components of $\mathbf{w}$ is $S = N \times M_1 \times M_2 \times \cdots\cdots \times M_L \times K$. We thus set

$$\nabla_{\mathbf{w}} E = \frac{\partial E}{\partial \mathbf{w}} = \mathbf{0}. \quad (9.44)$$

In equation 9.44, $\nabla_{\mathbf{w}}$ stands for the gradient operator with respect to the network weight vector as defined by equation 9.43. Now from equation 9.20, we can write

$$\begin{aligned} \nabla_{\mathbf{w}} E &= \frac{\partial E}{\partial \mathbf{w}} \\ &= \frac{1}{P} \sum_{p=1}^{P} \sum_{k=1}^{K} (t_{pk} - y_{pk}) \frac{\partial \left(\sum_{p=1}^{P} y_{pk} \right)}{\partial \mathbf{w}} \\ &= \mathbf{J} \frac{1}{P} \sum_{p=1}^{P} \sum_{k=1}^{K} (t_{pk} - y_{pk}). \end{aligned} \quad (9.45)$$

Like our previous discussion on the local optimization methods in Chapter 7, $\mathbf{J}$ in equation 9.45 is the Jacobi (Fréchet derivative) matrix whose components are derivatives of the network output with respect to each component of the weight vector $\mathbf{w}$. Because the network output is of size K (Figure 9.15) and the weight vector is of size $S = N \times M_1 \times M_2 \times \cdots\cdots \times M_L \times K$, $\mathbf{J}$ is a matrix of size $K \times S$ and given as

$$\mathbf{J} = \begin{pmatrix} \frac{\partial\left(\sum_{p=1}^{P} y_{p1}\right)}{\partial w_{11}^0} & \frac{\partial\left(\sum_{p=1}^{P} y_{p1}\right)}{\partial w_{12}^0} & \cdots & \frac{\partial\left(\sum_{p=1}^{P} y_{p1}\right)}{\partial w_{M_L K}^L} \\ \frac{\partial\left(\sum_{p=1}^{P} y_{p2}\right)}{\partial w_{11}^0} & \frac{\partial\left(\sum_{p=1}^{P} y_{p2}\right)}{\partial w_{12}^0} & \cdots & \frac{\partial\left(\sum_{p=1}^{P} y_{p2}\right)}{\partial w_{M_L K}^L} \\ \cdot & \cdot & \cdots & \cdot \\ \cdot & \cdot & \cdots & \cdot \\ \frac{\partial\left(\sum_{p=1}^{P} y_{pK-1}\right)}{\partial w_{11}^0} & \frac{\partial\left(\sum_{p=1}^{P} y_{pK-1}\right)}{\partial w_{12}^0} & \cdots & \frac{\partial\left(\sum_{p=1}^{P} y_{pK-1}\right)}{\partial w_{M_L K}^L} \\ \frac{\partial\left(\sum_{p=1}^{P} y_{pK}\right)}{\partial w_{11}^0} & \frac{\partial\left(\sum_{p=1}^{P} y_{pK}\right)}{\partial w_{12}^0} & \cdots & \frac{\partial\left(\sum_{p=1}^{P} y_{pK}\right)}{\partial w_{M_L K}^L} \end{pmatrix}. \tag{9.46}$$

Note that for a general activation and bias (equations 9.16 and 9.18), log-sigmoid (equations 9.17 and 9.19), and many activation functions, the Jacobi matrix can be analytically computed. Here, we derive the analytical expression for the log-sigmoid activation and leave deriving similar analytical forms for a general activation and bias, and other functions as exercises.

To derive the analytical form of the Jacobi matrix for log-sigmoid activation, let us first note that any component w_{jk}^i of the network weight vector $\mathbf{w}$ is the connection weight for the jth node of layer i to the kth node of layer $i+1$. For any training pattern p and log-sigmoid activation, we write from equation 9.40 the output at node k of layer $i+1$ as

$$h_{pk}^{i+1} = \frac{1}{1 + e^{\frac{\sum_{j=1}^{M_i} w_{jk}^i h_{pj}^i + b_i}{d_i}}}. \tag{9.47}$$

Consequently, the connection weight w_{jk}^i only contributes to h_{pk}^{i+1} for the entire network output. Now using the derivative formula for the ratio of functions

$$\frac{d}{dx}\left\{\frac{u(x)}{v(x)}\right\} = \frac{u'(x)v(x) - u(x)v'(x)}{\{v(x)\}^2}, \tag{9.48}$$

or the chain rule of the derivatives

$$\frac{d}{dx}\{u(x)v(x)\} = u'(x)v(x) + u(x)v'(x), \tag{9.49}$$

we obtain, from equation 9.47

$$\frac{\partial h_{pk}^{i+1}}{\partial w_{jk}^i} = h_{pk}^{i+1}\{1 - h_{pk}^{i+1}\}h_{pj}^i. \tag{9.50}$$

Therefore, the components of the Jacobi matrix can be explicitly written as

$$J_{ij} = \sum_{p=1}^{P} y_{pi} h_{pj}^{i+1}\{1 - h_{pj}^{i+1}\}h_{pj}^i, \tag{9.51}$$

in which $i = 1, 2, \ldots, K$ and $j = 1, 2, \ldots, S$. Once we have the Jacobi matrix, we can use any local optimization method to iteratively estimate the weights.

We discussed above gradient-based methods for ML optimization. There are, however, many other approaches, such as genetic algorithms and other Bayesian algorithms, which we discussed in Chapter 8. Please see Tahmasebi and Hezarkhani (2012) for further details of these different ML-based optimizations.

9.3.6 Different Flavors of DNNs

To this date, there are many flavors of DNNs, and the list is so broad that discussing them all is beyond the scope of this book. Here, we limit our discussion to the ones which have been successfully used to solve seismic and other geophysical problems.

Convolutional Neural Network

In MLP networks, each neuron (node) of any layer (input and hidden) is connected to all neurons (nodes) of the next layer (hidden and output) like the one illustrated in Figure 9.7. This, in turn, leads to data overfitting, which we discussed in connection with the FFBP network, and the same issues are true with the DNN architecture of Figure 9.15.

In contrast with MLP, the concept of a *convolutional neural network* (*CNN*) follows from the biological processes in which the connectivity pattern of the neurons resembles the organization of the *cortical neurons*, which belong to the *visual cortex* of animals. These individual cortical neurons respond to stimuli only in a restricted region of the visual field known as the *receptive field*. The receptive fields of different neurons partially overlap with one another in such a way that they cover the entire visual field.

CNN has been successfully used for seismic interpretation (Shi et al., 2019; Sen et al., 2019; Jia et al., 2022; Di et al., 2020; Di et al., 2021; Di et al., 2022; Feng et al., 2021; Gao et al., 2022; Jing et al., 2023; Liu et al., 2020a; Mosser and Naeini, 2022; Wu et al., 2019a; Wu et al., 2019c; Mukhopadhyay and Mallick, 2019; Wu et al., 2020; Ziong et al., 2018; Xu et al., 2022; Zhgralu et al., 2022; among others), seismic data processing and denoising (Almuteri and Sava, 2023; Liu et al., 2021; Sun et al., 2020; Vrolijk and Blacquière, 2021; Wu et al., 2019b; among others), microseismic event location and first arrival picking (Vinard et al., 2022; Wang and Alkhalifah, 2021; Yuan et al., 2020; Zhang et al., 2020; among others), seismic imaging (Dhara and Bagaini, 2020; Liu et al., 2020b; Sheng and Zhao, 2022; Sun et al., 2023; Zhanh et al., 2021; among others), and inversion problems (Biswas et al., 2019; Aleardi and Salusti, 2021; Cui et al., 2021; Das and Mukerji, 2020; Das et al., 2019; Liu et al., 2023; Park and Sacchi, 2020; Ren et al., 2021; Wu and McMechan, 2019; Yao and Wang, 2022; Yildirim et al., 2022; among others).

CNN uses convolution as its fundamental building block to mimic the operation of the cortical neurons responding to restricted and overlapping receptor fields. We previously defined convolution between two continuous function in Chapter 2 (equation 2.55). When synthetic seismic responses are computed in frequency domain using the methods discussed in Chapter 5, they are multiplied by the band-limited source term in frequency domain before Fourier transforming into time domain. And, from the convolution theorem, also discussed in Chapter 2, multiplication in frequency domain is equivalent to convolution in time domain. Thus, convolution is, in theory, equivalent to applying a filter to any input into a desired output, and in CNN the kernel of this filter is the set of weights.

Figure 9.16 illustrates the process of convolution that CNN uses. On the left is the input data shown as a matrix, in the middle is the convolution *kernel* with the network weights, and on the right is the result of convolution, called the output or *feature map*. Each element of the output is computed from the input and the weight kernel using the following operation:

$$h_{mn} = \sum_j \sum_k x_{jk} w_{(m-j)(n-k)}. \tag{9.52}$$

Figure 9.16 illustrates how equation 9.52 is used. To compute the first component, h_{11}, we just do a component-by-component multiplication and summation between the portion

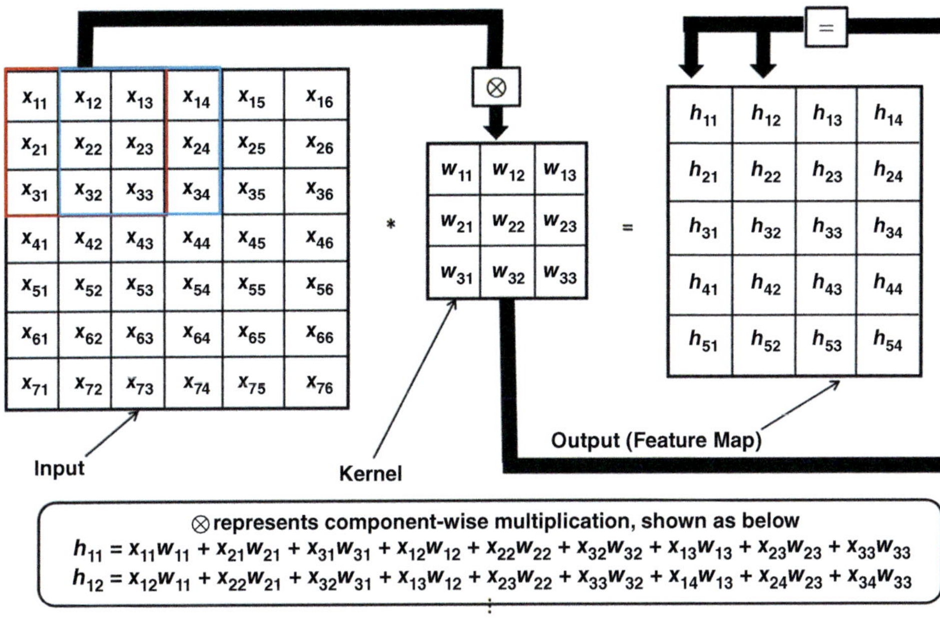

Figure 9.16 Illustration of the convolution operation used in CNNs.

of the input within the red box in Figure 9.16 with the kernel and write

$$h_{11} = x_{11}w_{11} + x_{21}w_{21} + x_{31}w_{31} + x_{12}w_{12} + x_{22}w_{22} + x_{32}w_{32} + x_{13}w_{13} + x_{23}w_{23} + x_{33}w_{33}. \tag{9.53}$$

Similarly, to compute h_{12}, we do the same component-by-component multiplication and summation between the portion of the input data within the blue box with the kernel and write

$$h_{12} = x_{12}w_{11} + x_{22}w_{21} + x_{32}w_{31} + x_{13}w_{12} + x_{23}w_{22} + x_{33}w_{32} + x_{14}w_{13} + x_{24}w_{23} + x_{34}w_{33}. \tag{9.54}$$

And, by repeating this process, we compute the entire output or the feature map. Note that each element h_{mn} is connected to restricted and overlapping regions of the input x_{ij} like the function of the cortical neurons. Additionally, the connection is achieved using the same set of weights, as defined by the kernel.

Convolution in place of multiplication is the fundamental building block of CNN in which the input data are filtered by the weight kernel. Most CNNs, however, not only use convolutional layers, but they also contain fully connected layers like a traditional MLP network discussed before. Before diving deep into CNN and discussing its architecture, it is, however, important to define a few additional concepts.

Stride In Figure 9.16, we explained convolution by component-wise multiplication and summation over overlapping regions of the input data. In generating these overlapping regions and creating the output (feature map), we shifted them by one both along the rows and columns. There is, however, no harm in computing the output by shifting row-wise and column-wise by other than one. Of course, if the kernel of weighting matrix is $M \times N$, by shifting more than M rows will not generate overlapping regions vertically and more than N columns will not generate overlapping regions horizontally. This amount of shift in vertical or horizontal direction is the *stride*. In the example of Figure 9.16, stride is therefore one in both directions.

The choice of stride depends on the underlying physics of the problem. Restricting to seismic inversion, assume that our input layer matrix is the prestack seismic data in plane-wave (τ – p) domain, in which each column represents the ray-parameter p and each row represents the intercept-time τ. Assuming a dominant seismic frequency of 30 Hz and a minimum subsurface P-wave velocity of 2000 m/s, the minimum dominant wavelength is then $2000/30 \approx 67$ m and a quarter wavelength of about 17 m. Two-way travel time for P-waves over a 17-m interval is about 16 ms. Because it is hard to resolve the subsurface below a resolution of quarter wavelengths of the dominant frequency, if $\tau - p$ domain seismic data are sampled at 2 ms time interval, we may safely use a stride of about 6 or 7 along the vertical (row) direction without losing any feature. And, according to the underlying physics, using a low value may lead to overfitting where the predictions may have spurious and unwanted oscillations below seismic resolution. In the horizontal (ray-parameter) direction, on the other hand, it is important to accurately resolve the variation of the reflection amplitudes with p and therefore a low value (1 or 2) for stride is advisable. The reason for this discussion is that given the training data and their labels (target output), any ML method would find an answer. However, because these methods tend to approximate the inverse physics via universal approximation, the answer from the method may not be correct unless the underlying physics is understood, and the algorithm is correctly designed.

The number of rows (or columns) in the output matrix is controlled by those of the input and kernel, and the value of the stride and is given as

$$o = \frac{n - f}{s} + 1. \tag{9.55}$$

In equation 9.55, o is the number of rows (or columns) in the output, n is the number of rows (or columns) in the input, f is the number of rows (or columns) in the weighting matrix kernel, and s is the stride. Using the example shown in Figure 9.15 with $s = 1$, we can readily verify from equation 9.55 that the number of rows and columns of the output to be 5 and 4.

Padding Depending on the sizes of the input matrix and kernel, and stride, the size of the output matrix after convolution is always less than the input matrix. In Figure 9.16, for example, the input matrix is 7×6, and with 3×3 weight kernel and $s = 1$, the output matrix is 5×4. Now, if we add another convolutional layer using a 3×3 weight kernel, and $s = 1$, then the resulting output will be a matrix of size 3×2. After being reduced to a matrix of size 3×2, we cannot add another convolutional layer with a 3×3 weighting matrix kernel.

Successive reduction in the size of the output is a problem for CNN because it loses the information content from one convolutional layer to the next. The way CNN avoids size reduction is by padding, as illustrated in Figure 9.17. As can be seen in this figure, the original input is padded with zeros, and by doing so, the output retains the original size of the input. With padding, we can modify equation 9.55 to

$$o = \frac{n + 2p - f}{s} + 1. \tag{9.56}$$

In equation 9.56, o, n, f, and s have the same meaning as they are in equation 9.55 and p represents the number of paddings. In Figure 9.16, $p = 1$, and we can immediately verify from equation 9.56 that the output is of the same size of the input.

Channels and Depth In Figures 9.16 and 9.17, we described CNN using a single input, weight kernel, and an output. In general, however, CNNs can have multiple channels.

As illustrated in Figure 9.18, for image processing and analysis, the input layer can have three different channels, red, green, and blue in which values in each channel represent a pixel value between 0 and 255, and all of them combined represents an image. In such a case, we may have multiple channels in both kernel and the output.

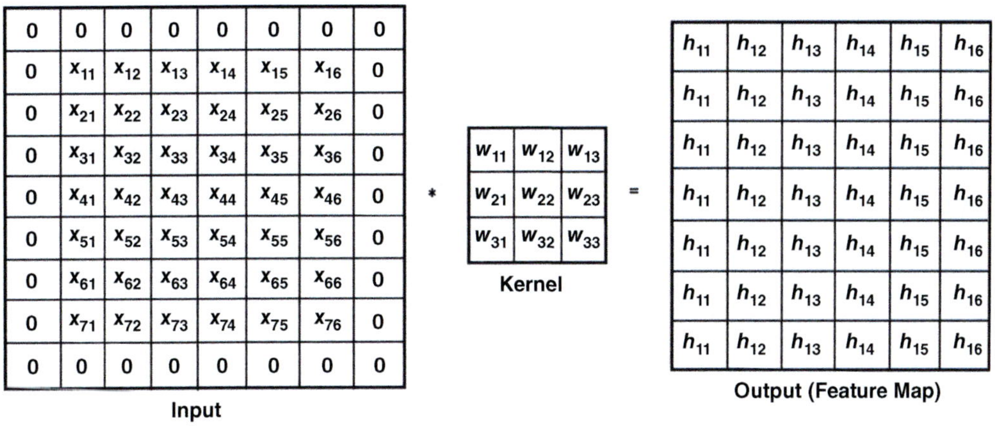

Figure 9.17 Illustration of padding used in CNNs.

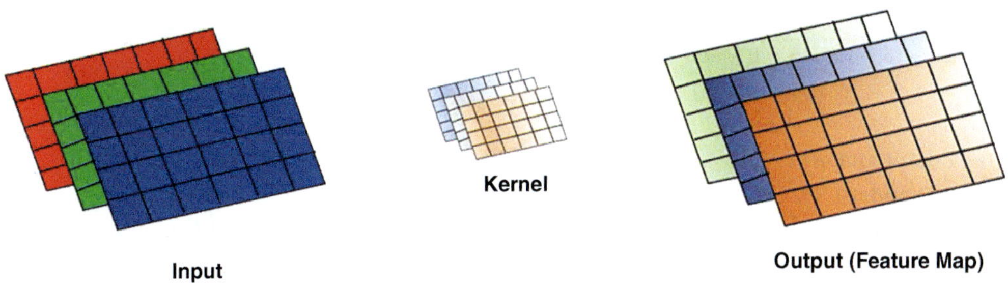

Figure 9.18 The concept of multiple channels used in CNNs.

For seismic inverse problems, the input is typically a single channel, representing the seismic data. However, if we want to invert multicomponent seismic data, there will be multiple channels in the input layer. Irrespective of single- or multi-channel input, the convolutional layers for seismic inverse problems can be multi-channel. Restricting to a single (vertical) component isotropic seismic inversion, for example, we know that the parameters (or features) controlling the reflection amplitudes in the input seismic data are (1) P-wave velocity (V_P), (2) S-wave velocity (V_S), and (3) density (ρ). So, we may use three different kernels (weighting matrices in Figures 9.16 and 9.17) and extract feature maps of V_P, V_S, and ρ. Consequently, the output for such a case becomes multichannel. The term *depth* in CNN is referred to as the number of kernels used to process the input.

Activation Functions In addition to convolution, CNN is also associated with activation and bias. Using the log-sigmoid function, we can write the computed value y_{mn} at each neuron as

$$y_{mn} = sigmoid(h_{mn}) = \frac{1}{1 + e^{-\frac{h_{mn}+b}{d}}}, \tag{9.57}$$

where h_{mn} is from convolution (equation 9.52), b is the bias, and d is the sigmoid shaping factor. Besides log-sigmoid, another popular activation function used in the neural networks is the hyperbolic tangent (*tanh*) or sigmoid activation, which can be written as

$$y_{mn} = \tanh(h_{mn}) = \frac{e^{\frac{h_{mn}+b}{d}} - e^{-\frac{h_{mn}+b}{d}}}{e^{\frac{h_{mn}+b}{d}} - e^{-\frac{h_{mn}+b}{d}}}. \tag{9.58}$$

Like log-sigmoid, b and d in equation 9.58 respectively represent the bias and shaping factor. Figure 9.19 shows behavior of the tanh activation function with different values of b and d. In Figure 9.19a, b is set to a value of 2, and d is varied from 2 to 12. In Figure 9.19b,

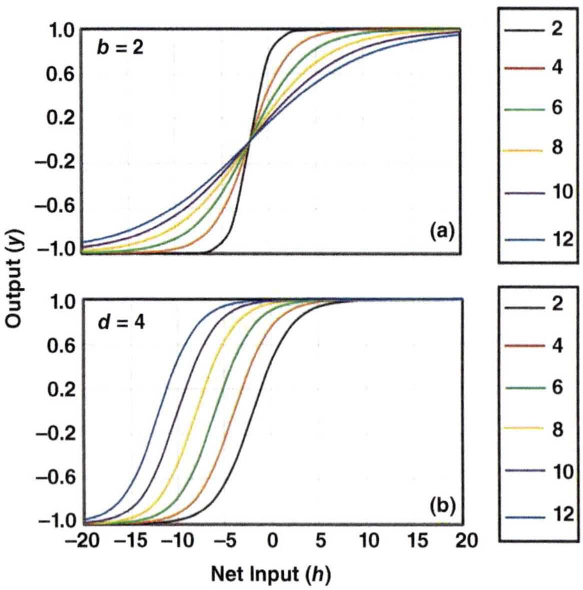

Figure 9.19 Influence of the hyperbolic tangent (tanh) or sigmoid activation function on the choices of the shaping factor d and the bias b. (a) Shapes of the activation function for $b = 2$ and d varying from 2 to 12, in steps of 2. (b) Same as (a) but for $d = 4$ and b varying from 2 to 12, in steps of 2.

on the other hand, d is set to 4, and b is varied between 2 and 12. By comparing Figure 9.19 with the one shown earlier for the log-sigmoid activation (Figure 9.3). we immediately see that besides lower and upper limits of the computed output (y), the tanh activation behaves like the log-sigmoid activation with the choice of d controlling the slope of the function and b influencing the bias of the function toward the positive axis of network output h. Like log-sigmoid activation, it is straightforward to analytically find the derivative of the tanh activation, leading to a fast optimization for network training.

Although log-sigmoid and tanh activations are easy to implement, using them can lead to numerical instability. From Figures 9.3 and 9.19, we can see that for large absolute values of the network input h, the gradient of both functions approaches zero. This is called the *vanishing gradient problem* for these activation functions and can lead to numerical instability. Therefore, in using these activations, it is important to appropriately scale the network input so that it stays outside the range of the vanishing gradients.

Most popular activation function used in CNNs is the *ReLU* (Fukushima, 1969), which can be mathematically described as

$$y = h_{mn}^{+} = \max(0, h_{mn}) = \begin{cases} h_{mn}; h_{mn} > 0 \\ 0; h_{mn} \leq 0 \end{cases}, \tag{9.59}$$

in which h_{mn} represents each component of the output (feature map) after convolution (see Figure 9.16 and equations 9.53 and 9.54).

Figure 9.20 shows the ReLU activation function. Note that this function is called ReLU for two reasons: (1) by setting the values for all $h \leq 0$ as analogous to half-wave rectification in electrical engineering and (2) for all $h > 0$ this function is a linear (ramp) function. Also note

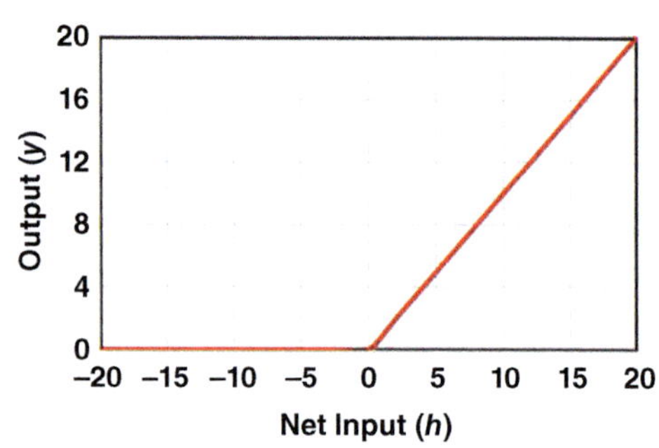

Figure 9.20 Behavior of the ReLU activation function.

from equation 9.59, computing the derivative of the ReLU function is straightforward.

For seismic inverse problems, the input layer will be seismic data with both positive and negative values. Because they both are important in mapping seismic data onto the respective elastic earth properties, using the ReLU activation function in its original form (equation 9.59 and Figure 9.19) is not appropriate because it will rectify all negative amplitudes to zeros. However, this problem can be easily avoided by adding a bias to the ReLU activation.

In addition to the ReLU defined by equation 9.59, there are other linear and nonlinear variants of ReLU. The list is so huge that it is beyond the scope of this book to discuss them. In addition, these variants can only provide marginal improvements over the ReLU activation function given by equation 9.59.

CNN Architecture *Neocognitron*, introduced by Fukushima (1980) and Fukushima and Miyake (1982), is the first known CNN example used for pattern recognition. This method introduced the fundamental building blocks of CNN and is the predecessor of the modern CNN architecture. The true power of CNN for using not only as a classifier system for image analysis, but also as regression to solve inverse problems, was not, however, realized until the introduction of the graphic processing unit in the computing hardware and the subsequent development of the massively parallel computing technology.

To this date, there are many variations in the CNN architecture such as LeNet (LeCun et al., 1998), AlexNet (Krizhevsky et al., 2017), ZFNet (Zeiler and Fergus, 2013), VGGNet (Simonyan and Zisserman, 2014), GoogLeNet (Szegedy et al., 2014), ResNet (He et al., 2015), etc. Although the implementation details of these different methods vary, they all are based on the *encoder–decoder* architecture with the following basic components:

1) Input layer (or layers).
2) Convolutional layers.
3) Pooling layers.
4) Fully connected layers.
5) Output layer.

General CNN architecture based on the above components is illustrated in Figure 9.21. The input layer (or layers) is followed by convolutional layers in which the feature maps are generated via convolution and activation as discussed above. Convolutional layers are followed by the pooling layers where the information content of the convolutional layers is subsampled. This subsampling is performed for two specific reasons:

1) The information content of the convolutional layers is far too many and may lead to overfitting issues and subsampling avoids potential overfitting of the network and performs poorly on the unseen data.
2) Subsampling reduces computational burden.

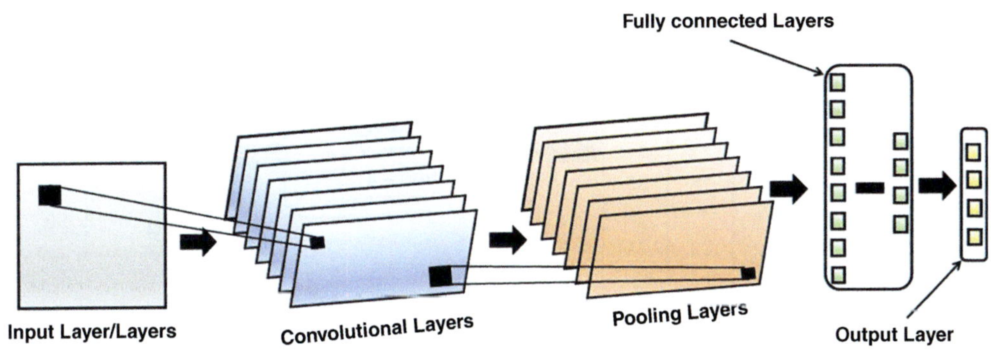

Figure 9.21 General CNN architecture.

Pooling layers are then followed by fully connected layers whose architecture is like the DNN architecture (Figure 9.15). Finally, the fully connected layers are followed by the output layer.

In CNN, there are three different ways to perform pooling: (1) maxpooling, (2) mean or average pooling, and (3) sum pooling. Out of these three ways, maxpooling is most popular. Irrespective of how this pooling is performed, there are two parameters that control pooling: (1) window size and (2) stride. Window size means the subset of the original convolutional layer to subsample, and stride is explained earlier. Figure 9.22 illustrates how the information content of a 6×6 convolutional layer is subsampled to size 3×3 via maxpooling using a window size of 2×2 and stride of 2. From each 2×2 submatrix of the convolutional layer, coded in different colors, only the maximum value is transferred in the pooling layer. In average pooling, the average value of each submatrix is transferred to the pooling layer, and in sum pooling, their sum is transferred. The entire propagation from the input layer(s) to convolutional layers and then to the pooling layers is referred to as the *encoder–decoder* architecture.

Another popular variation of CNN architecture is the U-Net, which is specifically useful for handling image segmentation problems. In seismic interpretation, U-Net has been used to interpret features such as faults and salt bodies from depth migrated images (Shi et al., 2019; Sen et al., 2019; Jia et al., 2022; Di et al., 2020; Di et al., 2021; Di et al., 2022; Feng et al., 2021; Gao et al., 2022; Jing et al., 2023; Liu et al., 2020a; Mosser and Naeini, 2022; Wu et al., 2019a; Wu et al., 2019c; Wu et al., 2020; Ziong et al., 2018; Xu et al., 2022; Zhgralu et al., 2022; among others). U-Net is also based on encoder–decoder architecture, which was originally developed by Ronneberger et al. (2015) and Shelhamer et al. (2017). Important characteristics of U-Net as compared with other CNN architectures are that U-Net does not have fully connected layers and the pooling layers are replaced by up-sampling operators so that they increase resolution of the output image, which is then followed by down-sampling with fully convolutional layers without padding. Many feature channels in the up-sampling part allow the network to propagate context information to high-resolution layers in a way that this expansive (up-sampling) part is approximately symmetric to the contracting (down-sampling) part and the network has a U-shaped architecture. Without using fully connected layers, U-net uses the valid part of each convolution. To predict the pixels in the border regions of the image, the missing context is extrapolated by mirroring the input image via augmentation. Although U-Net is useful for image processing problems, such as interpreting salts, faults,

Figure 9.22 Illustration of maxpooling used in CNN.

etc., from the depth images, its use to solve seismic inverse problems is limited and therefore we do not discuss U-Net architecture any further.

Recurrent Neural Network

In using a word processing software such as Microsoft Word, or any email platform like Gmail or Outlook, we note that as we start typing a sentence, these software tend to predict or suggest the sequences of words for us. Although these suggestions are not always correct, in many cases they are. Most of these predictions or suggestions are, in fact, made by a *recurrent neural network* or *RNN*. The way RNN attempts to make such predictions ahead of typing is by creating a network memory by feeding the output from the previous step to the input of the current step.

RNN is a special type of neural network, designed for time series data. The architecture of the neural networks we discussed so far is designed to work for data points that are independent of one another. However, in the case of data points that are in a sequence in which their values depend on the values at previous data points, like the above example of predicting words ahead of typing, then it is necessary to modify the network architecture in a way that such dependence is incorporated into the algorithm. And, the RNN is designed in this way with a memory that stores the states or information of previous inputs to generate the next output of the sequence.

As shown in Figure 9.23, RNN consists of recurrent layers and one feedforward layer, in which each recurrent layer corresponds to one time step. Figure 9.23, for example, is shown with k time steps. Each time step i consists of the two inputs, (1) the input data $\mathbf{x}_i$ and (2) $\mathbf{h}_{i-1}$, the hidden units containing the state of information from the previous time step $i - 1$ and they output $\mathbf{h}_i$, the state of information from the current time step using the following formula:

$$\mathbf{h}_i = f(\mathbf{W}_x\mathbf{x}_i + \mathbf{W}_h\mathbf{h}_{i-1} + \mathbf{b}_h). \qquad (9.60)$$

In equation 9.60, f is an appropriate activation function (log-sigmoid, tanh, ReLU, etc.), $\mathbf{W}_x$ and $\mathbf{W}_h$ are respectively the weights associated with the inputs and the hidden units in the recurrent layer, and $\mathbf{b}_h$ is the bias associated with the recurrent layer. These weights and bias are temporally shared between the recurrent layers. Also, for time step 1, $\mathbf{h}_0$ is set to zero.

After k such time steps outputs $\mathbf{h}_k$, from which the network output is calculated from the feedforward layer as

$$\mathbf{y}_k = f(\mathbf{W}_y\mathbf{h}_k + \mathbf{b}_y), \qquad (9.61)$$

in which f has the same meaning as in equation 9.60, and $\mathbf{W}_y$ and $\mathbf{b}_y$ are respectively the weights and bias associated with

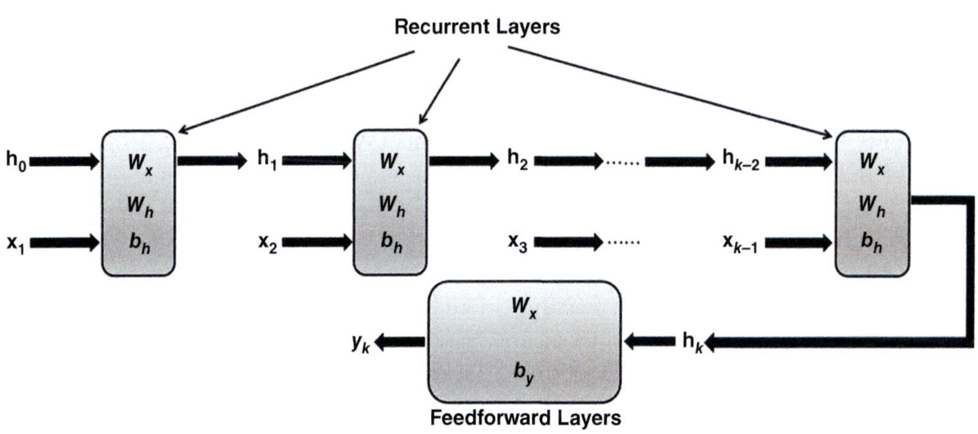

Figure 9.23 A general RNN architecture with k time steps.

the output layer. Like other neural networks, the network error is computed as discussed before and back propagated in an optimization scheme to adjust weights $\mathbf{W}_x$, $\mathbf{W}_h$, and $\mathbf{W}_y$ for network training.

Figure 9.23 shows the general (unwrapped) RNN architecture after k time steps. Although it may appear from this figure that to obtain the output $\mathbf{y}_k$ at time step k, the network uses input $\mathbf{x}_i$ for all $i = 1, 2, \ldots, k$, and the network output $\mathbf{h}_j$ for all $j = 0, 1, \ldots, k-1$, this is not true. Figure 9.23 requires further explanation on how RNN is implemented in practice, and we use a seismic example to do so. As a mitigation step to reduce greenhouse gas emissions from fossil-fuel consumptions, carbon dioxide (CO_2) sequestration into deep saline aquifers and depleted hydrocarbon reservoirs is now an acceptable technique. CO_2 sequestration is, however, incomplete unless the subsurface is remotely monitored using geophysical (primarily seismic) methods to ensure the containment of the injected fluids within the target formations and the overlying seal integrity. This is done using time-lapse seismology, where seismic data are acquired at different times and the subsurface fluids within the target formations and the integrity of the overlying seal (cap) rocks are characterized from these repeated surveys. For example, assume that we have three sets of seismic data over an area, which were acquired at three different times: (1) before sequestration, which we call survey 1 or the baseline data; (2) survey 2, acquired 5 years after sequestration; and (3) survey 3, acquired 10 years after sequestration. Time-lapse seismic analysis attempts to find the difference between these repeated surveys to monitor fluid movements and ensure seal integrity. It assumes that all parts of the seismic data acquired at different times are same, except in the target formation and possibly in the overlying seal (if the injected fluid pressure caused damage to the seal forming leakage pathways for CO_2 to escape). However, this assumption is not correct. Even if there is no CO_2 injection and it is impossible to reproduce the same seismic data acquired at different times. This is the issue of repeatability. Even if we use the same acquisition parameters, seismic data are not repeatable. There are many reasons for this. First, noise levels on data can be different. Second, there are always seasonal variations in the water table. Third, the earth is dynamic and there are always natural fluid movements in the subsurface, and there could be some tectonic activity in-between the two surveys, which may modify some subsurface features.

In seismic processing, the issue of non-repeatability is handled by applying a matching or cross-equalization filter, which attempts to equalize time-lapse data everywhere but the target regions where the change is expected. This cross-equalization filter is a standard procedure for time-lapse analysis. However, one must be careful in applying such a filter because it can potentially change the relative amplitudes. Another, and possibly more elegant, approach is to replace cross-equalization by RNN, as illustrated in Figure 9.24. Using the baseline data ($\mathbf{x}_1$) and the network output from the previous time step ($\mathbf{h}_0 = \mathbf{0}$) as input, we feed it into an RNN to compute the output $\mathbf{y}_1$ via a feedback loop (Figure 9.24a). We store the network output ($\mathbf{h}_1$) and the adjusted weights and biases ($\mathbf{W}_{x1}$, $\mathbf{W}_{h1}$, $\mathbf{W}_{y1}$, $\mathbf{b}_{h1}$, $\mathbf{b}_{y1}$). Now, for survey 2 (Figure 9.24b), we feed RNN the input data ($\mathbf{x}_2$) and the network output ($\mathbf{h}_1$) from the baseline survey to compute the output $\mathbf{y}_2$ via feedback loop. For the weights and biases, we can start from their previous values from the baseline ($\mathbf{W}_{x1}$, $\mathbf{W}_{h1}$, $\mathbf{W}_{y1}$, $\mathbf{b}_{h1}$, $\mathbf{b}_{y1}$) and via feedback loop we can compute and store the network output ($\mathbf{h}_2$) and the adjusted weights ($\mathbf{W}_{x2}$, $\mathbf{W}_{h2}$, $\mathbf{W}_{y2}$, $\mathbf{b}_{h2}$, $\mathbf{b}_{y2}$). And, we do the same for survey 3 (Figure 9.24c). Note that for any time step i, RNN uses the input data $\mathbf{x}_i$ for that time step and the output of the RNN layers ($\mathbf{h}_{i-1}$) from the previous time step. Additionally, each time step estimates a model $\mathbf{y}_i$ as output and comparing them with one another would provide the fluid movements in the target formation. Such an RNN-based time-lapse seismic application has been carried out by Alali et al. (2022). We postpone the

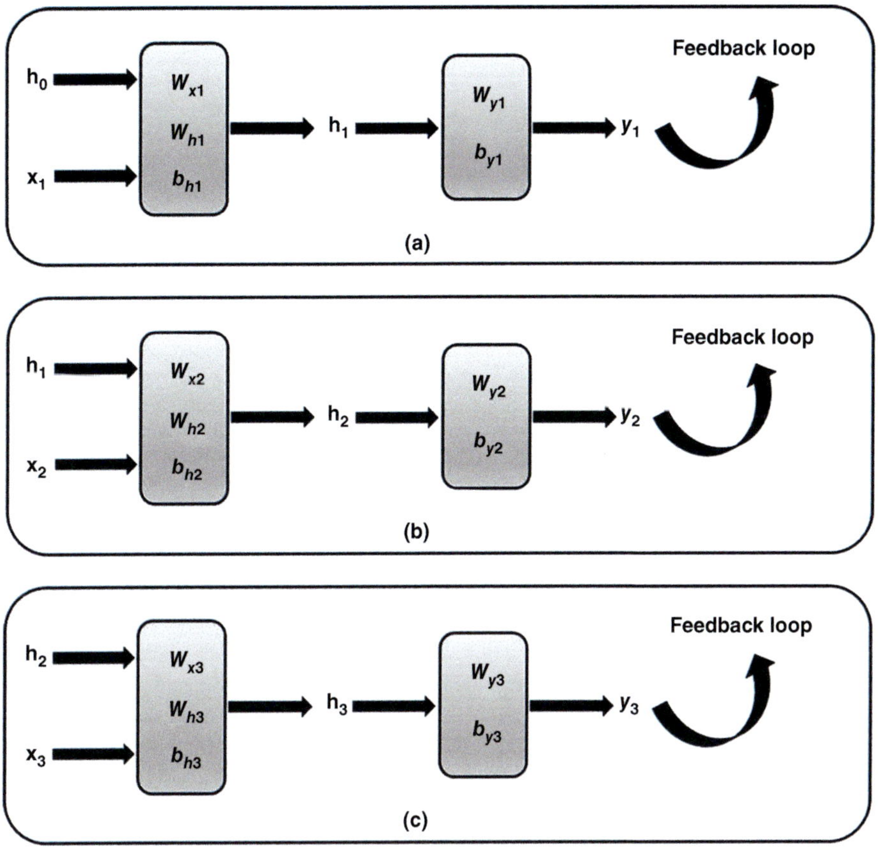

Figure 9.24 Implementation of RNN for subsurface characterization using time-lapse seismic data. (a) Survey 1 (baseline). (b) Surve 2 (5 years after CO_2 sequestration). (c) Survey 3 (10 years after CO_2 sequestration).

discussion of the work until later part of this chapter.

Different Types of RNNs Figures 9.23 and 9.24 show an input vector **x** of any size producing an output vector **y** whose size can be different form that of **x**. Depending on the specific problem, RNN can be of different types. The simplest one is *one to one* in which both input and output are scalars, and the most general one is *many to many* where both input and output are vectors of different sizes. Additionally, we can have intermediate types such *as one to many* and *many to one*. Figure 9.25 illustrates these different types of RNNs.

Variations of RNN Architecture Like CNNs, RNNs also have different variations in their architectures. Some of the popular ones are the following:

1) *Bidirectional Recurrent Neural Network (BRNN)*: In BRNN, inputs from future time steps are used to improve the accuracy of the network. This architecture is specifically useful in word processing where the network attempts to predict the middle words of a sentence from the knowledge of the first and the last word.
2) *Gated Recurrent Unit (GRU)*: This architecture is designed to handle the vanishing gradient problem, which we discussed in relation with the log-sigmoid and tanh activation functions. Using a reset and an update gate, GRU determines which information is to be retained for future predictions.

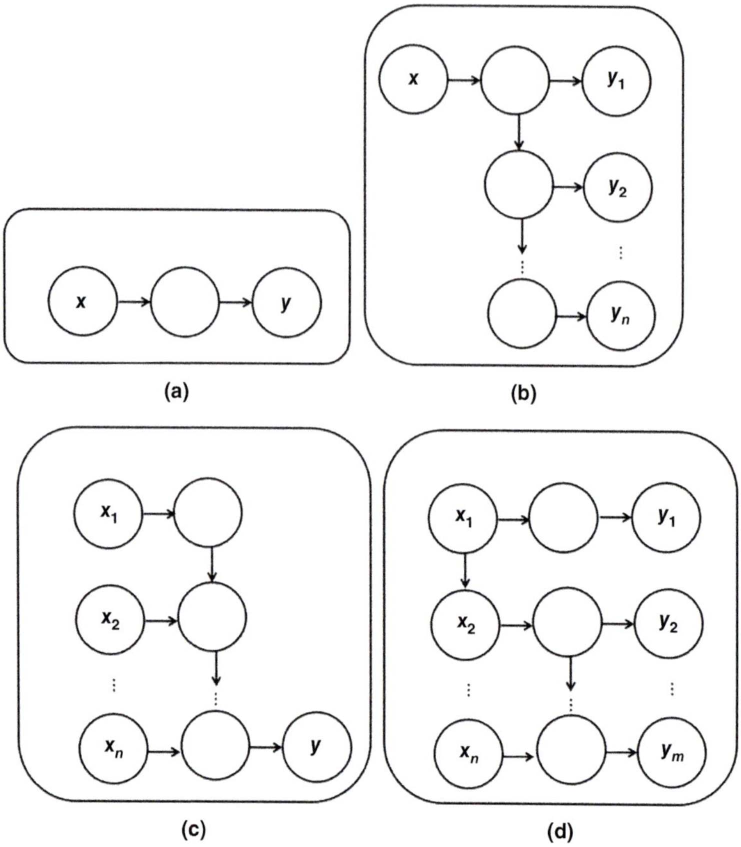

Figure 9.25 Different types of RNN: (a) one to one, (b) one to many, (c) many to one, and (d) many to many.

3) *Long–Short-Term Memory (LSTM)*: Like GRU, LSTM is also specifically designed to handle the vanishing gradient issue. This architecture uses three gates: (1) input gate, (2) output gate, and (3) forget gate to decide which information to retain.

Further details on RNNs can be found in Goodfellow et al. (2015) and Di et al. (2018).

Seismic Applications of RNN Compared with CNN, the application of RNN in seismology is limited. This, however, does not mean that RNN is not useful for seismic applications. Specialty of RNN is that it can efficiently handle time sequence data, where the output $\mathbf{h}_t$ of a given time step t is predicted not only from the input $\mathbf{x}_t$ for the time step but also from the output $\mathbf{h}_{t-1}$ from the previous time step $t - 1$. Consequently, as previously stated, RNN is potentially useful for subsurface characterization from time-lapse (4D) seismic data. One such successful application was reported by Alali et al. (2022) in monitoring the plume movement from the baseline and time-lapse seismic data after CO_2 injection. Although the results were based on synthetic data, the method showed a lot of promise in using RNN for time-lapse analysis instead of the cross-equalization using matching filters, which is the current state of the art. In fact, Alali et al. (2022) claimed that RNN outperformed cross-equalization in most parts of their analysis. Yuan et al. (2020) used a two-step neural network for first break picking from noisy seismic data. In this approach, they first cast the problem as image segmentation and used CNN with U-Net architecture to identify the segmentation boundary, which was then followed by RNN for picking first arrivals. Fabien-Ouellet and Sarkar (2020) showed an RNN-based velocity estimation procedure. Liu et al. (2021) combined seismic interferometry with RNN to invert vehicle-induced passive seismic data for estimating subsurface S-wave velocities. Finally, Birnie and Hansteen (2022) used BRNN for event detection from passive seismic data.

Hopfield Neural Network

Hopfield neural network (*HNN*) is named after Hopfield (1982), which was further developed by Hopfield and Tank (1985) and Tank and Hopfield (1986). In a strict sense, HNN is a variation of RNN. However, the variations are large enough to consider HNN as a different neural network class.

The primary characteristic feature of HNN is its multi-directional nature. This means that the flow of data from any neuron is not in a single direction, instead the network receives data from all neurons except itself. This is illustrated in Figure 9.26 using four neurons, N_1, N_2, N_3, and N_4. Like other networks, these connections are via weights, which are equal in the forward and reverse directions. Referring to Figure 9.26, for example, if W_{12} is the connection weight from N_1 to N_2, and

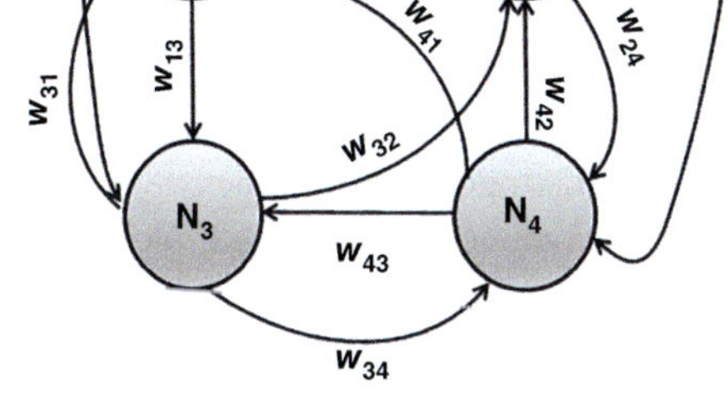

Figure 9.26 A simplified architecture of the Hopfield neural network with four neurons.

W_{21} is the connection weight from N_2 to N_1, then $W_{12} = W_{21}$. This condition ensures that the data entered in the network has the same effects in all neurons. The sum of the squared error of HNN is called the energy function and is given as

$$E(v) = \frac{1}{2}\sum_{i=1}^{n}\sum_{j\neq i=1}^{n} W_{ij}v_iv_j - \sum_{i=1}^{n} I_iv_i. \tag{9.62}$$

In equation 9.62 $E(v)$ is the energy function of HNN, n is the total number of neurons, W_{ij} is the connection weight between neurons i and j, v_i is the ith component of the input vector, v_j is the jth component of the input vector, and I_i is the ith input, also known as the threshold. Because $W_{ij} = W_{ji}$, the weights in HNN form a symmetric matrix, and the energy function $E(v)$ will never increase as the state of the neurons change (Hopfield, 1982; Hopfield and Tank, 1985). This, in turn, means that the network will converge to a state where $E(v)$ is locally minimized (Wang and Mendel, 1992). Therefore, HNN can invert any set of measured geophysical data to the corresponding model parameters and can be used like conventional inversion techniques without any training or supervised learning. The major challenge in a successful HNN implementation to solve an inverse problem is formulating an appropriate cost function between the measured data and those theoretically calculated based on the model parameters. Once this formulation is established, any method, local or global, can be used to minimize the cost function and estimate the model.

In geophysics, HNN has been used for deconvolution and multiple attenuation (Wang and Mendel, 1992; Calderón-Macías et al., 1997), horizon picking (Huang, 1997), bright-spot interpretation (Huang et al., 1989), seismic velocity picking (Huang and Yang, 2015), prestack seismic inversion (Phan and Sen, 2019), and depth migration (Vamaraju and Sen, 2018, 2019; Vamaraju et al., 2019). Using simulated annealing approaches to minimize the cost function in HNNs is formally called the *Boltzmann Machines*.

9.4 Other Machine-Learning Methods

Our discussion so far has been limited to different flavors of ANNs. At the beginning of this chapter, we also mentioned that ANNs are a subset of a broader class of learning algorithms called the machine-learning methods. We must note that ANNs are not the only ML methods, but there are many other ML methods. In this section, we discuss two such methods, (1) *support vector machine* (*SVM*) and (2) *gradient boosting*, which were successfully used in seismic inversion.

9.4.1 Support Vector Machine

The *support vector machine* (*SVM*) belongs to the class of supervised learning, whose original concept was developed in a Russian paper by Vapnik and Chervonenkis (1964) and was further advanced by Boser et al. (1992). These works developed this supervised learning for classification. Later Cortes and Vapnik (1993) extended SVM to solve regression problems, which is called the *support vector regression* (*SVR*). A very good account of SVM and SVR can be found in Christianini and Shawe-Taylor (2000). Additionally, the classic book on numerical recipes by Press et al. (2007) provides an excellent overview of SVM in Chapter 16, pp. 893–898, and parts of the discussion below are motivated by this numerical recipes textbook.

SVR is of particular interest for seismic and other geophysical inverse problems. However, to understand how SVR works, it is first necessary to understand the fundamental concepts of SVM as a classifier. We first discuss below SVM as a classifier before discussing SVR.

SVM as a Classifier

Assume a simple case of supervised learning for classification, in which we are given a set of m training data given as

$$(\mathbf{x}_i, y_i), i = 1, 2, \ldots, m. \tag{9.63}$$

In equation 9.63, we assume that each $\mathbf{x}_i$ is an n-dimensional feature vector, describing the data point, and their corresponding y_i has the value of either +1 or −1. This, in turn, means that each feature vector belongs to two classes: (1) those belonging to y_i values of +1 and (2) those belonging to y_i values of −1. Our objective is to train the algorithm using these given m data points. After successful training, we want to predict these two classes from other data points, which were not used for training. One may wonder, if each n-dimensional feature vector can only take two values, then why do we want this classification? The reason is the fact that we would typically use this classification to recognize the existence of certain patterns from images. Taking seismic as an example, seismic interpreters may want to identify the locations of faults from 3D depth images. Identifying these faults is of extreme importance in all aspects of seismic exploration, and it is impossible for an interpreter or even a team of interpreters to manually identify all of them from large data volumes. So, we want to generate a set of training dataset, in which the feature vectors $\mathbf{x}_i$ are portions of depth images and their corresponding y_i belong to two classes, (1) fault and (2) no-fault. We may assign +1 for the fault class and −1 to the no-fault class. Although we can prepare these training data with a large number of images, some with fault and some with no-fault and label them as +1 or −1, note that no two images in such are going to be same, and such hard labeling will lead to overfitting which will learn to fit the training data perfectly but poorly perform on other data. So, to recognize these two classes, we must define a decision rule in a way that these two classes form two different clusters of points, and we train the system to differentiate one cluster from the other.

A decision rule like above, once established to distinguish between two classes, can also be generalized for multiple classes. Figure 9.27, for example, shows the probability of three lithology classes, (1) oil shale, (2) silty mudstone, and (3) dolostone from a reservoir characterization case study from Northeast China, reported by Yuan et al.

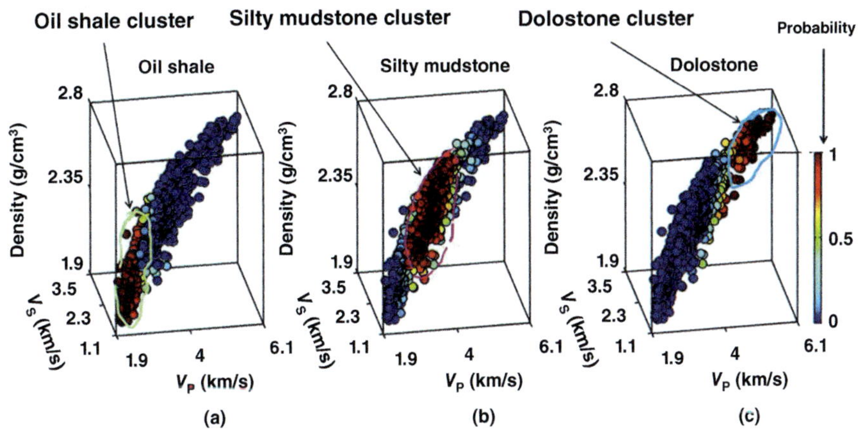

Figure 9.27 Probability of (a) oil shale, (b) silty mudstone, and (c) dolostone in a three-dimensional cross-plot with V_P and V_S as horizontal axes and density as the vertical axis. The plotted points are in the probability color scale, shown at the far right. Yuan et al. (2016)/Society of Exploration Geophysicists.

(2016). Three-dimensional cross-plots with the P- and S-wave velocity (V_P and V_S) as the horizontal axes, density (ρ) as the vertical axis, and the color-coded probabilities for each lithology type are shown in this figure. It is evident that the highly probable regions of oil shale, silty mudstone, and dolostone respectively cluster at the bottom left corner, middle, and upper right corner of these cross-plots. These clusters are also circled and marked in light green in Figure 9.27a, in pink in Figure 9.27b, and in cyan in Figure 9.27c. Therefore, if we can define a learning rule to identify these different regions, we then have an efficient method to predict these three different lithologies from the inverted volumes of V_P, V_S, and ρ without any human intervention. Going beyond V_P, V_S, and ρ, we may even add additional geophysical data such gravity and electromagnetic, define learning rule in a hyperplane with more than three dimensions to identify clusters, and attempt to predict multiple lithology and different fluid content within them. So, in essence, classification is establishing a learning rule in a multidimensional hyperplane and identifying regions or clusters within this hyperplane and classifying them.

Before discussing further on SVM as a classifier, we must note that the ANNs, especially CNNs, have been extensively and successfully used as a classifier system. In the discussion of ANNs in the previous section, although we have mentioned that they can be used for classification, such as fault/salt identification, denoising, interpretation, etc., we deliberately omitted detailed discussion of their usage as a classifier and emphasized their use to solve seismic inverse problems for three specific reasons:

1) Because the primary objective of this book is computational seismology and optimization, we emphasized how ANNs can be used to solve seismic inverse rather than classification problems.
2) Although ANNs are very good in solving classification problems, the way they work is not very intuitive. On the other hand, it will be seen from the discussion below, the way SVM works as a classifier system can be easily understood.
3) Although discussing SVM as regression rather than classification is important for seismic inversion, it is first important to describe its use as a classifier before we can discuss its use to solve regression problems.

Linearly Separable Data We restrict to the simple case of two classes from where we started and discuss how to develop a decision rule to identify one class from the other. As a simple case for such a decision rule, let us assume that there exists an $(n - 1)$-dimensional surface in the n-dimensional hyperplane, which can be defined as

$$f(\mathbf{x}) = \mathbf{w} \cdot \mathbf{x} + b = 0, \quad (9.64)$$

which completely separates the training data, given by equation 9.63. In other words, this means that all training points with $y_i = +1$ lie on one side of this hyperplane where $f(\mathbf{x}_i) > 0$, and the training points with $y_i = -1$ lie on the other side where $f(\mathbf{x}_i) < 0$. For such simple case, if we can find $\mathbf{w}$, the n-dimensional vector normal to the $(n - 1)$-dimensional hyperplane, and the scalar offset b, then $f(\mathbf{x})$ defined by equation 9.64 is the decision rule for the training data given by equation 9.63.

However, more than one hyperplane can separate linearly separable data. Therefore, to further improve the decision rule, we can pick the hyperplane that has the largest margin, which can be found by maximizing the perpendicular distance to the points that are nearest to the hyperplane on both sides. This, in other words means that given a hyperplane that separates the data, we can always scale the normal vector $\mathbf{w}$ and appropriately adjust the scalar b so that

$$\mathbf{w} \cdot \mathbf{x}_i + b \begin{cases} \geq +1; y_i = +1 \\ \leq = 1; y_i = -1 \end{cases}. \quad (9.65)$$

As shown in Figure 9.28, equation 9.65 represents the parallel bounding hyperplanes separating the data. In SVM, this bounding hyperplane is called the *fat plane* or the *maximum margin SVM*. It can be easily shown that the perpendicular distance between these bounding hyperplanes (that is two times the margin shown in Figure 9.28) is equal to $2(\mathbf{w}.\mathbf{w})^{-1/2}$. In addition, two equations shown in equation 9.65 can be written as a single equation as

$$y_i(\mathbf{w} \cdot \mathbf{x}_i + b) \geq 1. \quad (9.66)$$

Based on above discussions, the maximum margin SVM can be obtained by solving the following *quadratic programming* problem:

$$\left.\begin{array}{l} \text{minimize} \rightarrow \frac{1}{2}\mathbf{w} \cdot \mathbf{w} \\ \text{subject to} \rightarrow y_i(\mathbf{w} \cdot \mathbf{x}_i + b) \geq 0, \forall i = 1, 2, \ldots m \end{array}\right\}. \quad (9.67)$$

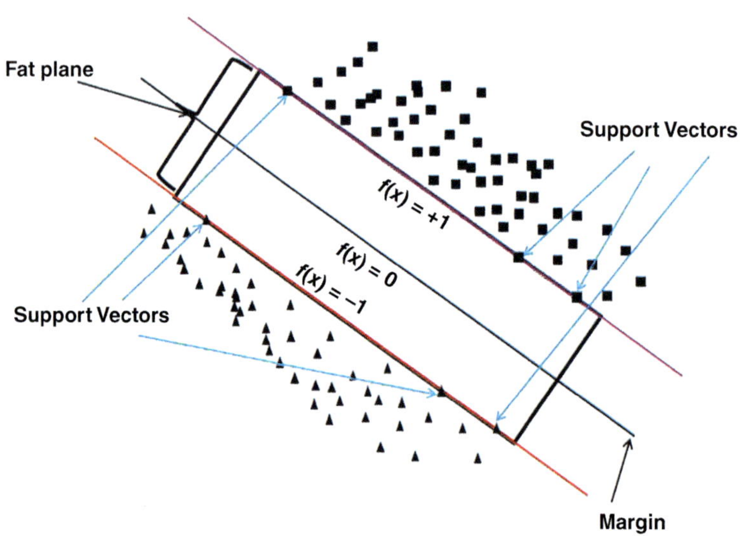

Figure 9.28 Concept of SVM in the context of linearly separable data. The objective is to classify the regions of the plane consisting of black triangles and black squares. The fat plane, defined as $-1 \leq f(\mathbf{x}) \leq +1$, is chosen to maximize the margin as shown. The points lying on the bounding planes are called the support vectors as marked. Press et al. (2007)/with Permission of Cambridge University Press.

The symbol ∀ in equation 9.67 represents "for all."

Before going into further details, note that in solving this problem a few data points must fall exactly on the bounding hyperplane, otherwise the fat plane would be fatter (see Figure 9.28). These data points with $f(\mathbf{x}) = \pm 1$ are called the *support vectors* of the solution, which are marked in Figure 9.28.

Primal and Dual Problems in Quadratic Programming The general or *primal* problem in quadratic programming can be stated as follows:

$$\left.\begin{aligned} \text{minimize} \rightarrow f(\mathbf{w}) \\ \text{subject to} \rightarrow g_j(\mathbf{w}) \leq 0 \\ h_k(\mathbf{w}) = 0 \end{aligned}\right\}. \quad (9.68)$$

In equation 9.68 $f(\mathbf{w})$ is a quadratic form in $\mathbf{w}$; $g(\mathbf{w})$ and $h(\mathbf{w})$ are affine (linear plus a constant) in $\mathbf{w}$; and j and k indices are respectively the sets of inequality and equality constraints. Note that the maximum margin SVM problem (equation 9.67) is a subset of the primary problem shown by equation 9.68 which does not have the equality constraints.

For every primal problem (equation 9.68), there is an associated *dual* problem, which can be considered to be an alternative way to solve the problem. To go from primal to dual, we write a Lagrangian that incorporates both the quadratic form and with Lagrange multipliers, all the constraints as

$$L = \frac{1}{2} f(\mathbf{w}) + \sum_j \alpha_j g_j(\mathbf{w}) + \sum_k \beta_k h_k(\mathbf{w}). \quad (9.69)$$

We then write this subset of conditions for an extremum as

$$\left.\begin{aligned} \frac{\partial L}{\partial w_i} = 0 \\ \frac{\partial L}{\partial \beta_k} = 0 \end{aligned}\right\}, \quad (9.70)$$

and use the resulting equations to eliminate $\mathbf{w}$ from L to write L as a function of the vectors $\boldsymbol{\alpha}$ and $\boldsymbol{\beta}$, $L(\boldsymbol{\alpha}, \boldsymbol{\beta})$, in which α_j and β_k are the components of $\boldsymbol{\alpha}$ and $\boldsymbol{\beta}$. This leads to the following dual problem, which is equivalent to the original primal problem (equation 9.68):

$$\left.\begin{aligned} \text{maximize} \rightarrow L(\alpha, \beta) \\ \text{subject to} \rightarrow \alpha_j \leq 0, \forall j \end{aligned}\right\}. \quad (9.71)$$

The derivation of equation 9.71 follows from the *strong duality* and *Kuhn–Tucker* theorems, details of which can be found in Press et al. (2007), Chapter 10.

It can be shown that the result shown by equation 9.71 is approximately true for any convex $f(\mathbf{w})$. In addition, if $\hat{\mathbf{w}}$ is the optimal solution to the primal problem (equation 9.68) and $\hat{\boldsymbol{\alpha}},\hat{\boldsymbol{\beta}}$ are the solutions to the corresponding dual problem (equation 9.71), then it can also be shown that

$$\left.\begin{aligned} f(\hat{\mathbf{w}}) = L(\hat{\boldsymbol{\alpha}}, \hat{\boldsymbol{\beta}}) \\ \hat{\alpha}_j g_j(\hat{\mathbf{w}}) = 0, \forall j \end{aligned}\right\}. \quad (9.72)$$

The second equation in equation set 9.72, saying that at least one of $\hat{\alpha}_j$ and $g_j(\hat{\mathbf{w}})$ must be zero for each j, is called the *Kraush–Kuhn–Tucker complimentary condition.*

Dual Formulation of the Maximum Margin SVM The procedure of the last subsection can easily be extended to the quadratic programming of the maximum margin SVM (equation 9.67). Note that unlike the primal form shown in equation 9.68, the primal form of the maximum margin SVM shown in equation 9.67 does not have the third (equality) condition $h_k(\mathbf{w}) = 0$. Consequently, there are no β_k's. Therefore, the Lagrangian becomes

$$L = \frac{1}{2} \mathbf{w} \cdot \mathbf{w} + \sum_{i=1}^{m} \alpha_i [1 - y_i(\mathbf{w} \cdot \mathbf{x}_i + b)]. \quad (9.73)$$

With the corresponding conditions for the extremum, we obtain

$$\left.\begin{aligned} 0 = \frac{\partial L}{\partial \mathbf{w}} = \mathbf{w} - \sum_{i=1}^{m} \alpha_i y_i \mathbf{x}_i; \\ \hat{\mathbf{w}} = \sum_{i=1}^{m} \hat{\alpha}_i y_i \mathbf{x}_i \end{aligned}\right\} \quad (9.74)$$

and

$$\frac{\partial L}{\partial b} = \sum_{i=1}^{m} \alpha_i y_i = 0. \quad (9.75)$$

We now substitute equations 9.74 and 9.75 into equation 9.73 and write

$$L(\boldsymbol{\alpha}) = \sum_{j=1}^{m} \alpha_j - \frac{1}{2}\sum_{j=1}^{m} \alpha_j y_j \sum_{k=1}^{m} (\mathbf{x}_j \cdot \mathbf{x}_k) y_k \alpha_k$$
$$= \mathbf{e} \cdot \boldsymbol{\alpha} - \frac{1}{2}\boldsymbol{\alpha} \cdot \text{diag}(\mathbf{y}) \cdot \mathbf{G} \cdot \text{diag}(\mathbf{y}) \cdot \boldsymbol{\alpha}. \tag{9.76}$$

In the far right-hand side of equation 9.76, some convenient matrix and vector notations are introduced as follows:

1) The vector $\mathbf{e}$ is an m-dimensional vector whose all components are 1, i.e., $e_i = 1$, $i = 1, 2, \ldots, m$.
2) diag($\mathbf{y}$) means an $m \times m$ diagonal matrix, formed from the vector $\mathbf{y}$.
3) $\mathbf{G}$ is the Gram matrix of dot products of all $\mathbf{x}_j$, i.e., $G_{ij} = \mathbf{x}_i \cdot \mathbf{x}_j$. Note that subscripts of $\mathbf{x}$ are the indices of which data points are referenced, not the rows and columns of the matrix $\mathbf{G}$. Also note that this matrix is of size $n \times n$ with n being the dimensionality of the feature vector $\mathbf{x}$.

From the above discussion, the dual problem becomes

$$\left.\begin{aligned} \text{minimize} \rightarrow \frac{1}{2}\boldsymbol{\alpha} \cdot \text{diag}(\mathbf{y}) \cdot \mathbf{G} \cdot \text{diag}(\mathbf{y}) \cdot \boldsymbol{\alpha} - \mathbf{e} \cdot \boldsymbol{\alpha} \\ \text{subject to} \rightarrow \alpha_i > 0, \forall i \\ \boldsymbol{\alpha} \cdot \mathbf{y} = 0 \end{aligned}\right\}. \tag{9.77}$$

Additionally, we also have the Karush–Kuhn–Tucker relation

$$\hat{\alpha}_i[y_i(\hat{\mathbf{w}} \cdot \mathbf{x}_i + b) - 1] = 0. \tag{9.78}$$

Equation 9.74 shows the way to obtain the optimal solution $\hat{\mathbf{w}}$ of the primal problem from the optimal solution $\hat{\boldsymbol{\alpha}}$ of the corresponding dual problem (equation 9.77). Once $\hat{\mathbf{w}}$ is found, we can then use equation 9.78 for the optimal solution $\hat{b}$. To do so, we can just find any nonzero α_i and the corresponding y_i, $\mathbf{x}_i$, and $\hat{\mathbf{w}}$, and solve equation 9.78 to solve for $\hat{b}$. Alternatively, we can choose to average out the roundoff error by taking a weighted average

$$\hat{b} = \frac{\sum_{i=1}^{m} \hat{\alpha}_i (y_i - \hat{\mathbf{w}} \cdot \mathbf{x}_i)}{\sum_{i=1}^{m} \alpha_i}. \tag{9.79}$$

Once we find $\hat{\mathbf{w}}$ and $\hat{b}$, we are then all set to use the decision rule $f(\mathbf{x}) = \hat{\mathbf{w}} \cdot \mathbf{x} + \hat{b}$.

Before proceeding further, we must note the following important points:

1) All data points with nonzero $\hat{\alpha}_i$ satisfy constraints as equalities, not as inequalities; therefore, they are support vectors.
2) The only place the data $\mathbf{x}_i$ appear in the dual formulation (equation 9.77) is the matrix $\mathbf{G}$.
3) The only place in the entire calculation where n, the dimensionality of the feature vector, comes into play is in the computation of $\mathbf{G}$, the Gram matrix. And for the rest other parts of the calculation, the number of data points (m) is needed.

So, going from primal to dual is equivalent to mapping the problem that approximately scales the dimensionality n of the feature matrix onto the dimensionality m of the training data points. It may sound strange for the case where $m \gg n$. However, when $m \ll n$, meaning the situation where the dimensionality of the training data is much smaller than the dimensionality of the feature vectors, SVM may outperform ANNs.

Soft-Margin SVM and Its Dual In our above discussions, we assumed the existence of a hyperplane that completely separates the training data (Figure 9.28). However, in practice, the situation may not be so ideal. Take the example illustrated in the 3D cross-plot of $V_P - V_S - \rho$, as shown in Figure 9.27. Although we can identify three different lithologies in this figure, they are not completely separable from one another and there is overlap in-between each lithology class. For the simpler two-class problem we are dealing here, an equivalent situation is illustrated in Figure 9.29, where there is no complete separation between the two classes, and yet we must identify a zone with least overlap and the zones belonging to each cluster (cluster 1 and cluster 2).

The problem illustrated in Figure 9.29 is called the *soft-margin SVM*, which requires

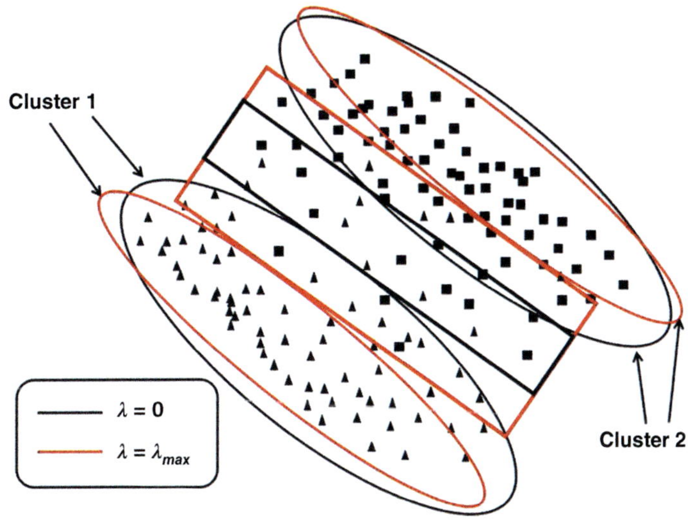

Figure 9.29 Illustration of the soft-margin SVM. Note that there is no clear separation between the two classes. We must, however, find a fat plane with the least overlap between the classes and identify them as two separate clusters, cluster 1 and cluster 2. The trade-off curves for $\lambda = 0$ and $\lambda = \lambda_{max}$ are shown respectively in red and black.

generalization of the concepts we just discussed. This generalization can be made using a slack variable, say ξ_i, such that for each data point $\mathbf{x}_i$ if the data point can be separated by a fat plane (maximum margin SVM), we then set $\xi_i = 0$, otherwise we set $\xi_i > 0$, as a measure of the amount of discrepancy. Based on this slack variable, we now modify the original inequality (equation 9.66) to

$$y_i(\mathbf{w} \cdot \mathbf{x}_i + b) \geq 1 - \xi_i. \tag{9.80}$$

In our optimization, we must make ξ_i as small as possible and preferably zero when possible. In our problem, we therefore have the trade-off between making ξ's small and making the fat plane fat. This leads our problem to not only optimization but also regularization. Therefore, to our Lagrangian L (equations 9.73 and 9.76), we need to add a regularization term like we regularized the objective in Chapters 7 and 8. With this regularization term, our primal problem (equation 9.67) would thus change to

$$\left.\begin{aligned} \text{minimize} \rightarrow \frac{1}{2}\mathbf{w} \cdot \mathbf{w} + \lambda \sum_{i=1}^{m} \xi_i \\ \text{subject to} \rightarrow \xi_i \geq 0 \\ y_i(\mathbf{w} \cdot \mathbf{x}_i + b) \geq 1 - \xi_i, i = 1, 2, \ldots, m \end{aligned}\right\}. \tag{9.81}$$

In equation 9.81, λ is a positive number, varying which within a range, say $0 < \lambda < \lambda_{max}$, allows exploring the regularization trade-off curve. Equation 9.81 shows the 1-norm soft-margin SVM. A variant of this 1-norm soft-margin SVM is the 2-norm soft-margin SVM where the regularization term is replaced by $\lambda \sum \xi_i^2$. We only discuss the 1-norm here.

The trade-off curve obtained varying $0 < \lambda < \lambda_{max}$, we can find a range of solutions to the problem. On one extreme of this range ($\lambda = 0$), we find the solution that prefers a real fat plane, irrespective of how many points are inside the plane or on the wrong side of it. In Figure 9.29, this extreme is illustrated in red with the fat plane shown as a square and the resulting clusters in ovals. Note that it makes the fat plane as fat as possible, so the resulting clusters do not contain any points from the other class. On the other extreme for $\lambda = \lambda_{max}$, illustrated in black in Figure 9.29, the solution does not allow many discrepancies and settles for a narrow flat plane. Clearly, as $\lambda_{max} \rightarrow \infty$, the resulting fat plane collapses to a line. The result from a low value of λ may not work well on training data but may be more robust on new data (i.e., the data not used for training) than the result from a high value of λ. We discuss below on how to optimally choose λ. However, its choice is problem dependent.

Interestingly, the dual of the 1-norm soft-margin SVM is not very different from the dual of the maximum margin SVM (equation 9.77) and can be written as

$$\left.\begin{array}{r}\text{minimize} \rightarrow \frac{1}{2}\boldsymbol{\alpha}\cdot \text{diag}(\mathbf{y})\cdot \mathbf{G}\cdot \text{diag}(\mathbf{y})\cdot \boldsymbol{\alpha} - \mathbf{e}\cdot\boldsymbol{\alpha} \\ \text{subject to} \rightarrow 0 \le \alpha_i \le \lambda, \forall i \\ \boldsymbol{\alpha}\cdot\mathbf{y} = 0\end{array}\right\}. \tag{9.82}$$

Only way equation 9.82 differs from equation 9.77 is its constrain on the α_i's. The formula for $\widehat{\mathbf{w}}$ remains the same as in equation 9.74, and the Karush–Kuhn–Tucker conditions become

$$\left.\begin{array}{r}(\widehat{\alpha}_i - \lambda)\widehat{\xi}_i = 0 \\ \widehat{\alpha}_i\{y_i(\widehat{\mathbf{w}}\cdot\mathbf{x}_i - \widehat{b}) - 1 + \xi_i\} = 0\end{array}\right\}. \tag{9.83}$$

It follows that for $\widehat{\alpha}_i = 0$ the data point i is on the correct side of the fat plane, for $0 < \widehat{\alpha}_i < \lambda$ the data point i is a support vector lying on the fat plane boundary, and for $\widehat{\alpha}_i = \lambda$ the data point i is inside or on the wrong side of the fat plane. Reducing λ toward zero would pin more α_i's to the value of λ, and we obtain increasing number of wrong points and fatter fat planes as illustrated in Figure 9.29.

The average estimator for $\widehat{b}$ is similar to equation 9.79 and is given by

$$\widehat{b} = \frac{\sum_{i=1}^{m}\widehat{\alpha}_i(\lambda - \widehat{\alpha}_i)(y_i - \widehat{\mathbf{w}}\cdot\mathbf{x}_i)}{\sum_{i=1}^{m}\widehat{\alpha}_i(\lambda - \widehat{\alpha}_i)}. \tag{9.84}$$

Nonlinearly Separable Data Points and the Kernel Trick Maximum margin SVM and its soft-margin variation can be used if the data points are (approximately) linearly separable. But what happens if they are not? Taking the three-class illustration shown in Figure 9.27, for example, we can clearly see that not only there are overlaps between the three lithology classes, but also these classes are not linearly separable. Such a nonlinear behavior is, in general, true for all classification problems of practical interest. The kernel trick gives the generalization to SVM that not only allows handling such data points for classification, but it also allows extension of SVM as the SVR.

To understand how to extend linearly separable data into a nonlinear regime, let us recall the functionality of the ANNs, discussed in the previous section. Irrespective of the network architecture, the input vector ($\mathbf{x}$) is mapped onto the output ($\mathbf{y}$) vector via a series of linear functions with weights, activation, and bias through the hidden layers. The relation between $\mathbf{x}$ and $\mathbf{y}$ can be nonlinear of any arbitrary order, but this nonlinear relation is approximated by a series of linear relations. We also discussed the universal approximation theorem to justify the validity of such mapping. However, mathematically speaking, this assumes the original functional relationship $\mathbf{x}\leftrightarrow\mathbf{y}$, when mapped onto a higher-dimensional space, we can use linear relations in this dimension, which approximates any nonlinear relation in the original dimension. In other words, we assume the existence of an embedding function $\boldsymbol{\varphi}$ that maps the n-dimensional feature vector $\mathbf{x}$ in the original space into a much higher N-dimensional embedding space, i.e.,

$$\mathbf{x} \rightarrow (n\text{-dimensional}) \rightarrow \boldsymbol{\varphi}(\mathbf{x}) \rightarrow (N\text{-dimensional}), \tag{9.85}$$

where the relationships are linear, even if they are nonlinear in the original n-dimensional space. And, higher are the number of hidden layers in the network design, higher is dimensionality of such mapping from $n \rightarrow N$.

Figure 9.30 illustrates the basic idea behind this mapping onto higher dimensions. A nonlinear separating surface in a two-dimensional original space shown on left can map onto linear planes in a 3D embedding space, shown on right.

As an example of this mapping to higher dimensions, consider mapping from two to five dimensions:

$$(x_0, x_1) \rightarrow \boldsymbol{\varphi} \rightarrow (x_0^2, x_0x_1, x_1^2, x_0, x_1). \tag{9.86}$$

If a two-to-five-dimensional mapping shown by equation 9.86 maps the original decision rule $f(\mathbf{x})$ as linear in the embedding space, it would then include all linear and quadratic forms such as lines, ellipses, hyperbolas, etc., in the original feature space. In other words,

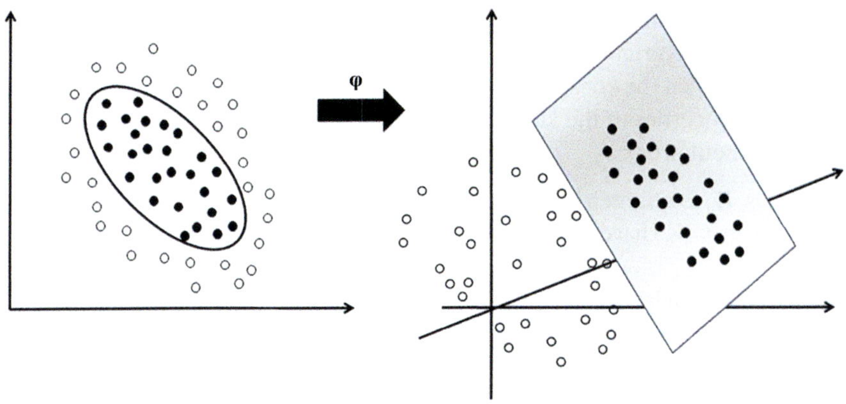

Figure 9.30 Concept of higher-dimensional mapping. When feature vectors in the two-dimensional original space (left) are mapped onto a three-dimensional embedding space (right), nonlinear separation surfaces can become approximated by linear ones.

we have

$$f(\mathbf{x}) = F[\boldsymbol{\varphi}(x)] = \mathbf{W} \cdot \boldsymbol{\varphi}(\mathbf{x}) + B, \qquad (9.87)$$

in which the uppercase letters indicate quantities in the embedding space. In the example given above, $N = 5$. There is, however, no limit to the value of N and we can make N to million or even billion. We see how to do such mapping in much higher-dimensional embedding space from a lower-dimensional original space. But the bottom line of the idea is if we can do such a successful mapping onto much higher-dimensional embedding space, then a linear decision rule, given by equation 9.87 in the embedding space, will include all nonlinearity in the original space.

Now, given our data, how do we find **W** and B in the embedding space? Well, because the decision rule is linear in the embedding space, if we use 1-norm soft-margin SVM, we can easily find the primal problem from equation 9.81 as

$$\left.\begin{array}{r} \text{minimize} \rightarrow \dfrac{1}{2}\mathbf{W} \cdot \mathbf{W} + \Gamma \displaystyle\sum_{i=1}^{m} \Xi_i \\ \text{subject to} \rightarrow \Xi_i \geq 0 \\ y_i(\mathbf{W} \cdot \boldsymbol{\varphi}(\mathbf{x}_i) + B) \geq 1 - \Xi_i, i = 1, 2, \ldots, m \end{array}\right\}. \qquad (9.88)$$

Following our conventions of the embedding space, in equation 9.88, we used the uppercase Greek letters Γ and Ξ_i instead of their lowercase equivalents λ and ξ_i, used in equation 9.81.

The quadratic problem, shown by equation 9.88, is in much higher (millions or billions) dimensional embedding space and is not solvable even using the fastest available computers. Following the steps as before, the dual problem of equation 9.88 however, turns out to be (see equation 9.82)

$$\left.\begin{array}{r} \text{minimize} \rightarrow \dfrac{1}{2}\boldsymbol{\alpha} \cdot \text{diag}(\mathbf{y}) \cdot \mathbf{K} \cdot \text{diag}(\mathbf{y}) \cdot \boldsymbol{\alpha} - \mathbf{e} \cdot \boldsymbol{\alpha} \\ \text{subject to} \rightarrow 0 \leq \alpha_i \leq \Gamma, \forall i \\ \boldsymbol{\alpha} \cdot \mathbf{y} = 0 \end{array}\right\}. \qquad (9.89)$$

Equation 9.89 is the same as equation 9.82, except that the Gram matrix G_{ij} is replaced by the *kernel* K_{ij} and is given as

$$K_{ij} = K(x_i, x_j) = \boldsymbol{\varphi}(\mathbf{x}_i) \cdot \boldsymbol{\varphi}(\mathbf{x}_i). \qquad (9.90)$$

Quadratic problem, given by equation 9.89, is therefore the same as the original problem of equation 9.82 as they both are in m-dimensional space, where m is the number of data points and they both require a matrix of size $n \times n$, the Gram matrix $G_{ij} = \mathbf{x}_i \cdot \mathbf{x}_j$ in equation 9.82, and the Kernel $K_{ij} = \boldsymbol{\varphi}(\mathbf{x}_i). \boldsymbol{\varphi}(\mathbf{x}_j)$ in equation 9.89.

In essence, we never need to know the mapping $\boldsymbol{\varphi}(\mathbf{x})$—all we need to know is how to compute the kernel **K** that could come from some mapping $\boldsymbol{\varphi}(\mathbf{x})$ onto a higher-dimensional embedded space from the original space where the behavior of the feature vectors are linear. Once we find such a kernel, equation 9.89

will incorporate any nonlinear behavior of the feature vectors in the original space. This procedure is what is called the *kernel trick*.

Some important properties of the kernel K_{ij} are listed below without proof:

1) $K_{ij} = K_{ji}$, that is, K_{ij} is symmetric in i and j.
2) The kernel matrix $\mathbf{K}$ have nonnegative eigenvalues.
3) Combining kernel functions by multiplication, addition, and scaling is also a kernel function.
4) $K[\boldsymbol{\varphi}(\mathbf{x}_i).\ \boldsymbol{\varphi}(\mathbf{x}_j)]$ is a kernel function if $K(,)$ is one for any $\boldsymbol{\varphi}$. This property generalizes the original idea of the embedding space.
5) For any function g, $g(\mathbf{x}_i)g(\mathbf{x}_j)$ is always a kernel.

There are many kernels that are in use in SVM. Some popular ones are given below:

$$\left.\begin{aligned} &\text{Linear} \rightarrow K(\mathbf{x}_i, \mathbf{x}_j) = \mathbf{x}_i \cdot \mathbf{x}_j \\ &\text{Power} \rightarrow K(\mathbf{x}_i, \mathbf{x}_j) = (\mathbf{x}_i \cdot \mathbf{x}_j)^p \\ &\text{Polynomial} \rightarrow K(\mathbf{x}_i, \mathbf{x}_j) = (a\mathbf{x}_i \cdot \mathbf{x}_j + b)^p \\ &\text{Tanh} \rightarrow K(\mathbf{x}_i, \mathbf{x}_j) = \tanh(a\mathbf{x}_i \cdot \mathbf{x}_j + b) \\ &\text{Radial basis function} \rightarrow K(\mathbf{x}_i, \mathbf{x}_j) \\ &\qquad = \exp\left(-\frac{\frac{1}{2}|\mathbf{x}_i - \mathbf{x}_j|^2}{\sigma^2}\right) \end{aligned}\right\}. \tag{9.91}$$

The appropriate choice of kernel is problem dependent. In practice, it is usually necessary to try a few (small) test runs to choose the appropriate kernel. In general, however, the radial basis function kernel is well known to be appropriate for many problems, specifically for SVR.

Once we settle on a kernel and solve equation 9.89, then the final decision rule for any new feature vector $\mathbf{x}$ becomes

$$f(\mathbf{x}) = \sum_i \hat{\alpha}_i y_i K(\mathbf{x}_i, \mathbf{x}) + \hat{b}, \tag{9.92}$$

in which $\hat{b}$ can be obtained by averaging as before, which can be written as

$$\hat{b} = \frac{\sum_i \hat{\alpha}_i(\Gamma - \hat{\alpha}_i)\left[y_i - \sum_j \hat{\alpha}_j y_j K(\mathbf{x}_j, \mathbf{x}_i)\right]}{\sum_i \hat{\alpha}_i(\Gamma - \hat{\alpha}_i)}. \tag{9.93}$$

Thus, the kernel trick is not different from a universal approximator used by the neural networks, where the input and output that may be nonlinearly related are approximated as linear combinations of the weights, activation, and bias in a high-dimensional space via hidden layers.

From Classifier to Regression: The Support Vector Regression

Now that we have discussed SVM as a classifier, we are now ready to discuss its extension as the SVR. In the classifier problem, we assumed that for every feature vector $\mathbf{x}_i \in \Re^n, i = 1, 2, \ldots, m$, there is a scalar output y_i, $i = 1, 2, \ldots, m$. In SVR, on the other hand, the output is a vector $\mathbf{y} \in \mathcal{R}^p, i = 1, 2, \ldots, m$. For seismic inversion under an isotropic assumption for example, each feature vector $\mathbf{x}_i$ is seismic data with $n = N_T N_X$, where N_T and N_X may respectively represent the number of time and offset (or ray-parameter) samples. And, for corresponding output $\mathbf{y}$ for each $\mathbf{x}_i$, $p = 3N_D$ where N_D is the number of depth samples on output. In this output vector, samples $1 - N_D$ represent P-wave velocity (V_P), samples $N_D + 1 - 2N_D$ represent S-wave velocity (V_S), and samples $2N_D + 1 - 3N_D$ represent density (ρ). The primal problem for SVR can thus be modified from equation 9.88 as

$$\left.\begin{aligned} &\text{minimize} \rightarrow \frac{1}{2}\mathbf{W} \cdot \mathbf{W} + \Gamma \sum_{i=1}^{m} (\Xi_i + \Xi_i^*) \\ &\text{subject to} \rightarrow y_i - \mathbf{W} \cdot \boldsymbol{\varphi}(\mathbf{x}_i) - b \le \mathrm{E} + \Xi_i, \\ &\qquad \forall i = 1, 2, \ldots, p \\ &\mathbf{W} \cdot \boldsymbol{\varphi}(\mathbf{x}_i) + b - y_i \le \mathrm{E} + \Xi_j^*, \forall i = 1, 2, \ldots, p \\ &\Xi_i, \Xi_i^* \ge 0, \forall i = 1, 2, \ldots, p \end{aligned}\right\}, \tag{9.94}$$

with the corresponding dual problem being

$$\left.\begin{aligned} &\text{minimize} \rightarrow \frac{1}{2}[\boldsymbol{\alpha} - \boldsymbol{\alpha}^*]^T \mathbf{K}[\boldsymbol{\alpha} - \boldsymbol{\alpha}^*] \\ &\quad + \mathrm{E}[\boldsymbol{\alpha} - \boldsymbol{\alpha}^*] - \mathbf{y}^T[\boldsymbol{\alpha} - \boldsymbol{\alpha}^*] \\ &\text{subject to} \rightarrow \mathbf{e}^T[\boldsymbol{\alpha} - \boldsymbol{\alpha}^*] = 0 \\ &0 \le \alpha_i, \alpha_i^* \le \Gamma, \forall i = 1, 2, \cdots, p \end{aligned}\right\}. \tag{9.95}$$

In the above equation, **e** is a vector of all ones, $\Gamma > 1$ is the upper bound, and **K** is the kernel with $K_{ij} \equiv K(\mathbf{x}_i, \mathbf{x}_j) = \boldsymbol{\varphi}(\mathbf{x}_i) \cdot \boldsymbol{\varphi}(\mathbf{x}_j)$. The input training vectors $\mathbf{x}_i$ are implicitly mapped onto higher-dimensional space by the function $\boldsymbol{\varphi}$, with the cost function $\sum_{i=1}^{p}(\alpha_i - \alpha_i^*)K(x_i, x_j) + P$, where P is a regularization parameter; see Cortes and Vapnik, (1993) and Cristianini and Shawe-Taylor (2000) for further details on SVR.

Geophysical Applications of SVM

SVM has been used in geophysics for noise elimination (Deng et al., 2010), fracture characterization (Li et al., 2018), group to phase mapping for anisotropic medium (Mallick et al., 2019), multiple elimination (Li, 2020), and migration noise suppression (Chen et, al., 2020).

9.4.2 Gradient Boosting

Gradient boosting is an ML method, which, like SVM and neural networks, is used for both regression and classification tasks (Friedman, 2001). This method gives a prediction model in the form of an ensemble of weak prediction models, which are considered as decision trees.

Easiest way to explain gradient boosting is using the least-squares regression example where the primary objective is to teach a model, say F, to predict values from $\hat{y} = F(x)$ by minimizing the sum of the squared errors and given as

$$\varepsilon = \frac{1}{n}\sum_i (\hat{y}_i - y_i)^2, \tag{9.96}$$

in which $\hat{y}_i$ is the predicted value from the model $F(x_i)$, y_i is the target (true) value, and n is the number of sample points. Instead of doing least-squares minimization, let us assume that we make the approximation of the model $\hat{y} = F(x)$ in M steps. Assume that at any step m where $1 \le m \le M$, we have an imperfect model F_m. So, to improve F_m, we must add some new estimator $h_m(x)$ such that $F_{m+1}(x_i) = F_m(x_i) + h_m(x_i) = y_i$ or

$$h_m(x_i) = y_i - F_m(x_i). \tag{9.97}$$

The gradient boosting method fits the correction term h_m to the residual $y_i - F_m(x_i)$ in each of the M steps. Now note that in equation 9.96, $\hat{y}_i = F_m(x_i)$ at any step m. Therefore, for any step m, we may write

$$-\frac{\partial \varepsilon}{\partial F_m(x_i)} = \frac{2}{n}\{y_i - F_m(x_i)\} = \frac{2}{n}h_m(x_i). \tag{9.98}$$

Or, in other words, the correction term h_m is proportional to the negative gradient of the MSE ε, and that is why the method is called gradient boosting.

Extreme Gradient Boosting

Extreme gradient boosting (*XGBoost*, also known as *Random Forest*), developed by Chen and Guestrin (2016), is an ensemble ML method that uses a set of additive functions as above (called trees) to fit the relationship between the input and output. The tree ensemble model can be described as

$$y_i = \varphi(\mathbf{x}_i) = \sum_{k=1}^{K} f_k(\mathbf{x}_i), \tag{9.99}$$

where $\mathbf{x}_i$ represents the input features of the ith sample, f_k corresponds to an independent regression tree that contains a continuous score on each of the leaf, and $\hat{y}_i$ is the predicted output from the ensemble of trees. The learning process used in the set given by equation 9.99 is obtained through minimizing the regularized objective function given as

$$L = \sum_i l(\hat{y}_i, y_i) + \sum_k \Omega(f_k), \tag{9.100}$$

with

$$\Omega(f_k) = \gamma T + \frac{1}{2}\lambda \|\mathbf{W}\|^2. \tag{9.101}$$

The first term on the right-hand side of equation 9.100 measures the difference between the prediction $\hat{y}_i$ and the target y_i, and the second term is the regularization term that penalizes the complexity of the model and is given by equation 9.101. As we see from equation 9.99, the tree ensemble model includes functions as parameters and cannot be trained using traditional optimization methods. To solve the problem, the gradient tree boosting method is used to train the model in an additive manner discussed as above,

$$L^{(t)} = \sum_{i=1}^{n} l(y_i, \hat{y}_i^{(t-1)}) + f_t(\mathbf{x}_i) + \Omega(f_t), \tag{9.102}$$

where $\hat{y}_i^t$ is the prediction of the ith sample at the tth iteration and f_t is the added correction like h_m of equation 9.97 that can mostly improve the model according to equation 9.100. In gradient boosting machines, many (hundreds) regression trees are usually needed to build an accurate model, which can be compute-intense. For this reason, XGBoost is designed with an optimized-distributed gradient boosting library that solves large-scale data science problems in a fast and efficient manner.

Geophysical Applications of Gradient Boosting

Gradient Boosting methods have been used in characterizing unconventional reservoirs (Gupta et al., 2019; Mohamed et al., 2019), group to phase mapping for anisotropic medium (Mallick et al., 2019), and inversion (Zhao et al., 2021; Zou et al., 2023).

9.5 Physics-Informed Machine Learning

Physics-informed learning is a relatively new development in ML. In all methods discussed so far, we assumed the existence of a set of m training (labeled) data set with input data given as $\mathbf{x}_i$, $i = 1, 2, \ldots, n$, and their corresponding target output $\mathbf{y}_j$, $j = 1, 2, \ldots, N$. Via and iterative function approximation $\hat{\mathbf{y}}_j = F_k(\mathbf{x}_i)$, these methods then define the error or mismatch between $\mathbf{y}_j$ and $\hat{\mathbf{y}}_j$ and refine the accuracy of prediction F_k by minimizing the error. In physics-informed learning (Raissi et al., 2017, 2018, 2019, 2020; Biswas et al., 2019; Stefano, 2021; Dwivedi et al., 2021; De Florio et al., 2021; Mishra and Roberto, 2022; Ali and Mukerji, 2022; Ryck et al., 2023; Fu, et al., 2023; among others), labeled data are not used and instead they use the underlying physics of the problem to iteratively approximate F_k for model prediction.

The concept of physics-informed ML is illustrated in Figure 9.31 using an example of a seismic inverse problem. Like any other method, the input seismic data are fed to the algorithm to produce the output model. For

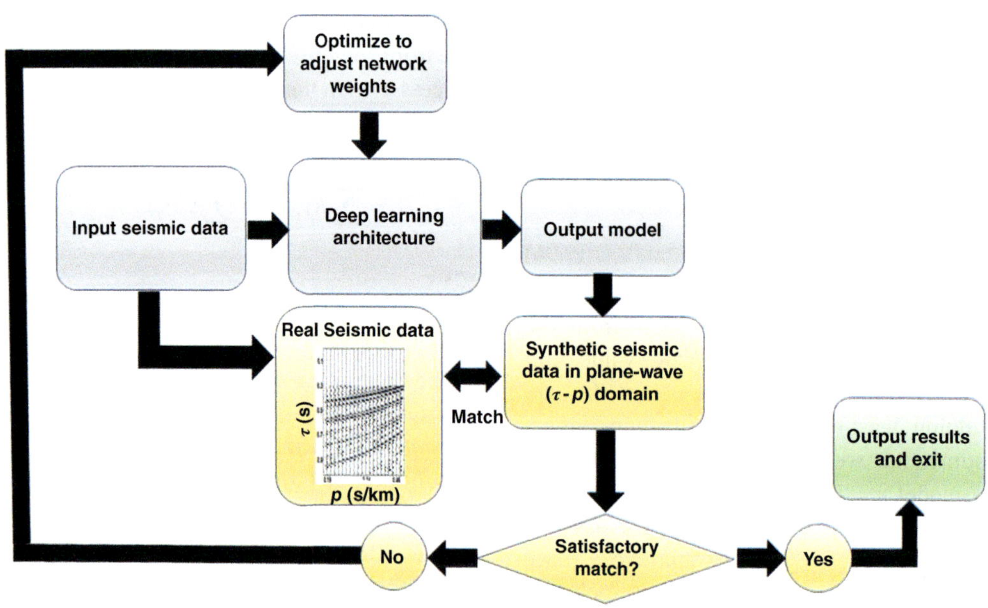

Figure 9.31 Concept of the physics-informed learning illustrated using a seismic inversion example.

isotropic seismic inversion, this output model can be the depth-domain model of V_P, V_S, and ρ or equivalent. Now, instead of matching the output model with the target (true) models, synthetic seismic data are computed using the output model and matched with the real seismic data in an iterative manner for back propagation (see Figure 9.31).

9.6 Multi-task Learning

Multi-task learning (MTL) systems where multiple tasks are jointly solved are equivalent to multi-objective optimization (MOO) methods in ML (Sener and Koltun, 2018). In Chapter 8, we thoroughly discussed MOO and the concept of Pareto-optimality, and the same concepts are used in MTL.

MOO has also been used for MTL in recent years. Igel (2005) provides a comprehensive review for model selection process for SVMs for two conflicting objectives. Multi-objective reinforcement learning (MORL) policy via a gradient approach to learn a continuous approximation of the Pareto fronts in multi-objective Markov decision problems was reviewed by Parisi et al. (2016). Drugan (2015) used MORL paradigm to estimate Pareto front but used a scalar objective and a single-objective optimization. Sener and Koltun (2018) cast MTL as MOO and use a gradient-based method to simultaneously optimize each objective and estimate the Pareto-optimal solutions. A very good overview of different approaches to Pareto-based multi-objective ML including applications of multi-objective evolutionary algorithms is given by Jin and Sendhoff (2008).

MTL is a relatively new concept, and its applications have been limited so far. The primary reason for this is the high computing cost for solving any MTL problem of practical interest. With further improvements in computing technology, it is, however, expected the MTL will be used more in future. We postpone further discussion of MTL to the next chapter (Chapter 10).

9.7 Machine Learning in a Bayesian Framework

It is now obvious from the above discussions that ML is also an optimization process that we discussed in the previous chapters. Consequently, these methods can be cast in a Bayesian framework in the same fashion as discussed in Chapter 8. Here, we omit these discussions and refer to Mukhopadhyay and Mallick (2019), Grana et al. (2020), and Feng et al. (2021) for further details.

9.8 Examples

Machine learning has been widely used in recent years in all fields of geophysics, and the examples are far too many to list them all. Here, we restrict two examples related to seismic inversion.

Our first example, shown in Figure 9.32, is from Wu and McMechan (2019). This is a synthetic acoustic example for estimating V_P. We omit the details of the CNN architecture used and show that the results on two models with the true models in Figure 9.32a,c are the corresponding CNN-estimated models in Figures 9.32b,d.

The second example is a physics-informed CNN on real seismic data from Biswas et al. (2019). Six angle-domain traces from 3° to 28° in 5° intervals were extracted from the offset-domain seismic data, and Figure 9.33 shows one such angle gather.

Low-frequency V_P and V_S models were extracted from six wells and were interpolated over four interpreted horizons on the stacked seismic data to obtain a low-frequency initial model for one arbitrary seismic line with 2669 common midpoint (CMP) locations. Using convolution as the primary modeling engine, a physics-informed CNN was used to train the network using 10% of the CMP locations, randomly chosen from the entire line. After successfully training the network, the adjusted network weights were used to predict V_P and

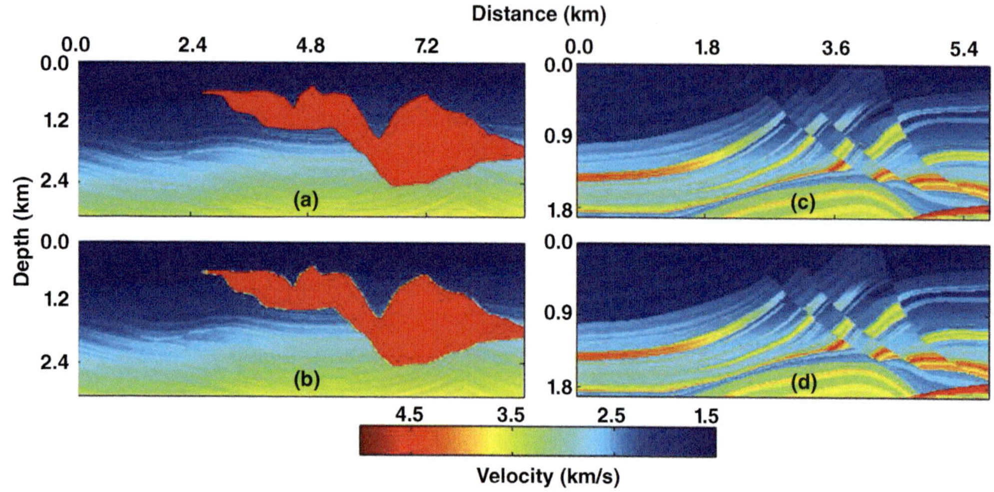

Figure 9.32 CNN example on synthetic seismic data: (a)–(c) True V_p models; (b)–(d) V_p models estimated by CNN. Wu and McMechan (2019)/Society of Exploration Geophysicists.

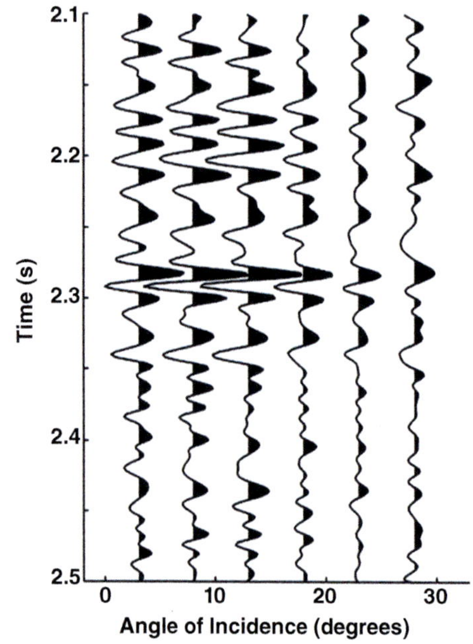

Figure 9.33 Single angle gather used for physics-informed CNN. Biswas et al. (2019)/Society of Exploration Geophysicists.

V_S for the entire line. Figure 9.34 shows the result for this physics-informed CNN on real seismic data.

9.9 Chapter Summary

In this chapter, we discussed the applications of ML methods for seismic inversion problems. We first discussed different variations of ANNs, such as CNN, HNN, RNN, etc., and then moved on to other methods such as SVM and gradient boosting. Finally, we ended the chapter with two examples.

9.10 Exercises

Problem 9.1 Derive equation 9.50.

Problem 9.2 Derive the analytical expression for the Jacobi matrix in MLP network for a general activation and bias (equation 9.16).

Problem 9.3 Repeat problem 9.2 for the tanh (sigmoid) activation (equation 9.58).

Problem 9.4 Repeat problem 9.2 for the ReLU function (equation 9.59).

Problem 9.5 Recall the maxpooling shown in Figure 9.21. Repeat this process with average

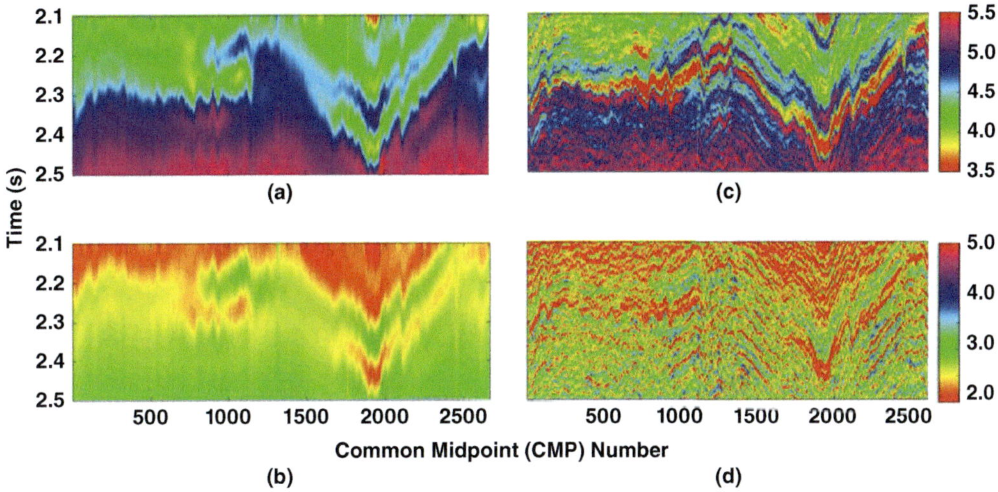

Figure 9.34 Example of the physics-informed CNN on real seismic data. (a, b) Low-frequency V_P and V_S; (c, d) V_P and V_S estimated from the physics-informed CNN. Biswas et al. (2019)/Society of Exploration Geophysicists.

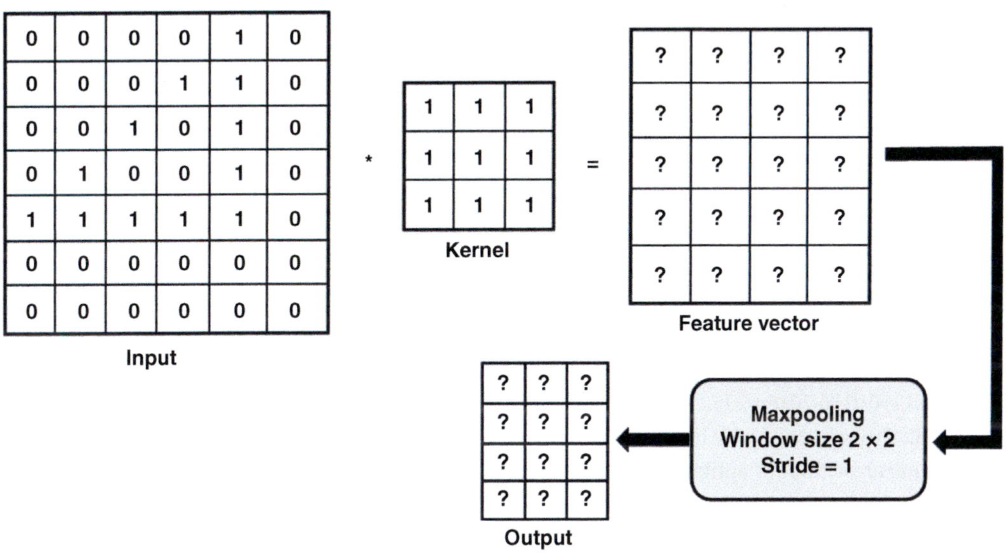

Figure 9.35 Convolution and maxpooling exercise.

pooling and sum pooling and write down the output matrices for each.

Problem 9.6 Repeat the maxpooling exercise shown in Figure 9.21 with stride = 1.

Problem 9.7 Refer to Figure 9.35.

The input and the kernel matrices are shown. Compute the feature vector matrix and the output matrix after maxpooling with a window size of 2×2 and stride = 1.

References

Alali, A., Kazei, V., Sun, B., and Alkhalifah, T., 2022, Time-lapse data matching using a recurrent neural network approach, *Geophysics*, *87*, V405–V417, https://doi.org/10.1190/GEO2021-0487.1.

Aleardi, M., and Salusti, A., 2021, Elastic prestack seismic inversion through discrete cosine transform reparameterization and convolutional neural network, *Geophysics*, *86*, R129–R146, https://doi.org/10.1190/GEO2020-0313.1.

Ali, K., and Mukerji, T., 2022, Physics-informed PointNet: A deep learning solver for steady-state incompressible flows and thermal fields on multiple sets of irregular geometries, *Journal of Computational Physics*, *468*, 111510, https://doi.org/10.1016/j.jcp.2022.111510.

Almuteri, K., and Sava, P., 2023, Seismic deghosting using convolutional neural networks, *Geophysics*, *88*, V113–V125, https://doi.org/10.1190/GEO2021-0521.1.

Birnie, C., and Hansteen, F., 2022, Bidirectional recurrent neural networks for seismic event detection, *Geophysics*, *87*, KS97–KS111, https://doi.org/10.1190/GEO2020-0806.1.

Biswas, R., Sen, M.K., Das, V., and Mukerji, T., 2019, Prestack and poststack inversion using a physics-guided convolutional neural network, *Interpretation*, *7*, SE161–SE174, https://doi.org/10.1190/INT-2018-0236.1.

Boser, B.E., Guyon, I.M., and Vapnik, V.N., 1992, A training algorithm for optimal margin classifiers, Proceedings of the fifth annual workshop on Computational learning theory – COLT '92. p. 144, https://doi.org/10.1145/130385.130401.

Calderón-Macías, C., Sen, M.K., and Stiffa, P.L., 1997, Hopfield neural networks, and mean field annealing for seismic deconvolution and multiple attenuation, *Geophysics*, *62*, 992–1002, https://doi.org/10.1190/1.1444205.

Chen, T. and Guestrin, C., 2016, Xgboost: A scalable tree boosting system, Proceedings of the 22nd ACM SIGKKD international conference on knowledge discovery and data mining, 785–794, https://doi.org/10.1145/2939672.2939785.

Chen, Y., Huang, Y., and Huang, L., 2020, Suppressing migration image artifacts using a support vector machine method, *Geophysics*, *85*, S255–S268, https://doi,org/10.1190/GEO2019-0157.1.

Cristianini, N., and Shawe-Taylor, J., 2000, *An Introduction to Support Vector Machines and Other Kernel-based Learning Methods*, Cambridge University Press, ISBN 0-521-78019-5.

Cortes, C., and Vapnik, V.N., 1993, Support-vector networks, *Machine Learning*, *20*(3), 273–297, https://doi.org/10.1007/BF00994018.

Cui, R., Cao, D., Liu, Q., Zhu, Z., and Jia, Y., 2021, V_P and V_S prediction from digital rock images using a combination of U-Net and convolutional neural networks, *Geophysics*, *86*, MR27–MR37, https://doi.org/10.1190/GEO2020-0162.1.

Cybenko, G., 1989, Approximation by superpositions of a sigmoidal function, *Mathematics of Control, Signals, and Systems*, *2*(4), 303–314, https://doi.org/10.1007/BF02551274.

Das, V., and Mukerji, T., 2020, Petrophysical properties prediction from prestack seismic data using convolutional neural networks, *Geophysics*, *85*, N41–N55, https://doi.org/10.1190/GEO2019-0650.1.

Das, V., Pollack, A., Wollner, U., and Mukerji, T., 2019, Convolutional neural network for seismic impedance inversion, *Geophysics*, *84*, R869–R880, https://doi.org/10.1190/GEO2018-0838.1.

De Florio, M., Schiassi, E., Ganapol, B.D., and Roberto, 2021, Physics-informed neural networks for rarefied-gas dynamics: Thermal creep flow in the Bhatnagar–Gross–Krook approximation, *Physics of Fluids*, *33*(4), 047110, https://doi.org/10.1063/5.0046181.

Deng, X., Yang, D., Peng, J., Guan, X., and Yang, B., 2010, Noise reduction and drift removal using least-squares support vector regression with the implicit bias term, *Geophysics*, *75*, V119–V127, https://doi.org/10.1190/1.3506602.

Dhara, A., and Bagaini, C., 2020, Seismic image registration using multiscale convolutional neural networks, *Geophysics*, *85*, V425–V441, https://doi.org/10.1190/GEO2019-0724.1.

Di, H., Li, C., Smith, S., Li, Z., and Abubakar, A., 2021, Imposing interpretational constraints on a seismic interpretation convolutional neural network, *Geophysics*, *86*, IM63–IM71, https://doi.org/10.1190/GEO2020-0449.1.

Di, H., Li, Z., and Abubakar, A., 2022, Using relative geologic time to constrain convolutional neural network-based seismic interpretation and property estimation, *Geophysics*, *87*, IM25–IM35, https://doi.org/10.1190/GEO2021-0257.1.

Di, H., Li, Z., Maniar, H., and Abubakar, A., 2020, Seismic stratigraphy interpretation by deep convolutional neural networks: A semisupervised workflow, *Geophysics*, *85*, WA77–WA86, https://doi.org/10.1190/GEO2019-0433.

Di, W., Bhardwaj, A., and Wei, J., 2018, *Deep Learning Essentials*, Packt Publishing, ISBN 978-1-78588-036-0.

Dorrington, K.P., and Link, C.A., 2004, Genetic-algorithm/neural-network approach to seismic attribute selection for well-log prediction, *Geophysics*, *69*, 212–221, https://doi.org/10.1190/1.1649389.

Drugan, M.M., 2015, Multi-objective optimization perspectives on reinforcement learning algorithms using reward vectors, ESANN 2015 proceedings, European Symposium on Artificial Neural Networks, Computational Intelligence and Machine Learning. Bruges (Belgium), 22–24 April 2015, http://www.i6doc.com/en/.

Dwivedi, V., Parashar, N., and Srinivasan, B., 2021, Distributed learning machines for solving forward and inverse problems in partial differential equations, *Neurocomputing*, *420*, 299–316, https://doi.org/10.1016/j.neucom.2020.09.006.

Fabien-Ouellet, G., and Sarkar, R., 2020, Seismic velocity estimation: A deep recurrent neural-network approach, *Geophysics*, *85*, U21–U29, https://doi.org/10.1190/GEO2018-0786.1.

Feng, R., Grana, D., and Balling, N., 2021, Uncertainty quantification in fault detection using convolutional neural networks, *Geophysics*, *86*, M41–M48, https://doi.org/10.1190/GEO2020-0424.1.

Friedman, J.H., 2001, Greedy function approximation: A gradient boosting machine, *Annals of Statistics*, *29*, 1189–1232, http://www.jstor.org/stable/2699986.

Fu, J., Xiao, D., Fu, R., Li, C., Zhu, C., Arcucci, R., and Navon, I.M., 2023, Physics-data combined machine learning for parametric reduced-order modelling of nonlinear dynamical systems in small-data regimes, *Computer Methods in Applied Mechanics and Engineering*, *404*, 115771, https://doi.org/10.1016/j.cma.2022.115771.

Fukushima, K., 1969, Visual feature extraction by a multilayered network of analog threshold elements, *IEEE Transactions on Systems Science and Cybernetics*, *5*, 322–333, https://doi.org/10.1109/TSSC.1969.300225.

Fukushima, K., 1980, Neocognitron: A self-organizing neural network model for a mechanism of pattern recognition unaffected by shift in position, *Biological Cybernetics*, *36*, 193–202, https://doi.org/10.1007/BF00344251.

Fukushima, K., and Miyake, S., 1982, Neocognitron: A self-organizing neural network model for a mechanism of visual pattern recognition, in S. Amari, and M.A. Arbib, eds., *Competition and Cooperation in Neural Nets*, Lecture Notes in Biomathematics, Vol. *45*, Springer, Berlin, Heidelberg https://doi.org/10.1007/978-3-642-46466-9_18.

Funahashi, K.-I., 1989, On the approximate realization of continuous mappings by neural networks, *Neural Networks*, *2*(3), 183–192, https://doi.org/10.1016/0893-6080(89)90003-8.

Gao, L., Huang, L., Zheng, Y., Lin, R., Hu, H., and Cladohous, T., 2022, Automatic fault detection on seismic images using a multiscale attention convolutional neural network, *Geophysics*, *87*, N13–N29, https://doi.org/10.1190/GEO2020-0945.1.

Goodfellow, I., Bengio, Y., and Courville, A., 2015, *Deep Learning*, MIT Press, ISBN 978-0-26203-356-56-13.

Grana, D., Azevedo, L., and Liu, M., 2020, A comparison of deep machine learning and Monte Carlo methods for facies classification from seismic data, *Geophysics*, *85*, WA41–WA52, https://doi.org/10.1190/GEO2019-0405.1.

Gupta, I., Devegowda, D., Jayaram, V., Rai, C., and Sondergeld, C., 2019, Machine Learning Regressors and their Metrics to predict Synthetic Sonic and Brittle Zones, Unconventional Resources Technology Conference (URTeC), https://doi.org/10.15530/urtec-2019-0148.

Günaydın, K., and Günaydın, A., 2008, Peak ground acceleration prediction by artificial neural networks for Northwestern Turkey, *Mathematical Problems in Engineering, 2008*, 919420, https://doi.org/10.1155/2008/919420.

Hajian, A., and Styles, P., 2018, *Application of Soft Computing and Intelligent Methods in Geophysics*, Springer, ISBN 978-3319665313.

Hajian, A., Zomorrodian, H., and Styles, P., 2012, Simultaneous estimation of shape factor and depth of subsurface cavities from residual gravity anomalies using feed-forward back-propagation neural networks, *Acta Geophysica*, *60*(4), 317–336, https://doi.org/10.2478/s11600-012-0049-1.

He, K., Zhang, X., Ren, S., and Sun, J., 2015, Deep residual learning for image recognition, in *2016 IEEE Conference on Computer Vision and Pattern Recognition (CVPR)*, 770–778 arXiv:1512.03385v1, https://doi.org/10.48550/arXiv.1512.03385.

Hopfield, J.J., 1982, Neural networks and physical systems with emergent collective computational abilities, *Proceedings of the National Academy of Sciences*, *79*, 2554–2558, https://doi.org/10.1073/pnas.79.8.2554.

Hopfield, J.J., and Tank, D.W., 1985, "Neural" computation of decisions in optimization problems, *Biological Cybernetics*, *52*, 141–152, https://doi.org/10.1007/BF00339943.

Hornik, K., 1991, Approximation capabilities of multilayer feedforward networks, *Neural Networks*, *4*(2), 251–257, https://doi.org/10.1016/0893-6080(91)90009-T.

Hornik, K., Stinchcombe, M., and White, H., 1989, Multilayer feedforward networks are universal approximators, *Neural Networks*, *2*(5), 359–366, https://doi.org/10.1016/0893-6080(89)90020-8.

Huang, K.-Y., 1997, Hopfield neural network for seismic horizon picking, in *SEG Technical Program Expanded Abstracts*, 562–565, Society of Exploration Geophysicists, https://doi.org/10.1190/1.1885963.

Huang, K.-Y., Liu, W.H., and Chang, I.C., 1989, Hopfield model of neural networks for detection of bright spots, in *SEG Technical Program Expanded Abstracts*, 444–446, Society of Exploration Geophysicists, https://doi.org/10.1190/1.1889690.

Huang, K.-Y., and Yang, J.-R., 2015, Seismic velocity picking using Hopfield neural network, in *SEG Technical Program Expanded Abstracts*, 5317–5321, Society of Exploration Geophysicists, https://doi.org/10.1190/segam2015-5815718.1.

Huang, X., Alkhalifah, T., and Song, C., 2021, A modified physics-informed neural network with positional encoding, in *First International Meeting of Applied Geoscience and Energy Expanded Abstracts*, 2480–2484, Society of Exploration Geophysicists, https://doi.org/10.1190/segam2021-3584127.1.

Igel, C., 2005, Multi-objective model selection for support vector machines, in C.A. Coello Coello, A. Hernández Aguirre, and E. Zitzler, eds., *Proceedings of the Third International Conference on Evolutionary Multi-Criterion Optimization (EMO 2005)*, Lecture Notes in Computer Science, Vol. *3410*, 534–546, Springer, Verlag https://doi.org/10.1007/978-3-540-31880-4-37.

Islam, K.M., and Noor, R., 2020, Pattern Recognition Using Neuromorphic Computing, Thesis for B.Sc in EEE, https://doi.org/10.13140/RG.2.2.16339.81447.

Jia, L., Sen, S., and Mallick, S., 2022, Improvement of generalization capability of

2D salt segmentation via iterative semi-supervised learning, *Interpretation, 10*, 1–10, https://doi.org/10.1190/INT-2021-0089.1.

Jin, Y., and Sendhoff, B., 2008, Pareto-Based Multiobjective Machine Learning: An Overview and Case Studies, *IEEE Transactions on Systems, Man and Cybernetics- Part C: Applications and Reviews, 38*, 397–415, https://doi.org/10.1109/TSMCC.2008.919172.

Jing, J., Yan, Z., Zhang, Z., Gu, H., and Han, B., 2023, Fault detection using a convolutional neural network trained with point-spread function-convolution-based samples, *Geophysics, 87*, IM1–IM14, https://doi.org/10.1190/GEO2021-0824.1.

Kingma, D.P., and Ba, J.L., 2014, ADAM: A method for stochastic optimization, International Conference on Learning Representations (ICLR), Ithaca, arXiv:1412.6980, https://doi.org/10.48550/arXiv.1412.6980.

Krizhevsky, A., Sutskever, I., and Hinton, G.E., 2017, ImageNet classification with deep convolutional neural networks, *Communications of the ACM, 60*, 84–90, https://doi.org/10.1145/3065386.

LeCun, Y., Bottou, L., Bengio, Y., and Haffner, P., 1998, Gradient-based learning applied to document recognition, *Proceedings of the IEEE, 86*, 2278–2324, https://doi.org/10.1109/5.726791.

Levenberg, K., 1944, A method for the solution of certain non-linear problems in least squares, *Quarterly of Applied Mathematics, 2*, 164–168, https://doi.org/10.1090/qam/10666.

Li, T., Wang, R., Wang, Z., Zhao, M., and Li, L., 2018, Prediction of fracture density using genetic algorithm support vector machine based on acoustic logging data, *Geophysics, 83*, D49–D60, https://doi,org/10.1190/GEO2017-0229.1.

Li, Z., 2020, Adaptive multiple subtraction based on support vector regression, *Geophysics, 85*, V57–V69, https://doi.org/10.1190/GEO2018-0245.1.

Liu, M., Jervis, M., Li, W., and Nivlet, P., 2020a, Seismic facies classification using supervised convolutional neural networks and semisupervised generative adversarial networks, *Geophysics, 85*, O47–O58, https://doi.org/10.1190/GEO2019-0627.1.

Liu, Q., Fu, L., and Zhang, M., 2021, Deep-seismic-prior-based reconstruction of seismic data using convolutional neural networks, *Geophysics, 86*, V131–V142, https://doi.org/10.1190/GEO2019-0570.1.

Liu, S., Ni, W., Fang, W., and Fu, L., 2023, Absolute acoustic impedance inversion using convolutional neural networks with transfer learning, *Geophysics, 88*, R163–R174, https://doi.org/10.1190/GEO2022-0045.1.

Liu, Z., Chen, Y., and Schuster, G., 2020b, Deep convolutional neural network and sparse least-squares migration, *Geophysics, 85*, WA241–WA253, https://doi.org/10.1190/GEO2019-0412.1.

Mallick, S., Rath, P., Liu, M., Darhower, S.A., Ha, C.D.M., 2019, Machine-learning based prediction of phase velocities and phase angles from group velocities and group angles in an anisotropic elastic medium- a feasibility study, *SEG Technical Program Expanded Abstracts*, 2443–2447, Society of Exploration Geophysicists, https://doi.org/10.1190/segam2019-3214496.1.

Marquardt, D.W., 1963, An algorithm for least-squares estimation of nonlinear parameters, *SIAM Journal on Applied Mathematics, 11*, 431–441, https://doi.org/10.1137/01110.

Mishra, S., and Roberto, M., 2022, Estimates on the generalization error of physics-informed neural networks for approximating a class of inverse problems for PDEs, *IMA Journal of Numerical Analysis, 42*, 981–1022, https://doi.org/10.1093/imanum/drab032.

Mohamed, I.M., Mehta, D., Mohamed Salah, M., Ibrahim, M., and Ozkan, E., 2019, Advanced Machine Learning Methods for Prediction of Fracture Closure Pressure, Closure Time, Permeability and Time to Late Flow Regimes From DFIT, Unconventional Resources Technology Conference (URTeC), https://doi.org/10.15530/urtec-2020-2762.

Mosser, L., and Naeini, E.Z., 2022, A comprehensive study of calibration and uncertainty quantification for Bayesian convolutional neural networks – An application to seismic data, *Geophysics*, *87*, IM157–IM176, https://doi.org/10.1190/GEO2021-0318.1.

Mukhopadhyay, P., and Mallick, S., 2019, Bayesian deep learning for seismic facies classification and its uncertainty estimation, in *SEG Technical Program Expanded Abstracts*, 2488–2492, Society of Exploration Geophysicists, https://doi.org/10.1190/segam2019-3216870.1.

Parisi, S., Pirotta, M., and Restelli, M., 2016, Multi-objective Reinforcement Learning through Continuous Pareto Manifold Approximation, *Journal of Artificial Intelligence Research*, *57*, 187–227, https://doi.org/10.1613/jair.4961.

Park, M.J., and Sacchi, M.D., 2020, Automatic velocity analysis using convolutional neural network and transfer learning, *Geophysics*, *85*, V33–V43, https://doi.org/10.1190/GEO2018-0870.1.

Phan, S.D.T., and Sen, M.K., 2019, A Boltzmann machine for high-resolution prestack seismic inversion, *Interpretation*, *7*, SE215–SE224, https://doi.org/10.1190/INT-2018-0234.1.

Pinkus, A., 1999, Approximation theory of the MLP model in neural networks, *Acta Numerica*, *8*, 143–195, https://doi.org/10.1017/S0962492900002919.

Pomerat, J., Segev, A., and Datta, R., 2019, On Neural Network Activation Functions and Optimizers in Relation to Polynomial Regression, 2019 IEEE International Conference on Big Data (Big Data), Los Angeles, CA, USA, 6183–6185, https://doi.org/10.1109/BigData47090.2019.9005674.

Press, W.H., Teulolsky, S.A., Vetterling, W., and Flannery, B.P., 2007, Numerical recipes, in *The Art of Scientific Computing*, Third Edn, Cambridge University Press, ISBN ISBN 978-0-521-88068-8.

Raissi, M., Perdikaris, P., and Karniadakis, G.E., 2017, Physics Informed Deep Learning (Part I): Data-driven Solutions of Nonlinear Partial Differential Equations, arXiv:1711.10561v1, https://doi.org/10.48550/arXiv.1711.10561.

Raissi, M., Perdikaris, P., and Karniadakis, G.E., 2019, Physics-informed neural networks: A deep learning framework for solving forward and inverse problems involving nonlinear partial differential equations, *Journal of Computational Physics*, *378*, 686–707, https://doi.org/10.1016/j.jcp.2018.10.045.

Raissi, M., Yazdani, A., and Karniadakis, G.E., 2018, Hidden fluid mechanics: A Navier–Stokes informed deep learning framework for assimilating flow visualization data, *Computational Engineering, Finance, and Science*, arXiv:1808.04327v1, https://doi.org/10.48550/arXiv.1808.04327.

Raissi, M., Yazdani, A., and Karniadakis, G.E., 2020, Hidden fluid mechanics: Learning velocity and pressure fields from flow visualizations, *Science*, *367*(6481), 1026–1030, https://doi.org/10.1126/science.aaw4741.

Ren, Y., Liu, B., Yang, S., Li, D., and Jiang, P., 2021, Seismic data inversion with acquisition adaptive convolutional neural network for geologic forward prospecting in tunnels, *Geophysics*, *86*, R659–R670, https://doi.org/10.1190/GEO2020-0370.1.

Ronneberger, O., Fischer, P., and Brox, T., 2015, U-Net: Convolutional Networks for Biomedical Image Segmentation, In Medical image computing and computer-assisted intervention–MICCAI 2015: 18th international conference, Munich, Germany, October 5–9, 2015, proceedings, part III 18, pp. 234–241. Springer International Publishing, https://doi.org/10.48550/arXiv.1505.04597.

Ryck, T.D., Jagtap, A.D., and Mishra, S., 2023, Error estimates for physics informed neural networks approximating the Navier–Stokes equations, *IMA Journal of Numerical Analysis*, *44*(1), 83–119, https://doi.org/10.1093/imanum/drac085.

Shelhamer, E., Long, J., and Darrell, T., 2017, Fully convolutional networks for semantic segmentation, *IEEE Transactions on Pattern Analysis and Machine Intelligence*, *39*,

640–651, https://doi.org/10.1109/TPAMI.2016.2572683.

Sen, S., Kainkaryam, S., Ong, C., and Sharma, A., 2019, Regularization strategies for deep-learning-based salt model building, *Interpretation*, *7*, T911–T922, https://doi.org/10.1190/INT-2018-0229.1.

Sener, O. and Koltun, V., 2018, Multi-task learning as multi-objective optimization, 32nd Conference on Neural Information Processing Systems (NeurIPS 2018), Montréal, Canada, 525–536, https://dl.acm.org/doi/pdf/10.5555/3326943.3326992.

Sheng, T., and Zhao, J., 2022, Separation and imaging of diffractions using a dilated convolutional neural network, *Geophysics*, *87*, S117–S127, https://doi.org/10.1190/GEO2021-0260.1.

Shi, Y., Wu, X., and Fomel, S., 2019, SaltSeg: Automatic 3D salt segmentation using a deep convolutional neural network, *Interpretation*, *7*, SE113–SE122, https://doi.org/10.1190/INT-2018-0235.1.

Simonyan, K., and Zisserman, A., 2014, Very deep convolutional networks for large-scale image recognition, International Conference on Learning Representations, arXiv:1409.1556v6, https://doi.org/10.48550/arXiv.1409.1556.

Stefano, M., 2021, Physics-Informed Deep-Learning for Scientific Computing, arXiv:2103.09655v2, https://doi.org/10.48550/arXiv.2103.09655.

Sun, J., Slang, S., Elboth, T., Greiner, T.L., McDonald, S., and Gelius, L.-J., 2020, A convolutional neural network approach to deblending seismic data, *Geophysics*, *85*, WA13–WA26, https://doi.org/10.1190/GEO2019-0173.1.

Sun, J., Yang, J., Li, Z., Huang, J., Xu, J., and Zhuang, S., 2023, Reflection and diffraction separation in the dip-angle common-image gathers using convolutional neural network, *Geophysics*, *88*, WA281–WA291, https://doi.org/10.1190/GEO2022-0157.1.

Szegedy, C., Liu, W., Jia, Y., Sermanet, P., Reed, S., Anguelov, D., Erhan, D., Vanhoucke, V., and Rabinovich, A., 2014, Going Deeper with Convolutions, 2015 IEEE Conference on Computer Vision and Pattern Recognition (CVPR), 1–9, arXiv:1409.4842v1, https://doi.org/10.48550/arXiv.1409.4842.

Tahmasebi, P., and Hezarkhani, A., 2012, A hybrid neural networks-fuzzy logic-genetic algorithm for grade estimation, *Computers & Geosciences*, *42*, 18–27, https://doi.org/10.1016/j.cageo.2012.02.004.

Tank, D.W., and Hopfield, J.J., 1986, Simple "neural" optimization networks: An A/D converter, signal decision circuit, and a linear programming circuit, *IEEE Transactions on Circuits and Systems*, *33*, 533–541, https://doi.org/10.1109/TCS.1986.1085953.

Vamaraju, J., and Sen, M.K., 2018, Mean-field Boltzmann machines for high-resolution Kirchhoff migration, in *SEG Technical Program Expanded Abstracts*, 2006–2010, Society of Exploration Geophysicists, https://doi.org/10.1190/segam2018-2997793.1.

Vamaraju, J., and Sen, M.K., 2019, Unsupervised physics-based neural networks for seismic migration, *Interpretation*, *7*, SE189–SE200, https://doi.org/10.1190/INT-2018-0230.1.

Vamaraju, J., Zhao, Z., and Sen, M.K., 2019, Pre-stack target-oriented least squares RTM in image domain using conjugate Hopfield networks and the Boltzmann machine, in *SEG Technical Program Expanded Abstracts*, 2308–2312, Society of Exploration Geophysicists, https://doi.org/10.1190/segam2019-3216032.1.

Vapnik, V.A., and Chervonenkis, A.Y., 1964, A class of algorithms for pattern recognition learning, *Avtomatika i Telemekhanika*, *25*(6), 937–945, English Translation from the original paper in Russian, https://www.mathnet.ru/eng/at/v25/i6/p937.

Vinard, N.A., Drijkoningen, G.G., and Verschuur, D.J., 2022, Localizing microseismic events on field data using a U-Net-based convolutional neural network trained on synthetic data, *Geophysics*, *87*, KS33–KS43, https://doi.org/10.1190/GEO2020-0868.1.

Vrolijk, J.-W., and Blacquière, G., 2021, Source deghosting of coarsely sampled common-receiver data using a convolutional neural network, *Geophysics*, *86*, V185–V196, https://doi.org/10.1190/GEO2020-0186.1.

Wang, H., and Alkhalifah, T., 2021, Direct microseismic event location and characterization from passive seismic data using convolutional neural networks, *Geophysics*, *86*, KS109–KS121, https://doi.org/10.1190/GEO2020-0636.1.

Wang, L., and Mendel, J.M., 1992, Adaptive minimum prediction-error deconvolution and source wavelet estimation using Hopfield neural networks, *Geophysics*, *57*, 670–679, https://doi.org/10.1190/1.1443281.

Wu, H., Zhang, B., Lin, T., Cao, D., and Lou, Y., 2019a, Semiautomated seismic horizon interpretation using the encoder–decoder convolutional neural network, *Geophysics*, *84*, B403–B417, https://doi.org/10.1190/GEO2018-0672.1.

Wu, H., Zhang, B., Lin, T., Li, F., and Liu, N., 2019b, White noise attenuation of seismic trace by integrating variational mode decomposition with convolutional neural network, *Geophysics*, *84*, V307–V317, https://doi.org/10.1190/GEO2018-0635.1.

Wu, X., Geng, Z., Shi, Y., Pham, N., Fomel, S., and Caumon, G., 2020, Building realistic structure models to train convolutional neural networks for seismic structural interpretation, *Geophysics*, *85*, WA27–WA39, https://doi.org/10.1190/GEO2019-0375.1.

Wu, X., Liang, L., Shi, Y., and Fomel, S., 2019c, FaultSeg3D: Using synthetic data sets to train an end-to-end convolutional neural network for 3D seismic fault segmentation, *Geophysics*, *84*, IM35–IM45, https://doi.org/10.1190/GEO2018-0646.1.

Wu, Y., and McMechan, G., 2019, Parametric convolutional neural network-domain full-waveform inversion, *Geophysics*, *84*, R881–R896, https://doi.org/10.1190/GEO2018-0224.1.

Xu, M., Zhao, L., Gao, S., Zhu, X., and Geng, J., 2022, Joint use of multiseismic information for lithofacies prediction via supervised convolutional neural networks, *Geophysics*, *87*, M151–M162, https://doi.org/10.1190/GEO2021-0554.1.

Yao, J., and Wang, Y., 2022, Building a full-waveform inversion starting model from wells with dynamic time warping and convolutional neural networks, *Geophysics*, *87*, R223–R230, https://doi.org/10.1190/GEO2021-0168.1.

Yildirim, I.E., Alkhalifah, T., and Yildirim, E.U., 2022, Machine learning-enabled traveltime inversion based on the horizontal source-location perturbation, *Geophysics*, *87*, U1–U8, https://doi.org/10.1190/GEO2020-0735.1.

Yuan, C., Li, J., Chen, X., and Rao, Y., 2016, Quantitative uncertainty evaluation of seismic facies classification: A case study from northeast China, *Geophysics*, *81*, B87–B99, https://doi.org/10.1190/GEO2015-0228.1.

Yuan, P., Wang, S., Hu, W., Wu, X., Chen, J., and Van Nguyen, H., 2020, A robust first-arrival picking workflow using convolutional and recurrent neural networks, *Geophysics*, *85*, U109–U119, https://doi.org/10.1190/GEO2019-0437.1.

Zeiler, M.D., and Fergus, R., 2013, Visualizing and Understanding Convolutional Networks, arXiv:1311.2901v3, https://doi.org/10.48550/arXiv.1311.2901.

Zhang, G., Lin, C., and Chen, Y., 2020, Convolutional neural networks for microseismic waveform classification and arrival picking, *Geophysics*, *85*, WA227–WA240, https://doi.org/10.1190/GEO2019-0267.1.

Zhang, W., Gao, J., Yang, T., Jiang, X., and Sun, W., 2021, Least-squares reverse time migration using convolutional neural networks, *Geophysics*, *86*, R959–R971, https://doi.org/10.1190/GEO2021-0006.1.

Zhang, Z., 2018, Improved Adam Optimizer for Deep Neural Networks, 2018 IEEE/ACM 26th International Symposium on Quality of Service (IWQoS), Banff, AB, 1–2, https://doi.org/10.1109/IWQoS.2018.8624183.

Zhao, L., Zou, C., Chen, Y., Shen, W., Wang, Y., Chen, H., and Geng, J., 2021, Fluid and lithofacies prediction based on integration of well-log data and seismic inversion: A machine-learning approach, *Geophysics*, *86*, M151–M165, https://doi.org/10.1190/GEO2020-0521.1.

Zou, C., Zhao, L., Hong, F., Wang, Y., Chen, Y., and Geng, J., 2023, A comparison of machine learning methods to predict porosity in carbonate reservoirs from seismic-derived elastic properties, *Geophysics*, *88*, B101–B120, https://doi.org/10.1190/GEO2021-0342.1.

10

The Future of Seismic Inversion and Machine Learning

10.1 Introduction

This book first provides the mathematical background needed to understand the elastic wave theory and a complete overview of seismic wave propagation in homogeneous and heterogeneous media and different methods to compute forward synthetic seismograms. This is then followed by detailed discussions on the optimization of functions and various local and global optimization methods for seismic inversion. Finally, current advances on machine learning (ML) are reviewed, and how they are used to solve seismic inverse problems are discussed.

The primary objective of this book is to provide details of seismic modeling and inversion in a single textbook. As can be seen from the previous chapters, seismic modeling and inversion are complimentary to one another. Although it may appear that the ML methods do not require any understanding of the underlying physics, this is not at all true. This is specifically emphasized in almost all sections of Chapter 9, in the discussion of ML methods. Therefore, this textbook attempts to provide both modeling and inversion methods in a single place.

Most of the theories on how to compute forward synthetic seismograms, different local and global optimization methods, and most of the ML algorithms discussed in this book are not new. However, because computers were not fast enough until the mid-1980s, their applications were limited to solving simple problems. And it is the current and continuing advances in high-performance, massively parallel computing technology that revolutionized the applications of seismic inversions to solve complex problems. As computation technology advances further, we expect to see attempts to solve more complex problems in future using optimization and ML methods. The scope of these new advances is too broad, and in this chapter, we outline some such future possibilities and make some concluding remarks.

10.2 The Road Ahead

We expect that the application optimization methods to solve complex problems will continue in future. The ML methods, specifically the physics-informed ML where the underlying physics of the problem is used for back propagation (Figure 9.30) instead of supervised learning from labeled data, have many potential future applications. In addition, no optimization method, including ML, is perfect, and these imperfections are due to two factors: (1) noise in the observed data and (2) limitations in the underlying physics used to solve the forward problem. Consequently, it is not only important to estimate the model from optimization and ML, but it is also important to quantify the uncertainties associated with their estimates. Additionally, in

Computation, Optimization, and Machine Learning in Seismology, Advanced Textbook 8, First Edition. Subhashis Mallick.

the previous chapters, we mostly restricted to seismic data. To solve complex problems of practical interest, seismic data require to be combined with other geophysical, engineering, and maybe other data types in a joint optimization/ML framework. Solving such problems requires multi-objective optimization (or multi-task learning or MTL for ML). In addition, because these different data types are described by different underlying physics that may resolve subsurface properties at different scales (resolutions), these problems are not only multi-objective but also multi-physics and multi-scale. Although the concepts of these multi-objective, multi-physics, and multi-scale problems are already in place, and we discussed them in Chapter 8, their applications are still limited. As computing technology advances further, it is expected that more of such problems will be tackled in future.

The scope of future applications of optimization and ML is broad. We discuss below a selected few applications.

10.2.1 Carbon Capture, Utilization, and Storage

To mitigate global warming from fossil-fuel consumptions, carbon capture, utilization, and storage (CCUS) is now a well-accepted method where carbon dioxide (CO_2) from fossil-fuel consumptions is captured and sequestered into deep saline aquifers or depleted hydrocarbon reservoirs for permanent underground storage. Such CCUS is, however, incomplete until the injected fluids are remotely monitored to ensure containment, and there are no leakage pathways for CO_2 to escape into the atmosphere or contaminate groundwater. Such a monitoring, in turn, requires time-lapse seismic and possibly other geophysical data to be combined with reservoir simulation, geomechanical modeling, and rock physics. Although substantial research and developments are already in place and CCUS is being attempted in different parts of the world, it is still an emergent field of research and require further developments, specifically for accurately quantifying the totality of the injected CO_2 volume as a function of time, estimating in situ stress fields for the seal integrity to ensure there are no fracture initiations causing pathways for CO_2 leakage, and quantifying uncertainties associated with these estimates. We briefly discuss below how optimization and ML can be used to further advance future CCUS technology.

In a recent work, Padhi et al. (2014) carried out a synthetic study to demonstrate that as CO_2 migrates, the fluid saturation in the sequestered formation becomes patchy and tends to behave seismically anisotropic. Figures 10.1 and 10.2 summarize these results.

Using the replica of an eolean sandstone of about 66 m in thickness, Padhi et al. (2014) synthetically sequestered CO_2 at the bottom and ran reservoir simulation to map the behavior of the fluids in the formation. Then by using the rock-physics-based fluid substitutions given by the Gassmann equations (Mavko et al., 2010), they computed the isotropic elastic properties of the sandstone at different times after sequestration. Figure 10.1a–c shows the P-wave velocity (V_P), S-wave velocity (V_S), and density (ρ) of the sandstone 5 years after CO_2 sequestration and Figure 10.1d–f shows the same 55 years after CO_2 sequestration. As can be noted from Figure 10.1, movements of CO_2 plumes cause substantial change in both V_P and ρ. However, because S waves are relatively insensitive to fluids, V_S remains nearly constant.

To quantify how does the change in V_P and V_S shown in Figure 10.1 manifest on the seismic data, Padhi et al. (2014) produced a simple four-layer subsurface model with two isotropic overburden layers on top, the 66.3 m eolean sandstone as the third layer, and another isotropic half-space below it. Figure 10.2a shows this model. Then, by using a time-domain isotropic elastic finite-difference modeling, they computed two synthetic seismograms for the model. For the first, they used the properties of the eolean sand from 5 years after CO_2 injection (Figure 10.1a–c); while for the second,

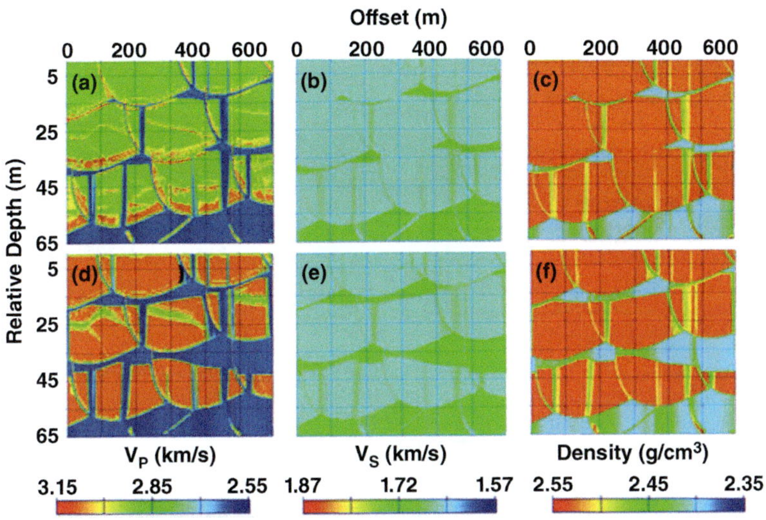

Figure 10.1 Time-dependent behavior of the elastic properties in an eolean sandstone caused by CO_2 injection at the bottom of the sandstone. (a–c) P-wave velocity, S-wave velocity, and density 5 years after injection. (d–f) Same as (a–c) but 55 years after injection. Padhi et al. (2014)/with permission of Elsevier.

they used the properties from 55 years after CO_2 injection (Figure 10.1d–f). The acquisition geometry for computing these synthetic data is shown in Figure 10.2a. Then, they used the equivalent medium theory (Schoenberg and Douma, 1988; Schoenberg and Sayers, 1995) to replace the laterally and vertically heterogeneous eolean sand by a homogeneous but anisotropic medium and used the anisotropic reflectivity modeling method to compute another set of two synthetic seismic data for the same acquisition geometry and source wavelet. The theory of computing time-domain finite-difference synthetic seismograms and anisotropic synthetic seismograms using reflectivity method is discussed in Chapter 5, and the concepts of an equivalent medium are discussed in Chapter 4. Figure 10.2b shows the comparison between the isotropic finite-difference and anisotropic reflectivity synthetic data for one near and one far source-to-receiver offset where the eolean sand properties are from 5 years after sequestration, and Figure 10.2c shows the same comparison in which the eolean sand properties are from 55 years after sequestration, demonstrating that the overall characteristics of the isotropic finite-difference and anisotropic reflectivity synthetics are comparable. Although synthetics for only two offsets are shown in Figure 10.2b,c for clarity, Padhi et al. (2014) computed them for a range of offsets from 200 to 3100 m and verified that the synthetic data for the entire range of these offsets computed using both modeling methods are comparable. Furthermore, Padhi et al. (2014) developed a method to predict CO_2 saturation from the properties of the equivalent anisotropic medium, and Figure 10.2d compares true CO_2 saturation from the reservoir simulation with that estimated from the anisotropic properties of the resulting equivalent medium.

The takeaway messages from Figures 10.1 and 10.2 are the following:

1) As CO_2 migrates, the sequestered formation exhibits a patchy saturation, which becomes seismically anisotropic.
2) If this anisotropy can be quantified from time-lapse seismic data and combined with reservoir simulation and rock-physics modeling and inversion via Gassmann equation or equivalent, it is possible to estimate the saturation of the injected fluid in the formation at different times and ensure containment of the injected fluid.

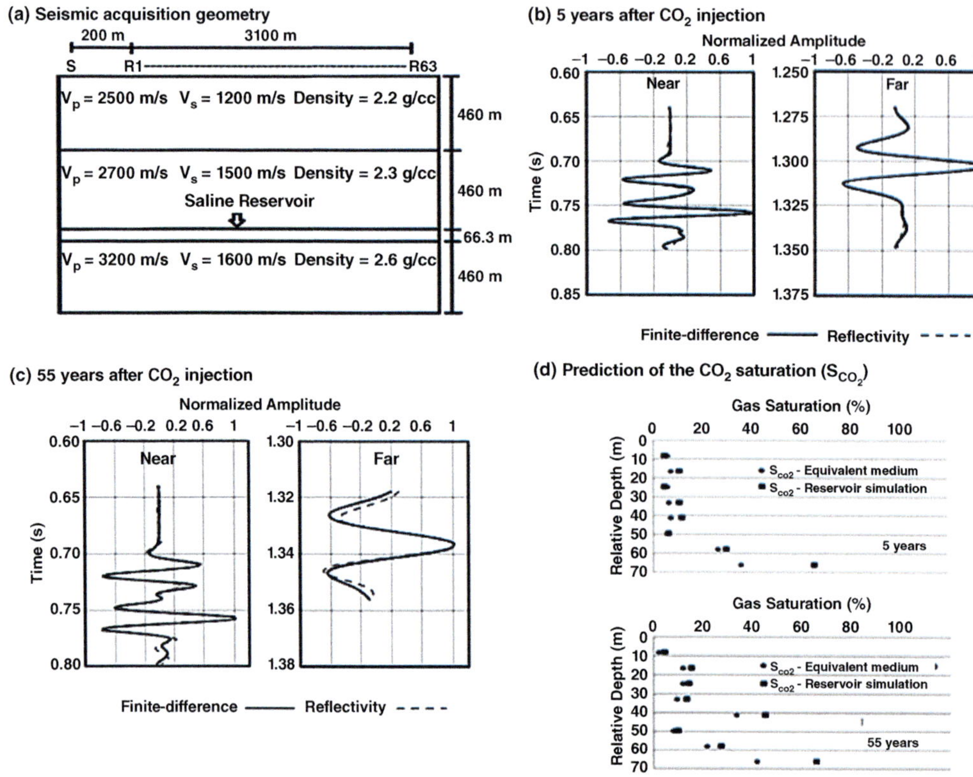

Figure 10.2 Seismic characteristics of CO_2 injection. (a) Model and the synthetic seismogram computation geometry. The eolean sand is the third layer. (b) Near (left) and far (right) offset seismograms in which the properties of the eolean sand are taken from 5 years after the CO_2 injection. Isotropic finite-difference seismograms are in solid and anisotropic reflectivity seismograms are in dashed curves. (c) Same as (b) except that the properties of the eolean sand are from 55 years after the CO_2 injection. (d) Comparison between the true (black squares) and estimated (black circles) CO_2 saturations. The top figure is for 5 years, and the bottom figure is for 55 years after injection. Padhi et al. (2014)/with permission of Elsevier.

To implement the above synthetic study in practice, we must combine seismic, rock-physics, and reservoir simulation, into a single optimization workflow. Furthermore, to characterize seal integrity and ensure that there are no open fractures as conduits of the injected fluids to leak, it is also necessary to use geomechanics that would allow estimating stress fields within the neighborhood of the sequestered formation. There are many fluid-flow simulations coupled with geomechanical modeling systems (for example, Settari and Mounts, 1998; Settari and Walters, 2001; Thomas et al., 2003) that can be used for such.

As shown in Figure 10.3a, the baseline seismic data (seismic data acquired prior to CO_2 sequestration) can be inverted using a classical inversion/optimization, local or global, and combined with rock physics to obtain the static reservoir model of porosity, permeability, fluid saturation, and pressure. Using real seismic data, Grana et al. (2017) showed the way the inverted model from seismic data can be used in a rock-physics inversion to estimate such a static model. As an Alternative to such classical optimization, we can also use a physics-informed ML approach of Biswas et al. (2019) on a portion of the input seismic data and combine with rock physics and obtain such static model, which is shown in Figure 10.3b.

Once a static model is obtained, Figure 10.4 shows a possible workflow for classical optimization of using this model and combine

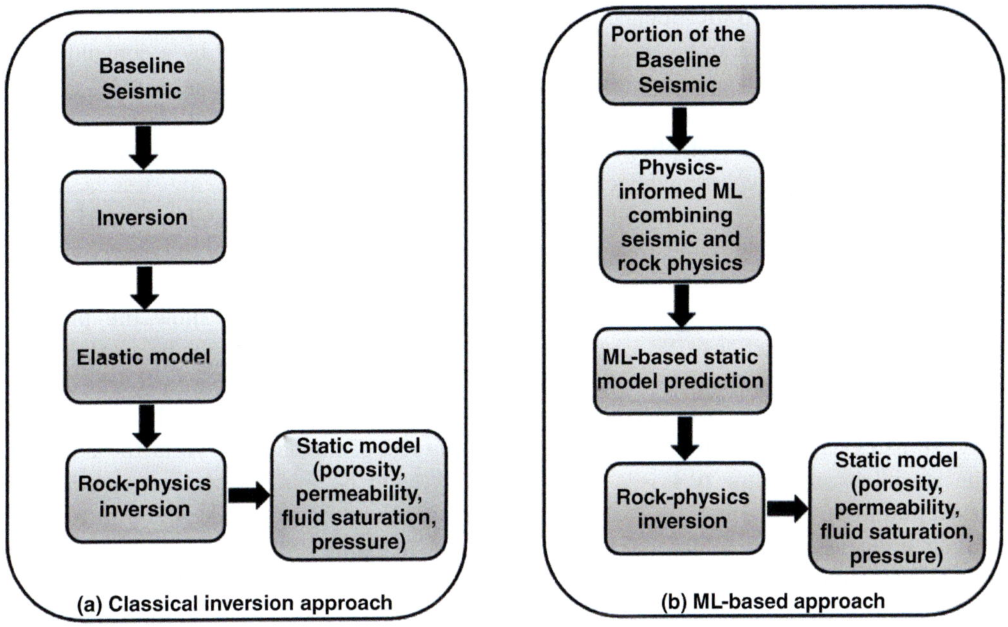

Figure 10.3 Estimation of the static model from the baseline seismic data using (a) classical inversion/optimization approach. (b) ML-based approach.

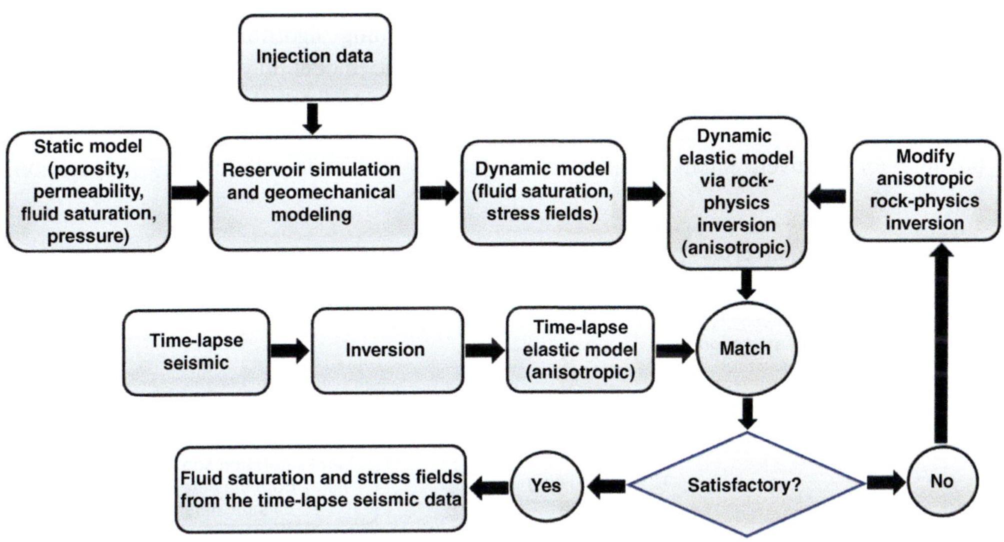

Figure 10.4 Inversion of the time-lapse model for estimating fluid saturation and stress fields.

with the time-lapse seismic data for estimating fluid saturation and stress fields at different times after CO_2 injection. As shown in this figure, the static model of porosity, permeability, fluid saturation, and pressure can be combined with the injection data (injection volume, injection pressure, etc.) into a coupled simulation–geomechanics system which will provide the dynamic model of saturation and stress fields at any time after CO_2 injection. From this dynamic model, we can use an iterative rock-physics/seismic inversion tool to predict fluid saturation and stress fields. In this iterative approach, the anisotropic model from the inversion of the time-lapse seismic data and the dynamic model from simulation–geomechanics–anisotropic–rock physics could be matched until they are reasonably close to one another. This iterative process would then provide the way to predict the dynamic model of fluid saturation and stress fields, directly from the time-lapse seismic data. Additionally, the entire such approach can be cast in a Bayesian framework to obtain the probability density functions and the associated uncertainty estimates, as discussed in Chapter 8.

The workflow of Figure 10.4 is a multi-objective, multi-physics, and multi-scale optimization problem as it combines time-lapse seismic with fluid-flow simulation, geomechanics, and rock physics. Because such problems are nonunique with multiple (Pareto-optimal) solutions, they require global optimization, such as NSGA-II or SPEA2, discussed in Chapter 8. Although plausible, such multi-objective, multi-physics, and multi-scale optimization problems are computationally challenging. Using the present-day computers, applying the workflow of Figure 10.4 is possible for relatively small datasets, or for large datasets, if the physics of the forward problem is replaced by a convolutional amplitude-variation-with-angle (AVA) type of synthetic modeling instead of a wave-equation-based modeling. As we discuss later in this chapter, AVA synthetics are less accurate than the wave equation synthetics and fail to describe intricate details of the observation as wave equation synthetics can.

Figures 10.5 and 10.6 show the physics-informed ML alternative to the workflow of Figure 10.4. Using only a portion of the data to derive a prediction model and then applying it to the rest of the data shown in these figures is likely to be much more computationally efficient than the workflow of Figure 10.4. As shown in Figure 10.5, we can select a portion of the static model, injection data, and time-lapse seismic data in a learning algorithm and derive the model. Figure 10.6 provides how such algorithm can be implemented in practice. As shown in Figure 10.6, the injection and static model from the input layer is fed to a deep learning algorithm to output dynamic anisotropic model in the output layer. We then compute forward synthetic seismograms from this dynamic model and match with the time-lapse seismic data for back propagation.

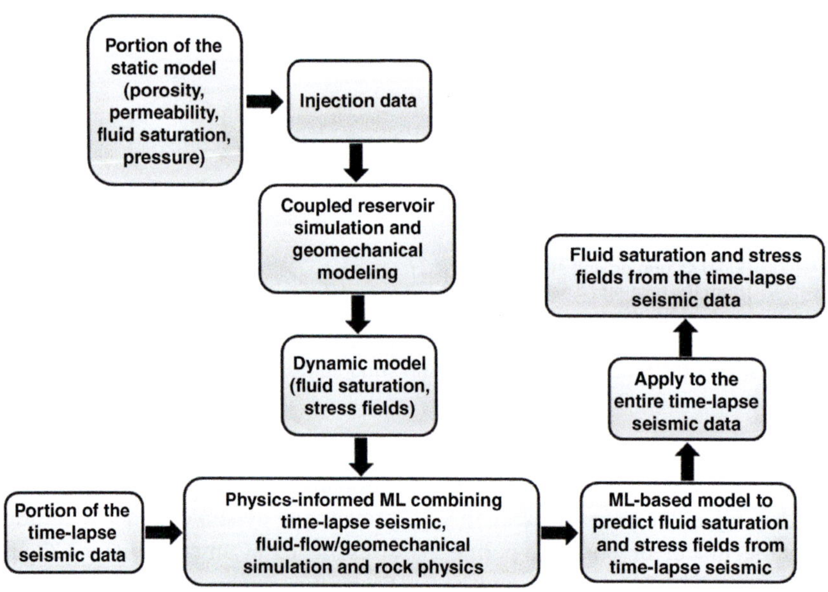

Figure 10.5 A machine-learning alternative to the workflow shown in Figure 10.4.

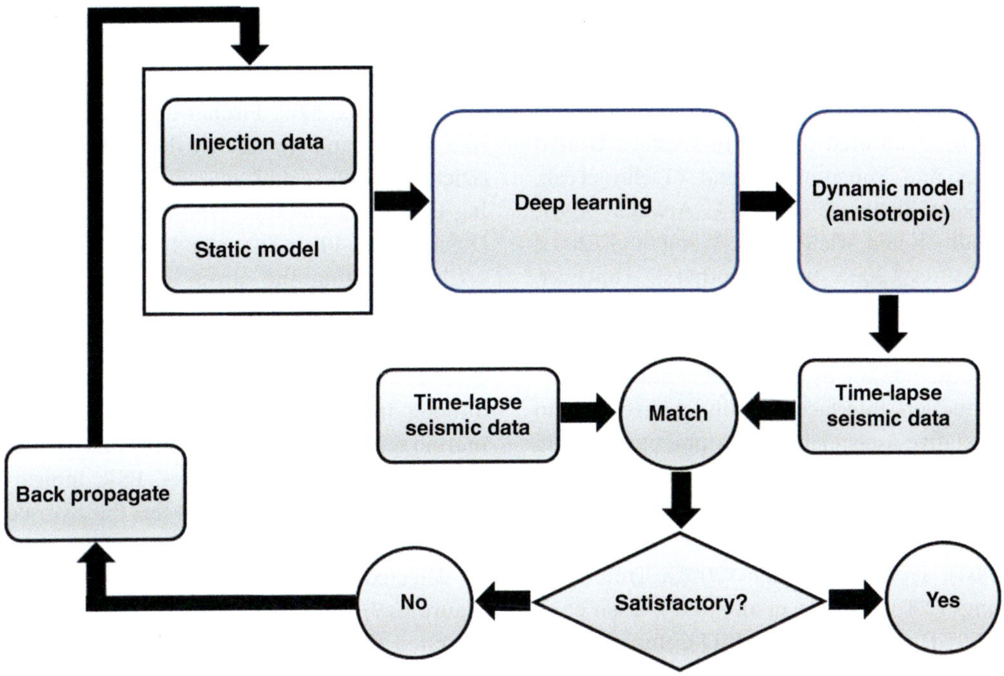

Figure 10.6 Detailed steps for a physics-informed approach to combine the injection data, static (baseline) model, and time-lapse seismic data.

We should also note here that to predict fluid movements in the injected formation, such ML approach must be repeated for a series of time-lapse seismic data, acquired at different times after the CO_2 injection, in which the prediction from a given time-lapse data can use the prediction from the previous time-lapse data. Consequently, the recurrent neural network architecture, discussed in Chapter 9, may be effective for such predictions. It is also important in these ML methods to quantify the uncertainties associated with all estimations. We must also emphasize that like Figure 10.4 the ML method outlined by Figures 10.5 and 10.6 is also a multi-objective, multi-physics, and multi-scale problem and therefore requires the implementation of a MTL process.

10.2.2 Hydrogen Storage Systems

With the developments of fuel-cell technology, hydrogen is likely to be the most important future energy resource. For such future use of hydrogen as the source of energy, it is important to find ways to store hydrogen, and the most economically viable options for hydrogen storage are depleted hydrocarbon reservoirs, saline aquifers, and salt caverns (Lindblom, 1985; Hassanpouryouzband et al., 2021; Heinemann et al., 2021). These underground hydrogen storage systems face the same set of challenges as CCUS. Therefore, the same multi-physics, multi-scale, and multi-objective approach outlined above for CCUS can be effectively used to remotely monitor the hydrogen storage systems.

10.2.3 Geohazards and Related Environmental Impacts

Beyond CCUS and hydrogen storage, the future use of optimization and ML is applicable to many other problems. One such future potential application is in the field of geohazards and the related environmental impacts. Cascadia subduction zone (CSZ), offshore Washington, USA, for example, poses a major seismic and environmental hazard to the population centers of the northwestern United States and western Canada with 10,000-year record of magnitude 9 or greater earthquakes occurring in every 500-year intervals that ruptured the entire Western United States and Vancouver Island (Tréhu et al., 1995; Williams et al., 2011; Atwater, 1987; Hyndman and Wang, 1995; Wang et al., 2003; McCaffrey, 2007; among others). However, key details of the geologic structure and fault mechanisms of the CSZ are poorly understood. In particular, the position and character of the subduction interface remain elusive due to its relative aseismicity and segmented forearc structure (Fortin, 2015). Additionally, seismic analyses by MacKay et al. (1994), Tréhu et al. (1995, 1999, 2002, 2004), Torres et al. (2004), Tréhu and Flueh (2001), Tréhu and Bangs (2001), Spence et al. (1995), Dash et al. (2009), Dash and Spence (2011), among others reported wide presence of bottom-simulating reflector (BSR) in the Cascadia Ocean margin indicating the presence of methane hydrates and free gas underneath. In most places, these BSRs appear stronger in the uplifted strata on regional highs suggesting active expulsion of methane-charged fluids (Hyndman and Davis, 1992; Hyndman et al., 1994; MacKay et al., 1994; Tréhu et al., 2002). Besides seismic, the presence of methane hydrates in the Cascadia margin using electromagnetic (EM) methods was also identified and analyzed by Weitemeyer et al. (2006, 2011). Although anaerobic oxidation of methane does not permit large quantities of methane to leak into the ocean waters, small-to-moderate quantities of methane are continuously seeping into the ocean along the Cascadia margin (Suess, 2014), and in the event of a major earthquake, the hydrate stability field is likely to be disrupted to release large quantities of methane into the atmosphere, which may cause major impact to the climate. Therefore, mapping the regional distribution of methane hydrates in the Cascadia margin and other subduction zones is important not only for developing constraints on the overall fluid budget of the subducting plate but also for other environmental issues such as the impact to the climate change forced by the hydrate eruptions in regional and global scales (McDonald, 1990; Dickens et al., 1997; Kennett et al., 2000; Fischer et al., 2013), the role of methane hydrate in the carbon cycle (Dickens, 1999), and possibility of exploiting them as a potential future energy source (Kvenvolden, 1993; Thena and Mohan, 2016).

Characterizing the stress and strain conditions within the CSZ plate boundary and studying the associated earthquake-related geohazards from marine seismic data alone is difficult. In relatively shallow depths, however, marine seismic and EM data have been used to characterize methane hydrates. BSR, indicating the phase boundary between the hydrate and free gas phase underneath, can easily be detected from the seismic depth images. Figure 10.7, for example, shows a reverse-time migrated (RTM) depth image from the CSZ, reported by Jia and Mallick (2018) with a clearly visible ocean-bottom reflection and BSR, being marked with arrows. Additionally, two possible faults within the BSR are also marked. Although these reflections and faults can be identified from the seismic depth images, the depth at which the BSR can be present is controlled by the hydrate stability field, which, in turn, is a function of the subsurface temperature and pressure distributions (Motiee, 1991). Therefore, to characterize gas hydrates and free gas for environmental impacts, and possibility of methane leakage via faults, we must combine geophysical data with heat-flow and pressure data in an optimization or ML workflow as shown in Figure 10.8.

As can be seen from Figure 10.8, properly characterizing methane hydrates is a multi-objective, multi-physics, and multi-scale problem that requires implementing multi-objective optimization or MTL methodologies.

10.2.4 Use of Seismology for Weather Prediction, Climate Modeling, and Marine Biology Research

Studying climate change and the associated environmental impacts is of utmost importance

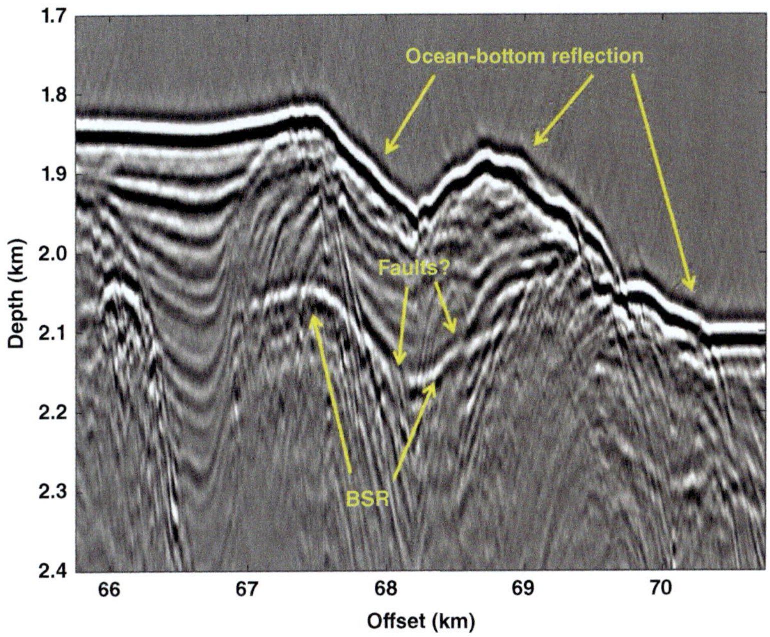

Figure 10.7 RTM depth image of one seismic line from the Cascadia subduction zone marine seismic data with the ocean-bottom reflection, an identifiable BSR, and two probable faults within the BSR marked with arrows. Jia and Mallick (2018)/Society of Exploration Geophysicists.

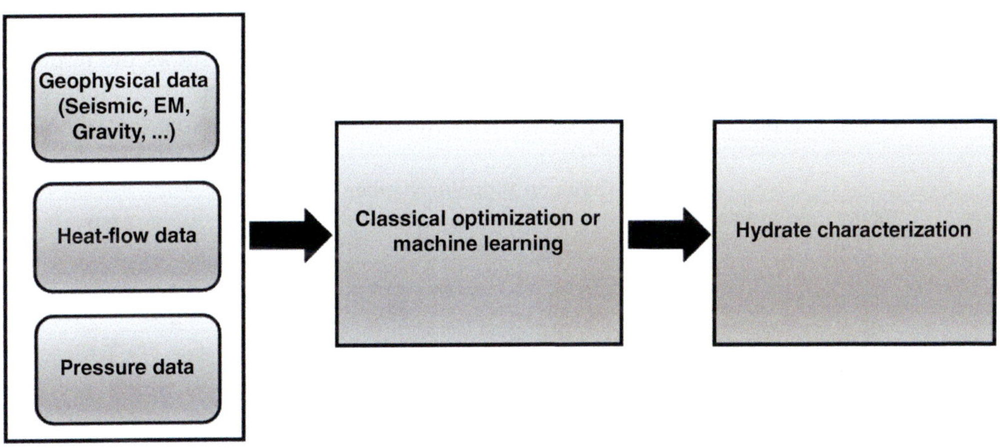

Figure 10.8 A simplified workflow to characterize the gas hydrates by combining geophysical, pressure, and heat-flow data.

for the future of society, where optimization and ML can play major roles. Weather prediction and climate modeling require a detailed analysis of the atmosphere–hydrosphere interactions in which three key parameters are the atmospheric temperature distributions and the temperature and salinity distributions within the oceans (Warner et al., 2010). Geophysics has no role to play in estimating the atmospheric temperature distributions. However, subtle sound-speed variations showing visible reflections within the water columns in the marine seismic data can provide accurate estimates of the oceanic sound speeds. These sound speeds are, in turn, functions of temperature, salinity, and pressure via an equation of state (Chen and Millero, 1977; Fofonoff and Millard, 1983), which have been used in an inversion/optimization scheme to predict temperature and salinity from sound speeds (for example, Padhi et al., 2015). Considering that the seismic data are of much higher lateral resolution than the resolution at which temperature and salinity are measured in the oceans, using salinity and temperature from the seismic can potentially improve the accuracy of climate modeling and weather prediction. In addition, global climate change and variations of temperature and salinity control the marine ecosystem and the production and migration of fish population in marine, lake, and river environments (Castillo et al., 1996; Rijnsdorp et al., 2009; Brucet et al., 2012; Mahadevan, 2016; Sugie et al., 2020; among others). Consequently, seismically derived temperature and salinity with high lateral resolution may also have future use in marine biology and fishery research.

Padhi et al. (2015) used both global optimization using a genetic algorithm (GA) and a local optimization using a nonlinear least-squares (NLS) method based on a Gauss–Newton algorithm and showed that they both can extract accurate sound-speed models from real marine seismic data. Padhi et al. (2015) also demonstrated that the GA inversion results are superior to the NLS in matching the subtle features of the real data. Although the results of Padhi et al. (2015) are accurate, the global and local methods they used needed careful inversion runs to obtain sound-speed models at sparse locations, and interpolating them via horizons, interpreted from the stacked seismic data for an initial model for inversion. These sparse inversion runs, and horizon-based interpolations, are labor-intensive and prone to human error and bias.

Mallick and Chakraborty (2022) modified the GA approach of Padhi et al. (2015) and introduced the concept of an attribute-guided inversion. This attribute-guided inversion does not require any careful inversion runs at sparse locations, nor does it require any manual horizon interpretation and interpolation. Yet, the method finds an accurate sound-speed model consistent with the reflections on the stacked seismic data. Figure 10.9 shows the result of the attribute-guided GA inversion along with the stacked seismic data. Note that although

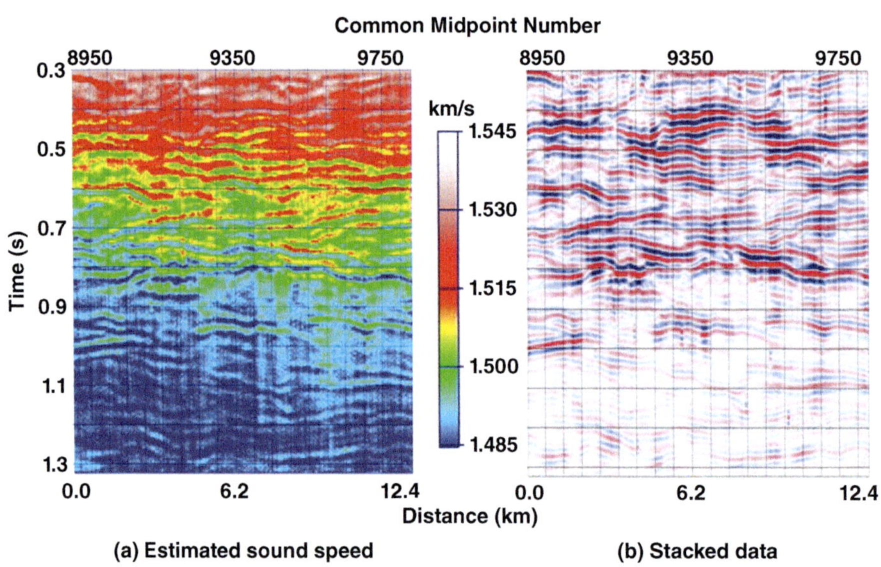

Figure 10.9 Inversion of water-column reflection data from the South China Sea. (a) Estimated sound speed, (b) stacked seismic data. Mallick and Chakraborty (2022)/Society of Exploration Geophysicists.

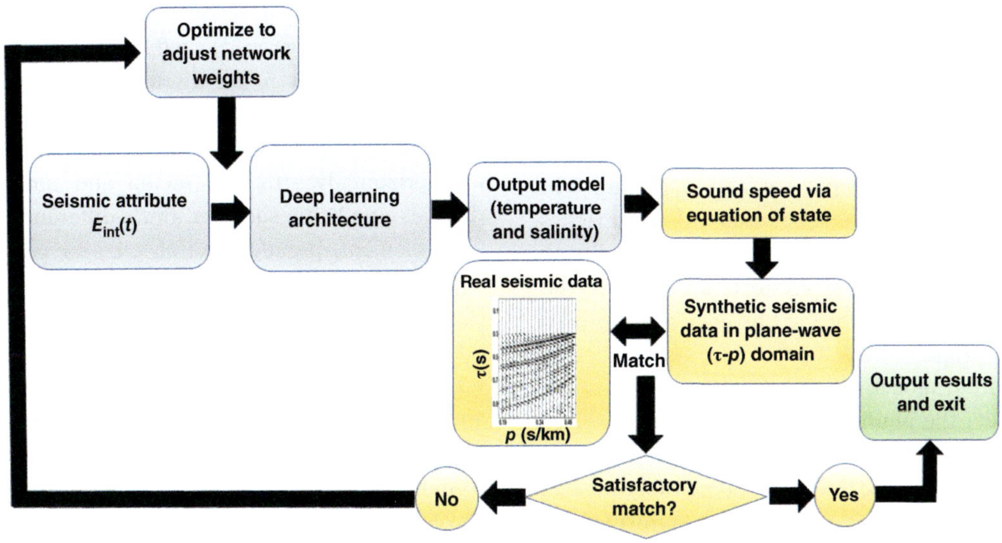

Figure 10.10 A machine-learning approach to estimate temperature and salinity from marine seismic data for weather prediction, climate modeling, and marine biology research.

no horizon-guided interpolation was provided as an initial model, the sound-speed model shown in Figure 10.9a is consistent with the reflections on the stacked data (Figure 10.9b).

Figure 10.10 shows a possible ML approach to predict temperature and salinity from marine seismic data for weather prediction, climate modeling, and marine biology research. Seismic attributes in the input layer are fed into a deep learning algorithm to output temperature and salinity. From temperature and salinity, we can then compute sound speed via the equation of state (Chen and Millero, 1977; Fofonoff and Millard, 1983). We then compute synthetic data from the sound speed and iteratively match with observed seismic data to drive ML algorithm. This physics-informed ML using both the physics of seismic wave propagation and the equation of state is again a multi-physics and multi-objective problem and therefore requires MTL.

10.2.5 The Overall Picture for Future Developments and Applications

Therefore, irrespective of whether we would like to use optimization/ML for CCUS, hydrogen storage, gas hydrates, climate modeling, etc., they all require combining multiple data types into a single workflow. This, in turn, is not only a multi-objective optimization problem, but it is also multi-physics because each data type is described by different physics. Additionally, these problems are also multi-scale because each data type samples the subsurface at different resolutions (scales). As discussed in Chapter 8, these multi-objective, multi-physics, and multi-scale problems are, in general, nonunique with multiple (Pareto-optimal) solutions. Solving these problems using multi-objective optimization methods, discussed in Chapter 8, is computationally demanding. However, MTL, the ML alternatives to these multi-objective optimizations, can potentially solve these problems in a compute-efficient way. We briefly discussed MTL in Chapter 9. MTL is an emergent field of research, and it is expected that more efficient MTL methods will be developed in future to solve many complex problems. One new development in this field is quantum computing, which allowed solving many large-scale computational challenges (Sahimi and Tahmasebi, 2022). Routine applications of these new concepts are expected in the future.

10.3 Few Aspects of Practical Importance

Classical optimization and physics-informed ML requires solving the forward problem using the underlying physics and matching with observation. For supervised ML approaches, where labeled data are used, do not directly require solving the physics of the forward problem. However, for the universal approximator to properly map input data onto the desired output model, all such ML designs must consider the physics of the problem being solved. If not, they will lead to an overfitting where ML will fit the training (and possibly the validation) data extremely well but perform poorly on other data that are unseen during training. Therefore, a thorough understanding of the underlying physics of the problem and matching synthetic and real data are vital elements to solve any optimization and ML problem. In this section, we discuss these elements and some consequences of using an inappropriate physics in solving problems via classical optimization and ML in seismic inversion.

In Chapter 4, we discussed computing synthetic seismograms using ray theory and via convolution using both exact and approximate (linearized) reflection coefficients. In Chapter 5, we discussed different methods of computing wave equation synthetic seismograms such as the reflectivity method for horizontally stratified media and more rigorous methods such as finite-difference, finite-element, pseudo-spectral, etc., for geologically complex media. In addition, in Chapter 5, we made comparisons between the convolution-based AVA synthetic with wave-equation-based synthetic seismograms, and argued that as the subsurface becomes increasingly complex, AVA fails to model complex wave effects, and therefore using wave-equation-based modeling is more appropriate than AVA-based modeling (see Figures 5.1–5.3). These conclusions were, however, purely based on a synthetic study, and here we use real seismic data for further analysis with a primary focus on the practical implementations of using optimization and ML to solve real seismic inverse problems.

The black curves, shown in Figure 10.11a,b, shows real well-log data of P- and S-wave velocity (V_P and V_S) and density (ρ). Note that the well-log data starts from a depth of

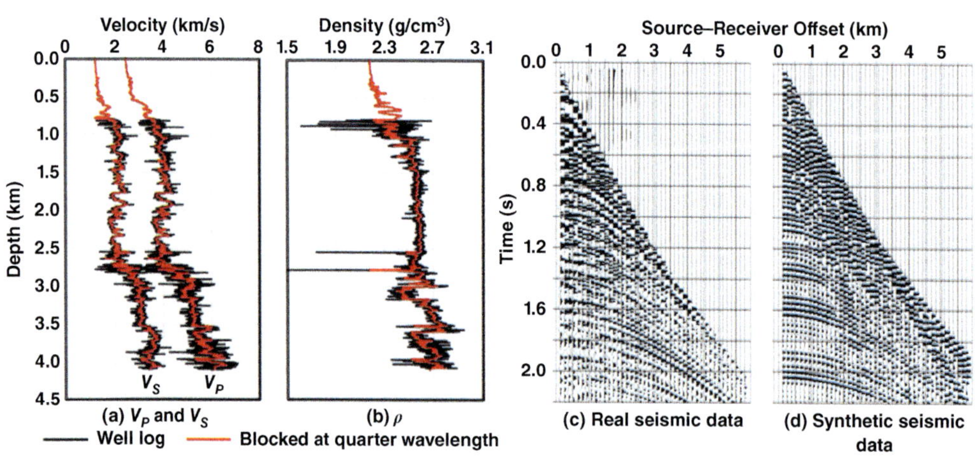

Figure 10.11 Example of well-log, real seismic, and synthetic seismic data. (a) Well-log V_P and V_S shown in black and the corresponding V_P and V_S blocked at the dominant seismic wavelength shown in red. (b) Same as (a) but for the density (ρ). (c) Real prestack seismic data that corresponds to the location of the well-log data shown in (a) and (b). (d) Wave-equation-based synthetic seismograms computed using the blocked well-log data shown in (a) and (b) and the same range of source–receiver offsets as they are in the real seismic data.

about 600 m. Because the location of the well corresponds to real seismic data, shown in Figure 10.11c, the portion of the well log from 0–600 m is not the measured well-log data, but derived from well-to-seismic tie such that the reflection times in the seismic data correspond to the formation boundaries in the measured well log.

The prestack data shown in Figure 10.11c were carefully processed through prestack Kirchhoff time migration (Yilmaz, 1987) and relative amplitude preserved processing (Resnick, 1993). The dominant frequency of the seismic data shown in Figure 10.11c was found to be 30 Hz, and the red curves shown in Figure 10.11a,b are the well logs blocked at the dominant wavelengths of the P waves. The area of these well-log and seismic data was from the Rock-Springs Uplift (RSU), Wyoming, USA, where the geology is simple with dips less than 1°–2°, and therefore after prestack migration, the subsurface can be adequately represented by a stack of horizontally stratified layers (Mallick and Adhikari, 2015; Pafeng et al., 2017). So, an approximate wavelet was estimated from the seismic and well-log data, and synthetic seismic data were computed using the reflectivity method (Kennett, 1983) using the blocked V_P, V_S, and ρ (red curves in Figure 10.11a,b) and the estimated wavelet. These synthetic seismograms, computed for the same range of source–receiver offsets as they are in real seismic data, are shown in Figure 10.11d. From a visual comparison between Figure 10.11c,d, it is evident that although they are similar, there are considerable differences. The primary reason for this difference is the noise in the real data. Additionally, the blocked well-log model may not be the true model represented by the seismic data and require minor adjustments.

In practice, attempting to directly match real and synthetic seismic data in time and offset $(t-x)$ domain shown in Figure 10.11c,d is usually not preferable because of computational expenses. As discussed in Chapter 5, reflectivity method computes synthetics in frequency–ray-parameter $(\omega - p)$ domain and then converts them to $x-t$ domain. For an azimuthally anisotropic medium, it must compute the response in $\omega - p_x - p_y$ domain and then use triple inverse Fourier transform (equation 5.150) to compute response in time (t) and horizontal offsets (x and y). For azimuthally isotropic (isotropic and transversely isotropic with vertical symmetry axis or VTI) media, however, the p_x and p_y combine as $p = \sqrt{(p_x^2 + p_y^2)}$, and the $t-x$ domain response is computed from the $\omega - p$ domain response using the inverse Fourier–Hankel transform, which for vertical response and pressure or explosion source can be written as (Aki and Richards, 2002)

$$R(t,x) = \frac{1}{2\pi} \int_{-\infty}^{+\infty} e^{i\omega t} \int_{0}^{+\infty} R(\omega, p) J_0(\omega p x) d\omega dp. \tag{10.1}$$

In equation 10.1, $R(\omega, p)$ is the response computed in $\omega - p$ domain, $R(t, x)$ is the computed response in time and offset, and J_0 is the Bessel function of order zero. As with any integral transforms, computing the integrals shown in equation 10.1 requires seismic response to be finely sampled in both ω and p, which is computationally demanding. And for seismic inversions, we must compute them not just once but many times in an iterative manner, which is not practical.

For the horizontally stratified subsurface as above, a practical way to model seismic data is the plane-wave domain instead of the time-offset domain. In Chapter 5, we discussed how to compute the plane-wave synthetic response in $\tau - p$ domain (equation 5.151). Because computing the $\tau - p$ domain response requires computing them only for a specific set of p values, they are compute-efficient and practical to use.

So, we can convert real seismic data from $x-t$ domain to $\omega - p$ domain

$$R(\omega, p) = \int_{-\infty}^{+\infty} e^{-i\omega t} \int_{0}^{+\infty} R(x, t) J_0(\omega p x) dt dx, \tag{10.2}$$

and then compute the $\tau - p$ response using a Fourier transform

$$R(\tau, p) = \frac{1}{2\pi} \int_{-\infty}^{+\infty} e^{i\omega\tau} R(\omega, p) d\omega. \qquad (10.3)$$

Although using equations 10.2 and 10.3 is the theoretical way to compute plane-wave response in $\tau - p$ domain, it requires input seismic data to be finely sampled in both time and offset, otherwise the computed response will be temporally and spatially aliased. Typically, real seismic data are finely sampled in time but not in offset. The real data shown in Figure 10.11c, for example, are sampled finely in time at 2 ms sampling interval but have only 43 source–receiver offsets from 134 to 5628 m at intervals of 134 m. A $\tau - p$ transform of such data is likely to be spatially aliased and unusable for inversion.

A practical alternative to using equations 10.2 and 10.3 is slant stacking along constant p or constant incident angle θ on the normal moveout (NMO) corrected data. Figure 10.12 illustrates the process along constant θ for offset-to-angle transformation, developed by Todd and Backus (1985). The procedure to stack along constant p to obtain NMO-corrected $\tau - p$ transform is similar and can be found in Brysk and McCowan (1986). Figure 10.12a shows the P-wave velocity (V_P) and the root-mean-square (RMS) velocity, computed from V_P. The V_P, shown as a solid curve in Figure 10.12a, is the same as the blocked V_P of Figure 10.11a but shown in two-way P-wave time instead of depth. Figure 10.12b shows the NMO-corrected real prestack data of Figure 10.11c using the velocity field of Figure 10.12a. Figure 10.12c shows the three constant P-wave incidence-angle trajectories (15°, 35°, and 55°) computed using the V_P field shown in Figure 10.12a. Finally, Figure 10.12d shows the offset-to-angle transform from 5° to 55° in steps of 5°. To compute the angle-domain data, shown in Figure 10.12d, the traces lying within the Fresnel radius on either side of the of the constant angle trajectory (angle mute) on the NMO-corrected gather shown in Figure 10.12b for each angle are partially stacked to obtain the angle-domain trace for that angle. The equations to compute constant angle trajectories or angle mutes and the corresponding radii

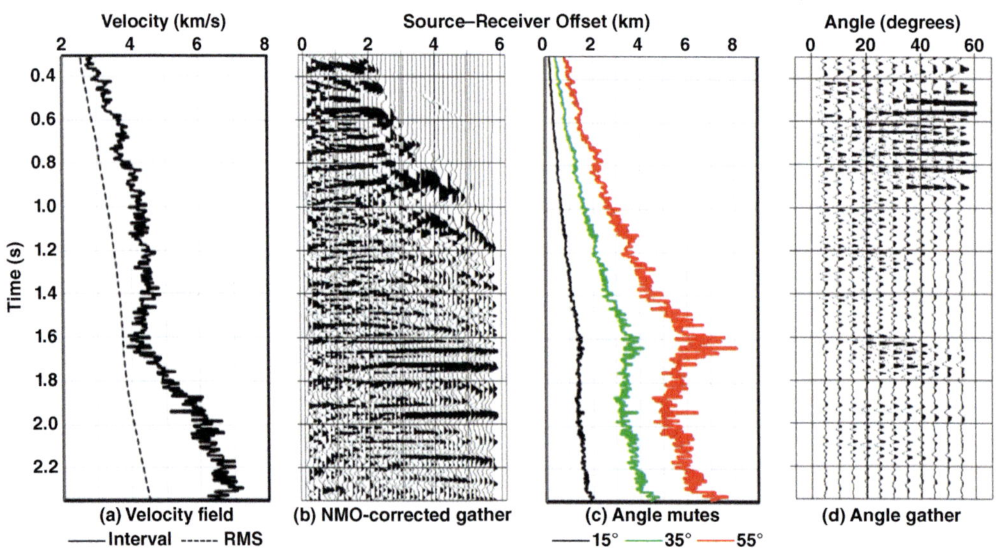

Figure 10.12 Procedure for offset-to-angle transform. (a) Interval and RMS velocity field. (b) NMO-corrected real prestack gather. (c) Angle mutes (constant incidence-angle trajectories) for the angle θ= 15°, 35°, and 55°. (d) Computed angle gather for θ values of 5°–55° in steps of 5°. Jia et al. (2021)/Society of Exploration Geophysicists.

of the Fresnel zone can be found in Resnick (1993). And, the equivalent expressions for slant staking to obtain NMO-corrected $\tau - p$ transform is similar and can be found in Brysk and McCowan (1986).

Note that because of partial stacking, angle gathers (Figure 10.12d) are of much higher signal-to-noise ratio than the original $x-t$ domain gather (Figure 10.12b).

So, we can use slant stacking to compute NMO-corrected plane-wave data either in incidence-angle (θ) or ray-parameter (p) domain. To compute synthetic data in either domain and matching them iteratively with observation, we have the following three options:

1) Compute convolutional synthetics (equation 4.99) by convolving the wavelet with the linearized P–P reflection coefficient (R_{PP}) formula (equation 4.91).
2) Same as 1 but using the exact (Zoeppritz) R_{PP} formula (equation 4.65).
3) Compute synthetic seismograms, either in angle or in ray-parameter domain using the wave equation-based reflectivity method for computing complete seismograms, as discussed in Chapter 5.

Figure 10.13 compares the real angle gather from the offset-to-angle transform (Figure 10.13a) with the angle-domain synthetic seismograms computed using the blocked well-log data (red curves shown in Figure 10.11a,b) and (1) wave-equation-based reflectivity method (Figure 10.13b), (2) convolution with exact R_{PP} (Figure 10.13c), and (3) convolution with the approximate (linearized) R_{PP}.

Note that the first 0.4 s of synthetic data shown in Figure 10.13b–d should be ignored because it comes from the contribution of the overburden model between 0 and 600 m, used for well-to-seismic tie and is not really from the measured V_P, V_S, and ρ from the well. For times greater than 0.4 s, the

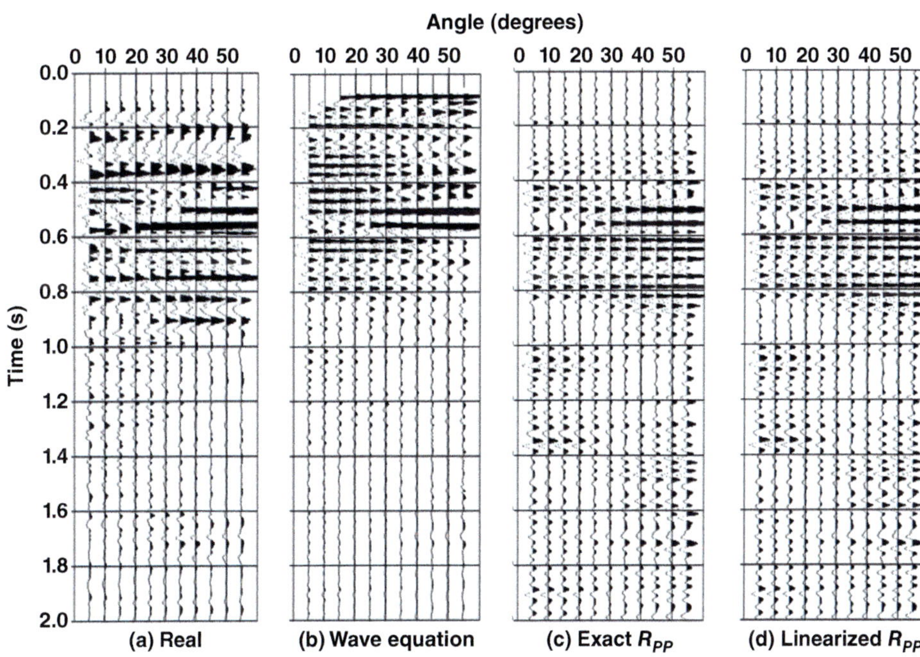

Figure 10.13 Comparison between the real and synthetic data in time angle domain. (a) Real angle gather (same as Figure 10.12d). (b) Synthetic angle gather computed using the wave-equation-based reflectivity method. (c) Synthetic convolutional angle gather, computed using the exact R_{PP}. (d) Same as (c) but with the approximate linearized R_{PP}.

convolutional synthetics reasonably match real data in the shallow parts (less than 1 s) but poorly for times greater than 1 s. Although not perfect, the wave-equation-based synthetics (Figure 10.13b) match the real data with much better accuracy than any of the convolution-based synthetics.

To quantify the visual comparisons of Figure 10.13, Figure 10.14 compares the real angle gather with two convolution-based and the wave-equation-based synthetic angle gathers for θ values of 5° and 55° and for 0.4–2 s time window. This quantitative comparison again demonstrates that the wave-equation-based methods match real seismic data with much better accuracy than any of the convolutional methods. For small incidence angle (5°), the convolution-based synthetics match real data to only 0.6 s (Figure 10.14a,c) while the wave-equation-based synthetics match reasonably for the entire time window besides a small part in the 0.6–0.7 s time window (Figure 10.14e). For large incidence angle (55°), the convolution-based synthetics are again accurate only in the 0.4–0.6 s window (Figure 10.14b,d) while the wave-equation-based synthetics are reasonably accurate for the entire time window.

Therefore, as previously demonstrated using synthetic data in Chapter 5 (Figures 5.1–5.3), here we used real seismic data to demonstrate the value of using wave equation for the underlying physics instead of convolution. As the subsurface model becomes complex, wave propagation effects, such as transmission loss, interbed reflections, and mode conversions, effects due to ray bending, etc., which are not accounted for in the convolution-based synthetics, result in the synthetic data to match poorly with real data. Wave equation methods, on the other hand, honor these effects, and therefore are considered as more appropriate method to use. This is vital, especially for any classical optimization and physics-informed ML methods. Note from Figures 10.13 and 10.14 that if the real data do not match with synthetic data, even using an (almost) correct model, as is the case for convolutional

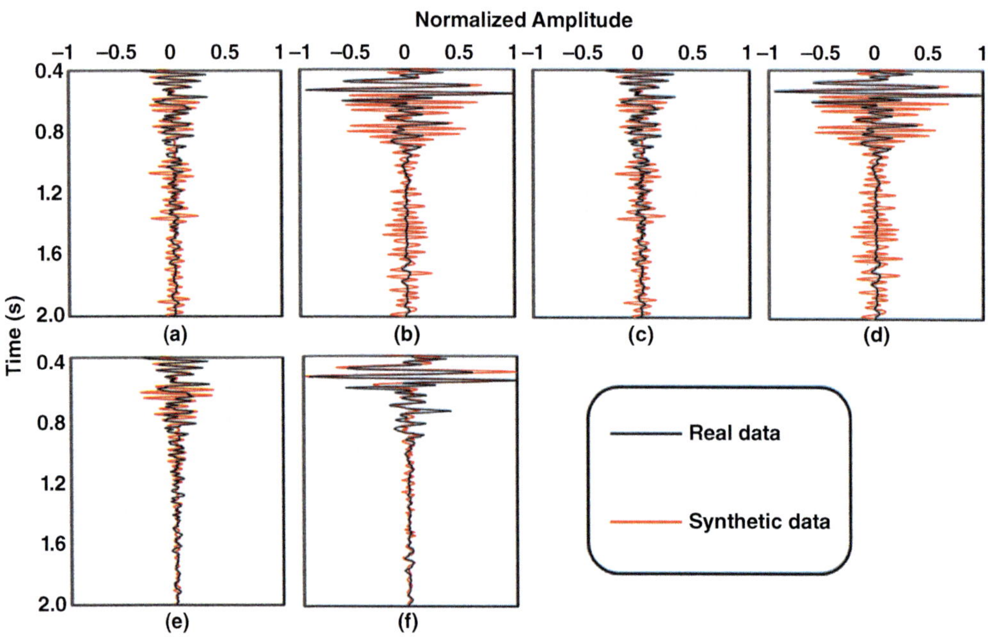

Figure 10.14 Quantitative comparison of real seismic data with wave-equation-based and convolution-based synthetics. (a, b) Comparison between the real angle-domain data (black) with convolution-based synthetic angle-domain data using approximate (linearized) R_{PP} for θ values of 5° and 55°. (c, d) Same as (a, b) but using the exact R_{PP}. (e, f) Same as (a–d), except the synthetic angle-domain data were computed using a wave-equation-based method.

synthetics, the optimization method will attempt to find a model that will match the data. But unfortunately, the model it will find may not be the correct model. On the other hand, if wave-equation-based method is used instead of convolution, it will find the correct model in a few iterations. This is the consequence of using incorrect physics to model data. So, for any optimization problem, it is important to use appropriate physics, and if not, we must make some quantitative analyses as above. In recent years, there have been many developments in the field of using Bayesian approaches for uncertainty quantification via optimization and ML for reservoir characterization applications in the field of exploration geophysics. Although elegant, most of these approaches use convolution as the basis for forward modeling, and the above analysis is meant to point out that unless a correct method is used for forward modeling, the results from these analyses are questionable irrespective of how sophisticated these methods are.

In the above discussion, we used the reflectivity method as the basis for computing wave-equation-based synthetic seismograms. As discussed in Chapter 5, the reflectivity method is valid for horizontally stratified models and is adequate for areas with simple geology like RSU. For relatively complex subsurface, computing prestack common image gathers from RTM would allow seismic data at each image location to be approximated as a stack of horizontal layers, and we can still use the reflectivity method as the basis for forward modeling (Jia and Mallick, 2018). In geologically complex areas, such as near salt bodies, subducting plate boundaries, etc., using reflectivity method as forward modeling engine is inadequate and it is necessary to use other numerical methods, such as finite-difference, finite-element, pseudo-spectral, etc. We have discussed these numerical methods in Chapter 5. However, the main problem in using these methods is their computational expense. The current approach for practically implementing these methods in any optimization/ML is to use a few selected frequencies in a frequency domain finite-difference, spectral-element, or pseudo-spectral modeling method for forward modeling (for example, see Pratt et al., 1998; Pratt, 1999; Guo and Alkhalifah, 2017; Guitton and Alkhalifah, 2017; Zhao and Sen, 2022;, among others). With further advances in computing technology, we expect more of such applications of rigorous numerical modeling approaches in future applications of optimization and ML.

Finally, we must point out that irrespective of how sophisticated or numerically rigorous of the forward modeling we use, none of them are perfect in matching every feature of the observed seismic data besides noise. The reflectivity method is an analytical solution to the wave equation, but its accuracy depends upon how finely we discretize the subsurface as a stack of horizontal layers. Numerical approaches (finite-difference, finite-element, spectral-element, etc.) approximate the partial derivatives as differences, and their accuracy depends upon the grid or element size. Typically, for the features on the observed data which we cannot model, we assume them to be the noise on data. This fact must be always kept in mind in implementing any optimization or ML method in practice.

10.4 Conclusions

Optimization and ML methods are powerful tools for solving complex problems. To solve any practical problem, it is however important to completely understand the underlying physics of the problem and understand its limitations. This book provides an overview of the physics of seismic wave propagation and different flavors of local optimization, global optimization, and ML methods to solve seismic inverse problems.

As computing technology advances further, we expect that solving more complex problems, especially multi-objective, multi-physics, and multi-scale problems, will become common. This, however, requires seamless integration of different data types. Current continuing advances in the field of data science and artificial intelligence methods are the key to the success of these future attempts.

References

Aki, K., and Richards, P.G., 2002, *Quantitative Seismology*, Second Edn, University Science Books, ISBN 0-978-1891389634.

Atwater, B.F., 1987, Evidence for great Holocene earthquakes along the outer coast of Washington state, *Science*, *236*(4804), 942–944, https://doi.org/10.1126/science.236.4804.942.

Biswas, R., Sen, M.K., Das, V., and Mukerji, T., 2019, Prestack and poststack inversion using a physics-guided convolutional neural network, *Interpretation*, *7*, SE161–SE174, https://doi.org/10.1190/INT-2018-0236.1.

Brucet, S., Boix, D., Nathansen, L.W., Quintana, X.D., Jensen, E., Balayla, D., Meerhoff, M., and Jeppesen, E., 2012, Effects of temperature, salinity and fish in structuring the macroinvertebrate community in Shallow lakes: implications for effects of climate change, *PLoS One*, *7*(2), e30877, https://doi.org/10.1371/journal.pone.0030877.

Brysk, H., and McCowan, D.W., 1986, A slant-stack procedure for point-source data, *Geophysics*, *51*, 1370–1386, https://doi.org/10.1190/1.1442187.

Castillo, J., Basrbieri, M.A., and Gonzalez, A., 1996, Relationships between sea surface temperature, salinity, and pelagic fish distribution off northern Chile, *ICES Journal of Marine Science*, *53*, 139–146, https://doi.org/10.1006/jmsc.1996.0014.

Chen, C.T., and Millero, F.J., 1977, Speed of sound in seawater at high pressures, *Journal of the Acoustical Society of America*, *62*(5), 1129–1135, https://doi.org/10.1121/1.381646.

Dash, R., and Spence, G., 2011, P-wave and S-wave velocity structure of northern Cascadia margin gas hydrates, *Geophysical Journal International*, *187*, 1363–1377, https://doi.org/10.1111/j.1365-246X.2011.05215.x.

Dash, R., Spence, G., Hyndman, R., Grion, S., Wang, Y., and Ronen, S., 2009, Wide-area imaging from OBS multiples, *Geophysics*, *74*, Q41–Q47, https://doi.org/10.1190/1.3223623.

Dickens, G.R., 1999, Carbon cycle: the blast in the past, *Nature*, *401*, 752–753, https://doi.org/10.1038/99999.

Dickens, G.R., Paull, C., Wallace, P., and ODP Leg 164 Scientific Party, 1997, Direct measurement of in situ methane quantities in a large gas-hydrate reservoir, *Nature*, *385*, 426–428, https://doi.org/10.1038/385426a0.

Fischer, D., Mogollón, J.M., Strasser, M., Pape, T., Bohrmann, G., Fekete, N., Spiess, V., and Kasten, S., 2013, Subduction zone earthquake as potential trigger of submarine hydrocarbon seepage, *Nature Geoscience*, *6*, 647–651, https://doi.org/10.1038/ngeo1886.

Fofonoff, P. and Millard, R.C., Jr, 1983, Algorithms for computation of fundamental properties of seawater, UNESCO Technical Paper in Marine Sciences, No. 44, 53 pp.

Fortin, W.F.J., 2015, Extracting physical parameters from marine seismic data: New methods in seismic oceanography and velocity inversion, ProQuest Dissertations And Theses; Thesis (Ph.D.) – University of Wyoming, 2015.; Publication Number: AAT 3722992; ISBN: 9781339054759; Source: Dissertation Abstracts International, Volume: 77-02(E), Section: B.; 145 p.

Grana, D., Verma, S., Pafeng, J., Lang, X., Sharma, H., Wu, W., McLaughlin, F., Campbell, E., Ng, K., Alvarado, V., Mallick, S., and Kaszuba, J., 2017, A rock physics and seismic reservoir characterization study of the Rock Springs uplift, a carbon dioxide sequestration site in Southwestern Wyoming, *International Journal of Greenhouse Gas Control*, *63*, 296–309, https://doi.org/10.1016/j.ijggc.2017.06.004.

Guitton, A., and Alkhalifah, T., 2017, A parameterization study for elastic VTI full-waveform inversion of hydrophone components: synthetic and North Sea field data examples, *Geophysics*, *82*, R299–R308, https://doi.org/10.1190/GEO2017-0073.1.

Guo, Q., and Alkhalifah, T., 2017, Elastic reflection-based waveform inversion with a nonlinear approach, *Geophysics*, *82*, R309–R321, https://doi.org/10.1190/GEO2016-0407.1.

Hassanpouryouzband, A., Joonaki, E., Edlmann, K., and Haszeldine, R.S., 2021, Offshore geological storage of hydrogen: is this our best option to achieve net-zero? *ACS Energy Letters*, *6*(6), 2181–2186, https://doi.org/10.1021/acsenergylett.1c00845.

Heinemann, N., Alcalde, J., Miocic, J.M., Hangx, S.J.T., Kallmeyer, J., Christian Ostertag-Henning, C., Hassanpouryouzband, A., Thaysen, E.M., Strobel, G.J., Schmidt-Hattenberger, C., Edlmann, K., Wilkinson, M., Bentham, M., Haszeldine, R.S., Carbonell, R., and Rudloff, A., 2021, Enabling large-scale hydrogen storage in porous media – the scientific challenges, *Energy & Environmental Science*, *14*, 853–964, https://doi.org/10.1039/D0EE03536J.

Hyndman, R.D., and Davis, E.E., 1992, A mechanism for the formation of methane hydrate and sea floor bottom simulating reflectors by vertical fluid expulsion, *Journal of Geophysical Research*, *97*(H25), 7025–7041, https://doi.org/10.1029/91JB03061.

Hyndman, R.D., Spence, G.D., Yuan, T., and Davis, E.E., 1994, Regional geophysical and structure framework of the Vancouver Island margin accretionary prism, in Westbrook, G.K., Carson, B., Musgrave, R.J., et al., (Eds.), Proceedings of the Ocean Drilling Program, Initial Reports, 146(1), College Station, TX, 399–419, https://doi.org/10.2973/ODP.PROC.IR.146-1.002.1994.

Hyndman, R.D., and Wang, K., 1995, The rupture zone of Cascadia great earthquakes from current deformation and the thermal regime, *Journal of Geophysical Research – Solid Earth*, *100*(B11), 22133–22154, https://doi.org/10.1029/95JB01970.

Jia, L., and Mallick, S., 2018, Characterization of the Cascadia ocean margin methane hydrates using prestack waveform inversion and reverse time migration, SEG Technical Program, Expanded Abstracts, https://doi.org/10.1190/segam2018-2996629.1.

Jia, L., Mallick, S., and Wang, C., 2021, Data-driven prestack-waveform inversion using genetic algorithm: methodology and examples, *Interpretation*, *9*, T1069–T1088, https://doi.org/10.1190/INT-2020-0201.1.

Kennett, B.L.N., 1983, *Seismic Wave Propagation in Stratified Media*, Cambridge University Press.

Kennett, J.P., Cannariato, K.G., Hendy, I.L., and Behl, R.J., 2000, Carbon isotopic evidence for methane hydrate instability during quaternary interstadials, *Science*, *288*, 128–133, https://doi.org/10.1126/science.288.5463.128.

Kvenvolden, K.A., 1993, Gas hydrates – geological perspective and global change, *Reviews of Geophysics*, *31*, 173–183, https://doi.org/10.1029/93RG00268.

Lindblom, U.E., 1985, A conceptual design for compressed hydrogen storage in mined caverns, *International Journal of Hydrogen Energy*, *10*, 667–675, https://doi.org/10.1016/0360-3199(85)90006-0.

MacKay, M.E., Jarrard, R.D., Westbrook, G.K., Hyndman, R.D., and the Shipboard Scientific Party of ODP Leg 146, 1994, Origin of bottom simulating reflectors: geophysical evidence from the Cascadia accretionary prism, *Geology*, *22*, 459–462, https://doi.org/10.1130/0091-7613(1994)022>3c0459:OOBSRG<3e2.3.CO;2.

Mahadevan, A., 2016, The impact of submesoscale physics on primary productivity of plankton, *Annual Review of Marine Science*, *8*, 161–184, https://doi.org/10.1146/annurev-marine-010814-015912.

Mallick, S., and Adhikari, S., 2015, Amplitude-variation-with-offset and prestack waveform inversion: a direct comparison using a real data example from the Rock Springs uplift, Wyoming, USA, *Geophysics*, *80*, B45–B59, https://doi.org/10.1190/geo2014-0233.1.

Mallick, S., and Chakraborty, D., 2022, Prediction of the ocean water sound speeds via attribute-guided seismic waveform inversion, *Geophysics*, *87*, U67–U79, https://doi.org/10.1190/geo2021-0442.1.

Mavko, G., Mukerji, T., and Dvorkin, J., 2010, *Rock Physics Handbook*, Second Edn, Cambridge University Press, ISBN: 9780511626753, https://doi.org/10.1017/CBO9780511626753.

McCaffrey, R., 2007, The next great earthquake, *Science*, *315*(5819), 1675–1676, https://doi.org/10.1126/science.1140173.

McDonald, I.R., 1990, Role of methane clathrates in past and future climates, *Climate Change*, *16*, 247–281, https://doi.org/10.1007/BF00144504.

Motiee, M., 1991, Estimate possibility of hydrate, *Hydrocarbon Processing*, *70*, 98–99.

Padhi, A., Mallick, S., Behzadi, H., and Alvarado, V., 2014, Efficient modeling of seismic signature of patchy saturation for time lapse monitoring of carbon sequestrated deep saline reservoirs, *Applied Energy*, *114*, 445–455, https://doi.org/10.1016/j.apenergy.2013.10.016.

Padhi, A., Mallick, S., Fortin, W., Holbrook, W.S., and Blacic, T.M., 2015, 2-D ocean temperature and salinity images from pre-stack seismic waveform inversion methods: an example from the South China Sea, *Geophysical Journal International*, *202*, 800–810, https://doi.org/10.1093/gji/ggv188.

Pafeng, J., Mallick, S., and Sharma, H., 2017, Prestack waveform inversion of three-dimensional seismic data – an example from the Rock Springs uplift, Wyoming, USA, *Geophysics*, *82*(1), B1–B12, https://doi.org/10.1190/geo2016-0079.1.

Pratt, R.G., 1999, Seismic waveform inversion in the frequency domain. Part 1: theory and verification in a physical scale model, *Geophysics*, *64*, 888–901, https://doi.org/10.1190/1.1444597.

Pratt, R.G., Shin, C., and Hicks, G.H., 1998, Gauss-newton and full newton methods in frequency-space seismic waveform inversion, *Geophysical Journal International*, *133*, 341–362, https://doi.org/10.1046/j.1365-246X.1998.00498.x.

Resnick, J., 1993, Seismic data processing for AVO and AVA analysis, in J.E. Castagna, and M.M. Backus, eds., *Offset Dependent Reflectivity—Theory and Practice for AVO and AVA Analysis*, 175–189, Society of Exploration Geophysicists, https://doi.org/10.1190/1.9781560802624.ch3.

Rijnsdorp, A.D., Peck, M.A., Engelhard, G.H., Möllmann, C., and Pinnegar, J.K., 2009, Resolving the effect of climate change on fish populations, *ICES Journal of Marine Science*, *66*, 1570–1583, https://doi.org/10.1093/icesjms/fsp056.

Sahimi, M., and Tahmasebi, P., 2022, *Transport in Porous Media*, *145*, 367–387, https://doi.org/10.1007/s11242-022-01855-8.

Schoenberg, M., and Douma, J., 1988, Elastic-wave propagation in media with parallel fractures and aligned cracks, *Geophysical Prospecting*, *36*, 571–590, https://doi.org/10.1111/j.1365-2478.1988.tb02181.x.

Schoenberg, M., and Sayers, C., 1995, Seismic anisotropy of fractured rock, *Geophysics*, *60*, 204–211, https://doi.org/10.1190/1.1443748.

Settari, A., and Mounts, F.M., 1998, A coupled reservoir and geomechanical simulation system, *SPE Journal*, *3*, 219–226, https://doi.org/10.2118/50939-PA.

Settari, A., and Walters, D.A., 2001, Advances in coupled geomechanical and reservoir modeling with applications to reservoir compaction, *SPE Journal*, *6*, 334–342, https://dx.doi.org/10.2118/51927-MS.

Spence, G.D., Minshull, T.A., and Fink, I.C., 1995, Seismic studies of methane gas hydrate, offshore Vancouver Island, in Carson, B., Westbrook, G.K., Musgrave, R.J., and Suess, E. (Eds.), Proceedings of the Ocean Drilling Program, Scientific Results, 146 (1), 163–174, https://dx.doi.org/10.2204/iodp.proc.311.109.2006.

Suess, E., 2014, Legacy of Hydrate Ridge: An illustrated account, Proceedings of the 8th International Conference on Gas Hydrates (ICGH8-2014), Beijing, China, https://oceanrep.geomar.de/id/eprint/26036.

Sugie, K., Fugiwara, A., Nishino, S., Kameyayama, A., and Harada, N., 2020, Impacts of temperature, CO_2, and salinity on phytoplankton community composition in the Western Arctic Ocean, *Frontiers in Marine Science*, https://doi.org/10.3389/fmars.2019.00821.

Thena, T., and Mohan, K., 2016, Estimation of pore space methane hydrate in marine sediments of Cascadia margin, *Journal of Natural Gas Science and Engineering*, *35*, 956–963, https://doi.org/10.1016/j.jngse.2016.09.046.

Thomas, L.K., Chin, L.Y., Pierson, R.G., and Stylte, J.E., 2003, Coupled geomechanics and reservoir simulation, *SPE Journal*, *8*, 350–358, https://doi.org/10.2118/87339-PA.

Todd, C.P. and Backus, M.M., 1985, Offset-dependent reflectivity in a structural context, SEG International Meeting Expanded Abstracts, 586–588, https://doi.org/10.1190/1.1892805.

Torres, M.E., Wallmann, K., Tréhu, A.M., Bohrmann, G., Borowski, W.S., and Tomaru, H., 2004, Gas hydrate growth, methane transport, and chloride enrichment at the southern summit of hydrate ridge, Cascadia Margin off Oregon, *Earth and Planetary Science Letters*, *226*, 225–241, https://doi.org/10.1016/j.epsl.2004.07.029.

Tréhu, A.M. and Bangs, N., 2001, 3-D seismic imaging of an active margin hydrate system, Oregon continental margin, Report of cruise TTN112, Oregon State University Data Report 182, ref. 2001-2, Corvallis, Oregon, https://ir.library.oregonstate.edu/concern/technical_reports/02870x13k.

Tréhu A.M., Bangs, N.L., Arsenault, M.A., Bohrmann, G., Goldfinger, C., Johnson, J.E., Nakamura, Y., and Torres, M.E., 2002, Complex subsurface plumbing beneath southern Hydrate Ridge, Oregon continental margin, from high resolution 3-D seismic reflection and OBS Data, 4th Int. Conf. Gas Hydrates, Yokohama, Japan, 90–96.

Tréhu, A.M., Flemings, P.B., Bangs, N.L., Chevalier, J., Gracia, E., Johnson, J., Liu, C.-S., Liu, X., Riedel, M., and Torres, M.E., 2004, Feeding methane vents and gas hydrate deposits at south hydrate ridge, *Geophysical Research Letters*, *31*(23), L23310, https://doi.org/10.1029/2004GL021286.

Tréhu, A.M., and Flueh, E.R., 2001, Estimating the thickness of the free gas zone beneath hydrate ridge, Oregon continental margin, from seismic velocities and attenuation, *Journal of Geophysical Research*, *106*, 2035–2045, https://doi.org/10.1029/2000JB900390.

Tréhu, A.M., Lin, G., Maxwell, E., and Goldfinger, C., 1995, A seismic-reflection profile across the Cascadia subduction zone offshore Central Oregon – new constraints on methane distribution and crustal structure, *Journal of Geophysical Research – Solid Earth*, *100*(B8), 15101–15116, https://doi.org/10.1029/95JB00240.

Tréhu, A.M., Torres, M.E., Moore, G.F., Suess, E., and Bohrmann, G., 1999, Temporal and spatial evolution of a gas hydrate-bearing accretionary ridge on the Oregon continental margin, *Geology*, *27*(10), 939–942, https://doi.org/10.1130/0091-7613(1999)027>3c0939:TASEOA<3e2.3.CO;2.

Wang, K., Wells, R., Mazzotti, S., Hyndman, R.D., and Sagiya, T., 2003, A revised dislocation model of interseismic deformation of the Cascadia subduction zone, *Journal of Geophysical Research – Solid Earth*, *108*(B1), https://doi.org/10.1029/2001JB001227.

Warner, J.C., Armstrong, B., He, R., and Zambon, J.B., 2010, Development of a coupled-ocean-atmosphere-wave sediment transport (COAWST) modeling system, *Ocean Modeling*, *35*, 230–244, https://doi.org/10.1016/j.ocemod.2010.07.010.

Weitemeyer, K.A., Constable, S.C., Key, K.W., and Behrens, J.P., 2006, First results from a marine controlled-source electromagnetic survey to detect gas hydrates offshore Oregon, *Geophysical Research Letters*, *33*, L03304, https://doi.org/10.1029/2005GL024896.

Weitemeyer, K.A., Constable, S., and Tréhu, A.M., 2011, A marine electromagnetic survey to detect gas hydrate at hydrate ridge, Oregon, *Geophysical Journal International*, *187*, 45–62, https://doi.org/10.1111/j.1365-246X.2011.05105.x.

Williams, M.C., Tréhu, A.M., and Braunmiller, J., 2011, Seismicity at the Cascadia plate boundary beneath the Oregon continental shelf, *Bulletin of the Seismological Society of America*, *101*(3), 940–950, https://doi.org/10.1785/0120100198.

Yilmaz, O., 1987, *Seismic Data Processing, Investigations in Geophysics*, Vol. 2, Society of Exploration Geophysicists, ISBN: 978-0931830402.

Zhao, Z., and Sen, M.K., 2022, Uncertainty quantification for full waveform inversion with a mini-batch gradient based sampling method, Second International meeting of Applied Geoscience & Energy (Expanded Abstracts), https://doi.org/10.1190/image2022-3751538.1.

Index

Note: *Italic* page numbers refer to *figures*, and **bold** page numbers refer to **tables**.

Computation, Optimization, and Machine Learning in Seismology, Advanced Textbook 8, First Edition. Subhashis Mallick.

t

Printed and bound by CPI Group (UK) Ltd, Croydon, CR0 4YY
19/08/2025
14720465-0001